QD 262 TIE

QM Library

23 1335219 4

MAIN LIBRARY
QUEEN MARY, UNIVERSITY OF LONDON
Mile End Road, London E1 4NS
DATE DUE FOR RETURN

CANCELLED
NEW ACCESSIONS

D1422379

WITHDRAWN
FROM STOCK
QMUL LIBRARY

Lutz F. Tietze, Gordon Brasche, and Kersten M. Gericke
**Domino Reactions
in Organic Synthesis**

Related Titles

Berkessel, A., Gröger, H.

Asymmetric Organocatalysis

**From Biomimetic Concepts to
Applications in Asymmetric Synthesis**

2005
ISBN 3–527-30517–3

Zhu, J., Bienaymé, H. (eds.)

Multicomponent Reactions

2005
ISBN 3–527-30806–7

de Meijere, A., Diederich, F. (eds.)

**Metal-Catalyzed
Cross-Coupling Reactions**

2004
ISBN 3–527-30518–1

Beller, M., Bolm, C. (eds.)

**Transition Metals for Organic
Synthesis, 2 Vol.**

Building Blocks and Fine Chemicals

2004
ISBN 3–527-30613–7

Mahrwald, R. (ed.)

Modern Aldol Reactions, 2 Vol.

2004
ISBN 3–527-30714–1

Nicolaou, K. C., Snyder, S. A.

Classics in Total Synthesis II

More Targets, Strategies, Methods

2003
ISBN 3–527-30685–4

Eicher, T., Hauptmann, S.

The Chemistry of Heterocycles

**Structure, Reactions, Syntheses, and
Applications**

2003
ISBN 3–527-30720–6

Lutz F. Tietze, Gordon Brasche,
and Kersten M. Gericke

Domino Reactions
in Organic Synthesis

**WILEY-
VCH**

WILEY-VCH Verlag GmbH & Co. KGaA

The Authors

Prof. Dr. Dr. h.c. Lutz Tietze
Inst. f. Organische Chemie
Georg-August-Universität
Tammannstr. 2
37077 Göttingen

Dr. Gordon Brasche
Inst. f. Organische Chemie
Georg-August-Universität
Tammannstr. 2
37077 Göttingen

Dr. Kersten Matthias Gericke
Inst. f. Organische Chemie
Georg-August-Universität
Tammannstr. 2
37077 Göttingen

QM LIBRARY
(MILE END)

■ All books published by Wiley-VCH are carefully produced. Nevertheless, authors, editors, and publisher do not warrant the information contained in these books, including this book, to be free of errors. Readers are advised to keep in mind that statements, data, illustrations, pro cedural details or other items may inadvertently be inaccurate.

Library of Congress Card No.:
applied for

British Library Cataloguing-in-Publication Data
A catalogue record for this book is available from the British Library.

Bibliographic information published by Die Deutsche Bibliothek
Die Deutsche Bibliothek lists this publication in the Deutsche Nationalbibliografie; detailed bibliographic data is available in the Intern et at <http://dnb.ddb.de>.

© 2006 WILEY-VCH Verlag GmbH & Co. KGaA, Weinheim

All rights reserved (including those of translation into other languages). No part of this book may be reproduced in any form – by photoprinting, microfilm, or any other means – nor transmitted or translated in to a machine language without written permission from the publishers. Registered names, trademarks, etc. used in this book, even when not specifically marked as such, are not to be considered unprotected by law.

Typesetting primustype Robert Hurler GmbH
Printing betz-Druck GmbH, Darmstadt
Binding Litges & Dopf Buchbinderei GmbH, Heppenheim
Cover Design Adam Design, Weinheim

Printed in the Federal Republic of Germany
Printed on acid-free paper

ISBN-13: 978-3-527-29060-4
ISBN-10: 3-527-29060-5

Table of Contents

Preface

The ability to create complex molecules in only a few steps has long been the dream of chemists. That such thinking is not unrealistic could be seen from Nature, where complicated molecules such as palytoxin, maitotoxin and others are synthesized with apparent ease and in a highly efficient manner. Now, with the development of domino reactions, the dream has become almost true for the laboratory chemist – at least partly. Today, this new way of thinking represents a clear change of paradigm in organic synthesis, with domino reactions being frequently used not only in basic research but also in applied chemistry.

The use of domino reactions has two main advantages. The first advantage applies to the chemical industry, as the costs not only for waste management but also for energy supplies and materials are reduced. The second advantage is the beneficial effect on the environment, as domino reactions help to save natural resources. It is, therefore, not surprising that this new concept has been adopted very rapidly by the scientific community.

Following our first comprehensive review on domino reactions in 1993, which was published in *Angewandte Chemie*, and a second review in 1996 in *Chemical Reviews*, there has been an "explosion" of publications in this field. In this book we have included carefully identified reaction sequences and selected publications up to the summer of 2005, as well as details of some important older studies and very recent investigations conducted in 2006. Thus, in total, the book contains over 1000 citations!

At this stage we would like to apologize for not including *all* studies on domino reactions, but this was due simply to a lack of space. In this book, the term "domino" is used throughout to describe the reaction sequences used, and we seek the understanding of authors of the included publications if we did not use their terminology. Rather, we thought that for a better understanding a unified concept based on our definition and classification of domino reactions would be most appropriate. Consequently, we would very much appreciate if everybody working in this field would in future use the term "domino" if their reaction fulfills the conditions of such a transformation.

We would like to thank Jessica Frömmel, Martina Pretor, Sabine Schacht and especially Katja Schäfer for their continuous help in writing the manuscript and preparing the schemes. We would also like to thank Dr. Hubertus P. Bell for manifold ideas and the selection of articles, Dr. Sascha Hellkamp for careful overseeing of the manuscript and helpful advice, and Xiong Chen for controlling the literature. We also like to thank the publisher Wiley-VCH, and especially William H. Down, Dr. Romy Kirsten and Dr. Gudrun Walter, for their understanding and help in preparing the book.

Göttingen, summer 2006   *Lutz F. Tietze*
                                                                 Gordon Brasche
                                                                 Kersten M. Gericke

Abbreviations

)))	Sonification
18-C-6	18-Crown-6 ether
A-3CR (A-4CR)	Asinger three(four)-component reaction
Ac	acetyl
acac	acetylacetonato
ACCN	1,1'-azobis(cyclohexanecarbonitrile)
AcOH	acetic acid
Ac_2O	acetic anhydride
AIBN	2,2'-azobisisobutyronitrile
ALA	δ-amino levulinic acid
ALB	AlLibis[(S)-binaphthoxide] complex
All	allyl
Ar	aryl
BB-4CR	Bucherer–Bergs four-component reaction
BEH	bacterial epoxide hydrolase
$BF_3 \cdot OEt_2$	boron trifluoride–diethyl ether complex
BINAP	2,2'-bis(diphenylphosphino)-1,1'-binaphthyl
BINOL	2,2'-dihydroxy-1,1'-binaphthyl
[bmim]BF_4	1-butyl-3-methylimidazolium tetrafluoroborate
[bmim]PF_6	1-butyl-3-methylimidazolium hexafluorophosphate
BMDMS	bromomethyldimethylsilyl
Bn	benzyl
Boc	*tert*-butoxycarbonyl
BOM	benzyloxymethyl
BOXAX	2,2'-bis(oxazolin-2-yl)-1,1'-binaphthyl
BP	1,1'-biphenyl
BS	*p*bromophenylsulphonyloxy
BTIB	bis(trifluoroacetoxy)-iodobenzene
Bu	*n*-butyl
Bz	benzoyl
CALB	*Candida antarctica* lipase
CAN	ceric ammonium nitrate
cat.	catalytic; catalyst
Cbz	benzyloxycarbonyl

*c*Hx	cyclohexyl
CM	cross-metathesis
COD	cycloocta-1,5-diene
COX	cyclooxygenase
CuTC	copper thiophene-2-carboxylate
Cy	cyclohexyl
d	day(s)
DAIB	(diacetoxy)iodobenzene
DBU	1,8-diazabicyclo[5.4.0]undec-7-ene
DCM	dichloromethane
DCTMB	1,4-dicyano-tetramethylbenzene
DDQ	2,3-dichloro-5,6-dicyano-1,4-benzoquinone
de	diastereomeric excess
DIBAL	diisobutylaluminum hydride
diglyme	diethyleneglycol dimethylether
dimeda	N,N'-dimethylethylenediamine
DIPEA	diisopropylethylamine
DMA	dimethylacetamide
DMAD	dimethyl acetylenedicarboxylate
DMAP	4-(*N,N*-dimethylamino)pyridine
DME	1,2-dimethoxyethane
DMF	dimethylformamide
DMP	Dess–Martin periodinane
DMPU	*N,N*'-dimethylpropylene urea
DMSO	dimethyl sulfoxide
dppb	1,4-bis(diphenylphosphino)butane
dppe	1,2-bis(diphenylphosphino)ethane
dppf	1,2-bis(diphenylphosphino)ferrocene
dppp	1,3-bis(diphenylphosphino)propane
dr	diastereomeric ratio
DTBMP	2,6-di-*tert*-butyl-4-methylpyridine
EDDA	ethylenediamine-*N,N*'-diacetic acid
ee	enantiomeric excess
Et	ethyl
FVP	flash-vacuum pyrolysis
h	hour(s)
H-4CR	Hantzsch four-component reaction
HFIP	hexafluoroisopropanol
HIV	human immunodefficiency virus
HLE	human leukocyte elastase
HMG	hydroxymethylglutamate
HMPA	hexamethylphosphoric triamide
HOMO	highest occupied molecular orbital
HTX	histrionicotoxin
HWE	Horner–Wadsworth–Emmons or Horner–Wittig–Emmons
IBX	2-iodoxybenzoic acid

IMCR	isocyanide MCR
LDA	lithium diisopropylamide
LiHMDS	lithium hexamethyldisilazide
LiTMP	lithium 2,2',6,6'-tetramethylpiperidide
LUMO	lowest unoccupied molecular orbital
M-3CR	Mannich-three-component reaction
MCR	multicomponent reaction
Me	methyl
MEM	(2-methoxyethoxy)methyl
MeOH	methanol (methyl alcohol)
min	minute(s)
MOB	masked *o*-benzoquinones or *o*-benzoquinoid structures
MOM	methoxymethyl
MPEG	polyethyleneglycol monomethylether
MPV	Meerwein–Ponndorf–Verley
Ms	mesyl/methanesulfonyl
MS	molecular sieves
Mts	2,4,6-trimethylphenylsulfonyl
NBS	*N*-bromosuccinimide
NCS	*N*-chlorosuccinimide
NMO	*N*-methylmorpholine *N*-oxide
NMP	*N*-methyl-2-pyrrolidinone
NMR	nuclear magnetic resonance
N,N-DEA	*N,N*-Diethylamine
P-3CR	Passerini three-component reaction
PBG	porphobilinogen
PEG	poly(ethylene glycol)
PET	photo-induced electron transfer
PGE	prostaglandin E_1
PIDA	phenyliodine(III) diacetate
PLE	pig liver esterase
PMB	*p*-methoxybenzyl
PMP	*p*-methoxyphenyl
PNA	peptide nucleic acid
Pr	propyl
PrLB	(Pr = Praseodymium; L = lithium; B = BINOL)
PTSA	*p*-toluenesulfonic acid
Py	pyridine
r.t.	room temperature
RAMP	(*R*)-1-amino-2-(methoxymethyl)pyrrolidine
RCM	ring-closing metathesis
RDL	*Rhizopus delemar* lipase
RNR	ribonucleotide reductase
ROM	ring-opening metathesis
S-3CR	Strecker three-component reaction
SAMP	(*S*)-1-amino-2-(methoxymethyl)pyrrolidine

SAWU-3CR	Staudinger reduction/aza-Wittig/Ugi three component reaction
SEM	2-trimethylsilylethoxymethoxy
SET	single-electron transfer
SHOP	Shell Higher Olefin Process
TADDOL	(–)-(4*R*,5*R*)-2,2-dimethyl-α,α,α',α'-tetraphenyl-1,3-dioxolane-4,5-dimethanol
TBABr	tetrabutylammonium bromide
TBACl	tetrabutylammonium chloride
TBAF	tetrabutylammonium fluoride
TBS	*tert*-butyldimethylsilyl
TBSOTf	*tert*-butyldimethylsilyl trifluoromethanesulfonate
TDMPP	tri(2,6-dimethoxyphenyl)phosphine
TEMPO	tetramethylpiperidinyl-1-oxy
TES	triethylsilyl
tetraglyme	tetraethyleneglycol dimethylether
TFA	trifluoroacetic acid
TFAA	trifluoroacetic anhydride
TfOH	trifluoromethanesulfonic acid
THF	tetrahydrofuran
THP	tetrahydropyran-2-yl
TIPS	triisopropylsilyl
TMANO	trimethylamine-*N*-oxide
tmeda	N,N,N'N'-tetramethylethylenediamine
TMOF	trimethyl orthoformate
TMS	trimethylsilyl
TMSCl	trimethylsilyl chloride
TMSI	trimethylsilyl iodide
TMSOTf	trimethylsilyl trifluoromethanesulfonate
TPAP	tetrapropylammonium perruthenate
TPS	*tert*-butyldiphenylsilyl
triglyme	triethyleneglycol dimethylether
Ts	tosyl/*p*-toluenesulfonyl
TTMSS	tris(trimethylsilyl)silane
U-4CR	Ugi four-component reaction
UDC	Ugi/De-Boc/Cyclize strategy
UV	ultraviolet
μSYNTAS	miniaturized-SYNthesis and Total Analysis System

Introduction

During the past fifty years, synthetic organic chemistry has developed in a fascinating way. Whereas in the early days only simple molecules could be prepared, chemists can now synthesize highly complex molecules such as palytoxin [1], brevetoxine A [2] or gambierol [3]. Palytoxin contains 64 stereogenic centers, which means that this compound with its given constitution could, in principle, exist as over 10^{19} stereoisomers. Thus, a prerequisite for the preparation of such a complex substance was the development of stereoselective synthetic methods. The importance of this type of transformation was underlined in 2003 by the awarding of the Nobel Prize to Sharpless, Noyori and Knowles for their studies on catalytic enantioselective oxidation and reduction procedures [4]. Today, a wealth of chemo-, regio-, diastereo- and enantioselective methods is available, which frequently approach the selectivity of enzymatic process with the advantage of a reduced substrate specificity.

The past decade has witnessed a change of paradigm in chemical synthesis. Indeed, the question today is not only what can we prepare – actually there is nearly no limit – but how do we do it?

The main issue now is the efficiency of a synthesis, which can be defined as the increase of complexity per transformation. Notably, modern syntheses must obey the needs of our environment, which includes the preservation of resources and the avoidance of toxic reagents as well as toxic solvents [5]. Such an approach has advantages not only for Nature but also in terms of economics, as it allows reductions to be made in production time as well as in the amounts of waste products.

Until now, the "normal" procedure for the synthesis of organic compounds has been a stepwise formation of individual bonds in the target molecules, with work-up stages after each transformation. In contrast, modern synthesis management must seek procedures that allow the formation of several bonds, whether C–C, C–O or C–N, in one process. In an ideal procedure, the entire transformation should be run without the addition of any further reagents or catalysts, and without changing the reaction conditions. We have defined this type of transformation as a "domino reaction" or "domino process" [6]. Such a process would be the transformation of two or more bond-forming reactions under identical reaction conditions, in which the latter transformations take place at the functionalities obtained in the former-bond forming reactions.

Thus, domino processes are time-resolved transformations, an excellent illustration being that of domino stones, where one stone tips over the next, which tips the

Domino Reactions in Organic Synthesis. Lutz F. Tietze, Gordon Brasche, and Kersten M. Gericke
Copyright © 2006 WILEY-VCH Verlag GmbH & Co. KGaA, Weinheim
ISBN: 3-527-29060-5

next, and the next… such that they all fall down in turn. In the literature, although the word "tandem" is often used to describe this type of process, it is less appropriate as the encyclopedia defines tandem as "locally, two after each other", as on a tandem bicycle or for tandem mass spectrometers. Thus, the term "tandem" does not fit with the time-resolved aspects of the domino reaction type; moreover, if three or even more bonds are formed in one sequence the term "tandem" cannot be used at all.

The time-resolved aspect of domino processes would, however, be in agreement with "cascade reactions" as a third expression used for the discussed transformations. Unfortunately, the term "cascade" is employed in so many different connections – for example, photochemical cascades, biochemical cascades or electronic cascades – on each occasion aiming at a completely different aspect, that it is not appropriate; moreover, it also makes the database search much more difficult! Moreover, if water molecules are examined as they cascade, they are simply moving and do not change. Several additional excellent reviews on domino reactions and related topics have been published [7], to which the reader is referred.

For clarification, individual transformations of independent functionalities in one molecule – also forming several bonds under the same reaction conditions – are not classified as domino reactions. The enantioselective total synthesis of (–)-chlorothricolide **0-4**, as performed by Roush and coworkers [8], is a good example of tandem and domino processes (Scheme 0.1). In the reaction of the acyclic substrate **0-1** in the presence of the chiral dienophile **0-2**, intra- and intermolecular Diels–Alder reactions take place to give **0-3** as the main product. Unfortunately, the two reaction sites are independent from each other and the transformation cannot therefore be classified as a domino process. Nonetheless, it is a beautiful "tandem reaction" that allows the establishment of seven asymmetric centers in a single operation.

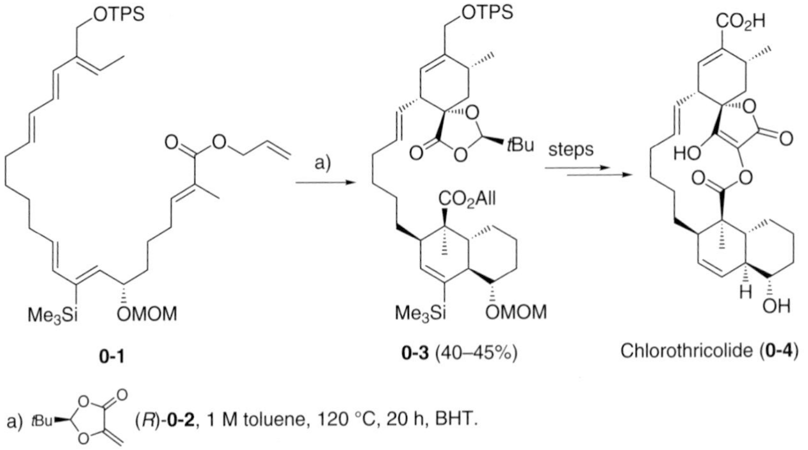

Scheme 0.1. Synthesis of chlorothricolide (**0-4**) using a tandem process.

Domino reactions are not a new invention – indeed, Nature has been using this approach for billions of years! However, in almost of Nature's processes different enzymes are used to catalyze the different steps, one of the most prominent examples being the synthesis of fatty acids using a multi-enzyme complex starting from acetic acid derivatives.

There are, however, also many examples where the domino process is triggered by only one enzyme and the following steps are induced by the first event of activation.

The term "domino process" is correlated to substrates and products without taking into account that the different steps may be catalyzed by diverse catalysts or enzymes, as long as all steps can be performed under the same reaction conditions.

The quality of a domino reaction can be correlated to the number of bond-forming steps, as well as to the increase of complexity and its suitability for a general application. The greater the number of steps – which usually goes hand-in-hand with an increase of complexity of the product, the more useful might be the process.

An example of this type is the highly stereoselective formation of lanosterol (**0-6**) from (*S*)-2,3-oxidosqualene (**0-5**) in Nature, which seems not to follow a concerted mechanism (Scheme 0.2) [9].

Knowledge regarding biosyntheses has induced several biomimetic approaches towards steroids, the first examples being described by van Tamelen [10] and Corey [11]. A more efficient process was developed by Johnson [12] who, to synthesize progesterone **0-10** used an acid-catalyzed polycyclization of the tertiary allylic alcohol **0-7** in the presence of ethylene carbonate, which led to **0-9** via **0-8** (Scheme 0.3). The cyclopentene moiety in **0-9** is then transformed into the cyclohexanone moiety in progesterone (**0-10**).

In the biosynthesis of the pigments of life, uroporphyrinogen III (**0-12**) is formed by cyclotetramerization of the monomer porphobilinogen (**0-11**) (Scheme 0.4). Uroporphyrinogen III (**0-12**) acts as precursor of inter alia heme, chlorophyll, as well as vitamin B_{12} [13].

The domino approach is also used by Nature for the synthesis of several alkaloids, the most prominent example being the biosynthesis of tropinone (**0-16**). In this case, a biomimetic synthesis was developed before the biosynthesis had been disclosed. Shortly after the publication of a more than 20-step synthesis of tropinone by Willstätter [14], Robinson [15] described a domino process (which was later improved by Schöpf [16]) using succinaldehyde (**0-13**), methylamine (**0-14**) and acetonedicarboxylic acid (**0-15**) to give tropinone (**0-16**) in excellent yield without isolating any intermediates (Scheme 0.5).

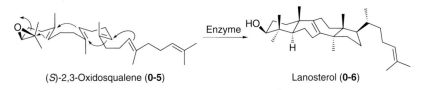

(*S*)-2,3-Oxidosqualene (**0-5**)   Lanosterol (**0-6**)

Scheme 0.2. Biosynthesis of lanosterol (**0-6**).

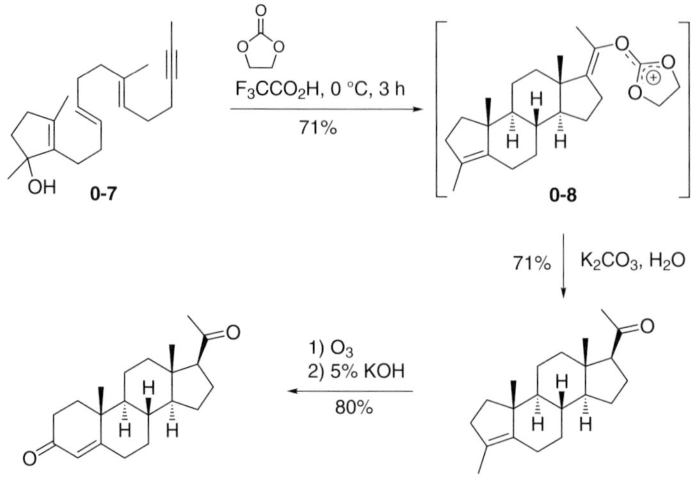

Scheme 0.3. Biomimetic synthesis of progesterone (**0-10**).

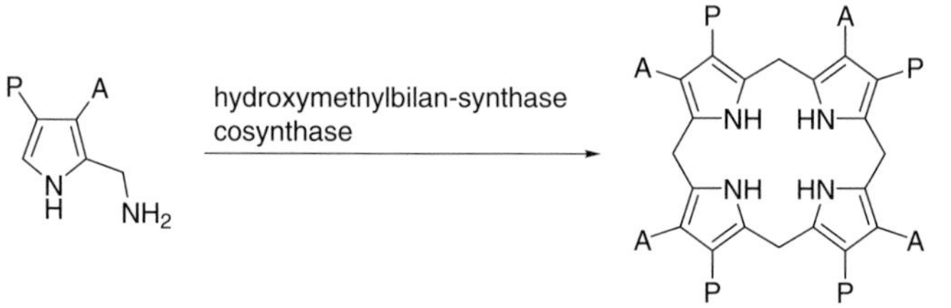

Porphobilinogen (**0-11**)
A = –CH$_2$–CO$_2$H
P = –(CH$_2$)$_2$–CO$_2$H

Uroporphyrinogen III (**0-12**)

Scheme 0.4. Biosynthesis of uroporphyrinogen III (**0-12**).

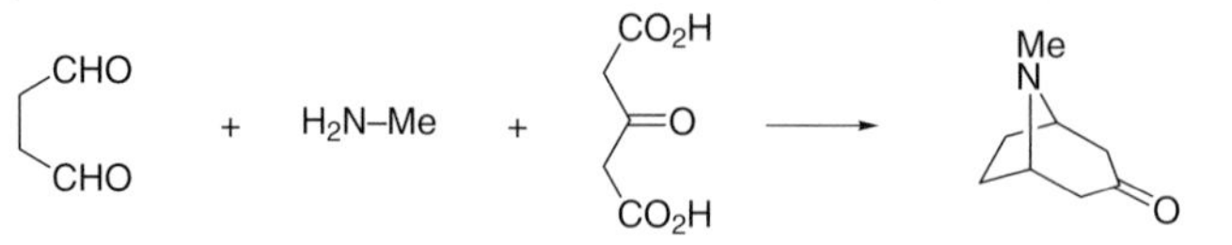

0-13   0-14   0-15   Tropinone (**0-16**)

Scheme 0.5. Domino process for the synthesis of tropinone (**0-16**).

Tropinone is a structural component of several alkaloids, including atropine. The synthesis is based on a double Mannich process with iminium ions as intermediates. The Mannich reaction in itself is a three-component domino process, which is one of the first domino reactions developed by humankind.

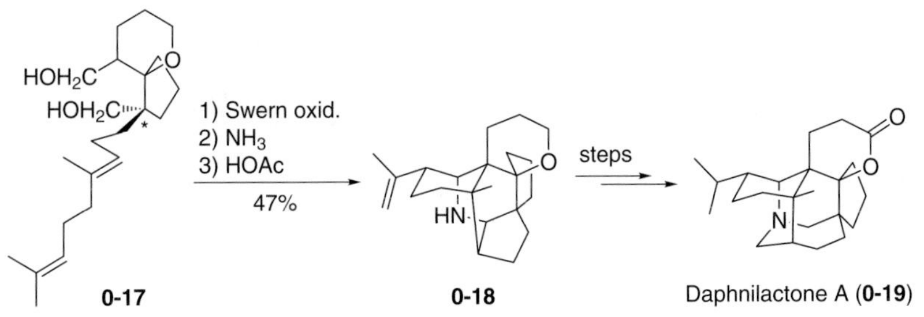

Scheme 0.6. Total synthesis of the daphnilactone A.

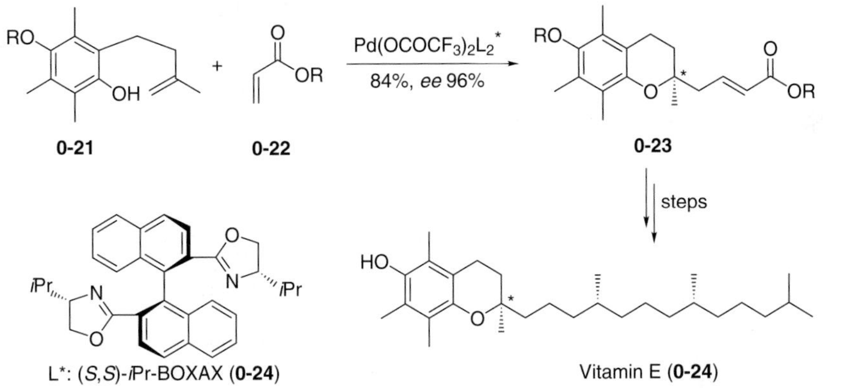

Scheme 0.7. Enantioselective Pd-catalyzed domino reaction for the synthesis of Vitamin E (**0-24**).

Another beautiful example of an early domino process is the formation of daphnilactone A (**0-19**), as described by Heathcock and coworkers [17]. In this process the precursor **0-17** containing two hydroxymethyl groups is oxidized to give the corresponding dialdehyde, which is condensed with methylamine leading to a 2-azabutadiene. There follow a cycloaddition and an ene reaction to give the hexacycle **0-18**, which is transformed into daphnilactone A (**0-19**) (Scheme 0.6).

One of the first enantioselective transition metal-catalyzed domino reactions in natural product synthesis leading to vitamin E (**0-23**) was developed by Tietze and coworkers (Scheme 0.7) [18]. This transformation is based on a Pd^{II}-catalyzed addition of a phenolic hydroxyl group to a C–C-double bond in **0-20** in the presence of the chiral ligand **0-24**, followed by an intermolecular addition of the formed Pd-species to another double bond.

One very important aspect in modern drug discovery is the preparation of so-called "substance libraries" from which pharmaceutical lead structures might be selected for the treatment of different diseases. An efficient approach for the preparation of highly diversified libraries is the development of multicomponent reactions, which can be defined as a subclass of domino reactions. One of the most

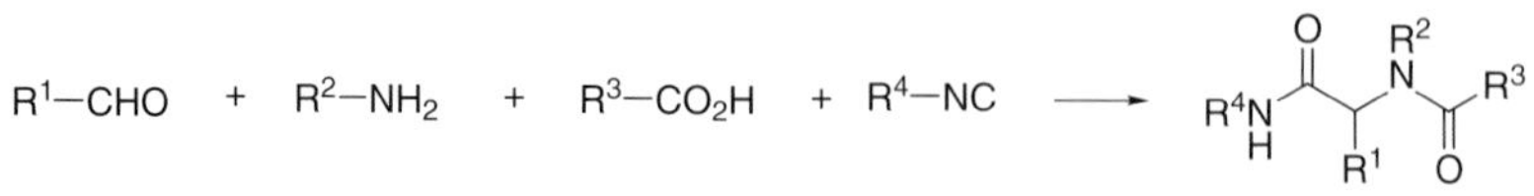

R^1—CHO + R^2—NH$_2$ + R^3—CO$_2$H + R^4—NC ⟶

0-25   0-26   0-27   0-28   0-29

Scheme 0.8. Ugi four-component (U-4CR) approach.

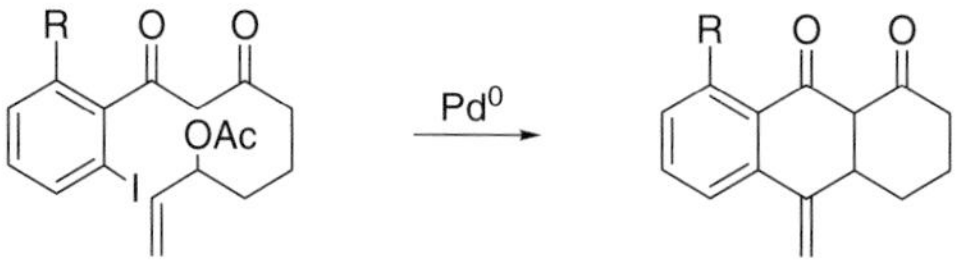

Pd^0

0-30a: R = H
0-30b: R = OMe

0-31a (23%)
0-31b (89%)

Scheme 0.9. Pd-catalyzed domino reaction.

widely used transformations of this type was described by Ugi and coworkers using an aldehyde **0-25**, an amine **0-26**, an acid **0-27**, and an isocyanide **0-28** to prepare peptide-like compounds **0-29** (Scheme 0.8) [7c]. This process could be even enlarged to an eight-component reaction.

As a requisite for all domino reactions, the substrates used must have more than two functionalities of comparable reactivity. They can be situated in one or two molecules or, as in the case of multicomponent domino reactions, in at least three different molecules. For the design and performance of domino reactions it is of paramount importance that the functionalities react in a fixed chronological order to allow the formation of defined molecules.

There are several possibilities to determine the course of the reactions. Thus, one must adjust the reactivity of the functionalities, which usually react under similar reaction conditions. This can be done by steric or electronic differentiation. An illustrative example of the latter approach is the Pd^0-catalyzed domino reaction of **0-30** to give the tricyclic compound **0-31**, as developed by the Tietze group (Scheme 0.9) [19]. In this domino process a competition exists between a Pd-catalyzed nucleophilic allylation (Tsuji–Trost reaction) and an arylation of an alkene (Heck reaction). By slowing down the oxidative addition as part of the latter reaction, through introducing an electronic-donating moiety such as a methoxy group, substrate **0-30b** could be transformed into **0-31b** in 89 % yield, whereas **0-30a** gave **0-31a** in only 23 % yield.

Another possibility here is to use entropic acceleration. In this way, it is possible to use a substrate that first reacts in an intramolecular mode to give an intermediate, which then undergoes an intermolecular reaction with a second molecule. An impressive older example is a radical cyclization/trapping in the synthesis of prostaglandin F$_{2\alpha}$, as described by the Stork group [20]. A key step here is the radical transformation of the iodo compound **0-32** using nBu$_3$SnH formed *in situ* from

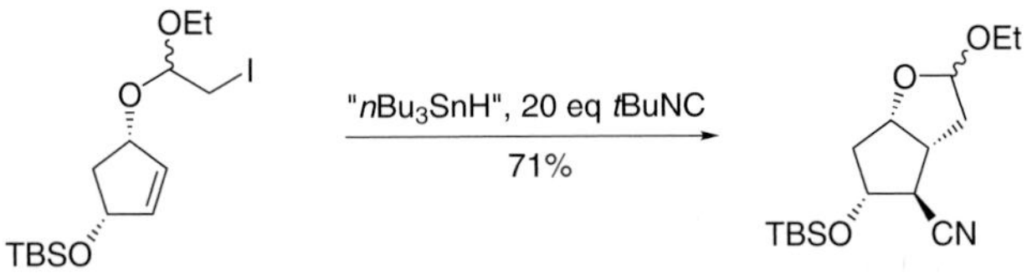

Scheme 0.10. Radical reaction in the synthesis of prostaglandine F$_{2\alpha}$.

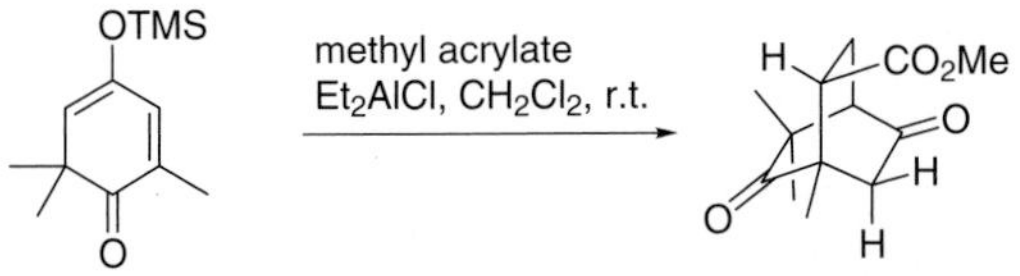

Scheme 0.11. Twofold Michael reaction in the synthesis of valeriananoid A.

nBu$_3$SnCl and NaBH$_3$CN in the presence of tBuNC and AIBN. The final product is the annulated cyano cyclopentane **0-33** (Scheme 0.10).

However, it is also possible to avoid an intramolecular reaction as the first step, for example if the cycle being formed in this transformation would be somehow strained, as observed for the formation of medium rings. In such a case, an intermolecular first takes place, followed by an intramolecular reaction.

On the other hand, many reactions are known where in a first intermolecular step a functionality is introduced which than can undergo an intramolecular reaction. A nice example is the reaction of dienone **0-34** with methyl acrylate in the presence of diethylaluminum chloride to give the bridged compound **0-35** (Scheme 0-11). The first step is an intermolecular Michael addition, which is followed by an intramolecular Michael addition. This domino process is the key step of the total synthesis of valeriananoid A, as described by Hagiwara and coworkers [21].

A different situation exists if the single steps in a domino process follow different mechanisms. Here, it is not normally adjustment of the reaction conditions that is difficult to differentiate between similar transformations; rather, it is to identify conditions that are suitable for both transformations in a time-resolved mode. Thus, when designing new domino reactions a careful adjustment of all factors is very important.

Classification

For the reason of comparison and the development of new domino processes, we have created a classification of these transformations. As an obvious characteristic, we used the mechanism of the different bond-forming steps. In this classification, we differentiate between cationic, anionic, radical, pericyclic, photochemical, transition metal-catalyzed, oxidative or reductive, and enzymatic reactions. For this type

Table 0.1 A classification of domino reactions.

I. Transformation	II. Transformation	III. Transformation
1. Cationic	1. Cationic	1. Cationic
2. Anionic	2. Anionic	2. Anionic
3. Radical	3. Radical	3. Radical
4. Pericyclic	4. Pericyclic	4. Pericyclic
5. Photochemical	5. Photochemical	5. Photochemical
6. Transition metal	6. Transition metal	6. Transition metal
7. Oxidative or reductive	7. Oxidative or reductive	7. Oxidative or reductive
8. Enzymatic	8. Enzymatic	8. Enzymatic

of classification, certain rules must be followed. Nucleophilic substitutions are always counted as anionic processes, independently of whether a carbocation is an intermediate as the second substrate. Moreover, nucleophilic additions to carbonyl groups with metal organic compounds as MeLi, silyl enol ethers or boron enolates are again counted as anionic transformations. In this way, aldol reactions (and also the Mukaiyama reaction) as well as the Michael addition are found in the chapter dealing with anionic domino processes. A related problem exists in the classification of radical and oxidative or reductive transformations, if a single electron transfer is included. Here, a differentiation according to the reagent used is employed. Thus, reactions of bromides with $n\mathrm{Bu_3SnH}$ follow a typical radical pathway, whereas reactions of a carbonyl compound with $\mathrm{SmI_2}$ to form a ketyl radical are listed under oxidative or reductive processes. An overview of the possible combinations of reactions of up to three steps is shown in Table 0.1.

Clearly, the list can be enlarged by introducing additional steps, whereas the steps leading to the reactive species at the beginning (such as the acid-catalyzed elimination of water from an alcohol to form a carbocation) are not counted.

The overwhelming number of examples dealing with domino processes are those where the different steps are from the same category, such as cationic/ cationic or transition metal/transition metal-catalyzed domino processes, which we term "homo domino processes". An example of the former reaction is the synthesis of progesterone (see Scheme 0.3), and for the latter the synthesis of vitamin E (Scheme 0.7).

There are, however, also many examples of "mixed domino processes", such as the synthesis of daphnilactone (see Scheme 0.6), where two anionic processes are followed by two pericyclic reactions. As can be seen from the information in Table 0.1, by counting only two steps we have 64 categories, yet by including a further step the number increases to 512. However, many of these categories are not – or only scarcely – occupied. Therefore, only the first number of the different chapter correlates with our mechanistic classification. The second number only corresponds to a consecutive numbering to avoid empty chapters. Thus, for example in Chapters 4 and 6, which describe pericyclic and transition metal-catalyzed reactions, respectively, the second number corresponds to the frequency of the different processes.

In our opinion, this approach provides not only a clear overview of the existing domino reactions, but also helps to develop new domino reactions and to initiate ingenious independent research projects in this important field of synthetic organic chemistry.

References

1 (a) R. W. Armstrong, J.-M. Beau, S. H. Cheon, W. J. Christ, H. Fujioka, W.-H. Ham, L. D. Hawkins, H. Jin, S. H. Kang, Y. Kishi, M. J. Martinelli, W. W. McWhorter, Jr., M. Mizuno, M. Nakata, A. E. Stutz, F. X. Talamas, M. Taniguchi, J. A. Tino, K. Ueda, J.-i. Uenishi, J. B. White, M. Yonaga, *J. Am. Chem. Soc.* **1989**, *111*, 7525–7530; (b) E. M. Suh, Y. Kishi, *J. Am. Chem. Soc.* **1994**, *116*, 11205–11206.

2 K. C. Nicolaou, Z. Yang, G. Q. Shi, J. L. Gunzner, K. A. Agrios, P. Gärtner, *Nature* **1998**, *392*, 264–269.

3 (a) I. Kadota, H. Takamura, K. Sato, A. Ohno, K. Matsuda, Y. Yamamoto, *J. Am. Chem. Soc.* **2003**, *125*, 46–47; (b) I. Kadota, H. Takamura, K. Sato, A. Ohno, K. Matsuda, M. Satake, Y. Yamamoto, *J. Am. Chem. Soc.* **2003**, *125*, 11893–11899.

4 (a) K. B. Sharpless, *Angew. Chem. Int. Ed.* **2002**, *41*, 2024–2032; (b) R. Noyori, *Angew. Chem. Int. Ed.* **2002**, *41*, 2008–2022; (c) W. S. Knowles, *Angew. Chem. Int. Ed.* **2002**, *41*, 1999–2007.

5 (a) R. A. Sheldon, *C. R. Acad. Sci., Ser. IIc: Chim.* **2000**, *3*, 41–551; (b) R. A. Sheldon, *Chem. Ind. (London, U. K.)* **1997**, 12–15; (c) R. A. Sheldon, *Pure Appl. Chem.* **2000**, *72*, 1233–1246; (d) R. A. Sheldon, *Russ. Chem. J.* **2000**, *44*, 9–20; (e) R. A. Sheldon, *Green Chem.* **2005**, *7*, 267–278.

6 (a) L. F. Tietze, U. Beifuss, *Angew. Chem. Int. Ed. Engl.* **1993**, *32*, 137–170; (b) L. F. Tietze, *Chem. Rev.* **1996**, *96*, 115–136; (c) L. F. Tietze, F. Haunert, Domino Reactions in Organic Synthesis. An Approach to Efficiency, Elegance, Ecological Benefit, Economic Advantage and Preservation of our Resources in Chemical Transformations, in: M. Shibasaki, J. F. Stoddart and F. Vögtle (Eds.), *Stimulating Concepts in Chemistry*, Wiley-VCH, Weinheim, **2000**, pp. 39–64; (d) L. F. Tietze, A. Modi, *Med. Res. Rev.* **2000**, *20*, 304–322; (e) L. F. Tietze, M. E. Lieb, *Curr. Opin. Chem. Biol.*

1998, *2*, 363–371; (f) L. F. Tietze, *Chem. Ind. (London, U. K.)*, **1995**, 453–457; (g) L. F. Tietze, N. Rackelmann, *Pure Appl. Chem.* **2004**, *76*, 1967–1983.

7 (a) H. Pellissier, *Tetrahedron* **2006**, *62*, 1619–1665; (b) A. Dömling, *Chem. Rev.* **2006**, *106*, 17–89; (c) A. Dömling, I. Ugi, *Angew. Chem. Int. Ed.* **2000**, *39*, 3168–3210; (d) J. Zhu, H. Bienaymé, *Multicomponent Reactions*, Wiley, Weinheim, **2005**; (e) G. H. Posner, *Chem. Rev.* **1986**, *86*, 831–844; (f) T.-L. Ho, *Tandem Organic Reactions*, Wiley, New York; **1992**; (g) R. A. Bunce, *Tetrahedron* **1995**, *51*, 13103–13159; (h) P. J. Parsons, C. S. Penkett, A. J. Shell, *Chem. Rev.* **1996**, *96*, 195–206; (i) H. Waldmann, *Nachr. Chem. Tech. Lab.* **1992**, *40*, 1133–1140

8 W. R. Roush, R. J. Sciotti, *J. Am. Chem. Soc.* **1998**, *120*, 7411–7419.

9 K. U. Wendt, G. E. Schulz, E. J. Corey, D. R. Liu, *Angew. Chem. Int. Ed.* **2000**, *39*, 2812–2833.

10 (a) E. E. van Tamelen, J. D. Willet, R. B. Clayton, K. E. Lord, *J. Am. Chem. Soc.* **1966**, *88*, 4752–4754; (b) E. E. van Tamelen, M. A. Schwartz, E. D. Hessler, A. Storni, *Chem. Commun.* **1966**, 409–411; (c) E. E. van Tamelen, *Acc. Chem. Res.* **1975**, *8*, 152–158; (d) E. E. van Tamelen, *J. Am. Chem. Soc.* **1982**, *104*, 6480–6481.

11 (a) E. J. Corey, W. E. Russey, P. R. Ortiz de Montellano, *J. Am. Chem. Soc.* **1966**, *88*, 4750; (b) E. J. Corey, S. C. Virgil, *J. Am. Chem. Soc.* **1991**, *113*, 4025–4026; (c) E. J. Corey, S. C. Virgil, S. Sashar, *J. Am. Chem. Soc.* **1991**, *113*, 8171–8172; (d) E. J. Corey, S. C. Virgil, D. R. Liu, S. Sashar, *J. Am. Chem. Soc.* **1992**, *114*, 1524–1525.

12 W. S. Johnson, *Angew. Chem. Int. Ed. Engl.* **1976**, *15*, 9–17.

13 (a) L. F. Tietze, H. Geissler, *Angew. Chem. Int. Ed. Engl.* **1993**, *32*, 1038–1040; (b) L. F. Tietze, H. Geissler, G. Schulz, *Pure Appl. Chem.* **1994**, *66*, 10–11.

14 (a) R. Willstätter, *Ber. Dtsch. Chem. Ges.*
1901, *34*, 129–144; (b) R. Willstätter, *Ber.
Dtsch. Chem. Ges.* **1901**, *34*, 3163–3165;
(c) R. Willstätter, *Ber. Dtsch. Chem. Ges.*
1896, *29*, 393–403; (d) R. Willstätter, *Ber.
Dtsch. Chem. Ges.* **1896**, 29, 936–947; (e)
R. Willstätter, *Justus Liebigs Ann. Chem.*
1901, *317* 204–265.

15 R. Robinson, *J. Chem. Soc.* **1917**, *111*,
762–768; *J. Chem. Soc.* **1917**, *111*, 876–
899.

16 C. Schöpf, G. Lehmann, W. Arnold,
Angew. Chem. **1937**, *50*, 779–787.

17 (a) C. H. Heathcock, *Angew. Chem. Int.
Ed. Engl.* **1992**, *31*, 665–681; (b) C. H.
Heathcock, J. C. Kath, R. B. Ruggeri, *J.
Org. Chem.* **1995**, *60*, 1120–1130.

18 L. F. Tietze, K. M. Sommer, J. Zinngrebe,
F. Stecker, *Angew. Chem. Int. Ed.* **2005**,
44, 257–259.

19 L. F. Tietze, G. Nordmann, *Eur. J. Org.
Chem.* **2001**, 3247–3253.

20 (a) G. Stork, P. M. Sher, H.-L. Chen, *J.
Am. Chem. Soc.* **1986**, *108*, 6384–6385; (b)
G. Stork, P. M. Sher, *J. Am. Chem. Soc.*
1986, *108*, 303–304.

21 H. Hagiwara, A. Morii, Y. Yamada, T.
Hoshi, T. Suzuki, *Tetrahedron Lett.* **2003**,
44, 1595–1597.

1
Cationic Domino Reactions

In this opening chapter, the class of domino reactions that covers processes in which carbocations are generated in the initial step will be discussed. In this context, it should be noted that it is of no relevance whether the carbocation is of formal or real nature. The formation of a carbocation can easily be achieved by treatment of an alkene or an epoxide with a Brønsted or a Lewis acid, by elimination of water from an alcohol or an alcohol from an acetal, or by reaction of carbonyl compounds and imines with a Brønsted or a Lewis acid. It is worth emphasizing that the reaction of carbonyl compounds and imines with nucleophiles or anionic process (e. g., in the case of an aldol reaction) is sometimes ambiguous. They could also be classified under anionic domino reactions. Thus, the decision between a cationic reaction of carbonyl compounds in the presence of a Brønsted or a Lewis acid will be discussed here, whereas reactions of carbonyl compounds under basic conditions as well as all Michael reactions are described in Chapter 2 as anionic domino processes. It is important to note that all transformations which are affiliated to a cationic initiation must be regarded as cationic processes, and those with an anionic initiation as anionic processes, as an alternation between these two classes would require an as-yet not observed two-electron transfer process. As just discussed for the cationic/anionic process, in examples for a cationic/radical domino process, an electron-transfer again must take place, although in this case it is a single electron transfer. Examples of these processes have been described, but the transfer of an electron is a synonym for a reduction process, and we shall discuss these transformations in Section 1.3, which deals with cationic/reductive domino processes. Furthermore, to date no examples have been cited in the literature for a combination of cationic reactions with photochemically induced, transition metal-catalyzed or enzymatic processes. Nevertheless, carbocations are feasible to act in an electrophilic process in either an inter- or intramolecular manner with a multitude of different nucleophiles, generating a new bond with the concomitant creation of a new functionality which could undergo further transformation (Scheme 1.1).

In most of the hitherto known cationic domino processes another cationic process follows, representing the category of the so-called homo-domino reactions. In the last step, the final carbocation is stabilized either by the elimination of a proton or by the addition of another nucleophile, furnishing the desired product. Nonetheless, a few intriguing examples have been revealed in which a succession

Domino Reactions in Organic Synthesis. Lutz F. Tietze, Gordon Brasche, and Kersten M. Gericke
Copyright © 2006 WILEY-VCH Verlag GmbH & Co. KGaA, Weinheim
ISBN: 3-527-29060-5

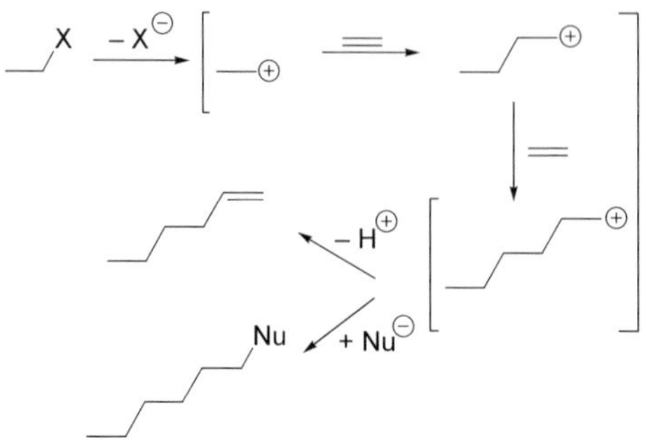

Scheme 1.1. General scheme of a cationic-cationic domino process.

of cationic (by a pericyclic step) or a reduction is also possible, these being catego-
rized as hetero-domino reactions. Furthermore, rearrangements, which traverse
several cationic species, are also quite common and of special synthetic interest.
Following this brief introduction, we enter directly into the field of cationic domino
reactions, starting with the presentation of cationic/cationic processes.

1.1
Cationic/Cationic Processes

The termination of cationic cyclizations by the use of pinacol rearrangements has
shown to be a powerful tool for developing stereoselective ring-forming domino reactions. During the past few years, the Overman group has invested much effort in the
design of fascinating domino Prins cyclization/pinacol rearrangement sequences
for the synthesis of carbocyclic and heterocyclic compounds, especially with regard
to target-directed assembly of natural products [1]. For example, the Prins/pinacol
process permits an easy and efficient access to oxacyclic ring systems, often occurring in compounds of natural origin such as the *Laurencia* sesquiterpenes (±)-*trans*-
kumausyne (1-1) [2] and (±)-kumausallene (1-2) [3] (Scheme 1.2). For the total synthesis of these compounds, racemic cyclopentane diol *rac*-1-3 and the aldehyde 1-4 were
treated under acidic conditions to give the oxocarbenium ion 1-5. Once formed, this
subsequently underwent a Prins cyclization affording the carbocationic intermediate 1-6 by passing through a chairlike, six-membered transition state. Further interception of carbocation 1-6 by pinacol rearrangement furnished racemic *cis*-hy-
drobenzofuranone *rac*-1-7 as the main building block of the natural products 1-1 and
1-2 in 69 % and 71 % yield, respectively.

The Prins/pinacol approach to ring formations is not limited to the assembly of
oxacyclic ring systems; indeed, carbocyclic rings can also be easily prepared [4, 5]. A
nice variant of this strategy envisages the Lewis acid-induced ring-expanding cyclopentane annulation of the 1-alkenylcycloalkanyl silyl ether 1-8 (Scheme 1.3) [1d].
Under the reaction conditions, the oxenium ion 1-9 produced performed a 6-*endo*
Prins cyclization with the tethered alkene moiety, giving cyclic carbocation 1-10.
Gratifyingly, the latter directly underwent a pinacol rearrangement resulting in the

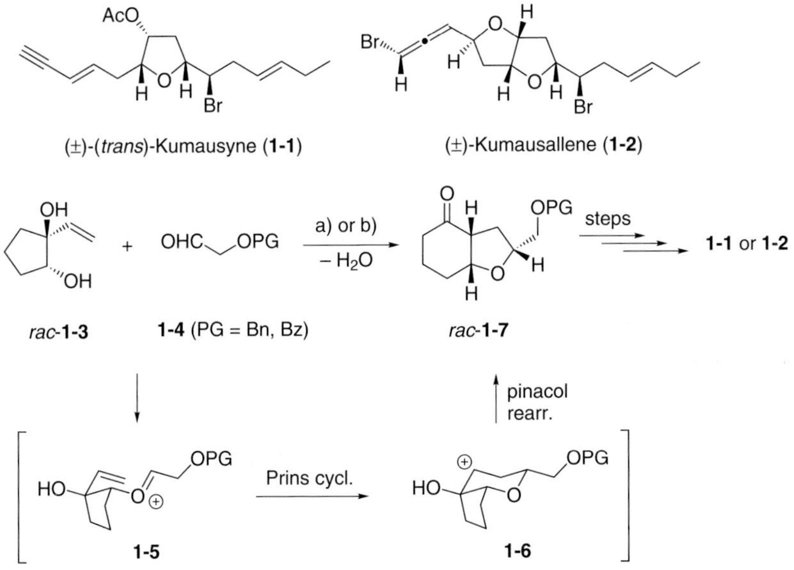

(±)-(*trans*)-Kumausyne (**1-1**)   (±)-Kumausallene (**1-2**)

a) or b) − H₂O

rac-**1-3**   **1-4** (PG = Bn, Bz)   *rac*-**1-7**

steps → **1-1** or **1-2**

pinacol rearr.

Prins cycl.

1-5   **1-6**

a) *p*TsOH, MgSO₄, CH₂Cl₂, 0 °C → r.t., 69%, PG = Bn.
b) BF₃•OEt₂, CH₂Cl₂, −23 °C, 71%, PG = Bz.

Scheme 1.2. Synthesis of annulated furans for an access to the terpenes kumausyne and kumausallene.

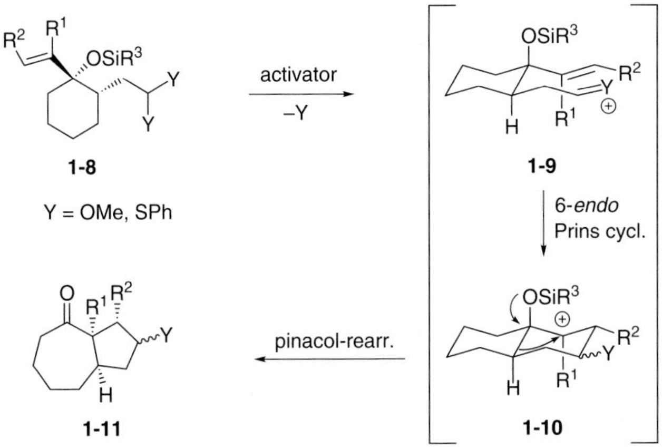

1-8   **1-9**

Y = OMe, SPh

activator −Y

6-*endo* Prins cycl.

pinacol-rearr.

1-11   **1-10**

Scheme 1.3. Domino Prins/pinacol rearrangement process.

formation of cycloalkanone **1-11**, which correlates to a one-carbon expansion of the substrate **1-8**.

This process allowed, for example, formation of the angular fused tricycle **1-13** containing a five-, six-, and eight-membered ring from precursor **1-12** in 64% yield (Scheme 1.4) [1d].

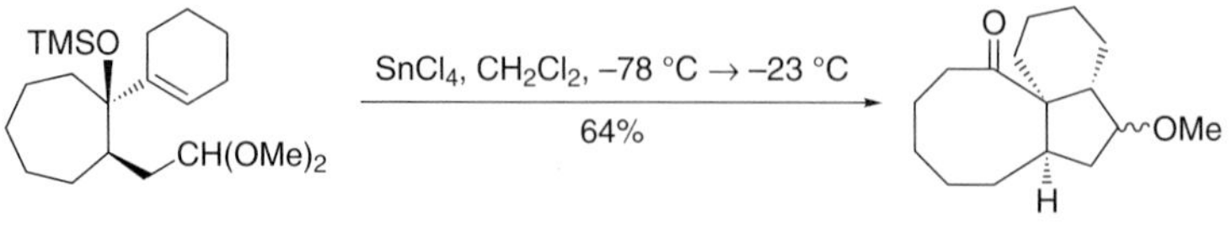

Scheme 1.4. Synthesis of annulated tricyclic compounds.

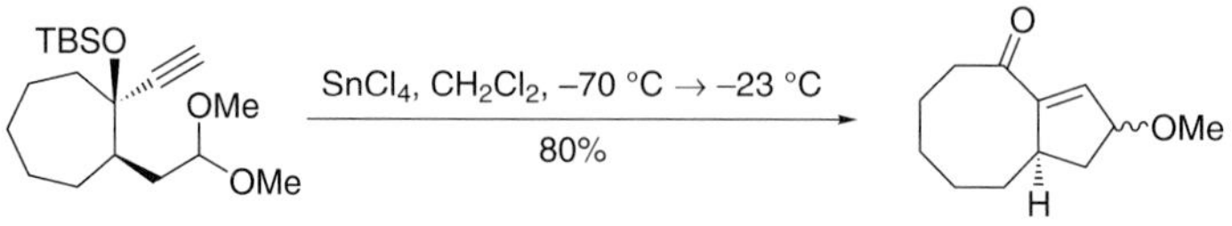

Scheme 1.5. Ring-enlarging cyclopentene annulation.

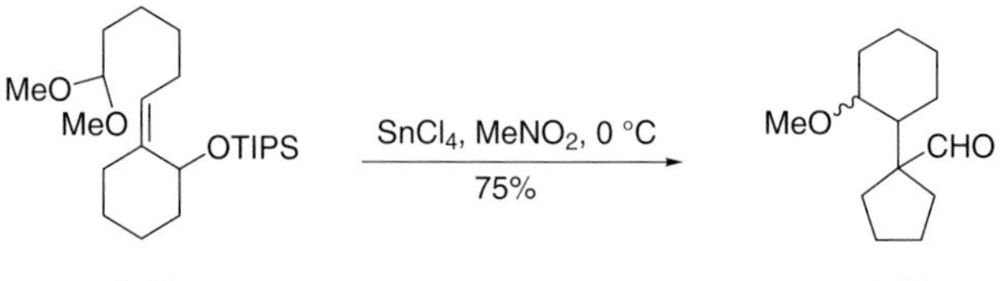

Scheme 1.6. Synthesis of cyclopentylcyclohexanes.

In a similar manner, terminal alkynes such as **1-14** participate in a Prins/pinacol reaction, resulting in a ring-expanding cyclopentene annulation to give compounds such as **1-15** in high yield (Scheme 1.5) [5].

The Prins cyclization can also be coupled with a ring-contraction pinacol rearrangement, as illustrated in Scheme 1.6. This allows a smooth conversion of alkylidene-cyclohexane acetal **1-16** to single bond-joined cyclohexane cyclopentane aldehyde **1-17** [1e].

It should be mentioned at this point that the strategy for ring construction is not restricted to being initiated by a Prins cyclization. The first step can also be triggered by preparing allylcarbenium ions from allylic alcohols. One virtue of using this initiator for cationic cyclization is the possibility of installing functionalities in the cyclopentane ring that can be employed readily to elaborate the carbocyclic products. Thus, treatment of precursor **1-18** with triflic anhydride led to a cyclization-rearrangement with concurrent protodesilylation, delivering hydroazulenone **1-19** in formidable 80% yield (Scheme 1.7) [6].

Finally, a carbocyclic ring formation initiated by a keteniminium cyclization is depicted in Scheme 1.8 [6]. In the presence of triflic anhydride and DTBMP, pyrrolidine amide **1-20** was converted into the keteniminium ion **1-22**, traversing inter-

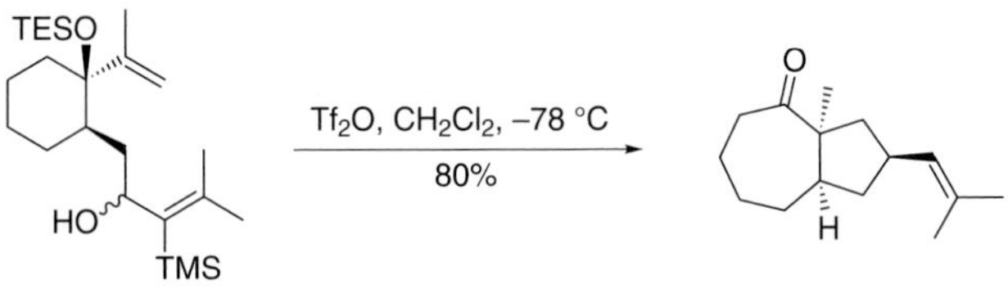

Scheme 1.7. Allyl cation-initiated cyclization/rearrangement.

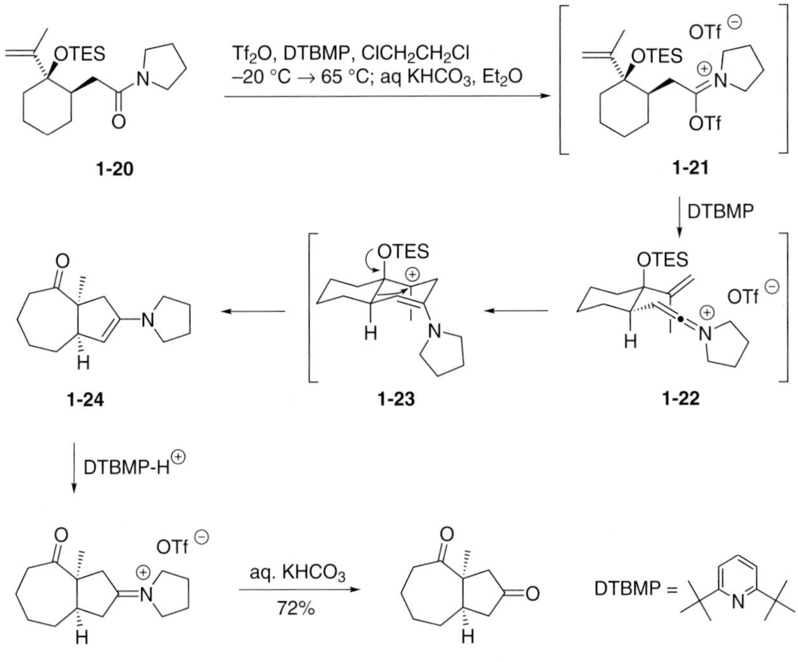

Scheme 1.8. Domino cyclization/rearrangements *via* iminium ions.

mediate **1-21**. Subsequently, cyclization through a chairlike transition state provided carbocation **1-23**, which was directly converted by way of a pinacol rearrangement to give enamine **1-24**. Due to the presence of a pyridinium triflate salt in the
reaction mixture the latter formed an iminium salt **1-25**. Ultimately, hydrolysis by
treatment with an aqueous base accomplished formation of the diketone **1-26** in
72 % yield.

In the following sections, we detail another functionality which is of major value
in the area of carbocationic domino processes, namely the epoxides. On the basis of
their high tendency to be opened in the presence of Lewis or Brønsted acids,
thereby furnishing carbocationic species, several challenging domino procedures
have been elaborated relatively recently.

An epoxide-triggered reaction involving two of these functionalities was developed by Koert and coworkers in 1994. Since 2,5-linked oligo(tetrahydrofurans) (THFs) are the key components of the *Annonaceous* acetogenins, a pharmaceutically promising class of natural products [7], their stereoselective synthesis is of great interest [8]. For this purpose, epoxy alcohols offer a perfect requisite to conduct a straightforward and convenient domino procedure furnishing oligo-THFs as products [9]. Treatment of monoepoxy-olefin **1-27** with *m*CPBA provided the diastereomeric oligo-THF precursors **1-28** and **1-29** as a 1:1 mixture (Scheme 1.9). The domino-epoxide cyclization was then initiated by addition of *p*-toluenesulfonic acid, leading to THF-trimers **1-30** and **1-31** (45 % each); these were subsequently separable using chromatography on silica gel.

Using this methodology, cyclic ethers of different ring-size can also be constructed. Due to the fact that many natural products of marine origin include fused polycyclic ether structural units, a convenient entry to this structural element is of continuing challenge.

In the following approach, the group of McDonald provided the first examples for the application of biomimetic regio- and stereoselective domino oxacyclizations of 1,5-diepoxides to yield oxepanes, as well as of 1,5,9-triepoxides to afford *trans*-fused bisoxepanes [10]. These authors observed that subjection of 1,5-diepoxides **1-32**, **1-34**, **1-36**, and **1-38** to BF₃·Et₂O at low temperatures, followed by acetylation, furnished the oxacycles **1-33**, **1-35**, **1-37**, and **1-39** in good yields (Scheme 1.10). Whereas the reaction of **1-32** and **1-34** are normal transformations, the carbonates **1-36** and **1-38** undergo a domino process leading to fused and spiro products, re-

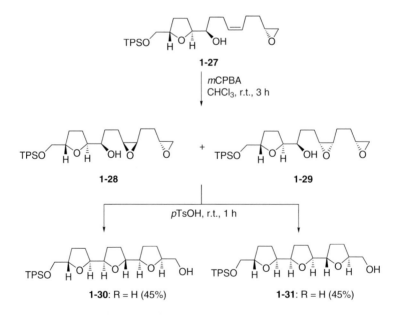

Scheme 1.9. Synthesis of poly-furans.

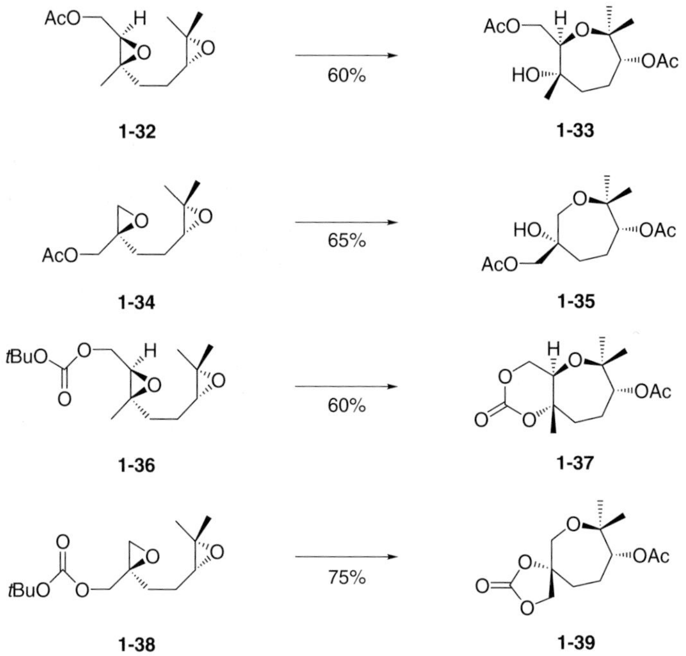

1-32   60%   1-33

1-34   65%   1-35

1-36   60%   1-37

1-38   75%   1-39

Reaction conditions: 1 eq BF$_3$•OEt$_2$, CH$_2$Cl$_2$, −40 °C, 10–30 min, H$_2$O quench; then Ac$_2$O, Et$_2$N, CH$_2$Cl$_2$.

Scheme 1.10. Synthesis of oxepanes.

spectively. Likewise, 1,5,9-triepoxides can be used as substrates, which react in a triple domino process; these reactions are described in Section 1.1.1.

In working towards the synthesis of nonracemic 3-deoxyschweinfurthin B (**1-42**), an analogue of the biological active schweinfurthin B, Wiemer and coworkers developed an acid-catalyzed cationic domino reaction to afford the tricyclic diol (*R*,*R*,*R*)-**1-41** from **1-40** in moderate yield (Scheme 1.11) [11].

The stereochemical outcome of the reaction of **1-43**, formed from **1-40** by desilylation, can be explained by assuming a pseudoequatorial orientation of the epoxide moiety in a pseudo-chair-chairlike transition state **1-44** which, after being attacked by the phenolic oxygen, furnishes the correct *trans*-fused stereoisomer **1-41** (Scheme 1.12). The conformation **1-45**, which would lead to **1-46** seems to be disfavored.

The tetrahydropyran moiety, another oxacycle, is also found in many biologically active natural products from marine and terrestrial origin. Consequently, an easy access to stereodefined tetrahydropyrans by inventive and reliable strategies has been shown to be an important issue [12].

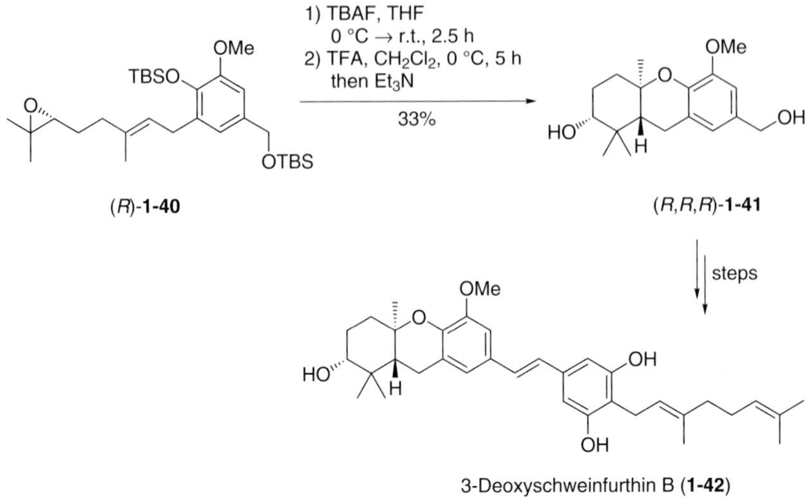

Scheme 1.11. Cationic domino poly-cyclization for the synthesis of 3-deoxyschweinfurthin B (**1-42**).

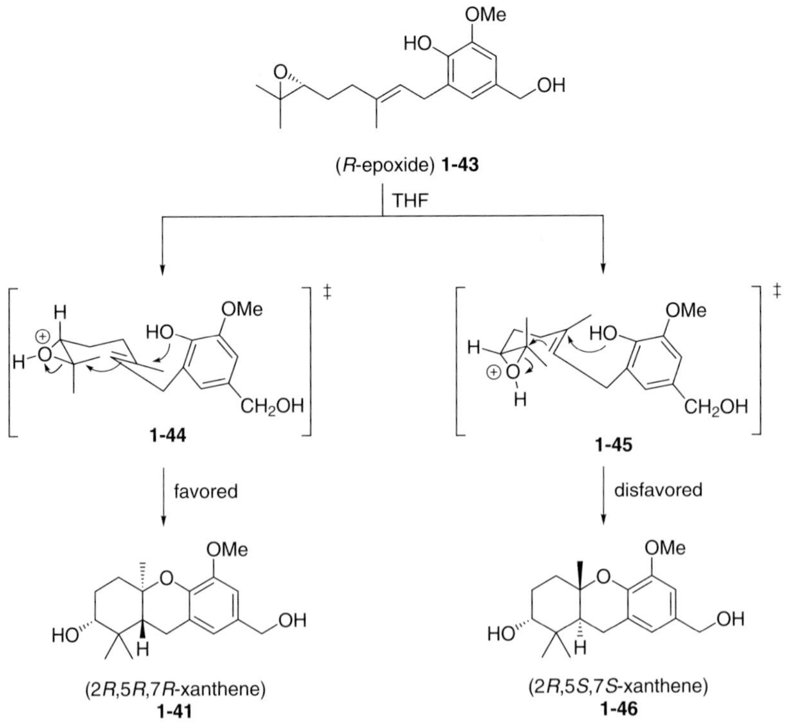

Scheme 1.12. Proposed mechanism of epoxide cyclization.

For their approach to the synthesis of pseudomonic acid C analogues, Markó and coworkers used a Sakurai reaction followed by formation of the tetrahydropyran moiety, starting from the allylsilane **1-47** and the acetal **1-48** to give **1-49** in 80% yield (Scheme 1.13) [13]. Pseudomonic acid C shows interesting activity against Gram-positive bacteria, as well as a high potency towards multiresistant *Staphylococcus aureus*. It is of interest that the reaction of **1-47** and **1-48** only occurs in propionitrile as solvent, and not in CH_2Cl_2. In contrast, the (Z)-isomer of **1-48** reacts smoothly in CH_2Cl_2.

Recently, a new multicomponent condensation strategy for the stereocontrolled synthesis of polysubstituted tetrahydropyran derivatives was re-published by the Markó group, employing an ene reaction combined with an intramolecular Sakurai cyclization (IMSC) (Scheme 1.14) [14]. The initial step is an Et_2AlCl-promoted ene reaction between allylsilane **1-50** and an aldehyde to afford the (Z)-homoallylic alcohol **1-51**, with good control of the geometry of the double bond. Subsequent Lewis acid-mediated condensation of **1-51** with another equivalent of an aldehyde provided the polysubstituted *exo*-methylene tetrahydropyran **1-53** stereoselectively and

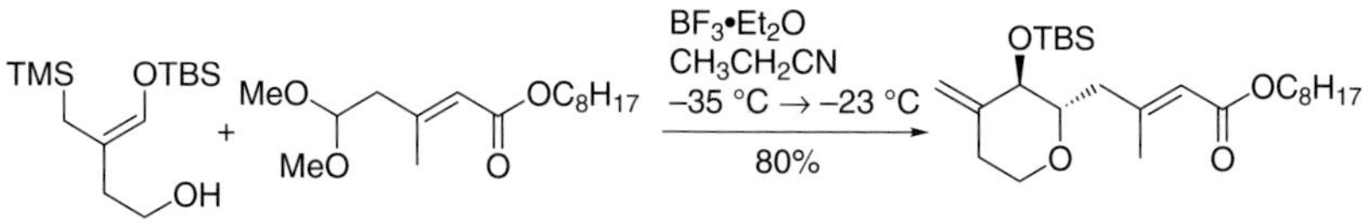

Scheme 1.13. Cationic domino process in the synthesis of pseudomonic acid C analogue.

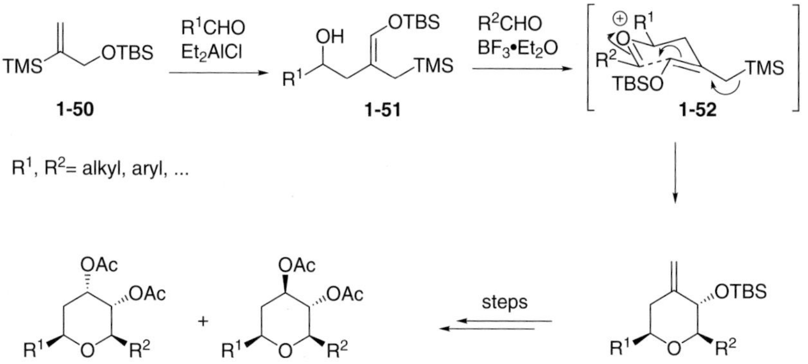

Scheme 1.14. Domino ene/Sakurai reaction for the synthesis of polysubstituted tetrahydropyrans.

in good yield. This IMSC reaction is thought to proceed *via* the formation of the oxenium cation **1-52** that undergoes an intramolecular addition of the allylsilane moiety through a chairlike transition state. Ozonolysis of the exocyclic double bond in **1-53** and subsequent stereoselective reduction of the formed carbonyl moiety allows the synthesis of the two diastereomeric diacetoxy-tetrahydropyrans **1-54** and **1-55**.

The use of (*E*)-enolcarbamates of type **1-56** allowed the generation of tetrahydropyrans **1-58** with complementary orientation of the carbamate functionality (Scheme 1.15). In all cases, the carbamate group adopts an axial orientation in the chairlike transition state **1-57**.

Loh and coworkers used a combination of a carbonyl-ene and an oxenium-ene reaction for the synthesis of annulated tetrahydropyrans **1-61**, using methylenecyclohexane **1-60** as substrates (Scheme 1.16) [15]. The most appropriate catalyst for this reaction with the aldehydes **1-59** turned out to be In(OTf)$_3$, which furnished the desired products in good to excellent yields and high stereoselectivity [16].

Another interesting cationic domino process is the acid-induced ring opening of α-cyclopropyl ketones and subsequent endocyclic trapping of the formed carboca-

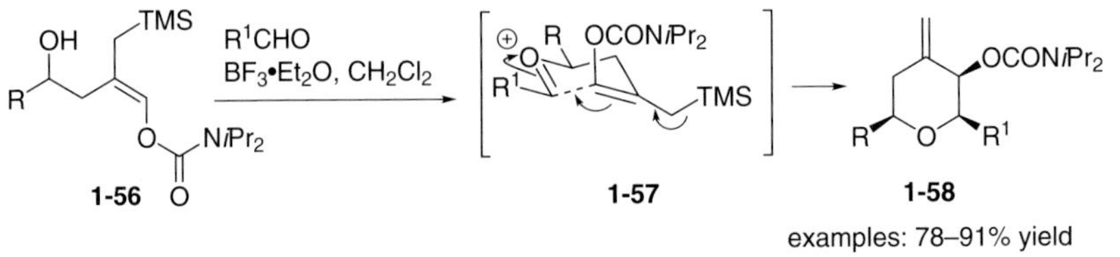

Scheme 1.15. Synthesis of tetrahydropyrans.

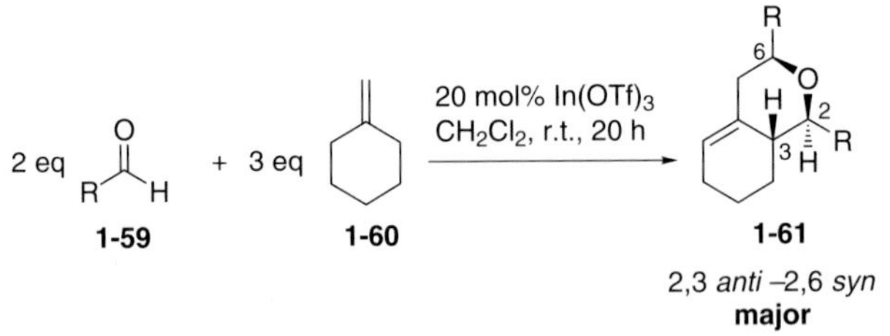

Scheme 1.16. Indium(III)-catalyzed domino carbonyl-ene reaction

Entry	Aldehydes (1-59)		R	Product (1-61)	Major:Minor[a]	Yield [%]
1	a		Ph	a	93:7	80
2	b		PhCH$_2$CH$_2$	b	94:6	91
3	c		CH$_3$(CH$_2$)$_4$	c	99:1	84
4	d		(CH$_3$)$_2$CH	d	95:5	75
5	e		CH$_3$(CH$_2$)$_7$	e	98:2	81

[a] The relative stereochemistry of the minor product was not determined.

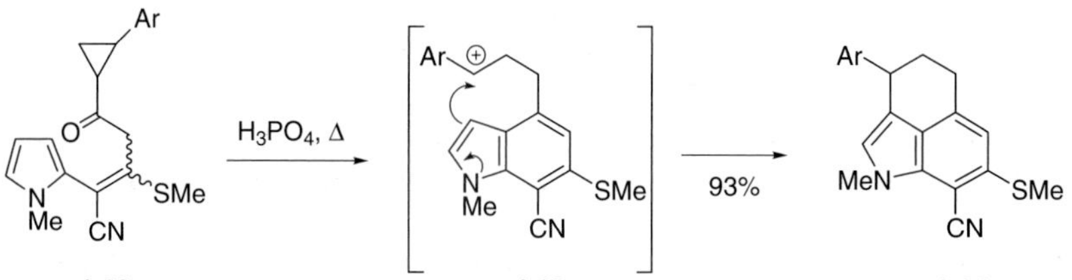

Scheme 1.17. Synthesis of tricyclic indoles.

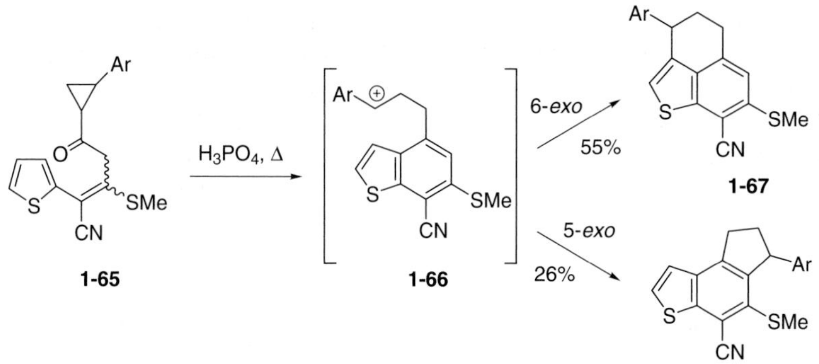

Scheme 1.18. Cationic domino rearrangement for the synthesis of tricyclic thiophenes.

tion by a double bond or an aryl group (Scheme 1.17) [17]. Ila and coworkers used this process to synthesize benzo-fused tricyclic arenes and heteroarenes [18.] For example, when cyclopropyl ketone **1-62** was heated in H_3PO_4 the tricyclic indole derivative **1-64** was obtained in 93% yield *via* the intermediate carbocation **1-63**. **1-64** contains a partial structural framework, which is found in several naturally occurring alkaloids.

Similarly, the thiophene-substituted cyclopropyl ketone **1-65** led to the fused benzothiophenes **1-67** and **1-68** as a 2:1 mixture. In this process the intramolecular interception of the cationic intermediate **1-66** took place at C-3 and C-5 of the benzothiophene moiety (Scheme 1.18).

Moreover, the authors were successful in extending this approach to a threefold domino procedure (for a discussion of this, see Section 1.1.1).

The ambivalent aptitude of sulfur [19] to stabilize adjacent anionic as well as cationic centers is a remarkable fact that has shown to be a reliable feature for the assembly of four-membered ring scaffolds utilizing cyclopropyl phenyl sulfides [20]. Witulski and coworkers treated the sulfide **1-69** with TsOH in wet benzene (Scheme 1.19) [21]. However, in addition to the expected cyclobutanone derivative **1-70**, the bicyclo[3.2.0]heptane **1-70** was also obtained as a single diastereoisomer, but in moderate yield. Much better yields of **1-71** were obtained using ketone **1-72**

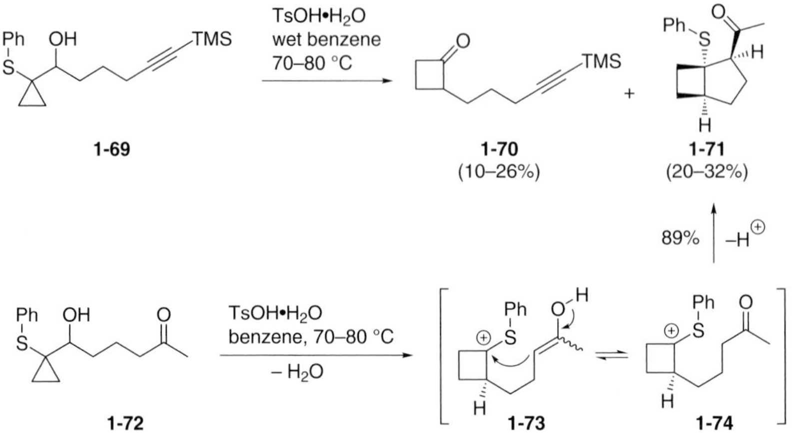

Scheme 1.19. Cationic domino ring-enlargement/annulation process for the synthesis of bicyclo[3.2.0]heptanes.

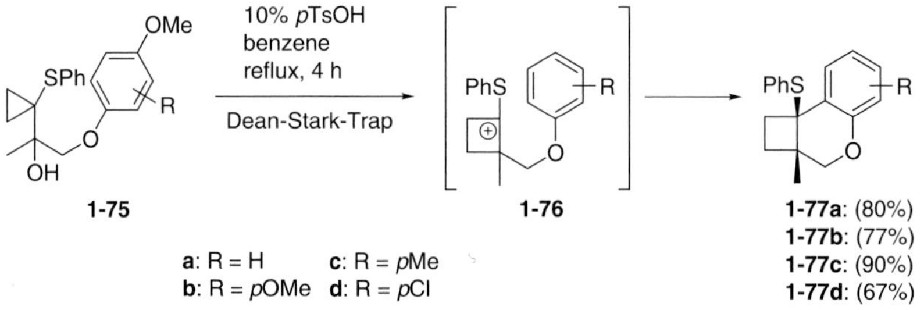

a: R = H   c: R = pMe
b: R = pOMe   d: R = pCl

Scheme 1.20. Cationic domino ring-enlargement/annulation process for the synthesis of chromanes.

as substrate, thereby developing a new domino reaction. The reaction was considered to be initiated by cationic ring enlargement, driven by the ability of the sulfur atom to stabilize the cationic center in the intermediates **1-73** and **1-74**, which undergo an intramolecular cyclization to give **1-71**. Due to entropic effects the cyclization occurs more rapidly than the competitive hydrolysis of the α-thiocarbocations.

A more recent approach, which also profits from the synthetic versatility of stabilized thionium ions, has been elaborated by Berard and Piras [22]. These authors observed that the cyclobutane thionium ions **1-76** obtained from the cyclopropyl phenyl sulfides **1-75** by treatment with pTsOH under anhydrous conditions can be trapped by an adjacent electron-rich aromatic ring to give the chromane derivatives **1-77** in good to excellent yields (Scheme 1.20). As expected, **1-77** were obtained as single diastereoisomers with a *cis*-orientation of the methyl and the phenylthio group as a consequence of steric constraints.

Furthermore, the authors showed that compounds of type **1-78**, easily accessible from **1-77**, can be used as starting materials for the production of a new cyclopropa[c]chromane framework **1-80** (Scheme 1.21). Oxidation of **1-77** to the corresponding sulfoxide and subsequent pyrolytic elimination generated the labile cyclobutene **1-78**, which was directly epoxidized leading to the desired tricyclic compound **1-80** in 46 % yield, probably *via* the epoxycyclobutane **1-79**, again in a domino fashion.

Cationic olefin polycyclizations represent important transformations in the field of domino-type reactions.

Such a polycyclization can be initiated by a Nazarov reaction, as described by West and coworkers [23]. The authors first observed that, in the presence of $BF_3 \cdot Et_2O$, aryltrienone **1-81** is transformed into the *exo*-methylene hydrindanone **1-83** in good yield (Scheme 1.22). As an intermediate an oxallylic cation can be assumed, which is trapped by a 6-*endo* cyclization to give the cationic intermediate **1-82**; this is stabilized by elimination of a proton. Using the stronger Lewis acid $TiCl_4$ at lower temperatures, it was possible to add a further C–C-bond forming step by reaction of the intermediate cation with an aromatic ring system to produce a tetracyclic framework. This threefold domino process is described in Section 1.1.1.

Using this procedure, however, the authors have also prepared tricyclic compounds such as **1-87** in high yield starting from 1,4-diene-3-ones **1-84** containing an electron-rich arylethyl side chain. Probable intermediates are **1-85** and **1-86** (Scheme 1.23) [24].

One of the most interesting structures of the past decade has been Taxol® [25]. This compound is important in the treatment of cancer, and has stimulated many

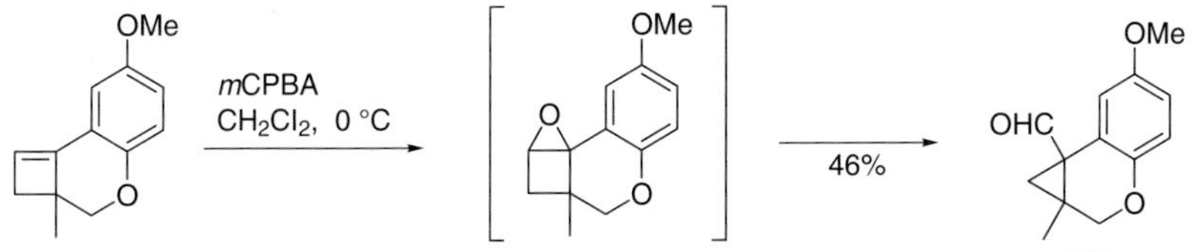

Scheme 1.21. Epoxidation/ring contraction domino process.

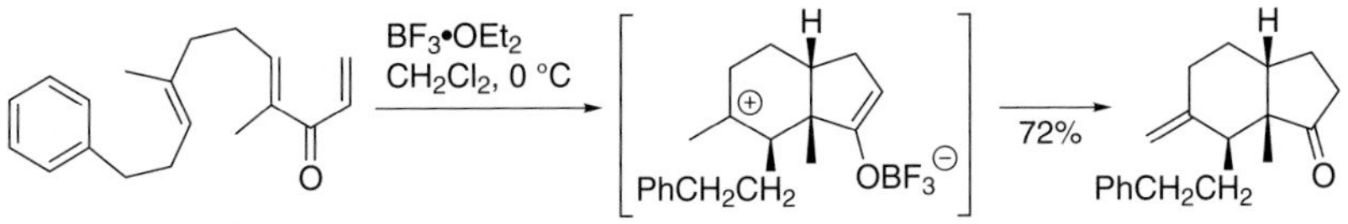

Scheme 1.22. Synthesis of indanone derivatives.

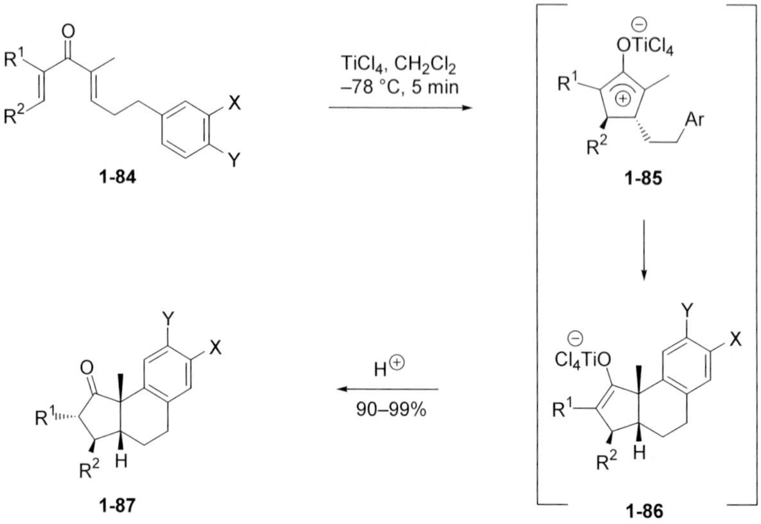

Scheme 1.23. Polycyclization initiated by a Nazarov reaction.

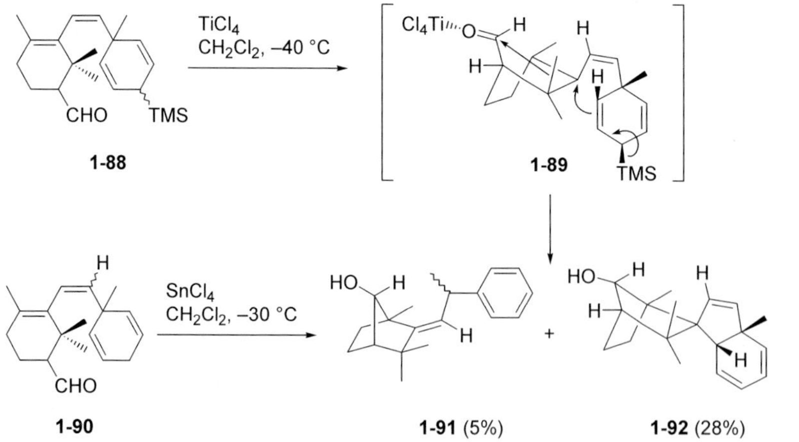

a: X = OMe, Y = H; R^1= Me, R^2= H
b: X = OMe, Y = H; R^1= Et, R^2= Me
c: X = OMe, Y = H; R^1= R^2= (CH$_2$)$_4$
d: X = Y = OCH$_2$O; R^1= Me, R^2= H
e: X = Y = OCH$_2$O; R^1= Et, R^2= Me
f: X = Y = OCH$_2$O; R^1= R^2= (CH$_2$)$_4$

Scheme 1.24. Lewis acid-catalyzed cationic domino cyclization to give bicyclo[2.2.1]heptane derivatives.

investigators to determine its total synthesis [26]. In working towards this goal, Fukumoto and Ihara [27] treated the allylsilane **1-88** (1:1 mixture of diastereomers) with TiCl$_4$. The reaction did not lead to the desired eight-membered ring, but rather yielded the bicyclo[2.2.1]heptane **1-92** through spirocyclization in a sterically encumbered environment. In addition to **1-92**, being formed in 28 % yield as a sole diastereoisomer, 5 % of the aromatized compound **1-91** as a mixture of two isomers was obtained (Scheme 1.24). With reference to the assumed transition state **1-89**, the authors argued that only one of the diastereoisomers of **1-88**, in which the silyl substituent is *trans*-oriented to the angular methyl group, is able to undergo this new cationic domino process, whereas the other isomer undergoes decomposition on treatment with the Lewis acid.

On further exploration it could be shown that the desilylated precursors **1-90** (Scheme 1.24) permit formation of the aromatized products **1-91** and its double bond isomer in up to 90 % yield, starting from the *E*-compound as a mixture of two diastereomers; with a (*Z*)-configuration of the double bond, **1-90** gave 50 % yield of **1-91**.

Cycloheptane-containing natural products occur widely in nature, and therefore the synthesis of such a carbocycle has captured the attention of many organic chemists. The Green group reported on the preparation of cycloheptyne hexacarbonyldicobalt complexes, which can be regarded as useful synthons for further transformations such as substitution or cycloaddition [28]. Treatment of the 1,4-diethoxy-alkyne-Co$_2$(CO)$_6$ complex **1-93** with allyltrimethylsilane in the presence of the Lewis acid BF$_3$·Et$_2$O allowed conversion into the cationic cycloheptyne complexes **1-96** *via* the intermediates **1-94** and **1-95**. This transformation represents a formal [4+3]-cycloaddition (Scheme 1.25). The cationic species **1-96** is trapped by fluoride, furnishing fluorocycloheptyne complexes **1-97** in good yields (67–75 %). The synthetic potential of this procedure can be expanded by utilizing different Lewis acids such as SnCl$_4$ and SnBr$_4$, leading to chloro- and bromo-derivatives in

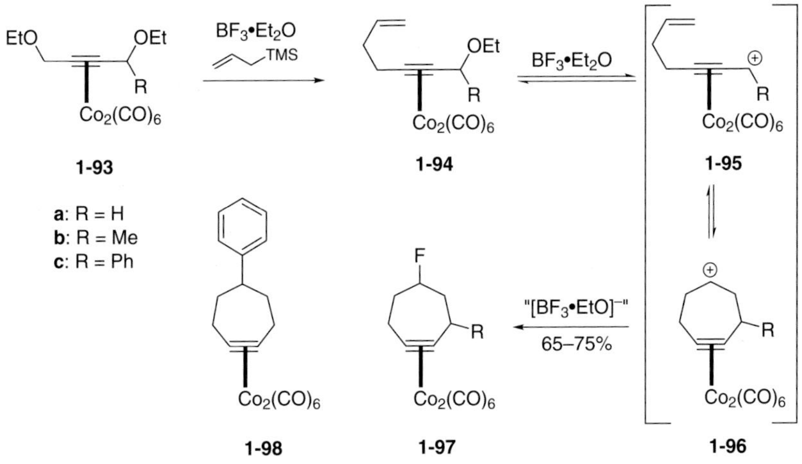

Scheme 1.25. Cationic [4+3]-cycloaddition/nucleophilic trapping domino reaction in the synthesis of halocycloheptynes.

good (SnCl$_4$) to rather low (SnBr$_4$) yields. Another feature of this process was revealed when benzene as solvent and B(C$_6$F$_5$)$_3$ as Lewis acid were used, since the formed cationic cycloheptyne intermediate **1-96** underwent a Friedel–Craft alkylation to provide **1-98** in 70% yield.

A general, efficient and diastereoselective approach to the marasmane scaffold (**1-104**) and, moreover, to the naturally occurring (+)-isovelleral (**1-103**), has been elaborated by Wijnberg and de Groot utilizing a MgI$_2$-catalyzed domino rearrangement/cyclopropanation reaction [29]. In this elegant study, the addition of MgI$_2$ to mesylate **1-99** resulted in an E-1-type elimination of the mesylate to give the carbocation **1-100** after a rearrangement (Scheme 1.27). As expected, this cation is then attacked in an intramolecular fashion by the TMS-enol ether, building up the cation **1-100** with a cyclopropane unit in **1-101**. Subsequent desilylation provided the ketone **1-102** in 82% yield, which was converted into (+)-isovelleral (**1-103**) in six steps.

An unusual cationic domino transformation has been observed by Nicolaou and coworkers during their studies on the total synthesis of the natural product azadirachtin (**1-105**) [30]. Thus, exposure of the substrate **1-106** to sulfuric acid in CH$_2$Cl$_2$ at 0 °C led to the smooth production of diketone **1-109** in 80% yield (Scheme 1.27). The reaction is initiated by protonation of the olefinic bond in **1-106**, affording the tertiary carbocation **1-107**, which undergoes a 1,5-hydride shift with concomitant disconnection of the oxygen bridge between the two domains of the molecule. Subsequent hydrolysis of the formed oxenium ion **1-108** yielded the diketone **1-109**.

Over the years, intensive studies in medicinal chemistry with regard to the structure–activity relationships of compounds being used in clinical praxis have revealed the exceptional position of heterocycles. Moreover, a multitude of bioactive natural products contain a heteroatom. Therefore, the development of reliable and efficient

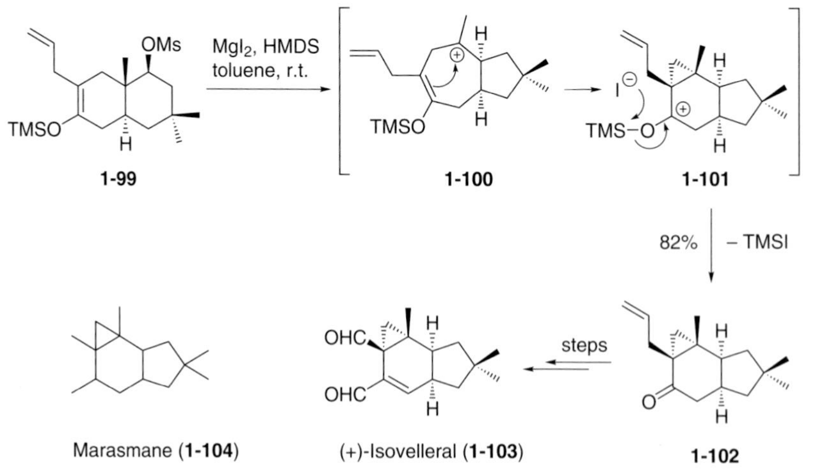

Scheme 1.26. Cationic domino rearrangement/cyclopropanation process for the total synthesis of (+)-isovelleral (**1-103**).

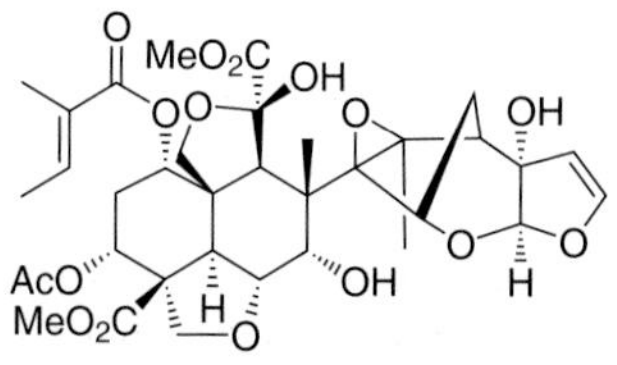

Azadirachtin (**1-105**)

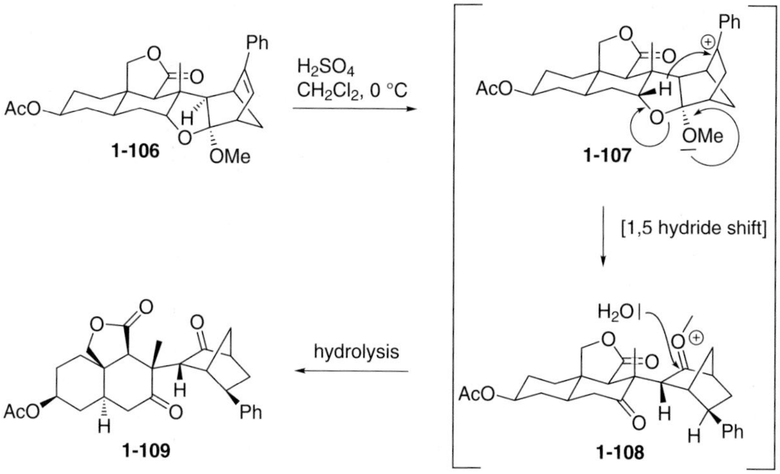

Scheme 1.27. Cationic domino transformation towards the synthesis of azadirachtin (**1-105**).

methods for constructing heterocyclic frameworks is of major importance, and it is of no wonder that several cationic domino processes have been designed in this field. Romero and coworkers prepared azasteroids based on an acyliminium ion-cyclization domino reaction [31]. Azasteroids display a broad variety of biological effects and thus are of continuing interest. When substrate **1-110** is submitted to acidic conditions, an acyliminium ion is formed by the elimination of ethanol; the ion then reacts with the C=C double bond in the molecule, performing the first ring closure with simultaneous formation of another carbocation (Scheme 1.28). Electrophilic aromatic substitution and demethylation of the methyl ether moiety led to the final product **1-111**. Despite the rather low yield of 30%, this reaction shows a new and elegant example of a domino reaction since four contiguous stereogenic centers are created in a single process, with high stereocontrol.

Another example of an intramolecular cyclization initiated by reactions of an acyliminium ion [32] with an unactivated alkene has been published by Veenstra and coworkers. In their total synthesis of CGP 49823 (**1-116**), a potent NK$_1$ antagonist [33], these authors treated the *N,O*-acetal **1-112** with 2 equiv. of chlorosulfonic acid in acetonitrile to afford acyliminium ion **1-113** (Scheme 1.29) [34]. This is qualified for a cyclization, creating piperidine cation **1-114**, which is then trapped by

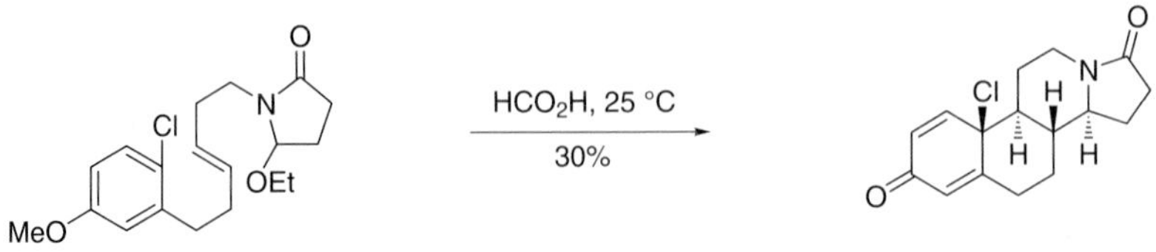

1-110 1-111

Scheme 1.28. Acyliminium ion-initiated domino cyclization in the synthesis of azasteroids.

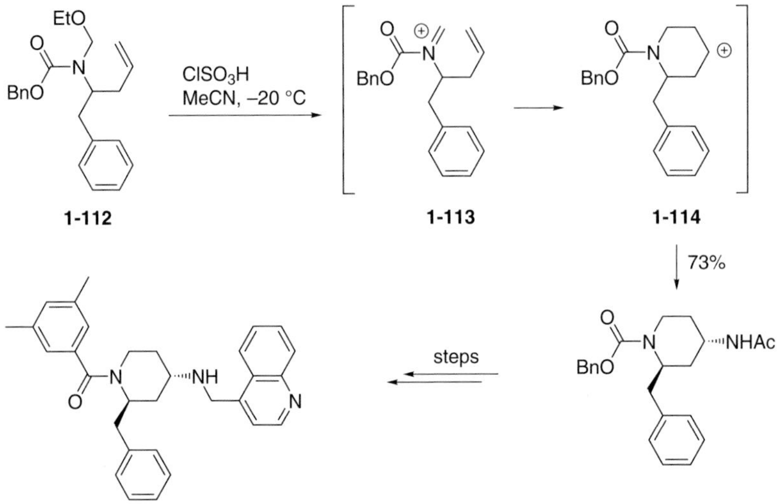

1-112 1-113 1-114

CPG 49823 (**1-116**) **1-115**

Scheme 1.29. Domino acyliminium ion cyclization/Ritter reaction procedure in the synthesis of CPG 49823 (**1-116**).

acetonitrile in a Ritter reaction to give **1-115** in 73% yield with high diastereoselectivity (*trans:cis*, 20:1). The interception of the cation **1-114** is assumed to occur for steric reasons from the opposite side of the benzyl moiety, leading to the *trans* product [35].

In an effort to assemble pyrrolidine heterocycles from acyclic precursors, the Rayner group developed a stereoselective domino cationic cyclization/aziridinium ion formation/nucleophilic ring-opening procedure [36]. The general process is displayed in Scheme 1.30, in which α-amino acetals **1-117** in the presence of a Lewis acid are converted into pyrrolidines **1-119** *via* the bicyclic aziridinium ions **1-118.** The aziridinium ions **1-118** can be isolated as a tetraphenylborate salt; however, when warming to room temperature, regioselective ring opening by a nucleophile occurs to give the pyrrolidines **1-119**.

Using the chiral α-amino acetal **1-120** ion and two different *N*-nucleophiles **1-122** and **1-123**, the pyrrolidines **1-124** and **1-125**, respectively, are obtained in good di-

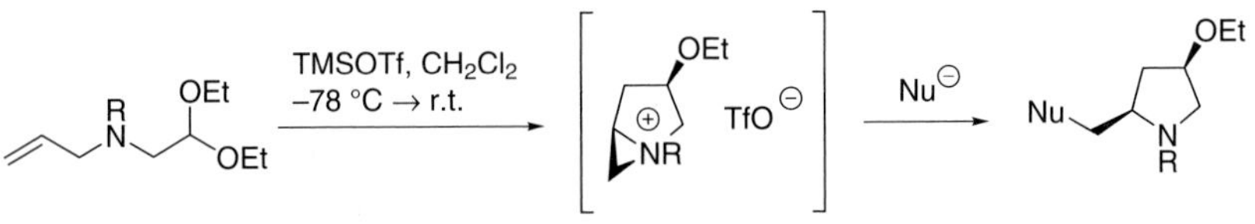

1-117 1-118 1-119

Scheme 1.30. Cationic cyclization/aziridinium ion formation/nucleophilic ring-opening procedure for the synthesis of pyrrolidines.

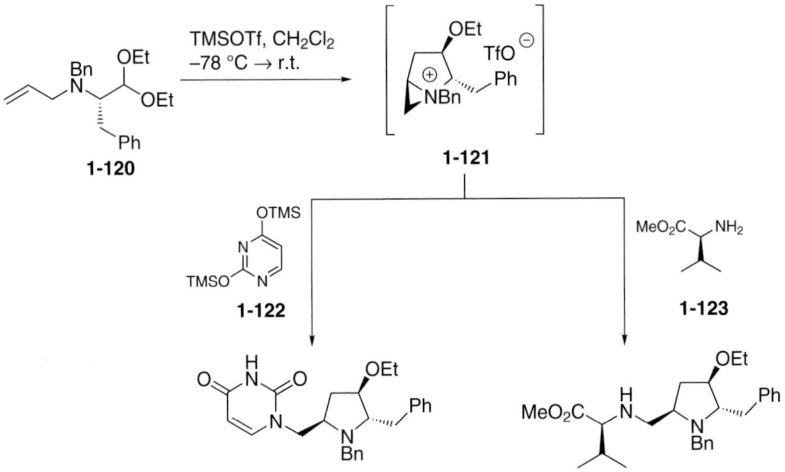

1-124 (78%, 87:13 *dr*) 1-125 (78%, 88:12 *dr*)

Scheme 1.31. Synthesis of enantiopure pyrrolidine derivatives.

astereoselectivity and yield (Scheme 1.31). The observed stereocontrol can be explained by assuming that all substituents adopt a pseudo-equatorial position in a chairlike transition state during the cyclization step.

Domino procedures are not limited to the construction of oxygen- and nitrogen-containing heterocycles. A novel *exo-endo*-cyclization of α,ω-diynesulfides mediated by bis(pyridyl)iodinium(I)tetrafluoroborate (IPy_2BF_4) has been revealed by Barluenga and coworkers [37]. Depending on the chain length of the diyne **1-126**, sulfur-containing tricycles **1-127** with different ring sizes were furnished in good to excellent yields (Scheme 1.32).

It is proposed that the electrophilic iodo ion first reacts with one triple bond in **1-126** to give the relative stable vinyl cation **1-128**. Ring closure leads to a seven-membered ring **1-129** containing another vinyl cation moiety, which then cyclizes to produce the final tricycle **1-127**.

Recently, a novel domino process for the synthesis of benzazepines with the concomitant formation of a C–N and a C–C bond has been published by the Basavaiah group [38]. These authors have shown that, under acidic conditions in the presence

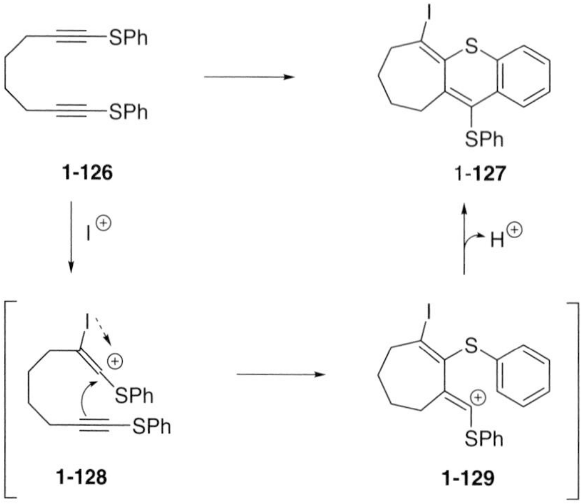

1-126   **1-127**

1-128   **1-129**

Entry	Substrate	Product	Yield [%]
1	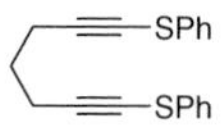	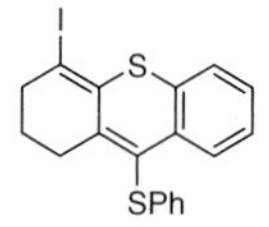	92
2	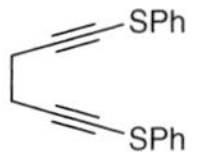	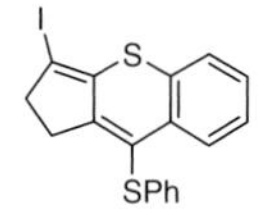	69
3	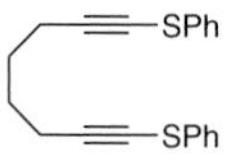	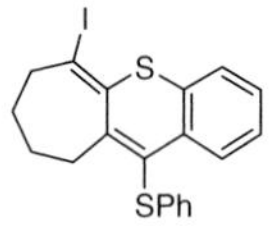	71
4	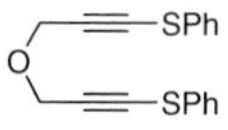	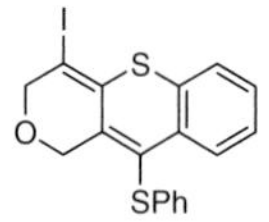	72

Reaction conditions: 1 eq IPy$_2$BF$_4$, 1 eq HBF$_4$, CH$_2$Cl$_2$, –80 °C.

Scheme 1.32. Domino cyclization of α-,ω-diynesulfides.

of nitriles, Baylis–Hillman adducts such as **1-130** can be easily transformed into 2-benzazepines **1-131**, a skeleton which is present in many bioactive molecules (Scheme 1.32). The process is assumed to proceed *via* a Ritter reaction of the primarily formed cation **1-132**, delivering the nitrilium ion **1-133**, which then undergoes ring closure – a so-called Houben–Hoesch reaction.

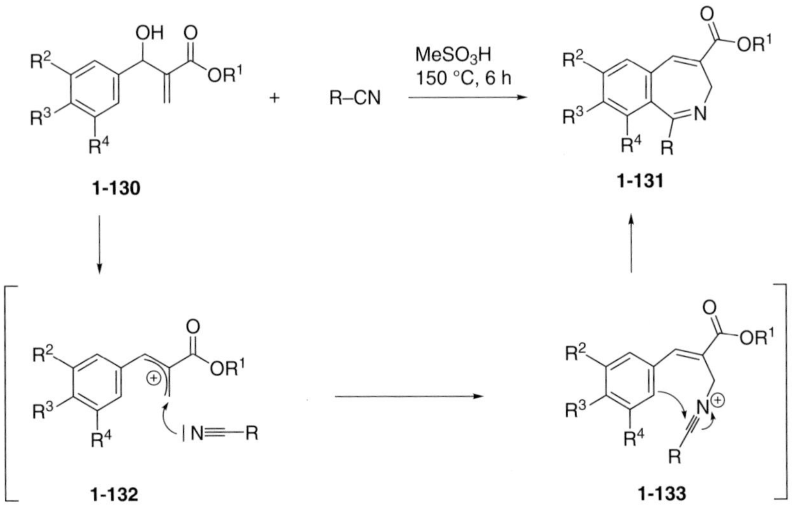

Scheme 1.33. Synthesis of benzazepines.

Entry	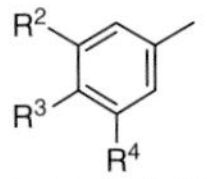 substrate **1-130**	R^1 (alkyl)	R (alkyl)	Product **1-131** Yield [%]
1	3-(MeO)Ph	Et	Me	55
2	3-(MeO)Ph	Et	Et	67
3	3-(MeO)Ph	Me	Et	44
4	3-(PrO)Ph	Et	Et	58
5	3-(PrO)Ph	Me	Et	65
6	3,5-(MeO)$_2$Ph	Et	Me	70
7	3,5-(MeO)$_2$Ph	Et	Et	74
8	3,5-(MeO)$_2$Ph	Me	Me	72
9	3,4,5-(OMe)$_3$Ph	Me	Et	33
10	3,4-(OCH$_2$O)Ph	Et	Et	48
11	3,4-(OCH$_2$O)Ph	Me	Et	46

An elimination/double Wagner–Meerwein rearrangement process has recently been developed by Langer and coworkers [39]. Treatment of compound **1-136**, obtained by reaction of **1-134** and **1-135**, with trifluoroacetic acid (TFA) led to the cationic species **1-137**, which then underwent a twofold Wagner–Meerwein rearrangement to give the bicyclic compound **1-139** *via* **1-138** (Scheme 1.34).

The construction of novel tetracyclic ring systems **1-142**, which can be considered as hybrids of the tetrahydropyrrolo[2,3-*b*]indole and tetrahydroimidazol[1,2-*a*]indole ring system, has been described by Herranz and coworkers [40]. The exposure of tryptophan-derived α-amino nitrile **1-140** to acidic conditions triggers a stereoselective tautomerization to give **1-142** in quantitative yield (Scheme 1.35).

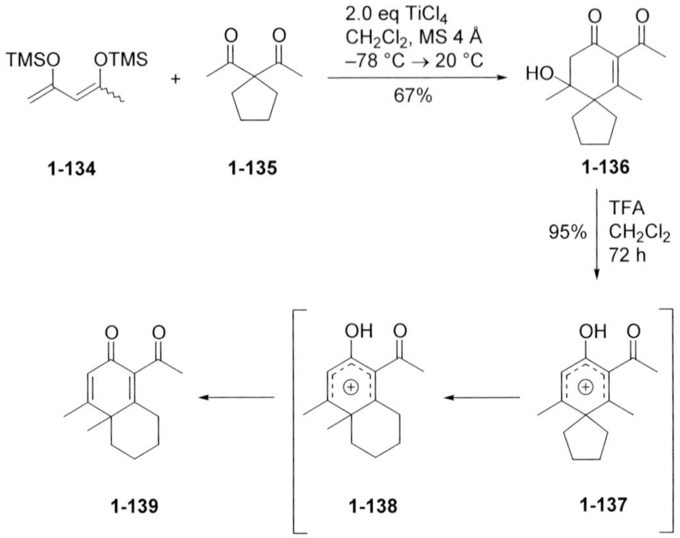

Scheme 1.34. Domino elimination/double Wagner–Meerwein rearrangement reaction.

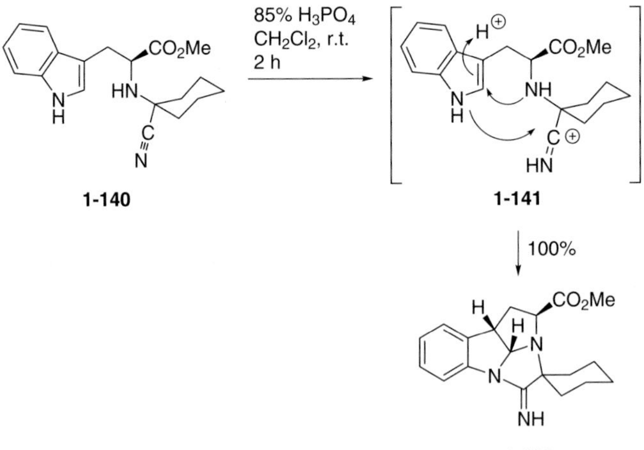

Scheme 1.35. Acid-mediated domino tautomerization.

1.1.1
Cationic/Cationic/Cationic Processes

One of the most fascinating aspects of domino reactions is the fact that the number of steps being included in one sequence is not really limited; therefore, the complexity of a domino process is simply a question of imagination and skill.

As mentioned earlier, the McDonald group was able to extend their epoxide-domino-cyclization strategy to 1,5,9-triepoxides [10]. Indeed, they were successful in converting precursor **1-143** into the tricyclic product **1-146** in 52 % yield after hydrolysis (Scheme 1.36) [41]. As a possible mechanism of this polyoxacyclization it can be assumed that, after activation of the terminal epoxide by BF_3, a sequence of intramolecular nucleophilic substitutions by the other epoxide oxygens takes place, which is induced by a nucleophilic attack of the carbonate oxygen, as indicated in **1-144** to give **1-145**.

As an alternative mechanistic assumption the initiation of the domino process by one molecule of water could be considered. However, this would lead to the corresponding enantiomer *ent*-**1-146**, which was excluded by crystallographic structure determination.

An impressive cationic domino polycyclization has been developed by Corey and coworkers in their short and efficient enantioselective total synthesis of aegicer-adienol (**1-150**), a naturally occurring pentacyclic nor-triterpene belonging to the β-amyrin family [42]. Thus, the treatment of the enantiopure monocyclic epoxy tetraene **1-147** with catalytic amounts of methylaluminum dichloride induces a cation-π-tricyclation by initial opening of the epoxide to form the tetracyclic ketone **1-148** in 52 % yield, and its C-14 epimer **1-149** in 23 % yield, after silylation and chromatographic separation (Scheme 1.37). Further transformations led to aegicer-adienol (**1-150**) and its epimer **1-151**.

Domino reactions are also used in the development of new and potent pharmaceuticals, which is an important target in organic synthesis. The following example by Katoh and coworkers presents an impressive approach to the tetracyclic core structure of the novel anti-influenza A virus agent, stachyflin (**1-152**), using as key feature a new Lewis acid-induced domino epoxide-opening/rearrangement/cy-

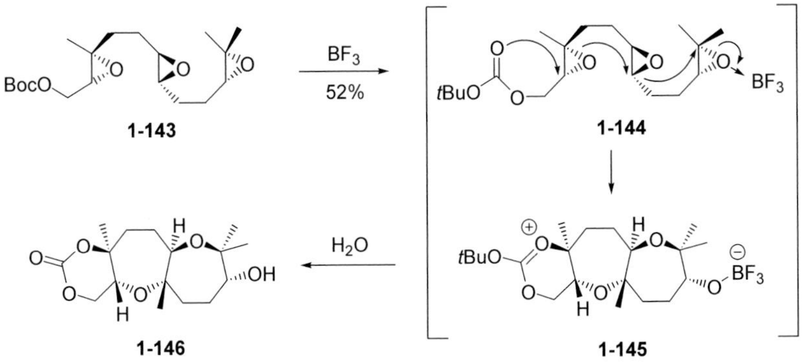

Scheme 1.36. Triple cyclization of triepoxides.

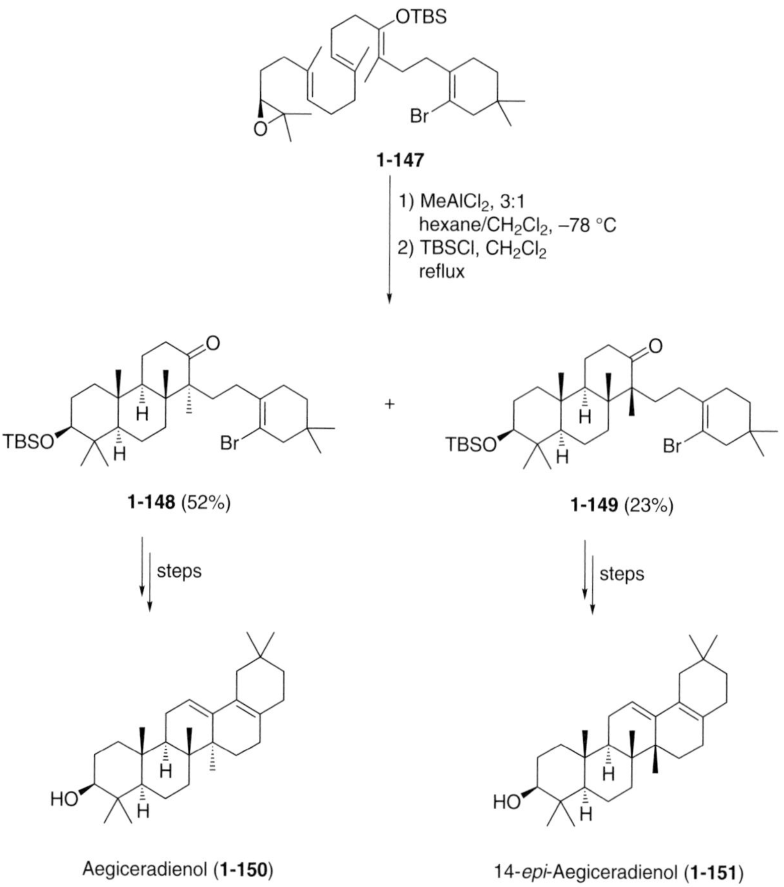

Scheme 1.37. Cationic domino polycyclization in the total synthesis of aegiceradienol (**1-150**).

clization reaction [43]. Thus, the α-epoxide **1-153** forms the tetracycle **1-157** on treatment with an excess of BF$_3$·Et$_2$O in one process in reasonable 41 % yield (Scheme 1.38). It is assumed that this domino transformation proceeds stepwise *via* the carbocations **1-154**, **1-155**, and **1-156**, with 1,2-methyl group migration, 1,2-hydride shift and trapping by nucleophilic attack of the phenolic hydroxyl group as the final step [44]. **1-157** could be further transformed into an analogue of **1-152** containing all stereogenic centers with the correct configuration.

During the course of a biomimetic synthesis of the indole-diterpene mycotoxin emindole SB (**1-158**), the Clark group has observed an unexpected domino reaction [45]. When epoxide **1-159** was treated with BF$_3$·Et$_2$O, the formation of three rings and four stereogenic centers in a highly regioselective and stereoselective single operation took place to give pentacycle **1-160** in 33 % yield (Scheme 1.39). Unfortunately, the desired emindole SB scaffold was not formed, this being due to unex-

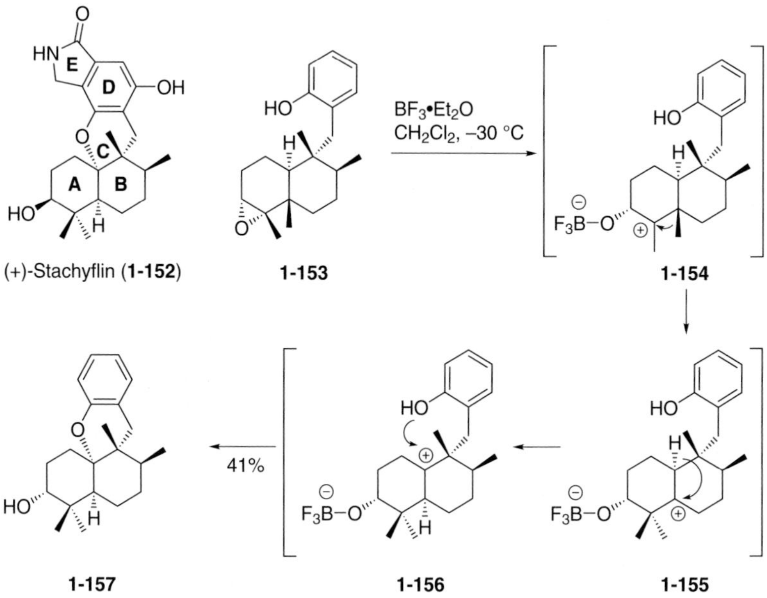

Scheme 1.38. Domino epoxide-opening/rearrangement/cyclization reaction towards the total synthesis of (+)-stachyflin (**1-152**).

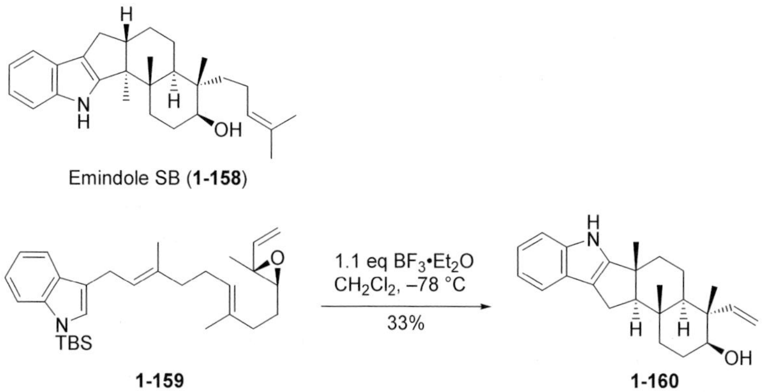

Scheme 1.39. Synthesis of an indole derivative by a three fold cationic polycyclization.

pected 6-*endo* cyclization instead of the anticipated 5-*exo* ring closure during assembly of the second ring. This led on the one hand to the assumption that the proposed biosynthetic pathway [46] is incorrect, and on the other hand to the assumption that the 5-*exo* ring closure event is controlled enzymatically.

An impressive number of eight steps within one cationic domino process was observed by Mulzer and coworkers, when treating the entriol derivative **1-161** with

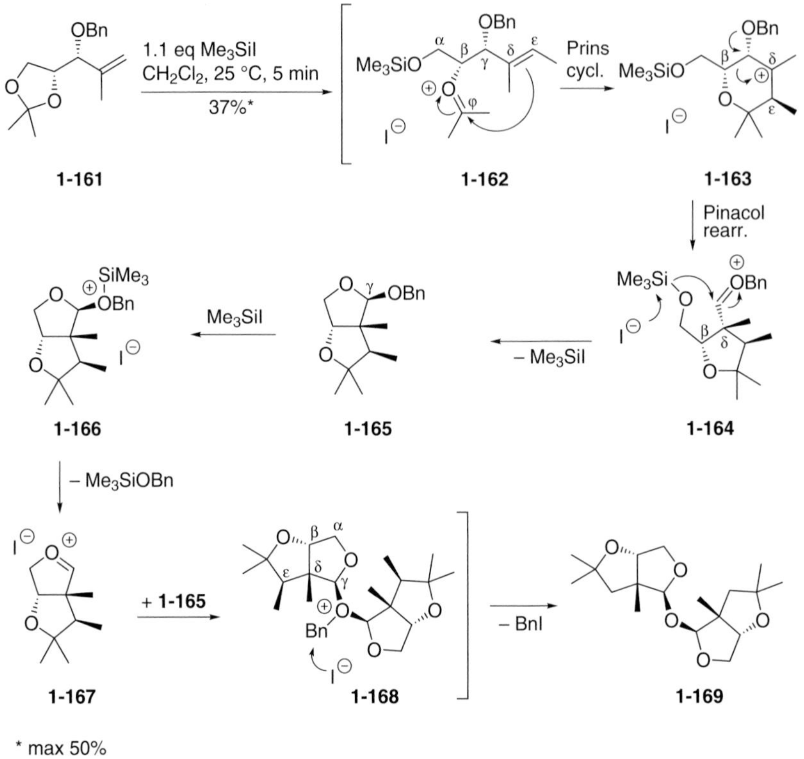

Scheme 1.40. Eight-step cationic domino process.

trimethylsilyliodide [47]. The transformation which provided the uniform bisacetal **1-169** in 37–50% yield included a silyliodide-promoted cleavage to give oxonium ion **1-162**, a Prins reaction, a pinacol rearrangement, and the formation of an oxenium ion **1-167** (Scheme 1.40). This reacts with the intermediate **1-165** to give the oxonium ion **1-168** and, after debenzylation, the bisacetal **1-169**.

As discussed earlier, Ila, Junjappa and coworkers used cyclopropyl units as cation-provider in cationic domino processes. Within their interesting approach, the indole derivatives **1-170** could be converted into the unexpected carbazoles **1-171** with 54–69% yield in a five-step transformation using SnCl₄ as reagent (Scheme 1.41) [48].

The domino reaction starts with a Lewis acid-promoted electrophilic ring opening of the cyclopropyl moiety in **1-170**. The resulting stable zwitterionic intermediate **1-171** undergoes an intermolecular enol capture by another zwitterionic species, leading to dimeric indole cation **1-172**. Two ring closures follow with **1-173** as an intermediate, and subsequently a Lewis acid-assisted elimination of indole and an oxidation through the solvent occur to give the carbazole **1-174**.

Another carbocationic domino process by the same group resulted in the formation of indane derivatives using cyclopropyl carbinol compounds as starting mate-

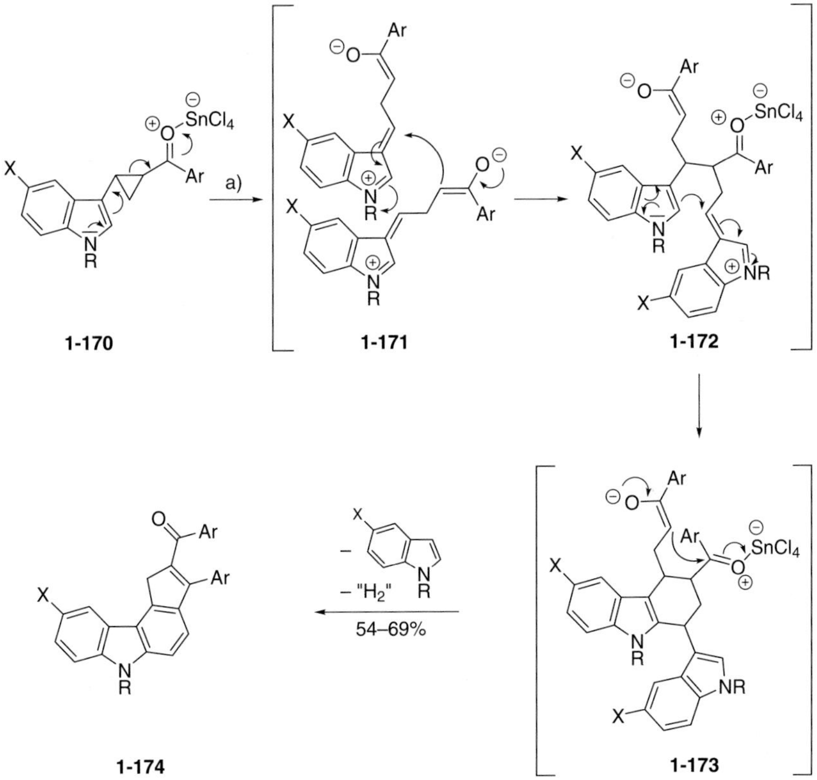

1-170 **1-171** **1-172**

1-174 **1-173**

54–69%

X = H, Br, OMe
Ar = pMeOC$_6$H$_4$, Ph, pClC$_6$H$_4$, 2-thienyl, 3,4-(MeO)$_2$C$_6$H$_3$
R = Me, Bn

a) 1.5 eq SnCl$_4$, MeNO$_2$, 0 °C: 0.5 h, r.t., 3–8 h.

Scheme 1.41. Cationic domino rearrangement in the synthesis of the 1H-cyclopenta[c]carbazole framework.

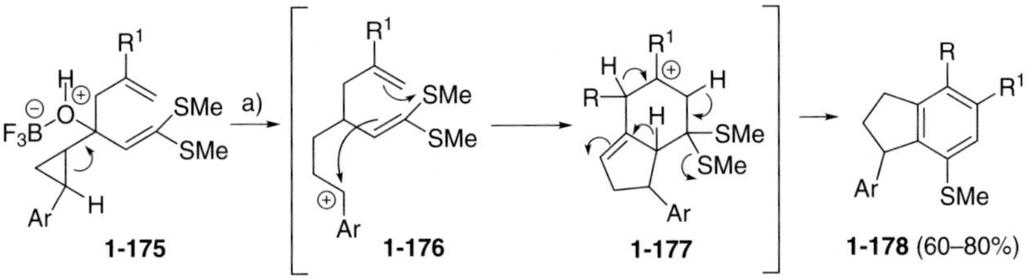

1-175 **1-176** **1-177** **1-178** (60–80%)

a) BF$_3$•Et$_2$O, MeNO$_2$, 0 °C → r.t..

a: R = R^1= H
b: R = Me, R^1= H

Scheme 1.42. Cationic domino reaction of cyclopropylcarbinols.

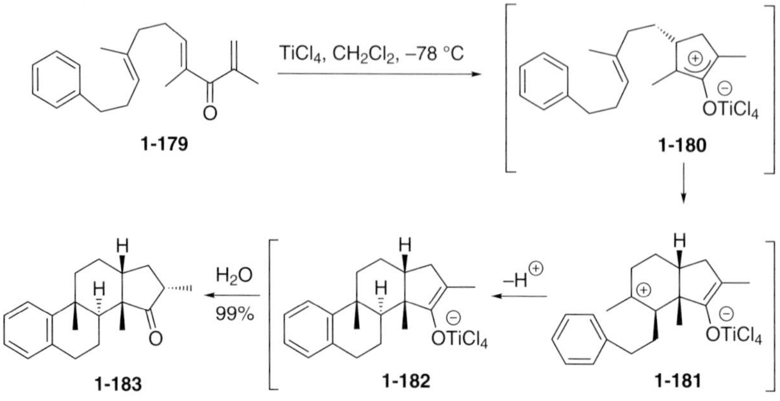

Scheme 1.43. Synthesis of steroid analogues.

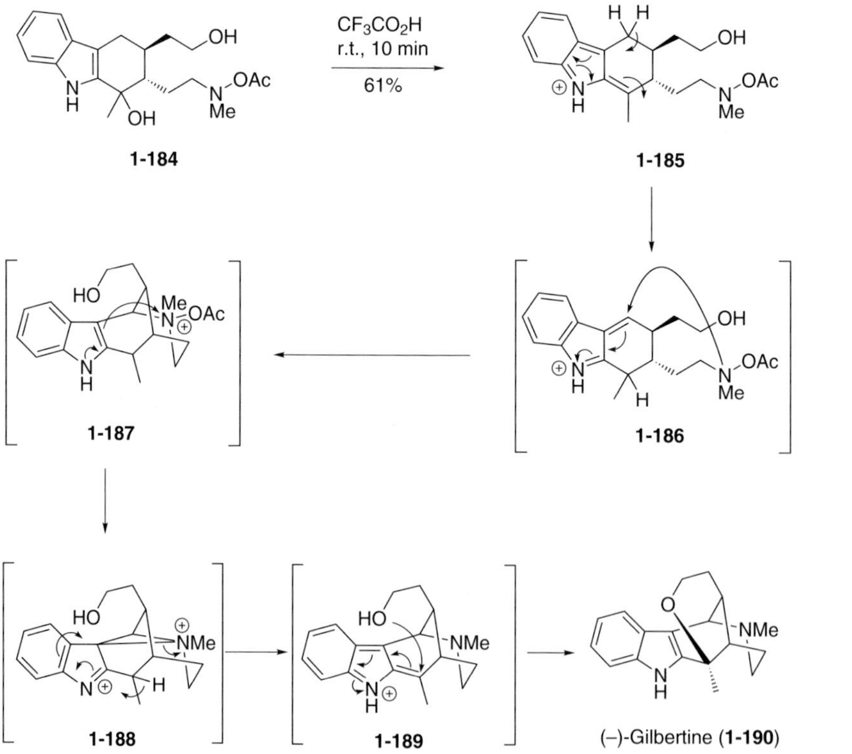

Scheme 1.44. Enantioselective synthesis of (–)-gilbertine (**1-190**) by a cationic domino cyclization.

rial [49]. Reaction of cyclopropyl carbinol **1-175** with $BF_3 \cdot Et_2O$ led to **1-178** in 60–80 % yield *via* **1-176** and **1-177** (Scheme 1.42). Using cyclopropyl carbinols with a different substitution pattern, isomers of **1-178** or monocyclic compounds are obtained.

As discussed previously, West and coworkers developed a two-step domino process, which is initiated by a Nazarov reaction. This can be extended by an electrophilic substitution. Thus, reaction of **1-179** with $TiCl_4$ led to **1-182** *via* the intermediate cations **1-180** and **1-181**. The final product **1-183** is obtained after aqueous workup in 99 % yield (Scheme 1.43) [23]. It is important to mention here that all six stereocenters were built up in a single process with complete diastereoselectivity; hence, the procedure was highly efficient.

A very impressive multiple cationic domino reaction was used in the enantioselective total synthesis of (–)-gilbertine (**1-190**), described by Blechert and coworkers [50]. When the tertiary alcohol **1-184** is treated with TFA, a carbocation is formed which undergoes a cascade of cyclizations to afford **1-190** in very good yield (61 %) (Scheme 1.44). The cations **1-185** to **1-189** can be assumed as intermediates.

1.2
Cationic/Pericyclic Processes

The first example of this rather under-represented class of domino processes depicts an attractive strategy to functionalized cycloheptenes which has been developed by West and coworkers [51]. As illustrated in Scheme 1.45, substrate **1-191a** is exposed at –30 °C to the Lewis acid $FeCl_3$ to induce a Nazarov electrocyclization to give carbocation **1-192**. Under the applied reaction conditions, this undergoes a [4+3]-cycloaddition to yield the two diastereomers **1-193a** and **1-194a** in good yield, but almost unselectively. Interestingly, the homologous substrate **1-191b** led

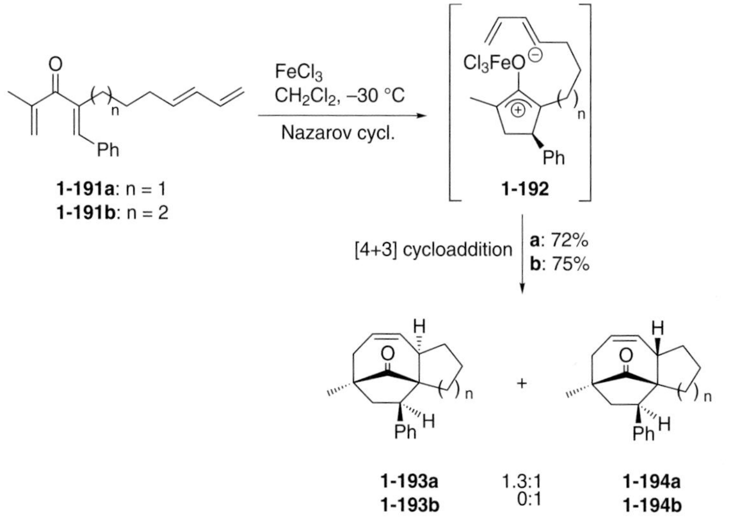

1-191a: n = 1
1-191b: n = 2

1-192

[4+3] cycloaddition | **a**: 72%
| **b**: 75%

| **1-193a** | 1.3:1 | **1-194a** |
| **1-193b** | 0:1 | **1-194b** |

Scheme 1.45. Cationic domino cycloisomerization of tetraenones.

to the formation of cycloadduct **1-194b** as a single diastereomer in 75 % yield. In this case, the additional methylene group in the tether causes an attack of the diene from below *anti* to the phenyl group in an *endo* mode. Until now, the origin of this surprising result has not been clear.

A combination of cationic processes and an ene reaction has been used by Kim and coworkers for the total synthesis of perhydrohistrionicotoxin (**1-202a**) using **1-198** as substrate (Scheme 1.46). Derivatives of histrionicotoxin **1-202b** have turned out to be attractive targets for synthetic chemists, as they demonstrate useful neurophysical properties [52]; **1-202b** has a unique structural feature and low natural abundance. For the synthesis of **1-198**, the ketone **1-195** was used as the starting material which, after epimerization, was treated with the Grignard reagent obtained from **1-197**. The domino process of **1-198** was initiated by using *p*TsOH, which impelled the pinacol rearrangement as well as the cleavage of the acetal unit, thus preparing the settings for the proximate carbonyl ene reaction. Ultimately, the tricyclic alcohol **1-201** and its epimer **1-200** were afforded *via* the transition state **1-199** in a

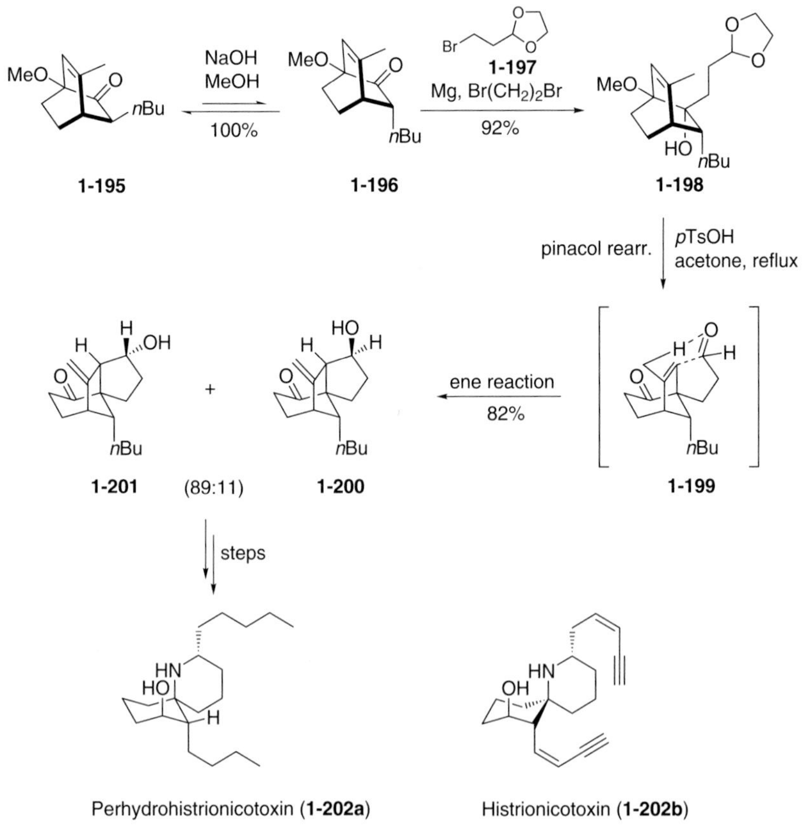

Scheme 1.46. Domino pinacol rerrangement/ene-reaction strategy in the total synthesis of perhydrohistrionicotoxin (**1-202a**).

89:11 ratio in 82 % total yield. **1-201** was further transformed into perhydrohistrioni-cotoxin (**1-202a**). It should be noted that ene reactions are classified as pericyclic reactions, although carbonyl ene reactions usually follow a cationic mechanism.

The Lewis acid-catalyzed 1,3-migration of divinyl esters allows the formation of 1,3-butadienes, which can undergo cycloaddition. In this respect, Dai and co-workers described a rearrangement of the divinyl alkoxyacetate **1-203** followed by a Diels–Alder reaction with a dienophile such as maleic anhydride **1-204** in the presence of catalytic amounts of Ln(fod)$_3$ to produce **1-205** in up to 61 % yield (Scheme 1.47) [53].

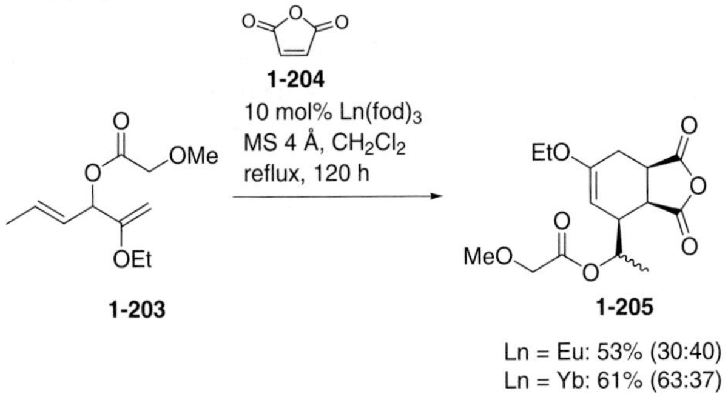

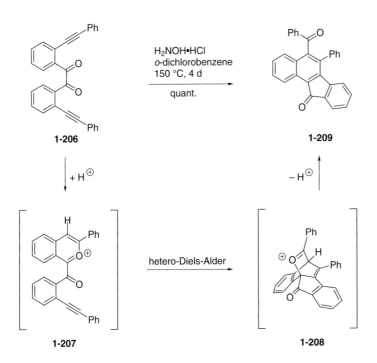

Scheme 1.47. Domino-1,3-rearrangement/Diels–Alder process.

Scheme 1.48. Cationic domino cyclization/hetero-Diels–Alder procedure.

The final example of a cationic/pericyclic domino process was discovered almost by accident. While preparing bisisoquinoline-bis-*N*-oxides, the Dyker group revealed an unexpected domino sequence when they treated the benzil derivative **1-206** with hydroxylamine hydrochloride at high temperature (Scheme 1.48) [54]. Rather than obtaining the desired bisoxime, 13 % of the tetracycle **1-209** was isolated. After further optimization of the reaction conditions, these authors were finally able to achieve **1-209** within four days as the sole product in a striking quantitative yield. For the mechanistic interpretation it was assumed that the reaction is initiated by protonation of one of the alkyne functions with ring closure under formation of the benzopyrylium cation **1-207**; this is followed by an intramolecular hetero-Diels–Alder reaction to give **1-208**. Finally, the latter undergoes a ring fragmentation accomplishing the observed product **1-209**.

1.3
Cationic/Reductive Processes

Cationic/reductive domino processes were first described in 2003, and are consequently among the "youngest" domino procedures described in this book. To date, only two (albeit very useful) examples typifying a combination of a cationic reaction with a reduction procedure have been identified.

Recently, it was discovered that THF-inactivated $AlEt_3$ has a rare double reactivity, namely Lewis acidity and reduction potential. By using this reagent, Tu and co-workers established a semipinacol rearrangement/reduction methodology of α-hydroxyepoxides **1-210** to give diols **1-214** with ring contraction (Scheme 1.49) [55]. In the transformation, **1-211** and **1-212** can be proposed as intermediates, with the Al-complex **1-213** as the product which is hydrolyzed to afford **1-214**.

In addition to hydroxyepoxides, which can be easily prepared in an enantiopure form, α-aminoepoxides can also be used to produce β-aminoalcohols (Scheme 1.50, entry 1). Moreover, α and β-hydroxyaziridines are also good substrates, leading again to aminoalcohols (Scheme 1.50, entries 2 and 3).

The same group has presented another highly diastereoselective domino procedure for the construction of quaternary 1,3-diols involving a samarium diiodide-catalyzed semipinacol/Tishchenko reduction sequence [56]. Treatment of α-hydroxy epoxides **1-212** with substoichiometric amounts of samarium diiodide in the presence of an excess of a reducing aldehyde (best results were obtained by using *p*-chlorobenzaldehyde) initiated a process in which at first a semipinacol-type rearrangement with C-1 to C-3 carbon migration and simultaneous formation of an intermediate aldehyde by epoxide opening takes place (Scheme 1.51). Furthermore, hetero-Tishchenko reduction of the formed aldehyde and the added aldehyde furnished the desired monoesters **1-216** in good yields (58–82 %) and high selectivity.

On the basis of these experimental results, a possible mechanism has been proposed for the reaction of **1-215** with SmI_2 (Scheme 1.52). After formation of the *syn*-complex **A**, a rearrangement occurs to give the aldehyde **B**, which coordinates to the added aldehyde RCHO to afford complex **C**. Subsequent samarium-catalyzed nucleophilic attack of the secondary alcohol to the carbonyl of RCHO generates a hemiacetal, **D**. There follows an irreversible intramolecular 1,5-hydride transfer *via*

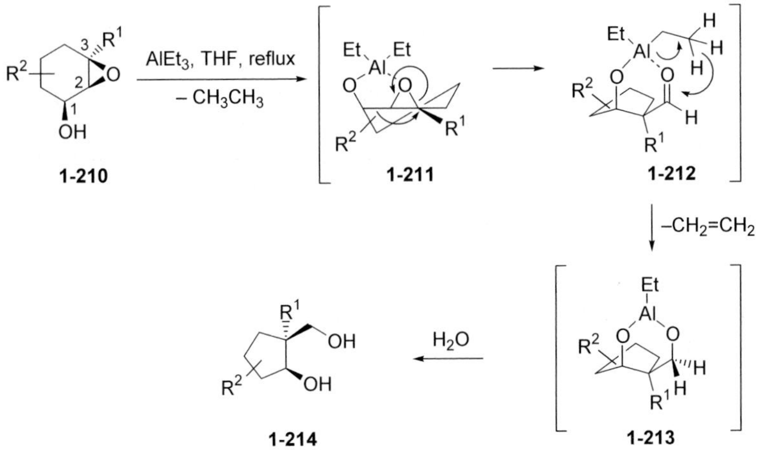

Scheme 1.49. Cationic domino rearrangement/reduction sequence of α-hydroxy epoxides.

Entry	Substrates	Products	Yield [%]
1	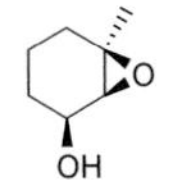	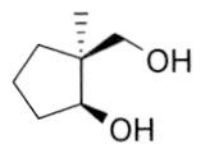	58
2	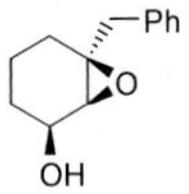	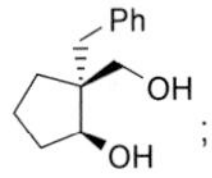 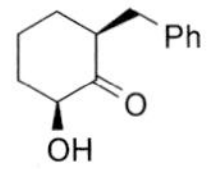	66; 31
3	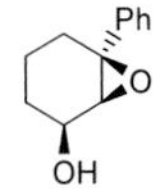	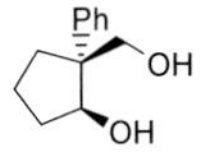	82
4	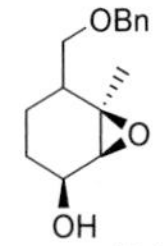	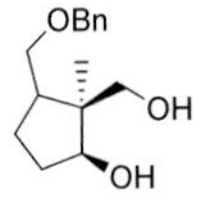	70
5	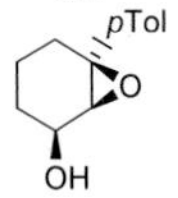	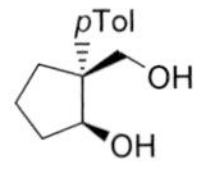	80
6	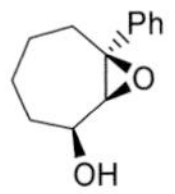	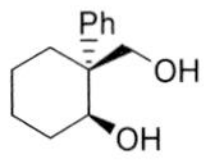	80
7	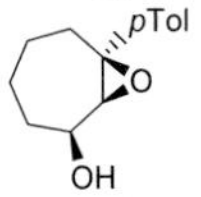	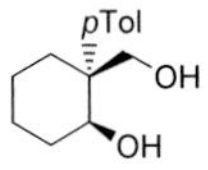	78

Entry	Substrates		Products	Yield [%]
1	(epoxide, Ph, NHTs)	AlEt₃, THF, reflux	(Ph, OH, NHTs)	80
2	(OH, Ph, NTs)	AlEt₃, THF, reflux	(Ph, OH, NHTs)	92
3	(HO, NTs)	AlEt₃, THF, reflux	(OH, NHTs)	60

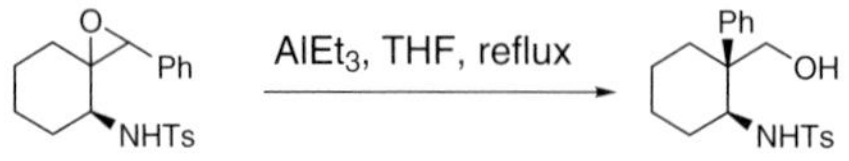

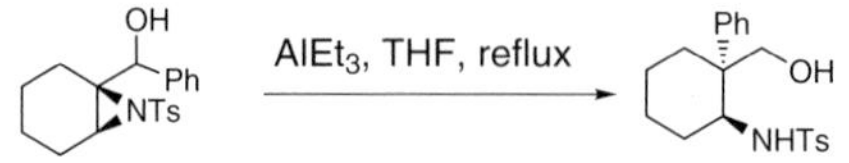

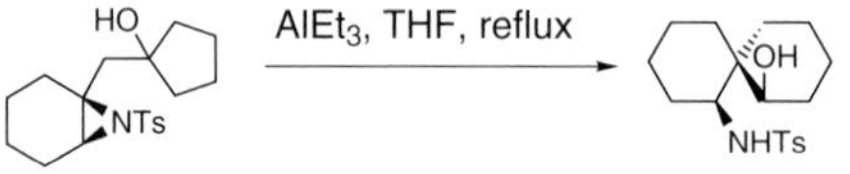

Scheme 1.50. Synthesis of β-amino alcohols.

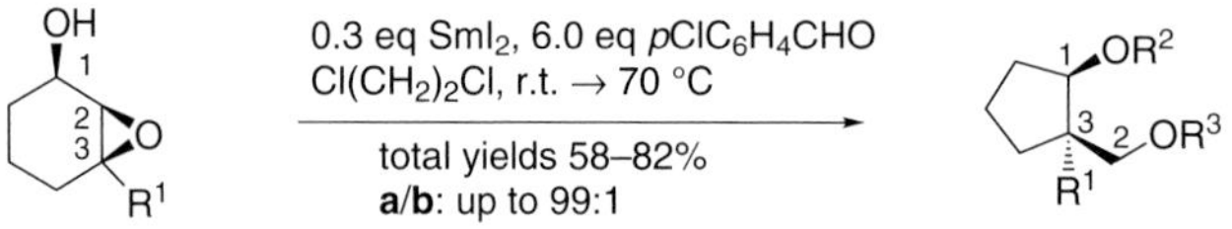

Scheme 1.51. Samarium-catalyzed domino reaction of secondary α-hydroxy epoxides.

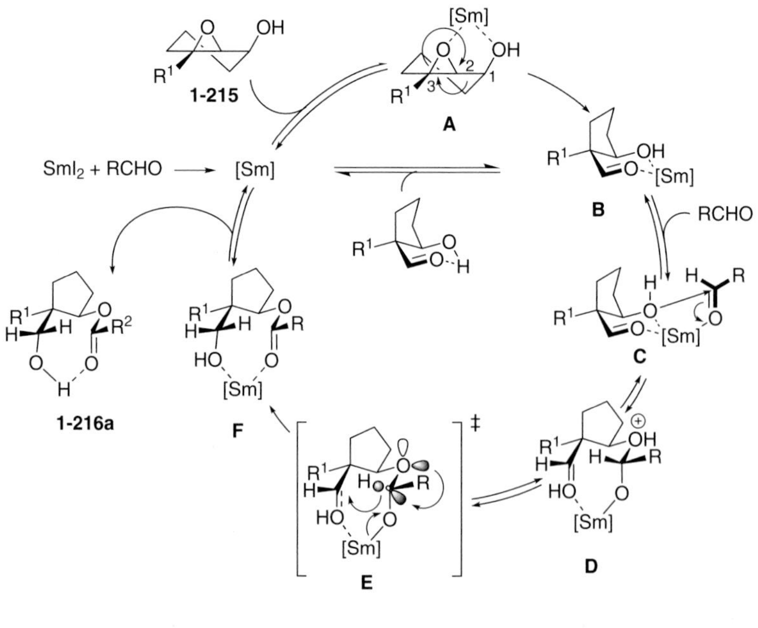

Scheme 1.52. Proposed catalytic cycle for the cationic domino rearrangement/hetero-Tishchenko reduction process of secondary α-hydroxy epoxides in the presence of SmI₂.

transition state **E** to afford the samarium-coordinated 1,3-diol monoester **F** that further accomplishes the production of monoesters **1-216a** by releasing the samarium species for the next cycle.

In an earlier publication by the same authors, a similar procedure – albeit without ring contraction – has already been described [57].

References

1 (a) L. E. Overman, *Aldrichim. Acta* **1995**, *28*, 107–120; (b) L. E. Overman, *Acc. Chem. Res.* **1992**, *25*, 352–359; (c) G. C. Hirst, P. N. Howard, L. E. Overman, *J. Am. Chem. Soc.* **1989**, *111*, 1514–1515; (d) T. Gahman, L. E. Overman, *Tetrahedron* **2002**, *58*, 6473–6483; (e) L. E. Overman, L. P. Pennington, *Can. J. Chem.* **2000**, *78*, 732–738; (f) K. P. Minor, L. E. Overman, *Tetrahedron* **1997**, *53*, 8927–8940.

2 M. J. Brown, T. Harrison, L. E. Overman, *J. Am. Chem. Soc.* **1991**, *113*, 5378–5384.

3 T. A. Grese, K. D. Hutchinson, L. E. Overman, *J. Org. Chem.* **1993**, *58*, 2468–2477.

4 (a) T. Sano, J. Toda, Y. Tsuda, *Heterocycles* **1984**, *22*, 53–58; (b) B. M. Trost, A. Brandi, *J. Am. Chem. Soc.* **1984**, *106*, 5041–5043; (c) M. Sworin, W. L. Neumann, *J. Org. Chem.* **1988**, *53*, 4894–4896; (d) B. M. Trost, D. C. Lee, *J. Am. Chem. Soc.* **1988**, *110*, 6556–6558; (e) T. Nakamura, T. Matsui, K. Tanino, I. Kuwajima, *J. Org. Chem.* **1997**, *62*, 3032–3033; (f) J.-H. Youn, J. Lee, J. K. Cha, *Org. Lett.* **2001**, *3*, 2935–2938.

5 T. O. Johnson, L. E. Overman, *Tetrahedron Lett.* **1991**, *32*, 7361–7364.

6 L. E. Overman, J. P. Wolfe, *J. Org. Chem.* **2002**, *67*, 6421–6429.

7 (a) J. K. Rupprecht, Y.-H. Hui, J. L. McLaughlin, *J. Nat. Prod.* **1990**, *53*, 237–278; (b) U. Koert, *Tetrahedron Lett.* **1994**, *35*, 2517–2520; (c) T. R. Hoye, P. R. Hanson, *Tetrahedron Lett.* **1993**, *34*, 5043–5046; (d) J.-C. Harmange, B. Figadere, *Tetrahedron Lett.* **1993**, *34*, 8093–8096.

8 (a) U. Koert, M. Stein, K. Harms, *Tetrahedron Lett.* **1993**, *34*, 2299–2302; (b) U. Koert, H. Wagner, U. Pidun, *Chem. Ber.* **1994**, *127*, 1447–1457.

9 U. Koert, H. Wagner, M. Stein, *Tetrahedron Lett.* **1994**, *35*, 7629–7632.

10 F. E. McDonald, X. Wang, B. Do, K. I. Hardcastle, *Org. Lett.* **2000**, *2*, 2917–2919.

11 J. D. Neighbors, J. A. Beutler, D. F. Wiemer, *J. Org. Chem.* **2005**, *70*, 925–931.

12 For excellent reviews, see: (a) T. L. B. Boivin, *Tetrahedron* **1987**, *43*, 3309–3362; (b) M. C. Elliott, *J. Chem. Soc. Perkin Trans. 1* **2000**, 1291–1318.

13 I. E. Markó, J.-M. Plancher, *Tetrahedron Lett.* **1999**, *40*, 5259–5262.

14 B. Leroy, I. E. Markó, *J. Org. Chem.* **2002**, *67*, 8744–8752.

15 T.-P. Loh, L.-C. Feng, J.-Y. Yang, *Synthesis* **2002**, *7*, 937–940.

16 (a) T.-P. Loh, Q.-Y. Hu, L.-T. Ma, *J. Am. Chem. Soc.* **2001**, *123*, 2450–2451; (b) T.-P. Loh, K.-T. Tan, Q.-Y. Hu, *Int. Ed. Engl. Angew. Chem.* **2001**, *40*, 2921–2922.

17 For a excellent review, see: H. N. C. Wong, M.-Y. Hon, C.-W. Tse, Y.-C. Yip, J. Tanko, T. Hudlicky, *Chem. Rev.* **1989**, *89*, 165–198.

18 S. Nandi, U. K. Syam Kumar, H. Ila, H. Junjappa, *J. Org. Chem.* **2002**, *67*, 4916–4923.

19 (a) P. C. B. Page, *Organosulfur Chemistry-Synthetic Aspects*, Vol. 1, Academic Press, New York, **1995**; (b) P. Metzner, A. Thuillier, *Sulfur Reagents in Organic Synthesis*, Academic Press, New York, **1994**.

20 (a) L. Fitjer, *Methoden Org. Chem. (Houben-Weyl)*, Vol. E17, **1997**, p. 251; (b) J. R. Y. Salaün, *Top. Curr. Chem.* **1988**, *144*, 1–71; (c) B. M. Trost, *Top. Curr. Chem.* **1986**, *133*, 3–82; (d) B. M. Trost, D. E. Keeley, H. C. Arndt, J. H. Rigby, M. J. Bogdanowicz, *J. Am. Chem. Soc.* **1977**, *99*, 3080–3087; (e) B. M. Trost, D. E. Keeley, H. C. Arndt, M. J. Bogdanowicz, *J. Am. Chem. Soc.* **1977**, *99*, 3088–3100.

21 B. Witulski, U. Bergsträßer, M. Gößmann, *Tetrahedron* **2000**, *56*, 4747–4752.

22 A. M. Bernard, E. Cadoni, A. Frongia, P. P. Piras, F. Secci, *Org. Lett.* **2002**, *4*, 2565–2567.

23 J. A. Bender, A. M. Arif, F. G. West, *J. Am. Chem. Soc.* **1999**, *121*, 7443–7444.

24 C. C. Browder, F. P. Marmsäter, F. G. West, *Org. Lett.* **2001**, *3*, 3033–3035.

25 For recent reviews on Taxol, see: (a) *The Chemistry and Pharmacology of Taxol and its Derivatives* (Ed.: V. Farina), Elsevier, Amsterdam, **1995**; (b) *TaxolÖ: Science and Applications* (Ed.: M. Suffness), CRC Press, Boca Raton, **1995**; (c) K. C. Nicolau, W.-M. Dai, R. K. Guy, *Angew. Chem. Int. Ed.* **1994**, *33*, 15–44; (d) G. I. George, T. T. Chen, I. Ojima, D. M. Vyas, *Taxane Anticancer Agents*, American Cancer Society, San Diego, **1995**.

26 (a) K. C. Nicolaou, Z. Yang, J. J. Liu, H. Ueno, P. G. Nantermet, R. K. Guy, C. F. Claiborne, J. Renaud E. A. Couladouros, K. Paulvannan, E. J. Sorensen, *Nature* **1994**, *367*, 630–634; (b) K. C. Nicolaou, H. Ueno, J.-J. Liu, P. G. Nantermet, Z. Yang, J. Renaud, K. Paulvannan, R. Chadha, *J. Am. Chem. Soc.* **1995**, *117*, 653–659; (c) R. A. Holton, H.-B. Kim, C. Somoza, L. Liang, R. J. Bieediger, P. D. Boatman, M. Shindo, C. C. Smith, S. Kim, H. Nadizadeh, Y. Suzuki, C. Tao, P. Vu, S. Tang, P. Zhang, K. K. Murthy, L. N. Gentile, J. H. Liu, *J. Am. Chem. Soc.* **1994**, *116*, 1599–1600; (d) S. J. Danishefsky, J. J. Masters, W. B. Young, J. T. Link, L. B. Snyder, T. V. Magee, D. K. Jung, R. C. A. Isaacs, W. G. Bornmann, C. A. Alaimo, C. A. Coburn, M. J. Di Grandi, *J. Am. Chem. Soc.* **1996**, *118*, 2843–2859; (e) P. A. Wender, N. F. Badham, S. P. Conway, P. E. Floreacig, T. E. Glass, J. B. Houze, N. E. Krauss, D. Lee, D. G. Marquess, P. I. McGrane, W. Meng, M. G. Natchus, A. J. Schuker, J. C. Sutton, R. E. Taylor, *J. Am. Chem. Soc.* **1997**, *119*, 2757–2758; (f) T. Mukaiyama, I. Shiina, H. Iwadare, M. Saitoh, T. Nishimura, N. Ohkawa, H. Sakoh, K. Nishimura, Y. Tani, M. Hasegawa, K. Yamada, K. Saitoh, *Chem. Eur. J.* **1999**, *5*, 121–161; (g) K. Morihira, R. Hara, S. Kawahara, T. Nishimori, N. Nakamura, H. Kusama, I. Kuwajima, *J. Am. Chem. Soc.* **1998**, *120*, 12980–12981.

27 (a) M. Ihara, S. Suzuki, Y. Tokunaga, K. Fukumoto, *J. Chem. Soc. Perkin Trans. 1* **1995**, 2811–2812; (b) V. J. Majo, S. Suzuki, M. Toyota, M. Ihara, *J. Chem. Soc. Perkin Trans. 1* **2000**, 3375–3381.

28 Y. Lu, J. R. Green, *Synlett* **2001**, *2*, 243–247.

29 R. P. L. Bell, J. B. P. A. Wijnberg, A. de Groot, *J. Org. Chem.* **2001**, *66*, 2350–2357.

30 K. C. Nicolaou, A. J. Roecker, H. Monenschein, P. Guntupalli, M. Follmann, *Angew. Chem. Int. Ed.* **2003**, *42*, 3637–3642.

31 A. G. Romero, J. A. Leiby, S. A. Mizsak, *J. Org. Chem.* **1996**, *61*, 6974–6979.

32 H. Hiemstra, W. N. Speckamp, in: *Comprehensive Organic Synthesis, Vol. 2* (Ed.: S. L. Schreiber), Pergamon Press, **1991**, pp. 1047–1082.

33 (a) D. Regoli, A. Boudon, J.-L. Fauchère, *Pharmacol. Rev.* **1994**, *46*, 551–599; (b) S. McLean, *Med. Res. Rev.* **1996**, *16*, 297–317.

34 S. J. Veenstra, K. Hauser, W. Schilling, C. Betschart, S. Ofner, *Bioorg. Med. Chem. Lett.* **1996**, *6*, 3029–3034.

35 (a) H. Paulsen, K. Todt, H. Ripperger, *Chem. Ber.* **1968**, *101*, 3365; (b) Y. L. Chow, C. J. Colon, J. N. S. Tam, *Can. J. Chem*, **1968**, *46*, 2821–2825.

36 M. A. Graham, A. H. Wadsworth, M. Thornton-Pett, C. M. Rayner, *Chem. Commun.* **2001**, 966–967.

37 J. Barluenga, G. P. Romanelli, L. J. Alvarez-García, I. Llorente, J. M. González, E. García-Rodríguez, S. García-Granda, *Angew. Chem. Int. Ed.* **1998**, *37*, 3136–3139.

38 D. Basavaiah, T. Satyanarayana, *Chem. Commun.* **2004**, 32–33.

39 G. Bose, E. Ullah, P. Langer, *Chem. Eur. J.* **2004**, *10*, 6015–6028.

40 J. A. González-Vera, M. T. García-López, R. Herranz, *Org. Lett.* **2004**, *6*, 2641–2644.

41 F. E. McDonald, F. Bravo, X. Wang, X. Wei, M. Toganoh, J. R. Rodríguez, B. Do, W. A. Neiwert, K. I. Hardcastle, *J. Org. Chem.* **2002**, *67*, 2515–2523.

42 A. X. Huang, Z. Xiong, E. J. Corey, *J. Am. Chem. Soc.* **1999**, *121*, 9999–10003.

43 M. Nakatani, M. Nakamura, A. Suzuki, M. Inoue, T. Katoh, *Org. Lett.* **2002**, *4*, 4483–4486.

44 M. Nakamura, A. Suzuki, M. Nakatani, T. Fuchikami, M. Inoue, T. Katoh, *Tetrahedron Lett.* **2002**, *43*, 6929–6932.

45 J. S. Clark, J. Myatt, C. Wilson, L. Roberts, N. Walshe, *Chem Commun.* **2003**, 1546–1547.

46 (a) K. Nozawa, M. Yuyama, S. Nakajima, K. Kawai, *J. Chem. Soc. Perkin Trans. 1*, **1988**, 2155–2160; (b) W. Acklin, F. Weibel, D. Arigoni, *Chimia*, **1977**, *31*, 63.

47 J. Mulzer, S. Greifenberg, J. Buschmann, P. Luger, *Angew. Chem. Int. Ed.* **1993**, *32*, 1173–1174.

48 C. Venkatesh, H. Ila, H. Junjappa, S. Mathur, V. Huch, *J. Org. Chem* **2002**, *67*, 9477–9480.

49 P. K. Mohanta, S. Peruncheralathan, H. Ila, H. Junjappa, *J. Org. Chem.* **2001**, *66*, 1503–1508.

50 J. Jiricek, S. Blechert, *J. Am. Chem. Soc.* **2004**, *126*, 3534–3538.

51 Y. Wang, A. M. Arif, F. G. West, *J. Am. Chem. Soc.* **1999**, *121*, 876–877.

52 (a) J. W. Daly, *J. Nat. Prod.* **1998**, *61*, 162–172; (b) E. X. Albuquerque, K. Kuba, J. W. Daly, *J. Pharmacol. Exp. Ther.* **1974**, *189*, 513–524.

53 W.-M. Dai, W. L. Mak, A. Wu, *Tetrahedron Lett.* **2000**, *41*, 7101–7105.

54 G. Dyker, W. Stirner, G. Henkel, M. Köckerling, *Tetrahedron Lett.* **1999**, *40*, 7457–7458.

55 X. Li, B. Wu, X. Z. Zhao, Y. X. Jia, Y. Q. Tu, D. R. Li, *Synlett* **2003**, 623–626.

56 C.-A. Fan, X.-D. Hu, Y.-Q. Tu, B.-M. Wang, Z.-L. Song, *Chem. Eur. J.* **2003**, *9*, 4301–4310.

57 C.-A. Fan, B.-M. Wang, Y.-Q. Tu, Z.-L. Song, *Angew. Chem. Int. Ed.* **2001**, *40*, 3877–3880.

2
Anionic Domino Reactions

Anionic domino processes are the most often encountered domino reactions in the chemical literature. The well-known Robinson annulation, double Michael reaction, Pictet–Spengler cyclization, reductive amination, etc., all fall into this category. The primary step in this process is the attack of either an anion (e. g., a carbanion, an enolate, or an alkoxide) or a "pseudo" anion as an uncharged nucleophile (e. g., an amine, or an alcohol) onto an electrophilic center. A bond formation takes place with the creation of a new "real" or "pseudo-anionic" functionality, which can undergo further transformations. The sequence can then be terminated either by the addition of a proton or by the elimination of an X$^-$ group.

Besides the numerous examples of anionic/anionic processes, anionic/pericyclic domino reactions have become increasingly important and present the second largest group of anionically induced sequences. In contrast, there are only a few examples of anionic/radical, anionic/transition metal-mediated, as well as anionic/reductive or anionic/oxidative domino reactions. Anionic/photochemically induced and anionic/enzyme-mediated domino sequences have not been found in the literature during the past few decades. It should be noted that, as a consequence of our definition, anionic/cationic domino processes are not listed, as already stated for cationic/anionic domino processes. Thus, these reactions would require an oxidative and reductive step, respectively, which would be discussed under oxidative or reductive processes.

2.1
Anionic/Anionic Processes

Domino transformations combining two consecutive anionic steps exist in several variants, but the majority of these reactions is initiated by a Michael addition [1]. Due to the attack of a nucleophile at the 4-position of usually an enone, a reactive enolate is formed which can easily be trapped in a second anionic reaction by, for example, another α,β-unsaturated carbonyl compound, an aldehyde, a ketone, an imine, an ester, or an alkyl halide (Scheme 2.1). Accordingly, numerous examples of Michael/Michael, Michael/aldol, Michael/Dieckmann, as well as Michael/S$_N$-type sequences have been found in the literature. These reactions can be considered as very reliable domino processes, and are undoubtedly of great value to today's synthetic chemist.

Domino Reactions in Organic Synthesis. Lutz F. Tietze, Gordon Brasche, and Kersten M. Gericke
Copyright © 2006 WILEY-VCH Verlag GmbH & Co. KGaA, Weinheim
ISBN: 3-527-29060-5

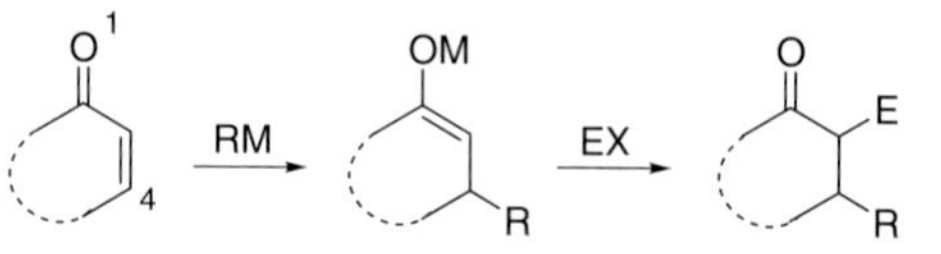

Scheme 2.1. Twofold anionic domino reaction initiated by a Michael addition.

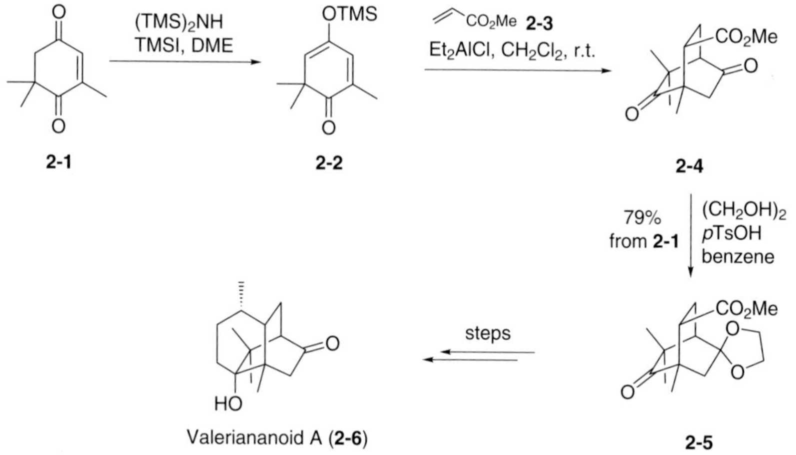

Scheme 2.2. Double Michael addition leading to bicyclo[2.2.2]octane **2-4**.

Twofold Michael additions have been utilized by the groups of Spitzner [2] and Hagiwara [3] to construct substituted bicyclo[2.2.2]octane frameworks. In Hagiwara's approach towards valeriananoid A (**2-6**) [4], treatment of trimethylsilyl-enol ether **2-2**, prepared from the corresponding oxophorone **2-1**, and methyl acrylate (**2-3**) with diethylaluminum chloride at room temperature (r.t.) afforded the bicyclic compound **2-4** (Scheme 2.2). Its subsequent acetalization allowed the selective protection of the less-hindered ketone moiety to provide **2-5**, which could be further transformed into valeriananoid A (**2-6**).

In addition, Hagiwara's group synthesized the solanapyrones D (**2-10**) and E (**2-11**), two members of the family of the solanapyrones (**2-7–2-11**) [5], using the same methodology. These compounds, which cause blight disease to tomatoes and potatoes, were isolated from the fungus *Altenaria solani* (Scheme 2.3).

The characteristic *trans*-decalin portion **2-15** of (–)-solanapyrone D (**2-10**) and E (**2-11**) was accessible from enolate **2-12** and methyl crotonate (**2-13**) via **2-14** (Scheme 2.4) [6].

Herein, the stereogenic center in **2-12** controls the stereochemistry in the way that the Michael addition occurs from the less-hindered α-face of the enolate to the *si*-side of the crotonate **2-13** according to transition structure **2-16**. The second Michael addition occurs from the same face, again under chelation control, followed by an axial protonation of the formed enolate to give the *cis*-compound **2-14a**. It should be noted that after the usual aqueous work-up procedure an inseparable

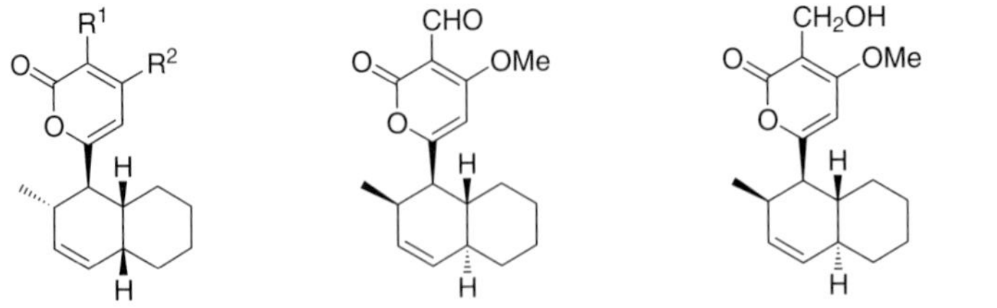

Solanapyrone A (**2-7**)
(R¹= CHO, R²= OMe)
Solanapyrone B (**2-8**)
(R¹= CH₂OH, R²= OMe)
Solanapyrone C (**2-9**)
(R¹= CHO, R²= NHCH₂CH₂OH)

Solanapyrone D (**2-10**)   Solanapyrone E (**2-11**)

Scheme 2.3. Natural products isolated from *Altenaria solani*.

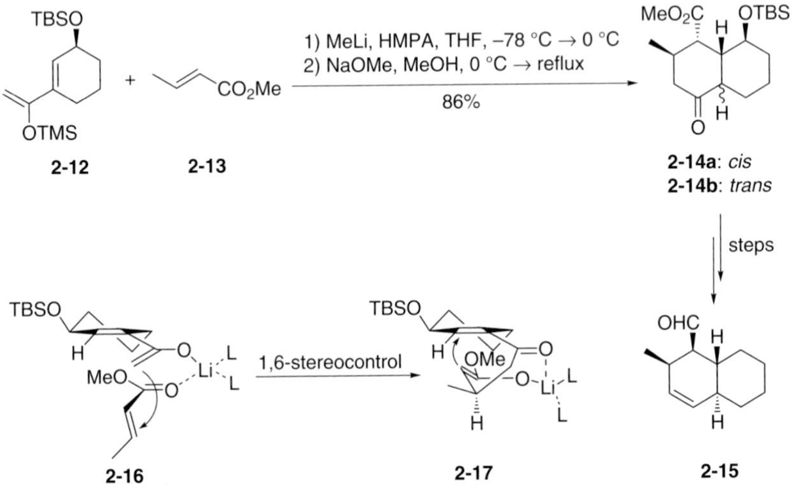

Scheme 2.4. Synthesis of the solanapyrones D (**2-10**) and E (**2-11**).

mixture of *cis*- and *trans*-decalones **2-14a** and **2-14b** is isolated. However, this can be circumvented by isomerizing the *cis*-compound **2-14a** to the *trans*-compound **2-14b** using NaOMe in MeOH at elevated temperatures.

An interesting domino Michael/Michael addition has also been reported by Carreño's group [7]. These authors started from chiral sulfur-containing quinamines and were able to build up enantiopure hydroindolones. Further sequences have been published by the research groups of Wicha [8], who elaborated a three-component, twofold Michael addition approach ensuring access to a valuable intermediate of vitamin D₃, Pollini [9], who designed a facile synthesis towards indolizidine as well as quinolizidine derivatives, and de Meijere [10]. By employing a domino Michael protocol, the latter group was able to prepare enantiopure β-amino acids (Scheme 2.5).

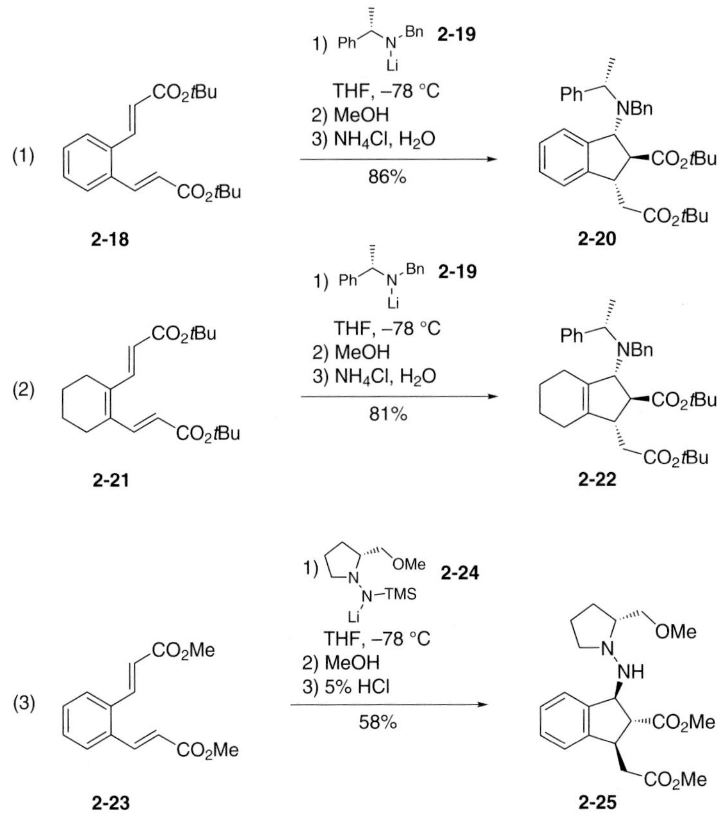

Scheme 2.5. Synthesis of β-amino acids.

For a facial selective assembly of the stereogenic centers and the introduction of the amino functionality, chiral nitrogen-containing reagents, such as benzyl(2-phenylethyl)amine (**2-19**) and trimethylsilyl RAMP derivative **2-24** were applied. Treatment of diacrylates **2-18**, **2–21**, and **2-23** with **2-19** and **2-24**, respectively, gave the protected amino acids **2-20**, **2–22**, and **2-25** in good yields as single isomers.

As the last example of this section, a conjugate addition sequence of Ikeda and coworkers will be discussed, in which a somewhat unusual Michael system **2-26** was used (Scheme 2.6) [11]. Reactions of 1-nitro-1-cyclohexene (**2-26**) and 4-hydroxybutynoates **2-27a–c** in the presence of the base KO*t*Bu gave *E/Z*-mixtures of substituted octahydrobenzofurans **2-28a–c** in excellent to good yield.

Domino Michael/aldol addition processes unquestionably represent the largest group of domino transformations. Numerous synthetic applications – for example, in natural product synthesis as well as for the preparation of other bioactive compounds – have been reported. Thus, the procedure is rather flexible and allows the use of many different substrates [12]. In this process it is possible, in theory, to establish up to two new C–C-bonds and three new stereogenic centers in a single step. For example, Collin's group developed a three-component approach.

2-26   **2-27**   **2-28**

a: $R^1= R^2= H$, $R^3=$ OMe
b: $R^1= nPr$, $R^2= H$, $R^3=$ OEt
c: $R^1= R^2= H$, $R^3=$ NMe$_2$

a: $R^1= R^2= H$, $R^3=$ OMe (100%)
b: $R^1= nPr$, $R^2= H$, $R^3=$ OEt (69%)
c: $R^1= R^2= H$, $R^3=$ NMe$_2$ (100%)

Scheme 2.6. Synthesis of octahydrobenzofurans.

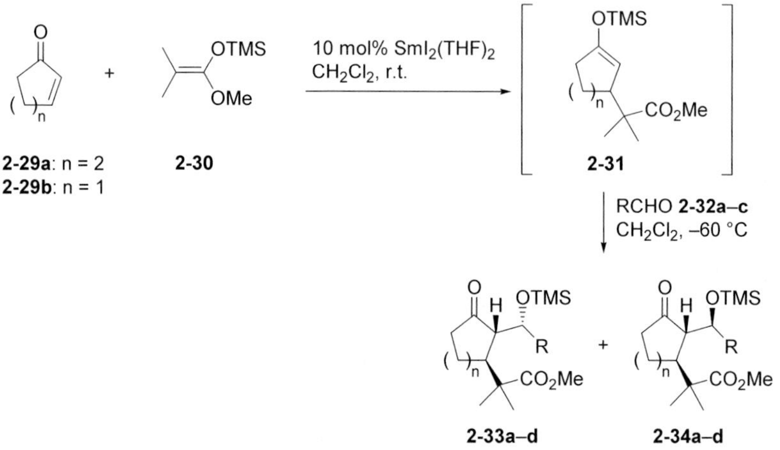

2-29a: n = 2   **2-30**   **2-31**
2-29b: n = 1

RCHO **2-32a–c**
CH$_2$Cl$_2$, –60 °C

2-33a–d   **2-34a–d**

Entry	α,β-Unsaturated ketone **2-29**	Aldehyde **2-32**	T [°C]	Major product **2-33**	**2-33/2-34**	Yield [%]
1	a	PhCHO a	–60	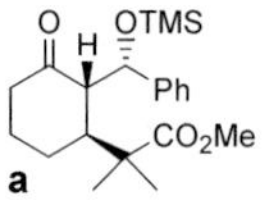 a	98:2	70
2	b	PhCHO a	–30	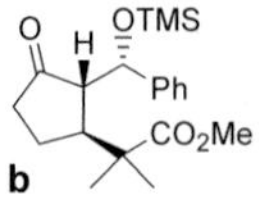 b	80:20	77
3	b	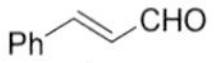 Ph⁓CHO b	–60	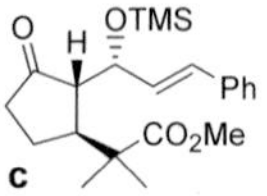 c	95:5	72
4	b	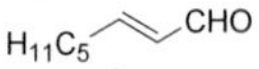 H$_{11}$C$_5$⁓CHO c	–60	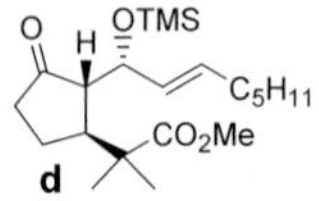 d	95:5	74

Scheme 2.7. Domino Mukaiyama/Michael/aldol reactions catalyzed by SmI$_2$(THF)$_2$.

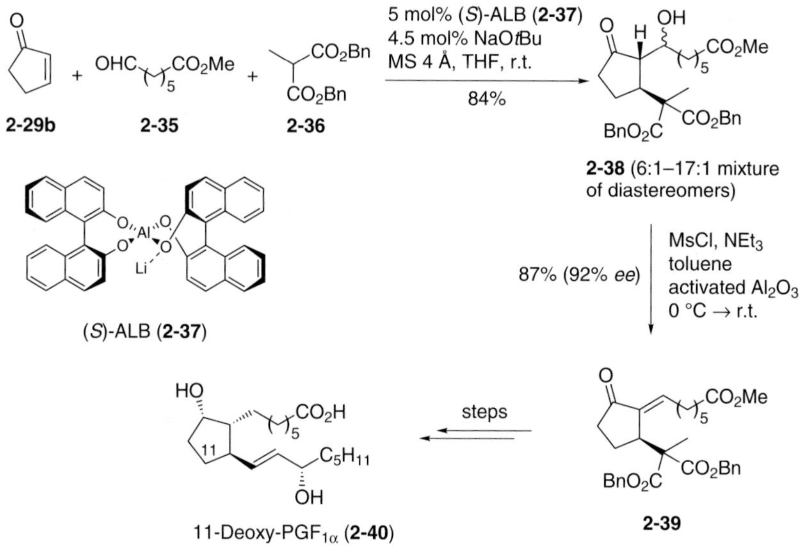

Scheme 2.8. Formation of 11-deoxy-PGF$_{1\alpha}$ (**2-40**).

Using a cyclic enone **2-29b** and an ester-TMS enolate **2-30** in the presence of catalytic amounts of SmI$_2$(THF)$_2$, the Michael addition and the Mukaiyama/aldol reaction with the added aldehyde **2-32** led to the diastereomeric adducts **2-33** and **2-34** via **2-31** with a $dr = 80{:}20$ to 98:2 and 70–77 % yield (Scheme 2.7) [13]. The major product is the *trans*-1,2-disubstituted cycloalkanone.

This finding is also in agreement with another three-component Michael/aldol addition reaction reported by Shibasaki and coworkers [14]. Here, as a catalyst the chiral AlLibis[(S)-binaphthoxide] complex (ALB) (**2-37**) was used. Such heterobimetallic compounds show both Brønsted basicity and Lewis acidity, and can catalyze aldol [15] and Michael/aldol [14, 16] processes. Reaction of cyclopentenone **2-29b**, aldehyde **2-35**, and dibenzyl methylmalonate (**2-36**) at r.t. in the presence of 5 mol% of **2-37** led to β-hydroxy ketones **2-38** as a mixture of diastereomers in 84 % yield. Transformation of **2-38** by a mesylation/elimination sequence afforded **2-39** with 92 % *ee*; recrystallization gave enantiopure **2-39**, which was used in the synthesis of 11-deoxy-PGF$_{1\alpha}$ (**2-40**) (Scheme 2.8). The transition states **2-41** and **2-42** illustrate the stereochemical result (Scheme 2.9). The coordination of the enone to the aluminum not only results in its activation, but also fixes its position for the Michael addition, as demonstrated in TS-**2-41**. It is of importance that the following aldol reaction of **2-42** is faster than a protonation of the enolate moiety.

Furthermore, the same methodology was used for an approach towards enantiopure PGF$_{1\alpha}$ (**2-46**) through a catalytic kinetic resolution of racemic **2-43** using (S)-ALB (**2-37**) (Scheme 2.10) [14]. Reaction of **2-35**, **2-36** and **2-43** in the presence of **2-37** led to **2-44** as a 12:1 mixture of diastereomers in 75 % yield (based on malonate **2-36**). The transformation proceeds with excellent enantioselectivity; thus, the enone **2-45** obtained from **2-44** shows an *ee*-value of 97 %.

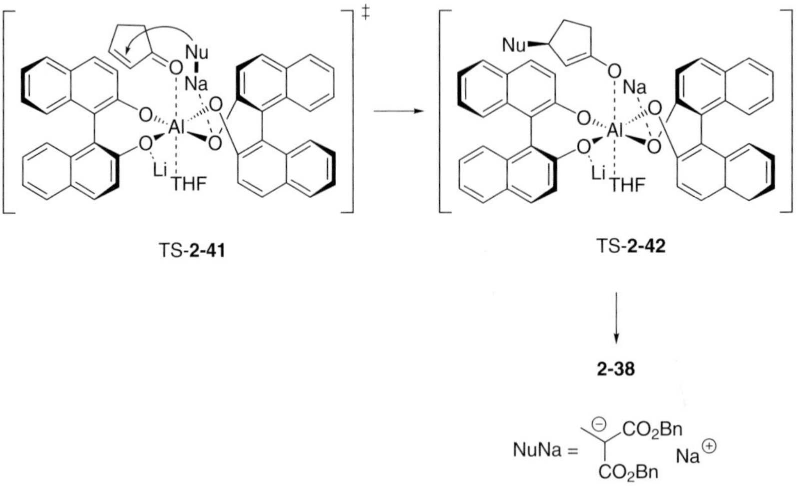

Scheme 2.9. Proposed transition states for the synthesis of **2-38**.

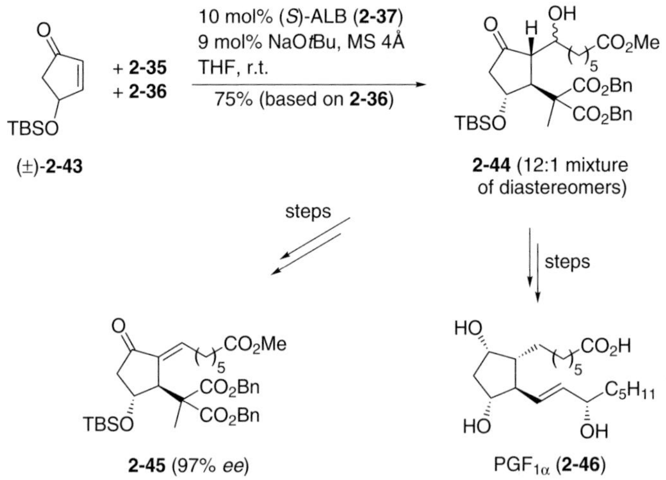

Scheme 2.10. Conversion of racemic **2-43** into enantiopure **2-45** using a kinetic resolution strategy.

Feringa's group has demonstrated that cyclopentene-3,5-dione monoacetals as **2-47** can also be successfully applied as substrates in an asymmetric three-component domino Michael/aldol reaction with dialkyl zinc reagents **2-48** and aromatic aldehydes **2-49** [17]. In the presence of 2 mol% of the *in-situ*-generated enantiomerically pure catalyst $Cu(OTf)_2$/phosphoramidite **2-54**, the cyclopentanone derivatives **2-51** were formed nearly exclusively in good yields and with high *ee*-values (Scheme 2.11).

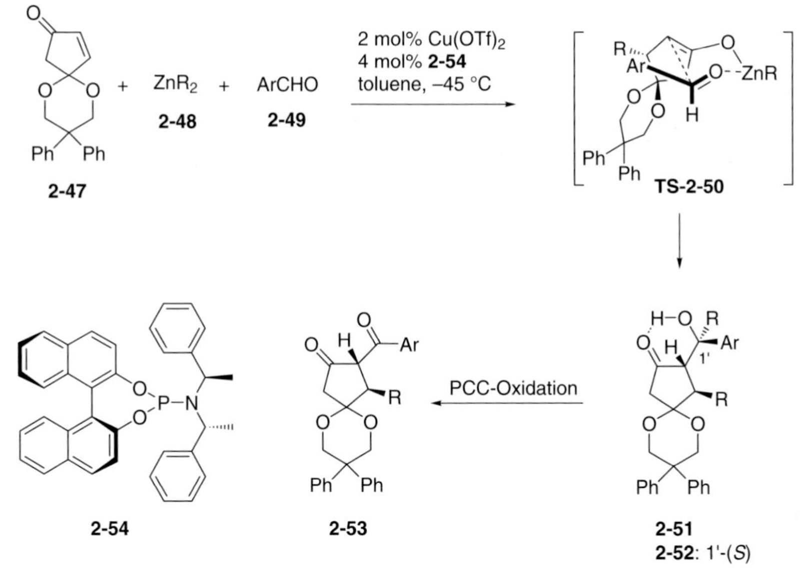

Scheme 2.11. Enantioselective Synthesis of **2-51** and **2-53**.

Entry	2-49	R	2-49	Ar	2-51	Yield [%]	2-53	ee [%]
1	a	Et	a	Ph	a	76	a	94
2	b	nBu	a	Ph	b	69	b	94
3	a	Et	b	pBrC$_6$H$_4$	c	69	c	96
4	b	nBu	b	pBrC$_6$H$_4$	d	64	d	97

The selectivity of the aldol addition can be rationalized in terms of a Zimmerman–Traxler transition-state model with TS-**2-50** having the lowest energy and leading to *dr*-values of >95:5 for **2-51** and **2-52** [18]. The chiral copper complex, responsible for the enantioselective 1,4-addition of the dialkyl zinc derivative in the first anionic transformation, seems to have no influence on the aldol addition. To facilitate the *ee*-determination of the domino Michael/aldol products and to show that **2-51** and **2-52** are 1'-epimers, the mixture of the two compounds was oxidized to the corresponding diketones **2-53**.

The method has been used for a short asymmetric synthesis of (–)-prostaglandin E$_1$ methyl ester (PGE$_1$) (**2-58**) starting from **2-47**, **2-55** and **2-56** (Scheme 2.12) [17]. The domino reaction provided **2-57** in 60 % yield as mixture of two diastereomers in reasonable stereoselectivity (*trans-threo:trans-erythro* ratio 83:17). Further transformations led to **2-58** in an overall yield of 7 % and 94 % *ee* in seven steps.

Another group of natural products, namely the biologically active lignans of the aryltetralin series – for example, isopodophyllotoxone (**2-59**), picropodophyllone (**2-60**), and podophyllotoxin (**2-61**) (Scheme 2.13) [19] – have also been synthesized using a domino Michael/aldol process.

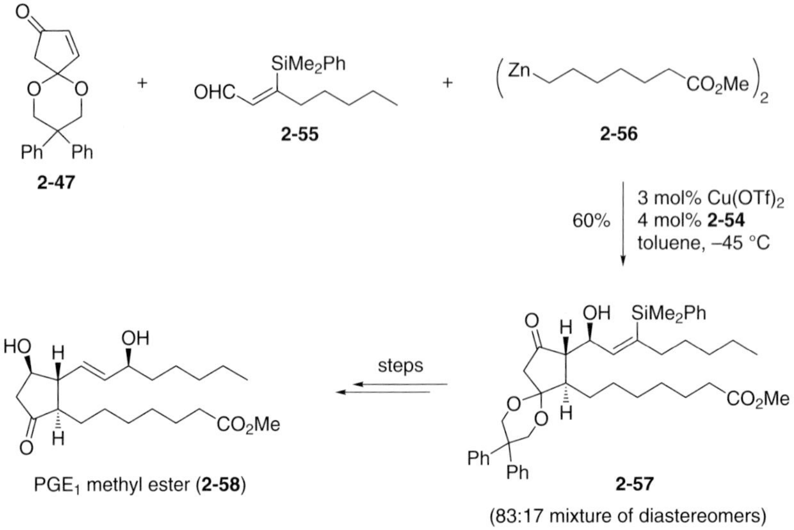

Scheme 2.12. Synthesis of PGE₁ methyl ester.

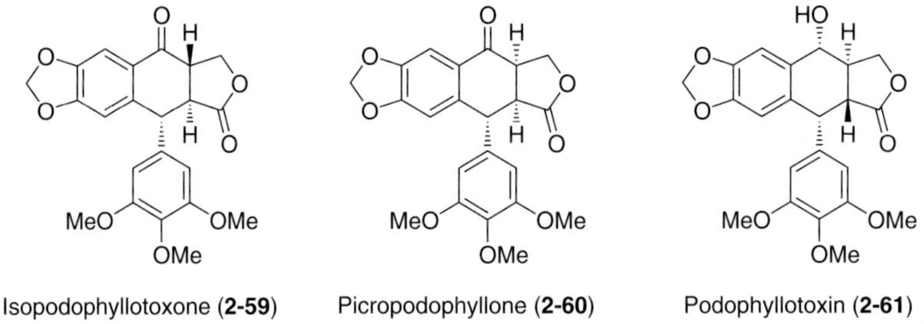

Isopodophyllotoxone (**2-59**)   Picropodophyllone (**2-60**)   Podophyllotoxin (**2-61**)

Scheme 2.13. Lignans of the aryltetralin series.

Thus, Pelter and Ward [20], as well as Ohmizu and Iwasaki [21], have shown independently that a conjugate addition of an acyl anion equivalent to a butenolide, followed by *in-situ* trapping of the resulting enolate with an aromatic aldehyde can be employed for the synthesis of podophyllotoxin derivatives. Reaction of TBS-protected cyanohydrin **2-62**, 2-butenolide (**2-63**) and 3,4,5-trimethoxybenzaldehyde (**2-64**) in the presence of LDA afforded **2-65**, which led to **2-66** on treatment with TFA in CH_2Cl_2 in 77 % yield over two steps (Scheme 2.14) [21]. Finally, the desired **2-59** was obtained in 92 % yield by unveiling the keto function with NH_4F in a slightly aqueous THF/DMF (10:1) solution. Interestingly, by treatment of **2-66** with TBAF in CH_2Cl_2/AcOH the isomerized diastereomeric product **2–60** was obtained.

Instead of using an aldehyde for trapping the primarily formed enolate, there are also a few examples which involve an imino acceptor in the second anionic step. The Collin group used a lanthanide iodide-mediated reaction of a ketene silyl acetal

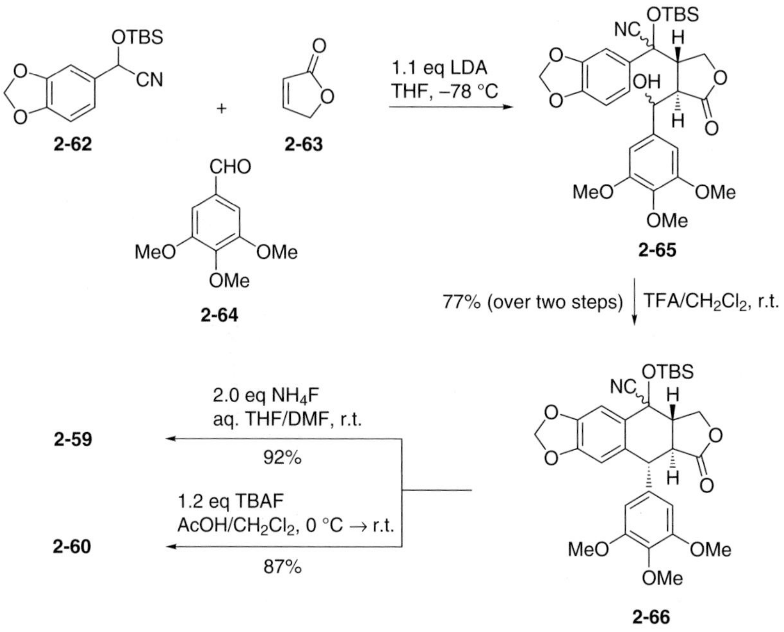

Scheme 2.14. Synthesis of isopodophyllotoxone (**2-59**) and picropodophyllone (**2-60**).

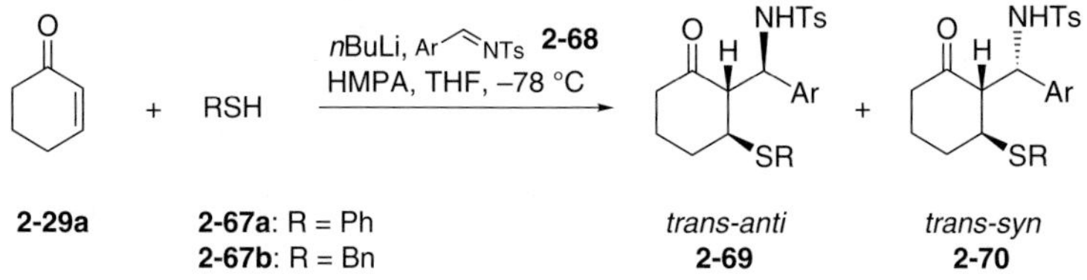

Entry	Thiol 2-67	Ar	Imine 2-68	Product 2-69/2-70	Ration	Yield [%]
1	a	pMePh	a	a	16:1	92
2	a	pClPh	b	b	17:1	92
3	a	pMeOPh	c	c	9:1	83
4	a	2-Furyl	d	d	23:1	82
5	b	Ph	e	e	4:1	96

Scheme 2.15. Three-component coupling of thiolate **2-67**, cyclohexenone (**2-29a**), and imines **2-68**.

and a cyclic α, β-unsaturated ketone, followed by the addition of a glyoxylic or aromatic imine to give 2,3-disubstituted cyclopentanones and cyclohexanones with an amine moiety in the side chain [22]. A further variant has been published by Hou and coworkers [23]. In a three-component reaction, cyclohexenone (**2-29a**), thiols **2-67**, as non-carbon Michael donors, and different N-tosylimines **2-68** as acceptors led to the β-amino ketones **2-69** and **2-70** in partly remarkable diastereoselectivity (Scheme 2.15).

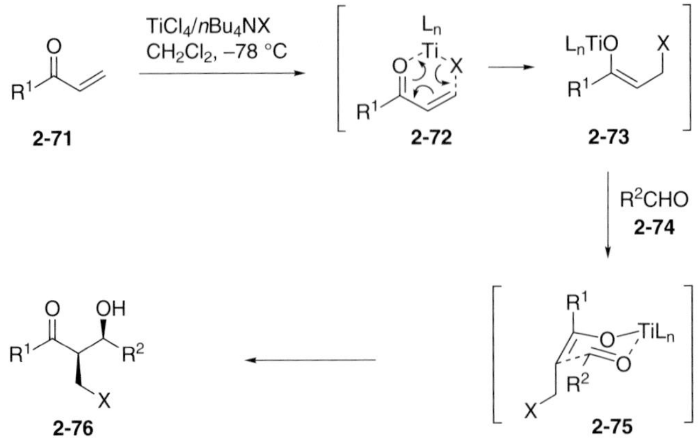

Scheme 2.16. Intermolecular domino Michael/aldol process initiated by the addition of halide to an enone.

Entry	2-71	R¹	2-74	R²	X	Product 2-76	syn/anti	Yield [%]
1	a	Me	a	Ph	Cl	a	>99:1	61
2	a	Me	b	$oNO_2C_6H_4$	Cl	b	>99:1	90
3	a	Me	a	Ph	Br	c	>99:1	90
4	b	Ph	c	nHex	Br	d	>99:1	66
5	b	Ph	d	nBu	I	e	>99:1	73
6	c	Me	c	nHex	I	f	>99:1	80

Domino Michael/aldol processes, which are initiated by the addition of a halide to an enone or enal, have found wide attention. They are valuable building blocks, as they can be easily converted into a variety of extended aldols *via* subsequent S_N2 reactions with nucleophiles or a halide/metal exchange. As an example, α-haloalkyl-β-hydroxy ketones such as **2-76** have been obtained in very good yields and selectivities by reaction of enones **2-71** with nBu_4NX in the presence of an aldehyde **2-74** and $TiCl_4$ as described by the group of Shinokubo and Oshima (Scheme 2.16) [24].

The selective generation of *syn*-isomers can plausibly be explained by the formation of (Z)-enolates **2-73** via **2-72** in the addition step and by a six-membered transition state **2-75** in the subsequent aldol reaction.

Further investigations in this field revealed that the described domino sequence can also be performed in an intramolecular mode to afford to 2-acyl-3-halocyclohexanols **2-78** from **2-77** (Scheme 2.17) [24e].

Enantiopure compounds of type **2-78** can easily be obtained, as elaborated by the group of Li and Headley, using an Evans auxiliary controlled conjugate addition of nucleophiles to α,β-unsaturated compounds **2-79** and subsequent trapping of the resulting enolates by aldehydes **2-81** to give **2-83** in high yield and excellent selectivity (Scheme 2.18) [25]. The proposed mechanism based on the primary formation of an (Z)-enolate **2-80** and an open transition state **2-82** accounts for the observed high *anti*-diastereoselectivity.

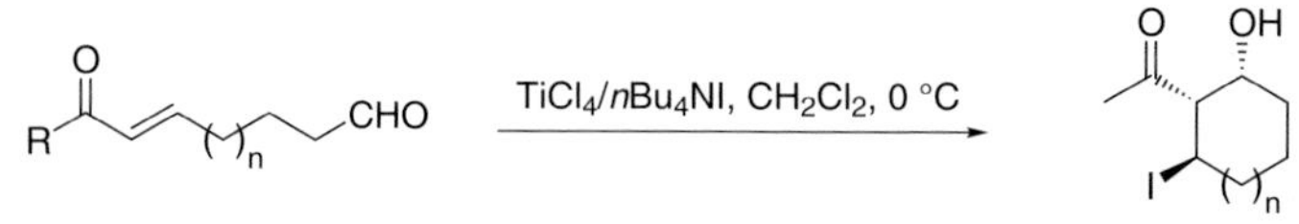

2-77 2-78

Entry	2-77	Product 2-78	*dr*	Yield [%]
1	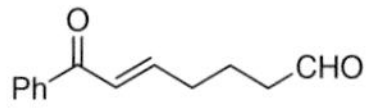	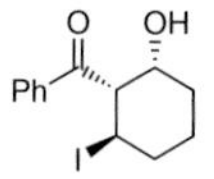	>99 : 1	85
2	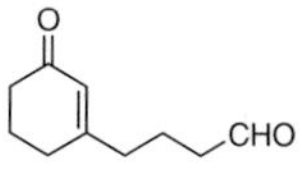	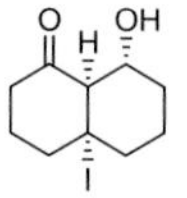	>99 : 1	33
3	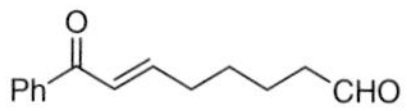	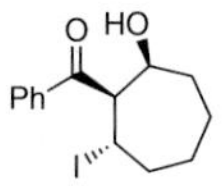	>99 : 1	70

Scheme 2.17. Intramolecular domino Michael/aldol process initiated by the addition of halide to an enone.

The halogenide-initiated domino Michael/aldol process can also be performed with ynones such as **2-84** and aldehydes **2-85** using $TiCl_4$, $TiBr_4$, $TiCl_4/(n\mathrm{Bu}_4)NI$, Et_2AlI, MgI_2 or TMSI as halogen sources to give β-halo Baylis–Hillman adducts such as **2-86**, as described by Li (Scheme 2.19) [26]. These compounds are of interest for Pd-mediated coupling reactions, for example. As a general trend, the use of $TiCl_4$, $TiBr_4$, as well as $TiCl_4/(n\mathrm{Bu}_4)NI$ leads to (*E*)-β-halo Baylis–Hillman adducts as the major product. In contrast, with MgI_2, Et_2AlI or TMSI the corresponding (*Z*)-vinyl isomers are formed preferentially. In a very recent paper, Paré and coworkers [27] have demonstrated this (see Scheme 2.19), within the synthesis of α-substituted-β-iodovinyl ketones **2-86** with a (*Z*)-configuration. Magnesium iodide in CH_2Cl_2 is the halogen source of choice.

An unusual two-component domino Michael/aldol process was described by Tomioka and coworkers in which the initiating step is the formation of an α-lithiated vinyl-phosphine oxide [28] or vinyl phosphate [29].

As shown in Scheme 2.20, selective lithiation of substrate **2-87** by treatment with LDA in THF at –78 °C triggers an intramolecular Michael/intermolecular aldol addition process with benzaldehyde to give a mixture of diastereomers **2-90** and **2-91**. **2-91** was afterwards transformed into **2-92**, which is used as a chiral ligand for Pd-catalyzed asymmetric allylic substitution reactions [29].

Besides this unique above-described process, there a numerous examples of inter- and intramolecular domino Michael/aldol processes in which the sequence is initiated by the addition of a metalorganic compound to an enone moiety. The Kamimura group [30] synthesized several five- to seven-membered thio- and hy-

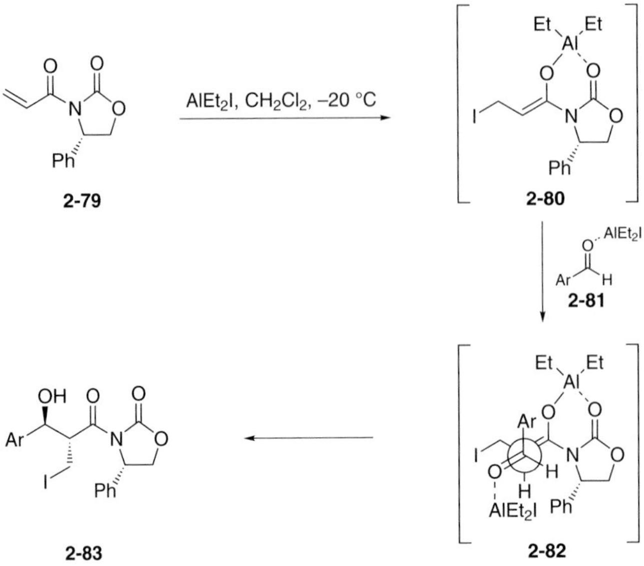

AlEt₂I, CH₂Cl₂, −20 °C

2-79

2-80

2-81

2-83

2-82

Entry	Substrate 2-82	Ar	Product 2-83	*de* [%]	Yield [%]
1	a	Ph	a	>95	91
2	b	2-naphthyl	b	>95	80
3	c	pMeC₆H₄	c	>95	85
4	d	m, p(Me)₂C₆H₄	d	>95	93
5	e	pPhC₆H₄	e	>95	91
6	f	pMeOC₆H₄	f	>95	82
7	g	pClC₆H₄	g	>95	89
8	h	pCF₃C₆H₄	h	>95	84

Scheme 2.18. Diastereoselective AlEt₂I-mediated domino Michael/aldol process.

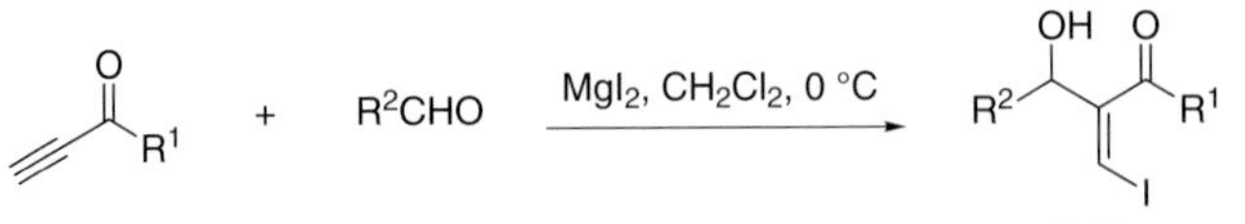

2-84   +   R²CHO   **2-85**   MgI₂, CH₂Cl₂, 0 °C   **2-86**

6 examples: 81–90% yield, (*Z*):(*E*) >98:2

Scheme 2.19. MgI₂-mediated synthesis of β-iodo Baylis–Hillman adducts.

droxy-functionalized carbocycles; Tomioka and coworkers [31] used the methodology in the total synthesis of the carbonucleoside (−)-neplanocin A [32], a naturally occurring compound showing (*S*)-adenosylhomocysteine hydrolase inhibitory activity [33]. Schneider and coworkers produced functionalized enan-

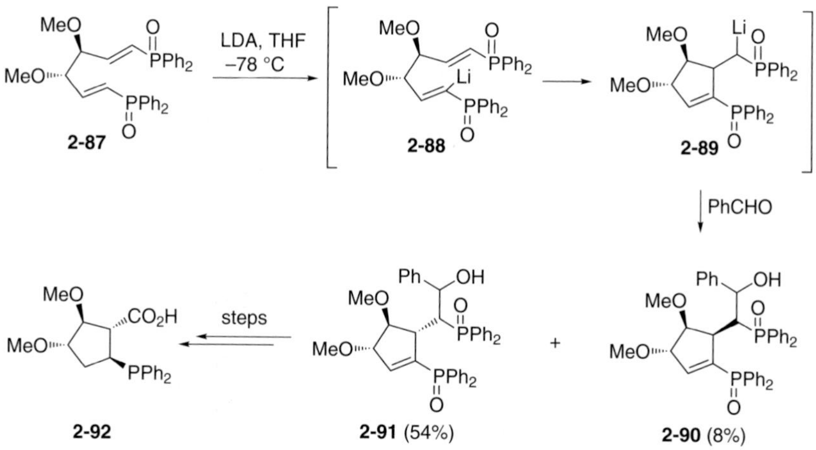

Scheme 2.20. Synthesis of substituted cyclopentenes and cyclopentanes.

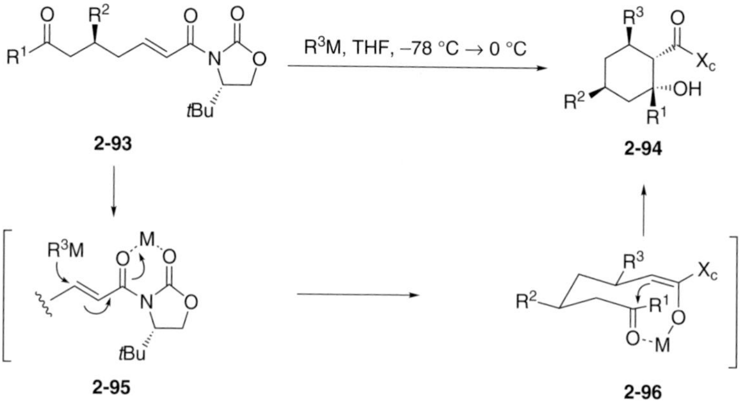

Entry	Substrate 2-93	R^1	R^2	R^3	M	Product 2-94	Yield [%]
1	a	H	Me	nBu	CuLiI/Me$_2$AlCl	a	41
2	b	H	Ph	nBu	CuLiI/Me$_2$AlCl	b	42
3	c	Me	Me	Et	CuMgBr$_2$	c	71
4	c	Me	Me	nBu	CuLiI/Me$_2$AlCl	d	83
5	c	Me	Me	allyl	CuMgBr$_2$	e	81
6	d	Me	Ph	nBu	CuLiI/Me$_2$AlCl	f	51

Scheme 2.21. Domino Michael/aldol reactions of 7-keto-2-enimides **2-93**.

tiopure cyclohexanes of type **2-94**, starting from **2-93** and a cuprate with excellent selectivity of >95:5 (Scheme 2.21) [34]. The stereochemical outcome can be explained taking the TS **2-95** and TS **2-96** into account.

Similarly, the addition of an amine to the enone moiety can initiate a domino process leading to substituted diaminocyclohexanes [34]. In this transformation an imino aldol reaction occurs. The observed stereoselectivity was again >95:5, and the yield between 51% and 69% in all cases.

The group of Terashima [35] developed an asymmetric domino Michael/aldol process using the chinchona alkaloid (–)-cinchonidine (**2-103**), to prepare an intermediate for the synthesis of the natural product (–)-huperzine A (**2-102**) [36] (Scheme 2.22).

Reaction of β-ketoester **2-97** and acrolein **2-98** in presence of stoichiometric amounts of **2-103** led to the desired product **2-100** in 45% yield. A transition-state model **2-99** may be postulated assuming an ion-pairing mechanism as reported for similar asymmetric transformations [37]. The diastereomeric mixture of **2-100** was transformed into **2-101** by mesylation and subsequent elimination. Despite the moderate 64% *ee* determined for **2-101**, it was possible to obtain optically pure **2-101** by recrystallization from hexane.

An impressive organocatalytic asymmetric two-component domino Michael/aldol reaction has been recently published by Jørgensen and coworkers (Scheme 2.23) [38].

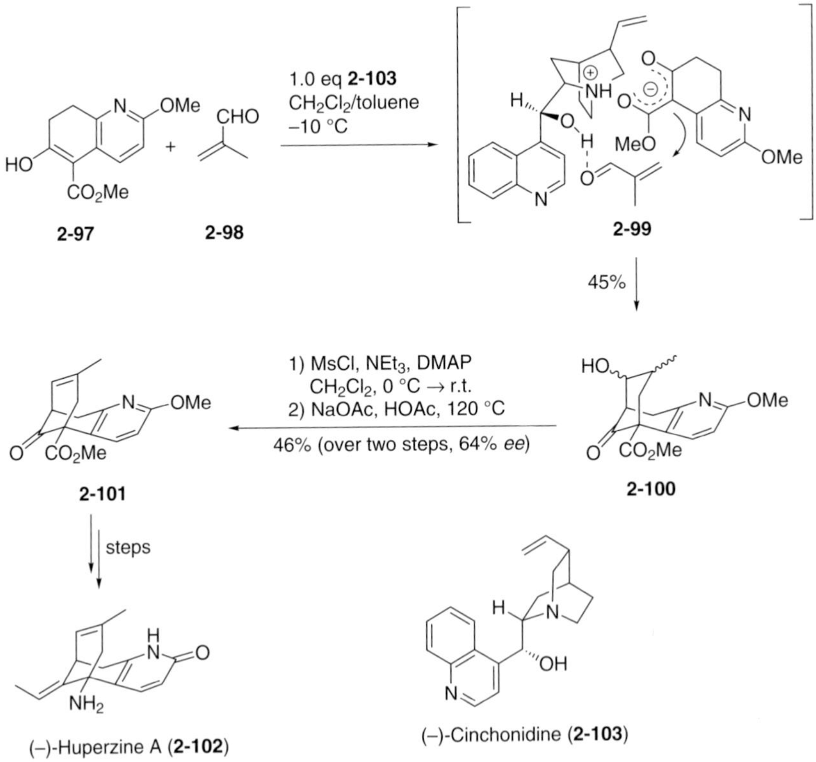

Scheme 2.22. Enantioselective domino Michael/aldol process.

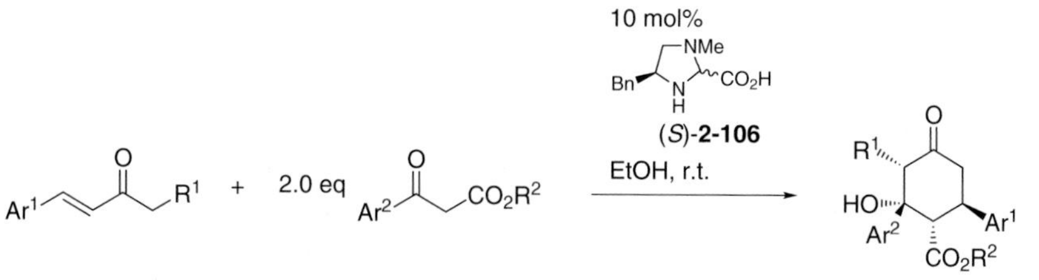

10 mol%

(*S*)-**2-106**

EtOH, r.t.

Ar¹$\overset{O}{\diagdown}$R¹ + 2.0 eq Ar²$\overset{O}{\diagdown}$CO₂R² ⟶ product **2-107**

2-104   **2-105**   **2-107**

Entry	2-104	Ar¹	R¹	Substrate 2-105	Ar²	R²	Product 2-107	dr	ee [%]	Yield [%]
1	a	2-Np	H	a	Ph	Bn	a	>97:3	91	85
2	b	*o*NO₂Ph	H	a	Ph	Bn	b	>97:3	96	56
3	c	2-furyl	H	a	Ph	Bn	c	>97:3	85	40
4	d	Ph	Me	a	Ph	Bn	d	>97:3	96	61
5	e	Ph	H	b	*p*FPh	Me	e	>97:3	92	44

Scheme 2.23. Domino Michael/aldol reaction of α,β-unsaturated ketones with β-ketoesters.

Inspired by the proline-catalyzed Robinson annulation pioneered by Wiechert, Hajos, Parrish and coworkers [39], they were able to construct cyclohexanones of type **2-107** with up to four stereogenic centers with excellent enantio- and diastereoselectivity from unsaturated ketones **2-104** and acyclic β-ketoesters **2-105** in the presence of 10 mol% phenylalanine-derived imidazolidine catalyst **2-106**. The final products can easily be converted into useful cyclohexanediols, as well as γ- and ε-lactones.

Alkenones were used by Rao and coworkers [40] to prepare cyclohexane derivatives which, for example, can be transformed into substituted arenes in a single step. Another interesting intermolecular Michael/intramolecular aldol reaction sequence for the construction of the highly substituted 2-hydroxybicyclo[3.2.1]octan-8-one framework has been described by Rodriguez' group [41]. This process can be extended to a three- and even a fourfold domino reaction [41a, 42, 43].

Krische and coworkers [44] developed a Rh-catalyzed asymmetric domino Michael/aldol reaction for the synthesis of substituted cyclopentanols and cyclohexanols. In this process, three contiguous stereogenic centers, including a quaternary center, are formed with excellent diastereo- and enantioselectivity. Thus, using an enantiopure Rh-BINAP catalyst system and phenyl boronic acid, substrates **2-108** are converted into the correspondding cyclized products **2-109** in 69–88 % yield and with 94 and 95 % *ee*, respectively (Scheme 2.24).

A proposed simplified mechanism for the conjugate addition/aldol cyclization, as depicted in Scheme 2.25, is based on detailed mechanistic studies performed on related Rh-catalyzed enone conjugate additions [45]. A model accounting for the observed relative stereochemistry invokes the intermediacy of a (*Z*)-enolate and a Zimmerman–Traxler-type transition state as shown in **2-110** to give **2-111**.

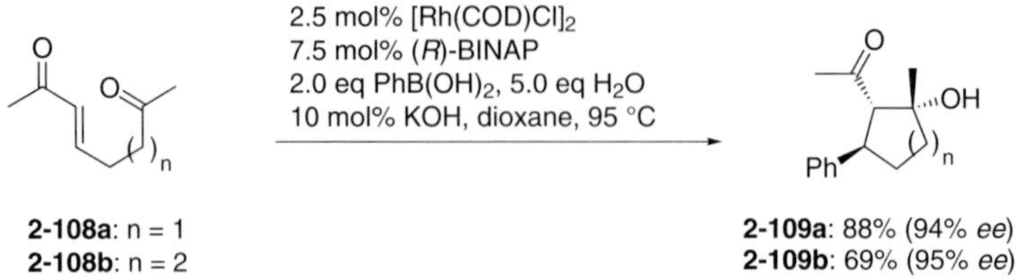
2.5 mol% [Rh(COD)Cl]2
7.5 mol% (R)-BINAP
2.0 eq PhB(OH)2, 5.0 eq H2O
10 mol% KOH, dioxane, 95 °C
OH
Ph
2-108a: n = 1
2-108b: n = 2
2-109a: 88% (94% ee)
2-109b: 69% (95% ee)

Scheme 2.24. Rh-catalyzed asymmetric domino Michael/aldol process.

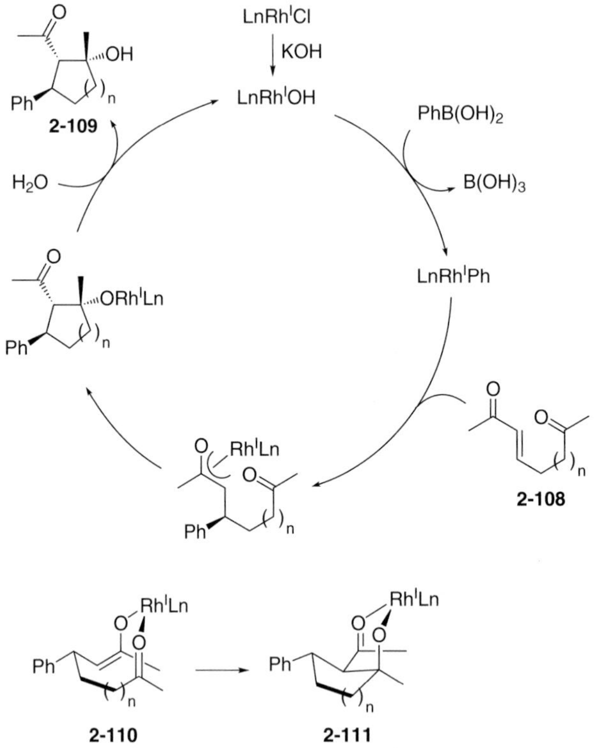
LnRhICl
KOH
LnRhIOH
PhB(OH)2
B(OH)3
LnRhIPh
2-108
2-109
H2O
ORhILn
RhILn
Ph
2-110
2-111

Scheme 2.25. Proposed catalytic cycle and stereochemical model.

In a more recent publication the same research group described a Cu(OTF)$_2$/(POEt)$_3$-catalyzed two-component Michael/aldol protocol of **2-112** and ZnEt$_2$ leading to annulated cyclopentanols [46]. They showed that the enolate formed in the 1,4-addition can be trapped not only by a keto moiety, but also by an ester (Dieckmann condensation) or a nitrile functionality present in the molecule. Thus, as depicted in Scheme 2.26, there is a broad product variety. Starting from **2-112**, compounds of type **2-114**, **2-115** and **2-116** can be obtained *via* the enolate **2-113**.

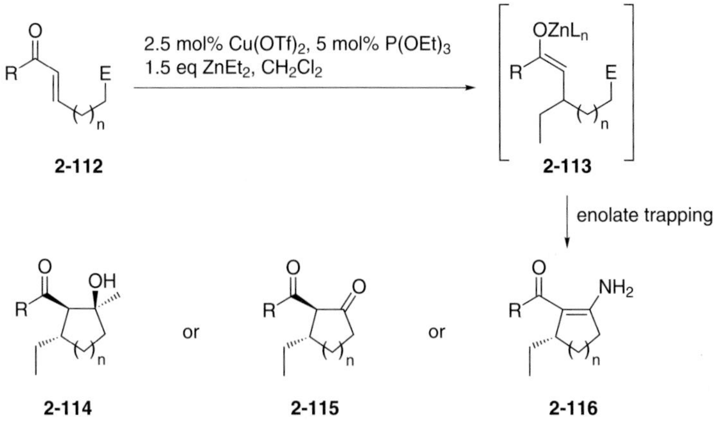

Scheme 2.26. Copper-mediated Michael addition/electrophilic trapping.

In the so-far described transformations at least one intermolecular process was involved. However, there are also examples where the necessary functionalities are all present in the starting material such as **2-117**, **2-119**, **2-121** and **2-123** [47]. These substrates can be transformed into polycyclic annulated cyclobutanes **2-118**, **2-120**, **2-122** and **2-124** using a base and a silylating agent (Scheme 2.27) [48]. The procedure is of great interest, since a number of biologically active polycyclic compounds incorporating a cyclobutane portion, for example endiandric acids [49], trihydroxydecipiadiene [50], italicene [51], lintenone [52] and filifolone [53], exist in nature. Fukumoto and coworkers reported that the domino process can be carried out under two different sets of conditions, either TBSOTf in the presence of Et$_3$N [48d] or TMSI in the presence of (TMS)$_2$NH [48c].

As already shown, domino Michael/Dieckmann processes are especially useful synthetic procedures with regard to the rapid, efficient assembly of complex organic molecules. This is particularly true for the construction of compounds containing a highly functionalized naphthoquinone or naphthalene unit as central element as found in napyradiomycin A1 (**2-125**) [54], bioxanthracene (–)-ES-242–4 (**2-126**) [55], dioxanthin (**2-127**) [56], and the bioactive compound S-8921 (**2-128**) [57] (Scheme 2.28).

The synthesis of these compounds using a domino Michael/Dieckmann process has been developed by Tatsuta [58], Müller [59], and Mori [60]. The reaction schemes are shown in Scheme 2.29.

A related domino Michael/Dieckmann process was used by Behar and coworkers [61] for the ABCD-ring system assembly of lactonamycin (**2-132**) [62].

Reaction of **2-129** with NaCN followed by treatment with NaH and TIPSCl led to the anthracene **2-130** in 68% yield (Scheme 2.30). The desired lactam **2-131** was then obtained by reduction of the cyano group and ring closure using CoCl$_2$/NaBH$_4$. Quite recently, this procedure has also been used for a formal total synthesis of tetracenomycin [63].

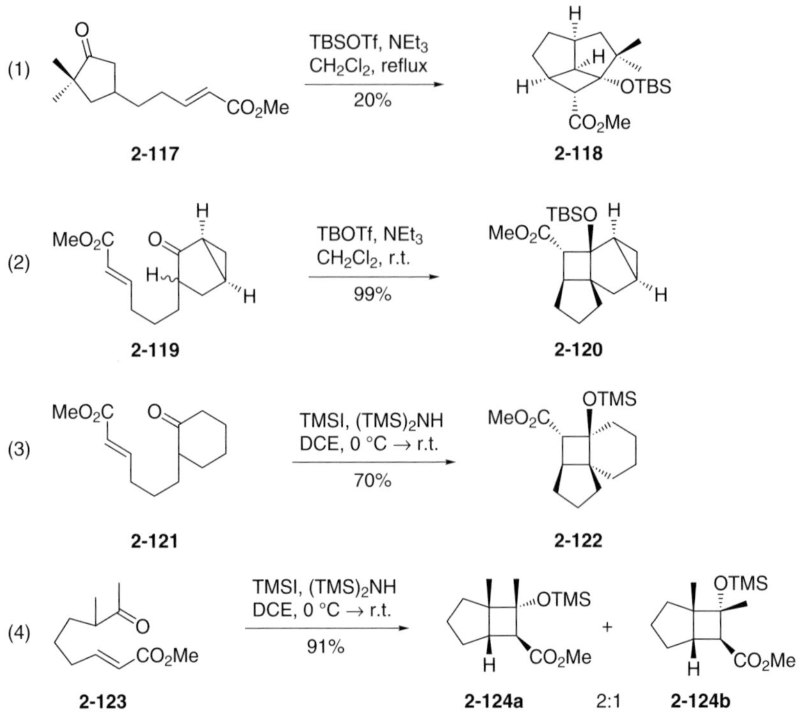

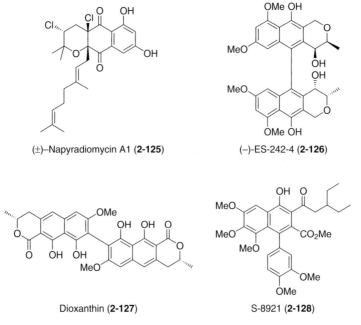

(1)

2-117

TBSOTf, NEt₃
CH₂Cl₂, reflux

20%

2-118

(2)

2-119

TBOTf, NEt₃
CH₂Cl₂, r.t.

99%

2-120

(3)

2-121

TMSI, (TMS)₂NH
DCE, 0 °C → r.t.

70%

2-122

(4)

2-123

TMSI, (TMS)₂NH
DCE, 0 °C → r.t.

91%

2-124a   2:1   2-124b

Scheme 2.27. Efficient synthesis of annulated cyclobutanes.

(±)–Napyradiomycin A1 (**2-125**)

(−)-ES-242-4 (**2-126**)

Dioxanthin (**2-127**)

S-8921 (**2-128**)

Scheme 2.28. Compounds containing a naphtoquinone or naphthalene unit as central element.

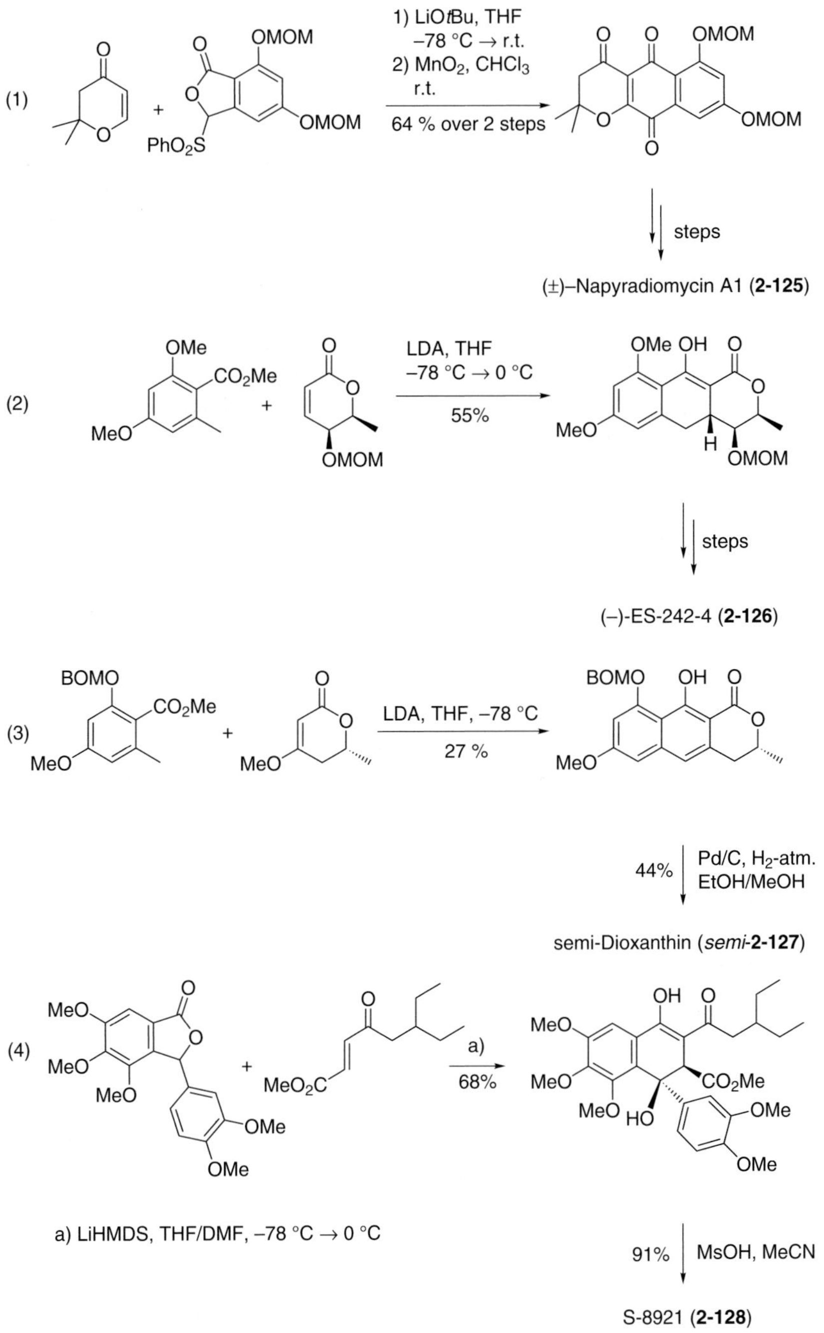

Scheme 2.29. Domino Michael addition/Dieckmann condensation sequences leading to precursors of natural products.

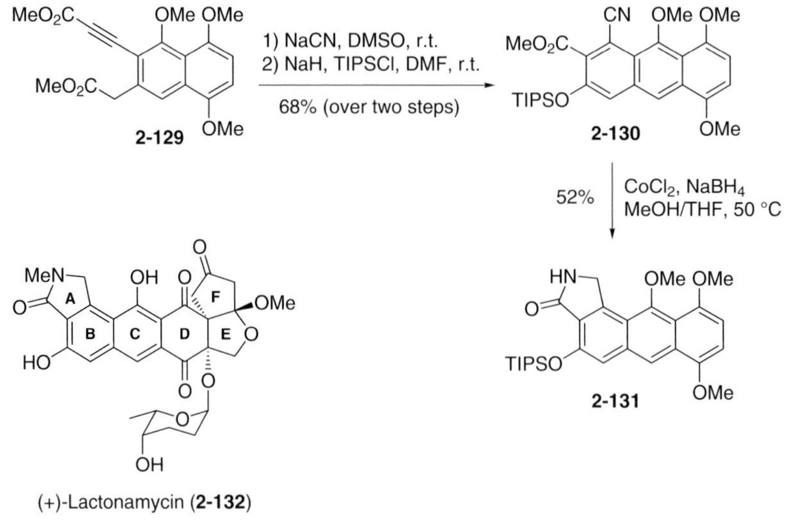

Scheme 2.30. Synthesis of the ABCD-ring skeleton of (+)-lactonamycin (**2-132**).

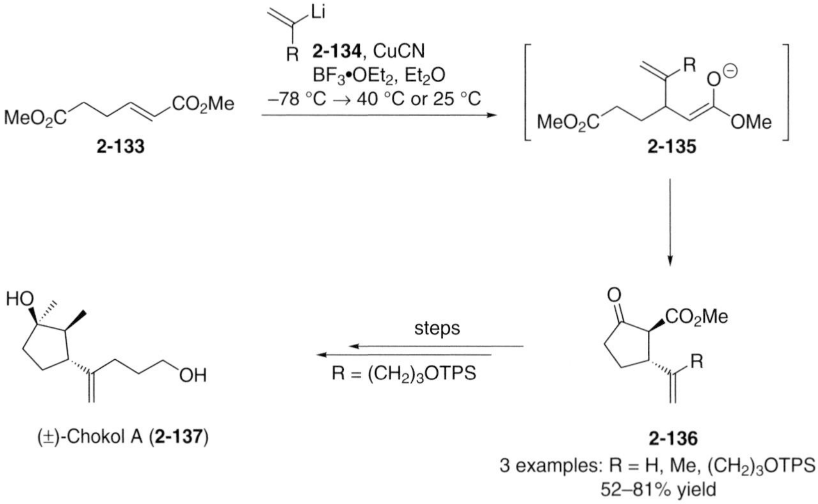

Scheme 2.31. Domino Michael addition/Dieckmann condensation process.

Groth and coworkers [64] used a Michael/Dieckmann combination for the synthesis of chokol (**2-137**) [65] starting from **2-133**; 1,4-addition employing a vinyl lithium compound **2-134** in the presence of CuCN led to the enolate **2-135** as intermediate, which was trapped by an intramolecular Dieckmann condensation to give **2-136** (Scheme 2.31). The domino process is highly selective affording only the *trans*-1,2-disubstituted cyclopentanones (dr>95 % according to gas chromatography) [66].

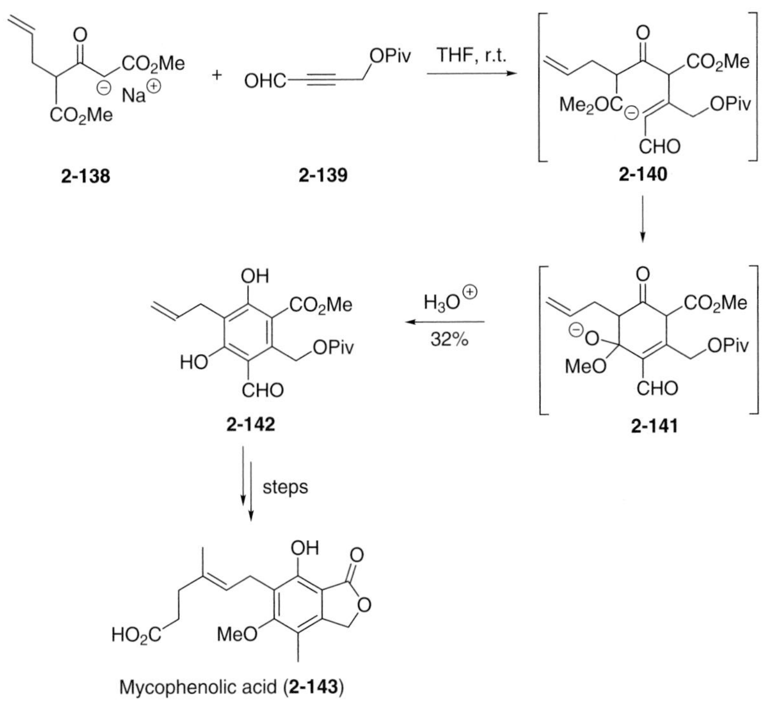

Scheme 2.32. Synthesis of mycophenolic acid (2-143).

Even fully substituted aromatic compounds can be prepared utilizing the Michael/Dieckmann strategy. As reported by Covarrubias-Zúñiga and coworkers (Scheme 2.32) [67], reaction of the anion of 1-allyl-1,3-acetonedicarboxylate (2-138) and the ynal 2-139 afforded the intermediate 2-140 which led to the resorcinol 2-142 with spontaneous aromatization under acidic conditions via 2-141 in an overall yield of 32%. 2-142 was transformed into mycophenolic acid (2-143) in only a few additional steps [68].

Tridachiahydropyrone belongs to the family of marine polypropionates [69]. Efforts towards its total synthesis have recently led to a revision of the structure with the new proposal 2-147 [70]. The construction of the highly substituted cyclohexenone moiety 2-146 which could be incorporated into this natural product [71] has been described by Perkins and coworkers (Scheme 2.33) [70, 72]. The conjugate addition/Dieckmann-type cyclization utilizing organocopper species as Michael donors afforded the enantiopure 2-145 in 68% yield. A further methylation of the β-ketoester moiety in 2-145 followed by an elimination led to the desired cyclohexenone 2-146.

Recently, the 2-substituted L-glutamate analogue (2R)-α-(hydroxymethyl)glutamate (HMG) (2-151) has been reported by the group of Kozikowski to serve as a potential bioactive compound [73]. Since the synthesis of such a small molecule should be rapid and practical in order to produce it on a multi-gram scale, a domino

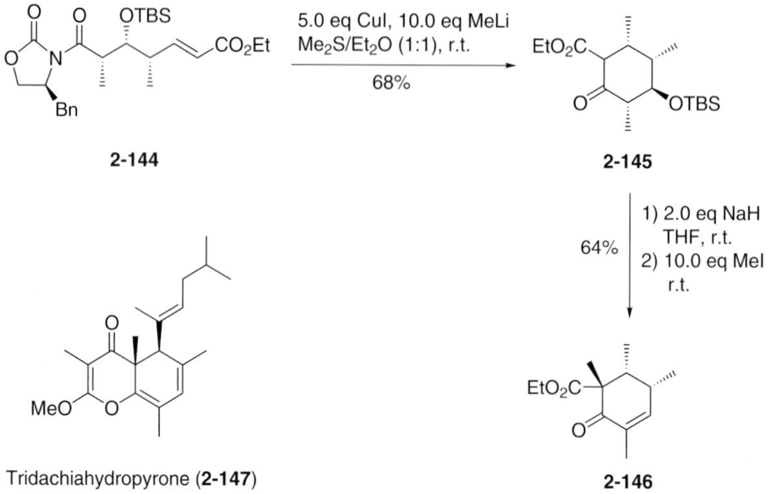

2-144   **2-145**

Tridachiahydropyrone (2-147)   **2-146**

Scheme 2.33. Formation of highly substituted chiral cyclohexenone derivatives using a domino conjugate addition/Dieckmann condensation.

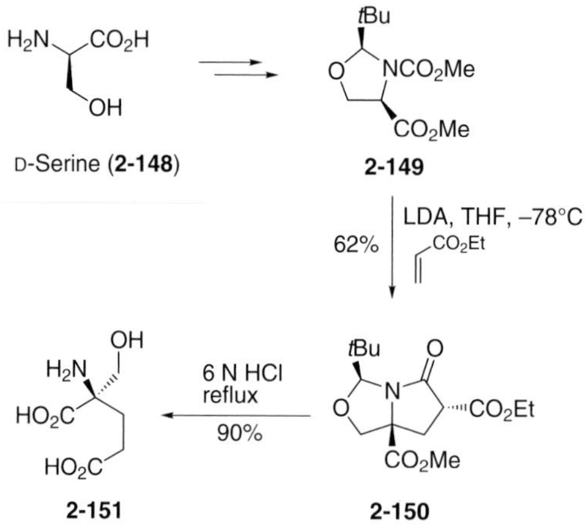

D-Serine **(2-148)**   **2-149**

2-151   **2-150**

Scheme 2.34. Synthesis of (2R)-α-(hydroxymethyl)glutamate (**2-151**).

Michael/Dieckmann strategy illustrated in Scheme 2.34 was envisaged [73]. Treatment of the oxazolidine **2-149**, derived from D-serine (**2-148**), with LDA followed by addition of acrylate gave the bicyclic compound **2-150** as a single stereoisomer in 62% yield after recrystallization. Finally, hydrolysis of **2-150** with 6 M HCl provided the desired HMG derivative(**2-151**) in 90% yield.

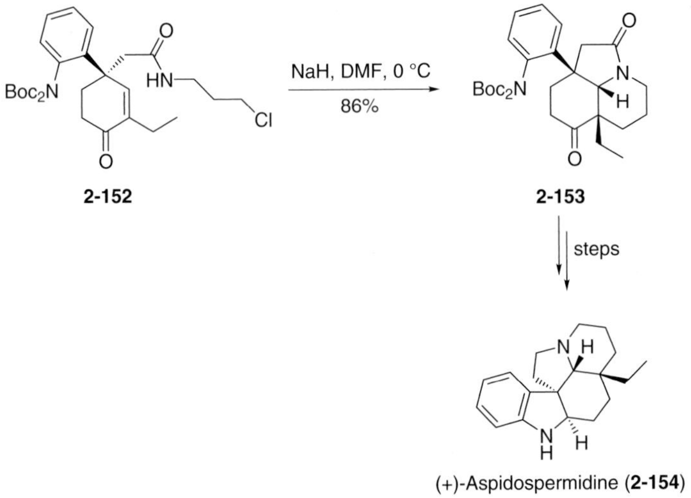

Scheme 2.35. Synthesis of (+)-aspidospermidine (**2-154**).

Enolates formed by a Michael addition can also be trapped by a S_N-type reaction. Applications of this approach have been found in almost every significant topic of organic chemistry. This includes the efficient construction of 2-substituted 3-aroyl-cyclohexanones and enantiopure β-amino acids by Enders' group [74] as well as the assembly of 3-azabicyclo[3.1.0]hexane skeletons by Braish and coworkers [75]. According to this scheme, key intermediates for the synthesis of natural products such as monamycin D_1 [76] and lignan substructures were also prepared by Hale's group [77] as well as by the groups of Mäkelä [78], Pelter and Ward [79] as an alternative to the Michael/aldol addition protocol [20, 21]. The groups of Marino [80] and Rubiralta [81] have independently developed synthetic routes towards the biologically interesting aspidosperma alkaloid (+)-aspidospermidine (**2-154**) [82]. The approach of Marino's group is depicted in Scheme 2.35. Treatment of substrate **2-152** with NaH at r.t. led to **2-153** as a single stereoisomer in 86 % yield. The stereogenic quaternary center in **2-152** was formed by a ketene lactonization involving a chiral vinyl sulfoxide [83].

Other applications of Michael/S_N-type domino sequences in natural product synthesis have been found in White's total synthesis [84] of (+)-kalkitoxin and Danishefsky's investigations into bicycloillicinone asarone acetal (**2-160**) (Scheme 2.36) [85]. The latter compound has been isolated from extracts of the wood *Illicium tashiroi*, and is reported to enhance the action of choline acetyltransferase, which catalyzes the synthesis of acetylcholine from its precursor [86]. Since one of the characteristic symptoms of Alzheimer's disease involves the degeneration of cholinergic neurons, resulting in markedly reduced levels of acetylcholine, compound **2-160** might serve as an agent in the treatment of such a disorder [87]. With the intention to construct the cage structure of **2-160**, a chromatographically inseparable 2:1 mixture of the epoxides **2-155** was treated with Et_2AlCN in THF at 0 °C to provide the enolate **2-156** in a regioselective 1,4-addition of cyanide, which subsequently at-

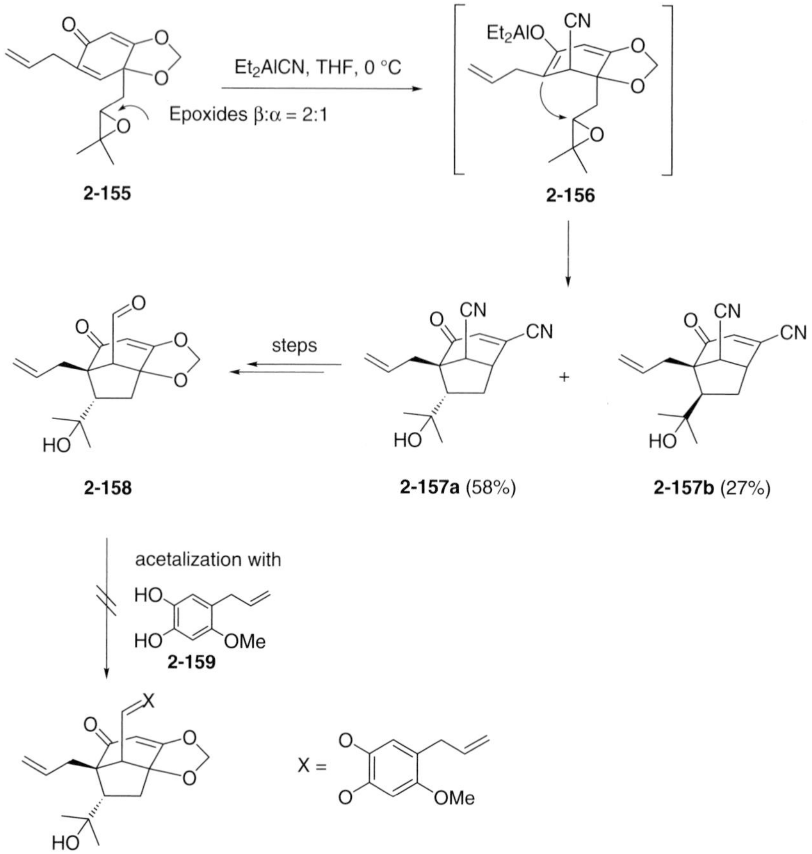

Bicycloillicinone asarone acetal (**2-160**)

Scheme 2.36. Synthesis of bicycloillicinone aldehyde (**2-158**) by a domino Michael addition/epoxide opening.

tacked the epoxide moiety. The products **2-157a** and **2-157b** were obtained after separation in 58% and 27% yields, respectively.

In only two further steps the aldehyde **2-158** was prepared, but this could not be transformed into the desired acetal **2-160** using catechol **2-159**.

The β-lactam antibiotic cephalosporin is one of the most important drugs for the treatment of bacterial infections. Recently, several new compounds such as cefadroxil (**2-161**), cephalexin (**2-162**), cefixime (**2-163**), and cefzil (**2-167**) have been isolated, which contain an alkyl or alkenyl substituent instead of an acetoxymethyl group at C-3 (Scheme 2.37 and 2.38) [88].

Kant and coworkers [89] synthesized cefzil (**2-167**) through a Normant cuprate addition to allene **2-164**, readily available from inexpensive penicillins, to give **2-165**, which cyclized to the cefzil precursor **2-166** in a S_N'-type reaction (Scheme 2.38). The conversion of **2-166** into **2-167** was already known [90].

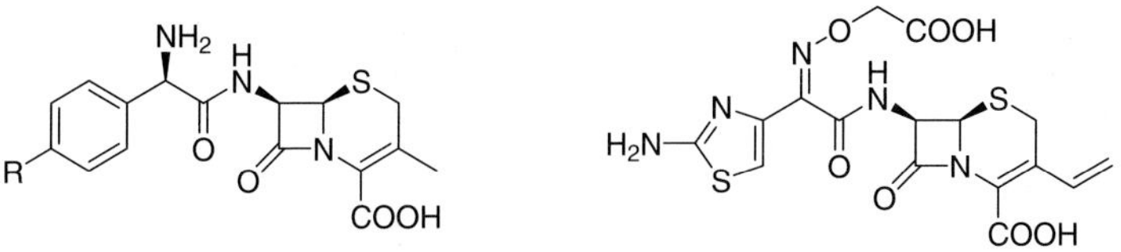

R = OH: Cefadroxil (**2-161**)
R = H: Cephalexin (**2-162**)

Cefixime (**2-163**)

Scheme 2.37. Novel cephalosporin antibiotics.

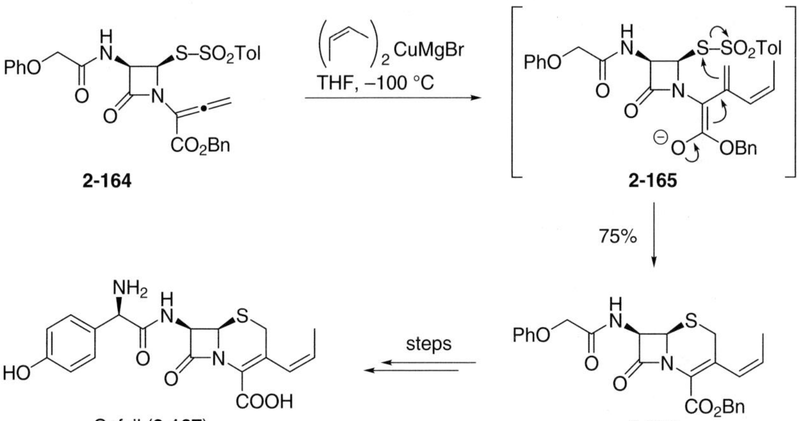

Scheme 2.38. Synthesis of cefzil (**2-167**).

The method developed for the assembly of **2-166** has a broad scope. Scheme 2.39 shows a variety of differently C-3-substituted cephalosporins obtained from allene **2-164** in yields ranging from 68 % to 95 %.

Another domino Michael addition/S_N sequence has been elaborated by the group of de Meijere. It was discovered that upon basic treatment of 2-chloro-2-cyclopropyl-idenacetates **2-168** with carboxamides **2-169** in MeCN, 4-spirocyclopropane-anne-lated oxazoline-5-carboxylates **2-172** are formed (Scheme 2.40) [91]. As intermediates, the carbanion **2-170** and **2-171** can be proposed.

In a similar way, thiocarboxamides can also be used to produce thiazoline-4-car-boxylates [92].

It is worth mentioning here that the spirocyclopropyl-substituted oxazoline-5-car-boxylates **2-172**, as well as the corresponding thiazoline-4-carboxylates, can be transformed into cyclopropyl-substituted amino acids, which might act as potential enzyme inhibitors [93] and interesting building blocks for peptidomimetics [94].

A novel cyclopropanation method based on a domino 1,4-addition/S_N sequence has recently been described by Florio and coworkers [95]. A diastereoselective Mi-chael reaction of lithiated aryloxiranes **2-174**, obtained from **2-173**, onto an α,β-un-

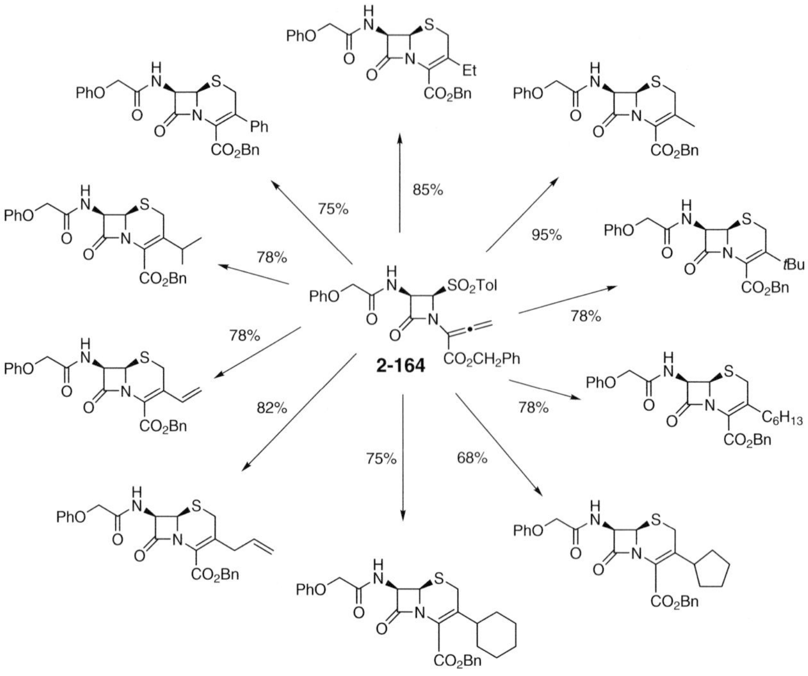

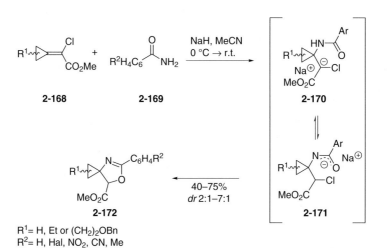

Scheme 2.39. Synthesis of cephalosporin analogues.

Scheme 2.40. Synthesis of oxazolinecarboxylates.

saturated Fischer carbene complex **2-175** followed by a nucleophilic ring opening of the epoxide moiety yielded the substituted cyclopropane carbene complexes **2-177** (Scheme 2.41). Subsequent treatment with pyridine *N*-oxide in THF at r.t. or expo-

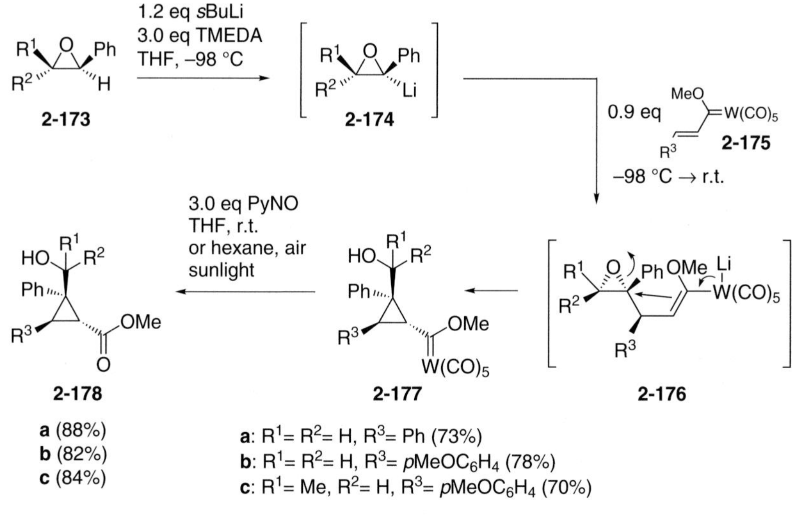

Scheme 2.41. Preparation of cyclopropanecarboxylates.

sure to air oxidation under sunlight in hexane furnished the corresponding cyclopropanecarboxylates **2-178**.

It should be noted that several domino reactions exist where the Michael/S_N-type processes are reversed. For instance, bicylo[3.3.1]nonane ring systems, applicable as synthons for the construction of various natural products [96], have been synthesized by Srikrishna and coworkers using a domino S_N/Michael process [97]. This type of domino reaction was also used by the group of Bunce to synthesize *N*-protected pyrrolidines and piperidines bearing functionalized side chains at C-2 [98]. For a domino alkylation/spiroannulation process to give homoerythrina alkaloids **2-177** [99], Desmaële/d'Angelo and coworkers have treated the 2-tetralones **2-179** with **2-180** in the presence of Cs_2CO_3 as base (Scheme 2.42) [100].

The obtained spiro keto esters **2-181** could be converted into **2-183**, employing among other transformations an intramolecular Schmidt rearrangement [101] of the azido ketones **2-182**.

Besides the domino Michael/S_N processes, domino Michael/Knoevenagel reactions have also been used. Thus, Obrecht, Filippone and Santeusanio employed this type of process for the assembly of highly substituted thiophenes [102] and pyrroles [103]. Marinelli and colleagues have reported on the synthesis of various 2,4-disubstituted quinolines [104] and [1,8]naphthyridines [105] by means of a domino Michael addition/imine cyclization. Related di- and tetrahydroquinolines were prepared by a domino Michael addition/aldol condensation described by the Hamada group [106]. A recent example of a domino Michael/aldol condensation process has been reported by Bräse and coworkers [107], by which substituted tetrahydroxanthenes **2-186** were prepared from salicylic aldehydes **2-184** and cycloenones **2-185** (Scheme 2.43).

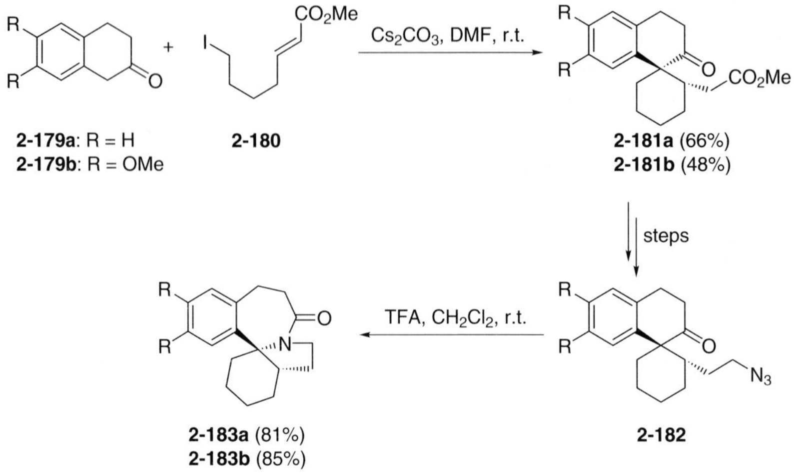
Cs2CO3, DMF, r.t.
2-179a: R = H
2-179b: R = OMe
2-180
2-181a (66%)
2-181b (48%)
steps
TFA, CH2Cl2, r.t.
2-183a (81%)
2-183b (85%)
2-182

Scheme 2.42. An access to the homoerythrina alkaloids.

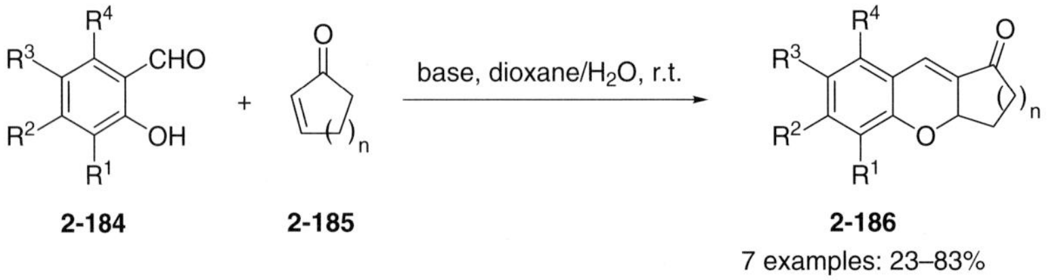
base, dioxane/H2O, r.t.
2-184
2-185
2-186
7 examples: 23–83%

Scheme 2.43. Domino Michael/aldol condensation.

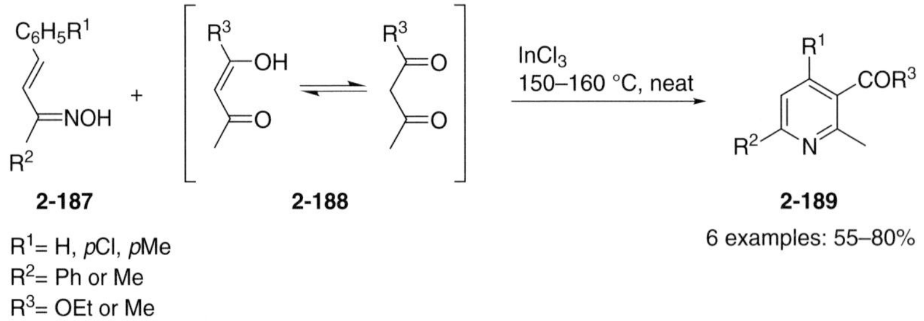
InCl3
150–160 °C, neat
2-187
2-188
2-189
R1= H, pCl, pMe
R2= Ph or Me
R3= OEt or Me
6 examples: 55–80%

Scheme 2.44. $InCl_3$-mediated synthesis of various tetrasubstituted pyridines.

Various methylenetetrahydrofurans were accessible by a combination of a Zn-promoted Michael addition and a cyclization using alkylidenemalonates and propargyl alcohol as substrates, as reported by Nakamura and coworkers [108]. Tetrasubstituted pyridines of type **2-189** have been obtained through a solvent-free $InCl_3$-promoted domino process of **2-187** and **2-188** (Scheme 2.44) [109].

Although the exact role of the indium in this reaction has not been clarified, its presence is necessary, according to a report of Prajapati and coworkers.

Laschat and coworkers described an interesting domino Michael addition/electrophilic-trapping process leading to **2-191** as a single diastereomer in 53 % yield, which was further transformed into enantiopure cylindramide (**2-192**) (Scheme 2.45) [110]. During the course of the process, substrate **2-190** was reacted with a TMS-protected propynyl cuprate and subsequently with orthoformate and $BF_3 \cdot OEt_2$.

Electron-rich aromatic systems can act efficiently as Michael donors, as shown recently by Krohn's group (Scheme 2.46) [111]. For example, reaction of the enone **2-193** with the resorcine derivative **2-194** in a domino Michael/acetalization process led to various naturally occurring xyloketals of type **2-196** in excellent yield.

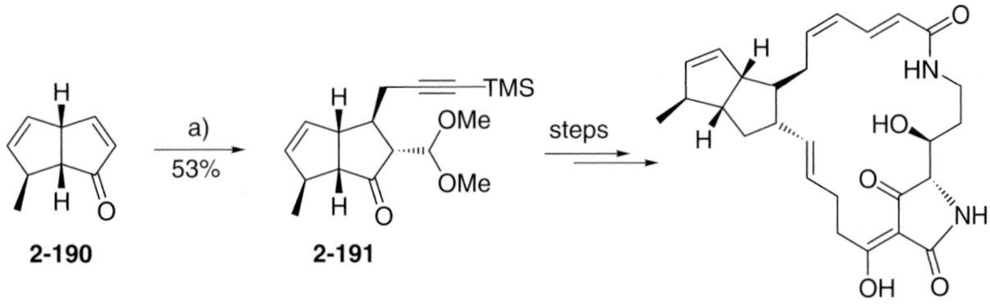

Cylindramide (**2-192**)

a) TMS——, *t*BuLi, TMEDA, THF, −40 °C, 1 h, CuI, TMSCl, THF, −78 °C,
2-190, 2 h, then BF₃•OEt₂, HC(OMe)₃, CH₂Cl₂, −20 °C, 1 h.

Scheme 2.45. Synthesis of cylindramide (**2-192**).

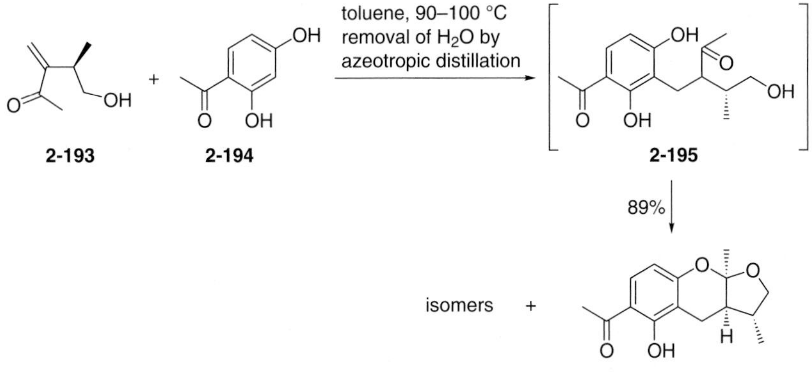

Xyloketal D (**2-196**)

Scheme 2.46. Synthesis of xyloketal D (**2-196**).

An unusual combination, namely a Michael addition and a Simmons–Smith reaction, has been elaborated by the group of Alexakis [112]. Using their protocol, 3/5- as well as 3/6-annulated systems can easily be obtained, which might be useful as precursors for the total syntheses of (–)-(S,S)-clavukerin A and (+)-(R,S)-isoclavukerin, both isolated from *Clavularia koellikeri* [113].

Various diphenylphosphine oxide-substituted seven-membered rings **2-200**, including also hydroazulenes, were synthesized by a Michael addition of **2-198**, obtained from **2-197**, and **2-199** followed by a Wittig reaction as developed by Fujimoto and coworkers (Scheme 2.47) [114].

This procedure can also be performed in a chiral auxiliary controlled version (Scheme 2.48) [114d]. Reaction of enantiopure α,β-unsaturated 8-phenyl-menthyl esters **2-201** with **2-197** yielded the corresponding cycloheptenyl menthyl ethers, which on cleavage with hydrochloric acid gave the corresponding cycloheptanone derivatives. Acetalization with chiral butane-1,3-diol then afforded the acetals **2-203** and **2-204**. The yields ranged from 46 to 81 %; however, the diastereoselectivity was good only for the reaction with the cinnamate **2-201a** as substrate.

Other useful building blocks, namely functionalized chromanes [115] are accessible through a process combining a Michael addition and a Friedel–Crafts-type cyclization, as introduced by the group of Rutjes and Jørgensen (Scheme 2.49) [116]. Interesting chromanes are the natural antioxidant α-tocopherol **2-205** [117] and compound **2-206**, which inhibits the multidrug transporter that decreases drug accumulation in resistant cells [118]. Other important chromanes are Ro 23-3544 (**2-207**), which is not only a potential peptidoleukotriene antagonist and biosynthesis

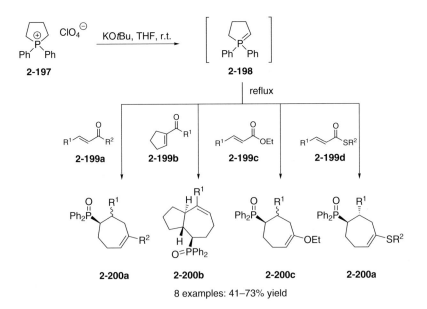

Scheme 2.47. Synthesis of cycloheptene derivatives **2-200**.

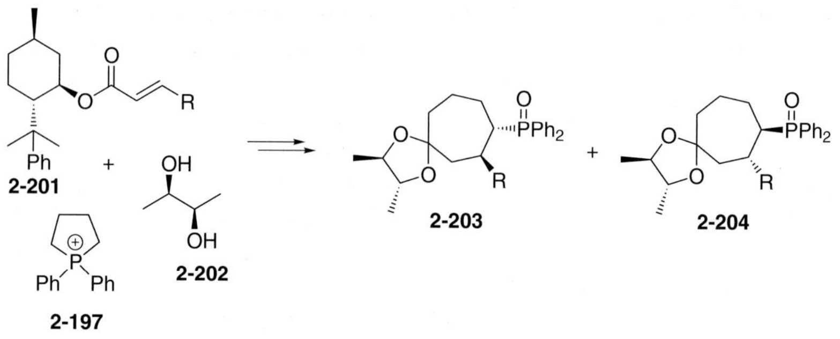
2-201
2-197
2-202
2-203
2-204
R
PPh2

Entry	Substrate 2-201	R	2-203:2-204	Overall yield [%]
1	**a**	Ph	94:6	61
2	**b**	Me	73:27	55
3	**c**	*i*Pr	65:35	81
4	**d**	*n*Pent	60:40	46

Scheme 2.48. Results of the chiral auxiliary supported domino Michael addition/ Wittig reactions.

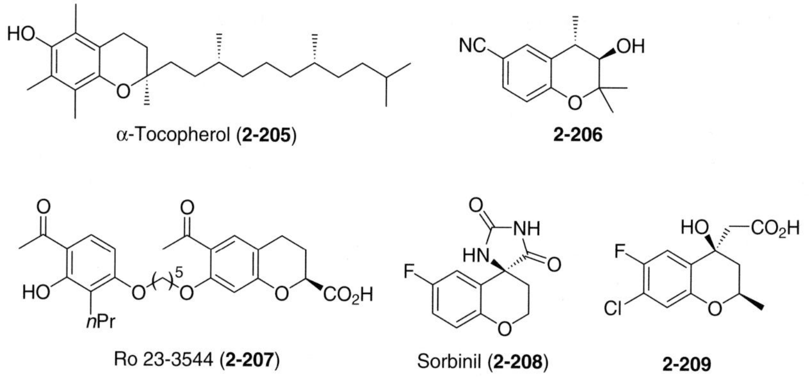
α-Tocopherol (2-205)
2-206
Ro 23-3544 (2-207)
Sorbinil (2-208)
2-209

Scheme 2.49. Examples of biologically active chromanes.

inhibitor [119], but also possesses a potential utility in the treatment of asthma; furthermore, the sorbinils (**2-208** and **2-209**) function as aldose reductase inhibitors [120].

For the enantioselective synthesis of chiral chromanes such as **2-213**, a chiral Lewis acid complex, formed *in situ* from $Mg(OTf)_2$ and **2-212**, is assumed to catalyze the domino transformation of the phenols **2-210** and the β,γ-unsaturated α-ketoesters **2-211** (Scheme 2.50). **2-213** was obtained in excellent diastereoselectivity, but only in mediocre enantioselectivity.

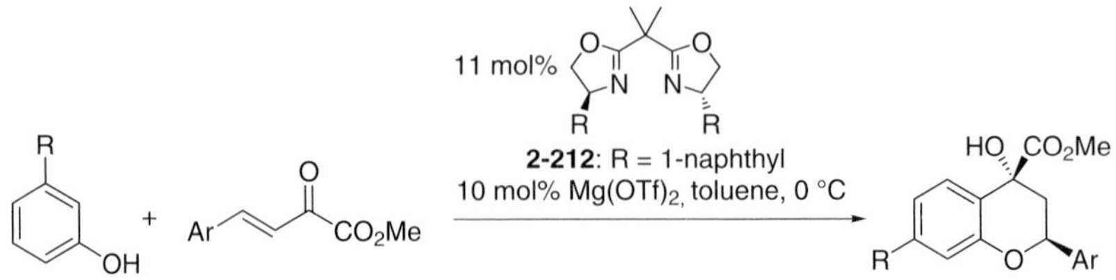

Scheme 2.50. Enantioselective synthesis of chromanes **2-213**.

Entry	2-210: R	2-211: Ar	2-213	ee [%]	Yield [%]
1	OMe	Ph	a	73	67
2[a]	OMe	Ph	b	80	77
3[a]	OMe	pFC$_6$H$_4$	c	74	43
4[a]	OMe	pBrC$_6$H$_4$	d	66	45
5	NMe$_2$	pClC$_6$H$_4$	e	<18	>95
6	NMe$_2$	Ph	f	18	>95

[a] The reaction was performed in the presence of pMe-N,N-dimethylaniline.

Furthermore, replacement of the aryl substituent by a simple methyl group resulted in the formation of a mixture of diastereomers (4:1) with low enantioselectivity. Remarkable results according to diastereoselectivity and yield were found with phenols, where the methoxy group was replaced by a dimethylamine substituent (entries 5 and 6). Thus, the corresponding chromanes were obtained as single diastereomers in >95 % yield, but the enantioselectivity was below 20 %. Most likely, due to the higher electron-donating character of the dimethylamine substituent (compared to the methoxy group) the nucleophile is sufficiently reactive to give an uncatalyzed, non-selective oxa-Michael addition resulting in a dramatic decrease of the enantioselectivity.

In a recently published report by MacMillan's group [121] on the enantioselective synthesis of pyrroloindoline and furanoindoline natural products such as (–)-flustramine B **2-219** [122], enantiopure amines **2-215** were used as organocatalysts to promote a domino Michael addition/cyclization sequence (Scheme 2.51). As substrates, the substituted tryptamine **2-214** and α,β-unsaturated aldehydes were used. Reaction of **2-214** and acrolein in the presence of **2-215** probably leads to the intermediate **2-216**, which cyclizes to give the pyrroloindole moiety **2-217** with subsequent hydrolysis of the enamine moiety and reconstitution of the imidazolidinone catalyst. After reduction of the aldehyde functionality in **2-217** with NaBH$_4$ the flustramine precursor **2-218** was isolated in very good 90 % ee and 78 % yield.

Besides the Michael addition-initiated domino reactions presented here, a multitude of other anionic domino reactions exist. Many of these take advantage of an incipient S$_N$-type reaction (for a discussion, see above). In addition to the presented S$_N$/Michael transformations [97, 98, 100], a S$_N$/retro-Dieckmann condensation was described by Rodriguez and coworkers, which can be used for the construction of substituted cycloheptanes as well as octanes [123]. Various twofold S$_N$-type domino

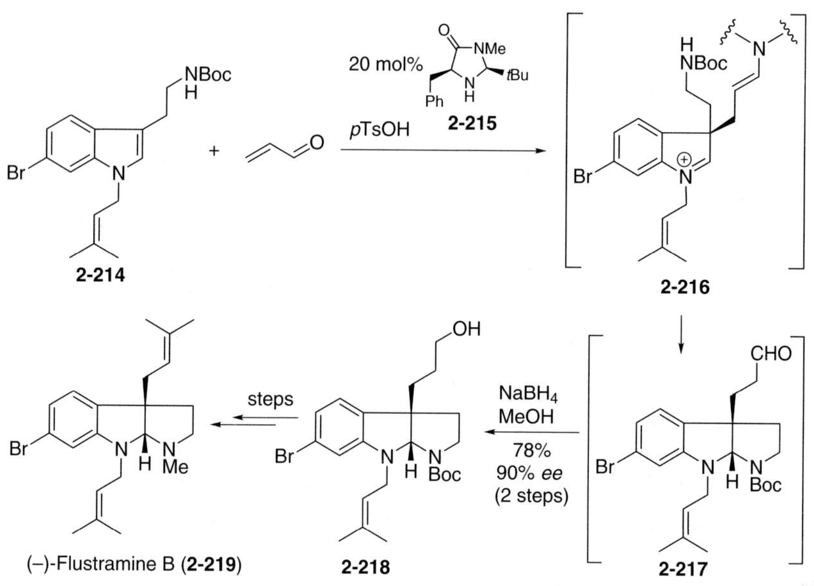

Scheme 2.51. Synthesis of the (−)-flusramine B (**2-219**).

reactions have been employed for the synthesis of natural products by Holton [124] and Takikawa [125], as well as by Barrero [126] and coworkers.

Langer and coworkers have utilized dianions of β-ketoesters or 1,3-diketones **2-220** with epibromohydrin (**2-222**) in S_N/S_N domino processes to give functionalized 2-alkylidene-5-hydroxymethyltetrahydrofurans **2-222** (Scheme 2.52) [127].

The process is assumed to take place by a chemoselective attack of the dianion **2-223** at the bromomethyl group of **2-221** and subsequent nucleophilic attack of the resultant monoanion **2-224** onto the epoxide moiety to give **2-225**. Use of the sodium-lithium-salt **2-223** of the dicarbonyl compound **2-220**, the reaction temperature as well as the Lewis acid LiClO$_4$, are crucial. The reaction seems to be quite general, since various 1,3-dicarbonyl compounds can be converted into the corresponding furans.

Moreover, a twofold S_N'-type domino reaction was reported by Krische and coworkers for the synthesis of γ-butenolides **2-229** (Scheme 2.53) [128]. Treatment of Morita–Baylis–Hillman acetates **2-226** with trimethylsilyloxyfuran (**2-227**) in the presence of triphenylphosphane in THF at 0 °C led to **2-229** in yields of up to 94 % and diastereoselectivities of >95:5.

Starting from salicylaldehydes **2-230** and triphenylchloroacetonylphosphorane (**2-231**), substituted 2,3-dihydro-1-benzoxepines **2-233** can be synthesized through a sequence that starts with a S_N2 Williamson ether synthesis followed by a Wittig olefination (Scheme 2.54) [129]. According to Huang and coworkers, treatment of **2-230** with NaOEt generates the corresponding sodium salt, which subsequently undergoes an O-alkylation with α-chloroketone **2-231** to give the intermediate **2-232**. Intramolecular ring formation by Wittig olefination then affords **2-233**.

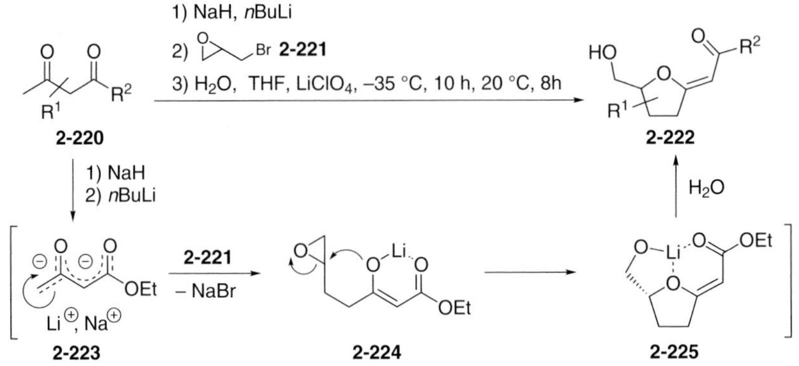

Reagents and conditions above the arrow from **2-220** to **2-222**:

1) NaH, *n*BuLi
2) (epoxide) Br **2-221**
3) H₂O, THF, LiClO₄, −35 °C, 10 h, 20 °C, 8h

Side pathway from **2-220**: 1) NaH 2) *n*BuLi → **2-223** → (**2-221**, −NaBr) → **2-224** → **2-225** → (H₂O) → **2-222**

Scheme 2.52. Reactions of epoxide **2-221** with 1,3-dicarbonyl dianions.

Entry	2-220	2-222	R¹	R²	Z/E	~dr	Yield [%]
1	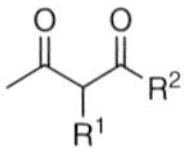	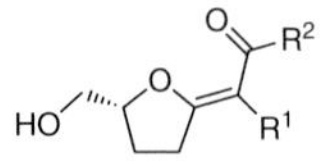	H	OEt	>98:2	–	74
2			H	O*i*Pr	10:1	–	92
3			H	O(CH₂)₂OMe	>98:2	–	57
4			H	NEt₂	>98:2	–	96
5			H	*t*Bu	<2:98	–	70
6			Me	Me	4:3	–	72
7			Bu	OEt	>98:2	–	72
8	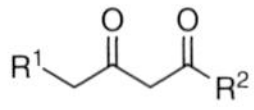	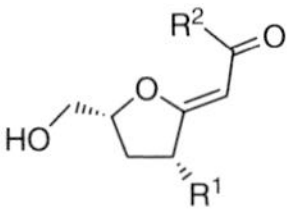	Et	OEt	>98:2	2:1	65
9			allyl	OEt	>98:2	4:3	78
10	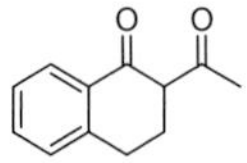	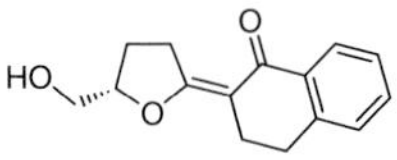	–	–	<2:98	–	71
11	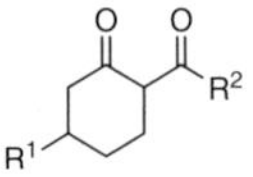	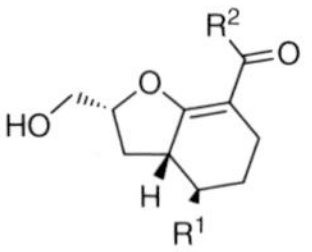	H	OEt	–	4:1	72
12			Me	OMe	–	4:1	30
13	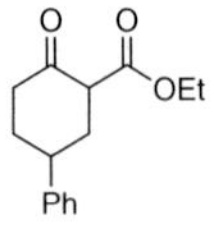	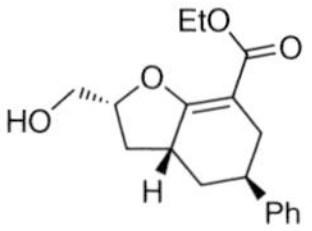	–	–	–	4:1	42
14	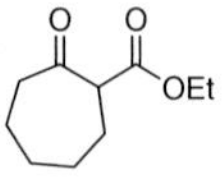	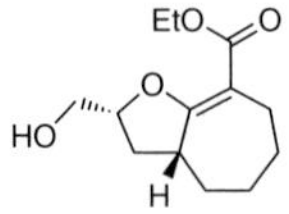	–	–	–	3:1	21

The process tolerates alkyl, methoxy, tertiary amino, and nitro groups at the salicylaldehyde. However, the yields of the nitro- and the amino-substituted salicylaldehydes are rather low. Best results were obtained with aldehydes containing electron-donating groups (entries 4, 6, and 7). Steric effects clearly also play an im-

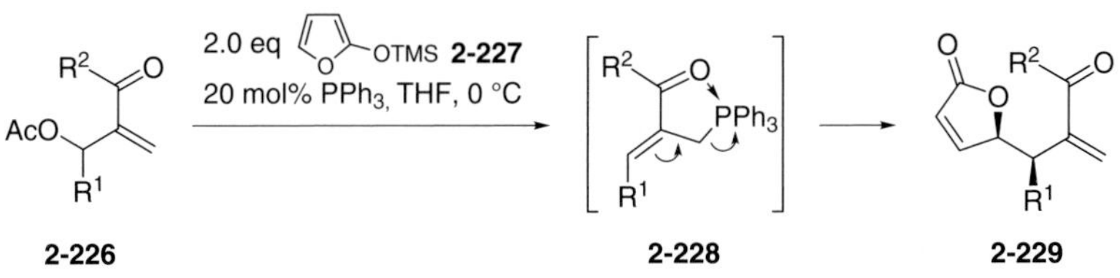
2.0 eq OTMS 2-227
20 mol% PPh3, THF, 0 °C
AcO
R2
R1
PPh3
2-226
2-228
2-229

Scheme 2.53. Synthesis of γ-butenolide through a twofold S_N′-process.

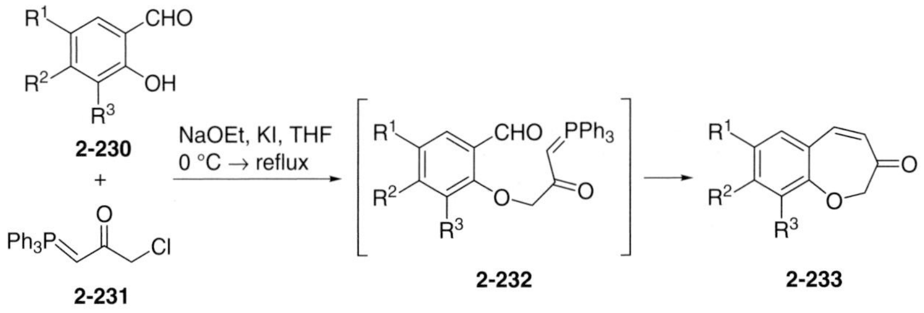
CHO
OH
R1
R2
R3
2-230
+
Ph3P
Cl
2-231
NaOEt, KI, THF
0 °C → reflux
CHO
PPh3
2-232
2-233

Entry	Aldehyde **2-230**	R^1	R^2	R^3	Product **2-233**	Yield [%]
1	**a**	H	H	H	**a**	53
2	**b**	Br	H	H	**b**	63
3	**c**	NO_2	H	H	**c**	34
4	**d**	MeO	H	H	**d**	65
5	**e**	H	Et_2N	H	**e**	21
6	**f**	Me	H	H	**f**	72
7	**g**	I	MOMO	H	**g**	68
8	**h**	H	H	Me	**h**	–

Scheme 2.54. Domino S_N/Wittig-sequence of salicylaldehydes **2-230** and phosphorane **2-231**.

portant role. Reaction of 3-methylsalicylaldehyde did not give any of the wanted benzoxepine (entry 8).

The versatility of the described domino S_N2/Wittig reaction is underlined by its application in the total synthesis of (*Z*)-pterulinic acid (**2-236**) [130], a natural product isolated from fermentations of the *Pterula* sp 82168 species (Scheme 2.55). Reaction of **2-234** and **2-231** led to **2-235** in 68% yield, which was then transformed into **2-236** [129].

Another attractive domino approach starts with an aldol reaction of preformed enol ethers and carbonyl compounds as the first step. Rychnovsky and coworkers have found that unsaturated enol ethers such as **2-237** react with different aldehydes **2-238** in the presence of $TiBr_4$. The process consists of an aldol and a Prins-type reaction to give 4-bromotetrahydropyrans **2-239** in good yields, and allows the formation of two new C–C-bonds, one ring and three new stereogenic centers (Scheme 2.56) [131]. In the reaction, only two diastereomers out of eight possible isomers were formed whereby the intermediate carbocation is quenched with a bromide.

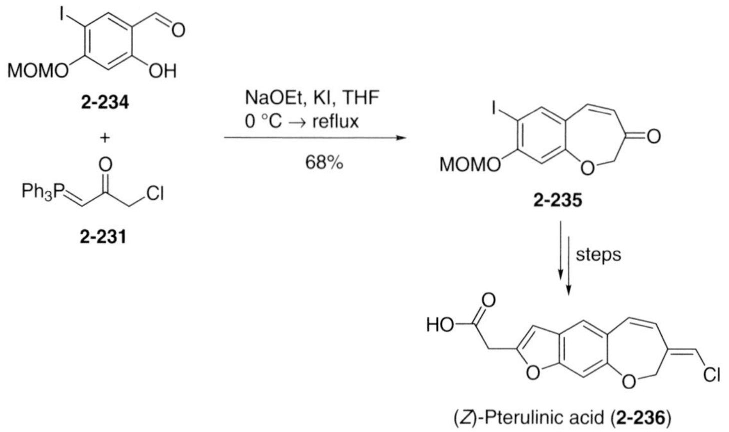

Scheme 2.55. Synthesis of (*Z*)-pterulinic acid (**2-236**).

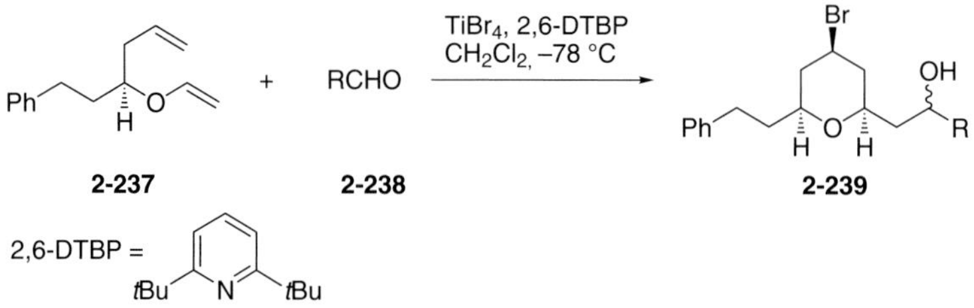

2,6-DTBP =

Entry	Aldehyde **2-238**	Product **2-239**	*dr*	Yield [%]
1	OHC / Ph	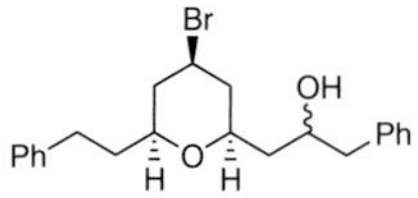	1.1:1	80
2	*i*PrCHO	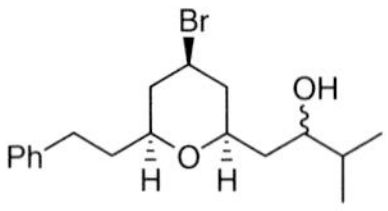	1.1:1	78
3	*c*HxCHO	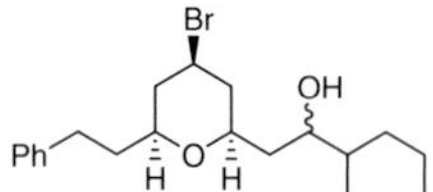	1:1	82
4	*t*BuCHO	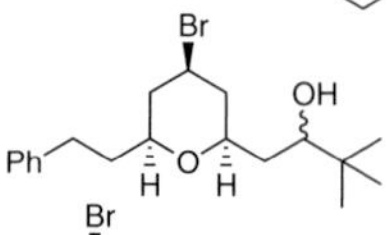	1.2:1	74
5	OHC / OTPS	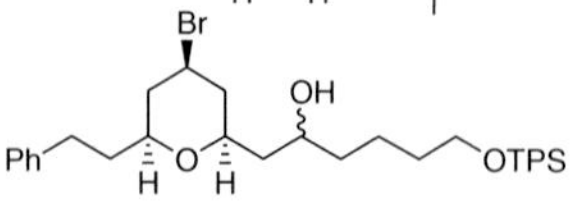	1.2:1	70
6	PhCHO	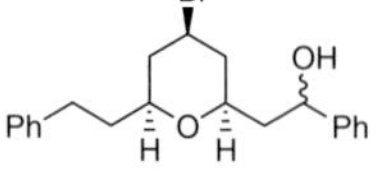	1.3:1	53

Scheme 2.56. Aldol/Prins-type/addition sequence.

In a similar way, enol ethers **2-240** containing an allylsilane moiety can also be treated with various aldehydes **2-238** (Scheme 2.57) [132]. In the presence of $BF_3 \cdot OEt_2$ **2-241** is formed with an *exo*-double bond in 72–98% yield.

This approach was used for the total synthesis of the macrolide leucascandrolide A (**2-245**) starting from the building blocks **2-242** and **2-243** [133]. The transformation led to **2-244** in 78% yield as a 5.5:1 mixture of the C-9-epimers (Scheme 2.58). The observed unusual high facial selectivity in the aldol reaction can apparently be traced back to the stereogenic center in β-position of the aldehyde **2-242**.

A combination of an aldol reaction and an elimination was used by Pandolfi's group to obtain access to the natural product prelunularin [134], and a domino aldol/aldol sequence was elaborated by the group of West for the synthesis of highly

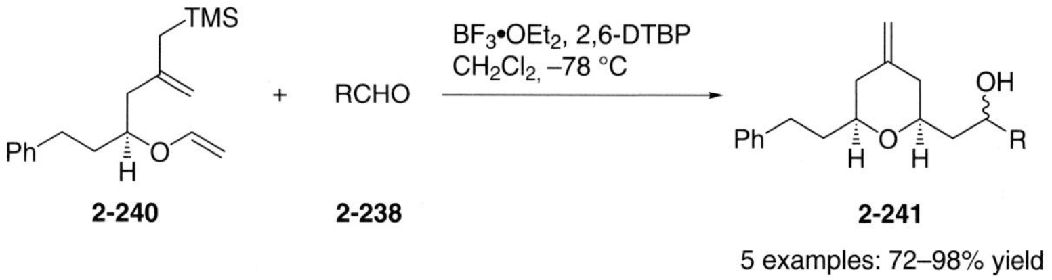

Scheme 2.57. Aldol/Prins cyclization of **2-240** and an aldehyde promoted by $BF_3 \cdot OEt_2$.

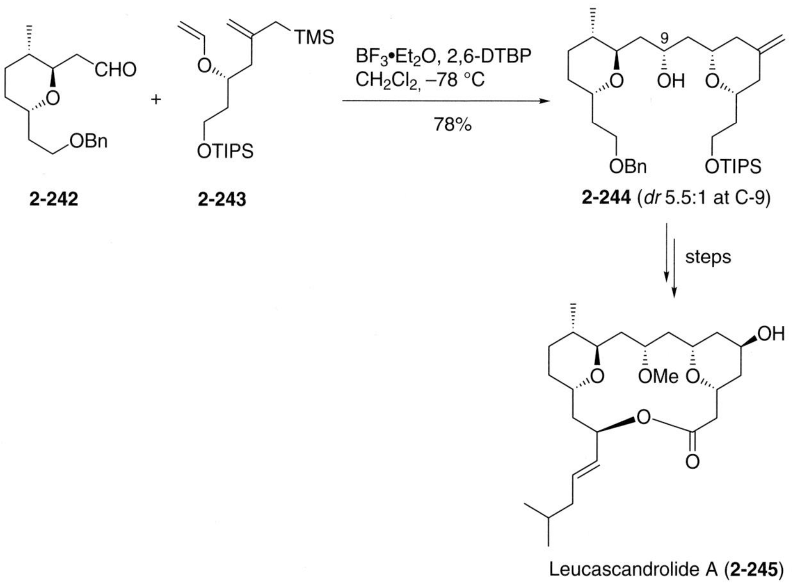

Scheme 2.58. Synthesis of leucascandrolide A (**2-245**).

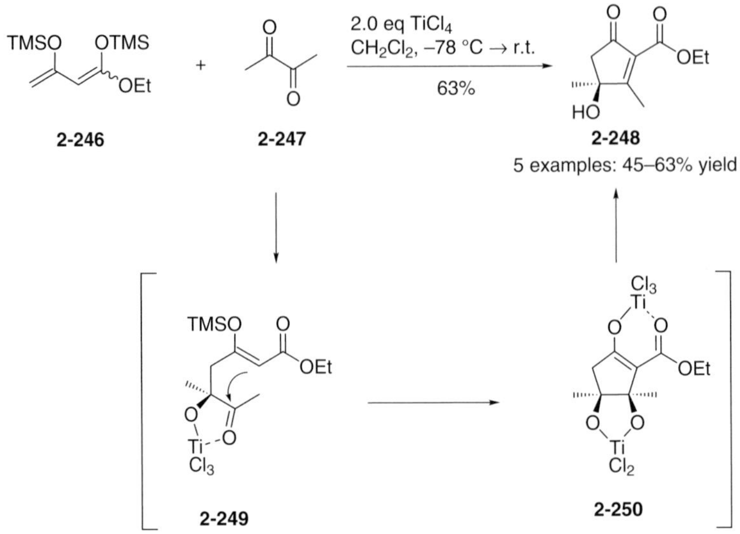

Scheme 2.59. Domino aldol addition/aldol condensation process for the synthesis of tetrasubstituted cyclopent-2-en-1-ones.

substituted triquinacenes [135]. Langer and coworkers reported on the Lewis acid-mediated Mukaiyama-aldol addition/aldol-condensation of 1,3-bis(trimethylsilyloxy)-1,3-diene **2-246** with 1,2-diketone **2-247** to provide a highly substituted cyclopent-2-en-1-one **2-248** (Scheme 2.59) [136]. The highest yield was obtained when the reaction was performed at −78 °C and the mixture then slowly warmed to room temperature. The success of this protocol can be explained by the assumption that the initial attack of the diene **2-246** onto the 1,2-diketone **2-247** takes place at low temperature, and that the cyclization step from **2-249** to **2-250** occurs only at elevated temperature.

A common procedure in C–C-bond formation is the aldol addition of enolates derived from carboxylic acid derivatives with aldehydes to provide the anion of the β-hydroxy carboxylic acid derivative. If one starts with an activated acid derivative, the formation of a β-lactone can follow. This procedure has been used by the group of Taylor [137] for the first synthesis of the 1-oxo-2-oxa-5-azaspiro[3.4]octane framework. Schick and coworkers have utilized the method for their assembly of key intermediates for the preparation of enzyme inhibitors of the tetrahydrolipstatin and tetrahydroesterastin type [138]. Romo and coworkers used a Mukaiyama aldol/lactonization sequence as a concise and direct route to β-lactones of type **2-253**, starting from different aldehydes **2-251** and readily available thiopyridylsilylketenes **2-252** (Scheme 2.60) [139].

Depending on the Lewis acid used, cis- [139a] or trans-substituted products [139b–d] are accessible. When $SnCl_4$ was employed at −78 °C as promoter, 2,3-cis-β-lactones were obtained almost exclusively, whereas the $ZnCl_2$-initiated reactions at r.t. afforded the corresponding trans-substituted compounds (Scheme 2.60).

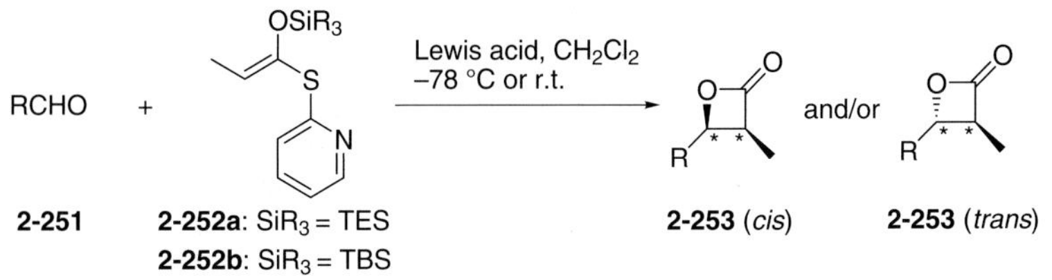

OSiR₃ ... Lewis acid, CH₂Cl₂ −78 °C or r.t.

RCHO +

2-251   **2-252a**: SiR₃ = TES   **2-253** (*cis*)   **2-253** (*trans*)
              2-252b: SiR₃ = TBS

and/or

Entry	SiR₃	Lewis acid	R	**2-253** *cis:trans*	Yield [%]
1	TES	SnCl₄	Ph(CH₂)₂	>19:1	62
2	TES	SnCl₄	pNO₂Ph	>19:1	66
3	TES	SnCl₄	nHep	>19:1	64
4	TBS	ZnCl₂	Ph(CH₂)₂	1:37	57
5	TBS	ZnCl₂	cHx	>1:19	16
6	TBS	ZnCl₂	nHep	>1:19	42
7	TBS	ZnCl₂	BnOCH₂	>1:19	74

Scheme 2.60. Synthesis of β-lactones.

In addition, enantiopure 2,3-*trans*-β-lactones could be obtained with good induced diastereoselectivity (1:5.3–1:22) when chiral α- or β-substituted aldehydes are employed [139d]

Notably, the Mukaiyama aldol/lactonization approach has been used in the total synthesis of panclicin D (**2-258**) [139b,c] and okinonellin B (**2-261**) (Scheme 2.61) [139d]. In the synthesis of **2-258**, aldehyde **2-254** and the ketene acetal **2-255** were used to prepare the β-lactone **2-256** with high simple and induced diastereoselectivity. There follows an esterification with the carboxylic acid **2-257**. For the synthesis of **2-261**, the aldehydes **2-259** and **2-252b** were employed as substrates leading initially to the β-lactone **2-260**.

An interesting double aldol domino approach has been reported by the group of Shibasaki [140], who showed that lanthanoide containing heterobimetallic asymmetric catalysts are able to promote an inter-/intramolecular nitro-aldol domino process of aldehyde **2-262** and nitromethane (**2-263**) (Scheme 2.62). As the best result, treatment of **2-262** with 30 equiv. of **2-263** in the presence of 5 mol% (*R*)-PrLB (Pr = Praseodymium; L = lithium; B = BINOL) in THF at −40 °C at r.t. gave **2-266** with 79 % *ee* in 41 % yield after crystallization from the reaction mixture. Recrystallization of this material gave the diastereomer **2-267** with increased optical purity of 96 % *ee* in 57 % yield. Mechanistic studies revealed that, at −40 °C, **2-266** is formed first *via* **2-264** and **2-265**. There follows a gradual transformation into **2-267** at r.t., indicating that this is the thermodynamically more stable product.

Yamamoto's group recently published a highly enantioselective chiral amine-catalyzed domino *O*-nitroso aldol/Michael reaction of **2-268** and **2-269** (Scheme 2.63) [141]. As products, the formal Diels–Alder adducts **2-271** were obtained with > 98 % *ee*, which is probably due to the selective attack of an enamine, temporarily formed from amine **2-270** and enone **2-268**, onto the nitroso functionality.

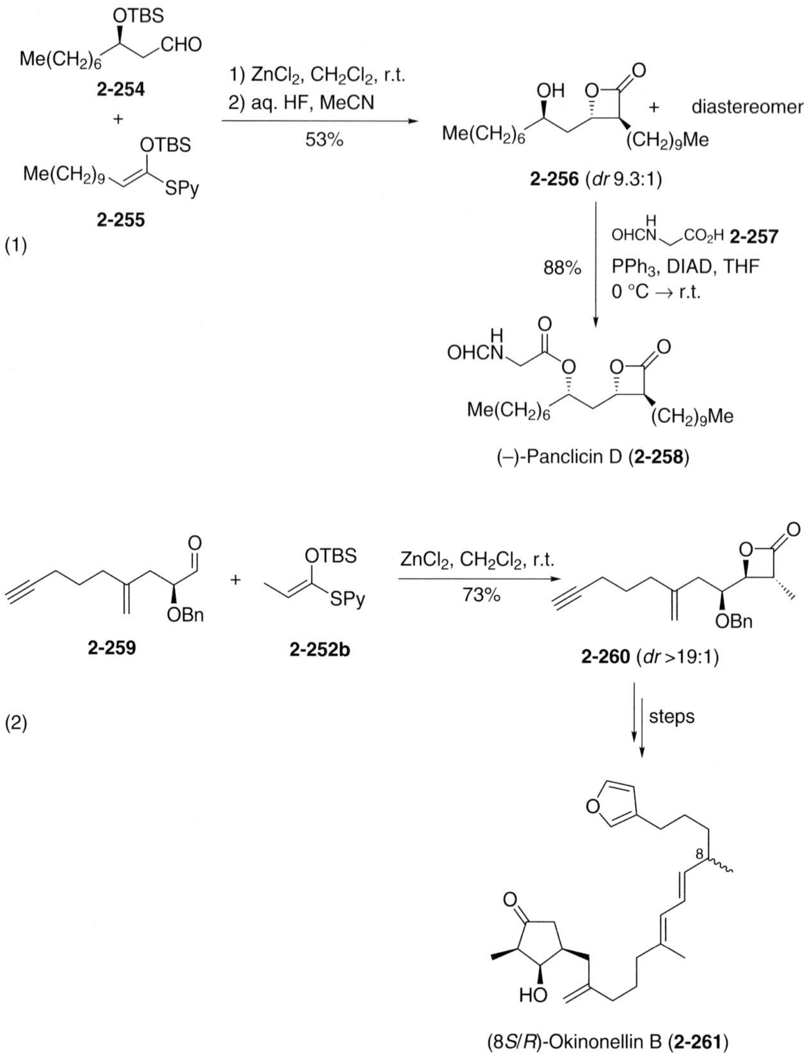

Scheme 2.61. Syntnesis of (–)-panclicin D (**2-258**) and okinonellin B (**2-261**) employing the domino Mukaiyama aldol/lactonization reaction sequence.

The following example shows that retro-aldol reactions can also be utilized in domino processes. Thus, Reißig and coworkers have successfully developed a route to cyclophanes based on a fluoride-initiated retro-aldol/Michael addition sequence [142]. Numerous examples varying the ring size and the number of functional groups were constructed, though usually with only moderate yield and selectivity. Thus, reaction of **2-272** in the presence of CsF and a phase-transfer catalyst provided 19 % of **2-274** together with 36 % of **2-273**. It should be noted, however, that in a few cases high yields and very good chemoselectivities were observed, as in the transformation of **2-275** to give **2-276** (Scheme 2.64).

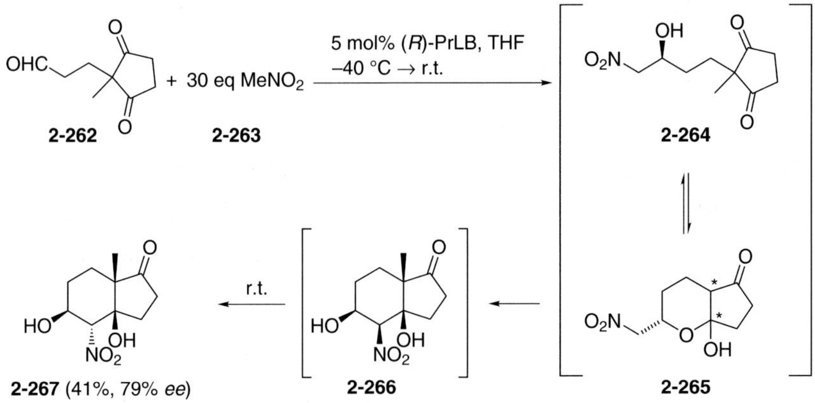

Scheme 2.62. Synthesis of an enantio-enriched indene by a twofold nitro-aldol reaction.

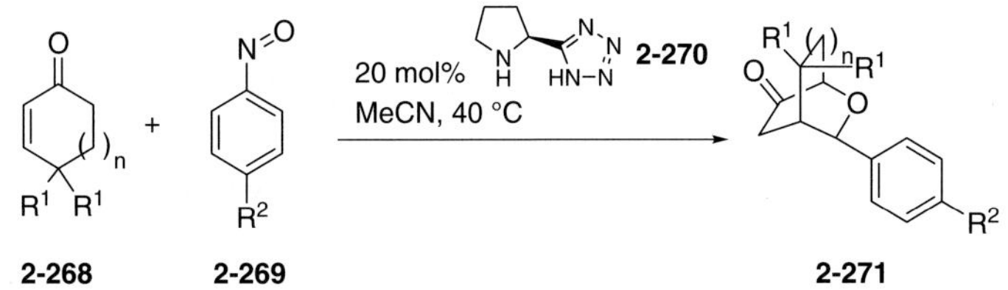

Entry	R^1	n	R^2	2-271	ee [%]	Yield [%]
1	Me	1	H	a	99	64
2	Ph	1	H	b	99	56
3	-(OCH$_2$CH$_2$O)-	1	H	c	98	61
4	Me	1	Me	d	98	47
5	Me	1	Br	e	99	50
6	H	2	H	f	99	51

Scheme 2.63. Organocatalytic synthesis of **2-271** by a domino O-nitroso-aldol/Michael reaction.

Domino processes involving Horner–Wadsworth–Emmons (HWE) reactions constitute another important approach. Among others, HWE/Michael sequences have been employed by the group of Rapoport for the synthesis of all-*cis*-substituted pyrrolidines [143], and by Davis and coworkers to access new specific glycoamidase inhibitors [144]. Likewise, arylnaphthalene lignans, namely justicidin B (**2-281**) and retrojusticidin B (**2-282**) [145], have been synthesized utilizing a domino HWE/aldol condensation protocol developed by Harrowven's group (Scheme 2.65) [146].

In these syntheses, **2-279** was prepared initially through a base-induced cyclization of ketoaldehyde **2-277** and the phosphonate **2-278** in 73 % yield, together with a small amount of the monoester **2-280** (14:1). To complete the total syntheses, the

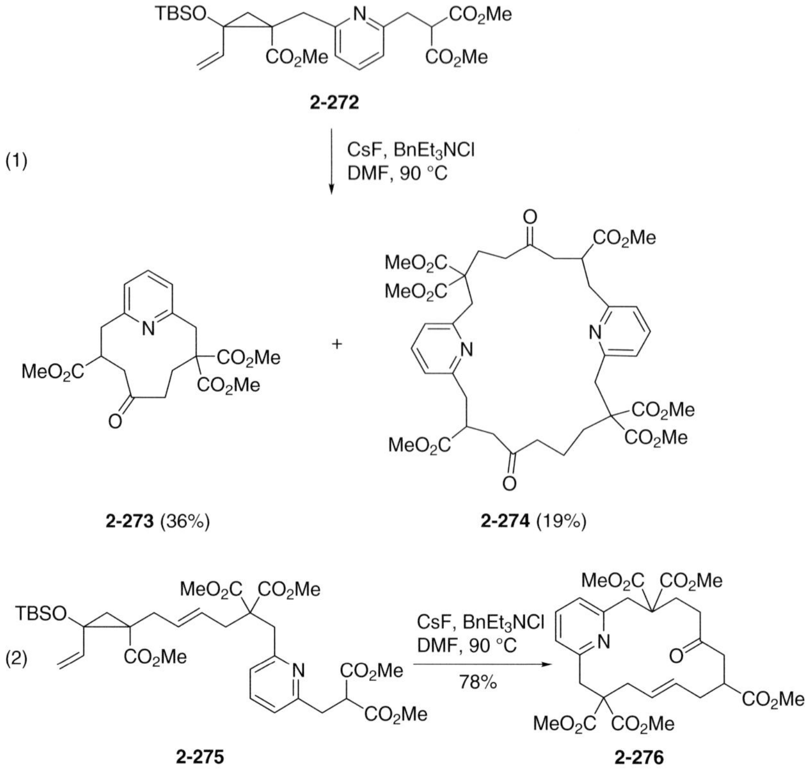

Scheme 2.64. CsF-initiated ring-expansion reactions of substituted pyridines **2-272** and **2-275**.

mixture was selectively hydrolyzed to give the monoacid **2-280** using potassium trimethylsilanolate. Reduction of **2-280** with borane dimethylsulfide complex gave justicidin B (**2-281**) almost exclusively, while reduction of the sodium salt of **2-281** with lithium borohydride afforded retrojusticidin B (**2-282**) as the major product together with some justicidin B (**2-281**).

The combination of a Corey–Kwiatkowski [147] and a HWE reaction efficiently furnishes α,β-unsaturated ketones of type **2-288** in good yields [148]. This unique domino reaction, developed by Mulzer and coworkers, probably proceeds *via* the intermediates **2-285** and **2-286** using the phosphonate **2-283**, the ester **2-284**, and the aldehyde **2-286** as substrates (Scheme 2.66).

Wittig reactions have also been employed in domino processes. For example, Schobert and coworkers developed an effective addition/Wittig reaction protocol which provides access to α,β-disubstituted tetronic acids, tetronates, as well as to five-, six- and seven-membered *O*-, *N*-, and *S*-heterocycles [149].

The group of Molina and Fresneda employed even two different types of anionic domino processes in their total synthesis of the novel marine alkaloid variolin B (**2-295**) (Scheme 2.67) [150]. It is generally accepted that marine organisms are among

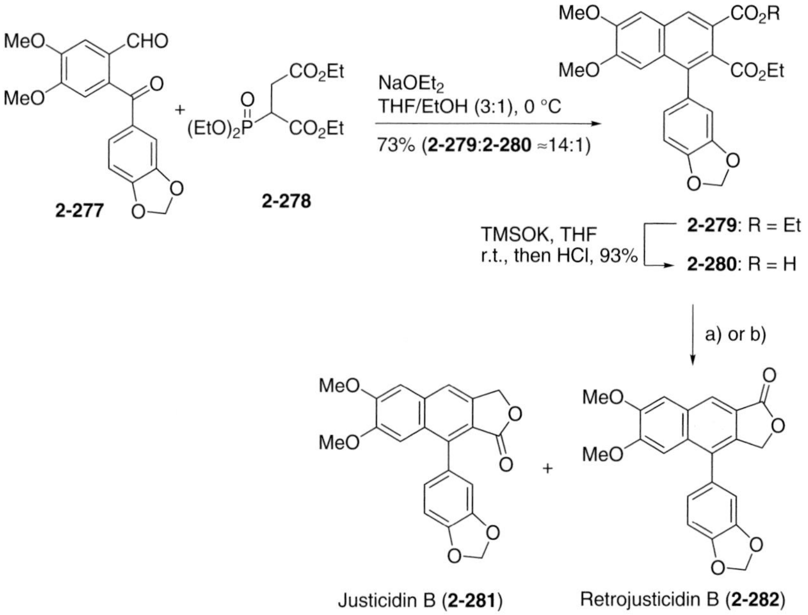

a) BH$_3$•SMe$_2$, THF, r.t., then ethanolic HCl: 76% of **2-281** and a trace of **2-282**.
b) NaH, LiBH$_4$, 1,4-dioxane, Δ, then 0.5 M HC: 28% of **2-281** and 67% of **2-282**.

Scheme 2.65. Synthesis of justicidin B (**2-281**) and retrojusticidin B (**2-282**).

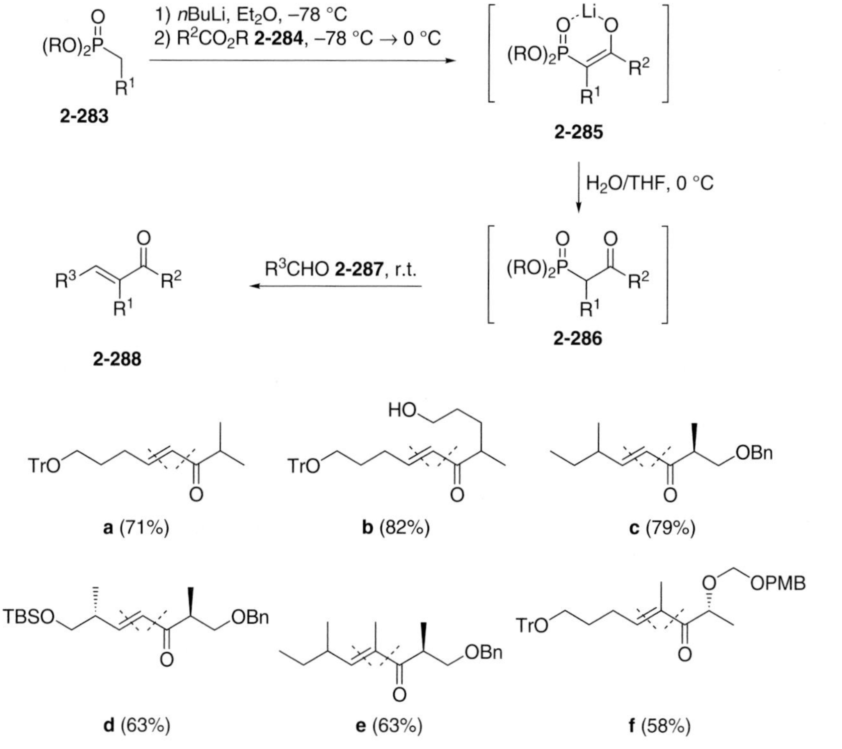

Scheme 2.66. Domino Corey–Kwiatkowski /Horner–Wadsworth–Emmons reaction.

the most promising new sources of biologically active molecules [151]. Indeed, the variolins, which are isolated from the Antarctic sponge *Kirkpatrickia varialosa* [152], are assumed to possess some pharmacological potential. Variolin B (**2-295**) is the most active compound of this family, having cytotoxic activity against P388 murine leukemia cells and also being effective against herpes simplex type I [152b]. All members have a pyridopyrrolopyrimidine ring in common, which has no precedent in other natural products. With regard to its synthesis, a domino aza-Wittig/carbodiimide-mediated cyclization provided the tricyclic core **2-291** in almost quantitative yield, starting from the known iminophosphorane **2-289** and the isocyanate **2-290** [153].

Completion of the total synthesis afforded only six further steps, including the installation of the second 2-aminopyrimidine ring *via* a second domino sequence. This process presumably involves a conjugate addition of guanidine (**2-293**) to the enone system of **2-292**, followed by a cyclizing condensation and subsequent aromatization. Under the basic conditions, the ethyl ester moiety is also cleaved and **2-294** is isolated in form of the free acid, in 89 % yield. Finally, decarboxylation and deprotection of the amino functionality yielded the desired natural product **2-295**.

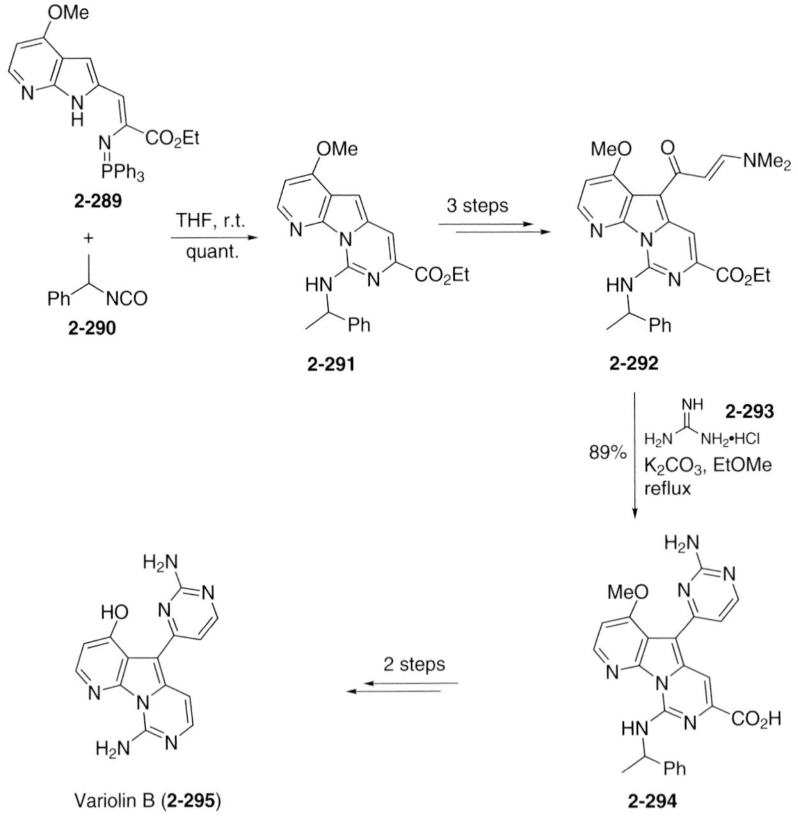

Scheme 2.67. Synthesis of variolin B (**2-295**).

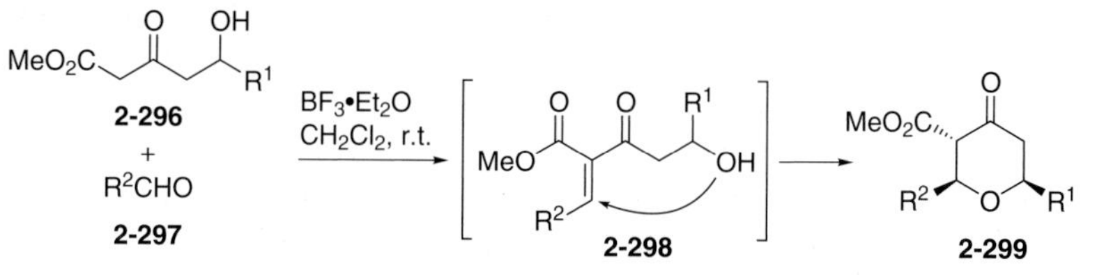

Scheme 2.68. Diastereoselective formation of substituted tetrahydropyran-4-ones.

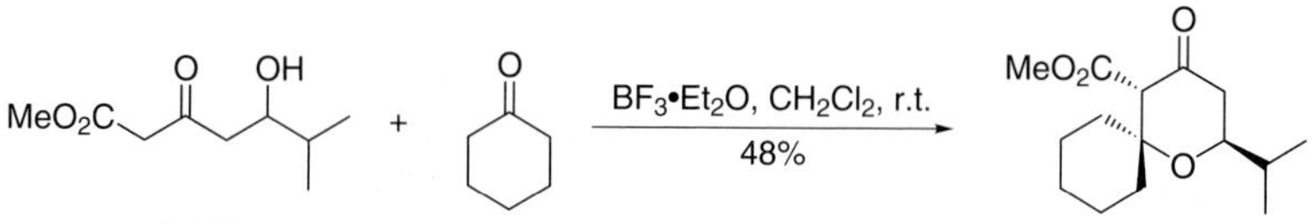

Scheme 2.69. Synthesis of a spirotetrahydropyran-4-one derivative.

Comparable with the Michael addition, the Knoevenagel condensation is also linked to a broad range of effective domino transformations, and many twofold anionic processes exploiting this approach have emerged during the past decade. For example, Daïch's group described an interesting Knoevenagel/amino-nitrile cyclization affording highly functionalized indolizines [154]. Moreover, domino Knoevenagel/Michael sequences have excelled as useful methods for the construction of indanones, according to a protocol of Sartori and coworkers [155], as well as for the generation of highly substituted tetrahydopyran-4-ones, as shown independently by the groups of Clarke [156] and Sabitha [157]. Dwelling on Clarke's approach, the reaction of 5-hydroxy-1,3-ketoesters **2-296** and aldehydes **2-297** under Lewis acidic conditions provided the desired tetrahydropyran-4-ones **2-299** in reasonable to good yields *via* intermediate **2-298** (Scheme 2.68). **2-299** exists as a 1:1 to 1:2 mixture of keto and enol forms, except for compounds with $R^1 = Ph$, $R^2 = i$Pr or Ph, where the keto form is found exclusively.

Interestingly, when cyclohexanone was used instead of aldehydes in the reaction with **2-296**, the diastereomerically pure spirocycle **2-300** was obtained in 48 % yield (Scheme 2.69) [156].

A Knoevenagel condensation/Michael addition sequence has been reported by Barbas III and coworkers (Scheme 2.70) [158] using benzaldehyde, diethyl malonate, and acetone in the presence of the chiral amine (*S*)-1-(2-pyrrolidinyl-methyl)-pyrrolidine (**2-301**). As the final product the substituted malonate **2-302** was isolated in 52 % yield with 49 % *ee*.

In the following section, anionic/anionic domino processes will be discussed which are, to date, much less common, although important structural elements may

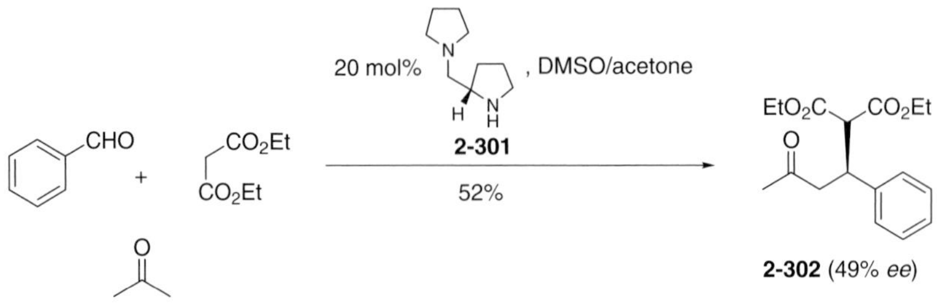

Scheme 2.70. Asymetric domino Knoevenagel/Michael addition reaction.

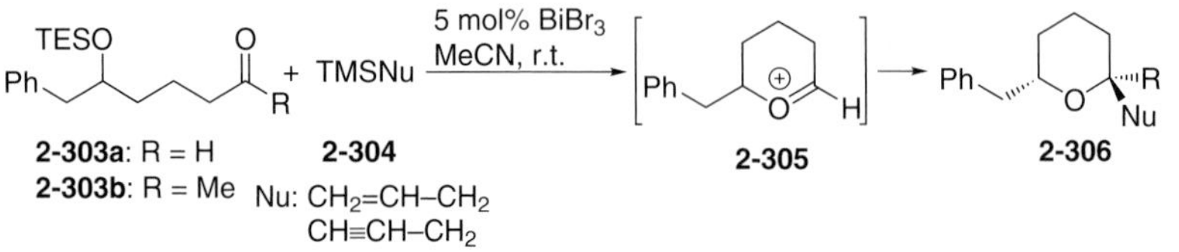

Entry	2-303	2-304	TMSNu	2-306	Nu	dr	Yield [%]
1	a	a	TMSCH₂CH=CH₂	a	–CH₂CH=CH₂	≥99:1	90
2	b	a	"	b	"	≥19:1	88
3	a	b	TMSCH₂C≡CH	c	–CH=C=CH₂	≥19:1	80
4	b	b	"	d	"	≥19:1	72
5	a	c	CH₂=C(OTMS)CH₃	e	–CH₂C(O)CH₃	≥19:1	73
6	b	c	"	f	"	≥19:1	80

Scheme 2.71. Domino etherification reaction.

be obtained using these approaches. The groups of Fry, Dieter and Flemming independently developed a S_N-type halide displacement cyclization to provide either substituted cyclic imines [159] *via* an incipient Grignard reaction or substituted tetrahydrofurans and pyrans [160] through an initial aldol addition. Moreover, Florio and coworkers have shown that an electrophilic addition/S_N-type aziridine ring opening can lead to useful building blocks [161]. An appealing tetrahydropyran synthesis has been reported by the group of Evans and Hinkle [162]. Products of type **2-306** were generated from aldehydes or ketones **2-303** and various trialkylsilyl nucleophiles **2-304**, in high yield and excellent diastereoselectivity (Scheme 2.71).

The use of the Lewis acid $BiBr_3$ gave the best results, with the formation of an oxenium ion **2-306** as intermediate.

Langer and coworkers constructed diverse *O*- and *N*-heterocyclic scaffolds, such as γ-alkylidene-α-hydroxybutenolides and pyrrolo[3,2-*b*]pyrrol-2,5-diones, exploiting the well-established cyclization strategy of bisnucleophiles with oxalic acid derivatives [163], while Stockman's research group reported in this context on a novel oxime formation/Michael addition providing the structural core of the alkaloid perhydrohistrionicotoxin [164].

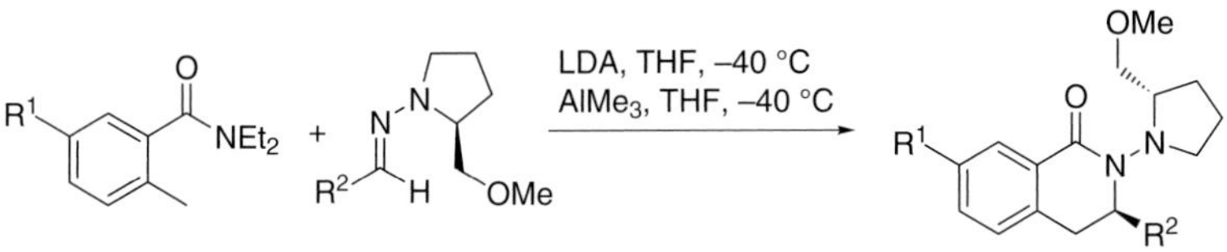

Scheme 2.72. Asymmetric domino 1,2-addition/lactamization.

Additions of stabilized carbanions to imines and hydrazones, respectively, have been used to initiate domino 1,2-addition/cyclization reactions. Thus, as described by Benetti and coworkers, 2-substituted 3-nitropyrrolidines are accessible *via* a nitro-Mannich (aza-Henry)/S_N-type process [165]. Enders' research group established a 1,2-addition/lactamization sequence using their well-known SAMP/RAMP-hydrazones **2-308** and lithiated *o*-toluamides **2-307** as substrates to afford the lactams **2-309** in excellent diastereoselectivity (Scheme 2.72) [166]. These compounds can be further transformed into valuable, almost enantiopure, dihydro-*2H*-isoquinolin-1-ones, as well as dihydro- and tetrahydroisoquinolines.

Shindo and coworkers have described anionic [2+2] cycloadditions of ynolates to carbonyl functionalities followed by either a Dieckmann condensation or a Michael addition [167]. These domino reactions allow the synthesis of five- and six-membered cyloalkenones and five- to seven-membered cycloalkenes. Exceptionally short as well as stereoselective routes to several tricyclic diterpenes have been designed by Ramana's group using novel types of domino acylation/cycloalkylation and alkylation/cycloacylation processes, respectively (Scheme 2.73) [168]. For instance, acid **2-310** was subjected to MsOH/P$_2$O$_5$ (10:1) with anisole **2-311** to give **2-312**, which was subsequently transformed into racemic ferruginol (**2-314**). In the present case, the desired product is likely to be formed *via* an acylation/cycloalkylation sequence. However, reaction of **2-310** and **2-311** in the presence of concentrated H$_2$SO$_4$ led to **2-313**, which could be transformed into totarol (**2-315**). Clearly, the use of H$_2$SO$_4$ causes a reversal of the reaction steps, resulting in an alkylation/cycloacylation domino process.

Beller and coworkers have elaborated a twofold hydroamination sequence employing halo styrenes which led to 2,3-dihydroindoles [169]. An aryne is postulated as the reactive intermediate, and although the expected products were obtained in somewhat moderate yields, the method is reported to be superior to conventional procedures by a factor of three. Another useful 2.2 domino process has been reported by Molander and coworkers [170]. These authors utilized a unique SmI$_2$-mediated Barbier-type cyclization/Grob fragmentation sequence which ultimately led to usually difficult accessible medium-sized carbocycles. However, the loss of two stereogenic centers during the reaction course is somewhat aggravating. A related fragmentation/rearrangement domino process with other reaction types has also been found in the literature. For example, Tu and coworkers developed a semipinacol rearrangement/alkylation of α-epoxy alcohols [171], with multifunctional and

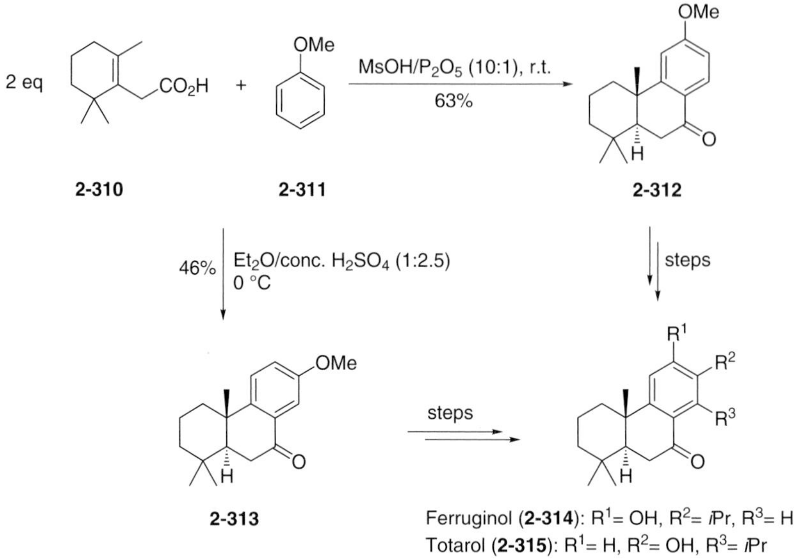

Scheme 2.73. Synthesis of diterpenes *via* domino acylation/cycloalkylation and alkylation/cycloacylation strategies.

diastereomerically pure 1,3-diols being obtained as products. In continuing with anionic/anionic domino reactions triggered by rearrangements, Oltra's group presented a novel enantioselective transannular cyclization/ring contraction approach for the synthesis of oxygen-bridged terpenoids [172]. The method has proved to be useful in the synthesis of the antimycobacterial (+)-dihydroparthenolide diol, as well as for other nine-membered ring systems.

Recently, a further unique domino methodology has been reported by Lu and coworkers (Scheme 2.74) [173]. Herein, a triphenyl phosphine-catalyzed umpolung addition/cyclization of allenes and alkynes containing an electron-withdrawing group **2-316**–**2-318** followed by reaction with a double nucleophile **2-319** is assumed to account for the production of a broad palette of various heterocycles **2-321** and **2-323** *via* **2-320** and **2-322**, respectively. Dihydrofurans, piperazines, morpholines and diazepanes were obtained during the process.

A proposed catalytic cycle for the formation of **2-321** or **2-323** is illustrated in Scheme 2.75. Accordingly, the reaction is started by a nucleophilic addition of a PPh_3 molecule to the electron-deficient multiple bond of compounds **2-316**–**2-318**. Next, the formed zwitterionic intermediate deprotonates the nucleophile **2-319**, thereby facilitating the following addition. Proton transfer and elimination of PPh_3 from the new zwitterionic intermediate affords the corresponding γ- or α-adduct. Finally, an intramolecular conjugate addition reaction gives rise to the desired heterocyclic products **2-321** and **2-323**.

Parsons and coworkers [174] have published a route to anatoxin-*a*, which was isolated from strains of freshwater blue-green algae *Anabaena flos aqua* and is responsible for the death of livestock, waterfowl, and fish [175]. The sequence started

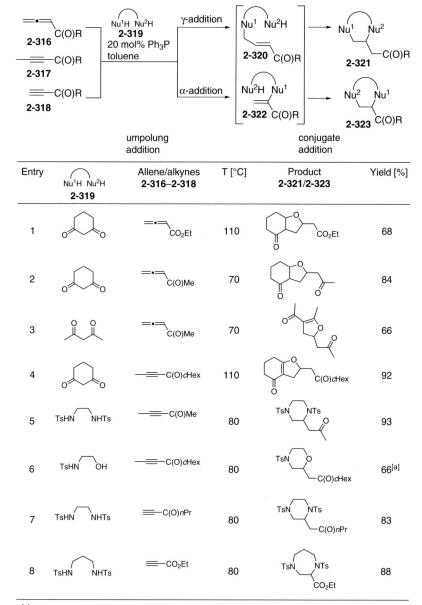

Entry	Nu¹H Nu²H **2-319**	Allene/alkynes **2-316–2-318**	T [°C]	Product **2-321/2-323**	Yield [%]
1		=•=\CO₂Et	110		68
2		=•=\C(O)Me	70		84
3		=•=\C(O)Me	70		66
4		—≡—C(O)cHex	110		92
5	TsHN⌒NHTs	≡—C(O)Me	80		93
6	TsHN⌒OH	≡—C(O)cHex	80		66[a]
7	TsHN⌒NHTs	≡—C(O)nPr	80		83
8	TsHN⌒NHTs	≡—CO₂Et	80		88

[a] MeCN was used as solvent.

Scheme 2.74. PPh₃-catalyzed domino nucleophilic additions.

with the addition of methyl lithium to the β-lactone moiety in **2-324**. There follows a rearrangement with a nucleophilic attack of the nitrogen at the epoxide moiety to give **2-325**, which was transformed into **2-326** (Scheme 2.76).

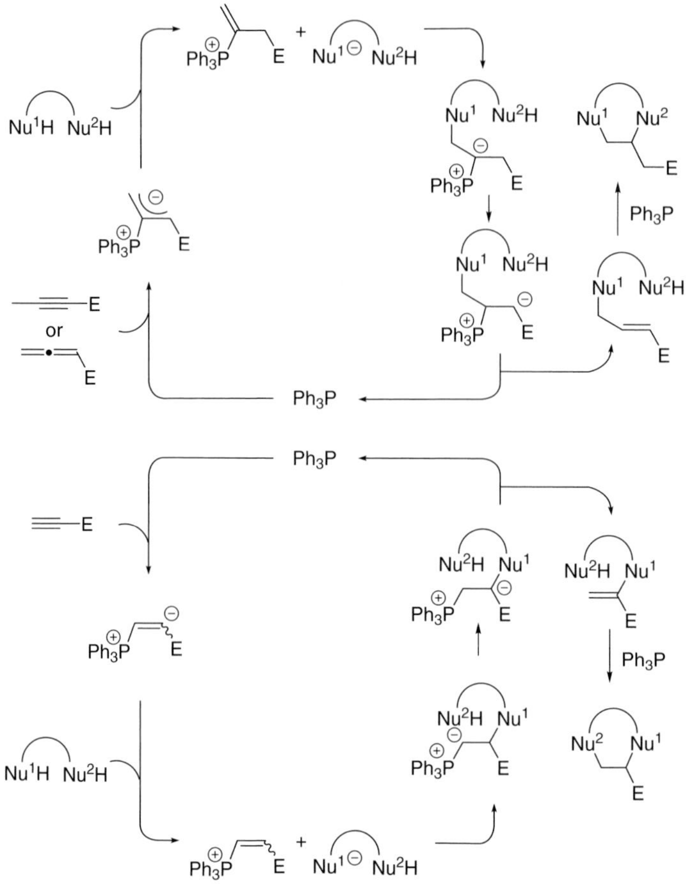

Scheme 2.75. Possible catalytic cycle leading to the isolated products.

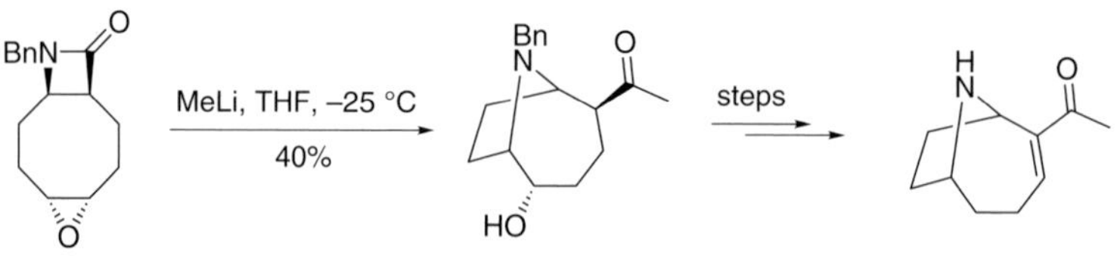

Scheme 2.76. MeLi-induced β-lactam ring-opening/intramolecular cyclization leading to anatoxin-*a* (**2-326**).

Several other twofold cyclization strategies have been developed by Smith and co-workers, ultimately to obtain access to the cyclic framework of penitrem D with the correct stereochemistry [176]. Williams' group has been interested in *Stemona* alkaloids, which are represented by approximately 50 structurally novel, polycyclic natu-

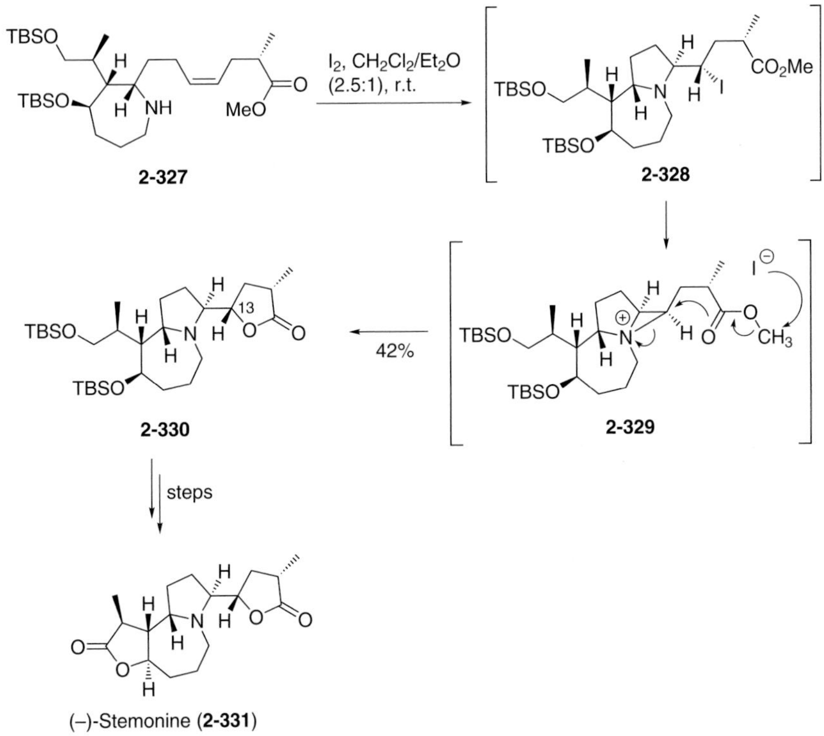

Scheme 2.77. Iodine-induced domino cyclization affording the pyrrolidino-butyro-lactone framework of (–)-stemonine (**2-331**).

ral products isolated from monocotyledonous plants comprising the genera *Stemona*, *Croomia*, and *Stichoneuron* [177]. Chinese and Japanese folk medicine has recorded the extensive use of extracts and herbal teas of Stemonaceae as remedies of respiratory diseases, including tuberculosis, and as anthelmintics [178]. Dried plant materials are utilized as powerful insecticidal resources for the treatment of livestock throughout East Asia [179]. (–)-Stemonine (**2-331**), an important secondary metabolite of *Stemona japonica* [180], is characterized by the presence of a unique 1-azabicyclo[5.3.0]decane as an integral part of the molecular architecture. The efficient construction of the bicyclic portion was achieved by an iodine-mediated twofold cyclization process of **2-327**, which provided **2-330** *via* **2-328** and **2-329** in a yield of 42%, along with 20% recovery of the starting material (Scheme 2.77) [181].

The transformation proceeds with excellent stereoselectivity by kinetic formation of the 2,5-*trans*-disubstituted pyrrolidine **2-328** [182]. The tertiary amine can now initiate a nucleophilic backside displacement of the vicinal iodide in **2-328**, leading to an aziridinium salt **2-329** [183]. This event ensures a net retention of the stereochemistry at C-13 in the following attack of the ester carbonyl in the butyrolactone ring closure to give **2-330**.

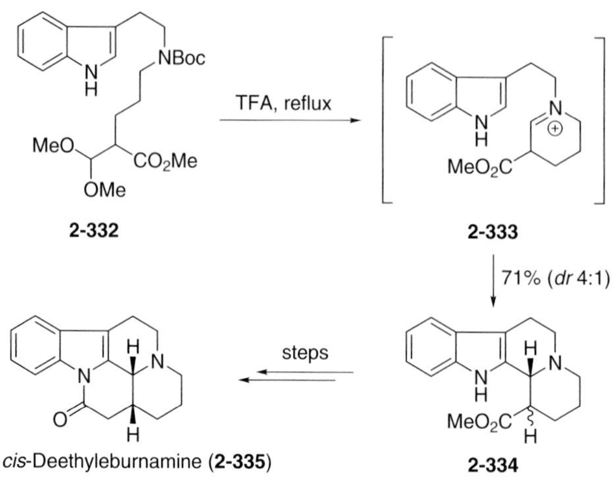

Scheme 2.78. Acid-mediated condensation/cyclization.

Domino sequences forming a reactive iminium ion intermediate in the first step by condensation of an amine or an amido moiety with a carbonyl functionality being trapped by a nucleophile in the subsequent step are well known (e. g., Pictet–Spengler-type reactions). An illustrative example is the work of Pátek and coworkers, who reported on a *N*-acyliminium ion-generation/cyclization process to produce various multifunctional heterocyclic scaffolds [184]. In addition, the twofold domino reaction can be extended by a further step which provides access to even more complex heterocycles. Several natural alkaloids and analogues as *cis*-deethyleburnamine (**2-335**) have been synthesized using iminium ion formation/cyclization processes (Scheme 2.78). The reaction of **2-332** (as described by Zard and coworkers) with TFA under reflux led to a deprotection of the secondary amine and the cleavage of the acetal moiety to give an aldehyde. Subsequently, the expected domino process involving the iminium ion **2-333** as intermediate led to the quadricyclic amine **2-334** in 71 % yield as a 4:1 *cis*/*trans* mixture [185]. Transformation of **2-334** into **2-335** can easily be performed *via* an already established four-step sequence [186].

The assembly of the benzazepane framework **2-338** (e. g., found in cephalotaxine [187]) starting from **2-336** has been achieved by the group of Schinzer [188] utilizing a combination of a Beckmann rearrangement to give the iminiumion **2-337** and an allylsilane cyclization (Scheme 2.79).

A highly efficient approach to cephalotaxine and its analogues with different ring sizes has been developed by Tietze and coworkers (Scheme 2.80) [189]. Reaction of the primary amine **2-339** with the ketoester **2-341** in the presence of AlMe₃ first led to the aluminum amide **2-340**, which afforded the spirocyclic lactam **2-343** *via* **2-342** by reaction with **2-341** in an amidation/Michael process in a very good yield of 81 %. A Pd⁰-catalyzed reaction of **2-343** then led to the pentacyclic skeleton **2-344** of cephalotaxine **2-345**. In a similar manner, ring size analogues with a six- and five-membered ring C were prepared using benzylamines or anilines instead of **2-339** as substrates.

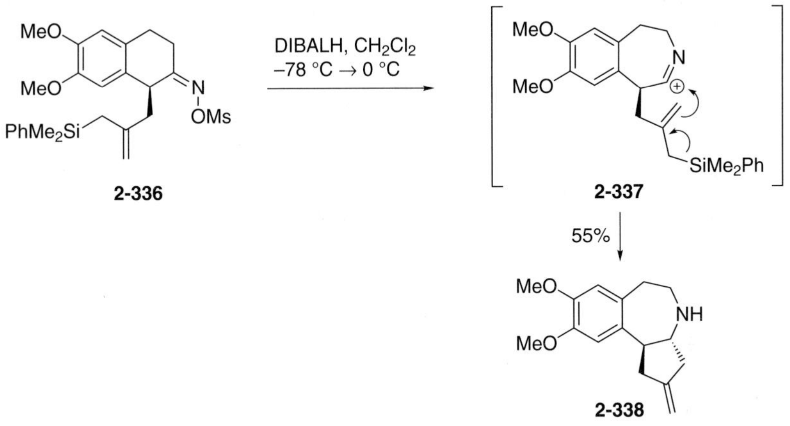

Scheme 2.79. Synthesis of benzazepanes *via* a Beckmann rearrangement/allylsilane-cyclization sequence.

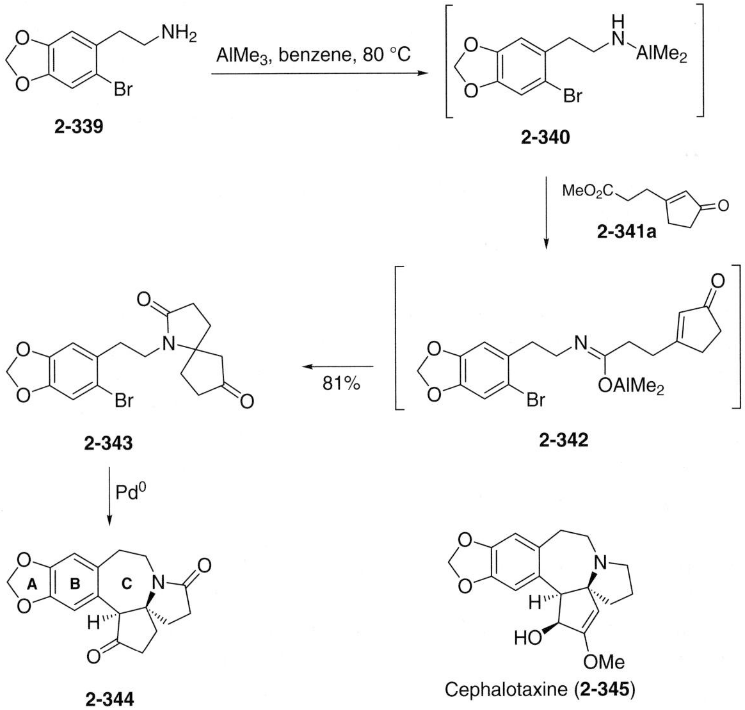

Scheme 2.80. Synthesis of cephalotaxine analogs.

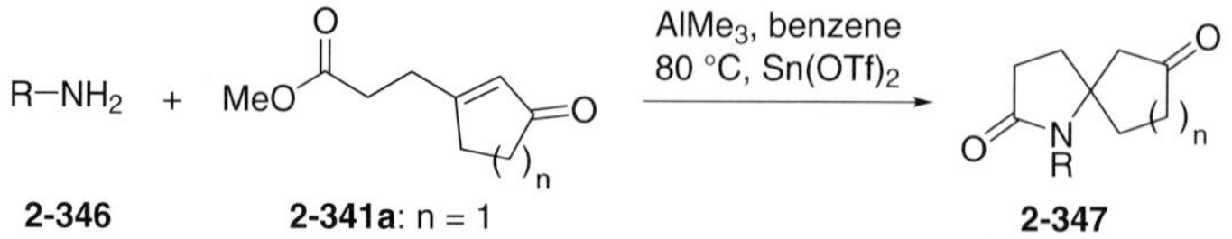

Scheme 2.81. Synthesis of spiro-azanonanediones.

Entry	R	Product 2-347	n	Yield [%]
1	Ph(CH₂)₂	a	1	71
2	Bn	b	1	79
3	nBu	c	1	53
4	sBu	d	1	21
5	HO(CH₂)₆	e	1	34
6	Bn	f	2	80
7	Ph(CH₂)₂	g	2	41

In order to explore the generality of this new domino reaction, the conversion of various primary amines with 2-341 and the cyclohexane analogue was investigated (Scheme 2.81). For example, the reaction proceeds with high yields when benzyl- or (2-phenylethyl)amine are used (entries 1 and 2). In comparison, sterically more hindered amines such as 2-butylamine produced much lower yield (entry 4). Furthermore, the reaction tolerates other functional groups, such as an unprotected hydroxyl group (entry 5), and variation of the enone ring size is possible (entries 6 and 7). More recent results have revealed that the addition of $Sn(OTf)_2$ or $In(OTf)_2$ makes the transformation more reliable.

As the final example in this section, a Li-mediated carboaddition/carbocyclization process will be described. Thus, Cohen and coworkers observed a 5-exo-trig-cyclization by reaction of the lithium compound 2-349 and α-methyl styrene 2-350 to give 2-352 via 2-351 (Scheme 2.82). Quenching of 2-352 with methanol then led to the final product 2-353 [189]. In this process, 2-349 is obtained by a reductive lithiation of the corresponding phenyl thioether 2-348 with the radical anion lithium 1-(dimethylamino)naphthalenide (LDMAN) (2-354). Instead of the homoallylic substance 2-348, bishomoallylthioesters can also be used to provide substituted six-membered ring compounds.

Of particular note here is that, besides the organolithium derivative, the solvent also plays an important role [191]; Me_2O was found to be a highly effective for the in-situ production of the radical anion 2-354 from its precursor N,N-dimethylamino naphthalene. For the reductive lithiation of the thioethers and the following steps, the addition of pre-cooled Et_2O combined with subsequent removal of the Me_2O under reduced pressure gave the best results.

For synthesis of the sesquiterpene (±)-cuparene (2-358) [192] using this method, 2-355 was reacted with 2-354 and 2-356 to give 2-358 via 2-357 in 45% overall yield.

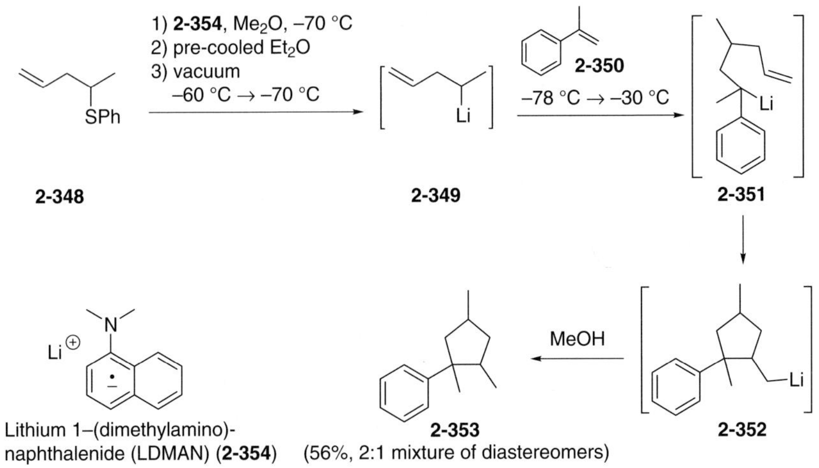

Scheme 2.82. Synthesis of substituted cyclopentanes.

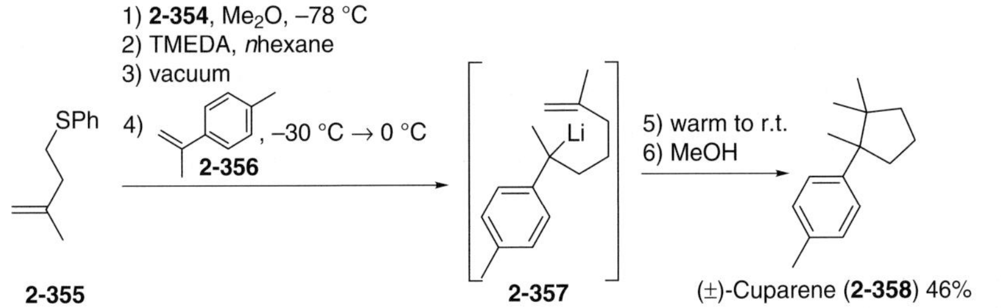

Scheme 2.83. Synthesis of (±)-cuparene (**2-358**).

Taylor and coworkers also developed similar twofold intramolecular Li-mediated carboadditions [193]. In contrast to the aforementioned sequences, they were able to utilize functionalized triple bonds and to trap the intermediate lithiated species such as **2-352** with other electrophiles besides a proton. Such a 2.2.2 process will be discussed in the following section, which relates to threefold anionic domino processes.

Quite recently, three interesting twofold anionic domino reactions have been published which we would like to include in this chapter. Thus, novel three-component domino Michael addition/electrophilic trapping protocols were independently developed by the groups of Jørgensen [194] and MacMillan [195]. In both approaches, two adjacent stereogenic centers could be formed with high *dr*- and *ee*-values by using organocatalysts. A Michael addition/alkylation has been reported by Pilli and coworker [196]. In this example, the azaspirodecane moiety of halichlorine [197] was assembled in a diastereoselective manner.

2.1.1
Anionic/Anionic/Anionic Processes

Threefold anionic domino processes clearly have more variants with regard to the possible combinations compared to twofold reaction sequences. In order to maintain some clarity, an attempt was made to classify and sort these, according to the preceding chapter. Correspondingly, the initiating reaction step is used as the primary identifying feature. Thus, we start with the Michael addition-initiated threefold anionic domino processes which are again quite common. Here, a highlight is a threefold Michael addition developed by the group of Carreño (Scheme 2.84) [198]. Treatment of enantiopure [(*S*)*R*]-*p*-[(*p*-tolylsulfinyl)methyl]quinamines **2-359a–c** and also their hydroxy analogues **2-359d,e** with 2-(trimethylsilyloxy)furan in the presence of TBAF gave the enantiopure heterocyclic cage compounds **2-360/2-361** in 56 to 67 % yield.

2–359a and **2-359d** led to a 1:1 mixture of the two diastereomers **2-360a,d** and **2-361a,d**, whereas under the same conditions the sulfoxides **2-359b,e** afforded exclusively diastereomer **2-360b,e**; as expected, **2-359c** gave the diastereomer **2-361c**. Mechanistically, as shown for the reaction of (4*R*)-**2-359b** in Scheme 2.85, Michael addition of 2-(trimethylsilyloxy)furan induced by TBAF takes place on the less-hindered and more electrophilic site of **2-359b**. The newly formed butenolide framework is then attacked by the amino moiety in a second Michael manner to give an enolate which, in a third Michael addition, then affords **2-360b**.

Another option is the twofold Michael addition/S_N-type sequence of which manifold versions have been published. Thus, Padwa's group reported on the diastereoselective synthesis of bicyclo[3.3.0]octenes [199], while in another approach by Hagiwara and coworkers various tricyclo[3.2.1.0]octane derivatives and similar bridged compounds have been constructed [200]. The group of Spitzner has also been engaged intensively in Michael/Michael/S_N-type processes [201]. One such ex-

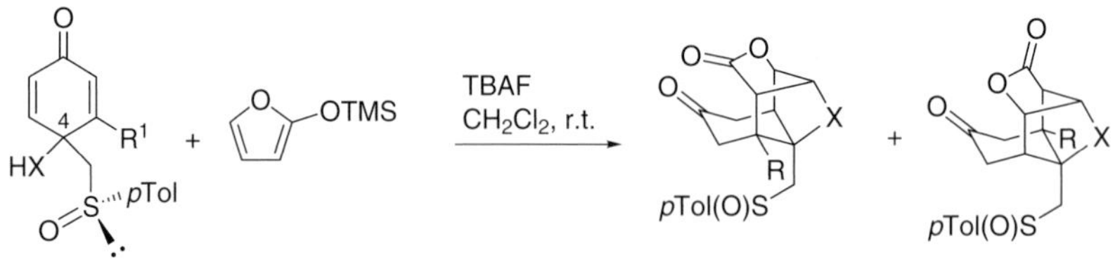

Entry	2-359	R¹	X	Products 2-360/2-361	Ratio	Yield [%]
1	a	H	NH	a	1:1	67
2	b (4*R*)	Me	NH	b	>99:1	56
3	c (4*S*)	Me	NH	c	>99:1	70
4	d	H	O	d	1:1	67
5	e (4*R*)	Me	O	e	>99:1	67

Scheme 2.84. Synthesis of **2-360/2-361**.

ample is the synthesis of enantiopure tricyclo[3.3.1.0]octenes, as illustrated in Scheme 2.86. Herein, treatment of the Li-enolate obtained from the enone **2-362** with the chloro esters (Z)-**2-363a** as well as (E)-**2-363b** gave the tricyclic compounds **2-364a** and **2-364b**, respectively, as single isomers.

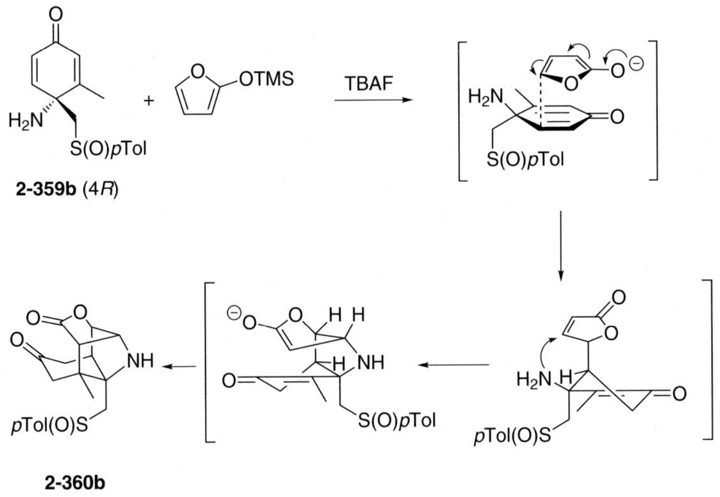

Scheme 2.85. Stereochemical and mechanistic course of the domino threefold conjugate addition.

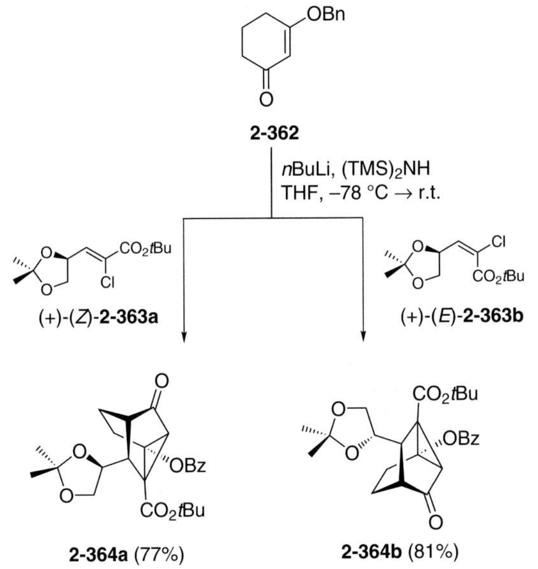

Scheme 2.86. Diastereoselective formation of tricyclic compounds **2-364**.

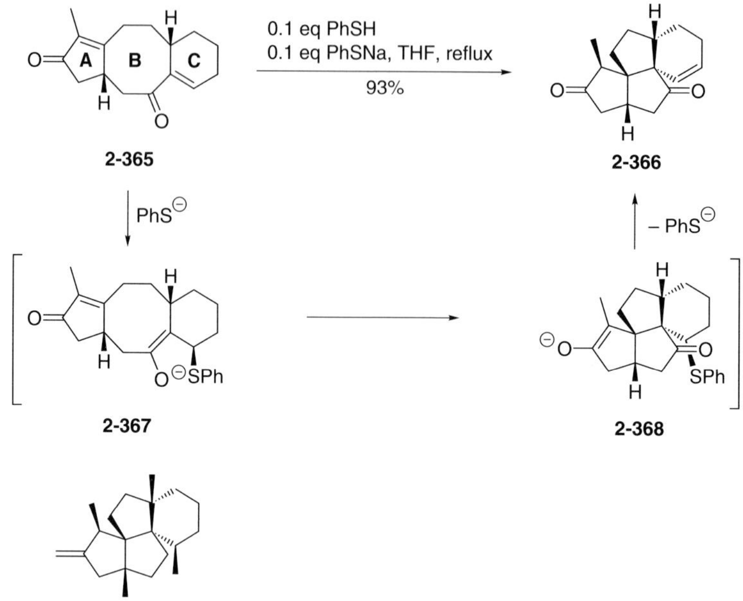

Scheme 2.87. Synthesis of the tetraquinanes **2-366**.

Generally, "push-pull" substituted cyclopropanes as **2-364** are flexible building blocks and represent an equivalent for 1,4-dicarbonyl compounds. They show a pronounced tendency to undergo ring opening [202].

Another interesting reaction sequence has been observed by Moore and co-workers when a THF solution of **2-365** and catalytic amounts of thiophenol as well as sodium thiophenolate were refluxed (Scheme 2.87) [203]. This resulted in a rearrangement to give the angularly fused tetraquinane **2-366**; this skeleton is found in the natural product waihoensene (**2-369**) [204]. It can be assumed that, primarily, a Michael addition of the thiolate from the β-face occurs to give **2-367**, which then undergoes a transannular Michael-type ring closure to provide **2-368**. The enolate moiety in **2-368** then induces an intramolecular E_2 *trans*-diaxial elimination to afford product **2-366** with regeneration of the thiophenolate catalyst. The last step seems to be favored due to the close proximity of the enolate moiety in **2-368** to the (pro-*S*)-hydrogen in 2-position of the thiophenol functionality being *trans* disposed to the leaving group.

This explanation is in accordance with the reaction of a ring size isomer of **2-365** containing a cyclopentene instead of a cyclohexene moiety as ring C. With catalytic amounts of thiophenolate and thiophenol, little reaction was observed; nevertheless, with stoichiometric amounts of thiophenol a tetraquinane, still containing the SPh-group, was obtained in 76% yield. Inspection of a molecular model revealed that the corresponding enolate is not proximately disposed to facilitate the E_2 elimination step.

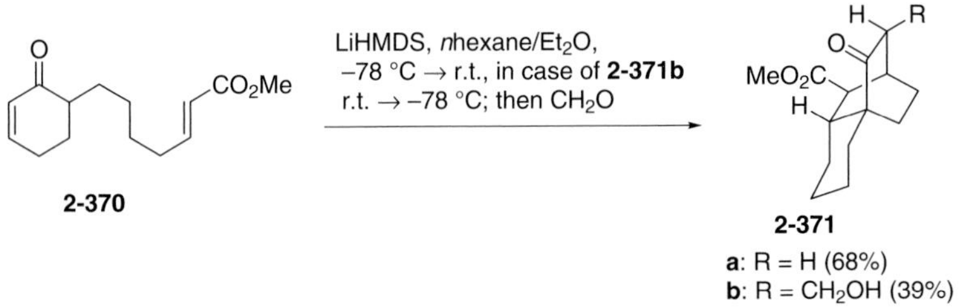

Scheme 2.88. Synthesis of bridged compounds **2-371**.

Domino processes forming tricyclo[6.2.2.0]dodecane and tricyclo[5.3.1.0]unde-cane systems have been described by the group of Fukumoto [205]. The sequences are based on two consecutive Michael additions which are followed by an aldol or a substitution reactions forming up to three new C–C-bonds and up to five stereo-genic centers in a single operation. For instance, when 2-cyclohexen-1-one **2-370**, containing an α,β-unsaturated ester moiety attached to C-2 through a tether, was treated with LiHMDS the double Michael product **2-371a** was obtained in 68 % yield as a single stereoisomer (Scheme 2.88). Moreover, on successive treatment of the intermediate enolate with gaseous CH_2O at –78 °C, the hydroxymethylated compound **2-371b** is obtained in 39 % overall yield.

Under the same conditions, tricyclo[5.3.1.0]undecanes are accessible from 5-sub-stituted 2-cyclohexen-1-one as **2-370** with a shorter tether by one CH_2-group. Recently, another Michael/Michael/aldol transformation was employed by Paulsen and coworkers to obtain access to the central aromatic core of compounds as **2-376** (Scheme 2.89) [206]. It is of value that such products are thought to act as cholesterol ester transfer protein (CETP) inhibitors, and the application of these drugs should prevent reduction of the HDL-cholesterol level and therefore reduce the risk of coronary heart diseases [207].

Under classical Mukaiyama conditions, silyl enol ether **2-372** and the Michael ac-ceptors **2-373** and **2-374** underwent a twofold 1,4-addition to form an enolate in which an ideal set-up exists for an intramolecular aldol reaction. This led to **2-375** with the desired structural core of **2-376** in an overall yield of 42 %.

Besides the described threefold anionic domino sequences involving three or two Michael reactions, other combinations with only an initial Michael addition are also well established. For instance, a Michael addition/iminium ion cyclization/elimi-nation process yielding naphthopyrandione derivatives [208] has been reported by K. Kobayashi and coworkers, and a Michael/imino aldol/ring closure procedure for the production of δ-lactams has been published by the group of S. Kobayashi [209]. Very recently, Zhai and colleagues have contributed a Michael-type addition/im-inium ion formation/cyclization to this field of domino reactions [210]. According to the illustration in Scheme 2.90, these authors treated the enaminone **2-377** with acrolein in the presence of the Lewis acid $BF_3 \cdot OEt_2$ as reaction promoter, and ob-served a smooth formation of the pentacycle **2-379**, probably *via* the iminium ion **2-378**. Since **2-379** is hardly soluble in most common organic solvents, it had to be transformed into its Boc-protected derivative in a subsequent step for the purpose

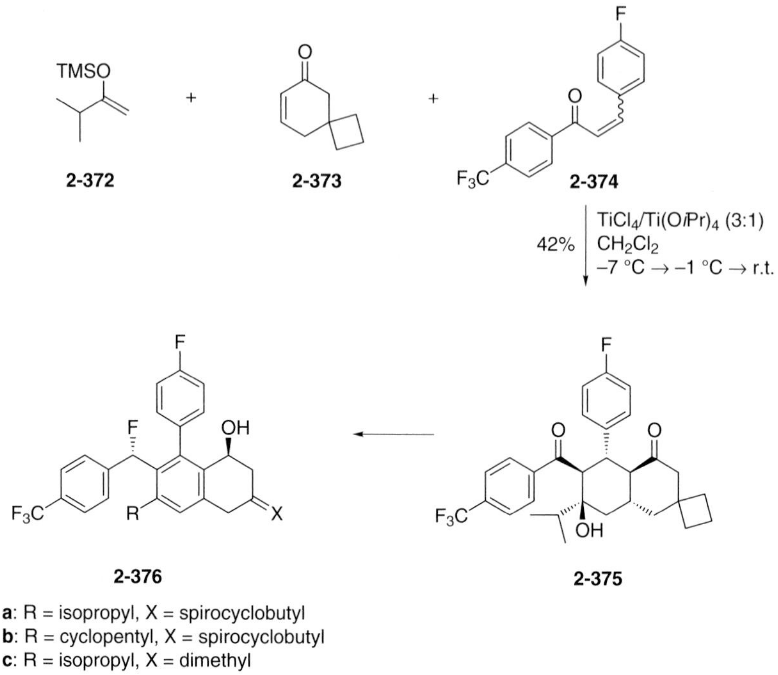

Scheme 2.89. Synthesis of CETP inhibitors.

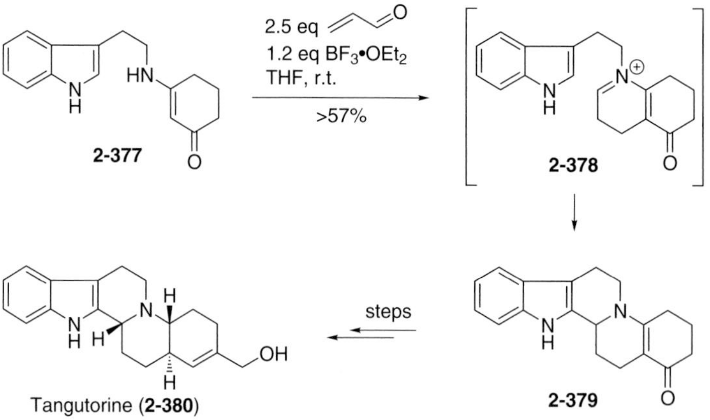

a: R = isopropyl, X = spirocyclobutyl
b: R = cyclopentyl, X = spirocyclobutyl
c: R = isopropyl, X = dimethyl

Scheme 2.90. Synthesis of tangutorine (**2-380**).

of purification. This was then used for total synthesis of the β-carboline alkaloid tangutorine (**2-380**) (Scheme 2.90) [211].

The groups of Palomo and Aizpurua also exploited this strategy to assemble compounds of type **2-381** (Scheme 2.91) [212]. These azetidin-2-ones have been found to

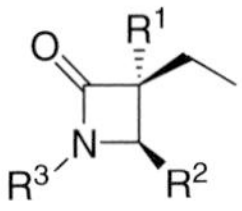

		R¹	R²	R³
	a:	H	CO₂Et	S(O)C₆H₄pNO₂
	b:	H	OPh	C(O)NHCH₂Ph
	c:	Et	OC₆H₄(CH₂COH)-P	(R)C(O)NHCH(nPr)C₆H₄pMe

Scheme 2.91. 3-Alkyl-4-alkoxy-carbonylazetidin-2-ones **2-381**.

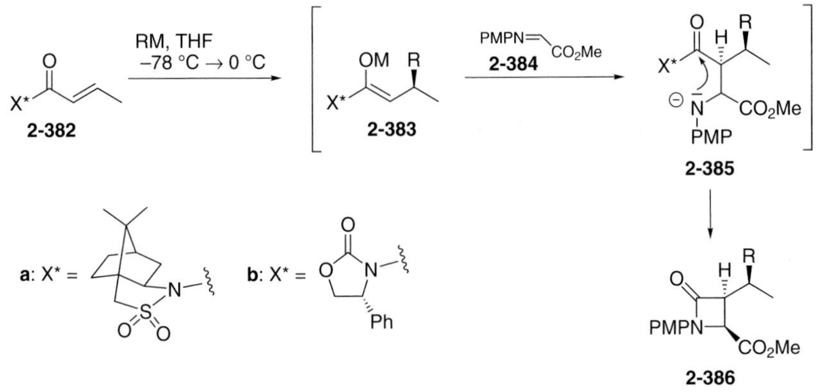

Scheme 2.92. Three-component asymmetric synthesis of 3-alkyl-4-methoxycarbonyl-azetidin-2-ones **2-386** from chiral crotonyl derivatives.

Entry	2-382	Cuprate RM	Product 2-386	dr	ee [%]	Yield [%]
1	a	Me₂CuLi	a	98:2	>99	57
2	a	Me₂CuCNLi₂	a	99:1	>99	40
3	b	Me₂CuLi	b	99:1	98	54
4	b	PhCuMgBr	b	92:8	>99	62

be effective inhibitors of the human leukocyte elastase (HLE), which is believed to be responsible for the enzymolytic degradation of a variety of proteins, including the structural proteins fibronectin, collagen, and elastin [213].

Their synthesis is based on a conjugate addition of organocuprates RM to chiral enantiopure α,β-unsaturated carboxylic acid derivatives **2-382** and subsequent condensation of the resulting enolates **2-383** with an imine **2-384** to give **2-385** which cyclize, affording **2-386** (Scheme 2.92). The best results with respect to reactivity and yield have been observed when organolithium or organomagnesium cuprates were used. Enolates formed from so-called "higher-order" cyanocuprates seemed not to be very reactive towards the imine **2-384**. Oppolzer's N-enoylsultams **2-382a** and Evans' N-enoyloxazolidinones **2-382b** were of similar efficiency to induce both, asymmetric conjugate addition and subsequent stereoselective enolate condensation.

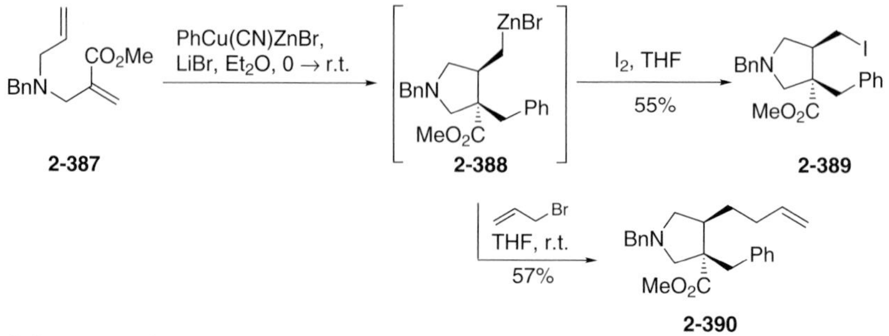

Scheme 2.93. Synthesis of pyrrolidines.

Another domino cuprate 1,4-addition-initiated threefold anionic domino sequence was developed by Chemla and coworkers (Scheme 2.93) [214]. Michael addition of the α,β-unsaturated ester **2-387** with PhCu(CN)ZnBr was followed by a carbocyclization to give the zinc species **2-388** which can be intercepted by iodine or an allyl bromide affording substituted pyrrolidines **2-389** and **2-390**, respectively.

Moreover, threefold anionic domino reactions where a Michael and an aldol reaction are combined with a third transformation such as an elimination, an electrophilic aromatic substitution, or a lactonization are attractive. Jauch used this strategy for the synthesis of enantiopure butenolides in which as the last step a BnBr/nBu₄NI-triggered *syn* elimination operates [215]. Its practicability was highlighted in the total synthesis of kuehneromycin A [216]. Further applications in this field have recently been reported by the groups of Casey [217] within their podophyllotoxin synthesis *via* a domino conjugate addition/aldol/electrophilic aromatic substitution reaction. D'Onofrio and Parlanti [218] were able to produce substituted γ-lactones by trapping the enolate formed in the aldol reaction in a ring-closing lactonization with a proximate methyl ester moiety. A related Michael addition/aldol/lactonization multi-step sequence reported by Thebtaranonth and coworkers gave access to various substituted spirocyclic lactones **2-397** (Scheme 2.94) [219].

Mechanistically, α-methylenecyclopentenone (**2-391**) reacts with ester enolate **2-392** in a Michael addition to give the enolate **2-393**, which is then trapped with an aldehyde **2-394** generating the alcoholate **2-396**. This eventually cyclizes through lactonization to afford **2-397** in good yield. The products **2-397** are obtained as single diastereomer; thus, it can be assumed that the aldol reaction proceeds via the six-membered chair-like transition state **2-395**.

A process by Li and coworkers with the aldol reaction as the last step is shown in Scheme 2.95. During a study towards enantiopure β-iodo Baylis–Hillman adducts (see Section 2.1), trace amounts of interesting side products **2-402** were isolated (Scheme 2.95) [220]. Their formation was assumed to arise from a α,β-conjugate addition of TMSI onto ethynyl alkyl ketone **2-398**, followed by a Lewis acid-mediated isomerization of the allenolates **2-399** and an aldol reaction of the resulting enolates **2-400** with an aldehyde **2-401**. In order to make this side reaction becoming the main one, a variety of different conditions were tested. In fact, BF₃·OEt₂ as Lewis acid in CH₂Cl₂ turned out to be the best catalyst, giving rise to the desired

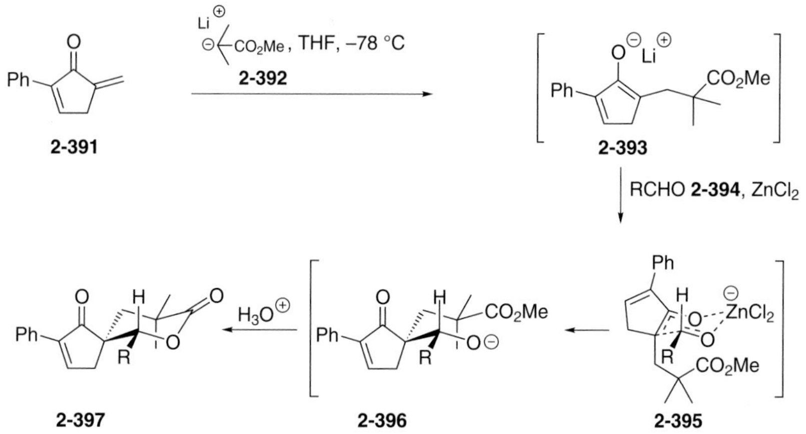

Entry	R–CHO 2-394	R	Product 2-397	Yield [%]
1	a	Et	a	72
2	b	Ph	b	60
3	c	mMeOC$_6$H$_4$	c	74
4	d	pMeOC$_6$H$_4$	d	69
5	e	(E)-MeCH=CH	e	56

Scheme 2.94. Stereoselective domino Michael/aldol/lactonization process.

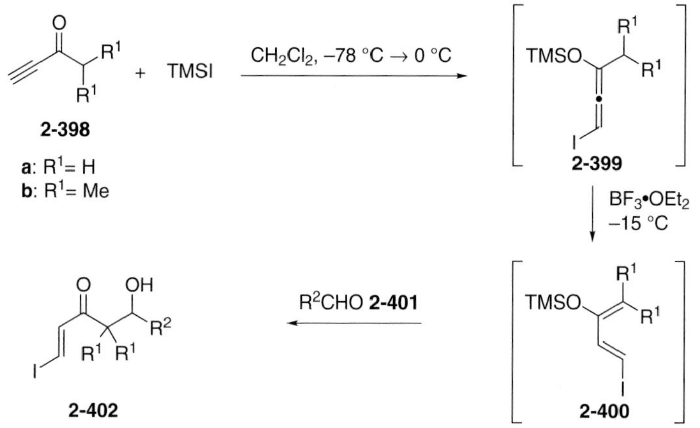

a: R^1= H
b: R^1= Me

Entry	2-398	R^{2}CHO 2-401	Product 2-402	Yield [%]
1	a	PhCHO	a	76
2	a	Ph⌒CHO	b	79
3	a	⌥CHO	c	73
4	b	PhCHO	d	82

Scheme 2.95. Results of the three-component reaction leading to compounds **2-402**.

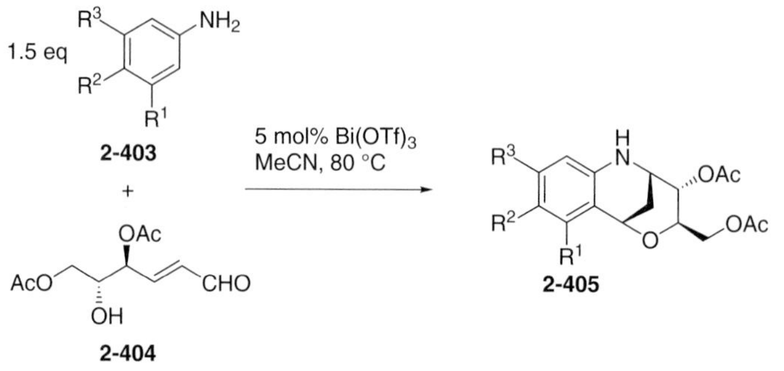

Scheme 2.96. Synthesis of bridged heterocycles **2-405** by a domino Michael/Friedel–Crafts/S$_N$-type cyclization.

Entry	2-403	R^1	R^2	R^3	2-405	Yield [%]
1	a	H	H	H	a	80
2	b	Me	H	H	b	86
3	c	H	H	Cl	c	85
4	d	H	Cl	F	d	78
5	e	CN	H	H	e	70

products **2-402** with (*E*)-geometry in yields of 73–82 % without formation of β-iodo Baylis–Hillman adducts [221].

Diastereoselective domino Michael addition/Friedel–Crafts/S$_N$-type cyclizations which allow the synthesis of bridged tetrahydroquinolines such as **2-405** have been observed by Yadav and coworkers when a mixture of anilines **2-403** and δ-hydroxy-α,β-unsaturated aldehydes **2-404** were treated with catalytic amounts of Bi(OTf)$_3$ or InCl$_3$ in MeCN at 80 °C (Scheme 2.96) [222].

Ballini and coworkers developed a domino Michael/base-promoted elimination/ hemiacetal process leading to substituted dihydropyranols from cheap starting materials [223]. Moreover, the compounds obtained contained at least three different functionalities for further manipulation. The group of Rodriguez observed a multicomponent domino reaction leading to fused cyclic aminals **2-409** when 1,3-dicarbonyls **2-406**, α,β-unsaturated carbonyl compounds **2-407**, and ω-functionalized primary amines **2-408** were heated together in toluene in the presence of 4 Å molecular sieves [224]. The products depicted in Scheme 2.97 are usually obtained in good yield and, in some cases, even as single diastereomers.

An interesting observation was made when o-aminophenol (**2-411**) was employed in the reaction with carbethoxypiperidone **2-410** and acrolein (Scheme 2.98). In this case, the spirocyclic scaffold **2-412** was exclusively formed in 67 % yield. This result can be explained by invoking a stereoelectronic control due to the presence of the aromatic ring which prevents the formation of the corresponding fused tetracyclic isomer. Moreover, both reactive sites can simultaneously be functionalized using 2-amino-1,3-propanediol (**2-413**) as partner in the multicomponent reaction. This leads to the formation of three new cycles

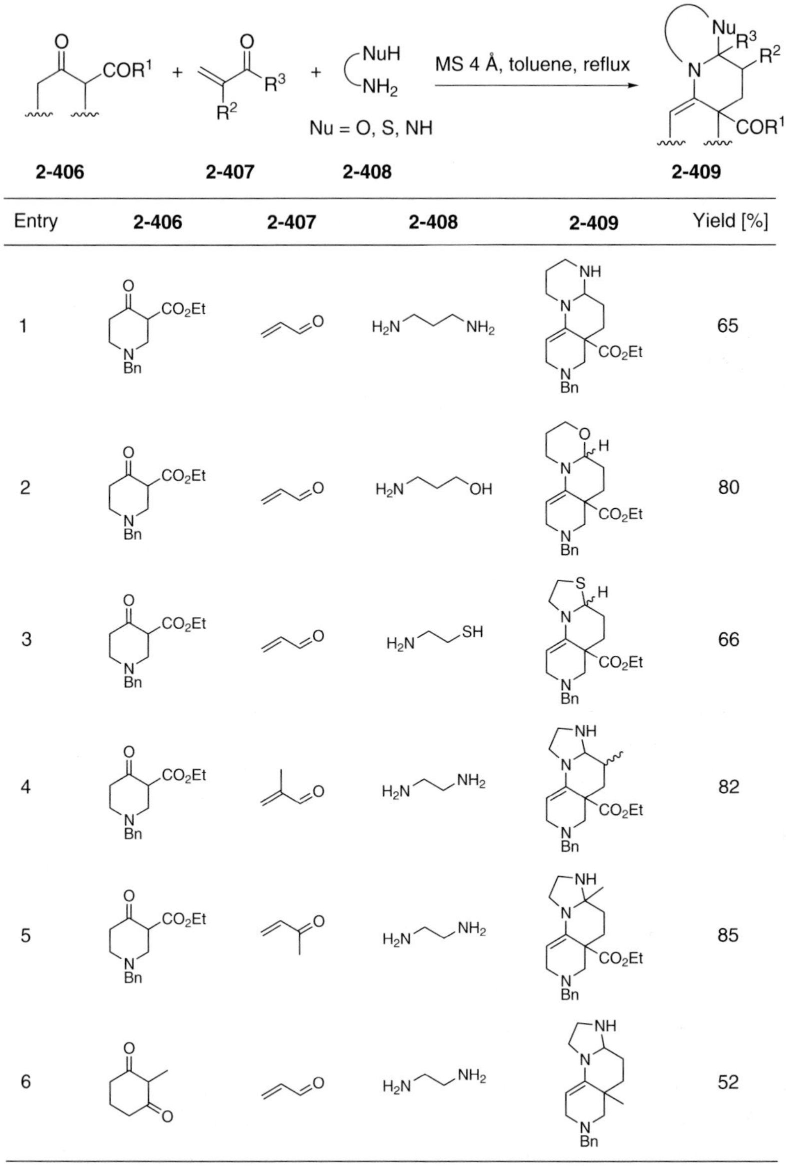

Scheme 2.97. Multicomponent reaction leading to polycyclic *N,N*-, *N,O*- and *N,S*-aminals **2-409**.

and five new bonds, ending with the tetracyclic structure **2-414** bearing six stereo-genic centers.

It is assumed that the overall process is initiated by a Michael addition of the 1,3-dicarbonyl compound onto the α,β-unsaturated carbonyl derivative. There follows the formation of either an aminal and an iminium intermediate which is followed by the formation of two *N,O*-acetals.

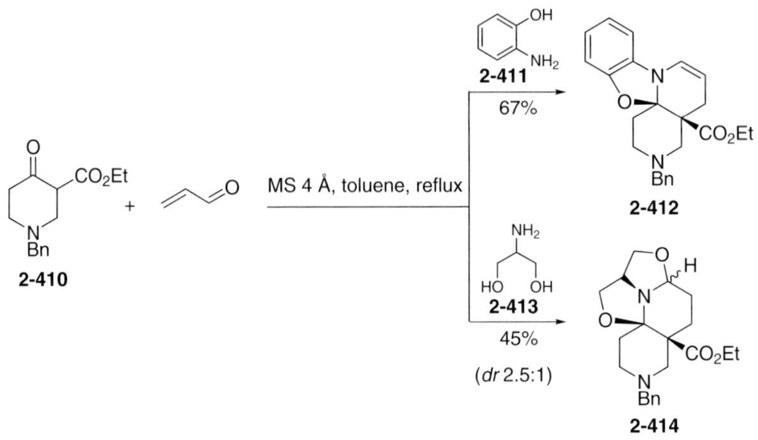

Scheme 2.98. Formation of tetracycles **2-412** and **2-414**.

In addition to the already discussed threefold domino processes starting with a Michael addition, many reactions are known in this field which are initiated by an aldol- or retro-aldol transformation. Langer and coworkers reported on a novel TiCl$_4$/TiBr$_4$-mediated twofold aldol-type condensation/S$_N$-type ring-opening converting 1,3-bis-silyl enol ethers with 1,1-diacetylcyclopropanes into substituted salicylates [225]. According to the process developed by Beifuss and coworkers, treatment of pyran-4-one (**2-415**) with a silyl triflate followed by addition of an α,β-unsaturated ketone **2-417** and 2,6-lutidine in CH$_2$Cl$_2$ furnished substituted tetrahydro-2H-chromenes of type **2-421** in up to 98% yield (Scheme 2.99) [226]. The sequence is assumed to start with an aldol-type reaction of oxenium ion **2-416** and enol ether **2-418**, being formed from **2-417**, to give **2-419**, which is followed by two 1,4-additions with **2-420** as proposed intermediate. In this domino process three new C–C-bonds and four stereogenic centers are built up in a highly efficient and selective manner, with the exception of the acyclic double bond.

Furthermore, as described by Mori and coworkers, the domino aldol/cyclization reaction of the β-keto sulfoxide **2-422** with succindialdehyde (**2-423**) in the presence of piperidine at r.t. afforded the chromone **2-424** which, on heating to 140 °C, underwent a thermal *syn*-elimination of methanesulfenic acid to provide **2-426** in 22% overall yield (Scheme 2.100) [227]. This approach was then used for the synthesis of the natural products coniochaetones A (**2-425**) and B (**2-427**) [228].

Cyclopentane ring systems have been constructed in a very elegant manner through a so-called [3+2] annulation strategy elaborated by Takeda and coworkers (Scheme 2.101) [229]. Thus, aldol addition of acylsilane **2-428** and enolate **2-429** led to the intermediate **2-430** which, in a 1,2-Brook rearrangement, generated a delocalized allylic anion **2-431**. This then cyclized to a mixture of the diastereomers **2-432** and **2-433**. Noteworthy, the ratio of **2-432** and **2-433** is not affected by the configuration of the double bond in the substrate **2-428**. This finding is in accordance with the proposed delocalized allylic anion **2-431** as an intermediate.

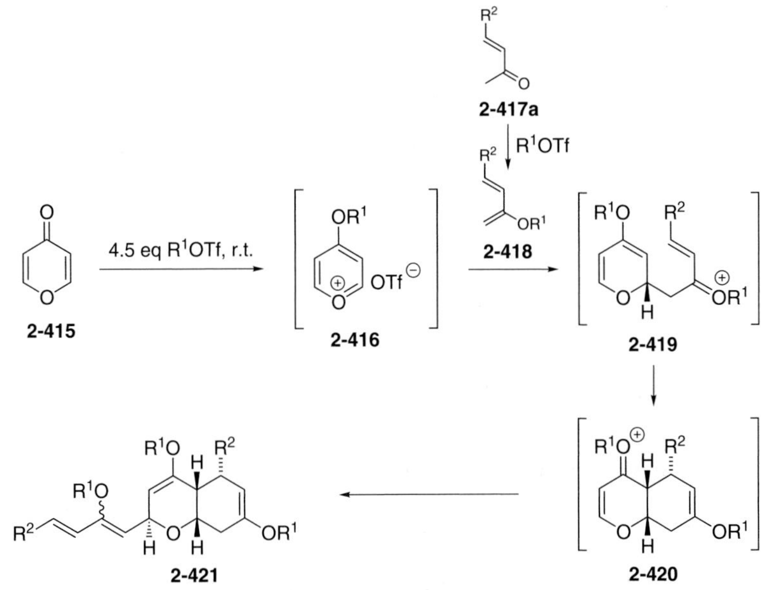

Scheme 2.99. Synthesis of tetrahydro-2*H*-chromenes **2-421**.

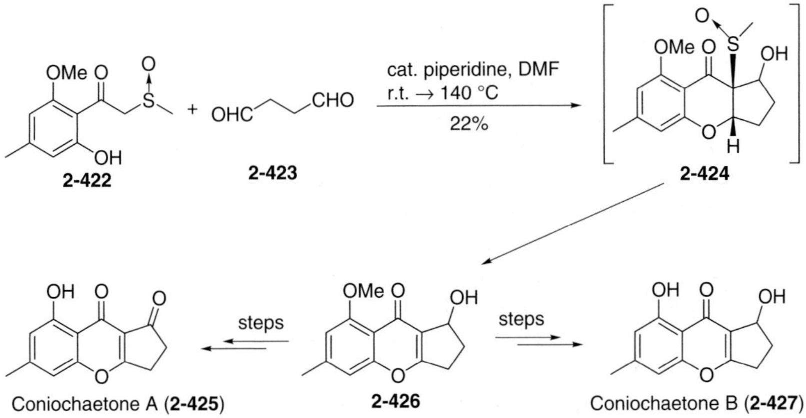

Entry	2-416/2-417		2-421	Ratio (*E/Z*)	Yield [%]
	R¹	R²			
1	TES	Ph	a	4.1:1.0	98
2	TBS	*p*CNC₆H₄	b	11.0:1.8	80
3	TBS	*p*ClC₆H₄	c	1.3:1.0	35
4	TMS	*p*CO₂MeC₆H₄	d	3.4:1.0	62
5	TMS	*p*NO₂C₆H₄	e	only *E*	56

Scheme 2.100. Synthesis of coniochaetone A (**2-425**) and B (**2-427**).

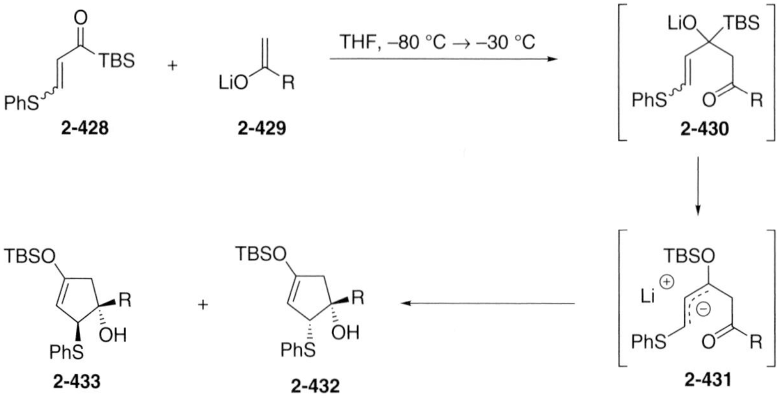

Scheme 2.101. [3+2] Annulation based on a Brook rearrangement.

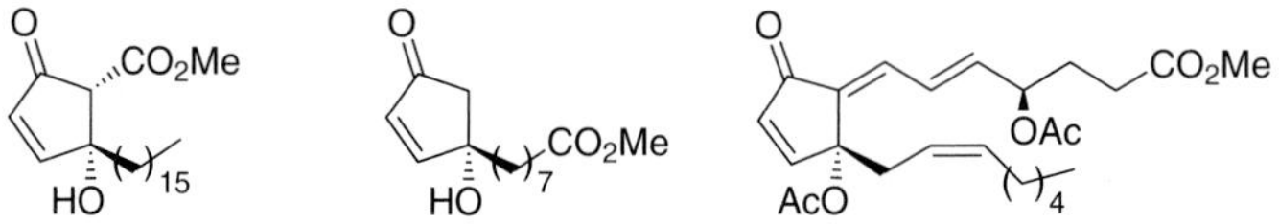

Entry	R	Yield [%]	
	2-429	2-432	2-433
1	*i*Pr	55	19
2	Et	70	5
3	*n*Pr	74	7

Untenone A (**2-434**)   Chromomoric acid D-II methyl ester (**2-435**)   Clavulone II (**2-436**)

Scheme 2.102. Natural products available through the [3+2] annulation strategy.

The concept was successfully applied in the synthesis of untenone A (**2-434**) [229b], chromomoric acid D-II methyl ester (**2-435**) [229a], and clavulone II (**2-436**) (Scheme 2.102) [229c]. Further studies allowed the development of a multifaceted [3+4] annulation approach [230]; moreover, numerous associated threefold anionic domino sequences such as cyanide addition/1,2-Brook rearrangement/alkylation [231] and epoxide opening/1,2-Brook rearrangement/alkylation processes [232] have been designed.

Ogasawara and coworkers have also published a complete series of threefold anionic domino reactions, all of which are based on an initial retro-aldol process. For instance, starting from chiral bicyclo[3.2.1]octenone **2-437**, a formal total synthesis of (–)-morphine (**2-445**) [233] has been successfully performed (Scheme 2.103) [234]. Transformation of **2-437** into the substrate **2-488**, necessary for the domino reaction, was achieved in seven linear steps. The domino process was then initiated by simply refluxing a solution of **2-438** in benzene in the presence of ethy-

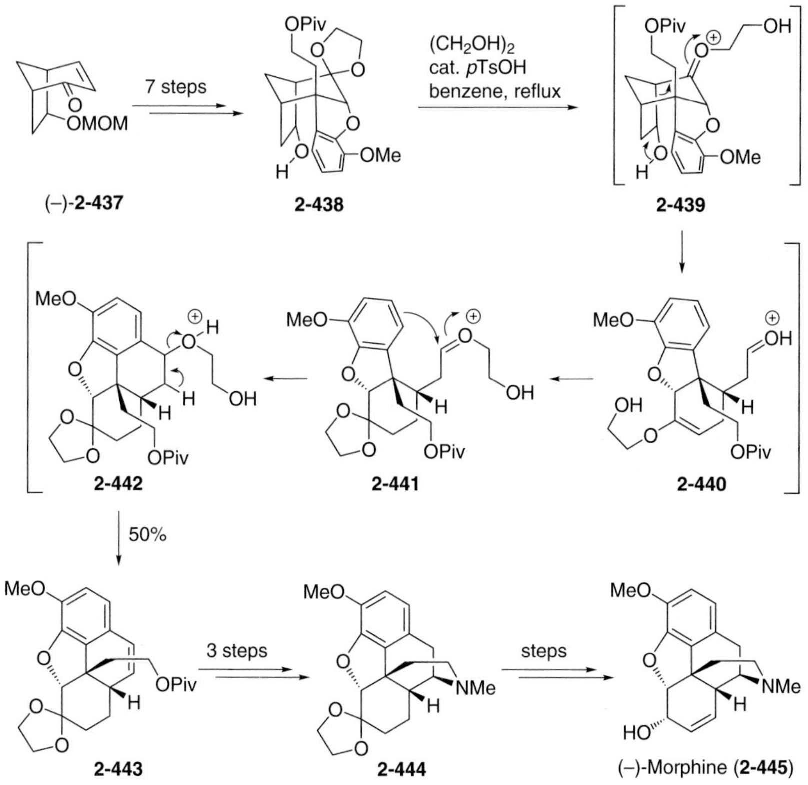

Scheme 2.103. Synthesis of (–)-morphine (**2-445**).

lene glycol and catalytic amounts of *p*TsOH to give the hydrophenanthrene **2-443** in 50 % yield as a single stereoisomer.

The course of the transformations can be explained by the initial formation of ox-enium ion **2-439**. This intermediate undergoes a retro-aldol cleavage leading to another oxenium ion **2-441** *via* the protonated aldehyde **2-440**. There follows an electrophilic aromatic substitution with subsequent elimination of a molecule of ethylene glycol to give the hydrophenanthrene **2-443** through a transient **2-442**. Worth noting here is that the addition of ethylene glycol seemed to be essential to accelerate the cyclization reaction. Conversion of **2-443** into the morphinan **2-444**, an intermediate of a previously published total synthesis of (–)-**2-445** [235], was per-formed in only three steps, thus affording **2-444** in an overall yield of 6 % in twelve steps from **2-437**. In addition, utilizing an identical domino retro-aldol/ring clo-sure/elimination strategy the hexahydrophenanthrene framework of the natural product (+)-ferruginol [236] was synthesized [237]. Ogasawara's group also estab-lished a retro-aldol/iminium ion formation/cyclization sequence usable for the construction of other naturally occurring compounds [238]. As depicted in Scheme 2.104, under acidic conditions the cyclic acetal and the MOM ether in **2-446**, ob-

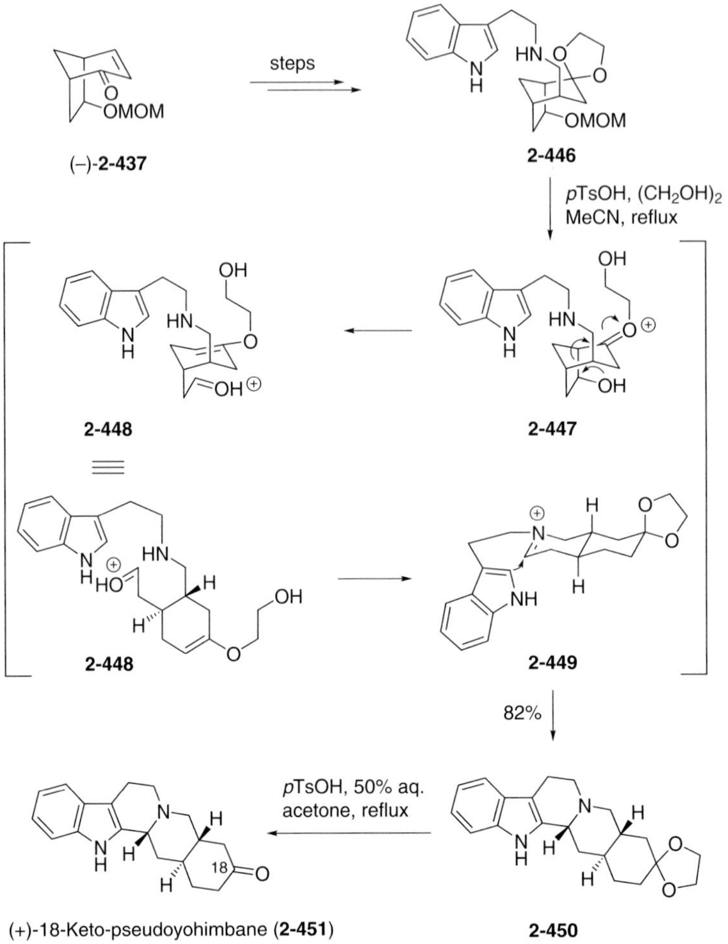

Scheme 2.104. A concise route to (+)-18-Keto-pseudoyohimbane (**2-451**) using bicyclo[3.2.1]octane **2-437** as chiral building block.

tained from (–)-**2-437** are simultaneously cleaved, providing the intermediate **2-447**. This undergoes a retro-aldol reaction, thus affording **2-448** with an aldehyde moiety. Its subsequent intramolecular condensation with the secondary amine group led to the iminium ion **2-449**, which cyclizes to give the desired pentacycle **2-450** in 82 % yield as a single isomer. Cleavage of the acetal moiety in **2-450** led to 18-keto-pseudoyohimbane **2-451** [239]. The exclusive generation of the pseudoyohimbane skeleton **2-451**, but not that of the C-3 epimeric yohimbane in the Pictet–Spengler reaction, can be explained by stereoelectronic effects [240].

2–**451** was subsequently transformed into the corynanthe-type indole alkaloid (–)-isocorynantheol [239], isolated from *Cichona ledgeriana* [241], by a simple three-step strategy.

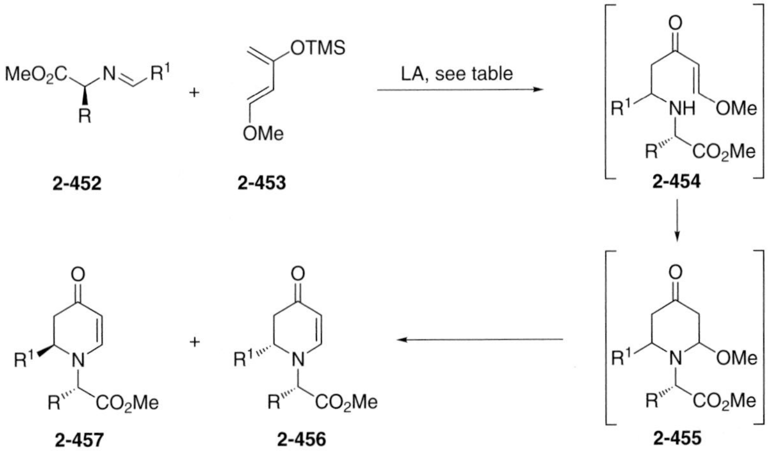

Scheme 2.105. Domino Mannich-type/Michael/elimination of **2-452** and diene **2-453**.

Entry	2-452	R¹	R (amino acid)	Temp [°C]	Lewis acid / solvent	2-457/2-456		Yield [%]
1	a	pMeOC₆H₄	Phg-OMe	−20	ZnCl₂ / THF	a	67:33	63
2	b	pNO₂C₆H₄	Phg-OMe	0	ZnCl₂ / THF	b	62:38	53
3	c	nPrl	Phg-OMe	0	ZnCl₂ / THF	c	71:29	69
4	d	pMeOC₆H₄	Val-OMe	0	ZnCl₂ / THF	d	92:8	54
5	e	pNO₂C₆H₄	Ile-OMe	−10	ZnCl₂ / THF	e	93:7	57
6	f	nPr	Ile-OMe	0	2 eq of ZnCl₂ / THF	f	15:85	11
7	g	iPr	Ile-OMe	−78 to −20	EtAlCl₂ / CH₂Cl₂	g	97:3	48
8	h	nBu	Ile-OMe	−78 to −20	Me₂AlCl / CH₂Cl₂	h	90:10	77
9	i	MeO₂C(CH₂)₃	Val-OBn	−78 to −20	EtAlCl₂ / CH₂Cl₂	i	93:7	46
10	j	EtO₂C(CH₂)₂	Val-OBn	−78 to −20	EtAlCl₂ / CH₂Cl₂	j	93:7	50

In this context, also mentionable are several publications by the groups of Díaz-de-Villegas [242], Guarna [243], Kunz [244] and Waldmann [245], which describe the formation of six-membered azaheterocycles *via* treatment of an imine with an appropriate substituted diene. For instance, as described by Waldmann and co-workers, reaction of the enantiopure amino acid-derived imines **2-452** with Danishefsky's diene **2-453** in the presence of equimolar amounts of a Lewis acid provided diastereomeric enaminones **2-456** and **2-457** (Scheme 2.105) [245a].

The mechanism of this transformation is a matter of debate, and may vary with the structure of the heteroanalogous carbonyl compound employed. Although a Diels–Alder-type process is conceivable [246], a Lewis acid-induced addition of the silyl enol ether moiety in **2-453** followed by a cyclization through a nucleophilic intramolecular attack of the amine and subsequent elimination of methanol is assumed in this case [247].

2,3-Didehydro-4-piperidinones of type **2-456** and **2-457** are useful intermediates for the synthesis of various substituted benzoquinolizines [243, 244, 245b–c]; this scaffold is found in many natural products [245b–d].

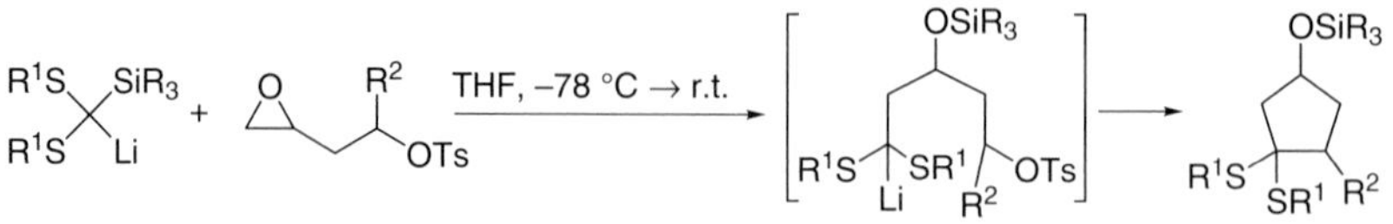

Entry	2-458	R^1	SiR$_3$	2-459	R^2	2-461	Yield [%]
1	a	Me	TMS	a	H	a	71
2	b	–(CH$_2$)$_3$–	TMS	a	H	b	70
3	c	Me	TBS	b	Me	c	69
4	d	–(CH$_2$)$_3$–	TBS	b	Me	d	41
5	c	Me	TBS	a	H	e	51
6	d	–(CH$_2$)$_3$–	TBS	a	H	f	75

Scheme 2.106. Synthesis of functionalized cyclopentanols.

Leaving the (retro-)aldol addition-initiated threefold anionic domino processes, we are now describing sequences which are initiated by a S_N-type transformation. In particular, domino reactions based on S_N/1,4-Brook rearrangement/S_N reactions are well known. For example, the group of Schaumann obtained functionalized cyclopentanols of type **2-461** by addition of lithiated silyldithioacetals **2-458** to epoxyhomoallyl tosylates **2-459** in 41–75 % yield (Scheme 2.106) [248].

The domino reaction is initiated by the chemoselective attack of the carbanion **2-458** on the terminal ring carbon atom of epoxyhomoallyl tosylate **2-459** to give the alkoxides **2-460** after a 1,4-carbon-oxygen shift of the silyl group. The final step to give the cyclopentane derivates **2-461** is a nucleophilic substitution. In some cases, using the TBS group and primary tosylates, oxetanes are formed as byproducts.

A highly useful twofold reaction of silyl dithioacetals with epoxides was described by Tietze and coworkers (Scheme 2.107) [249]. Treatment of 2.2 equiv. of enantiopure epoxides **2-463** with lithiated silyldithiane **2-458b** in the presence of a crown ether led to **2-467** after aqueous work-up. It can be assumed that by attack of the lithium compound **2-462** at the sterically less-hindered side of the epoxide **2-463**, the alkoxide **2-464** is formed which in a subsequent Brook rearrangement produces the lithium dithioacetal **2-465**. This reacts again with an epoxide to give **2-466** and furthermore **2-467**. Treatment with NaF then leads to the diol **2-468** which can be converted into the dihydroxy ketones **2-469** and the corresponding 1,3,5-triols, respectively.

The described procedure has been widely used by Smith III and coworkers [250] in the efficient total synthesis of natural products containing extended 1,3-hydroxylated chains. This architecture is often found as a structural element in polyene macrolide antibiotics [251] such as mycotoxin A and B, dermostatin, and roxaticin. The Smith group used the above-mentioned approach (e. g., as five-component coupling) for the synthesis of the pseudo-C_2-symmetric trisacetonide (+)-**2-471** [252], which was employed by Schreiber and coworkers [253] within the synthesis of (+)-mycotoxin A (**2-470a**) (Scheme 2.108). Thus, lithiation of 2.5 equiv. dithiane **2-462b** followed by treat-

nBuli, THF, –30 °C → 0 °C

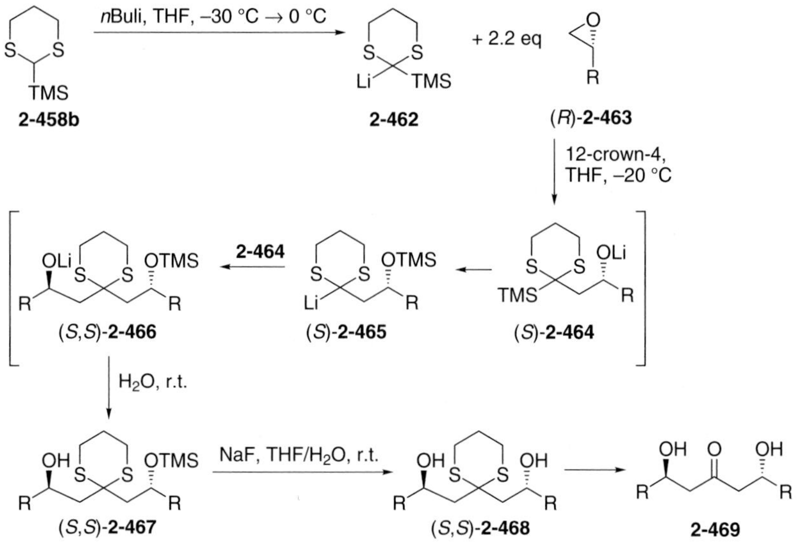

2-458b

2-462

+ 2.2 eq

(R)-2-463

12-crown-4,
THF, –20 °C

2-464

(S,S)-2-466 ← (S)-2-465 ← (S)-2-464

H₂O, r.t.

(S,S)-2-467

NaF, THF/H₂O, r.t.

(S,S)-2-468

2-469

Entry	Substrate 2-463	R	Product 2-468	Yield [%]
1	(R)-a	Ph	(S,S)-a	65
2	(R)-b	pClC₆H₄	(S,S)-b	41
3	(R)-c	pMeC₆H₄	(S,S)-c	54
4	(R)-d	pOMeC₆H₄	(S,S)-d	63

Scheme 2.107. Synthesis of enantiopure 1,5-diols.

ment with 2.3 equiv. (–)-benzyl glycidyl ether (2-473) and addition of diepoxypentane 2-474 in the presence of HMPA or DMPU furnished the protected diol 2-472 in 59 % yield.

Weigel and coworkers have found an interesting access to enantiopure benzo[f]quinolinones by utilizing a S$_N$-type metalloenamine alkylation/lactamization/Michael addition sequence [254]. To allow a stereoselective formation of the heterocycle an enantiopure amine was used as auxiliary. Moreover, as part of the investigation towards novel syntheses of the taxane framework, a domino epoxide-opening/retro-aldol addition/semiacetalization process for the formation of the A,B-ring part has been reported by Blechert's group (Scheme 2.109) [255]. Thus, when substrate 2-475 was treated with HCl, a ring enlargement takes place to give product 2-476 stereoselectively in an overall yield of 56 %.

Besides epoxides, the opening also of aziridines and cyclopropanes has been used as an initiating step in threefold anionic domino processes. Thus, the group of Shipman reported on an aziridine-opening/enamide alkylation/imine hydrolysis protocol ending with 1,3-disubstituted propanones [256]. A TMSI/(TMS)$_2$NH-induced cyclopropyl ring opening of compounds 2-477 and 2-481 to give the cyclobutane derivatives 2-480 and 2-482, respectively, has been observed by Fukumoto and

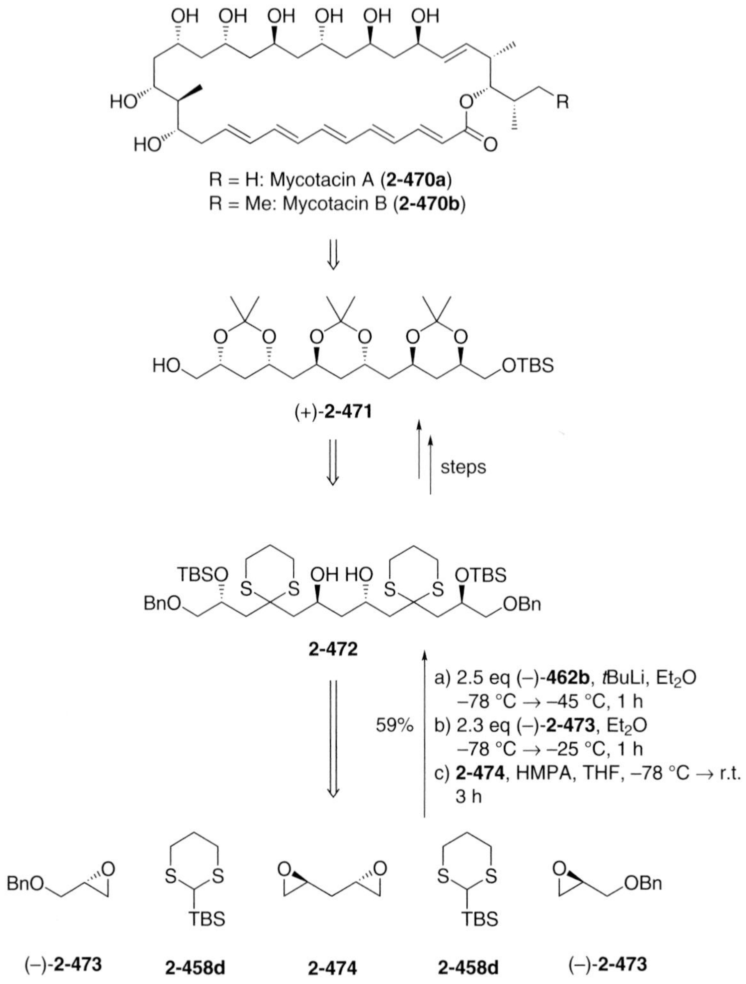

Scheme 2.108. Synthesis of a trisacetonide for the preparation of (+)-mycotacin A (**2-470a**).

coworkers (Scheme 2.110) [257]. The reactions likely proceed *via* transition state 2–478 in which the iodide anion attacks the cyclopropane moiety. The resulting ring-opening leads to the corresponding silyl enol ether **2-479** being suitable for a subsequent Michael/aldol-type domino reaction.

A S_N reaction-based domino route to clerodane diterpenoid tanabalin (**2-488**) [258] has been described by Watanabe's group (Scheme 2.111) [259]. This natural product is interesting as it exhibits potent insect antifeedant activity against the pink bollworm, *Pectinophora gossypiella*, a severe pest of the cotton plant. The domino sequence towards the substituted *trans*-decalin **2-487** as the key scaffold is induced by an intermolecular alkylation of the β-ketoester **2-484** with the iodoalkane **2-483** followed by an intramolecular Michael addition/aldol condensation (Robin-

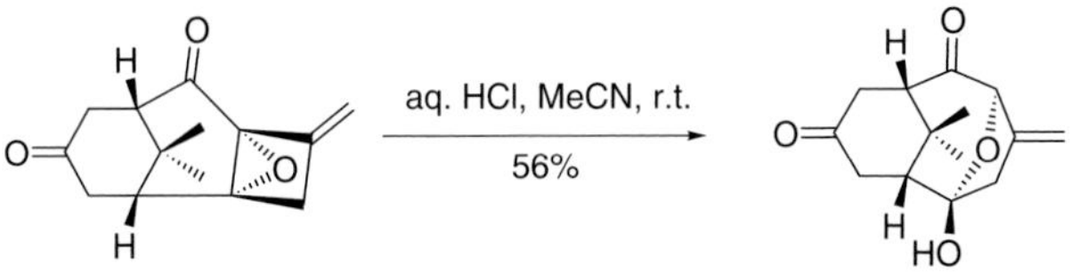

Scheme 2.109. Ring enlagement for the assembly of the portion of taxanes.

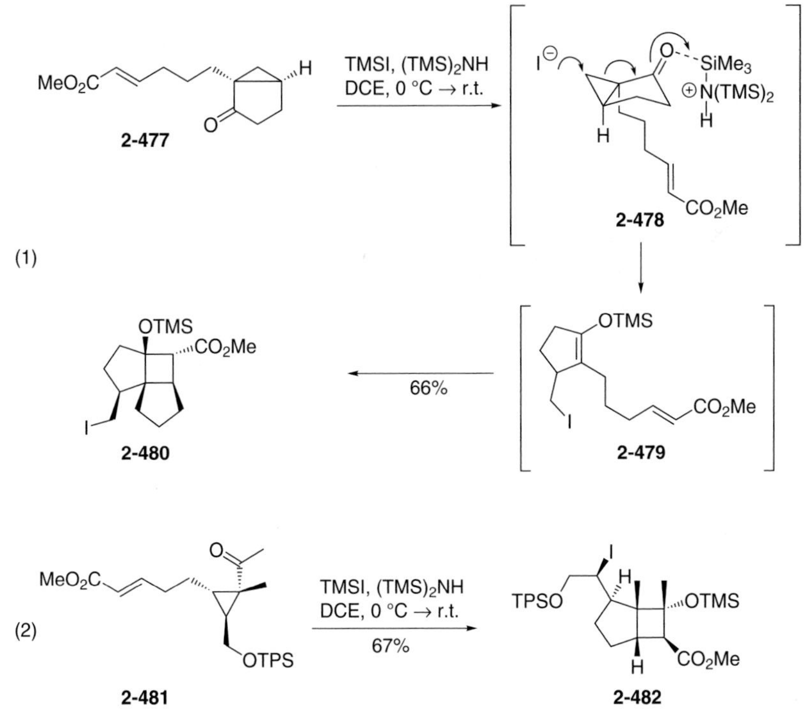

Scheme 2.110. Domino reaction of cyclopropyl ketones **2-477** and **2-481**.

son annelation) to give **2-487** *via* **2-485** and **2-486**. Overall, **2-487** is obtained in 82 % yield with formation of three new C–C-bonds and two new stereogenic centers. The natural product (–)-tanabalin (**2-488**) was accessible from **2-487** in a few more steps.

The following example completes the section of threefold anionic domino processes initiated by a S_N-type reaction. As discussed earlier in Section 2.2, the reaction of a five-membered cyclic phosphonium ylide with enones, α,β-unsaturated esters, and α,β-unsaturated thioesters provides cycloheptene or hydroazulene derivatives in a domino Michael/intramolecular Wittig reaction. This sequence

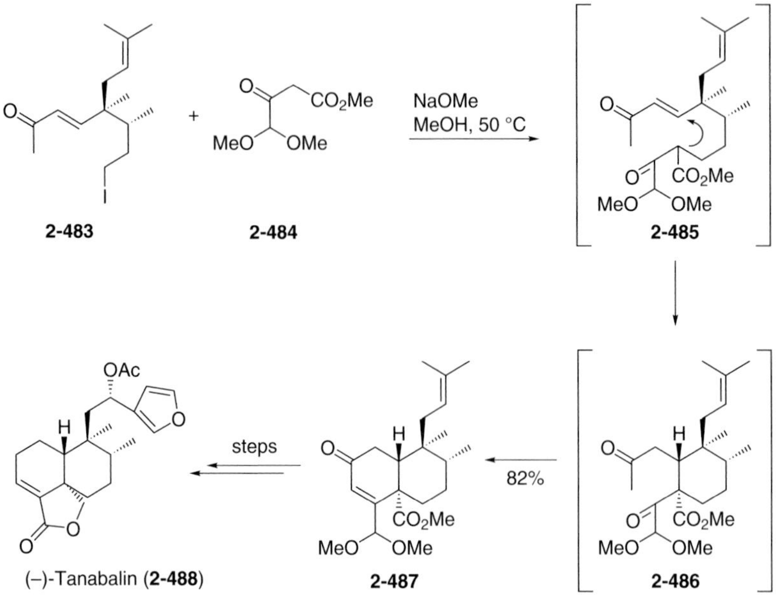

Scheme 2.111. Selective synthesis of (–)-tanabalin (**2-488**).

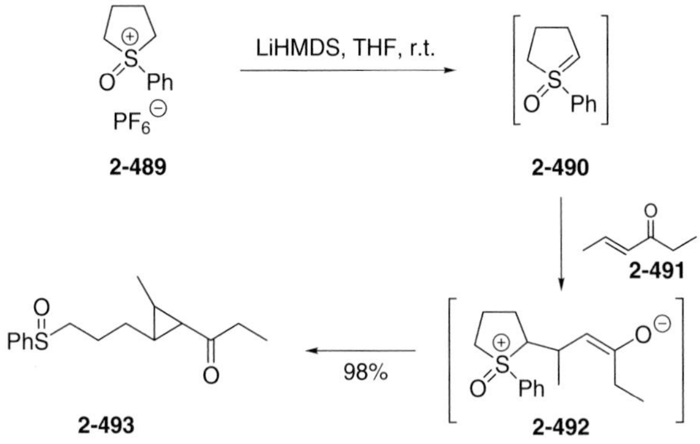

Scheme 2.112. Formation of the cyclopropane derivative **2-493**.

proceeds *via* a rigid phosphabicyclic or phosphatricyclic intermediate, so that the products are usually formed with good stereoselectivity. Based on these findings, Fujimoto and coworkers have developed a novel domino transformation employing the five-membered oxosulfonium ylide **2-490** obtained from **2-489** using a strong base (Scheme 2.112) [260].

However, as illustrated in Scheme 2.112, the reaction of ylide **2-490** with 4-hexen-3-one (**2-491**) did not lead to the expected cycloheptene, but to the cyclopropane derivative **2-493** in 98 % yield by a simple addition of the ylide to **2-491** to give **2-492**

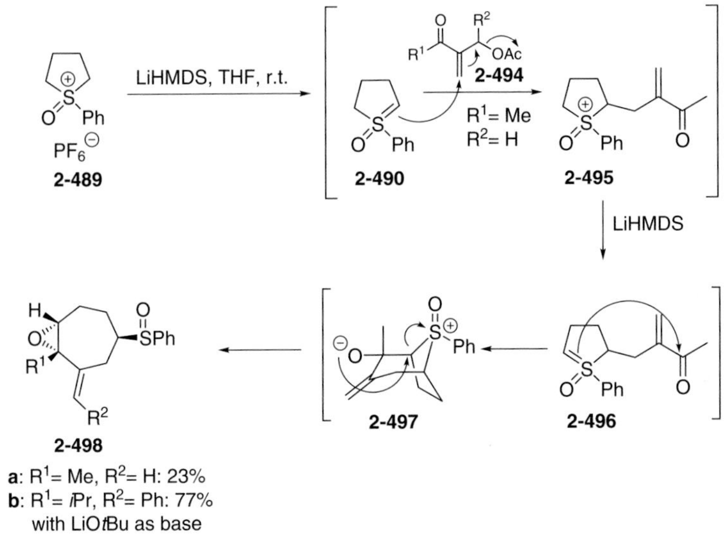

Scheme 2.113. Synthesis of cycloheptane derivatives.

followed by extrusion of the oxosulfonium group. In this reaction the intramolecular 1,3-nucleophilic substitution affording **2-493** is preferred over the regeneration of the corresponding ylide, which would be necessary for a reaction with the keto group. Nevertheless, if one uses substrates in which an enolate anion as **2-492** is not formed, the desired domino process can take place (Scheme 2.113). Thus, reaction of **2-489** and **2-494a** in the presence of 2 equiv. of LiHMDS led to **2-498a** *via* the proposed intermediates **2-490** and **2-495–2-497** in 23% yield as a single stereoisomer.

Using substituted α-methylene-β-acetoxy ketones **2-494** with R = Me, Et, *i*Pr, Ph and LiO*t*Bu as base, the yields of the cycloheptene oxide **2-498** could be greatly enhanced to up to 77%, as in the case of **2-498b**. In addition, cyclooctene oxides can be prepared (though in lower yield and stereoselectivity) starting from a six-membered oxosulfonium ylide.

Another group of threefold anionic domino processes involves an initial addition of nucleophiles other than enolates onto an electrophilic center such as a carbodiimide, an isocyanate, an aldehyde, or a related moiety. For example, the group of Volonterio and Zanda has observed that activated α,β-unsaturated carboxylic acids undergo a domino carboxyl group addition/aza-Michael ring closure/Dimroth-rearrangement with carbodiimides producing *N,N*-disubstituted hydantoins in good yield [261]. Tetronic acids, a subclass of β-hydroxybutenolides, can be prepared by a twofold addition/Michael sequences described by the group of García-Tellado [262]. Zercher and coworkers reported on the assembly of α-substituted-γ-keto esters from β-keto esters by a novel chain extension process which involves a domino cyclopropanation/rearrangement/aldol addition [263]. Furthermore, an efficient and simple domino protocol for the synthesis of 1,2,4-triazolo[1,5-*c*]qui-

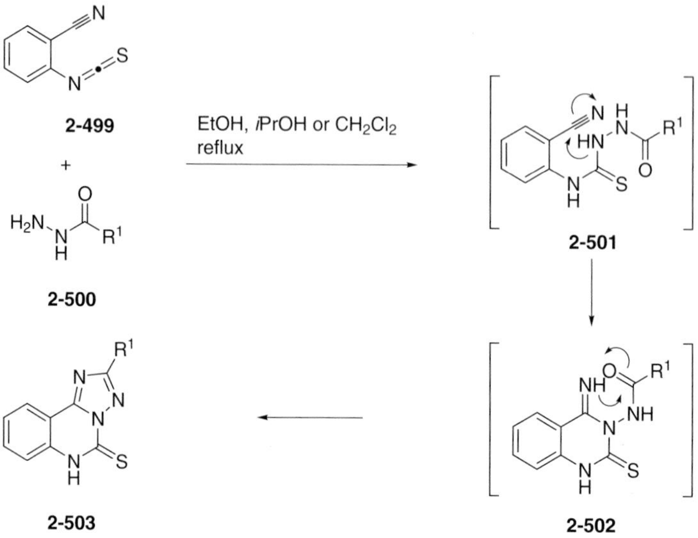

Scheme 2.114. Synthesis of 1,2,4-triazolo[1,5-*c*]quinazoline-5(6*H*)-thiones **2-503**.

Entry	2-500	R¹	2-503	Yield [%]
1	a	H	a	73
2	b	Me	b	88
3	c	CH₂CN	c	80
4	d	CH₂OPh	d	99
5	e	Ph	e	67
6	f	pBrC₆H₄	f	93
7	g	pNO₂C₆H₄	g	76
8	h	3-Pyridyl	h	95
9	i	2-Furyl	i	72

nazoline-5(6*H*)-thiones **2-503** has been reported by Langer and coworkers (Scheme 2.114) [264]. The development of efficient syntheses of such compounds is of interest due to their pronounced bioactivity. For example, 1,2,4-triazolo[5,1-*b*]quinazolines show antihypertonic activity [265]; antirheumatic and anti-anaphylactic activity has been recognized for 3-heteroaryl-1,2,4-triazolo[5,1-*b*]quinazolines [266], while 1,2,4-triazolo[1,5-*c*]quinazolines possess antiasthmatic, tranquilizing, and neurostimulating activity [267].

The formation of compounds **2-503** proceeds *via* the addition of a hydrazide **2-500** onto the central carbon of an isothiocyanate **2-499**. Subsequent cyclization by attack of the hydrazide nitrogen in the formed **2-501** onto the nitrile moiety gives intermediate **2-502**. A further attack of the newly formed amidine nitrogen onto the carbonyl group followed by extrusion of water affords the triazoloquinolines **2-503**.

Nucleophilic addition to an acylsilane followed by a 1,2-Brook rearrangement and final trapping of the resulting carbanion in either an acylation or intramolecular

Table 2.115. Synthesis of four- to six-membered carbocycles.

Entry	n	Yield [%]	
		2-507a	2-507b
1	1	49	30
2	2	53	29
3	3	47	23

Michael addition has been shown to lead to useful building blocks. According to a publication by Johnson and coworkers, highly functionalized unsymmetrical malonic acid derivatives are accessible in this way [268]. Moreover, as described by Takeda and coworkers, substituted four- to six-membered carbocycles 2-507 can be prepared starting from 2-504 by reaction with PhLi via the intermediates 2-505 and 2-506 (Scheme 2.115) [269].

Furthermore, Schaumann's group has described that lithiated silylthioacetals react not only with epoxides, but also with ω-bromoalkyl isocyanates to give lactams [270].

Threefold anionic domino processes can also be triggered by a condensation reaction between a nucleophile and a carbonyl or carboxyl moiety. Many different examples have been published using this approach. Ballini, Petrini and coworkers constructed 2,3-disubstituted (E)-4-alkylidenecyclopent-2-en-1-ones by exploiting an aldol condensation/Michael-addition/elimination sequence [271], while Tanabe's group prepared functionalized O-heterocycles using their Claisen condensation/aldol addition/lactonization combination [272]. The implementation of such condensation-triggered domino processes in natural product synthesis has been shown by Yoda's group with their HWE/Michael addition/lactonization reaction for the preparation of trilobatin [273], and furthermore by the group of Chamberlin with their lactam formation/iodination/S_N-type cyclization approach as part of the total synthesis of dysiherbaine [274]. Giorgi-Renault's and Husson's group reported on the design of a three-component Knoevenagel/Michael addition/cyclization

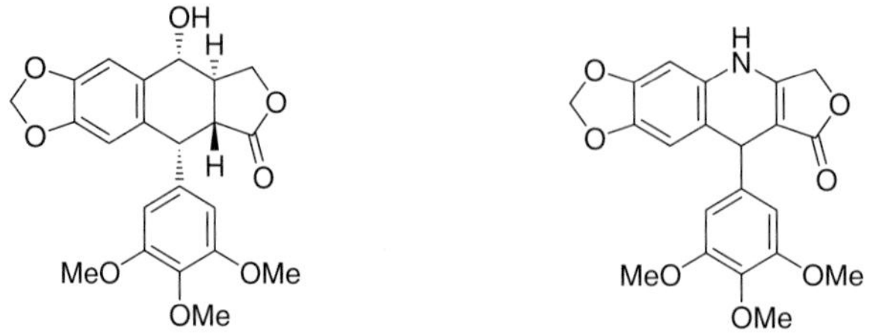

Podophyllotoxin (**2-508**)   4-Aza-2,3-didehydropodophyllotoxin (**2-509a**)

Scheme 2.116. Podophyllotoxin and aza-analog.

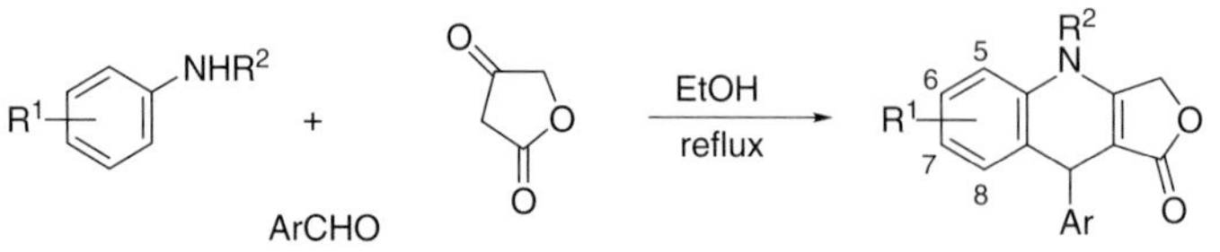

Entry	2-509	R^1	R^2	Ar	Yield [%]
1	a	6,7-methylenedioxy	H	3,4,5-trimethoxyphenyl	92
2	b	6,7-methylenedioxy	Me	3,4,5-trimethoxyphenyl	80
3	c	6,7-methylenedioxy	$Me_2N(CH_2)_3 \cdot HCl$	3,4,5-trimethoxyphenyl	45
4	d	6-hydroxy	H	3,4,5-trimethoxyphenyl	63
5	e	6,7-dimethoxy	H	3,4,5-trimethoxyphenyl	94
6	f	7,8-benzo	H	3,4,5-trimethoxyphenyl	59
7	g	6,7-methylenedioxy	H	2,3,4-trimethoxyphenyl	83
8	h	6,7-methylenedioxy	H	2-pyridyl	78

Scheme 2.117. Three-component domino reaction leading to aza-analogs of podophyllotoxin (**2-508**).

methodology [275]. Utilizing this process, rapid access to podophyllotoxin aza-analogues such as **2-509** is possible (Scheme 2.116).

The synthesis of analogues of podophyllotoxin (**2-508**) is important as the compound itself shows severe side effects in the treatment of human neoplasia. To date, no new analogues have been launched on the market, though for various reasons aza-analogues as **2-509** are of great interest [276]. The synthesis of these compounds is quite straightforward; simply heating a mixture of tetronic acid (**2-512**), an appropriate benzaldehyde **2-511**, and an aniline derivative **2-510** in MeOH for 10 min afforded the desired compounds **2-509** in good yields (Scheme 2.117).

It is assumed that in the formation of **2-509**, a Knoevenagel condensation of the benzaldehydes **2-511** and tetronic acid **2-512** initially takes place, and this is followed by the generation of a hemiaminal with the aniline **2-510** and an electrophilic substitution.

Very recently, a highly efficient synthesis of the erythrina and B-homoerythrina skeleton by an $AlMe_3$-mediated three-step domino condensation-type/iminium ion formation/iminium ion cyclization sequence has been reported by Tietze and co-

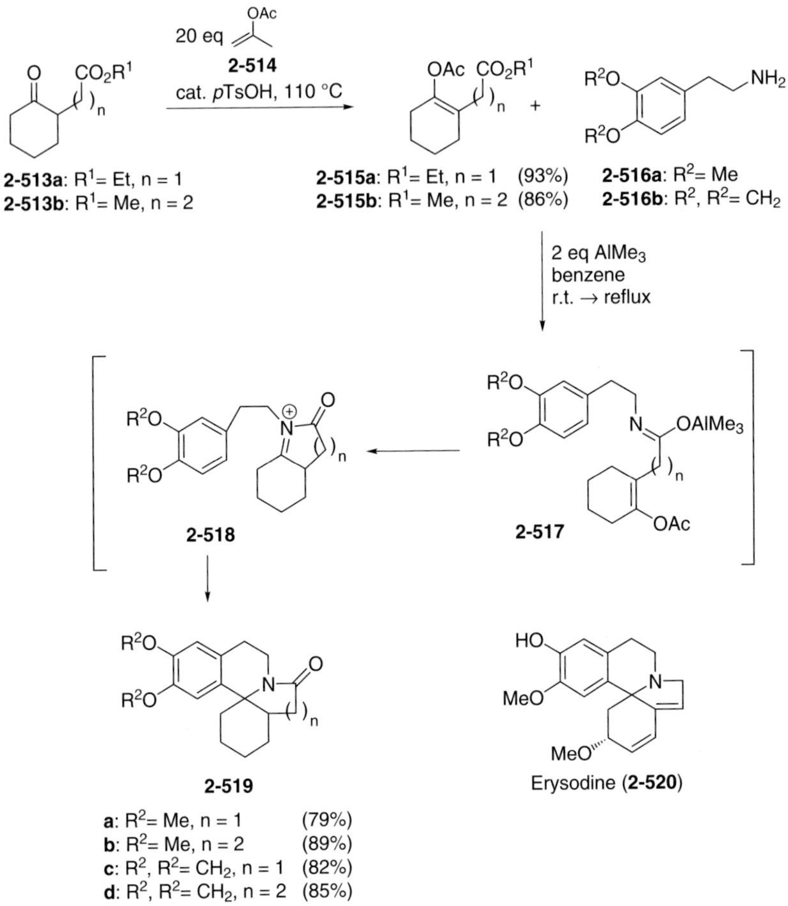

Scheme 2.118. Synthesis of the erythrina and homoerythrina skeleton.

workers (Scheme 2.118) [277]. The erythrina alkaloids [278] such as erysodine (2-520) are a widespread class of natural products with extensive biological activity [279]. Many compounds of this family exhibit curare-like activity as well as CNS-depressant properties [280]. As starting materials towards their structural core, primary amines **2-516** and enol acetates **2-515**, easily obtainable from the keto ester **2-513**, were employed.

For the synthesis of **2-519**, the amines **2-516** were first treated with AlMe₃ in benzene at r.t. and after addition of the enol acetates **2-515**, easily accessible from **2-513** and **2-514**, heated under reflux. Mechanistic investigations using on-line NMR spectroscopy, reveal that a metalated amide **2-517** is formed first. This then leads to a *N*-acyliminium ion **2-518** which undergoes an electrophilic substitution. Overall, three new bonds are formed selectively in the domino process, and the alkaloid scaffolds **2-519** are provided in very good yields of 79–89%. Interestingly, use of the keto esters **2-513** instead of **2-515** did not lead to the desired products **2-519**.

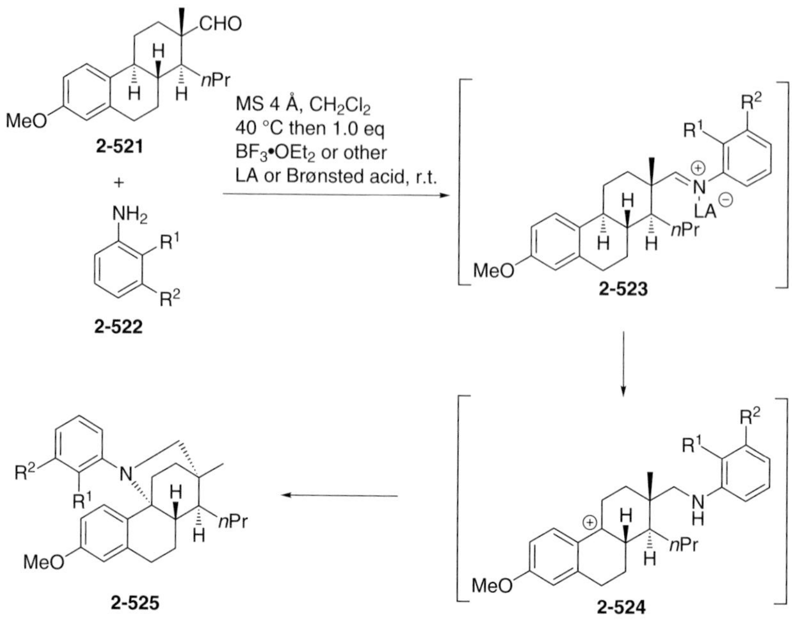

Scheme 2.119. Synthesis of steroid alkaloids.

Entry	2-525	R^1	R^2	Yield [%]
1	a	F	H	72
2	b	H	Cl	87
3	c	NO$_2$	H	69

A variety of unique 9,13-bridged D-secoestrone alkaloids of type **2-525** have been synthesized by treatment of 3-methoxy-16,17-secoestra-1,3,5(10)-trien-17-al (**2-521**) with aniline derivatives **2-522** in the presence of different Lewis and Brønsted acids in a joint study by the groups of Schneider, Wölfling, and Tietze (Scheme 2.119) [281]. In this transformation, an interesting 1,5-hydride shift after the formation of the iminium ion **2-523** takes place to give a carbocation **2-524** which reacts with the secondary amine, formed from the imine. The transformation gives good yields only if an electron-withdrawing group is present in the aniline component **2-522**.

The formation of an iminium ion as **2-530** is also proposed by Heaney and co-workers in the synthesis of a tetrahydro-β-carboline **2-531** (Scheme 2.120) [282]. Herein, heating a solution of tryptamine (**2-526**) and the acetal **2-527** in the presence of 10 mol% of Sc(OTf)$_3$ gives in the first step the *N,O*-acetal **2-528**, which then leads to the lactam **2-529** and further to the iminium ion **2-530** by elimination of methanol. The last step is a well-known Pictet–Spengler type cyclization to give the final product **2-531** in 91% yield.

Besides tryptamine **2-526**, tryptophan and 3,4-dimethoxy-β-phenylethylamine were also used. The latter led to a tetrahydroisoquinaline, but the yields were much lower.

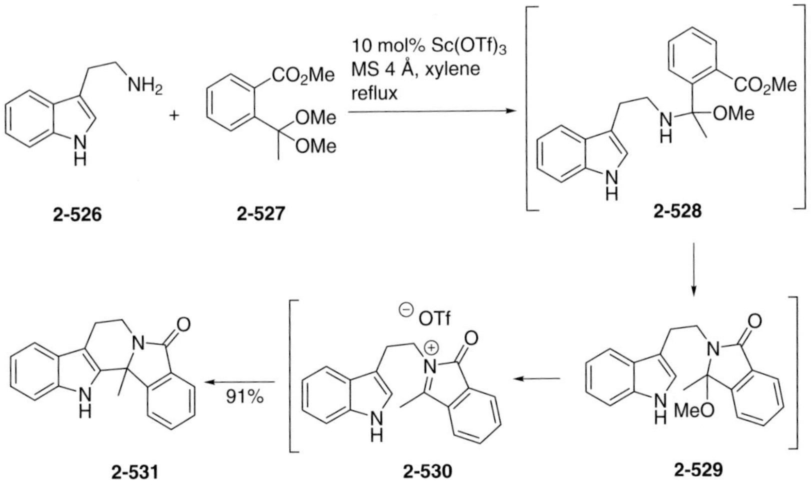

Scheme 2.120. Sc(OTf)$_3$-catalyzed domino process of **2-526** and **2-527**.

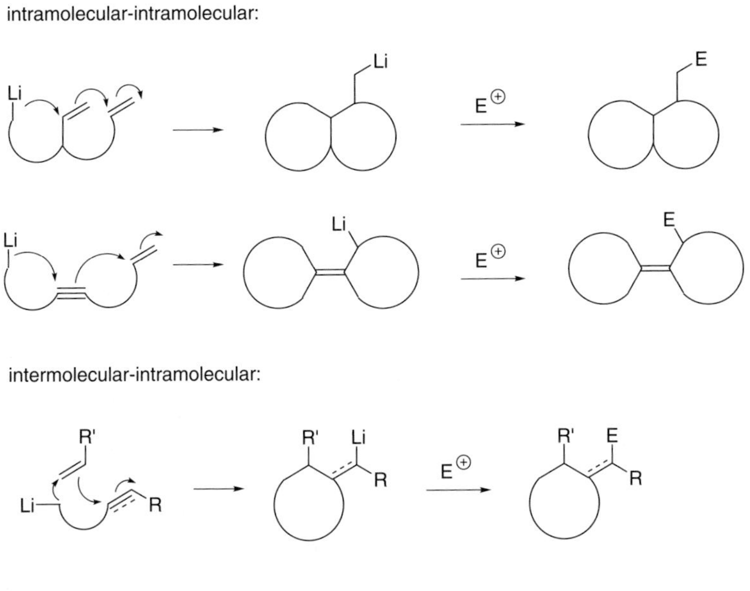

intramolecular-intramolecular:

intermolecular-intramolecular:

Scheme 2.121. General principle of inter-/intramolecular domino carbolithiations.

At the end of this section, carbolithiation-based domino processes will be discussed in which a bond and a new lithium organic moiety from an alkene and a starting lithium compound is produced. The new lithium compound can react with another C–C-double or -triple bond and finally with an electrophile, as depicted in Scheme 2.121 [283].

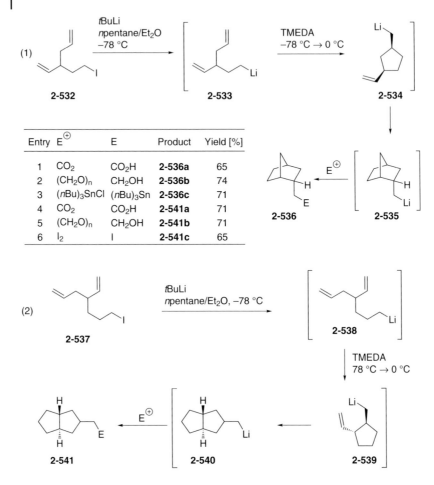

Scheme 2.122. Synthesis of bicyclo[2.2.1]heptanes **2-536** and *trans*-bicyclo[3.3.0]octanes **2-541**.

According to this scheme, Bailey and coworkers have reported on the domino cyclization of the alkadienyllithium compounds **2-533** and **2-538**, derived from the corresponding alkyl iodides **2-532** and **2-537** (Scheme 2.122) [284]. The reactions proceed through two stereo- and regioselective 5-*exo*-trig ring closures to deliver bicyclic alkyllithiums of type **2-535** and **2-540** via **2-534** and **2-539**, respectively. Subsequent trapping of these organolithium compounds by addition of an electrophile led to substituted bicyclo[2.2.1]heptanes **2-536** and *trans*-bicyclo[3.3.0]octanes **2-541** in 65–74% yield.

In a more recent contribution, O'Shea and coworkers described a related process leading to substituted indoles **2-544** and **2-545** by an intermolecular addition of alkyllithium to a styrene double bond and reaction of the formed intermediate **2-543** with an appropriate electrophile (Scheme 2.123) [285]. Using DMF, C-2 unsub-

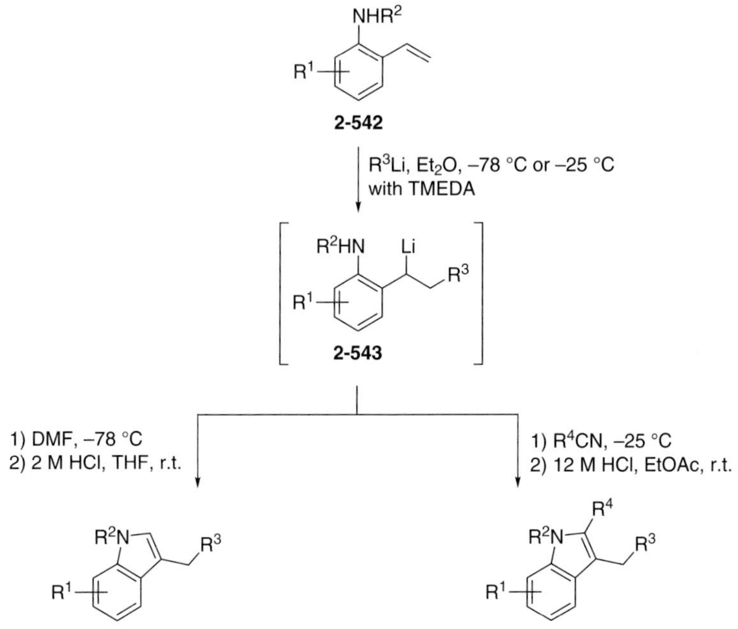

2-544a: R^1= H, R^2= Boc, R^3= *n*Bu (84%)
2-544b: R^1= 5-Me, R^2= Bn, R^3= *t*Bu (70%)
2-545a: R^1= H, R^2= H, R^3= *t*Bu, R^4= Ph (65%)
2-545b: R^1= 5-Me, R^2= Bn, R^3= *t*Bu, R^4= Ph (51%)

Scheme 2.123. Synthesis of indols *via* a domino carbolithiation/addition/cyclization process.

stituted indoles **2-544**, and with nitriles the corresponding substituted indoles **2-545** were formed.

The final group of threefold anionic domino processes described here includes transformations with an initiating elimination step which is either followed by two Michael additions or by substitutions. Thus, reaction of protected nitro alcohol **2-546** and α,β-unsaturated γ-aminoesters **2-547** in MeOH at r.t. afforded the pyrrolidines **2-548** as single isomers (except for entry 1) in good to very good yields, as described by Benetti and coworkers (Scheme 2.119) [286].

The first step in this domino process is the formation of an α,β-unsaturated nitro compound which reacts with the amino functionality in **2-547**.

Pyrrolidine **2-548e** was used as substrate for the synthesis of the natural product α-kainic acid [286, 287].

Another example belonging to the class of domino processes initiated by an elimination was described by Gurjar and coworkers in their synthesis of the 2,4-disubstituted tetrahydrofuran **2-554** [288]; this compound is a potent anti-asthmatic drug lead (Scheme 2.125) [289].

It can be assumed that the substrate **2-549** undergoes two elimination steps after lithiation to give **2-552** *via* **2-550** and **2-551**. The last step is an intramolecular substitution affording **2-553**, which was further transformed into the desired bioactive compound CMI-977 (also named LDP-977) (**2-554**).

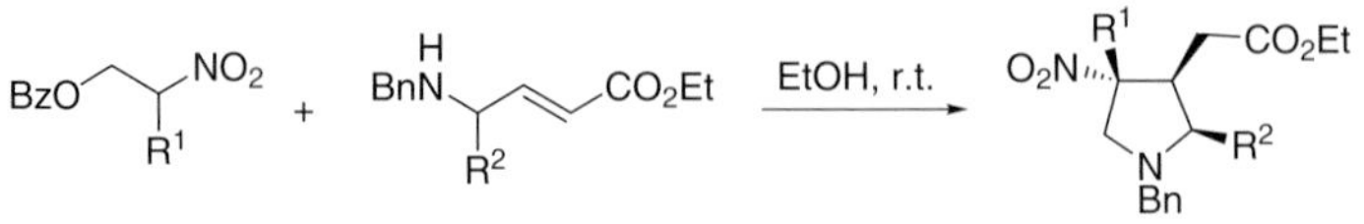

Entry	2-546	R^1	2-547	R^2	2-548	Yield [%]
1	a	H	a	H	a	90[a]
2	b	CH_2OH	b	CH_2OH	b	90
3	b	CH_2OH	c	CH_2OTBS	c	75
4	c	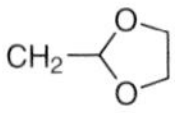 (CH₂-dioxolane)	c	CH_2OTBS	d	76
5	d	(isopropenyl)	b	CH_2OH	e	88

[a] *trans/cis* ratio 20:1

Scheme 2.124. Domino elimination/twofold Michael addition process.

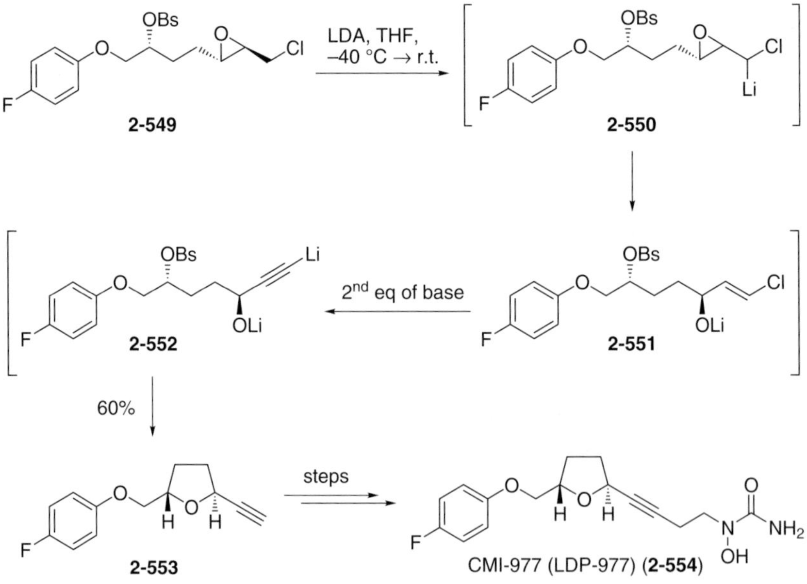

Scheme 2.125. Twofold elimination/nucleophilic substitution sequence for the synthesis of **2-554**.

2.1.2
Fourfold and Higher Anionic Processes

As mentioned in the Introduction of this book, the quality of a domino process can be correlated to the number of steps involved and the increase of complexity compared to the starting material. Indeed, there are anionic transformations which consist of four and even five separate steps under identical reaction conditions. Again, most of these transformations start with a Michael addition.

A fourfold anionic domino process consisting of a domino Michael/aldol/Michael/aldol process was used by Koo and coworkers for the synthesis of bicyclo[3.3.1]nonanes. They employed 2 equiv. of inexpensive ethyl acetoacetate and 1 equiv. of a simple α,β-unsaturated aldehyde [290]. Differently substituted dihydroquinolines were assembled in a Michael/aldol/elimination/Friedel–Crafts-type alkylation protocol by the Wessel group [291]. An impressive approach in this field, namely the construction of the indole moiety **2-557**, which represents the middle core of the manzamines, has been published by Markó and coworkers [292]. Manzamine A (**2-555**) and B (**2-556**) are members of this unique family of indole alkaloids which were isolated from sponges of the genus *Haliclona* and *Pelina* (Scheme 2.126) [293].

Coupling of the aminomethylindole **2-558** with acrolein in a Michael manner followed by a HWE reaction with phosphonates **2-560** afforded **2-561** which cyclized under basic conditions to give the desired product **2-566** *via* the intermediates **2-562–2-565** in 55 % yield (Scheme 2.127). The reaction sequence was also conducted in a stepwise manner, resulting in a greatly reduced yield; this clearly demonstrates again the advantage of domino strategies over conventional methods.

The reaction tolerates different *N*-protecting groups as well as a variety of substituted β-keto-phosphonates. In all cases, the tetracyclic structures were obtained as single diastereomers, however with respect to the natural product with the undesired *trans*-junction between the rings A and B.

In their synthesis of spirocyclopropanated oxazolines (see Section 2.1), the de Meijere group obtained initially unexpected cyclobutene-annelated pyrimidones **2-569** by reaction of the cyclopropylidene derivative **2-567** with the amidines **2-568**. In this fourfold anionic transformation a Michael addition takes place to furnish **2-570**, which is followed by an isomerization affording cyclobutenecarboxylates **2-572** and a final lactamization (Scheme 2.128) [294].

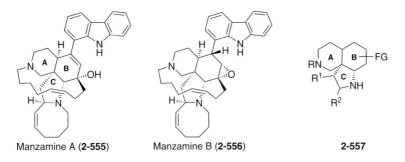

Manzamine A (**2-555**)   Manzamine B (**2-556**)   **2-557**

Scheme 2.126. Manzamine A (**2-555**) and B (**2-556**) – Members of a unique family of indole alkaloids.

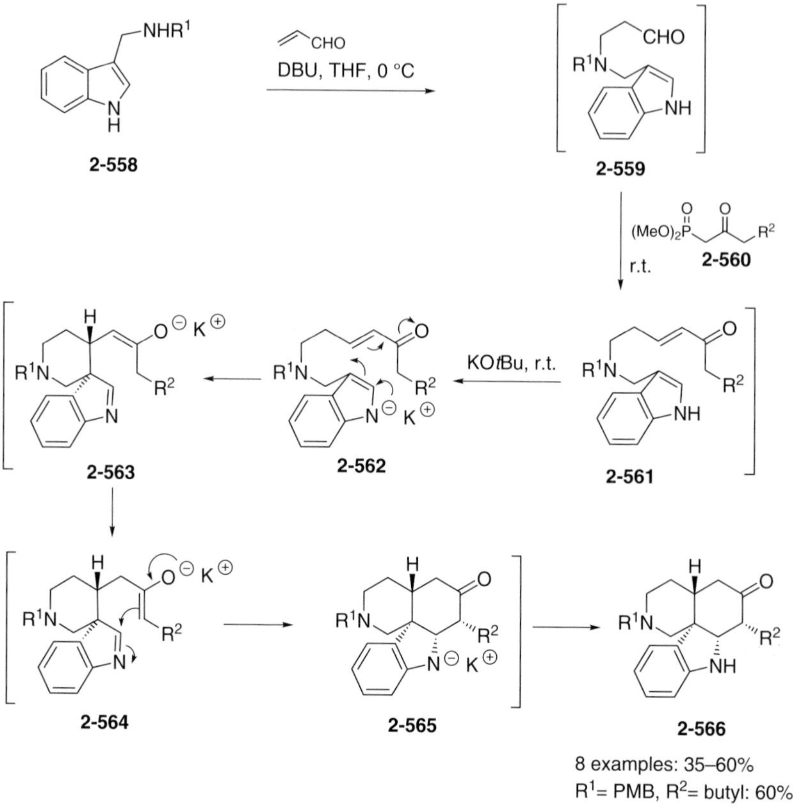

Scheme 2.127. Synthesis of the core of the manzamine alkaloids.

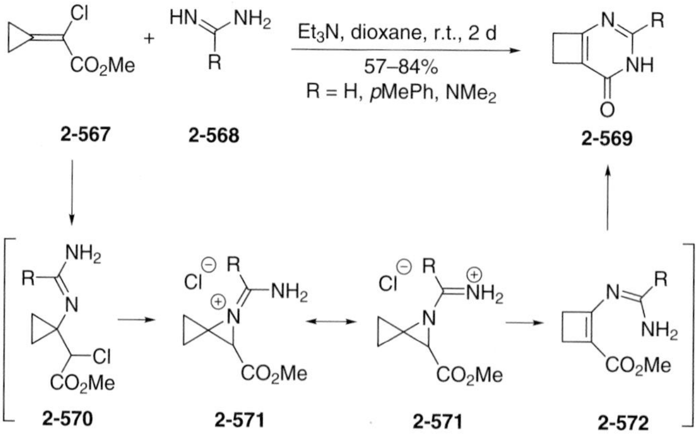

8 examples: 35–60%
R^1= PMB, R^2= butyl: 60%

Scheme 2.128. Synthesis of cyclobutene-annulated pyrimidinones.

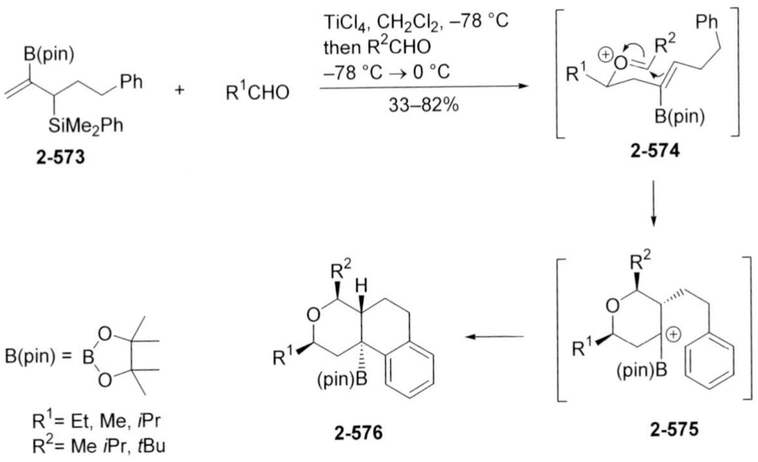

Scheme 2.129. Stereoselective synthesis of *trans*-1,2-benzoxadecalines.

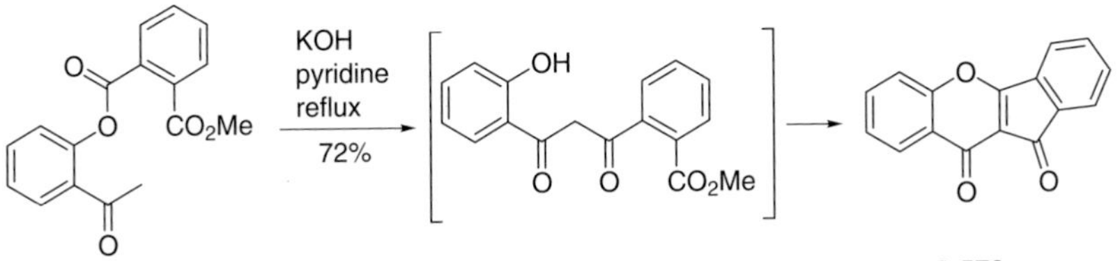

Scheme 2.130. Domino reaction involving a Baker-Venkataraman rearrangement to afford **2-579**.

A fourfold anionic sequence which is not initiated by a Michael but an aldol reaction has been reported by the group of Suginome and Ito (Scheme 2.129) [295]. In this approach, the borylallylsilane **2-573** reacts selectively in the presence of TiCl$_4$ with two different aldehydes which are added sequentially to the reaction mixture. First, a Lewis acid-mediated allylation of the aldehyde with **2-573** takes place to form a homoallylic alcohol which reacts with the second aldehyde under formation of the oxenium ion **2-574**. The sequence is terminated by a Prins-type cyclization of **2-574** and an intramolecular Friedel–Crafts alkylation of the intermediate **2-575** with formation of the *trans*-1,2-benzoxadecalines **2-576** as single diastereomers.

Furofuran lactones have been constructed by Enders and coworkers by an acid-mediated hydrolysis/lactone cleavage/hemiacetal formation/lactonization process [296]. Moreover, the group of Xu has reported on the efficient synthesis of 2-hydroxypyrano[3,2-c]quinolin-5-ones by utilizing a Knoevenagel condensation/Michael addition/cyclization/hydrolysis strategy [297]. A rapid access to the benz[b]indeno[2,1-e]pyran-10,11-dione framework (**2-579**) has been achieved by Ruchirawat's group [298], starting from the readily available precursor **2-577**, which under basic conditions undergoes a Baker–Venkataraman rearrangement to give intermediate **2-578** (Scheme 2.130). Cyclization then affords **2-579**. Noteworthy, a similar com-

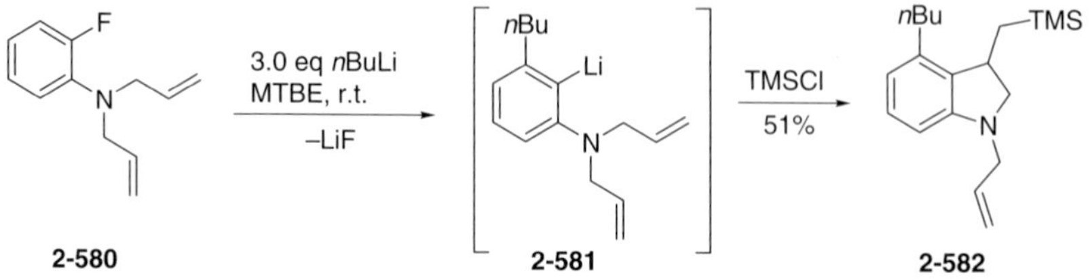

Scheme 2.131. Synthesis of indolines.

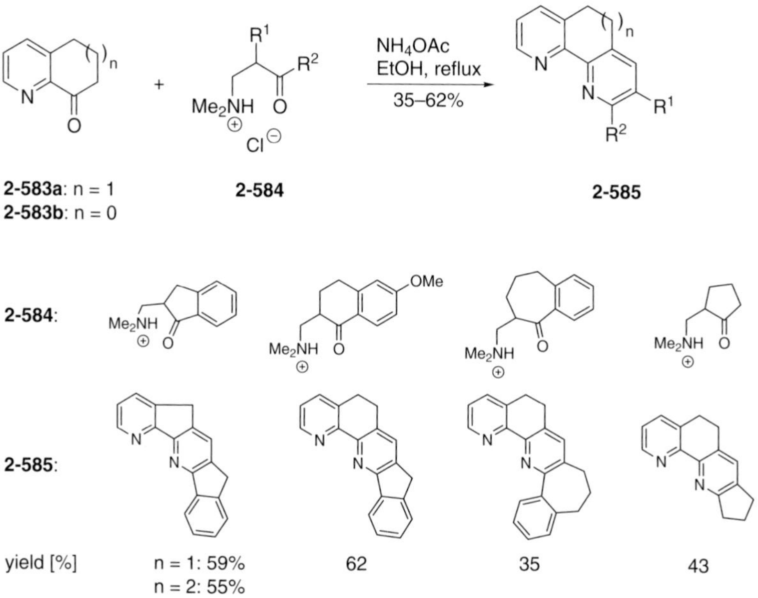

2-583a: n = 1
2-583b: n = 0

2-584

2-585

2-584:

Me₂NH

2-585:

yield [%]   n = 1: 59%   62   35   43
             n = 2: 55%

Scheme 2.132. Synthesis of annulated bipyridines.

pound as **2-579** has been reported to show weak activity against human immunodeficiency virus type I reverse transcriptase (HIV-1 RT) [299].

Bailey's group has elaborated a fourfold anionic domino approach leading to a *N*-allyl-3,4-disubstituted indoline **2-582** from **2-580** (Scheme 2.131) [300]. The central step is the formation of an aryne by treatment of 2-fluoro-*N,N*-diallylaniline (**2-580**) with *n*BuLi followed by a regioselective intermolecular addition of *n*BuLi to give **2-581**. This then cyclizes to afford a new lithiated species which is intercepted by added TMSCl.

Another elimination triggered tetrafold anionic domino process has been reported by Risch and coworkers [301]. These authors synthesized polycyclic bipyridine scaffolds as **2-585** (Scheme 2.132) which are potentially useful as ligands for host–guest systems or interesting building blocks for supramolecular applications [302]. The process combines an amine elimination from **2-584** and a Michael addi-

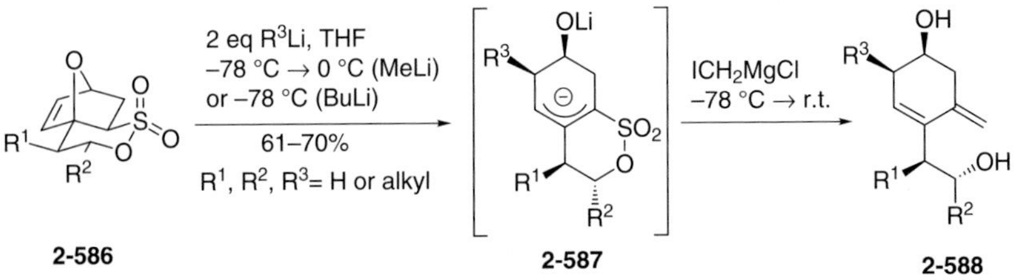

Scheme 2.133. Synthesis of methylenecyclohexenols.

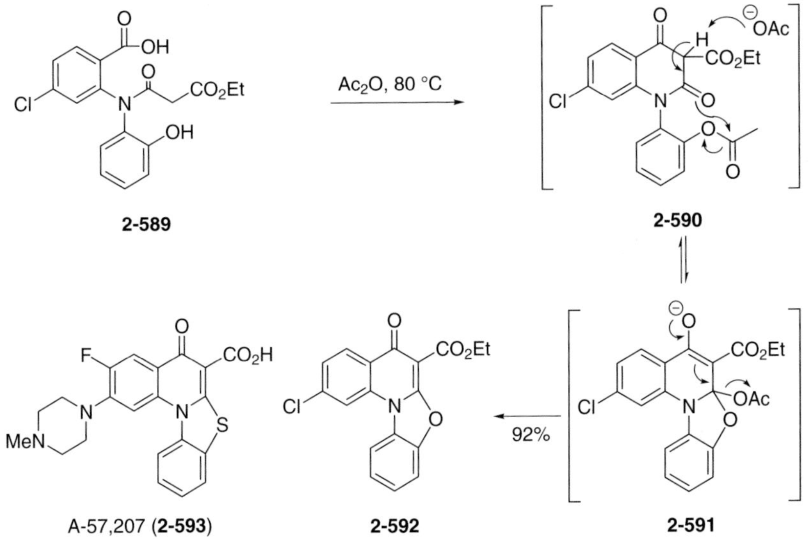

Scheme 2.134. Synthesis of a new analog of A-57,207 (**2-593**).

tion with the enol form of the second starting material **2-583** followed by condensation with NH_3, derived from added NH_4OAc, to give **2-585**.

Starting from the furan-derived substrates **2-586**, Metz and coworkers developed an unique lithium base-induced fourfold anionic domino ring-opening/alkoxide-directed 1,6-addition/alkylation/desulfurization process ending with the generation of methylenecyclohexenols **2-588** (Scheme 2.133) [303]. The reaction probably involves the formation of a cyclohexadienolate which reacts with a second molecule of R^3Li to give the allylic anion **2-587**; this is then able to react with the successively added ICH_2MgCl. The formed intermediate then undergoes an extrusion of SO_2 to yield the product **2-588**.

As the last example of a fourfold anionic domino process, the synthesis of a new analogue **2-592** of the antibacterially active quinoline derivative A-57,207 (**2-593**) [304] by the group of Kim will be described (Scheme 2.134) [305].

The reaction starts with the formation of a mixed anhydride and an acetate on treatment with an excess of acetic anhydride at 80 °C. There follows a Dieckmann condensation to give **2-590** and an intramolecular rearrangement/Michael addition/retro Michael addition to afford the desired tetracyclic compound **2-592** *via* **2-591** in an overall yield of remarkable 92 %.

Finally, we turn to processes which combine more than four anionic independent reaction steps. In this regard, Schäfer and coworkers have developed, besides several two- and one threefold imino aldol/lactamization sequences, a sextuple anionic approach by reaction of **2-594** and **2-595** to provide the tricyclic pyrrolidone derivative **2-596** as a single diastereomer in 17 % yield (Scheme 2.135) [306]. The C_6-building block aconitic acid trimethylester (**2-594**) to be used was obtained as a by-product in sugar manufacture, or by dehydration of citric acid, and seems very versatile in the context of domino reactions as it combines five functional groups: one potential donating group, namely the easily deprotonable methylene group, and four acceptor groups (one electrophilic double bond and three ester functionalities).

For the preparation of the benzo[*b*]cyclopropa[*d*]pyran **2-599**, Ohta and coworkers have treated 3-ethoxycarbonylcoumarin (**2-597**) with dimethylsulfoxonium methylide (**2-598**), derived from the corresponding trimethylsulfoxonium iodide and NaH (Scheme 2.136) [307]. The isolated product was not **2-599**, however, but an unexpected cyclopenta[*b*]benzofuran derivative **2-600** probably formed through **2-599**.

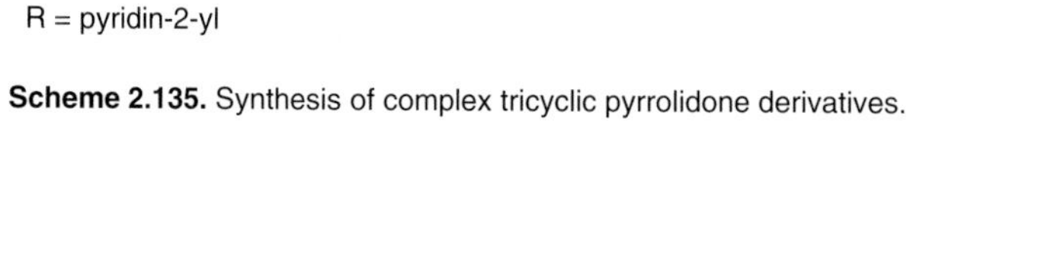

2-594   **2-595**   **2-596**

R = pyridin-2-yl

Scheme 2.135. Synthesis of complex tricyclic pyrrolidone derivatives.

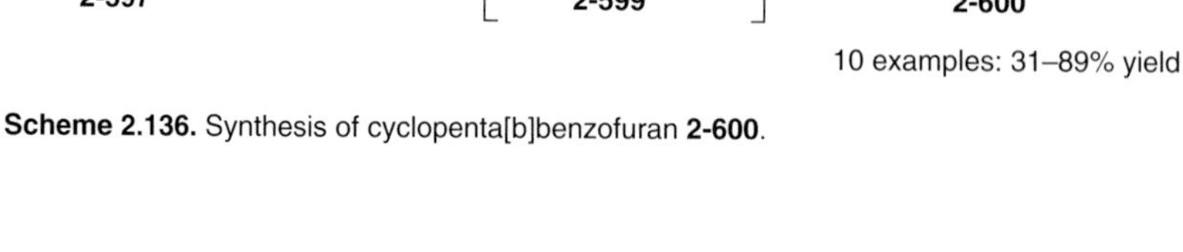

2-597   **2-599**   **2-600**

10 examples: 31–89% yield

Scheme 2.136. Synthesis of cyclopenta[b]benzofuran **2-600**.

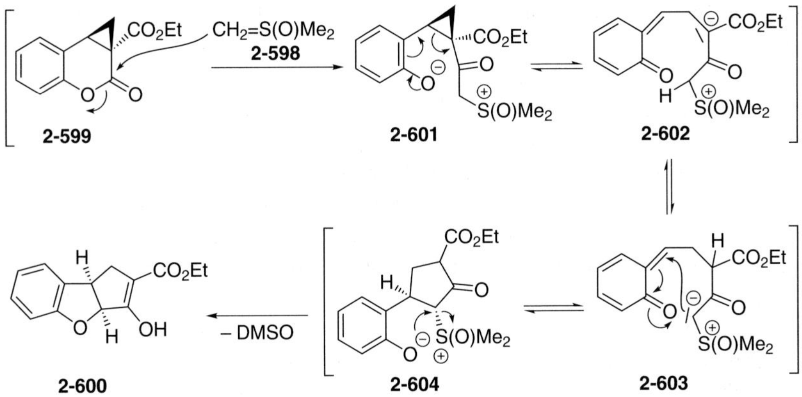

Scheme 2.137. Suggested mechanism for the formation of **2-600**.

Encouraged by this accidentally obtained result, and recognizing that the cyclopenta[*b*]benzofuran moiety is found in natural products such as aplysins [308] and rocaglamides [309], as well as in natural product analogues such as benzoprostacyclins [310], a whole series of coumarin derivatives all bearing a necessary electron-withdrawing group at the 3-position was subjected to the reaction conditions illustrated in Scheme 2.136. In most cases the obtained cyclopenta[*b*]benzofurans were isolated in useful to very good yields.

A possible reaction mechanism for the formation of **2-600** is depicted in Scheme 2.137. Reaction of the primarily formed cyclopropane derivative **2-599** with dimethylsulfoxonium methylide (**2-598**) leads to **2-601** which undergoes a ring opening and a proton shift to give **2-603** *via* **2-602**. There follows a Michael addition to give **2-604** and a S_N-type extrusion of DMSO to provide **2-600**. It must be assumed that the primarily formed cyclopropane derivative **2-599** is not stable under the reaction conditions due to the activation by two electron-withdrawing groups, and thus the cascade of reactions is started.

Very recently, the method has been used for the synthesis of racemic linderol A [311], a natural product isolated from the fresh bark of *Lindera umbellata*. This compound is reported to exhibit potent inhibitory activity on the melanin biosynthesis of cultured B-16 melanoma cells in guinea pigs, without causing any cytotoxicity in the cultured cell or skin irritation [312].

The final example in this section is the synthesis of a tristetrahydrofuran **2-606** described by the group of Rychnovsky [313]. Here, the tris(sulfate) **2-605** was converted into **2-606** by simply heating it in a mixture of MeCN and H_2O (Scheme 2.138). The domino reaction is most likely initiated by deprotection of the primary alcohol, which then attacks the adjacent sulfonate unit in a S_N2-type manner to afford the first furan moiety. Under the reaction conditions the formed acyclic sulfate is hydrolyzed affording a free secondary alcohol which then attacks the next adjacent cyclic sulfate unit. Overall, the S_N2/hydrolyzation sequence proceeds three times to finally provide the poly(tetrahydrofuran) **2-606** as a single isomer in 93% yield.

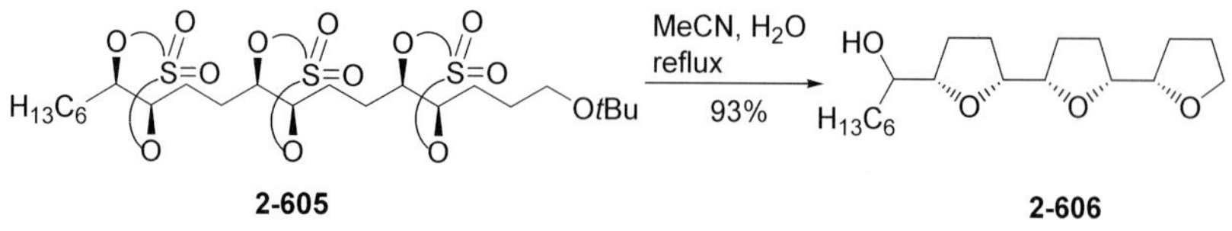

2-605   **2-606**

Scheme 2.138. Synthesis of a tris(tetrahydrofuran).

As the latest example in this section, the synthesis of clavolonine [314] by the group of Evans is mentioned [315]. Through the action of a multiple anionic domino reaction a functionalized linear carbon chain was converted into the polycyclic architecture of the natural product.

2.1.3
Two- and Threefold Anionic Processes Followed by a Nonanionic Process

The present section describes domino processes which combine two or three initiating anionic reaction steps with a following nonanionic transformation.

For domino reactions with three steps, only combinations with a pericyclic reaction have been found. In this regard, Schobert's group has reported on an addition/Wittig olefination/Claisen rearrangement to construct fully substituted butenolides [316]. A Corey–Kwiatkowski-type/HWE/Diels–Alder reaction route has been elaborated by the research group of Collignon (Scheme 2.139) [317]. Using this approach, they were able to prepare derivatives **2-611** of mikanecic acid (**2-612**), a terpenoid dicarboxylic acid isolable from *Mikanoidine* [318] or *Sarracine* [319]. Treatment of the allylic phosphonates **2-607** with an excess of LDA, followed by addition of ethyl chloroformate and then of formaldehyde, led to the transient **2-609** which underwent a spontaneous Diels–Alder dimerization to give **2-611**. The high regio- and stereoselectivity of this cycloaddition can be explained by the transition state **2-610**.

A combination of a transetherification and an intramolecular hetero-Diels–Alder reaction developed by Wada and coworkers allows the diastereoselective construction of fused hydropyranopyrans **2-616** (Scheme 2.140) [320]. The process involves a 1,4-addition of alcohols **2-614** containing a dienophile moiety to β-alkoxysubstituted α, β-unsaturated carbonyls **2-613** followed by elimination of MeOH to give an oxabutadiene **2-615** which undergoes a [4+2] cycloaddition. The formed cycloadducts **2-616** are generally isolated in good yields as single isomers [320a].

The domino process can be catalyzed by a Cu-complex with (*S,S*)-*t*Bu-bis(oxazoline) to give **2-616** with excellent enantioselectivity (97–98 % *ee*) [320b,c]. The use of 5 Å molecular sieves turned out to be obligatory. Wada and coworkers also reported on a related transetherification/1,3-dipolar cycloaddition procedure to give access to *trans*-fused bicyclic γ-lactones [321].

Li's group prepared a series of substituted 2-(hydroxyalkyl)tetrahydroquinoline derivatives **2-619** and **2-620** starting from anilines **2-617** and cyclic enol ethers **2-618** in the presence of catalytic amounts of InCl$_3$ (Scheme 2.141) [322]. Good yields of 73–90 % were obtained with an electron-donating or no substituent R at the aniline moiety.

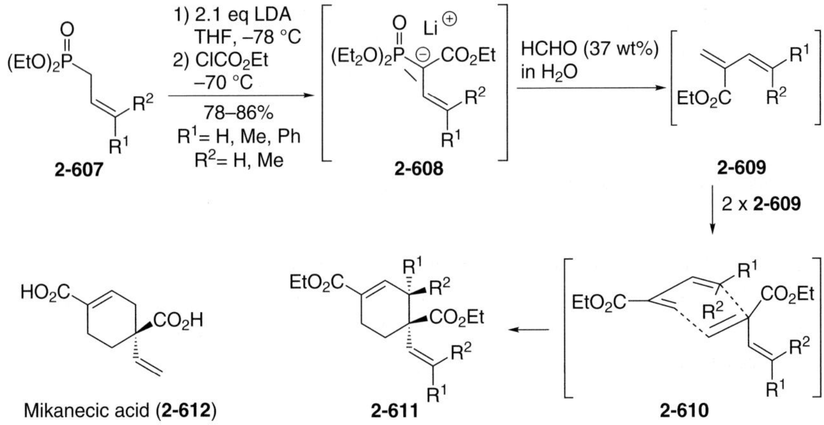

Scheme 2.139. Formation of derivatives of mikanecic acid.

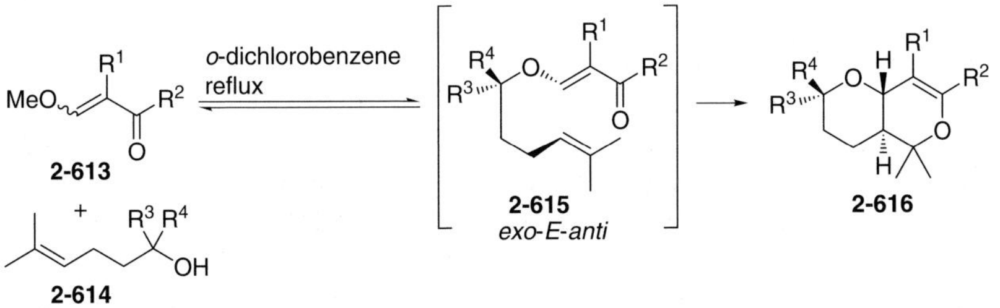

entry	2-613		2-614		2-616
	R^1	R^2	R^3	R^4	yield [%]
1	H	CO_2Me	H	H	70
2	C(O)Me	Me	H	H	76
3	C(O)Me	Me	Me	H	72
4	H	CO_2Me	Me	Me	63

Scheme 2.140. Domino transetherification/intramolecular hetero-Diels–Alder reaction.

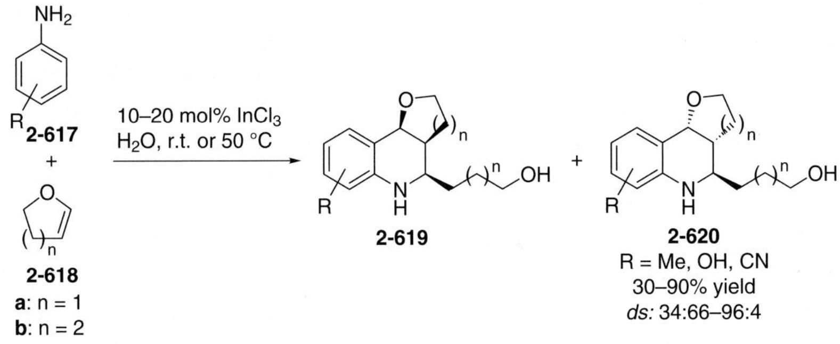

Scheme 2.141. Synthesis of tetrahydroquinolines.

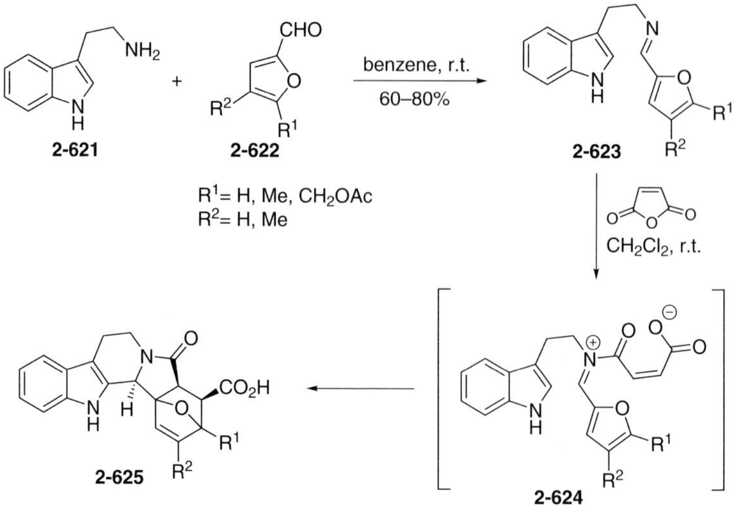

Scheme 2.142. Synthesis of hexacyclic tetrahydro-β-carbolines.

It can be assumed that, in the presence of $InCl_3$ and water, the cyclic enol ethers **2-618** form a hydroxy aldehyde which reacts with the aniline to give an aromatic iminium ion. This represents an electron-poor 1,3-butadiene which can undergo a hetero-Diels–Alder reaction [323] with another molecule of **2-618** to give a mixture of the diastereomeric tetrahydroquinolines **2-619** and **2-620**.

The described approach to this pharmaceutically important class of compounds [324] was also utilized by Bonnet-Delpon and coworkers one year later [325]. Interestingly, these authors employed hexafluoroisopropanol (HFIP) as solvent and were able to perform the domino process without adding any extra Lewis acid catalyst such as $InCl_3$ due to the acidic properties of HFIP ($pK_a = 9.3$) [326]. Besides dihydrofuran or dihydropyran, they have also used acyclic enol ethers.

Unusual hexacyclic tetrahydro-β-carbolines **2-625** have been assembled from tryptamine (**2-621**) and the furan **2-622** by the group of Paulvannan, using three anionic transformations followed by an intramolecular Diels–Alder reaction (Scheme 2.142) [327]. After condensation to give **2-623**, reaction with maleic anhydride leads to the iminium ions **2-624** which enter in an electrophilic aromatic substitution and an intramolecular Diels–Alder reaction to yield the β-carbolines **2-625**. These rigid oxa-bridged hexacyclic compounds **2-625** are obtained in good yields and as sole isomers. Overall, five stereogenic centers, including one or two quaternary centers (depending on R^1), and three rings are generated in this domino sequence.

It is well known that azomethine ylides, which are usually formed *in situ*, are very good substrates for 1,3-dipolar cycloadditions. The group of Novikov and Khlebnikov [328] generated such a 1,3-dipol by reaction of difluorocarbene formed from CBr_2F_2 (**2-626**) with the imine **2-627**. Cycloaddition of the obtained **2-629** with an ac-

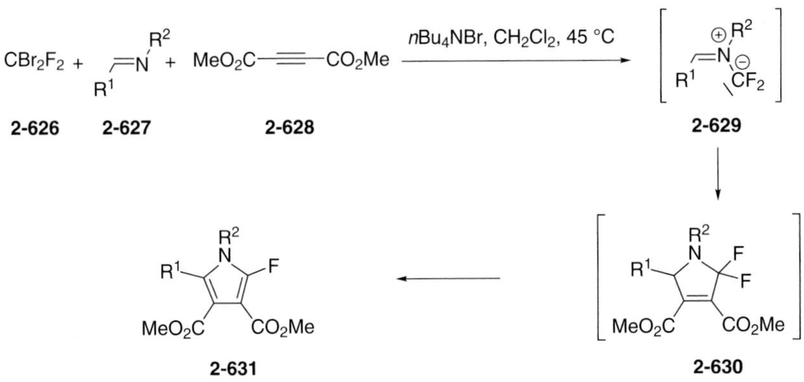

$$CBr_2F_2 + \underset{R^1}{\overset{R^2}{=}}N + MeO_2C\!\!=\!\!=\!\!CO_2Me \xrightarrow{\ nBu_4NBr,\ CH_2Cl_2,\ 45\ ^{\circ}C\ } \left[\ \underset{R^1}{\overset{\oplus\ R^2}{=}}N\overset{\ominus}{\underset{CF_2}{}}\ \right]$$

2-626   2-627   2-628   2-629

2-631

2-630

Entry	2-627	R^1	R^2	2-631	Yield [%]
1	a	Ph	$pMeOC_6H_4$	a	70
2	b	2-Furyl	Ph	b	69
3	c	$pBrC_6H_4$	Ph	c	78
4	d	$mNO_2C_6H_4$	Ph	d	32
5	e	$PhC\equiv C$	Ph	e	42
6	f	(E)-PhCH=CH	Ph	f	41

Scheme 2.143. Syntheses of 2-fluoropyrroles *via* domino carbene addition/1,3-dipolar cycloaddition.

tivated alkyne as DMAD (**2-628**) led to the adducts **2-630** from which, in a subsequent dehydrofluorination process, substituted fluoropyrroles **2-631** were formed. The imines used can contain aromatic, heteroaromatic, and unsaturated carbon compounds as substituents.

Stockman and coworkers [329] developed a straightforward synthesis of a tricyclic compound **2-636** which has some resemblance to the spirocyclic portion of the natural product halichlorine (**2-637**) [330]. On treatment of the symmetrical ketone **2-633**, accessible in five steps from alcohol **2-632**, with hydroxylamine hydrochloride the spiro piperidine **2-636** could be obtained in 62 % yield (Scheme 2.144). It is assumed that, after the initial formation of the oxime **2-634**, a Michael addition occurs to give **2-635** with formation of a nitrone moiety which then can undergo a 1,3-dipolar cycloaddition to give **2-636**.

Another anionic pericyclic domino process is a Pummerer-type rearrangement/ 1,3-dipolar cycloaddition/ring-opening sequence of **2-638** to give **2-639** (Scheme 2.145) [331]. Mechanistically, a very electrophilic α-acyl thienium ion **A**, generated from the imidosulfoxide **2-638**, rapidly reacts with the neighboring imido group and the resulting oxenium ion **B** undergoes subsequent deprotonation to produce isomünchnone **C**; this contains a carbonyl ylide dipole which can readily undergo a 1,3-dipolar cycloaddition with dipolarophiles. Exposure of the resulting cycloadducts **D** to Ac$_2$O in the presence of a trace of *p*TsOH results in a ring cleavage to give pyridones of type **2-639**.

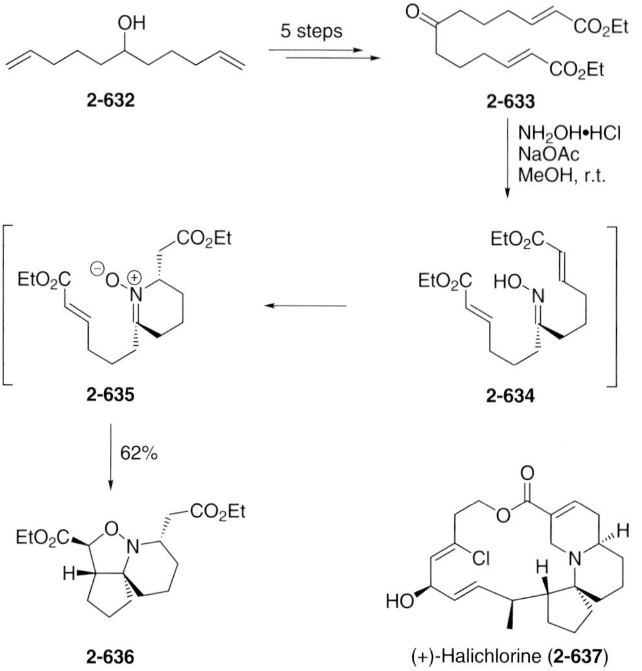

Scheme 2.144. Synthesis of the tricyclic spiro piperidine **2-636**.

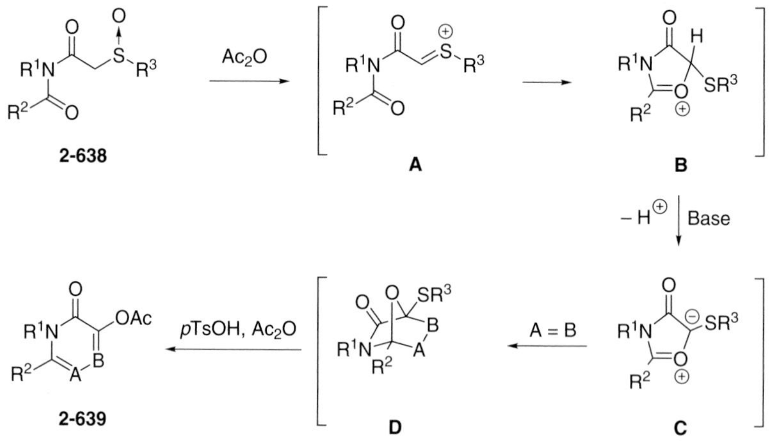

Scheme 2.145. General method for the construction of pyridones of type **2-639**.

According to this scheme, a wide variety of naturally occurring alkaloids [331], for example onychine (**2-640**) [332] dielsiquinone (**2-641**) [333], costaclavin (**2-642**) [334] and pumiliotoxin C (**2-643**) [335] have been prepared starting from **2-638a–c** *via* the cycloadducts **2-639a–c** (Scheme 2.146).

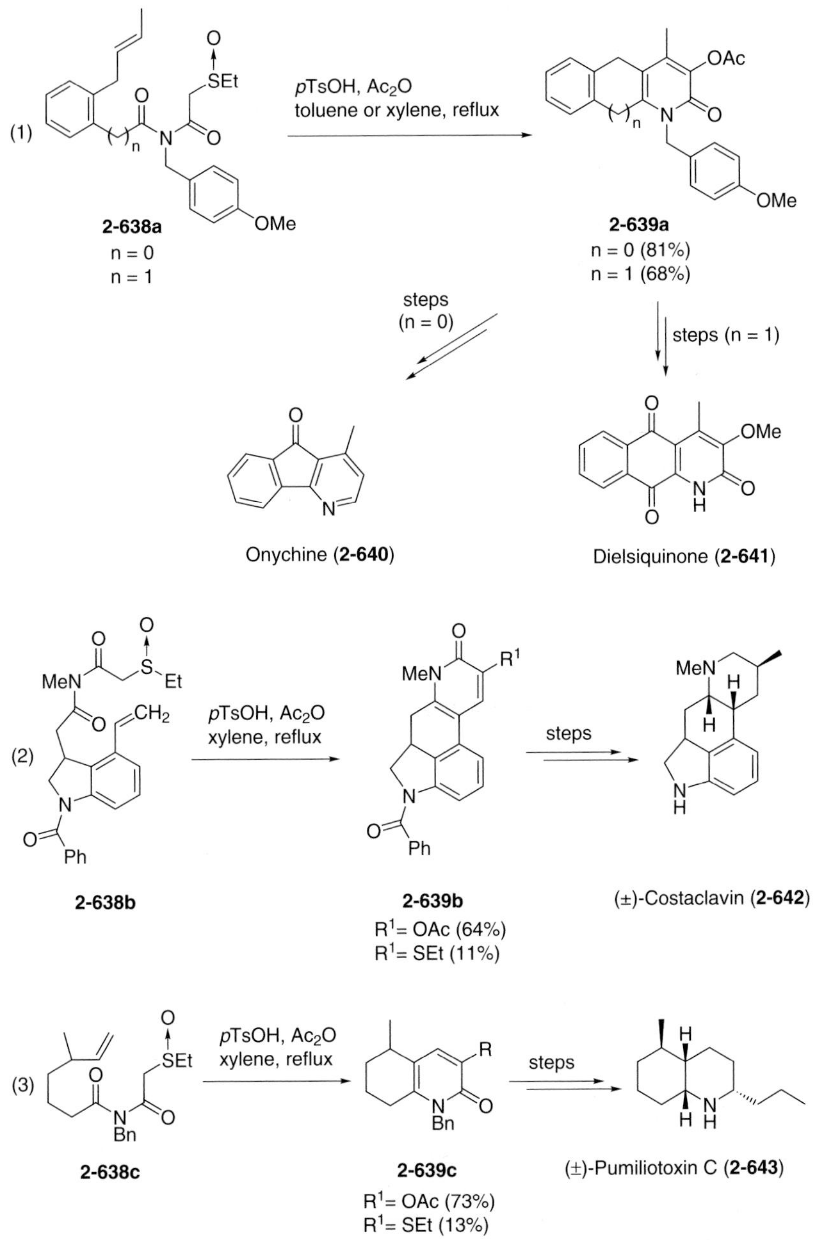

Scheme 2.146. Synthesis of several natural products *via* Pummerer-type rearrangement/
1,3-dipolar cycloaddition/ring-opening reactions.

A very attractive anionic/anionic/pericyclic/anionic fourfold domino sequence
was developed by Kuehne's group, as illustrated in Scheme 2.147 [336]. Herein, on
treatment of the enantiopure tryptophane-derived diester **2-644** with α,β-unsatu-
rated aldehydes **2-645** at 70 °C in benzene with benzoic acid and freshly activated

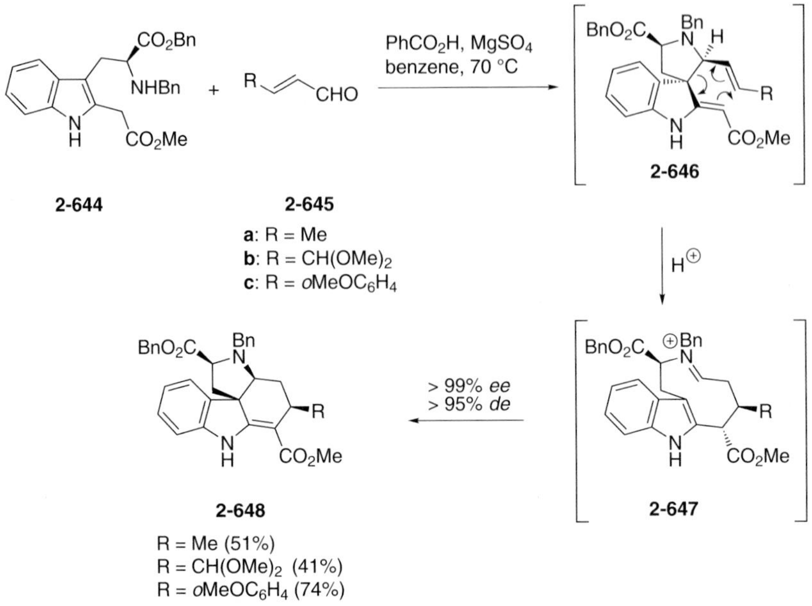

Scheme 2.147. Synthesis of complex indoles **2-648**.

MgSO$_4$, the tetracycles **2-648** were obtained with excellent diastereoselectivity in reasonable yield. The reaction presumably starts with a condensation of the aldehydes **2-645** with the benzyl-protected amine moiety of **2-644** to give an iminium ion which can subsequently cyclize to afford the spirocyclic intermediates **2-646**. A [3,3] sigmatropic Cope rearrangement then forms the nine-membered cyclic enamines **2-647** which, after protonation, act as the starting point for another indole iminium cyclization to provide the tetracycles **2-648** *via* **2-647**.

Recent reports have shown that this highly efficient domino sequence can also be employed for the synthesis of (–)-strychnine [337] and (–)-lochneridine [338].

A trifold anionic/pericyclic domino reaction was used for the synthesis of the dioxapyrrolidizine **2-655** combining a nitro aldol condensation, S$_N$-type cyclization, S$_N$-type etherification, and an intramolecular 1,3-dipolar cyclization as described by Rosini and coworkers (Scheme 2.148) [339].

As substrates, aliphatic aldehydes **2-649**, nitromethane **2-650** carrying an electron-withdrawing group, and chlorovinylsilane **2-653** were used. The reaction proceeds *via* the proposed intermediates **2-651**, **2-652**, and **2-654**. In this three-component transformation five new bonds and four new stereogenic centers are formed, leading to only two diastereomers out of eight possible isomers.

The novel [6+2] annulation approach developed by the Takeda group has also been included in a threefold anionic/pericyclic process (Scheme 2.149) [340]. The reaction leads to functionalized eight-membered rings **2-659** in a highly stereoselective manner, starting from acylsilanes **2-656** and β-(trimethylsilyl)vinyl-lithium (**2-657**). After 1,2-addition and 1,2-Brook rearrangement, the cyclobutane **2-**

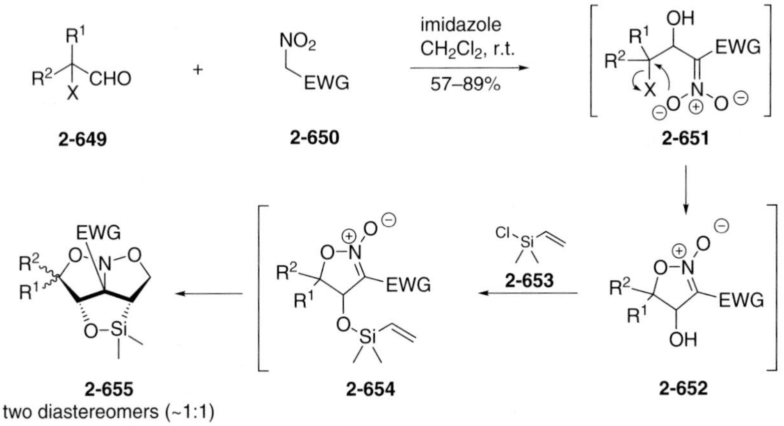
2-649
2-650
imidazole
CH2Cl2, r.t.
57–89%
2-651
2-652
2-653
2-654
2-655
two diastereomers (~1:1)
7 examples: EWG: CO2Et, SO2Ph; R1= H,Ph; R2= Alkyl; X = Br, OTs

Scheme 2.148. Synthesis of dioxapyrrolizidines **2-655**.

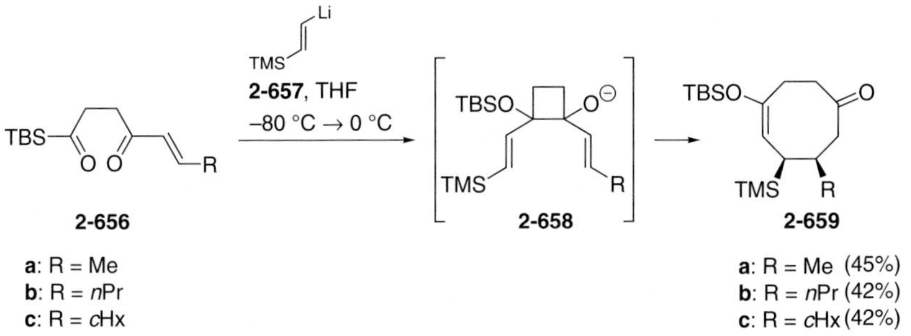
2-657, THF
−80 °C → 0 °C
2-656
2-658
2-659
a: R = Me
b: R = nPr
c: R = cHx
a: R = Me (45%)
b: R = nPr (42%)
c: R = cHx (42%)

Scheme 2.149. Synthesis of cyclooctenones.

658 is formed which undergoes an oxy-Cope process to give the cyclooctenenones **2-659** in 42–45 % yield.

Moreover, the tricyclic core **2-663** of the cyathins as erinacin E (**2-664**) and allocyathin B_2 (**2-665**) [341], isolated from bird nest fungi, has been synthesized by the same research group using a related approach with a [3+4] annulation reaction (Scheme 2.150) [342]. Addition of the enolate **2-660** to the acryloylsilane **2-661** gave the annulated tricyclic compound **2-663** as a single diastereomer in 60 % yield. The observed stereoselectivity can be rationalized assuming a concerted anionic oxy-Cope rearrangement of the *cis*-1,2-divinylcyclopropanediolate intermediate **2-662** which is selectively derived from the 1,2-adduct of **2-660** and **2-661** by a 1,2-Brook rearrangement followed by an internal trapping of the generated carbanion by the carbonyl group.

Besides morefold anionic domino processes with one pericyclic reaction, domino sequences combining two initiating anionic with two pericyclic steps have also been developed. For example, the group of Nesi and Turchi reported on the synthe-

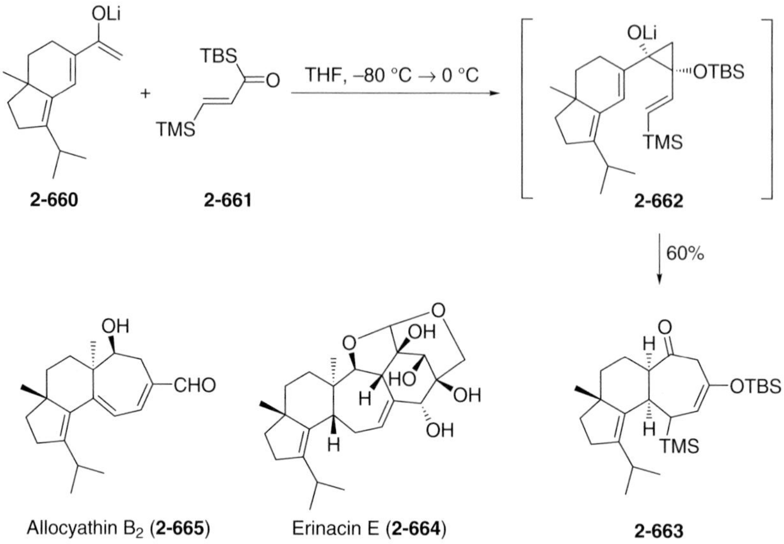

Scheme 2.150. Synthesis of the tricyclic skeleton of the cyathins.

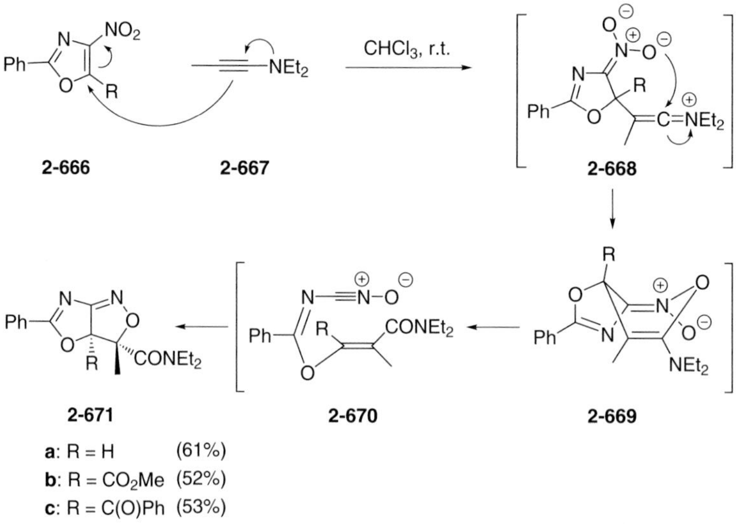

a: R = H   (61%)
b: R = CO₂Me (52%)
c: R = C(O)Ph (53%)

Scheme 2.151. Synthesis of compounds **2-671** by a fourfold domino process.

sis of substituted oxazolo[4,5-*c*]isoxazoles **2-671** in yields ranging from 52 to 61 %
(Scheme 2.151) [343]. Reaction of the oxazole **2-666** and the yneamine **2-667** in a Michael-type addition with subsequent cyclization of the zwitterionic intermediate **2-**
668 leads to **2-669**. It is assumed that the next step is a concerted retro-hetero-Diels–
Alder ring-opening due to the exclusive *cis* orientation observed for R and the CON-

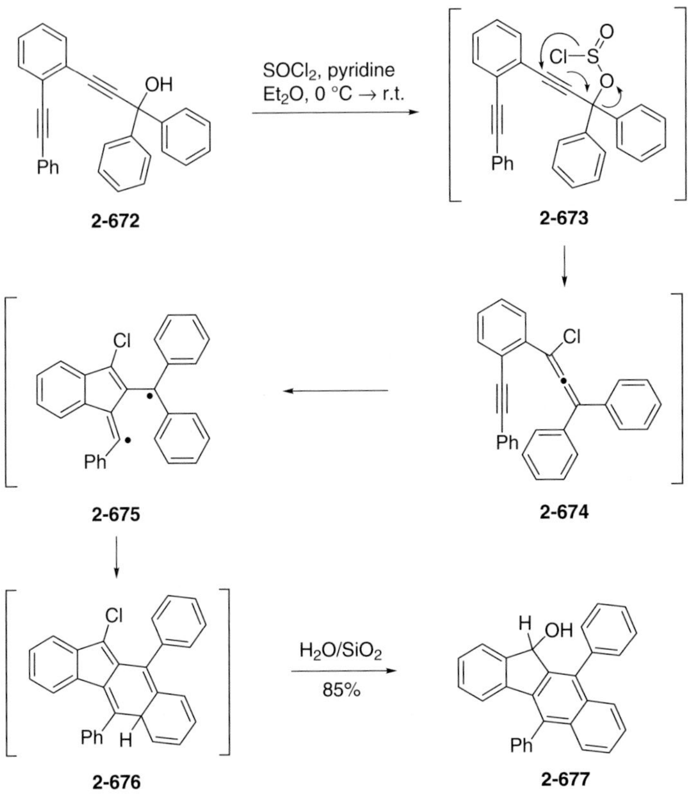

Scheme 2.152. Combination of anionic and radical transformations.

Et$_2$ group in **2-670**. Since **2-670** contains both, a 1,3-dipole as well as a 1,3-dipolarophile moiety, it immediately undergoes a 1,3-dipolar cycloaddition to form the heterocyclic compound **2-671**.

Further twofold anionic/twofold pericyclic sequences have been developed by the groups of Mikami and Grigg. The first group reported on an asymmetric domino addition/elimination (transetherification)/Claisen rearrangement/ene reaction sequence for the synthesis of a (+)-9(11)-dehydroestrone methyl ether [344], which is a valuable intermediate towards estrogen synthesis [345]. The second group examined the *in-situ* generation of azomethine ylides from an α-amino acid and a ketone *via* an iminium ion formation/cyclization/decarboxylation (decarboxylative methodology). The formed dipoles were trapped in a 1,3-dipolar cycloaddition to give complex molecular scaffolds [346].

There are also some rare domino sequences where two anionic and two radical reactions are combined (Scheme 2.152) [347]. According to a report of the Wang group, thionyl chloride is able to promote a succession of reactions by an initial formation of a chlorosulfite **2-673** of the tertiary alcohol **2-672**, followed by an S$_N$-type reaction to produce the chloroallene **2-674**. A Schmittel cyclization reaction [348] then generates

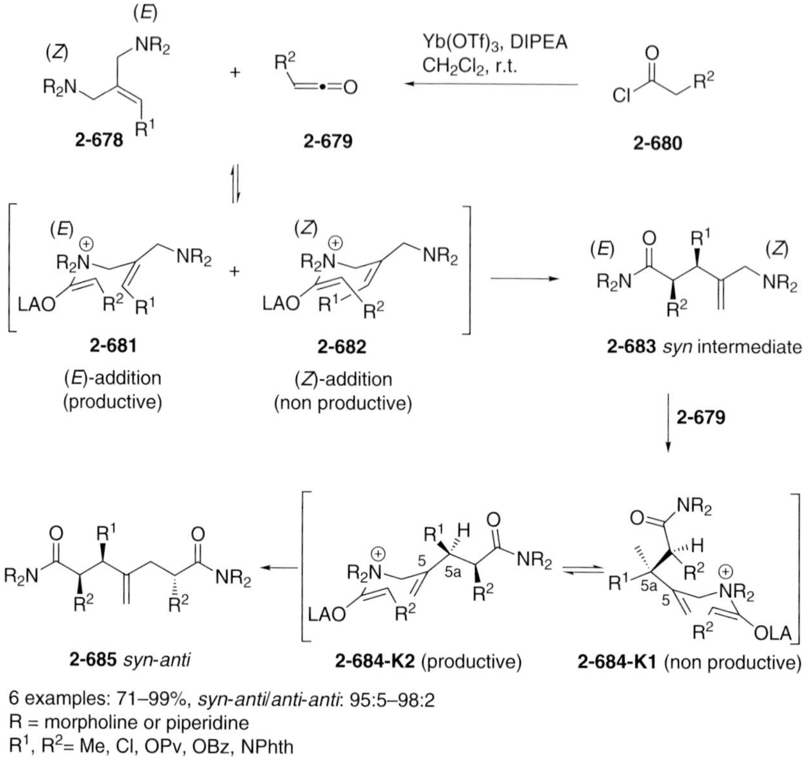

Scheme 2.153. Yb(OTF)$_3$-promoted domino ketene addition/acyl-Claisen rearrangement.

the biradical **2-675** which undergoes an intramolecular radical coupling to furnish the cycloadduct **2-676**. After rearomatization, the formed chloride is hydrolyzed during work-up to give the 11*H*-benzo[*b*]fluoren-11-ol **2-677** in 85 % yield.

A combination of three anionic and two pericyclic reactions has been elaborated by MacMillan's group to afford α,β-disubstituted-γ,δ-unsaturated amides (Scheme 2.153) [349]. As substrates, allyl diamines **2-678** and acid chlorides **2-680** were used. The domino process is initiated by the *in situ* generation of ketene **2-679**, which is then followed by a Lewis acid-mediated addition of **2-679** to either the (*Z*)- or (*E*)-amine moiety of the allyl diamine **2-678** to provide the regioisomeric allyl vinylam-monium complexes **2-681** and **2-682**. It is assumed that in the subsequent aza-Cope rearrangement only the (*E*)-1,5-diene **2-681** reacts. However, since the first steps seem to be reversible, a complete transformation is observed. The formed 2,3-disub-stituted intermediate **2-683** with *syn*-orientation of the two substituents contains again an allyl amine moiety which can react once more with ketene **2-679**. The formed 1,5-diene **2-684** underwent a second aza-Cope rearrangement *via* conforma-tion **2-684-K2** in which compared to **2-684-K1** the allylic 1,2-strain [350] about the C(5)–C(5a) bond is minimized. The final product is a 2,3,6-trisubstituted-1,7-di-amidoheptane **2-685** with mostly a 2,3-*syn*-3,6-*anti* orientation.

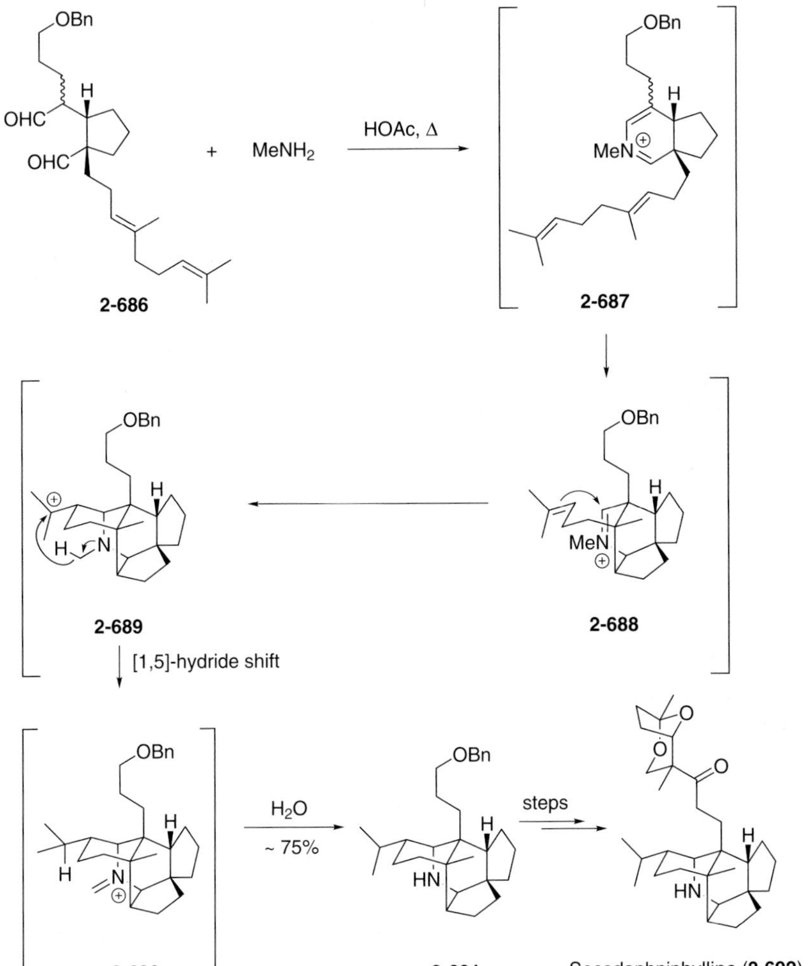

Scheme 2.154. Total synthesis of the *Daphniphyllum* alkaloid secodaphniphylline (**2-692**).

One very fascinating domino reaction is the fivefold anionic/pericyclic sequence developed by Heathcock and coworkers for the total synthesis of alkaloids of the *Daphniphyllum* family [351], of which one example was presented in the Introduction. Another example is the synthesis of secodaphniphylline (**2-692**) [352]. As depicted in Scheme 2.154, a twofold condensation of methylamine with the dialdehyde **2-686** led to the formation of the dihydropyridinium ion **2-687** which underwent an intramolecular hetero-Diels–Alder reaction to give the unsaturated iminium ion **2-688**. This cyclized, providing carbocation **2-689**. Subsequent 1,5-hydride shift afforded the iminium ion **2-690** which, upon aqueous work-up, is hydrolyzed to give the final product **2-691** in a remarkable yield of about 75 %. In a similar way, dihydrosqualene dialdehyde was transformed into the corresponding polycyclic compound [353].

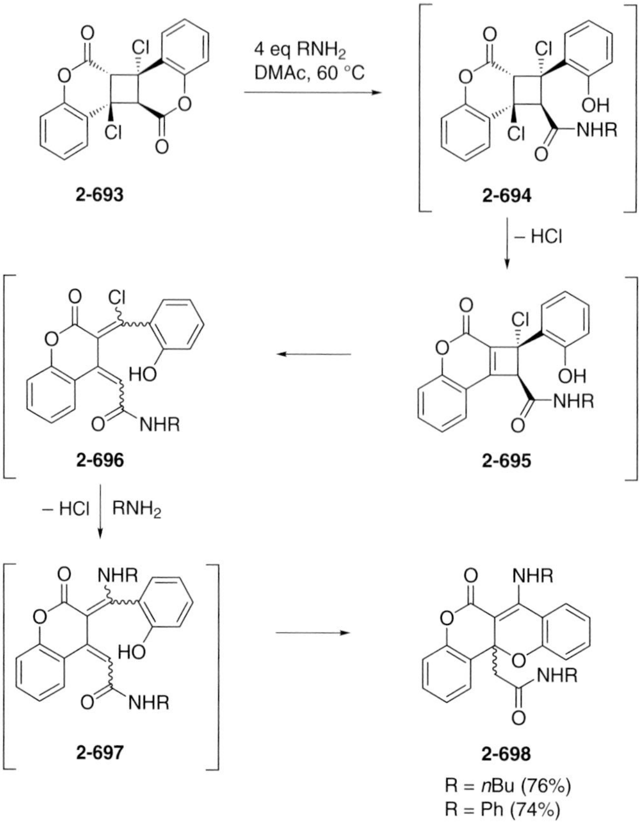

Scheme 2.155. Reaction of the chloro-coumarin dimer **2-693** with primary amines.

Another fivefold anionic/pericyclic sequence was developed by Hasegawa and co-workers, who had the original intention of producing polyamides by reaction of coumarin dimers with diamines [354]. Accordingly, the reaction of 4-chloro-coumarin dimer **2-693** with primary monoamines was investigated [355]. However, unexpectedly instead of obtaining the corresponding monoamides or diamides, benzopyranocoumarins **2-698** were produced in good yields (Scheme 2.155).

As a possible reaction mechanism the authors proposed the following. The dilactone **2-693** reacts with the amine to give the monoamide **2-694**; the remaining lactone ring in **2-694** seems to be less reactive due to a reduced ring strain as compared to **2-693**. Hydrogen chloride is then eliminated to give the cyclobutene **2-695**, which is immediately converted into the diene **2-696** through a cycloreversion of the cyclobutene ring. Afterwards, the chlorine atom is substituted by the amino group in an addition/elimination manner. Finally, a Michael addition of the phenolic hydroxy group takes place onto the enamide moiety in **2-697** to afford the benzopyranocoumarins **2-698**.

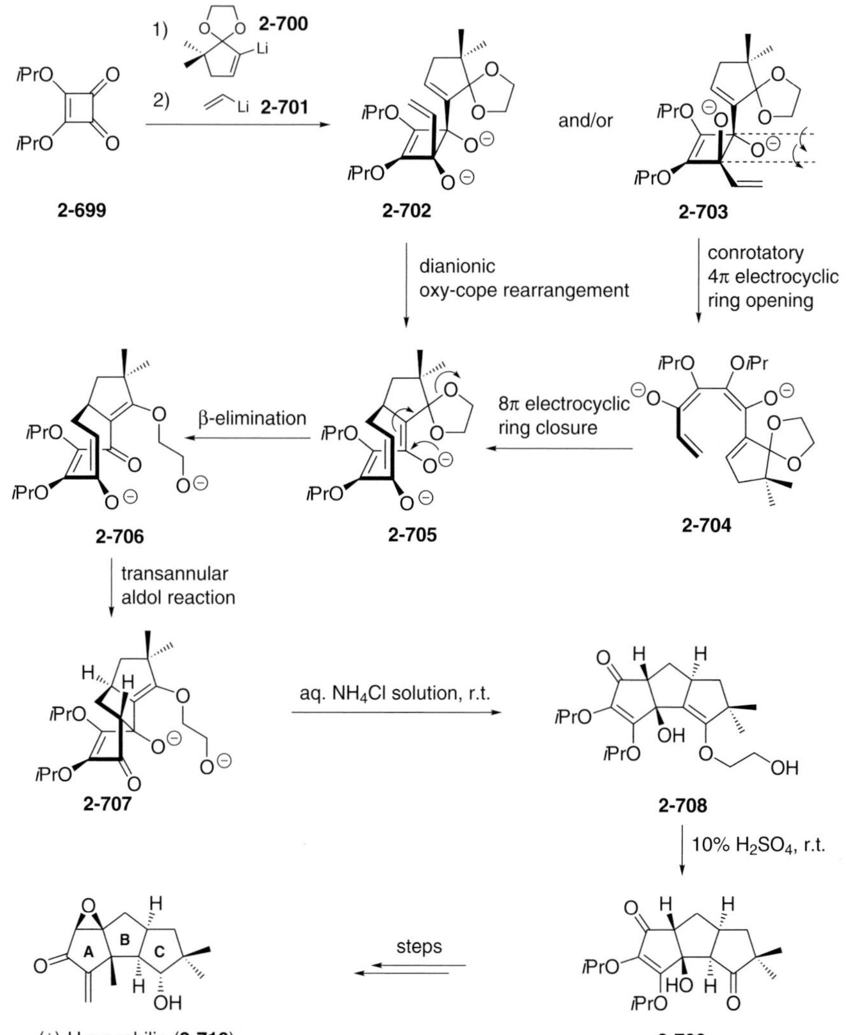

Scheme 2.156. Synthesis of (±)-hypnophilin (**2-710**).

More than a decade ago Paquette and coworkers investigated the sequential 1,2-addition of two molecules of an alkenyl anion (either the same or different) to *i*-propyl squarate **2-699** [356]. The reaction of **2-699** with **2-700** and vinyl lithium (**2-701**) is depicted in Scheme 2.156.

The nucleophilic attack with **2-700** and **2-701** can proceed in a *syn* or *anti* manner to provide either **2-702** or **2-703**, or both [357]. If **2-703** is formed, it follows a charge-driven conrotatory opening of the cyclobutene ring with generation of the coiled 1,3,5,7-octatetraene **2-704**. This intermediate is capable of a rapid helical equilibration [357] and a regioselective 8π electrocyclic ring closure to give **2-705** [358].

Nevertheless, arrival at **2-705** can also be realized somewhat more directly by a di-anionic oxy-Cope rearrangement starting from **2-702** [359]. These mechanistic options can be distinguished when additional stereochemical markers are present [360]. The proper positioning of the acetal moiety in **2-705** allows a β-elimination reaction with the formation of the ketone **2-706** which undergoes a subsequent trans-annular aldol ring closure. The final product is the highly functionalized linear triquinane **2-707** [361], which was transformed into the natural product (±)-hyp-nophilin **2-710** *via* **2-708** and **2-709** [362]. Of worth is the knowledge that this concept has also been used in the synthesis of ceratopicanol as well as other, similar compounds [362].

2.2
Anionic/Radical Processes

In an anionic/radical domino process an interim single-electron transfer (SET) from the intermediate of the first anionic reaction must occur. Thus, a radical is generated which can enter into subsequent reactions. Although a SET corresponds to a formal change of the oxidation state, the transformations will be treated as typical radical reactions. To date, only a few true anionic/radical domino transformations have been reported in the literature. However, some interesting examples of related one-pot procedures have been established where formation of the radical occurs after the anionic step by addition of TEMPO or Bu₃SnH. A reason for the latter approach are the problems associated with the switch between anionic and radical reaction patterns, which often do not permit the presence of a radical generator until the initial anionic reaction step is finished.

Jahn combined the formation of the enolate **2-713** resulting from an intermolecular Michael addition of **2-711** and **2-712** with a radical reaction (Scheme 2.157) [363]. The enolate **2-713** did not undergo any further transformations due to the lack of appropriate functionalities. However, after formation of a radical using a mixture of ferrocenium hexafluorophosphate (**2-714**) and TEMPO, a new reaction channel was opened which afforded the highly substituted cyclopentene **2-715a** diastereoselectively.

Moreover, by using only TEMPO without addition of **2-714** the Michael adduct **2-713** is transformed into the isopropenylcyclopentane **2-715b** as the major product. The process can also be extended by another radical reaction step [364].

Another anionic/radical one-pot sequence was developed by Guindon and co-workers for the stereoselective synthesis of substituted pentanoates **2-718** (Scheme 2.158) [365]. Such structures are found in polyketides and are, therefore, of great interest. The described approach offers a diastereoselective access to all four possible stereoisomers of **2-718** through a Mukaiyama aldol/radical defunctionalization sequence starting from **2-716** and **2-717** with addition of Bu₃SnH after completion of the first step.

The stereochemical outcome of the Mukaiyama reaction can be controlled by the type of Lewis acid used. With bidentate Lewis acids the aldol reaction led to the *anti* products through a Cram chelate control [366]. Alternatively, the use of a monodentate Lewis acid in this reaction led to the *syn* product through an open Felkin–Anh

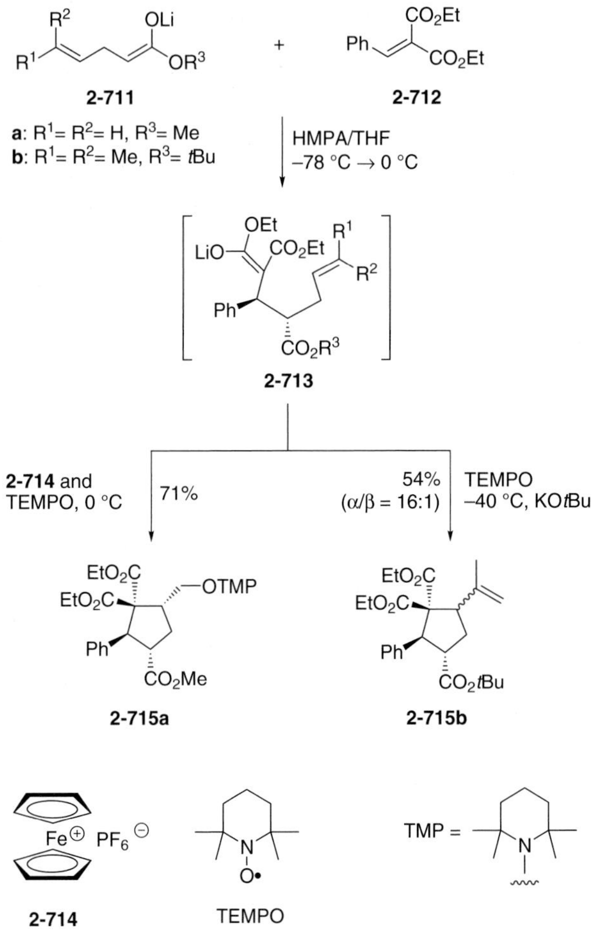

R² OLi
R¹ OR³

2-711

+

CO₂Et
Ph CO₂Et

2-712

a: R¹= R²= H, R³= Me
b: R¹= R²= Me, R³= *t*Bu

HMPA/THF
−78 °C → 0 °C

OEt R¹
LiO CO₂Et
R²
Ph
CO₂R³

2-713

2-714 and
TEMPO, 0 °C | 71%

54% | TEMPO
(α/β = 16:1) | −40 °C, KO*t*Bu

EtO₂C
EtO₂C OTMP
Ph
CO₂Me

2-715a

EtO₂C
EtO₂C
Ph
CO₂*t*Bu

2-715b

Fe⊕ PF₆⊖

N
O•

TMP =

N

2-714   TEMPO

Scheme 2.157. Michael addition/radical cyclization.

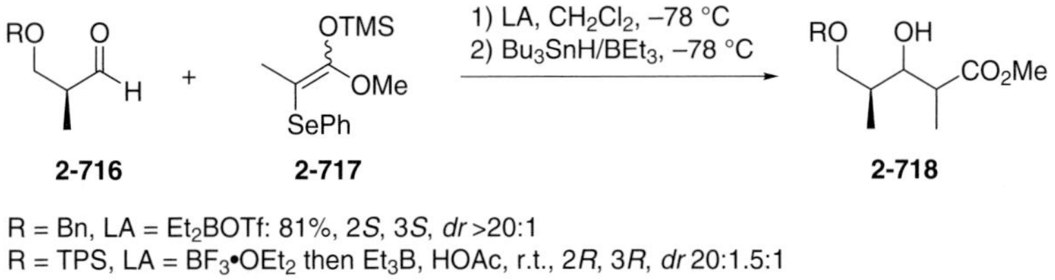

RO   O
H

2-716

+

OTMS
OMe
SePh

2-717

1) LA, CH₂Cl₂, −78 °C
2) Bu₃SnH/BEt₃, −78 °C

RO   OH
CO₂Me

2-718

R = Bn, LA = Et₂BOTf: 81%, 2*S*, 3*S*, *dr* >20:1
R = TPS, LA = BF₃•OEt₂ then Et₃B, HOAc, r.t., 2*R*, 3*R*, *dr* 20:1.5:1

Scheme 2.158. Mukaiyama aldol/radical defunctionalization process.

transition state [366]. The stereochemical outcome of the radical displacement of
the selenium substituent is again governed by the Lewis acid employed according
to their coordination as depicted in **2-719** and **2-720** (Scheme 2.159). The first tran-

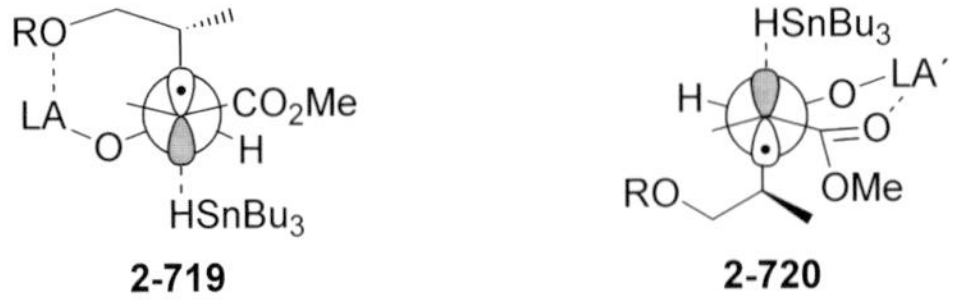
RO
LA
O
CO2Me
H
HSnBu3
2-719
HSnBu3
H
O
LA´
O
RO
OMe
2-720

Scheme 2.159. Transition states for the radical removal of a selenium substituent.

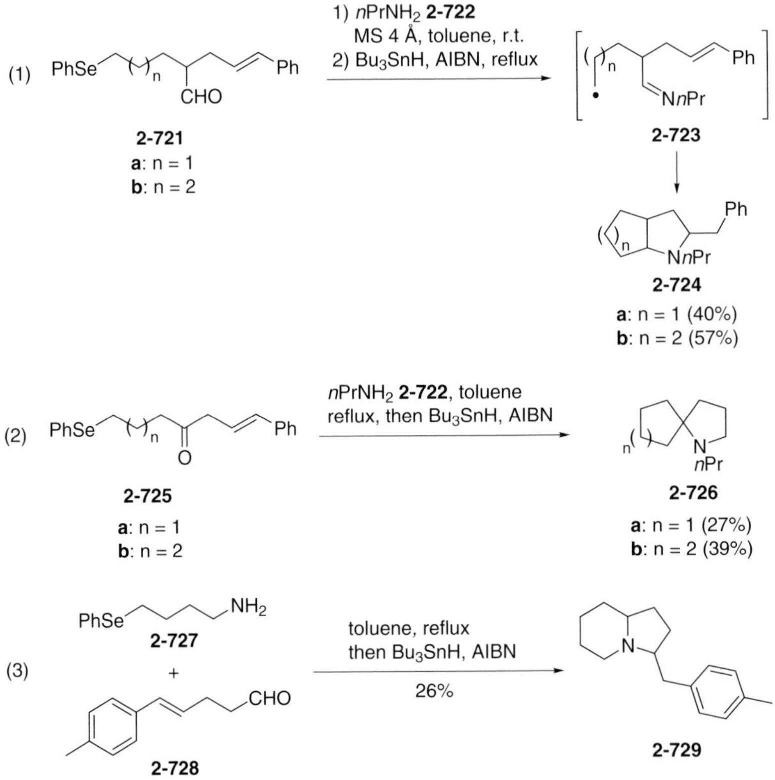
(1)
PhSe
n
CHO
Ph
2-721
a: n = 1
b: n = 2
1) nPrNH2 2-722
MS 4 Å, toluene, r.t.
2) Bu3SnH, AIBN, reflux
n
Ph
NnPr
2-723
n
Ph
NnPr
2-724
a: n = 1 (40%)
b: n = 2 (57%)
(2)
PhSe
n
O
Ph
2-725
a: n = 1
b: n = 2
nPrNH2 2-722, toluene
reflux, then Bu3SnH, AIBN
n
N
nPr
2-726
a: n = 1 (27%)
b: n = 2 (39%)
(3)
PhSe
NH2
2-727
+
CHO
2-728
toluene, reflux
then Bu3SnH, AIBN
26%
N
2-729

Scheme 2.160. Imin-formation/radical-cyclization for the synthesis of pyrrolidines.

sition state would lead to a 2,3-*anti*-orientation and the second transition state to a 2,3-*syn*-orientation in the product **2-718**.

The method can also be performed in an iterative manner [367], and the scope of the one-pot process can be broaden by replacing Bu_3SnH by its allylic derivative Bu_3SnAl-lyl, thus allowing the introduction of an allyl group instead of a hydrogen [365].

Bowman and coworkers described another nice approach in this field combining one anionic and two radical cyclization steps (Scheme 2.160) [368]. Thus, they were able to construct different types of nitrogen-containing heterocycles such as annulated pyrrolidines **2-724** and spiropyrrolidines **2-726** as well as indolizidines as **2-729**.

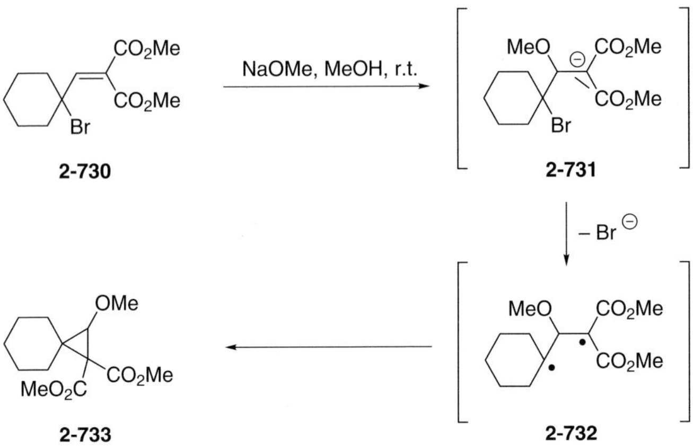

Scheme 2.161. Synthesis of spirocyclopropyl ether involving a SET process.

Reaction of the aldehyde **2-721** and the amine **2-722** gave an imine which, on treatment with Bu₃SnH, led to the radical **2-273**. This underwent a twofold cyclization to give **2-724**. In a similar way, **2-725** and **2-722** gave **2-726** whereas reaction of **2-728** with the selenoamine **2-727** afforded the indolizidine **2-729**. Although the yields are low, the transformations are simple and do not depend on complex starting materials.

The group of Walborsky probably has described one of the first true anionic/radical domino process in their synthesis of the spirocyclopropyl ether **2-733** starting from the tertiary allylic bromide **2-730** (Scheme 2.161) [369]. The first step is a Michael addition with methoxide which led to the malonate anion **2-731**. It follows a displacement of the tertiary bromide and a subsequent ring closure which is thought to involve a SET from the anionic center to the carbon–bromine antibonding orbital to produce the diradical **2-732** and a bromide anion. An obvious alternative S_N2 halide displacement was excluded due to steric reasons and the ease with which the reaction proceeded.

Another real anionic/radical domino sequence has been published by Molander and his group [370], who developed an efficient ring-building strategy utilizing SmI₂, initially in a metal iodide exchange reaction followed by a nucleophilic addition to a carbonyl moiety, and secondly as a radical generator. In this way, they were able to construct substituted bicyclic, tricyclic, and spirocyclic compounds from likewise readily available substrates (Scheme 2.162).

As an example, reactions of **2-734a** and **2-734b** with 4 equiv. of SmI₂ led to the annulated cyclopentanes **2-741a** and **2-741b** in good to excellent yields. However, the process is less suitable for the preparation of hydrindanes **2-741c**.

The transformation of **2-734** involves an initial generation of an organosamarium species **2-735** with subsequent nucleophilic addition to the lactone carbonyl. Presumably, a tetrahedral intermediate **2-736** is formed that collapses to yield the ketone **2-737**. This reacts with SmI₂ to give a ketyl radical **2-738**, which undergoes an intramolecular 5-*exo* radical cyclization reaction with the alkene moiety. The resultant

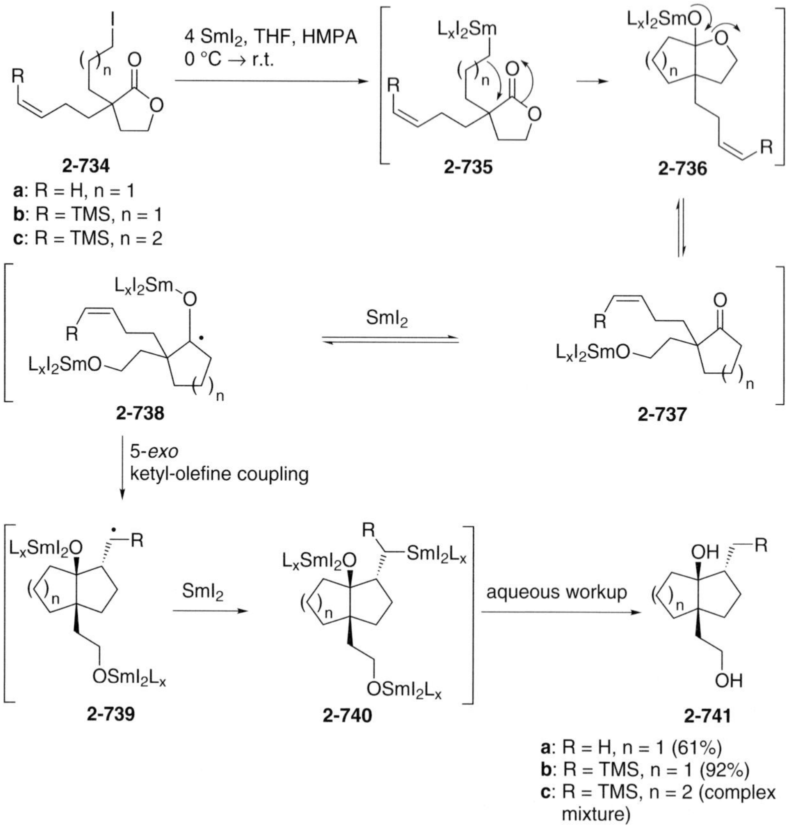

Scheme 2.162. SmI$_2$-mediated domino reaction leading to bicyclic compounds **2-741**.

carbon-centered radical **2-739** is rapidly reduced to an organosamarium compound **2-740**, generating the desired bicyclic products **2-741** after aqueous work-up. The major diastereomers in the *exo* cyclizations are those with the developing radical center *trans* to the alkoxy group. The formation of this isomer avoids unfavorable stereoelectronic interactions in the radical cyclization [371]. In the case of **2-734a** and **2-734b** the products **2-741a** and **2-742** were isolated as single isomers.

Moreover, alkynes instead of alkenes can be used as ketyl radical acceptors, and it was also demonstrated that the organosamarium species of type **2-740** can be trapped with an electrophile, such as acetone.

2.3
Anionic/Pericyclic Processes

The combination of an anionic and a pericyclic process has found broad application in synthetic organic chemistry. In these transformations, a primary addition is frequently followed by an elimination to give a reactive intermediate, which is then

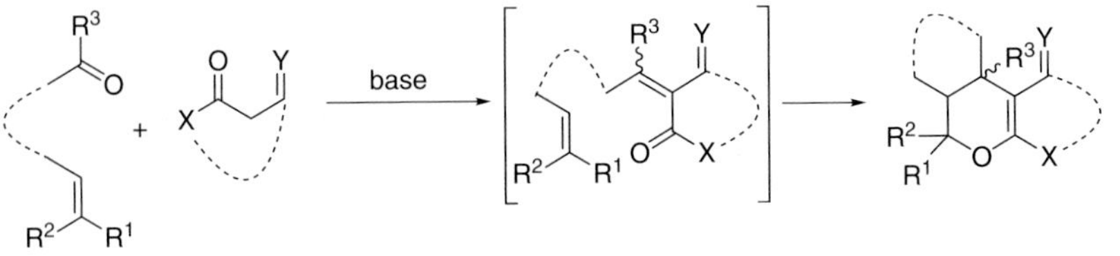

Scheme 2.163. The domino Knoevenagel/hetero-Diels–Alder reaction.

trapped in a cycloaddition, sigmatropic rearrangement, electrocyclization or ene re-action. In most cases reported, a C–C-double bond is formed in the anionic reaction step affording a reactive 1,3-diene, which undergoes a Diels–Alder cyclization or an ene reaction. Such an *in-situ* procedure is usually synthetically of advantage com-pared to a two-step protocol, since the formed dienes are often unstable and diffi-cult to purify. One of most useful protocols in this respect is the domino Knoevenagel/hetero-Diels–Alder reaction, which has been developed by Tietze and coworkers. Domino processes using an initial HWE or Wittig olefination followed by a pericyclic reaction are also of great interest. Moreover, the formation of 1,3-dipoles followed by a 1,3-dipolar cyclization reactions has been widely applied.

The domino Knoevenagel/hetero-Diels–Alder reaction is a prominent example of the great advantage of domino processes as it not only allows the efficient synthe-sis of complex compounds such as natural products starting from simple sub-strates, but also permits the preparation of highly diversified molecules. Due to the vast number of reports that have been made, only a few recent publications can be discussed here, although several excellent reviews on this topic have been produced that provide a more detailed insight into this useful method [372].

As the name implies, the first step of this domino process consists of a Knoevenagel condensation of an aldehyde or a ketone **2-742** with a 1,3-dicarbonyl compound **2-743** in the presence of catalytic amounts of a weak base such as ethy-lene diammonium diacetate (EDDA) or piperidinium acetate (Scheme 2.163). In the reaction, a 1,3-oxabutadiene **2-744** is formed as intermediate, which undergoes an inter- or an intramolecular hetero-Diels–Alder reaction either with an enol ether or an alkene to give a dihydropyran **2-745**.

In the Diels–Alder reaction with inverse electron demand, the overlap of the LUMO of the 1-oxa-1,3-butadiene with the HOMO of the dienophile is dominant. Since the electron-withdrawing group at the oxabutadiene at the 3-position lowers its LUMO dramatically, the cycloaddition as well as the condensation usually take place at room or slightly elevated temperature. There is actually no restriction for the aldehydes. Thus, aromatic, heteroaromatic, saturated aliphatic and unsaturated aliphatic aldehydes may be used. For example, α-oxocarbocylic esters or 1,2-dike-tones for instance have been employed as ketones. Furthermore, 1,3-dicarbonyl compounds cyclic and acyclic substances such as Meldrum's acid, barbituric acid and derivates, coumarins, any type of cycloalkane-1,3-dione, β-ketoesters, and 1,3-diones as well as their phosphorus, nitrogen and sulfur analogues, can also be ap-

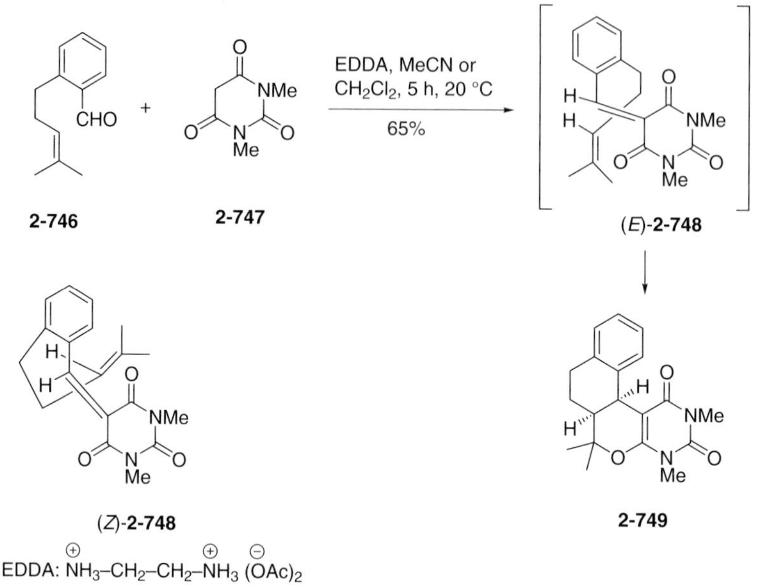

Scheme 2.164. Domino Knoevenagel/hetero-Diels–Alder reaction with aromatic aldehydes

plied. Hetero-analogue 1,3-dicarbonyl compounds such as the aromatic pyra-zolones and isoxazolones can also be put to reaction. However, depending on the substrates used, a domino Knoevenagel/ene process might occur as a side reaction, or it can become the main process [373]. The Tietze group has widely investigated the scope and limitation of this method. The most appropriate dienophiles are enol ethers, while enamines are more difficult to handle. Simple alkenes are also suitable as dienophiles, but usually only if the Diels–Alder reaction takes place in an intramolecular mode. In these cases, excellent stereochemical control is also possible, and it has been observed that the use of aromatic and aliphatic α,β-unsaturated aldehydes usually provides the *cis*-fused products with *dr* > 98%, whereas in the cases of simple aliphatic aldehydes the *trans*-annulated products are predominantly formed [374].

For the preparation of enantiopure products, chiral aldehydes, chiral 1,3-dicarbonyl compounds as well as chiral Lewis acids [375, 376] can be used.

A wide range of solvents can be employed. The most appropriate are acetonitrile, dichloromethane and toluene, but alcohols and water are also suitable. In those cases depending on the substrates an additional reaction such as the cleavage of formed lactones or acetals can be induced.

The *cis*-selectivity of the Diels–Alder reaction using aromatic aldehydes is demonstrated in the reaction of aldehydes such as **2-746**, which contain a dienophile moiety. Treatment with *N*,*N*-dimethylbarbituric acid (**2-747**) in the presence of ethylene diammonium diacetate at 20 °C led to the *cis*-fused product **2-749** exclusively in 95% yield (Scheme 2.164) [377]. As an intermediate, the benzylidene-1,3-dicarbonyl compound (*E*)-**2-748** is formed, which can be identified using online NMR-

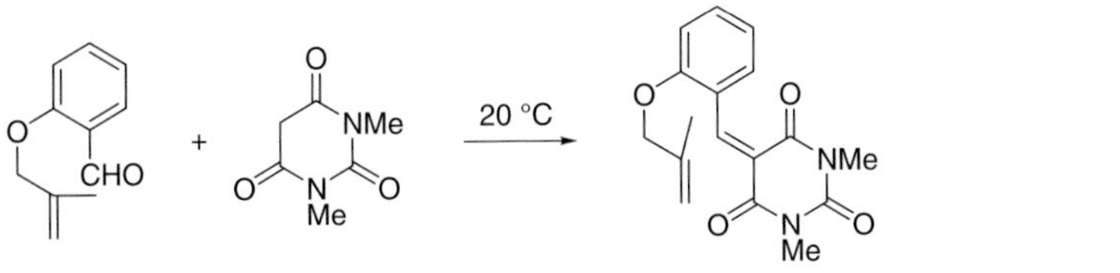

2-750   **2-747**   **2-751**

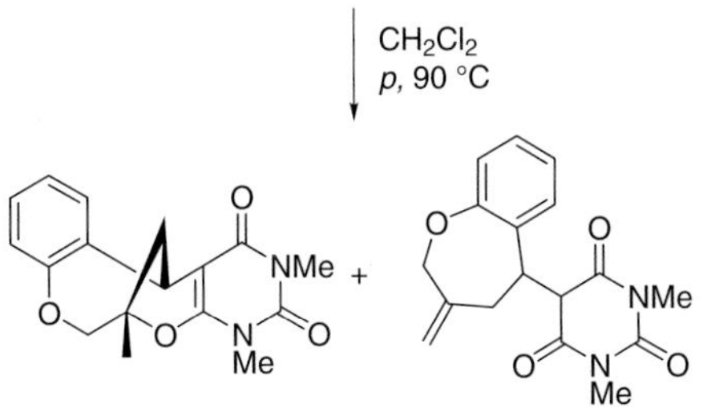

Pressure [MPa]	Selectivity (2-752:2-753)
75	19.5 : 1
100	23.5 : 1
320	40.7 : 1
550	76.3 : 1

$\Delta\Delta V^{\neq} = -(10.7 \pm 1.9)\ cm^3\ mol^{-1}$
$\Delta\Delta H^{\neq} = -(32.4 \pm 7.2)\ kJ\ mol^{-1}$

2-752   **2-753**

Scheme 2.165. Influence of pressure on the regioselectivity of the reaction of **2-751** in dichloromethane at 90°C.

spectroscopy. In the transition state an *endo-E-syn*-orientation is assumed; an *exo-Z-syn*-transition structure (Z)-**2-748** would lead to the same product, but this seems less likely due to steric interference.

The regioselectivity is controlled by the coefficients at the intermediately formed dienophile moiety. Thus, aldehydes of type **2-746** favor the formation of annulated compounds. However, with aldehyde **2-750** the bridged cycloadduct **2-752** is formed predominantly, in addition to small amounts of the ene product **2-753** *via* the 1-oxa-1,3-butadiene **2-751**. Interestingly, the preference for **2-752** can be improved by applying high pressure (Scheme 2.165) [378].

When using aromatic aldehydes such as **2-754** and pyrazolones such as **2-755** as 1,3-dicarbonyl compounds, a higher reaction temperature is necessary. The selectivity in these reactions depends on the substituent at the heteroaromatic compound and the substituents at the dienophile moiety (Scheme 2.166) [379].

In the Knoevenagel reaction using a pyrazolone with a bulky substituent at C-3, a (Z)-benzylidene-moiety is formed first due to a steric interaction of the substituent at the formed double bond and the substituent at C-3 of the pyrazolone. It could be proposed that the (Z)-1-oxa-1,3-butadiene undergoes a cycloaddition *via* an *exo-Z-syn* transition structure. However, it seems that this is less appropriate than the *endo-E-syn* transition structure. Thus, the Knoevenagel product first undergoes an (Z/E)-isomerization before the cycloaddition takes place to allow the formation of a *cis*-fused cycloadduct **2-756** *via* the *endo-E-syn* transition structure **2-758**. Thus, the reaction of **2-754a** and **2-755d** led to the isolable 1-oxa-1,3-butadiene **2-757** with a Z-configuration, which after isomerization at higher temperature yielded **2-756d**

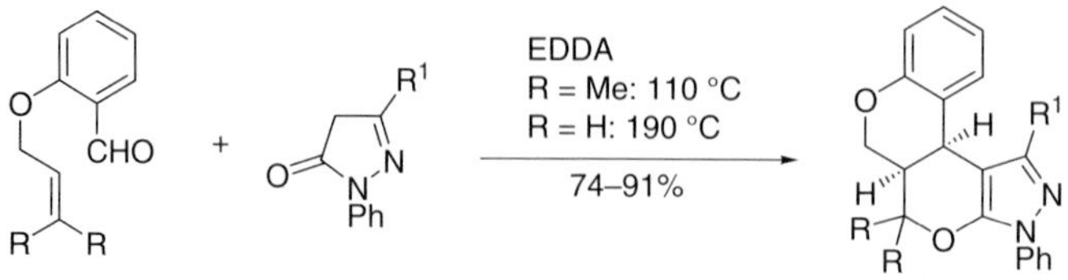

2-754 **2-755** **2-756**

a: R = Me
b: R = H

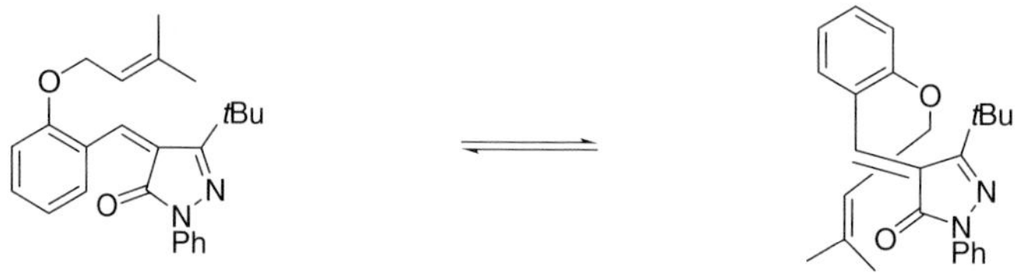

2-757 **2-758**

TS: endo-E-syn

2-754a + **2-755**	R^1	**2-756**: R = Me *cis*:*trans*
a	H	4.62:1
b	Me	16.75:1
c	Ph	4.49:1
d	*t*Bu	44.5:1

2-754b + **2-755**	R^1	**2-756**: R = H *cis*:*trans*
a	H	3.16:1
b	Me	23.2:1
c	Ph	10.2:1
d	*t*Bu	10.3:1

Scheme 2.166. Domino Knoevenagel/hetero-Diels–Alder reaction of pyrazolones

(R = Me) *via* **2-758** with an *E*-configuration. The transformation could be performed as a domino-process at 110 °C; however, under irradiation with UV-light, which facilitates the double bond isomerization, the cycloadduct was formed at only 40 °C.

By using a different length of tether between the aldehyde and the dienophile moiety in the aromatic or heteroaromatic substrates, a broad variety of different highly diversified heterocyclic compounds can be prepared. Thus, reaction of **2-759** and **2-755b** led to **2-760** containing a new 5,6-ring system, whereas reaction of **2-761** and **2-755b** gave **2-762** with a 7,6-ring system (Scheme 2.167) [380].

As examples of the influence of a stereogenic center in the aldehydes used on the induced diastereoselectivity, the reactions of **2-763a–d** with **2-746** to give the dihydropyrans **2-764a–d** were investigated. Moreover, the reactions show that the *trans*-cycloadducts are formed almost exclusively using aliphatic aldehydes (Scheme 2.168).

Chiral 1,3-dicarbonyl compounds such as **2-765** and **2-766** have also been used for the preparation of enantiopure products [381, 382]. In addition, chiral mediators such as **2-767** have been employed with great success (Scheme 2.169) [375, 376].

Moreover, an effective extension of the aforementioned procedure is the use of aldehydes derived from carbohydrates to yield polyhydroxylated condensed dihydropyrans, as shown by Yadav [383] and Gallos [384]. According to the latter ap-

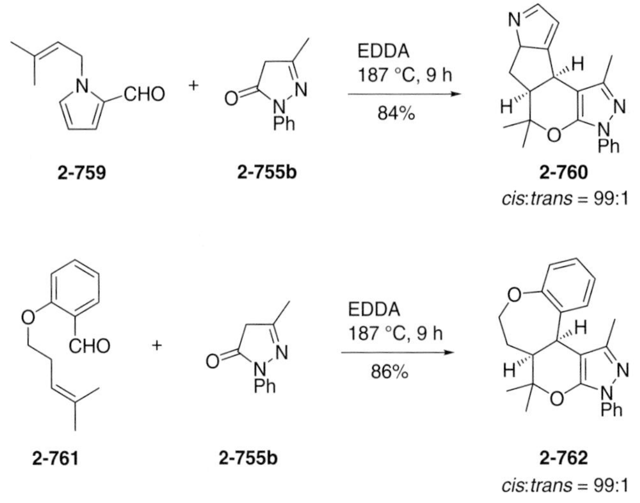

Scheme 2.167. Synthesis of diversified heterocycles by domino Knoevenagel/hetero-Diels–Alde reaction.

proach, good results were obtained when *N,N*-dimethylbarbituric acid (**2-747**) as when 1,3-dicarbonyl component was used (Scheme 2.170). As an example, the required aldehyde **2-769** was obtained from D-ribose *via* **2-768**. Condensation of **2-769** with *N,N*-dimethylbarbituric acid led to the enantiopure tricyclic compound **2-771** as a single diastereomer in 43 % yield based on the alcohol **2-768**.

It has been assumed that formation of the *cis*-fused product **2-771** in the domino reaction of aldehyde **2-769** is due to a strongly favored *exo-Z-syn* transition state **2-770**. The *endo-E-syn* structure is prohibited by the rigidity of the acetonide existing in **2-769**, whereas the proximity of the same moiety to the benzyloxymethyl substituent at the double bond disfavors the *exo-E-anti* transition state, which would be responsible for the formation of the *trans*-fused diastereomer.

Another attractive two-component domino Knoevenagel/hetero-Diels–Alder process employing the enantiopure aldehyde **2-773** and the hydroxypyridones **2-772** has been reported by the group of Snider (Scheme 2.171) [385]. The product, tricyclic compound **2-775**, was obtained in 35 % yield, and transformed into leporin A (**2-776**) [386].

4-Hydroxycoumarins and 4-hydroxyquinolinones have also been applied as 1,3-dicarbonyl compounds. Using these compounds, Raghunathan and coworkers prepared pyrano[3,2-*c*]coumarins [387] and pyranoquinolinones [388] under traditional conditions, while the group of Yadav synthesized similar pyrano[3,2-*c*]coumarins employing ionic liquids as solvents [389].

Other valuable substrates for the domino Knoevenagel/hetero-Diels–Alder reaction are chiral oxathiolanes such as **2-778**, which are easily accessible by condensation of 2-thioacetic acid and a ketone in the presence of *p*TsOH, followed by oxidation with hydrogen peroxide [390]. As described by Tietze and coworkers, the Knoevenagel condensation of **2-778** with aldehydes as **2-777** can be performed in

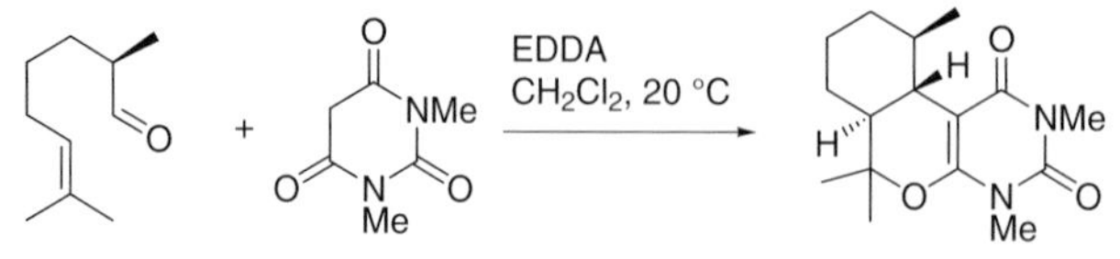

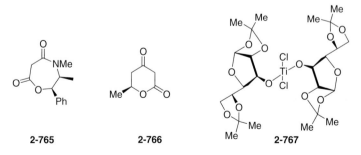

2-763a + **2-747** → EDDA CH$_2$Cl$_2$, 20 °C → **2-764a**
(*trans:trans:cis:cis =*
97.9:0:2.1:0)

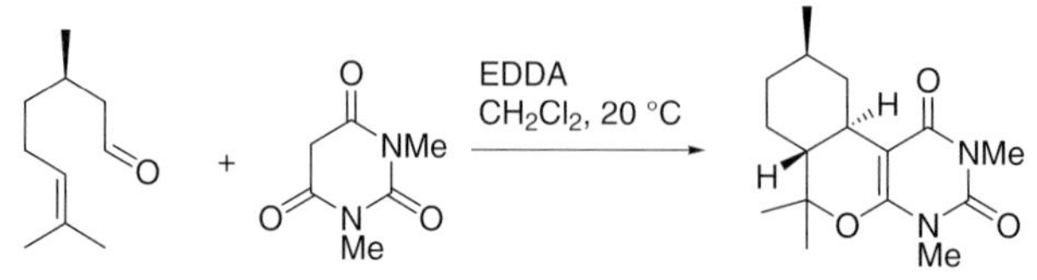

2-763b + **2-747** → EDDA CH$_2$Cl$_2$, 20 °C → **2-764b**
(*trans:trans:cis:cis =*
95.2:3.6:0.5:0.8)

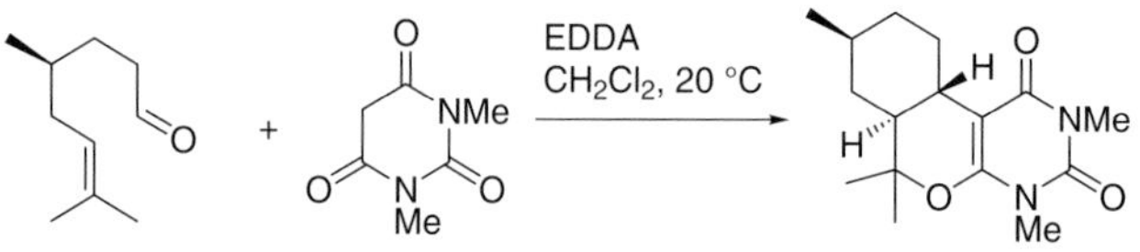

2-763c + **2-747** → EDDA CH$_2$Cl$_2$, 20 °C → **2-764c**
(*trans:trans:cis:cis =*
94.1:4.7:1.2:0)

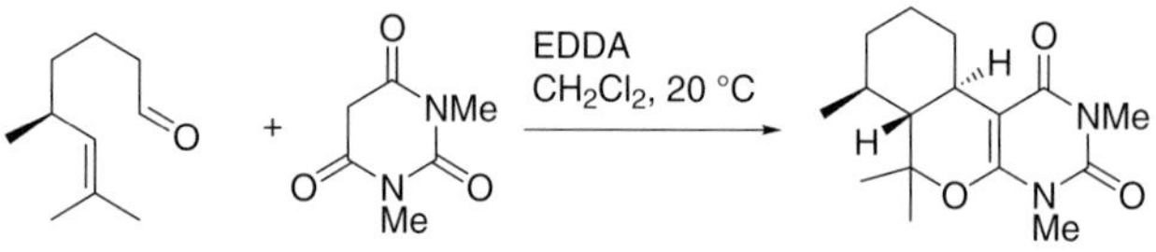

2-763d + **2-747** → EDDA CH$_2$Cl$_2$, 20 °C → **2-764d**
(*trans:trans:cis:cis =*
98.3:0:0.9:0.8)

Scheme 2.168. Diastereoselective domino Knoevenagel/hetero-Diels–Alder reaction with chiral aliphatic aldehydes.

2-765 **2-766** **2-767**

Scheme 2.169. Chiral 1,3-dicarbonyl compounds and chiral Lewis acids for the domino Knoevenagel/hetero-Diels–Alder reaction.

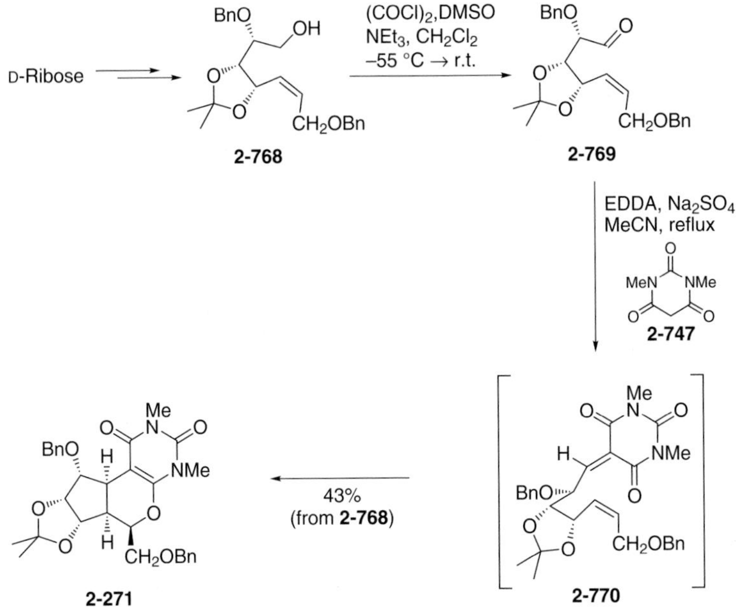

Scheme 2.170. Domino Knoevenagel/hetero-Diels–Alder reaction of aldehydes derived from carbohydrates.

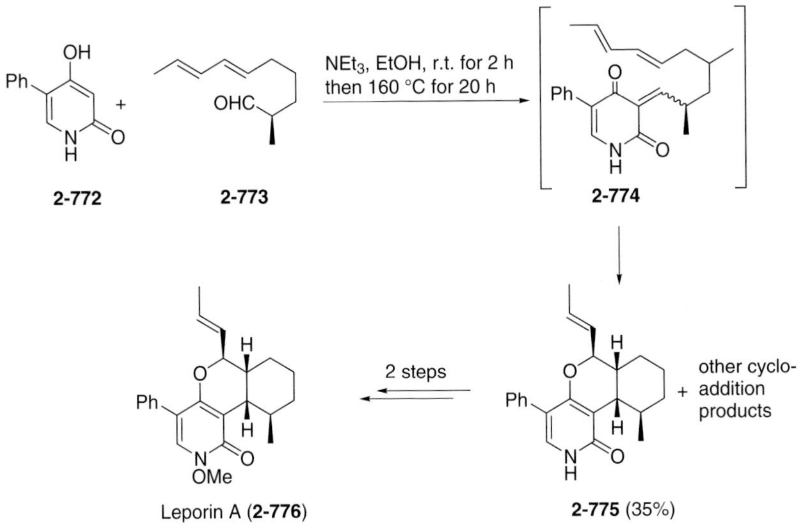

Scheme 2.171. Synthesis of leporin A (**2-776**).

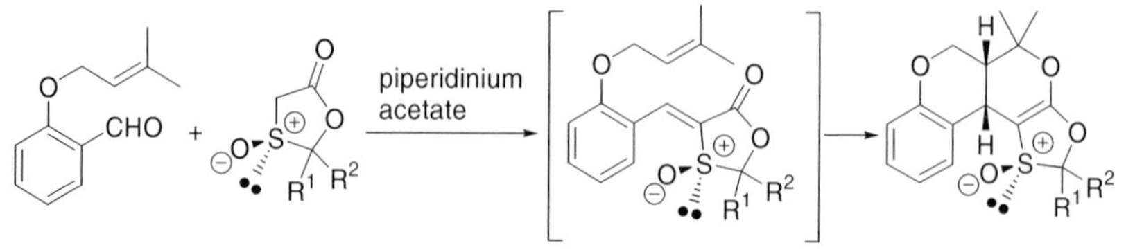

Scheme 2.172. Domino Knoevenagel/hetero-Diels–Alder reaction of oxathiolanes.

dichloromethane in the presence of catalytic amounts of piperidinium acetate with azeotropic removal of water to give the benzylidene compound **2-779** in good yields and high *Z*-selectivity (Scheme 2.172). The cycloaddition takes place at 82 °C or even with better selectivity at room temperature by addition of ZnBr$_2$. In the latter case, a single compound **2-780** with *cis*-annulation of the ring system and *anti*-orientation of the aryl moiety to the oxygen of the sulfoxide was obtained in 78 % yield starting from acetone for the preparation of **2-778** (R^1 = R^2 = Me). The transition structures have been calculated [390, 391].

So far, only those domino Knoevenagel/hetero-Diels–Alder reactions have been discussed where the cycloaddition takes place at an intramolecular mode; however, the reaction can also be performed as a three-component transformation by applying an intermolecular Diels–Alder reaction. In this process again as the first step a Knoevenagel reaction of an aldehyde or a ketone with a 1,3-dicarbonyl compound occurs. However, the second step is now an intermolecular hetero-Diels–Alder reaction of the formed 1-oxa-1,3-butadiene with a dienophile in the reaction mixture. The scope of this type of reaction, and especially the possibility of obtaining highly diversified molecules, is even higher than in the case of the two-component transformation. The stereoselectivity of the cycloaddition step is found to be less pronounced, however.

The group of Cravotto and Palmisano have accessed substituted coumarin derivatives of type **2-781** using chiral enol ethers **2-783** as dienophiles [392]. Such compounds are widely distributed in Nature and are reported to have various biological activities such as anticoagulant, insecticidal, anthelmintic, hypnotic, antifungal, and phytoalexin properties, and are also known to be HIV protease inhibitors [393]. For example, warfarin (**2-785a**) is today the dominant coumarin anticoagulant owing to its excellent potency and good pharmacokinetic profile (Scheme 2.173). While its marketed form is the racemic sodium salt (Coumadin®), the anticoagulant activity of the (*S*)-(–)-enantiomer is known to be six times higher than that of the (+)-enantiomer [394]. For the synthesis of enantiopure warfarin (**2-785a**), isopropenyl ether **2-783** derived from commercially available (–)-(1*R*,2*S*,5*R*)-menthol, hydroxycoumarin (**2-781**) and benzaldehyde (**2-782a**) were used. The cycloadduct **2-784** was obtained with an *endo/exo*-selectivity of 4.1:1 and an induced diastereoselectivity of 88:12. Treatment of **2-784** with trifluoroacetic acid/water (19:1) provided (*S*)-warfarin (**2-785a**) in an overall yield of 61 %, referred to 4-hydroxycoumarin (**2-781**) and an enantiomeric excess of 76 % (by HPLC). This could be increased to 95 % *ee* by recrystallization using the purified *endo*-product as substrate

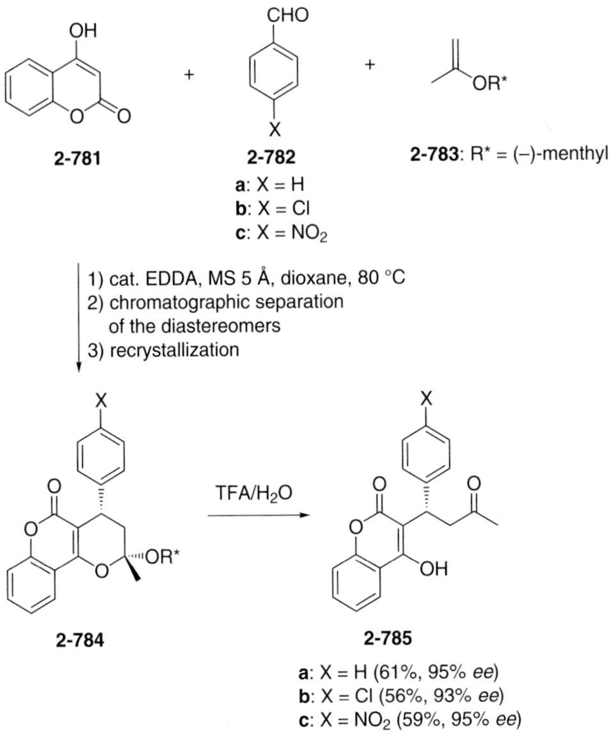

2-781 + 2-782 + 2-783: R* = (–)-menthyl

a: X = H
b: X = Cl
c: X = NO₂

1) cat. EDDA, MS 5 Å, dioxane, 80 °C
2) chromatographic separation
   of the diastereomers
3) recrystallization

2-784   TFA/H₂O   2-785

a: X = H (61%, 95% ee)
b: X = Cl (56%, 93% ee)
c: X = NO₂ (59%, 95% ee)

Scheme 2.173. Asymmetric synthesis of coumarin anticoagulants.

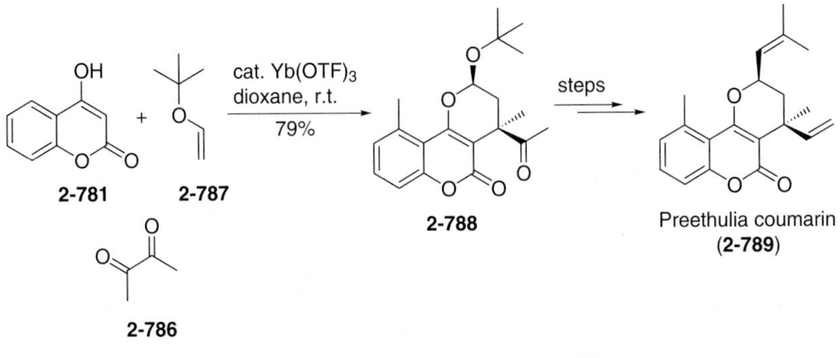

2-781 + 2-787   cat. Yb(OTF)₃
dioxane, r.t.
79%

2-786

2-788   steps   Preethulia coumarin
(2-789)

Scheme 2.174. Synthesis of a preethulia coumarin precursor 2-788.

for the hydrolysis (Scheme 2.173). In the same manner, (S)-coumachlor (2-785b) and (S)-acenocoumarol (2-785c) were obtained with 56 % yield and 93 % ee, or 59 % yield and 95 % ee, respectively.

In a similar manner, reaction of enol ether 2-787, hydroxycoumarin (2-781), and α-diketone 2-786 led to the cycloadduct 2-788 in 79 % yield using Yb(OTf)₃ as catalyst (Scheme 2.174). 2-788 could be transformed into the natural product preethulia coumarin (2-789).

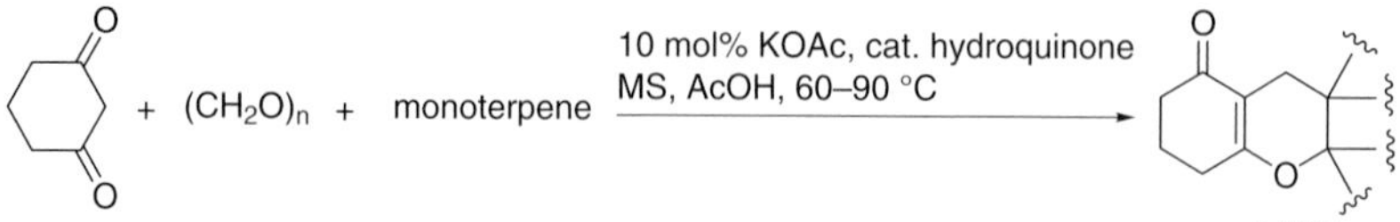

Entry	Monoterpene	Product **2-790**	dr	Yield [%]
1	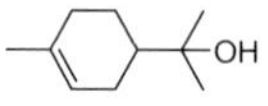 α-Terpineol	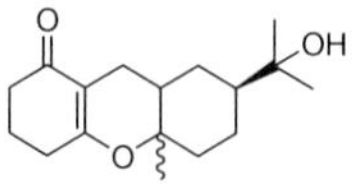	2.7:1	69
2	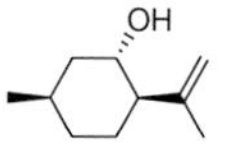 (−)-Isopulegol	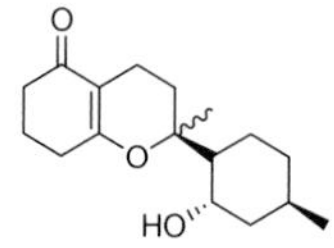	1:1	50
3	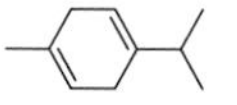 γ-Terpinene	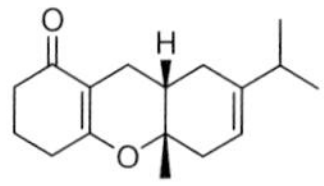	single isomer	47
4	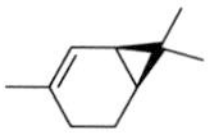 (+)-2-Carene	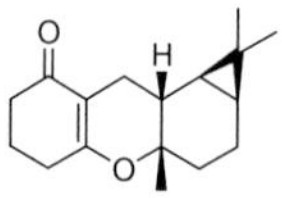	single isomer	51
5	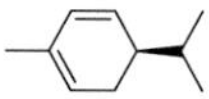 (+)-α-Phellandrene	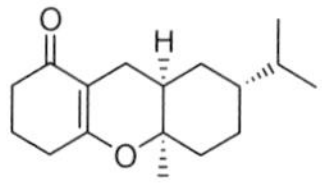	single isomer	49

Scheme 2.175. Domino Knoevenagel/hetero-Diels–Alder reaction for the construction of "polyketide" terpenes.

Hoffmann and coworkers have elaborated an appealing three-component dom-ino Knoevenagel/hetero-Diels–Alder procedure [395]. These authors converted cy-clohexane-1,3-dione, formaldehyde and several as dienophiles acceptor serving monoterpene into a variety of "polyketide" terpenes of type **2-790** (Scheme 2.175).

In addition to aldehydes and α-diketones, α-ketoesters can also be used in the domino process, as shown by Tietze and coworkers [396]. Reaction of methyl pyru-vate **2-791** with dimethylbarbituric acid (**2-747**) and the enol ether **2-792** in the pre-sence of trimethyl orthoformate (TMOF) and a catalytic amount of EDDA gave the cycloadduct **2-793** in 84 % yield (Scheme 2.176).

In a similar transformation using 4-hydroxycoumarin (**2-781**) as the 1,3-dicar-bonyl compound the cycloadduct **2-794** was obtained also in good yield. In order to demonstrate the general applicability of this process, a small library using substi-tuted pyruvate was prepared without optimizing the reaction conditions for the single transformations. α-Ketonitrile can also be used, though with a much lower yield.

The three-component domino Knoevenagel/hetero-Diels–Alder-reaction is espe-cially fruitful if one uses aldehydes containing a protected amino function. In such

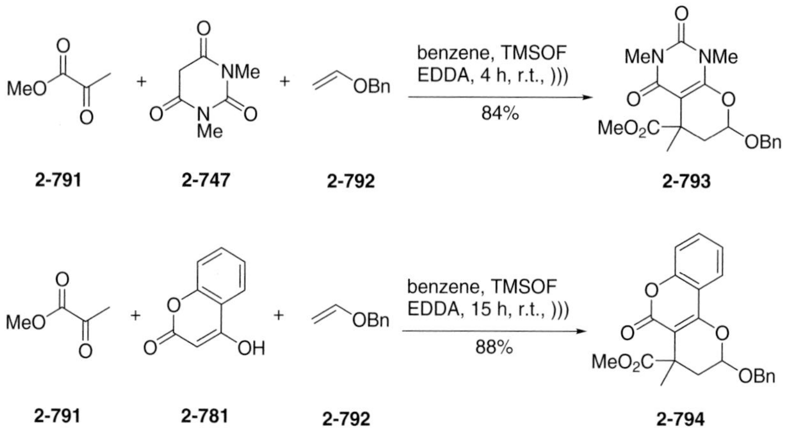

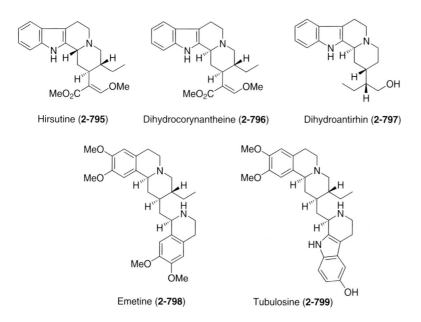

Scheme 2.176. Domino Knoevenagel/hetero-Diels–Alder reaction with pyruvate.

Hirsutine (**2-795**)   Dihydrocorynantheine (**2-796**)   Dihydroantirhin (**2-797**)

Emetine (**2-798**)   Tubulosine (**2-799**)

Scheme 2.177. Alkaloids synthesized by a three-component domino
Knoevenagel/hetero-Diels–Alder reaction.

a case the formed dihydropyranyl ether moiety can be used as a source of an aldehyde moiety, which can undergo a condensation with the amino group after deprotection. Thus, several alkaloids such as hirsutine (**2-795**), dihydrocorynantheine (**2-
796**), dihydroantirhin (**2-797**), emetine (**2-798**) and tubulosine (**2-799**) have been
synthesized using this approach (Scheme 2.177). In addition, two new concepts in

combinatorial chemistry were developed based on this type of domino Knoevenagel/hetero-Diels–Alder-reaction.

Hirsutine (**2-795**), which belongs to the corynanthe subgroup of the indole alkaloids, was isolated from the plant *Uncaria rhynchophylla* MIQ and used for the preparation of the old Chinese folk medicine "Kampo" [397]. It is of pharmacological interest as it shows a strong inhibitory effect on the influenza A virus (subtype H3N2) with an $EC_{50} = 0.40–0.57\ \mu g\ mL^{-1}$, which is about 11- to 20-fold higher than that of the clinically used ribavirin [398].

Reaction of the enatiopure aldehyde **2-800**, obtained from the corresponding imine by enantioselective hydrogenation, with Meldrum's acid (**2-801**) and the enol ether **2-802a** ($E/Z = 1:1$) in the presence of a catalytic amount of ethylene diammonium diacetate for 4 h gave **2-805** in 90% yield with a 1,3 induction of >24:1. As intermediates, the Knoevenagel product **2–803** and the primarily produced cycloadduct **2-804** can be supposed; the latter loses CO_2 and acetone by reaction with water formed during the condensation step (Scheme 2.178).

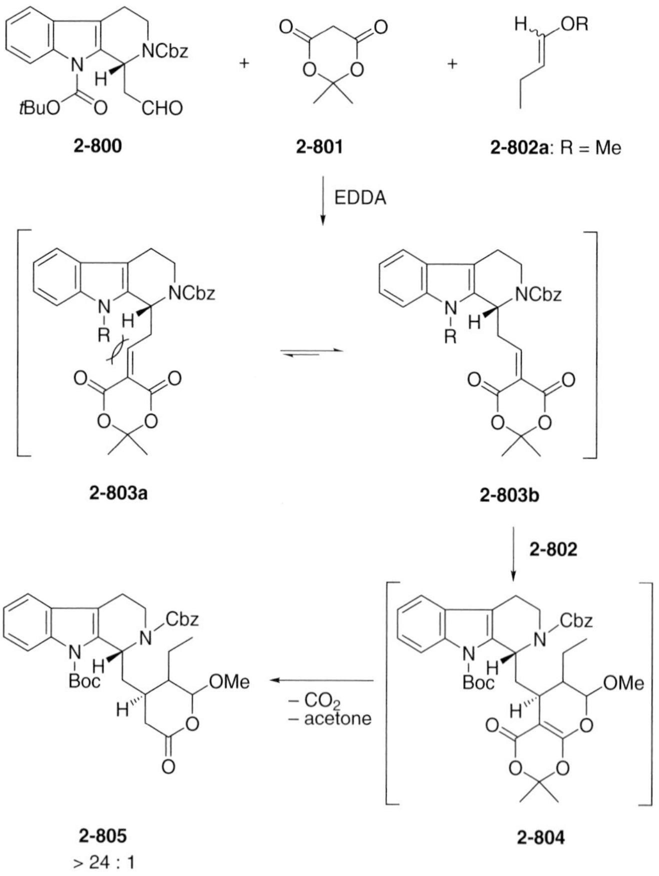

2-800   2-801   2-802a: R = Me

2-803a   2-803b

2-805   2-804
> 24 : 1

Scheme 2.178. Domino Knoevenagel/hetero-Diels–Alder reaction of **2-800**, **2-801** and **2-802a**.

Solvolysis of crude **2-805** with methanol in the presence of K_2CO_3 led to an opening of the lactone moiety with the formation of a methyl ester and a hemiacetal, which loses methanol to give **2-806a** containing an aldehyde. Under the following hydrogenolytic conditions the carbobenzoxy group at N-4 is removed to form the secondary amine **2-806b**, which reacts with the aldehyde moiety to give an enamine. Under these reaction conditions the enamine is hydrogenated to produce the indoloquinolizidine **2-807** as a single diastereomer in enantiopure form in a stereoelectronically controlled reaction *via* a chair-like transition state (Scheme 2.179).

The synthesis of (–)-hirsutine (**2-795**) was concluded by removal of the Boc-group, condensation with methyl formate, and methylation of the formed enol moiety. In a similar manner as was described for **2-795**, (+)-dihydrocorynantheine (**2-797**) [399] with the (3S)- and (15R)-configuration was synthesized from *ent*-**2-800**.

The described approach also allows a simple access to indole alkaloids of the vallesiachotamine type. In this process, the Cbz-protected secondary amino function in the formed cycloadducts such as *ent*-**2-805** is deprotected by hydrogenolysis. There follows an attack at the lactone moiety to form a lactam. In this way, the indole alkaloid (–)-dihydroantirhin (**2-797**) was prepared [400, 401].

Another class of alkaloids recently synthesized by Tietze and coworkers using a three-component domino Knoevenagel/hetero-Diels–Alder reaction included the *Ipecacuanha* alkaloids such as emetine (**2-798**) [402], and the *Alangium* alkaloids such as tubulosine (**2-799**) [403]. Both types belong to the group of tetrahydroisoquinoline alkaloids, and are formed in Nature from dopamine and the monoter-

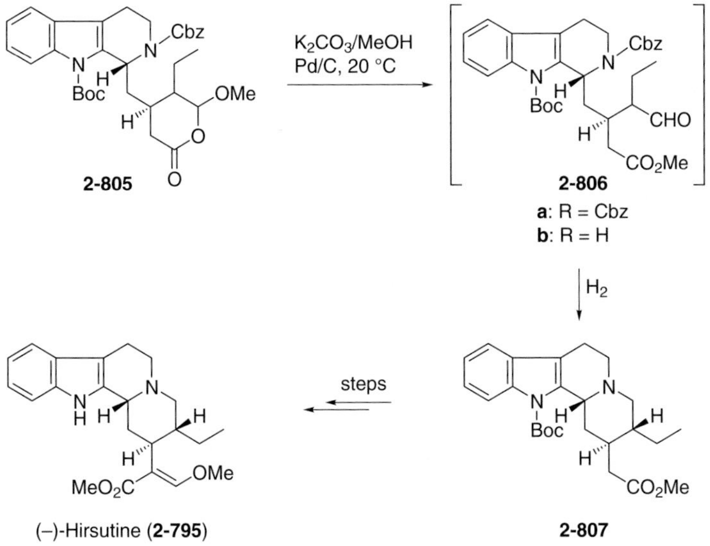

Scheme 2.179. Synthesis of the indole alkaloid (–)-hirsutine (**2-795**).

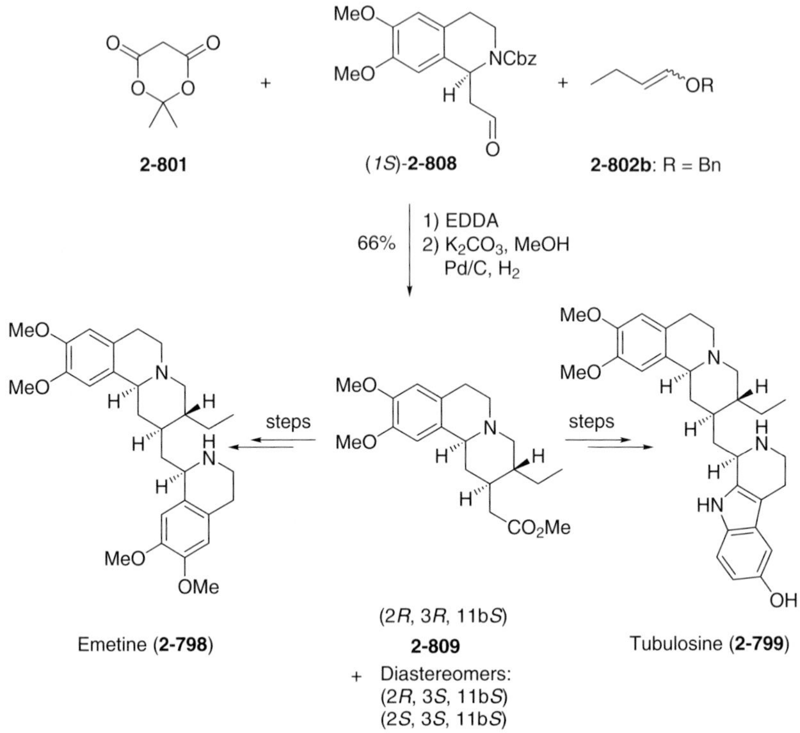

Scheme 2.180. Domino process for the synthesis of the benzoquinolizidine **2-809**.

pene secologanin. Emetine (**2-798**) was isolated from *Radix ipecacuanha* and the roots of *Psychotria ipecacuanha* and *Cephalis acuminata*, and possesses many interesting biological activities [404]. Emetine shows antiprotozoic properties and activity in the treatment of lymphatic leukemia; furthermore, as its name suggests, it was utilized as an emetic. Tubulosine (**2-799**) was isolated from the dried fruits of *Alangium lamarckii* and the sap of *Pogonopus speciosus*. It is remarkably active against several cancer cell lines, and has been studied for various other biological activities, such as the inhibition of protein biosynthesis and HIV reverse transcriptase inhibitory activities [405].

The domino reaction of (1*S*)-**2-808**, Meldrum's acid (**2-801**) and enol ether **2-802b** in the presence of EDDA, followed by treatment with K$_2$CO$_3$/MeOH and a catalytic amount of Pd/C in methanol under a nitrogen atmosphere for 50 min and subsequently under a H$_2$-atmosphere for 2 h at r.t. gave the benzoquinolizidine **2-809** with the correct stereochemistry at all stereogenic centers as in emetine (**2-798**) and tubulosine (**2-799**), together with two diastereomers (Scheme 2.180) [406]. Further manipulations of **2-808** led to emetine (**2-798**) and tubulosine (**2-799**).

Instead of the usual 1,3-dicarbonyl compounds, heteroanalogues such as the corresponding α-carbonylated phosphonates **2-810** can also be used in the Knoevenagel/hetero-Diels–Alder process. As in the case of carbonyl groups at the

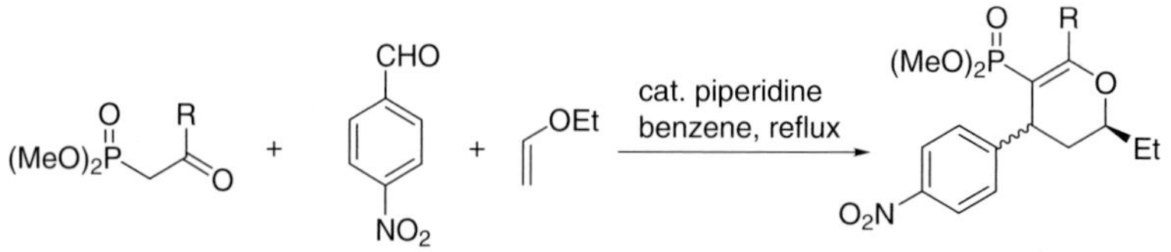

2-810a: R = CO₂Et
2-810b: R = CONEt₂

2-811

2-812a: R = CO₂Et
(87%, *cis/trans* 76:24)
2-812b: R = CONEt₂
(91%, *cis/trans* 78:22)

Scheme 2.181. Synthesis of dihydropyrans **2-812**.

3-position of 1-oxa-1,3-butadienes, a phosphono substituent also enhances the reactivity of the diene by lowering the energy of the LUMO. However, the transformations must be carried at about 80 °C, compared to r.t. when using oxabutadienes with a carbonyl compound [372b, 407]. Collignon and coworkers synthesized such 5-phosphono-substituted dihydropyrans **2-812** from **2-810**, the aldehyde **2-811** and ethyl vinyl ether (Scheme 2.181) [408]. However, using the phosphonopyruvate **2-810a**, the dienophile had to be added after complete formation of the intermediate Knoevenagel product in order to avoid the formation of byproducts. In contrast, the phosphonopyruvate **2-810b** could be used as a mixture with **2-811** and an enol ether. In both examples, the yields of the isolated cycloadducts **2-812a** and **2-812b** were high and much better than for a corresponding stepwise procedure.

Thioesters of 3-phosphonothio carboxylic acids can also be used with great success in this process to produce 5-phophono-dihydrothiopyrans in yields over 80 % with *dr* > 80:20.

Recently, a first example of an organocatalytic asymmetric domino Knoevenagel/Diels–Alder reaction was reported by Barbas and coworkers (Scheme 2.182) [409]. Spiro[5,5]undecane-1,5,9-triones of type **2-818/2-819** were obtained from commercially available 4-substituted-3-butene-2-ones **2-813**, aldehydes **2-814**, and Meldrum's acid (**2-801**) in the presence of 20 mol% of the amino acid **2-815** with 80–95 % *ee* and diastereoselectivity of *dr* > 12:1.

The domino process probably involves the chiral enamine intermediate **2-817** formed by reaction of ketone **2-813** with **2-815**. With regard to the subsequent cycloaddition step of **2-817** with the Knoevenagel condensation product **2-816**, it is interesting to note that only a "normal" Diels–Alder process operates with the 1,3-butadiene moiety in **2-817** and not a hetero-Diels–Alder reaction with the 1-oxa-1,3-butadiene moiety in **2-816**. The formed spirocyclic ketones **2-818/2-819** can be used in natural products synthesis and in medicinal chemistry [410]. They have also been used in the preparation of exotic amino acids; these were used to modify the physical properties and biological activities of peptides, peptidomimetics, and proteins [411].

Besides the combination of an anionic with a Diels–Alder reaction, a combination with a 1,3-dipolar cycloaddition is also possible. There are hundreds of examples of this type, since 1,3-dipoles are usually always prepared *in situ* in the presence of a dipolarophile. Here, only a few more recent publications are pre-

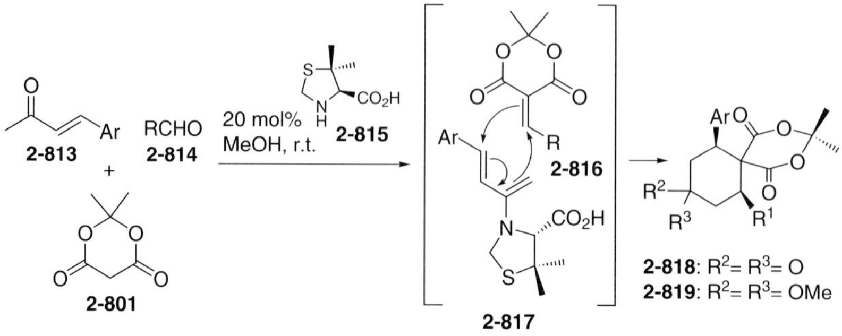

2-813 **2-814** **2-815** **2-816** **2-817**

2-801

2-818: $R^2 = R^3 = O$
2-819: $R^2 = R^3 = OMe$

20 mol% MeOH, r.t.

Entry	Enone 2-813: Ar	Aldehyde 2-814: R	Product ratio 2-818/2-819	ee [%]	Yield [%]
1	a: Ph	a: $pNO_2C_6H_4$	a: 13:1	86	95
2	a: Ph	b: $pCNC_6H_4$	b: 16:1	84	85
3	b: 1-Naphthalenyl	a: $pNO_2C_6H_4$	c: 100:1	99	93
4	c: 2-Furanyl	a: $pNO_2C_6H_4$	d: 12:1	88	92
5	d: 1-Thiophenyl	a: $pNO_2C_6H_4$	e: 15:1	99	80

Scheme 2.182. Amino acid-catalyzed asymmetric three-component Knoevenagel/Diels–Alder reactions.

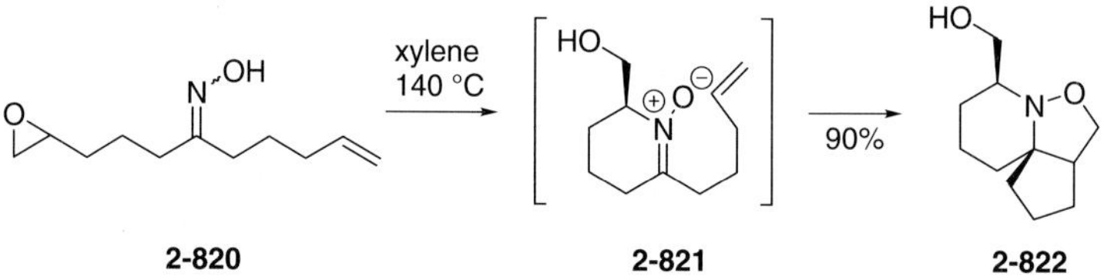

2-820 **2-821** **2-822**

xylene 140 °C

90%

Scheme 2.183. Examples of the domino nitrone formation/1,3-dipolar cycloaddition.

sented, which show the general approach. For example, Grigg's group has reported on regioselective nucleophilic ring-opening reactions of epoxides by oximes leading to nitrones which can be trapped by dipolarophiles to give isoxazolidines [412]. The ring-opening reaction and cycloaddition can be either inter- or intramolecular.

Heating of a racemic mixture of **2-820** in xylene at 140 °C led to the diastereomerically pure cycloadduct **2-822** *via* **2-821** in two intramolecular processes in a yield of 90 % (Scheme 2.183). Clearly, enantiopure isoxazolidines can also be obtained starting from enantiopure epoxides [412a].

Enantioenriched (–)-rosmarinecine, which belongs to the group of pyrrolizidine alkaloids [413], has been synthesized by Goti, Brandi and coworkers applying an intramolecular 1,3-dipolar cycloaddition as the key step [414]. The required nitrone was obtained *in situ* from L-malic acid. Moreover, 1,3-dienes as precursors for a cy-

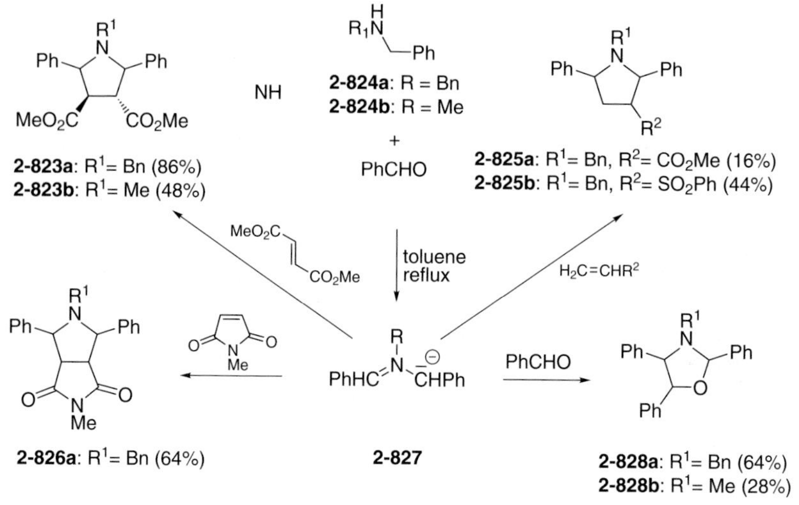

Scheme 2.184. Synthesis of diverse five-membered heterocycles *via* domino 1,3-dipole formation/cycloaddition.

cloaddition can also be obtained by a simple base-induced elimination of, for example, homoallylic mesylates such as **2-875** (Scheme 2.195).

In addition to nitrones, azomethine ylides are also valuable 1,3-dipoles for five-membered heterocycles [415], which have found useful applications in the synthesis of, for example, alkaloids [416]. Again, the groups of both Grigg [417] and Risch [418] have contributed to this field. As reported by the latter group, the treatment of secondary amines **2-824** with benzaldehyde and an appropriate dipolarophile leads to the formation of either substituted pyrrolidines **2-823**, **2-825** and **2-826** or oxazolidines **2–828** with the 1,3-dipole **2-827** as intermediate (Scheme 2.184). However, the yields and the diastereoselectivities are not always satisfactory.

Several other anionic/pericyclic domino processes use a Horner–Wadsworth–Emmons (HWE) or a Wittig olefination as the first step.

A sequence involving a Diels–Alder reaction as the second step has been used by Jarosz and coworkers for the synthesis of enantiomerically pure oxygenated perhydroindene derivatives such as **2-832** with a *trans* junction of the five- and the six-membered ring (Scheme 2.185) [419]. Olefination of the xylose derivative **2-829** with the mannose-derived keto-phosphonate **2-830**, having the (*S*)-configuration at the α-position led to the enone **2-831**, which underwent an intramolecular Diels–Alder reaction to give **2-832** in 80% yield as a single stereoisomer. It appears that the stereochemical outcome of the cycloaddition is mostly controlled by the stereogenic center at the α-position in the keto-phosphonate, as compounds with an (*R*)-configuration at that center mainly lead to products with the opposite stereochemistry at the newly formed stereogenic centers, despite the configuration of the other stereogenic centers in the substrates.

Interestingly, *cis*-annulated decalins can also be formed using a slightly modified protocol [420].

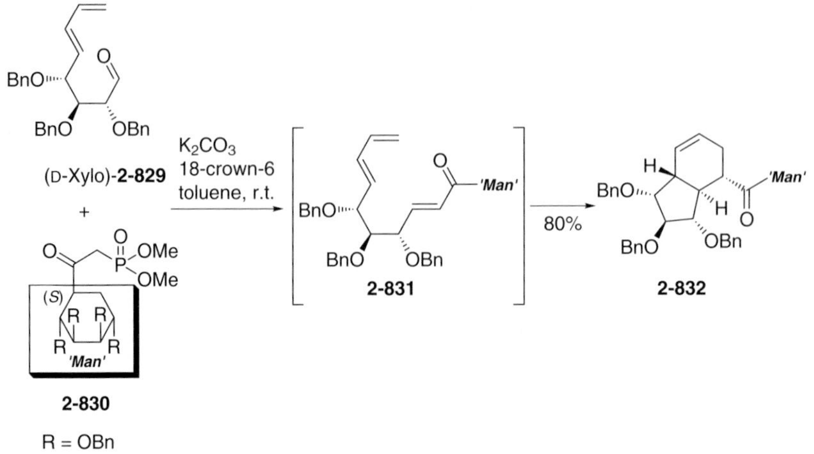

Scheme 2.185. Synthesis of enantiopure perhydroindenes **2-832** *via* domino HWE/Diels–Alder reaction.

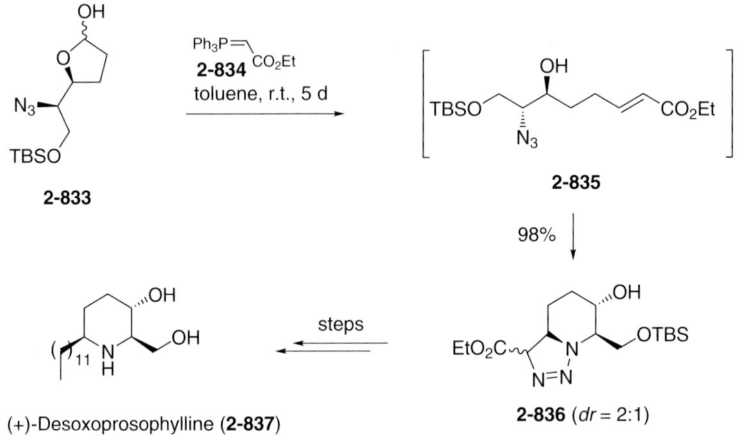

Scheme 2.186. Synthesis of (+)-desoxoprosophylline (**2-837**).

A combination of a Wittig reaction and a 1,3-dipolar cycloaddition was used by Herdeis and coworkers for the synthesis of hydroxy piperidine derivatives, again starting from an enantiopure natural product [421]. In a recent example, these authors prepared (+)-desoxoprosophylline (**2-837**) [422] using compound **2-833** as substrate which is easily accessible from L-ascorbic acid [2a]. Treatment of **2-832** with phosphorane **2-834** in dry toluene at r.t. afforded the triazoles **2-836** after 5 days in near-quantitative yield *via* the intermediate **2-835**. The long reaction is due to the sluggishness of the 1,3-dipolar cycloaddition, as the Wittig reaction is completed after only 80 min (Scheme 2.186).

The intramolecular cycloaddition of **2-835** led to a mixture of two diastereomers, both of which can be used for the synthesis of (+)-desoxoprosophylline (**2-837**) as the corresponding stereogenic center will be destroyed during the forthcoming steps.

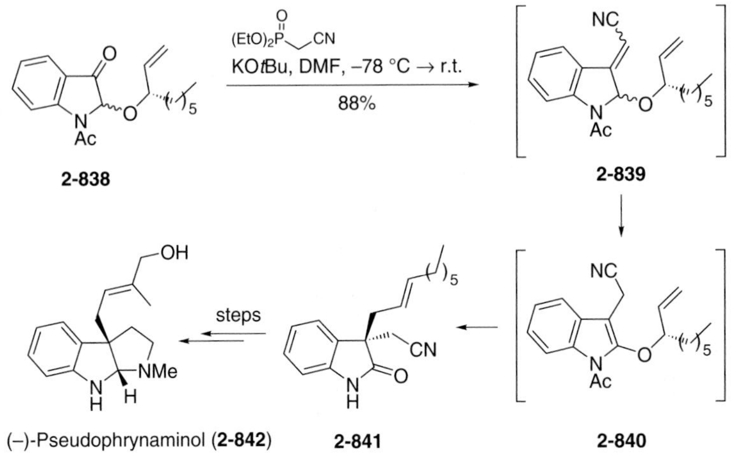

Scheme 2.187. Synthesis of (–)-pseudophrynaminol (**2-842**).

Kawasaki, Sakamoto and coworkers have utilized a HWE reaction in combination with a Claisen rearrangement to introduce an allylic moiety into hexahydropyrrolo[2,3-*b*]indole alkaloids [423]. Among this class of compounds are included amauromine [424], ardeemins [425], aszonalenin [426], flustramines [427], roquefortine [428], and pseudophrynamines [429]. As an example, pseudophrynaminol (**2-842**) has been prepared by olefination of a diastereomeric mixture of 2-allyloxy-indol-3-one **2-838** with diethyl cyanomethylphosphonate in the presence of KO*t*Bu. Initially, **2-839** is formed, which on warming leads to **2-840** with an allyl vinyl ether moiety. This is an ideal precursor for a Claisen rearrangement to give **2-841**, which was transformed into the desired natural product **2-842** (Scheme 2.187). The observed high selectivity of the Claisen rearrangement leading to **2-841** with 97 % *ee* can be attributed to a chair-like transition state in the reaction of enantiopure **2-840**.

Fused imidazopyridines have been prepared by Palacios and coworkers using a domino aza-Wittig/1,5-electrocyclic ring-closure process [430]. This heterocyclic moiety is found in alkaloids and in bioactive compounds [431]; thus, this structural element is part of potent short-acting neuromuscular blocking agents [432] of reversible inhibitors of the H⁺, K⁺-ATPase enzyme [433] with high antisecretory activity [434] and of sedative hypnotics of the nervous system [435]. The imidazo[1,5-*a*]pyridine skeleton is a basic element of the drug Pirmogrel, which has clinical applications as an effective inhibitor of platelet aggregation and thromboxane synthetase [436]. For the synthesis of imidazo[1,5-a]pyridines such as **2-845**, phosphazene **2-843** was treated with an aldehyde to induce an aza-Wittig olefination which leads to the intermediate **2-844** (Scheme 2.188). There follows a 1,5-electrocyclic ring-closure process to produce the desired imidazo-pyridine **2-845**, with good overall yields.

Of note, Quintela and coworkers used a related aza-Wittig/1,5-electrocyclic ring-closure process to obtain access to multiheterocyclic structural elements such as py-

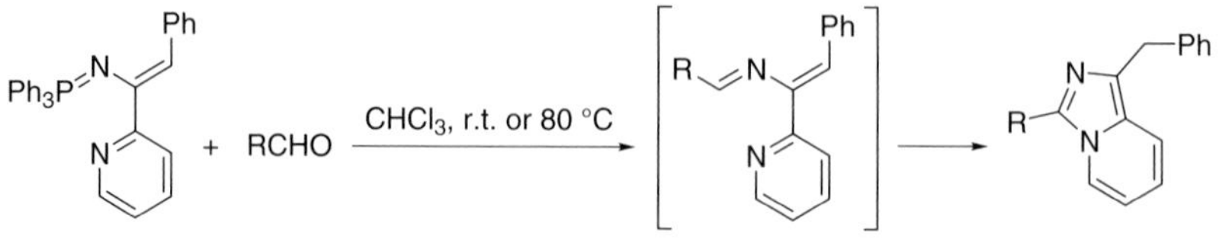

2-843 2-844 2-845

Scheme 2.188. Synthesis of imidazo[1,5-*a*]pyridines.

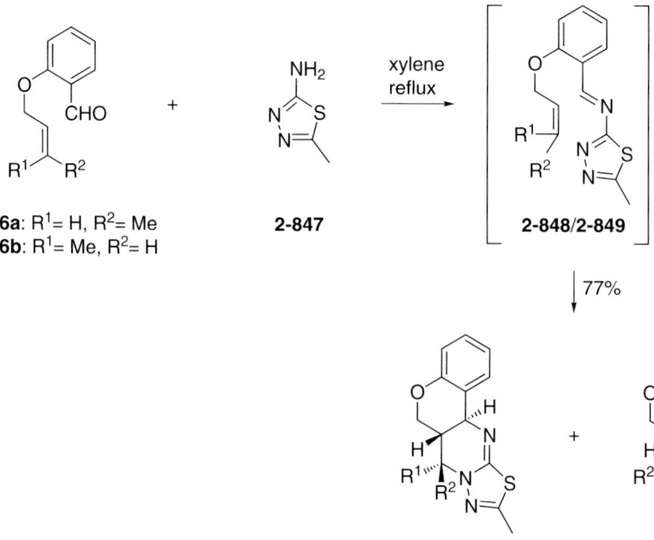

2-846a: R^1= H, R^2= Me
2-846b: R^1= Me, R^2= H

2-847 2-848/2-849 2-851 2-850

Scheme 2.189. Formation of a 2-aza-1,3-butadiene and cycloaddition.

ridothienopyridazines and pyrimidothienopyridazines [437]. In those reactions, the phosphazene was coupled with isocyanates, isothiocyanates as well as CO_2 and CS_2. Besides their formation by an aza-Wittig reaction imines can also be prepared by a simple reaction of aldehydes with primary amines. This was utilized by Tietze and coworkers to prepare 2-aza-1,3-butadienes which then can undergo a hetero-Diels–Alder reaction [438].

Thus, reaction of the aldehyde **2-846** with the aminothiadiazol **2-847** led to 1,3-diaza-1,3-butadienes **2-848** and **2-849** as intermediates which can undergo a cycloaddition to give *trans*-annulated products **2-849** and **2-850** (Scheme 2.189). The authors have shown that the cycloaddition does not proceed in a concerted manner, since starting from the (*E*)-aldehyde **2-846a** and the (*Z*)-aldehyde **2-846b**, respectively, gave the same product mixture. Laschat and coworkers have extended the process using anilines in the reaction with the aldehydes of type **2-846**, with good success [439].

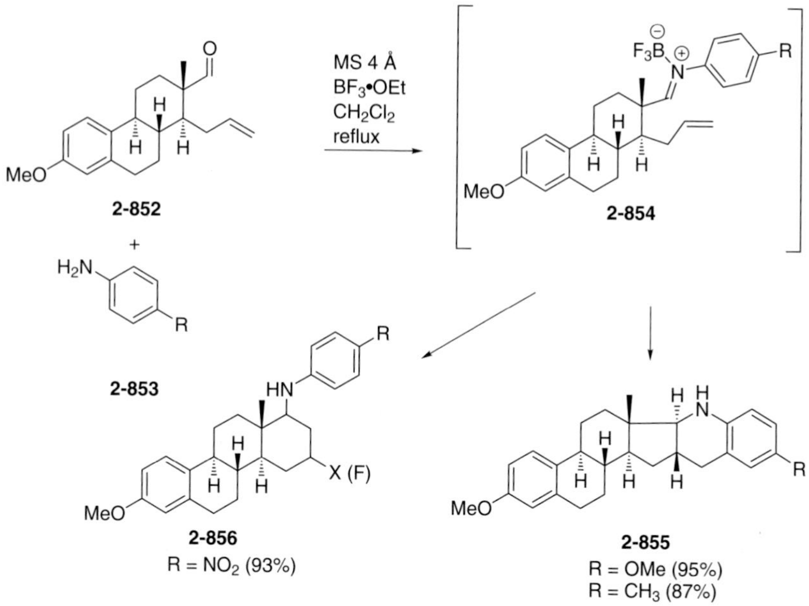

Scheme 2.190. Synthesis of steroid alkaloids.

Novel unusual heterocyclic steroids **2-855** and **2-856** have been prepared by the groups of Schneider, Wölfling and Tietze in a joint project (Scheme 2.190). Condensation of the secoestrone aldehyde **2-852** with aniline and derivatives **2-853** in the presence of a Lewis acid led to the iminium ion **2-854** which underwent cycloaddition to give the hexacyclic steroid alkaloids **2-855**. The cycloaddition, however, only occurs if aniline or aniline derivates **2-853** with electron-donating groups are used. With anilines **2-853**, which contain electron-withdrawing substituents, D-homosteroids of type **2-856** are obtained, where X depends upon on the Lewis acid applied; for example, in the case of $BF_3 \cdot OEt_2$, X is F.

A completely different course of reaction is found if a 16,17-dihydro-**2-852** with a propyl instead of a propenyl side chain is used. In this reaction, an unusual hydride shift takes place (see Section 2.1.1).

Interesting octahydroacridines **2-860** have been prepared by Beifuss and coworkers by combining the condensation step with a rare intramolecular polar $[4\pi^+ + 2\pi]$-cyclization of α-aryliminium ions **2-859**, obtained from anilines **2-857** by reaction with the ω-unsaturated aldehyde **2-858** (Scheme 2.191) [440]. The overall domino process seems to be stereoselective, since the formation of the two diastereomers **2-860** can be traced to the use of the substrate **2-858** as a diastereomeric mixture.

In a recent publication, Perumal and coworkers [441] described the condensation of an aldehyde **2-863** with an aniline **2-864** to give an imine which is trapped by a dienophile. However, when using this approach an intermolecular cycloaddition takes place as the reaction is performed as a three-component process using enol ethers or cyclopentadiene as dienophiles (Scheme 2.192). When using enol ether **2-**

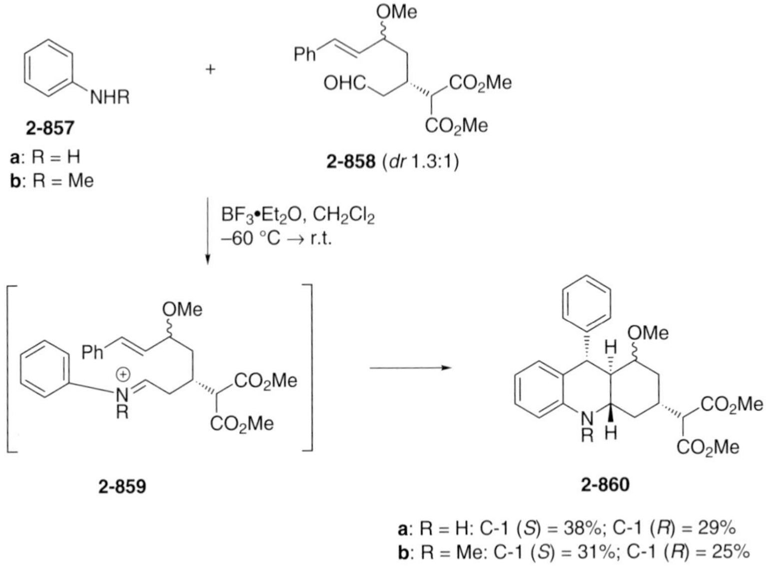

Scheme 2.191. Domino condensation/cycloaddition reaction leading to octahydroacridine 2-860.

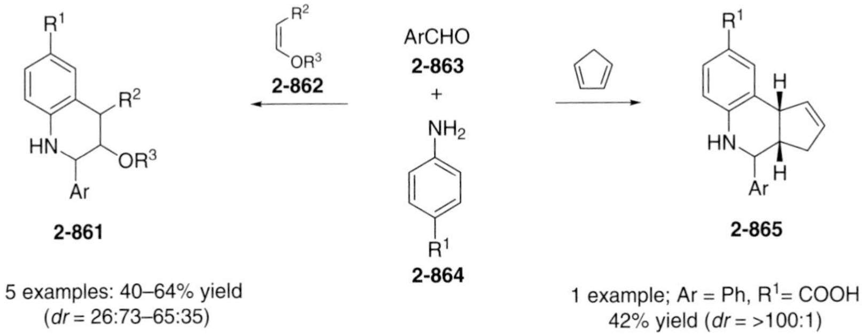

Scheme 2.192. Three-component imine formation/Diels–Alder reaction.

862, tetrahydroquinolines of type **2-861** are obtained, whereas with cyclopentadiene tricyclic aza-compounds of type **2-865** are formed. The yields are from mediocre to good, and this is also true for the diastereoselectivity.

Salicylaldehydes cannot be used in this process, as they form dihydropyrans, most likely *via* an intermediate *o*-quinomethide.

[2,3]-Wittig rearrangements can also be combined with a pericyclic transformation (Scheme 2.193). In a report by Hiersemann, treatment of esters **2-866** with LDA at –78 °C in THF followed by warming the reaction mixture to r.t., led initially to the enolate **2-867**, and then to the rearranged intermediate **2-868**, which undergoes an oxy-Cope rearrangement to afford α-keto esters **2-869** in moderate to high yields [442].

LDA, THF
−78 °C → r.t.
CO2iPr
OLi
OiPr
LiO
2-866
a: R1= H
b: R1= Ph
c: R1= Me
2-867
2-868
2-869
a: R1= H (84%)
b: R1= Ph (88%)
c: R1= Me (40%)

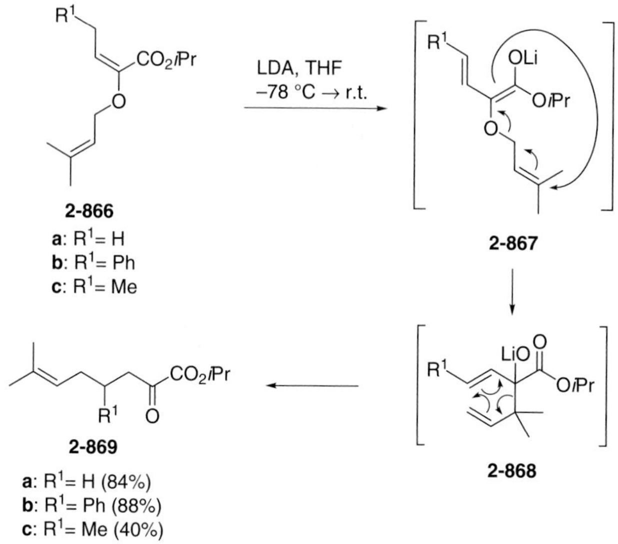

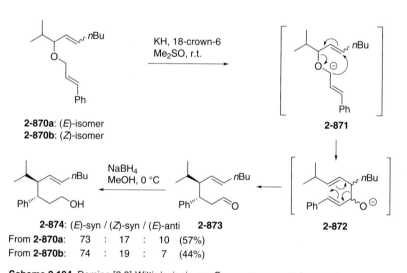

Scheme 2.193. Domino [2,3]-Wittig/anionic-oxy-Cope rearrangement process

KH, 18-crown-6
Me2SO, r.t.
NaBH4
MeOH, 0 °C
nBu
Ph
OH
2-870a: (E)-isomer
2-870b: (Z)-isomer
2-871
2-872
2-873
2-874: (E)-syn / (Z)-syn / (E)-anti
From 2-870a: 73 : 17 : 10 (57%)
From 2-870b: 74 : 19 : 7 (44%)

Scheme 2.194. Domino [2,3]-Wittig/anionic-oxy-Cope rearrangement process.

An earlier example of this type of domino reaction was reported by Greeves and coworkers (Scheme 2.194) [443]. Treatment of either the (*E*)- or (*Z*)-allyl vinyl ether **2-870** with NaH initiates the [2,3]-Wittig rearrangement to afford **2-872** via **2-871**. The subsequent oxy-Cope rearrangement led to the aldehyde **2-873**, which was reduced with $NaBH_4$ to give the alcohols **2-874**. Both isomers of **2-870** predominantly generated the (*E*)-*syn*-product **2-874** in comparable ratios as the main product.

Other types of anionic pericyclic domino reactions have been developed by the groups of Yamamoto and Bai. The former group described a unique α-methylenation/Diels–Alder sequence to give spiranones [444]. In the latter case, a novel Pummerer rearrangement/[4+3] cycloaddition process allowed the synthesis of a 5,7-membered ring system showing similarity to the key fragment of pseudolaric acid A [445].

In this way, tricyclic compounds of type **2-876** were obtained by Alcaide and coworkers in one process (Scheme 2.195) [446].

A combination of a Sakurai reaction [447] as the first step with an ene reaction has been developed by Tietze and coworkers for the synthesis of steroids [448]. These studies are discussed in Chapter 4.

As the final example in this section, a domino 5-*exo*-dig-cyclization/Claisen rearrangement process reported by Ovaska and coworkers will be described [449]. Reaction of the allylic alcohol **2-877** containing an alkyne moiety with MeLi induces the nucleophilic addition of the formed alkoxide to the triple bond to afford **2-878** with an allyl vinyl ether moiety (Scheme 2.196). This immediately undergoes a Claisen rearrangement to give **2-879**. Interestingly, if the alkyne moiety in **2-877** bears a silicon-containing substituent (TMS or TBS) at its end, the process is extended by a Brook rearrangement, providing a silyl enol ether **2-878** as the final product.

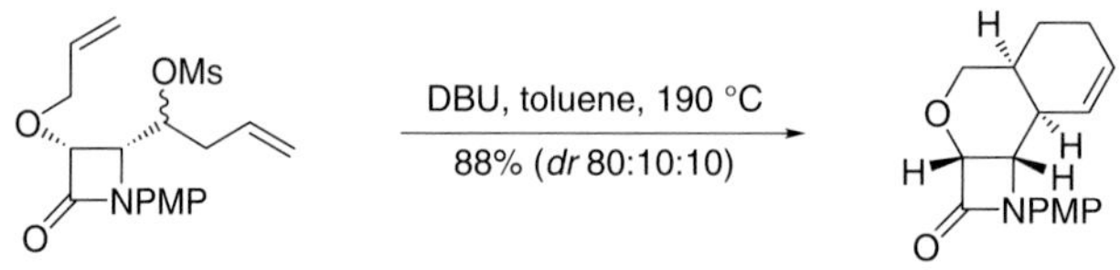

(+)-**2-875** (+)-**2-876**

Scheme 2.195. Domino elimination/Diels–Alder reactions.

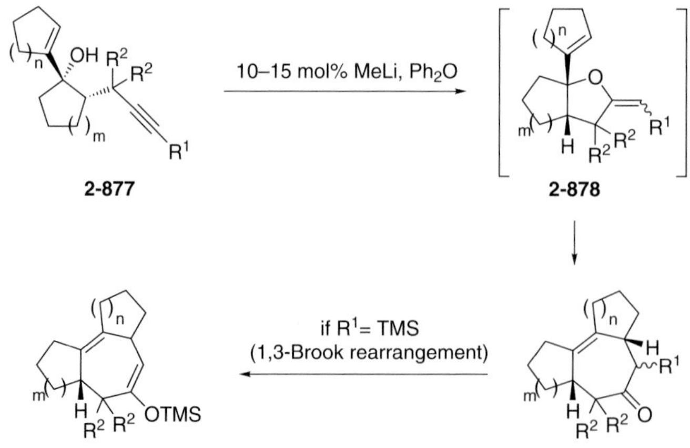

2-877 **2-878**

2-880 **2-879**

2 examples: 70 and 85% yield 5 examples: 76–81% yield

Scheme 2.196. 5-*exo*-dig-cyclization/Claisen rearrangement of compounds of type **2-877**.

The latter reaction belongs to the group of domino anionic/pericyclic reactions, which consists of three steps and will be discussed in the following section.

2.3.1
Anionic/Pericyclic Processes Followed by Further Transformations

As expected, some sequences also occur where a domino anionic/pericyclic process is followed by another bond-forming reaction. An example of this is an anionic/per-icyclic/anionic sequence such as the domino iminium ion formation/aza-Cope/imino aldol (Mannich) process, which has often been used in organic synthesis, especially to construct the pyrrolidine framework. The group of Brummond [450] has recently used this approach to synthesize the core structure **2-885** of the immu-nosuppressant FR 901483 (**2-886**) [451] (Scheme 2.197). The process is most likely initiated by the acid-catalyzed formation of the iminium ion **2-882**. There follows an aza-Cope rearrangement to produce **2-883**, which cyclizes under formation of the aldehyde **2-884**. As this compound is rather unstable, it was transformed into the stable acetal **2-885**. The proposed intermediate **2-880** is quite unusual as it does not obey Bredt's rule. Recently, this approach was used successfully for a formal total synthesis of FR 901483 **2-886** [452].

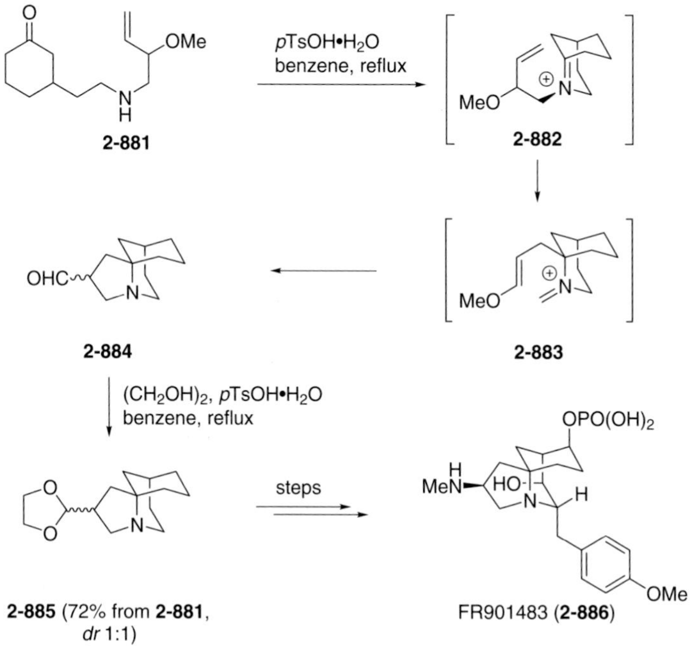

Scheme 2.197. Synthesis of the structural core **2-885** of the immunosuppressant FR 901483 (**2-886**).

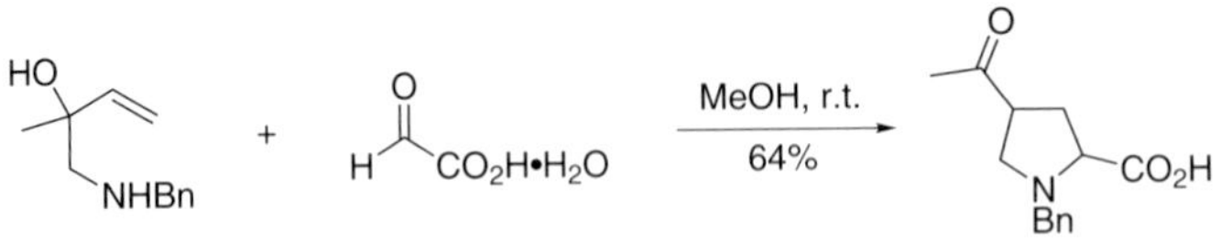

2-887 **2-888** **2-889** (*trans*:*cis* = 80:20 to 96:4)

Scheme 2.198. Domino condensation/aza-Cope/Mannich reaction sequence leading to **2-889**.

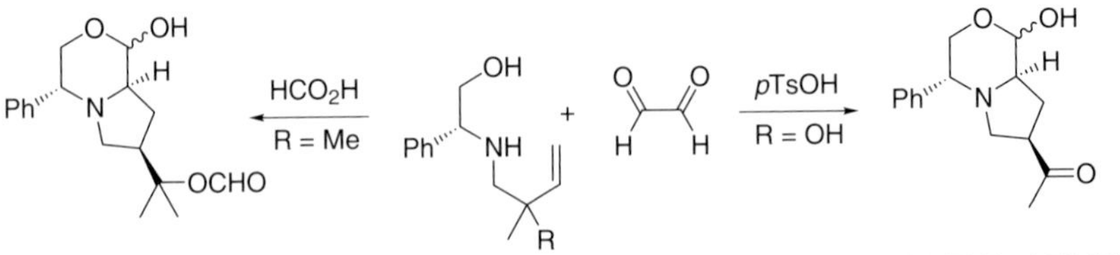

2-890 (76%, *dr* 77:23) **2-891a**: R = Me **2-891b**: R = OH **2-892** (50%, *dr* 53:47)

Scheme 2.199. Synthesis of diverse heterocycles **2-890** and **2-892** by varying the reaction conditions.

Cooke and coworkers reported on the synthesis of the amino acid *N*-benzyl-4-acetylproline (**2-889**) (Scheme 2.198) [453], as this might represent an interesting synthon for the preparation of bioactive compounds. These authors also used a domino iminium ion formation/aza-Cope/Mannich protocol. Thus, treatment of the secondary amine **2-885** with glyoxylic acid (**2-888**) primarily provided the corresponding iminium ion, which led to **2-889** in 64 % yield as a mixture of diastereomers.

According to the studies of the group of Agami and Couty, as depicted in Scheme 2.199, several nitrogen-containing heterocycles such as **2-890** and **2-892** can be obtained starting from substrates **2-891** and glyoxal using slightly different reaction conditions [454]. Moreover, these authors have also used this fruitful procedure for the synthesis of (–)-α-allokainic acid [454c].

A domino Claisen condensation/ene-type decarboxylation/aldol reaction was used by Braun's group [455] for the synthesis of spirocyclic compounds which have similarities with the structural core of the antitumor antibiotic fredericamycin A [456]. Moore used a 1,2-addition/oxy-Cope/aldol sequence for the synthesis of polyquinanes (Scheme 2.200) [450]. The addition of 2-lithiofuran (**2-894a**) or 2-lithiothiophene (**2-894b**) to the keto moiety of **2-893** gave the linearly fused polyquinanes **2-897a** and **2-897b**, respectively. As a first intermediate in this domino process, the alkoxide **2-895** can be assumed which undergoes an oxy-Cope ring expansion through a *cis*-boat conformational transition state. The subsequent vinylogous aldol addition then led to **2-897** under transannular ring closure.

This approach can also be used for the synthesis of angularly fused polyquinanes **2-898** [457].

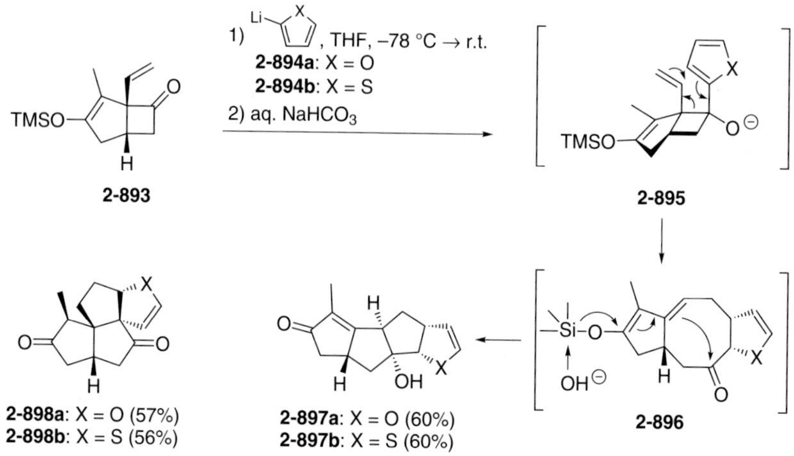

Scheme 2.200. Domino sequence for the construction of polyquinanes.

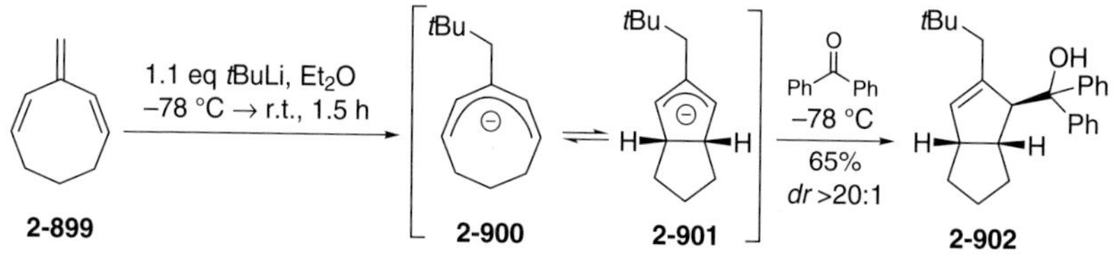

Scheme 2.201. Synthesis of *cis*-bicyclo[3.3.0]octenes *via* domino carbolithiation/electrocyclization/electrophile trapping.

A new sequence including a carbolithiation/electrocyclization/alkylation reaction was developed by Williams and coworkers for the synthesis of *cis*-bicyclo[3.3.0]octenes [458]. To give an example, to the readily available 3-methylene-1,4-cyclooctadiene (**2-899**) was added 1.1 equiv. *t*BuLi at –78 °C to give **2-900**. By raising the temperature to r.t., a disrotatory cyclization process was started and finally the resulting anion **2-901** quenched with benzophenone to yield the desired **2-902** in a good overall yield of 65 % and a very good diastereoselectivity of *dr* > 20:1 (Scheme 2.201). In other cases, transmetalation from lithium to copper after the electrocyclization step allowed the use of many electrophiles such as α,β-unsaturated ketones, epoxides, or others. Clearly, these transformations are not domino reactions by strict definition, but they are useful one-pot reactions and have the advantage that no isolation or separation of intermediates is necessary.

Bridged carbocyclic compounds such as **2-907** are interesting targets in organic synthesis [459]. A straightforward approach to these substances was developed by Shair and coworkers using a triple domino process [460]. Addition of a vinyl Grignard reagent **2-904** to the ketone **2-903** led to **2-907** *via* **2-905** and **2-906** in a highly stereoselective manner in reasonable to good yields (Scheme 2.202).

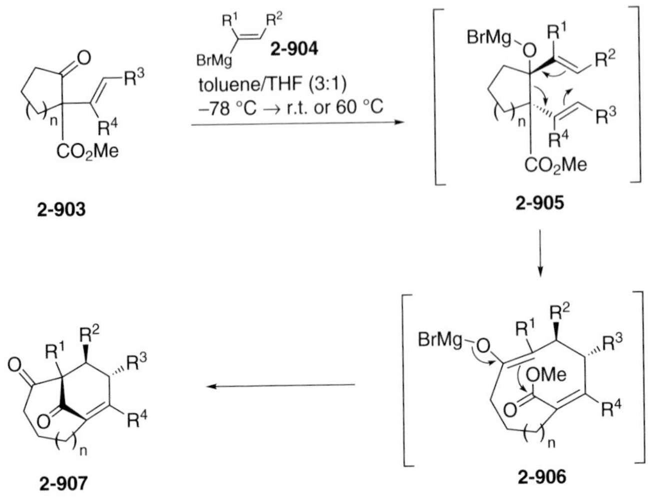

Scheme 2.202. Synthesis of bridged compounds.

In this domino process, the 1,2-addition is followed by an anion-accelerated oxy-Cope rearrangement and a transannular Dieckmann-type cyclization.

Moreover, the authors were able successfully to apply this process to the synthesis of the antibiotic (+)-CP-263,114 (**2-912**) (Scheme 2.203) [454]. Transformation of the bromide **2-908** into the vinyl Grignard reagent **2-909**, followed by addition to the ketone **2-910**, led to the bridged carbocyclic compound **2-911** which was then transformed into the target compound within a few steps.

Finally, a sequence consisting of thionium ion formation, thionium ion trapping, an aromatic substitution, and an alkylation, as elaborated by Padwa and coworkers, will be discussed [462]. Since several related sequences have already been mentioned in the foregoing sections, only the total synthesis of (±)-jamtine *N*-oxide (**2-916**) will be discussed here (Scheme 2.204). This natural product is an alkaloid from the shrub *Cocculus hirsutus*, which is commonly found in Pakistan [463].

The acid-catalyzed reaction of the substrate **2-913** gave the tetrahydroisoquinoline **2-914** in 88 % yield, though as a diastereomeric mixture of 5:2:1:1. Intramolecular alkylation led then to **2-915** which was transformed into jamtine *N*-oxide **2-916**.

The development of synthetic methods for the selective introduction of short-chain perfluoroalkyl groups into organic molecules is of interest in drug development [464]. Fluoromodifications often confer unique properties on a molecule, for example in terms of increased metabolic stability and lipophilicity and, as a consequence, the pharmacokinetic profiles are often improved [465]. Burger and coworkers developed a domino process consisting of a S_N reaction combined with a Claisen and a Cope rearrangement which allows the transformation of simple fluorinated compounds into more complex molecules with fluoro atoms [466]. Treatment of furan **2-917** with 2-hydroxymethyl thiophene (**2-918**) in the presence

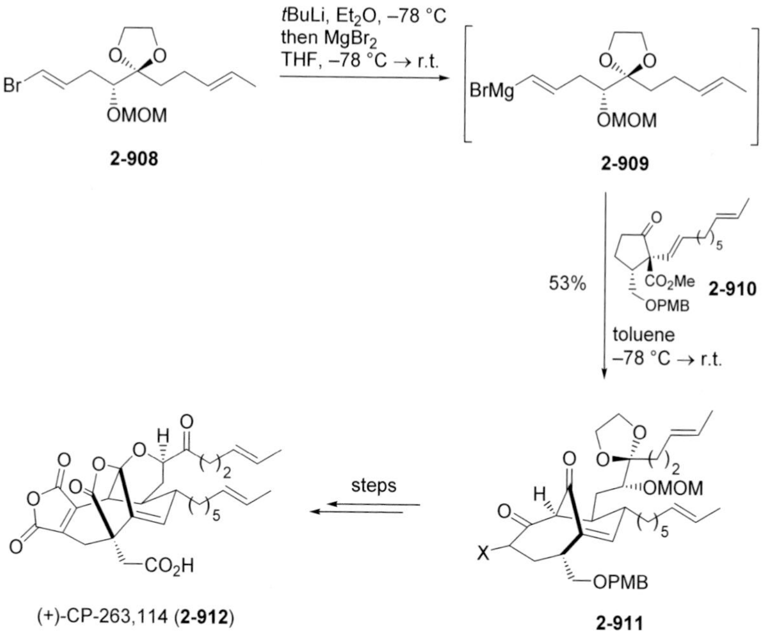

Scheme 2.203. Synthesis of the antibiotic (+)-CP-263,144 (**2-912**).

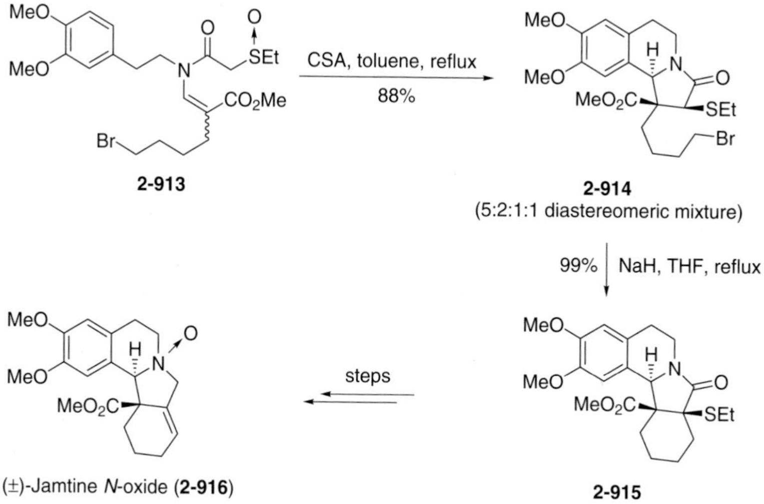

Scheme 2.204. Synthesis of (±)-jamtine *N*-oxide (**2-916**).

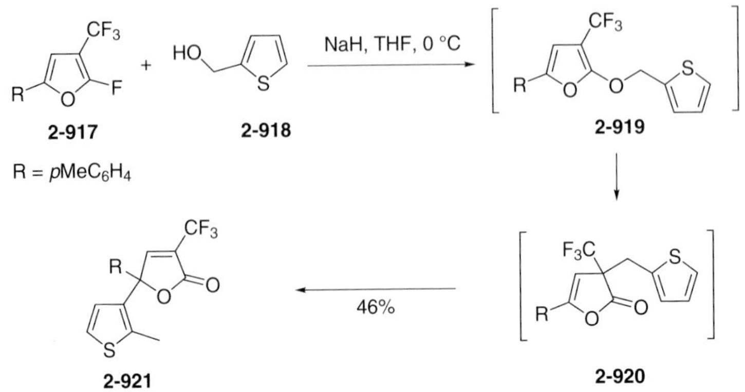

Scheme 2.205. Unusual domino reaction of a fluorinated furan **2-917**.

of NaH at 0 °C at first furnished the ether **2-919** by nucleophilic substitution (Scheme 2.205). Under the reaction conditions applied, **2-919** reacted further to give **2-920** through a [1,3]-thienyl migration. This compound proved to be unstable even at 0 °C. Thus, it slowly rearranges to give a Cope product which spontaneously underwent heteroaromatization. Overall, the final product **2-921** was isolated in 46 % yield.

Very recently, the Alcaide group constructed fused tricyclic systems from monocyclic precursors by a novel and elegant domino mesylation/[3,3] sigmatropic rearrangement/Diels–Alder reaction [467]. This domino process will be described in Chapter 4.

Anionic/pericyclic sequences with more than three and up to six independent reaction steps have also been reported. Thus, Florent and coworkers have published a domino alkylation/aza-Cope rearrangement/Mannich/elimination process [468] giving rise to essential parts of a 4-alkyl-4-hydroxy-2-cyclopentenone framework, for example as found in the central element of the punaglandins [469], clavulones [470] and chlorovulones [471]. Besides these examples, transformations with two or three additional pericyclic steps have also emerged. In this context, Taylor and coworkers developed an anionic/trifold pericyclic process giving access to nitrogen-containing polycycles [472] while, as depicted in Scheme 2.206, Mali and coworkers developed a route to naturally occurring coumarins **2-924** by employing two pericyclic steps [473]. Thus, the reaction of compounds **2-922** with phosphorane **2-291** in *N,N*-dimethylaniline at 200 °C afforded the natural 6-prenylcoumarins [474] suberosin (**2-924a**) and toddaculin (**2-994b**), as well as the analogue *O*-methylapigravin (**2-994c**) in 47–55 % yield.

Rychnovsky and coworkers very recently described a so-far unknown domino Michael addition/2-oxonia Cope/aldol-type reaction to give tetrahydropyran rings [475]. The transformation has as its origin an annulation process (see Section 2.1) which was discovered by the same group.

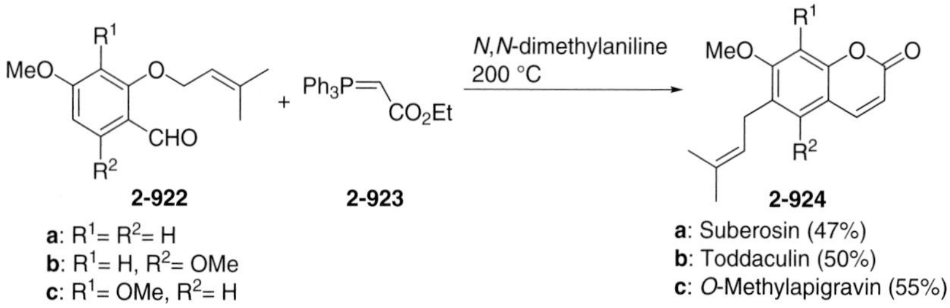

Scheme 2.206. Synthesis of naturally occurring coumarins and analogs.

2.4
Anionic/Transition Metal-Catalyzed Processes

To date, very few investigations have been made into the development of anionic/ transition metal-catalyzed domino processes. Nevertheless, a few examples have been published which admittedly in most of cases deal with the palladium-assisted assembly of simple heterocyclic structures. For instance, isoindolinones **2-931** are accessible through an approach employing the dihalide **2-926**, primary amines **2-927**, and carbon monoxide in the presence of palladium nanoparticles obtained from palladacycle **2-925**, as described by Grigg and coworkers (Scheme 2.207) [476]. In the first step, a nucleophilic substitution of the amine **2-927** at the benzyl-bromide moiety in **2-926** affords the benzylamine **2-928**, which undergoes an oxidative addition with Pd⁰ and subsequent CO insertion to give the intermediate **2-930**. This is trapped in a ring-closure reaction leading to the desired isoindolinones **2-931**.

Another sequence involving an anionic and a Pd-catalyzed step was described by the groups of Rossi and Arcadi [477]. These authors prepared substituted tetrahydro-2H-pyrrolo[3,2-c]pyrazolones **2-934** starting from hydrazones **2-932** and aryl-halides or alkenyl triflates **2-933** (Scheme 2.208). The first step is the formation of a pyrazolone. There follows cleavage of the urea moiety with piperidine and an inter- as well as an intramolecular Heck-type reaction with **2-933**.

Although aryl halides are the substrates of choice for Heck transformations, diazonium salts have also gained great interest, due mainly to reasons of economy [478]. With this in mind, the group of Beller explored a domino diazotization/Heck sequence of substituted anilines **2-935** with ethylene (Scheme 2.209) [479]. The reaction proceeds best in the presence of Pd(OAc)₂ and tBuONO as diazotization reagent in a mixture of acetic acid and CH₂Cl₂. Using these conditions, the desired styrenes **2-936** were isolated in yields up to 72%.

Balme and coworkers reported on a procedure for the preparation of highly functionalized furans of type **2-940** (Scheme 2.210) [480]. Their approach is based on a nucleophilic Michael addition of propargyl alcohols **2-937** to alkylidene or aryl-idenemalonates **2-938**, followed by a palladium-catalyzed cyclization *via* the carbanion **2-939**. The reactions with propargyl alcohol led to the formation of only one di-

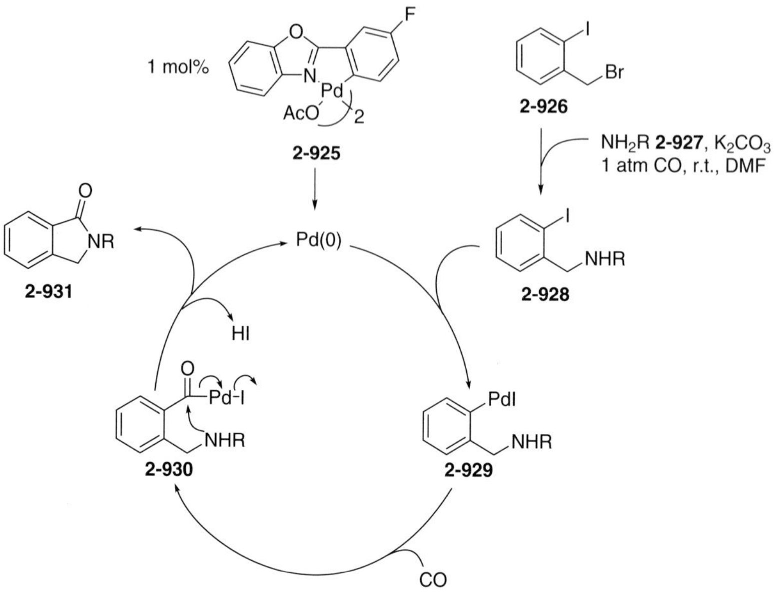

Entry	Amine **2-927**	Product **2-931**	Yield [%]
1	$BnNH_2$	**a**	72
2	L-phenylalanine methyl ester	**b**	33
3	propargyl amine	**c**	51

Scheme 2.207. Synthesis of isoindolinones.

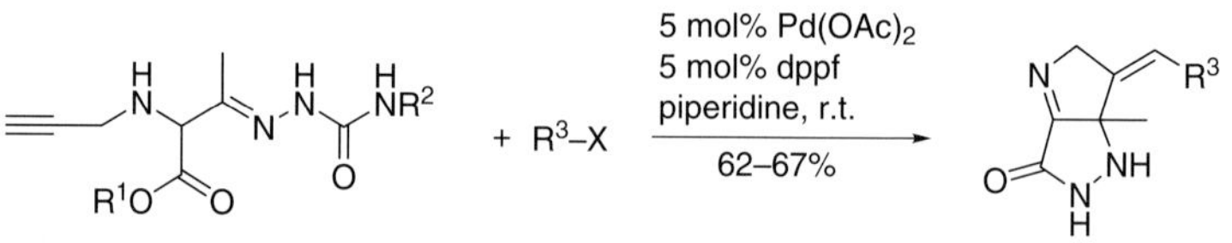

2-932 **2-933** **2-934**

Scheme 2.208. Synthesis of pyrrolopyrazolones.

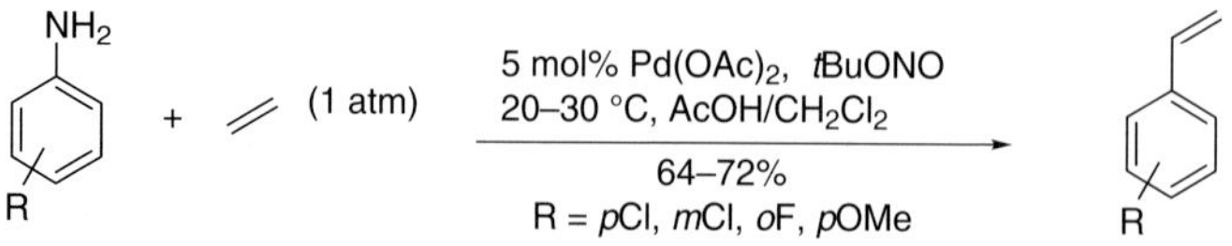

2-935 **2-936**

Scheme 2.209. Synthesis of styrenes **2-936** from anilines **2-935**.

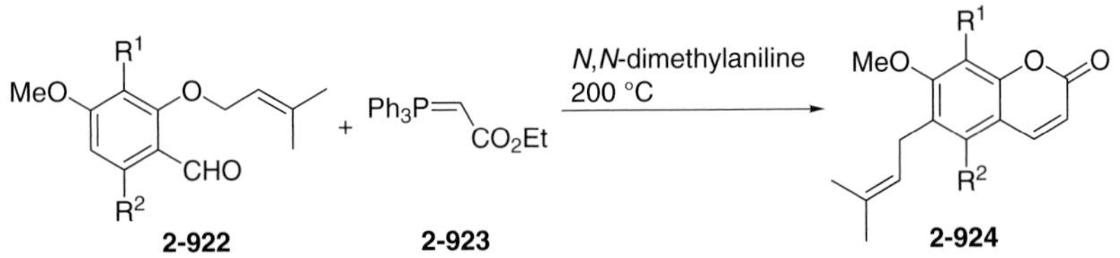

a: $R^1 = R^2 = H$
b: $R^1 = H$, $R^2 = OMe$
c: $R^1 = OMe$, $R^2 = H$

a: Suberosin (47%)
b: Toddaculin (50%)
c: O-Methylapigravin (55%)

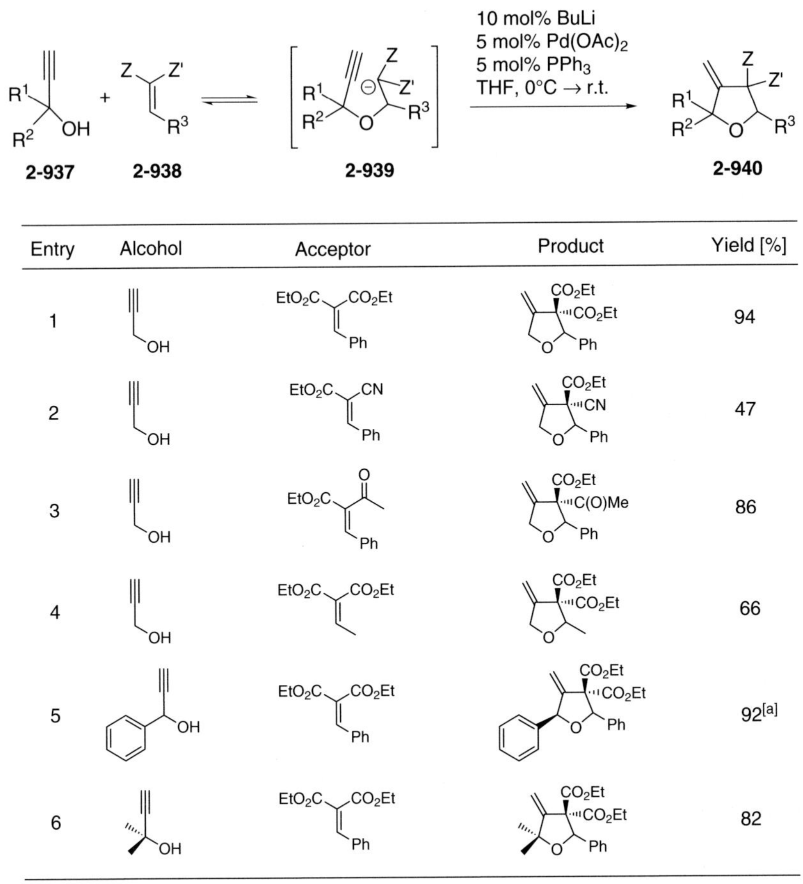

Entry	Alcohol	Acceptor	Product	Yield [%]
1		EtO₂C, CO₂Et, Ph	CO₂Et, CO₂Et, O, Ph	94
2		EtO₂C, CN, Ph	CO₂Et, CN, O, Ph	47
3		EtO₂C, O, Ph	CO₂Et, C(O)Me, O, Ph	86
4		EtO₂C, CO₂Et	CO₂Et, CO₂Et, O	66
5		EtO₂C, CO₂Et, Ph	CO₂Et, CO₂Et, O, Ph	92[a]
6		EtO₂C, CO₂Et, Ph	CO₂Et, CO₂Et, O, Ph	82

[a] Mixture of diastereomers (*cis/trans* 1:2).

Scheme 2.210. Synthesis of 3-methylene tetrahydrofurans **2-940**.

astereomer, whereas the use of an unsymmetrically substituted propargyl alcohol gave a mixture of two diastereomers.

In a similar way, substituted 3-methylene pyrrolidines were prepared in good to high yields using propargyl amines as substrates [481]. Herein, in contrast to the preliminarily utilized palladium salts, copper was the metal of choice, and was added as CuI at 3–20 mol% to the reaction mixture. Although the precise role of the metal is not clear, it is believed that a methylene copper(I) compound is formed as intermediate.

In continuation of the aforementioned reaction, Hiroya and coworkers used copper(II) acetate for the synthesis of indoles **2-943** in reasonable yields from the corresponding ethynylanilines **2-941** by a domino intermolecular Michael addition/copper-assisted nucleophilic tosylate displacement reaction *via* **2-942** (Scheme 2.211) [482].

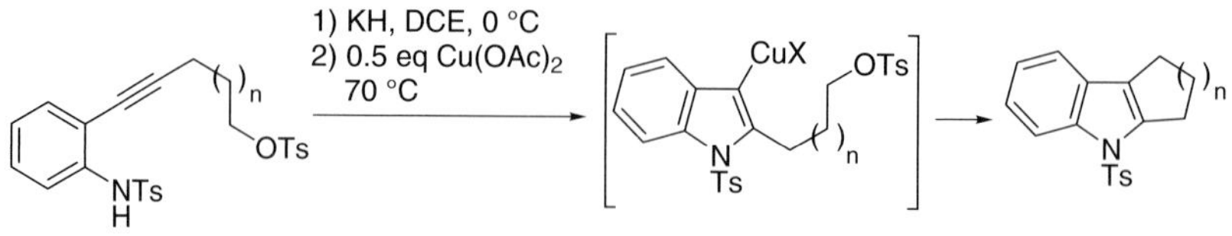

Scheme 2.211. Synthesis of indoles **2-943**.

As a final example in this section, the mechanistically interesting transformation of α,β-unsaturated aldehydes containing a chloro or bromo atom in the β-position into five-membered lactams or lactones is mentioned. In this transformation, which was developed by Rück-Braun and coworkers, an intermediate iron compound is formed by reaction with $[C_5H_5(CO)_2Fe]Na$, which yields the products either by adding a primary amine and $TiCl_4$ or a metalorganic alkane as RMgX or RLi [483].

2.5
Anionic/Oxidative or Reductive Processes

Among the anionic/reductive domino transformations, one of the most often encountered is the reductive amination of a carbonyl compound being followed by the formation of a lactam. As did many others before, Abdel-Magid and coworkers used this approach for the synthesis of γ- and δ-lactams as **2-949** from either ketone **2-944** and amine **2-945** or amine **2-946** and ketone **2-947** *via* **2-948**, employing sodium triacetoxyborane (Scheme 2.212) [484].

In a similar way, piperazinone derivatives can be prepared by either addition of the formed amine to a carboxylate moiety or by an intramolecular alkylation with a chloromethyl group present in the substrate [485].

Tietze and coworkers used this process to synthesize the monoterpene alkaloids bakankosin (**2-950**) and xylostosidine (**2-952**), as well as analogues starting from secologanin (**2-951**) (Scheme 2.213) [486].

A combination of a reductive amination and a Michael addition has been used to synthesize the anticancer alkaloid camptothecin (**2-955**) *via* **2-954**, starting from the quinoline carbaldehyde **2-953** and benzylamine (Scheme 2.214) [487].

Besides the domino reductive amination/lactamization reaction, the aldol/Tishchenko sequence is another important anionic/reductive domino reaction. In the classic aldol/Tishchenko variant, two enolizable aldehyde molecules undergo addition to form an aldol adduct which is subsequently reduced by a third aldehyde to yield 1,3-diol monoesters with the occurrence of an intramolecular hydride shift [488]. Nevertheless, the utility of this domino transformation is greatly enhanced upon reaction of aldehydes with ketones; by using these substrates, three adjacent stereogenic carbon centers with defined configuration can be synthesized in a single process [489]. As for the enolate part, preformed silyl enol ethers [489a], Li-enolates [489b] and Sm-enolates [490], as well as catalytic amounts of a metal

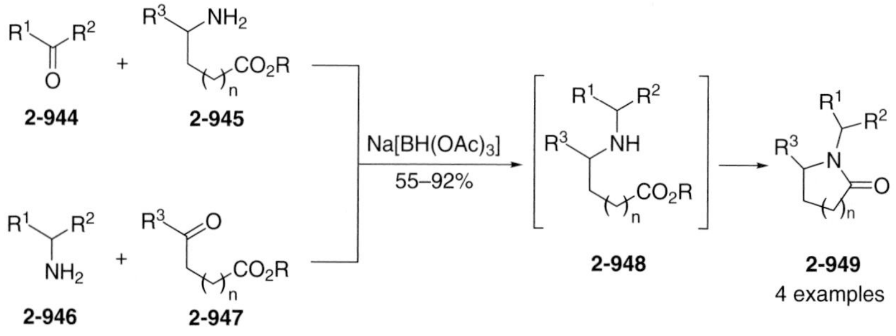

Scheme 2.212. Reductive amination/lactamization.

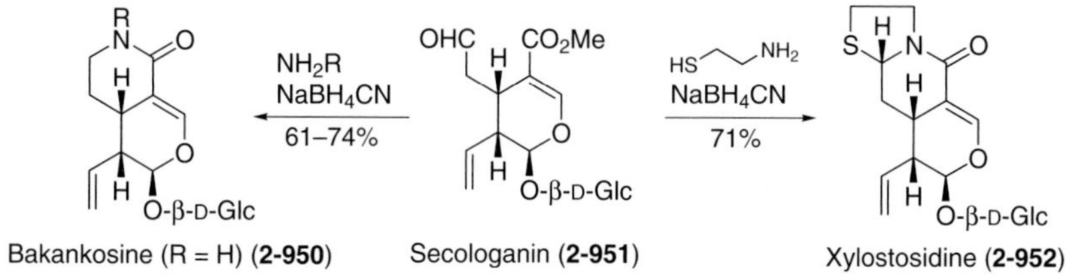

Bakankosine (R = H) (**2-950**)   Secologanin (**2-951**)   Xylostosidine (**2-952**)

Scheme 2.213. Reductive amination of secologanin (**2-951**).

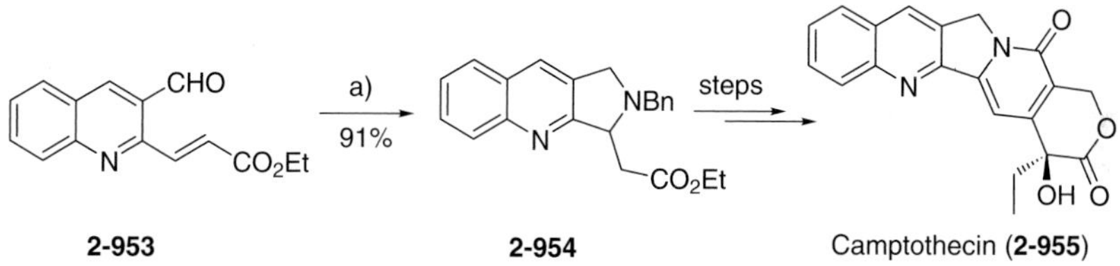

2-953   **2-954**   Camptothecin (**2-955**)

a) BnNH$_2$, MeOH, r.t. then NaBH$_4$, 0 °C to r.t..

Scheme 2.214. Synthesis of camptothecin (**2-955**).

alkoxide catalyst in combination with an appropriate ketone to form the reactive enolate [491] have been employed. Recently, it was reported by different research groups that aldol adducts of ketones such as acetone alcohol **2-957** can also be used for the *in-situ* generation of metal enolates through a retro-aldol process upon treatment with either Zr- or Al-based catalysts (Scheme 2.215) [492]. Schneider and coworkers have shown that these enolates can be trapped in an aldol/Tishchenko domino sequence with 2 equiv. of an aliphatic aldehyde **2-956** to furnish 1,3-*anti*-diol monoesters **2-958** in good to excellent yields, complete *anti*-diastereoselectivity, and with up to 57 % enantiomeric excess by using a Zr catalyst with TADDOL **2-959** as chiral ligand [493].

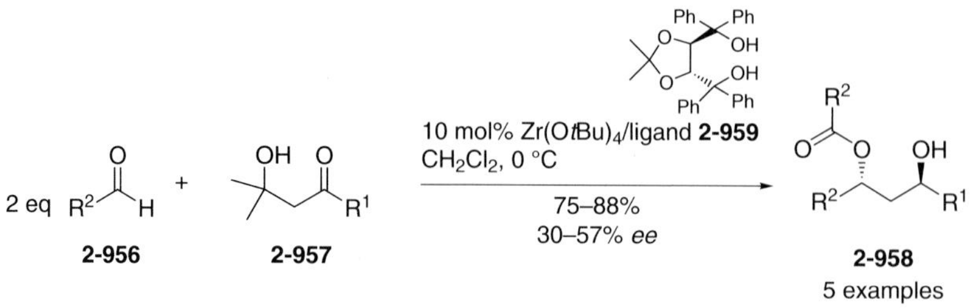

Scheme 2.215. Enantioselective aldol/Tishchenko reactions.

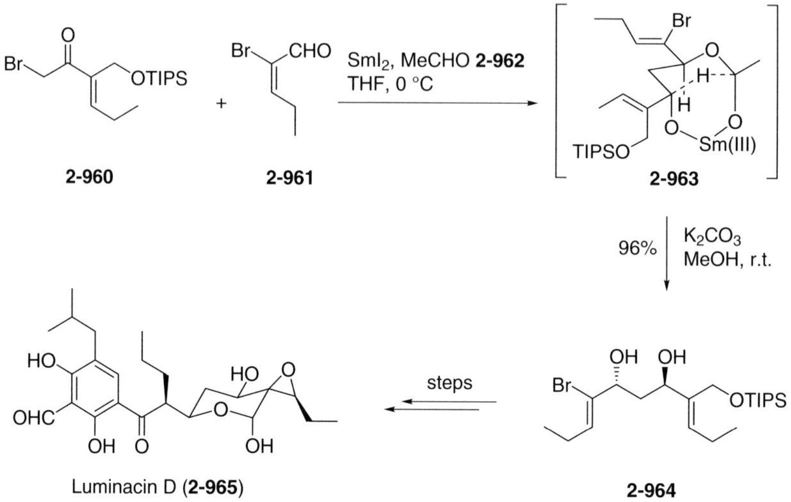

Luminacin D (**2-965**)

Scheme 2.216. Synthesis of luminacin D (**2-965**).

Furthermore, a domino aldol/Tishchenko process has been used by the group of Wood and Crews [494] to obtain rapid access to the natural product luminacin D (**2-965**). The luminacins belong to a growing class of anticancer agents which target angiogenesis (the process of neovascularization by which growing tumors establish their blood supply) [495]. The luminacins were isolated from the actinomycete *Streptomyces* sp. [496], and elicit their anti-angiogenic biological response by binding to an as-yet unknown intracellular receptor [497].

As illustrated in Scheme 2.216, generation of the Sm-enolate of **2-960** by treatment with SmI$_2$ in THF followed by sequential addition of (*E*)-2-bromo-2-pentenal (**2-961**) and acetaldehyde (**2-962**) furnished, after basic work-up, the diol **2-964** as a single diastereomer in excellent yield of 96%. The observed high degree of simple diastereoselectivity is assumed to arise from an organized eight-membered ring chelate **2-963** [498].

The Michael addition/Meerwein–Ponndorf–Verley (MPV) reduction sequence developed by Node and coworkers is another easily implementable strategy for the

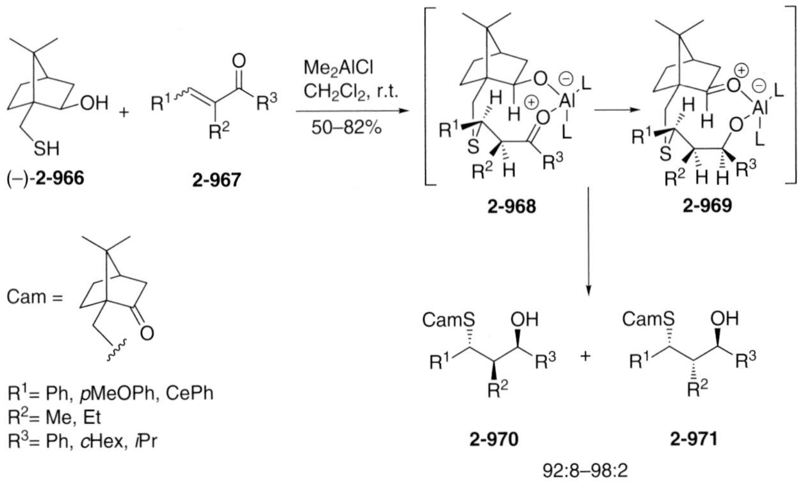

Scheme 2.217. Domino Michael/MPV reaction of the α,β-unsaturated ketones.

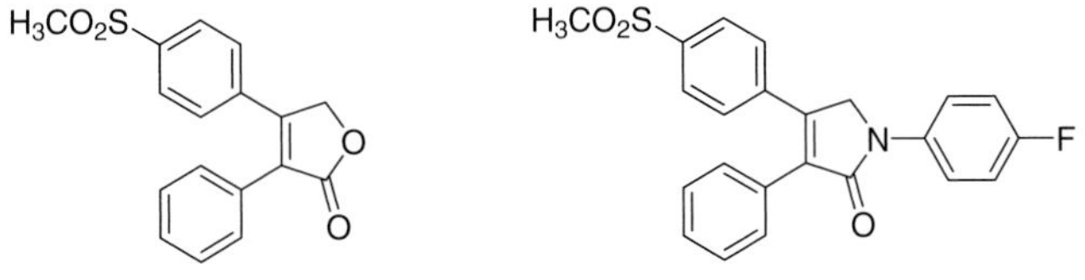

Rofecoxib (Vioxx) **(2-972)**   **2-973**

Scheme 2.218. Examples of tricyclic compounds as selective COX-2 inhibitor.

assembly of three contiguous stereogenic centers [499]. As for the reaction products, utilization of this methodology provides enantiopure mercapto alcohols of type **2-970/2-971** from α,β-unsaturated ketones **2-967** and (–)-**2-966** *via* **2-968** and **2-969** in moderate to good yields with very good diastereoselectivities (Scheme 2.217).

Anionic/oxidative reaction sequences have been developed in addition to the domino anionic/reductive processes. For example, with regard to the synthesis of novel diaryl heterocycles as COX-2 inhibitors [500], including rofecoxib (Vioxx) **2-972** [501] (which has recently been withdrawn from the market) or the pyrrolin-2-one derivative **2-973** [494], Pal and coworkers reported on a so-far unique domino aldol condensation/oxidation sequence (Scheme 2.218) [503].

The reaction of diaryl compounds **2-974** with 3 equiv. of DBU at r.t. results initially in an intramolecular aldol condensation to give **2-975**; this is followed by an oxidation to give the 3,4-diaryl disubstituted maleic anhydride or maleimide derivatives **2-976** when the reaction is conducted under atmospheric oxygen (Scheme 2.219).

Very recently, two examples involving an epoxidation as the oxidative step have been reported. Thus, Taylor and coworkers have described a S_N-type ring-opening/

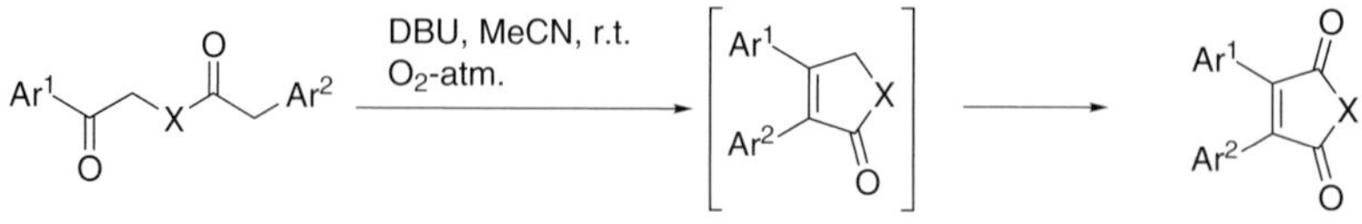

2-974 2-975 2-976

Entry	Substrate **2-974**	Ar^1	Ar^2	X	Product **2-976**	Yield [%]
1	**a**	Ph	*p*MeOPh	O	**a**	81
2	**b**	Ph	*m*FPh	O	**b**	88
3	**c**	*p*NO_2Ph	Ph	O	**c**	70
4	**d**	*p*MePh	Ph	O	**d**	71
5	**e**	Ph	Ph	N*p*FC_6H_4	**e**	61

Scheme 2.219. Base-promoted cyclization/oxidation reaction of compounds **2-974**.

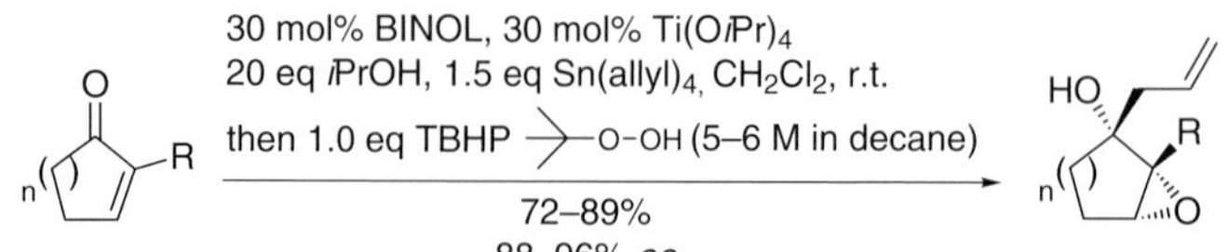

2-977
n = 1 or 2

2-978
4 examples
R = Me, C_5H_{11}, CH_2Ph, CH_2OTBS

Scheme 2.220. Synthesis of epoxy alcohols.

epoxidation of 1,2-dioxines [504], while the group of Walsh has developed a practical asymmetric allylation/epoxidation methodology employing α,β-unsaturated cyclic enones as substrates (Scheme 2.220) [505]. As products, epoxy alcohols of type **2-978** are formed in generally very good yields and with excellent *ee*-values.

This procedure is not a domino process in its strictest definition, but since the oxidant *tert*-butyl hydroperoxide is added after allylation is complete, it is a very impressive and useful transformation for the rapid assembly of three contiguous stereogenic centers, including a tertiary alcohol moiety.

References

1. H.-C. Guo, J.-A. Ma, *Angew. Chem. Int. Ed.* **2006**, *45*, 354–366.

2. (a) D. Spitzner, K. Oesterreich, *Eur. J. Org. Chem.* **2001**, 1883–1886; (b) R. A. Lee, *Tetrahedron Lett.* **1973**, *14*, 3333–3336; (c) D. Spitzner, *Tetrahedron Lett.* **1978**, *19*, 3349–3350.

3 H. Hagiwara, A. Morii, Y. Yamada, T. Hoshi, T. Suzuki, *Tetrahedron Lett.* **2003**, *44*, 1595–1597.

4 D. S. Ming, D. Q. Yu, Y. Y. Yang, C. H. He, *Tetrahedron Lett.* **1997**, *38*, 5205–5208.

5 (a) H. Oikawa, T. Yokota, A. Ichihara, S. Sakamura, *J. Chem. Soc., Chem. Com-*

mun. **1989**, 1284–1285; (b) H. Oikawa, Y. Suzuki, A. Naya, K. Katayama, A. Ichihara, *J. Am. Chem. Soc.* **1994**, *116*, 3605–3606; (c) A. Ichihara, H. Tazaki, S. Sakamura, *Tetrahedron Lett.* **1983**, *24*, 5373–5376.

6 (a) H. Hagiwara, K. Kobayashi, S. Miya, T. Hoshi, T. Suzuki, M. Ando, T. Okamoto, M. Kobayashi, I. Yamamoto, S. Ohtsubo, M. Kato, H. Uda, *J. Org. Chem.* **2002**, *67*, 5969–5976; (b) H. Hagiwara, K. Kobayashi, S. Miya, T. Hoshi, T. Suzuki, M. Ando, *Org. Lett.* **2001**, *3*, 251–254; (c) The same domino sequence was also applied to the synthesis of (+)-compactin, see: H. Hagiwara, T. Nakano, M. Kon-no, H. Uda, *J. Chem. Soc., Perkin Trans. 1* **1995**, 777–783.

7 (a) M. C. Carreño, M Ribagorda, G. H. Posner, *Angew. Chem.* **2002**, *114*, 2877–2878; (b) For further applications, see: M. C. Carreño, C. García Luzón, M. Ribagorda, *Chem. Eur. J.* **2002**, *8*, 208–216.

8 (a) B. Achmatowicz, E. Gorobets, S. Marczak, A. Przezdziecka, A. Steinmeyer, J. Wicha, U. Zügel, *Tetrahedron Lett.* **2001**, *42*, 2891–2895; (b) S. Marczak, K. Michalak, Z. Urbańczyk-Lipkowska, J. Wicha, *J. Org. Chem.* **1998**, *63*, 2218–2223; (c) K. Michalak, W. Stepanenko, J. Wicha, *Tetrahedron Lett.* **1996**, *37*, 7657–7658.

9 A. Barco, S. Benetti, C. De Risi, P. Marchetti, G. P. Pollini, V. Zanirato, *J. Comb. Chem.* **2000**, *2*, 337–340.

10 K. Voigt, A. Lansky, M. Noltemeyer, A. de Meijere, *Liebigs Ann.* **1996**, 899–911.

11 T. Yakura, T. Tsuda, Y. Matsumura, S. Yamada, M. Ikeda, *Synlett* **1996**, 985–986.

12 (a) For a recent review on some domino Michael/aldol reactions, see: T. Kataoka, H. Kinoshita, *Eur. J. Org. Chem.* **2005**, 45–48; (b)For an interesting previous example, see: M. Suzuki, H. Koyano, Y. Morita, R. Noyori, *Synlett* **1989**, *1*, 22–23.

13 N. Giuseppone, Y. Courtaux, J. Collin, *Tetrahedron Lett.* **1998**, *39*, 7845–7848.

14 K. Yamada, T. Arai, H. Sasai, M. Shibasaki, *J. Org. Chem.* **1998**, *63*, 3666–3672.

15 M. Shibasaki, H. Sasai, T. Arai, *Angew. Chem. Int. Ed.* **1997**, *36*, 1237–1256 and references cited herein.

16 T. Arai, H. Sasai, K. Aoe, K. Okamura, T. Date, M. Shibasaki, *Angew. Chem. Int. Ed.* **1996**, *35*, 104–106.

17 (a) L. A. Arnold, R. Naasz, A. J. Minnaard, B. L. Feringa, *J. Org. Chem.* **2002**, *67*, 7244–7254; (b) L. A. Arnold, R. Naasz, A. J. Minnaard, B. L. Feringa, *J. Am. Chem. Soc.* **2001**, *123*, 5841–5842.

18 H. E. Zimmerman, M. D. Traxler, *J. Am. Chem. Soc.* **1957**, *79*, 1920–1923.

19 (a) R. S. Ward, *Chem. Soc. Rev.* **1982**, *11*, 75–125; (b) R. S. Ward, *Tetrahedron* **1990**, *46*, 5029–5041; (c) R. S. Ward, *Synthesis* **1992**, 719–730; (d) S. Yamamura, *J. Synth. Org. Chem. Jpn.* **1985**, *43*, 583–593.

20 A. Pelter, R. S. Ward, N. P. Storer, *Tetrahedron* **1994**, *50*, 10829–10838.

21 T. Ogiku, S. Yoshida, T. Kuroda, M. Takahashi, H. Ohmizu, T. Iwasaki, *Bull. Chem. Soc. Jpn.* **1992**, *65*, 3495–3497.

22 (a) N. Jaber, J.-C. Fiaud, J. Collin, *Tetrahedron Lett.* **2001**, *42*, 9157–9159; (b) N. Jaber, M. Assié, J.-C. Fiaud, J. Collin, *Tetrahedron Lett.* **2004**, *60*, 3075–3083; (c) For a comparison, see also ref. [13].

23 X.-F. Yang, X.-L. Hou, L.-X. Dai, *Tetrahedron Lett.* **2000**, *41*, 4431–4431.

24 (a) Z. Han, S. Uehira, H. Shinokubo, K. Oshima, *J. Org. Chem.* **2001**, *66*, 7854–7857; (b) M. Shi, Y.-S. Feng, *J. Org. Chem.* **2001**, *66*, 406–411; (c) S. Uehira, Z. Han, H. Shinokubo, K. Oshima, *Org. Lett.* **1999**, *1*, 1383–1385; (d) M. Shi, J.-K. Jiang, S.-C. Cui, Y.-S. Feng, *J. Chem. Soc., Perkin Trans. 1* **2001**, 390–393; (e) K. Yagi, T. Turitani, H. Shinokubo, K. Oshima, *Org. Lett.* **2002**, *4*, 3111–3114.

25 G. Li, X. Xu, D. Chen, C. Timmons, M. D. Carducci, A. D. Headley, *Org. Lett.* **2002**, *21*, 3691–3693.

26 (a) H.-X. Wei, J. J. Gao, G. Li, P. W. Paré, *Tetrahedron Lett.* **2002**, *43*, 5677–5680; (b) G. Li, H.-X. Wei, J. J. Gao, J. Johnson, *Synth. Commun.* **2002**, *32*, 1765–1773; (c) H.-X. Wei, J. J. Gao, G. Li, *Tetrahedron Lett.* **2001**, *42*, 9119–9122; (d) G. Li, H.-X. Wei, B. S. Phelps, D. W. Purkiss, S. H. Kim, *Org. Lett.* **2001**, *3*, 823–826; (e) H.-X. Wei, S. H. Kim, T. D. Caputo, D. W. Purkiss, G. Li, *Tetrahedron* **2000**, *56*, 2397–2401.

27 (a) H.-X. Wei, J. Hu, D. W. Purkiss, P. W. Paré, *Tetrahedron Lett.* **2003**, *44*, 949–952; (b) The present sequence can even be extended by another step to provide related

products. For further details, see: H.-X. Wei, S. H. Kim, G. Li, *Org. Lett.* **2002**, *4*, 3691–3693.

28 (a) Y. Nagaoka, K. Tomioka, *Org. Lett.* **1999**, *1*, 1467–1469; (b) Y. Nagaoka, K. Tomioka, *J. Org. Chem.* **1998**, *63*, 6428–6429; (c) H. Inoue, H. Tsubouchi Y. Nagaoka, K. Tomioka, *Tetrahedron* **2002**, *58*, 83–90.

29 H. Inoue, Y. Nagaoka, K. Tomioka, *J. Org. Chem.* **2002**, *67*, 5864–5867.

30 (a) M. Ono, K. Nishimura, Y. Nagaoka, K. Tomioka, *Tetrahedron Lett.* **1999**, *40*, 6979–6982; (b) For related three-component Michael/aldol reactions, see: A. Kamimura, H. Mitsudera, Y. Omata, K. Matsuura, M. Shirai, A. Kakehi, *Tetrahedron* **2002**, *58*, 9817–9826 and A. Kamimura, H. Mitsudera, S. Asano, A. Kakehi, M. Noguchi, *Chem. Commun.* **1998**, 1095–1096.

31 M. Ono, K. Nishimura, H. Tsubouchi, Y. Nagaoka, K. Tomioka, *J. Org. Chem.* **2001**, *66*, 8199–8203.

32 (a) S. Yaginuma, N. Muto, M. Tsujino, Y. Sudate, M. Hayashi, M. Otani, *J. Antibiot.* **1981**, *34*, 359–366; (b) M. Hayashi, S. Yaginuma, H. Yoshioka, K. Nakatsu, *J. Antibiot.* **1981**, *34*, 675–680.

33 (a) P. M. Ueland, *Pharmacol. Rev.* **1982**, *34*, 223–253; (b) M. Inaba, K. Nagashima, S. Tsukagoshi, Y. Sakurai, *Cancer Res.* **1986**, *46*, 1063–1067; (c) M. S. Wolfe, R. T. Borchardt, *J. Med. Chem.* **1991**, *34*, 1521–1530; (d) S. Shuto, T. Obara, M. Toriya, M. Hosoya, R. Snoeck, G. Andrei, J. Balzarini, E. De Clercq, *J. Med. Chem.* **1992**, *35*, 324–331.

34 C. Schneider, O. Reese, *Chem. Eur. J.* **2002**, *8*, 2585–2594.

35 S. Kaneko, T. Yoshino, T. Katoh, S. Terashima, *Tetrahedron* **1998**, *54*, 5471–5484.

36 (a) J.-S. Liu, Y.-L. Zhu, C.-M. Yu, Y.-Z. Zhou, Y.-Y. Han, F.-W. Wu, B.-F. Qi, *Can. J. Chem.* **1986**, *64*, 837–839; (b) W. A. Ayer, L. M. Browne, H. Orszanska, Z. Valenta, J.-S. Liu, *Can. J. Chem.* **1989**, *67*, 1538–1540; (c) X.-C. Tang, P. D. Sarno, K. Sugaya, E. Giacobini, *J. Neurosci. Res.* **1989**, *24*, 276–285; (d) A. P. Kozikowski, *J. Heterocyclic Chem.* **1990**, *27*, 97–105.

37 (a) R. S. E. Corn, A. V. Lovell, S. Karady, L. M. Weinstock, *J. Org. Chem.* **1986**, *51*, 4710–4711; (b) K. Takagi, H. Katayama,

H. Yamada, *J. Org. Chem.* **1988**, *53*, 1157–1161.

38 N. Halland, P. S. Aburel, K. A. Jørgensen, *Angew. Chem. Int. Ed.* **2004**, *43*, 1272–1277.

39 (a) U. Eder, G. Sauer, R. Wiechert, *Angew. Chem. Int. Ed. Engl.* **1971**, *10*, 496–497; (b) Z. G. Hajos, D. R. Parrish, *J. Org. Chem.* **1974**, *39*, 1615–1621; see also: (c) T. Arai, H. Sasai, K.-I. Aoe, K. Okamura, T. Date, M. Shibasaki, *Angew. Chem. Int. Ed. Engl.* **1996**, *35*, 104–106; (d) T. Bui, C. F. Barbas, III, *Tetrahedron Lett.* **2000**, *41*, 6951–6954.

40 H. S. P. Rao, S. P. Senthilkumar, *J. Org. Chem.* **2004**, *69*, 2591–2594.

41 (a) J. Rodriguez, *Synlett* **1999**, 505–518; (b) M.-H. Filippini, R. Faure, J. Rodriguez, *J. Org. Chem.* **1995**, *60*, 6872–6882; (c) N. Ouvrard, P. Ouvrard, J. Rodriguez, M. Santelli, *J. Chem. Soc., Chem. Commun.* **1993**, 571–572.

42 M.-H. Filippini, J. Rodriguez, M. Santelli, *J. Chem. Soc.* **1993**, 1647–1648.

43 M.-H. Filippini, J. Rodriguez, *J. Org. Chem.* **1997**, *62*, 3034–3035.

44 D. F. Cauble, J. D. Gipson, M. J. Krische, *J. Am. Chem. Soc.* **2003**, *125*, 1110–1111.

45 T. Hayashi, M. Takahashi, Y. Takaya, M. Ogasawara, *J. Am. Chem. Soc.* **2002**, *124*, 5052–5058.

46 K. Agapiou, D. F. Cauble, M. J. Krische, *J. Am. Chem. Soc.* **2004**, *126*, 4528–4529.

47 (a) H. N. C. Wong, K.-L. Lau, K.-F. Tam, in: *Small Ring Compounds in Organic Synthesis I* (Ed.: A. de Meijere), Springer-Verlag, Berlin, **1986**, pp. 83–163; (b) R. D. Clark, K. G. Untch, *J. Org. Chem.* **1979**, *44*, 248–253; (c) R. D. Clark, K. G. Untch, *J. Org. Chem.* **1979**, *44*, 253–255.

48 (a) M. Ihara, T. Taniguchi, Y. Tokunaga, K. Fukumoto, *Synthesis* **1995**, 1405–1406; (b) M. Ihara, T. Taniguchi, Y. Tokunaga, K. Fukumoto, *J. Org. Chem.* **1994**, *59*, 8092–8100; (c) M. Ihara, T. Taniguchi, K. Makita, M. Takano, M. Onishi, N. Taniguchi, K. Fukumoto, C. Kabuto, *J. Am. Chem. Soc.* **1993**, *115*, 8107–8115; (d) M. Ihara, M. Ohnishi, M. Takano, K. Makita, N. Taniguchi, K. Fukumoto, *J. Am. Chem. Soc.* **1992**, *114*, 4408–4410.

49 (a) J. E. Banfield, D. S. Black, S. R. Johns, R. I. Willing, *Aust. J. Chem.* **1982**, *35*, 2247–2256; (b) K. C. Nicolaou, N. A. Petasis, R. E. Zipkin, J. Uenishi, *J. Am. Chem.*

Soc. **1982**, *104*, 5555–5557; (c) K. C. Nicolaou, N. A. Petasis, J. Uenishi, R. E. Zipkin, *J. Am. Chem. Soc.* **1982**, *104*, 5557–5558; (d) K. C. Nicolaou, R. E. Zipkin, N. A. Petasis, *J. Am. Chem. Soc.* **1982**, *104*, 5558–5560; (e) K. C. Nicolaou, N. A. Petasis, R. E. Zipkin, *J. Am. Chem. Soc.* **1982**, *104*, 5560–5562.

50 (a) E. L. Ghisalberti, P. R. Jefferies, P. Sheppard, *Tetrahedron Lett.* **1975**, *16*, 1775–1778; (b) K. D. Croft, E. L. Ghisalberti, P. R. Jeffries, D. G. Marshall, C. L. Raston, A. H. White, *Aust. J. Chem.* **1980**, *33*, 1529–1536; (c) M. L. Greenlee, *J. Am. Chem. Soc.* **1981**, *103*, 2425–2426; (d) W. G. Dauben, G. Shapiro, *J. Org. Chem.* **1984**, *49*, 4252–4258.

51 (a) J. Leimner, H. Maeschall, N. Meier, P. Weyerstahl, *Chem. Lett.* **1984**, *10*, 1769–1772; (b) T. Honda, K. Ueda, M. Tsubuki, T. Toya, A. Kurozumi, *J. Chem. Soc., Perkin Trans. 1* **1991**, *7*, 1749–1754.

52 E. Fattoarusso, V. Lanzotti, S. Magno, L. Mayol, M. Pansini, *J. Org. Chem.* **1992**, *57*, 6921–6924.

53 (a) J. J. Beereboom, *J. Am. Chem. Soc.* **1963**, *85*, 3525–3526; (b) J. J. Beereboom, *J. Org. Chem.* **1965**, *30*, 4320–4234; (c) R. B. Bates, M. J. Onore, S. K. Paknikar, C. Steelink, *Chem. Commun.* **1967**, 1037–1038.

54 (a) K. Shiomi, H. Iinuma, M. Hamada, H. Naganawa, M. Manabe, C. Matsuki, T. Takeuchi, H. Umezawa, *J. Antibiot.* **1986**, *39*, 487–493; (b) Y. Hori, Y. Abe, N. Shigematsu, T. Goto, M. Kohsaka, *J. Antibiot.* **1993**, *46*, 1890–1893.

55 (a) S. Toki, K. Ando, M. Yoshida, I. Kawamoto, H. Sano, Y. Matsuda, *J. Antibiot.* **1992**, *45*, 88–93; (b) S. Toki, K. Ando, I. Kawamoto, H. Sano, M. Yoshida, Y. Matsuda, *J. Antibiot.* **1992**, *45*, 1047–1054; (c) S. Toki, E. Tsukuda, M. Nazawa, H. Nonaka, M. Yoshida, Y. Matsuda, *Biol. Chem.* **1992**, *267*, 14884–14892.

56 A. Zeeck, P. Ruß, H. Laatsch, W. Loeffler, H. Wehrle, H. Zähner, H. Holst, *Chem. Ber.* **1979**, *112*, 957–978.

57 (a) T. Ichihashi, M. Izawa, K. Miyata, T. Mizui, K. Hirano, Y. Takagishi, *Pharmacol. Exp. Ther.* **1998**, *284(1)*, 43–50; (b) S. Hara, J. Higaki, K. Higashino, M. Iwai, N. Takasu, K. Miyata, K. Tonda, K. Nagata, Y. Go, T. Mizui, *Life Sci.* **1997**, *60(24)*, 365–370.

58 (a) Napyradiomycin A1: K. Tatsuta, Y. Tanaka, M. Kojima, H. Ikegami, *Chem. Lett.* **2002**, 14–15; (b) ES-242–4: K. Tatsuta, T. Yamazaki, T. Mase, T. Yoshimoto, *Tetrahedron Lett.* **1998**, *39*, 1771–1772.

59 D. Drochner, M. Müller, *Eur. J. Org. Chem.* **2001**, 211–215.

60 M. Mori, S. Takechi, S. Shimizu, S. Kida, H. Iwakura, M. Hajima, *Tetrahedron Lett.* **1999**, *40*, 1165–1168.

61 J. P. Deville, V. Behar, *Org. Lett.* **2002**, *4*, 1403–1405.

62 (a) N. Matsumoto, T. Tsuchida, M. Maruyama, R. Sawa, N. Kinoshita, Y. Homma, Y. Takahashi, H. Iinuma, H. Naganawa, T. Sawa, M. Hamada, T. Takeuchi, *J. Antibiot.* **1996**, *49*, 953–954; (b) N. Matsumoto, T. Tsuchida, M. Maruyama, N. Kinoshita, Y. Homma, H. Iinuma, T. Sawa, M. Hamada, T. Takeuchi, *J. Antibiot.* **1999**, *52*, 269–275; (c) N. Matsumoto, T. Tsuchida, H. Nakamura, R. Sawa, Y. Takahashi, H. Naganawa, H. Iinuma, T. Sawa, T. Takeuchi, *J. Antibiot.* **1999**, *52*, 276–280.

63 D. V. Kozhinov, V. Behar, *J. Org. Chem.* **2004**, *69*, 1378–1379.

64 U. Groth, W. Halfbrodt, T. Köhler, P. Kreye, *Liebigs Ann. Chem.* **1994**, 885–890.

65 T. Yoshihara, S. Togiya, H. Koshino, S. Sakamura, *Tetrahedron Lett.* **1985**, *26*, 5551–5554.

66 For related domino processes, see: U. Groth, W. Halfbrodt, A. Kalogerakis, T. Köhler, P. Kreye, *Synlett* **2004**, 291–294.

67 A. Covarrubias-Zúñiga, J. Diaz-Dominguez, J. S. Olguín-Uribe, *Synth. Commun.* **2001**, *31*, 1373–1381.

68 (a) B. Gosio, *Riv Igiene. Sanita Pubbl. Ann.* **1896**, *7*, 825; (b) B. Gosio, *Riv Igiene. Sanita Pubbl. Ann.* **1896**, *7*, 869; (c) B. Gosio, *Riv Igiene. Sanita Pubbl. Ann.* **1896**, *7*, 961.

69 C. M. Beaudry, J. P. Malerick, D. Trauner, *Chem. Rev.* **2005**, *105*, 4757–4778.

70 D. W. Jeffery, M. V. Perkins, J. M. White, *Org. Lett.* **2005**, *7*, 1581–1584.

71 M. Gavagnin, E. Mollo, G. Cimino, J. Ortea, *Tetrahedron Lett.* **1996**, *37*, 4259–4262.

72 (a) D. W. Jeffery, M. V. Perkins, *Tetrahedron Lett.* **2004**, *45*, 8667–8671; (b) D. W. Jeffery, M. V. Perkins, J. M. White, *Org. Lett.* **2005**, *7*, 407–409.

73 J. Zhang, J. L. Flippen-Anderson, A. P. Kozikowski, *J. Org. Chem.* **2001**, *66*, 7555–7559.

74 (a) D. Enders, J. Kirchhoff, D. Mannes, G. Raabe, *Synthesis* **1995**, 659–666; For related domino processes, see: (b) D. Enders, J. Wiedemann, *Liebigs Ann./Recueil* **1997**, 699–706; (c) D. Enders, W. Bettray, G. Raabe, J. Runsink, *Synthesis* **1994**, 1322–1326.

75 S. Chan, T. F. Braish, *Tetrahedron* **1994**, *50*, 9943–9950.

76 (a) K. Bevan, J. S. Davies, C. H. Hassall, R. B. Morton, D. A. S. Phillips, *J. Chem. Soc. (C)* **1971**, 514–521; (b) C. H. Hassall, Y. Ogihara, W. A. Thomas, *J. Chem. Soc. (C)* **1971**, 522–525; (c) C. H. Hassall, R. B. Morton, Y. Ogihara, D. A. S. Phillips, *J. Chem. Soc. (C)* **1971**, 526–531; (d) C. H. Hassall, W. A. Thomas, M. C. Moschidis, *J. Chem. Soc., Perkin Trans. 1* **1977**, 2369–2376.

77 K. J. Hale, N. Jogiya, S. Manaviazar, *Tetrahedron Lett.* **1998**, *39*, 7163–7166.

78 (a) T. H. Mäkelä, S. A. Kaltia, K. T. Wähälä, T. A. Hase, *Steroids* **2001**, *66*, 777–784; (b) T. H. Mäkelä, K. T. Wähälä, T. A. Hase, *Steroids* **2000**, *65*, 437–441.

79 A. Pelter, R. S. Ward, D. M. Jones, P. Maddocks, *Tetrahedron: Asymm.* **1992**, *3*, 239–242.

80 J. P. Marino, M. B. Rubio, G. Cao, A. de Dios, *J. Am. Chem. Soc.* **2002**, *124*, 13398–13399.

81 P. Forns, A. Diez, M. Rubiralta, *J. Org. Chem.* **1996**, *61*, 7882–7888.

82 J. E. Saxton, in: *The Alkaloids, Vol. 51* (Ed.: G. A. Cordell), Academic Press, New York, **1998**, Chapter 1.

83 (a) J. P. Marino, M. Neisser, *J. Am. Chem. Soc.* **1981**, *103*, 7687–7689; (b) J. P. Marino, A. D. Perez, *J. Am. Chem. Soc.* **1984**, *106*, 7643–7644; (c) J. P. Marino, M. Neisser, *J. Am. Chem. Soc.* **1988**, *110*, 966–968.

84 J. D. White, Q. Xu, C.-S. Lee, F. A. Valeriote, *Org. Biomol. Chem.* **2004**, *2*, 2092–2102.

85 T. R. R. Pettus, M. Inoue, X.-T. Chen, S. J. Danishefsky, *J. Am. Chem. Soc.* **2000**, *122*, 6160–6168.

86 (a) Y. Fukuyama, N. Shida, M. Kodama, H. Chaki, T. Yugami, *Chem. Pharm. Bull.* **1995**, *43*, 2270–2272; (b) Y. Fukuyama, Y. Hata, M. Kodama, *Planta Med.* **1997**, *63*, 275–277; c) H. Tsukui, I. Nihonmatsu, *Dev. Brain Res.* **1988**, *39*, 85–95.

87 (a) F. Hefti, *J. Neurobiol.* **1994**, 1418–1435; (b) F. Hefti, *Annu. Rev. Pharmacol. Toxicol.* **1997**, *37*, 239–267; (c) The cholinesterase inhibitors such as denepezil (Aricept) and tacrine (Cognex), which are capable of increasing acetylcholine levels, are now in use as Alzheimer's disease therapeutics. For a review, see: E. Giacobini, *Neurochem. Int.* **1998**, *32*, 413–419.

88 (a) E. P. Abraham, P. B. Loder, in: *Cephalosporins and Penicillins; Chemistry and Biology* (Ed.: E. H. Flynn), Academic Press, New York, **1972**, pp. 1–26; (b) S. Kukolja, R. R. Chauvette, in: *Chemistry and Biology of β-Lactam Antibiotics* (Eds.: R. B. Morin, M. Gorman), Academic Press, New York, **1982**, Vol. 1, pp. 93–198; (c) W. Dürckheimer, J. Blumbach, R. Lattrell, K. H. Scheunemann, *Angew. Chem., Int. Ed. Engl.* **1985**, *24*, 180–202.

89 J. Kant, J. A. Roth, C. E. Fuller, D. G. Walker, D. A. Benigni, V. Farina, *J. Org. Chem.* **1994**, *59*, 4956–4966.

90 K. Tomatsu, A. Shigeyuki, S. Masuyoshi, S. Kondo, M. Hirano, T. Miyaki, H. Kawaguchi, *J. Antibiot.* **1987**, *40*, 1175–1183.

91 M. W. Nötzel, M. Tamm, T. Labahn, M. Noltemeyer, M. Es-Sayed, A. de Meijere, *J. Org. Chem.* **2000**, *65*, 3850–3852.

92 M. W. Nötzel, T. Labahn, M. Es-Sayed, A. de Meijere, *Eur. J. Org. Chem.* **2001**, 3025–3030.

93 (a) M. C. Pirrung, J. Cao, J. Chen, *J. Org. Chem.* **1995**, *60*, 5790–5794; (b) M. C. Pirrung, H. Han, J. Chen, *J. Org. Chem.* **1996**, *60*, 4527–4531; (c) M. C. Pirrung, L. M. Kaiser, J. Chen, *Biochemistry* **1993**, *32*, 7445–7450.

94 K. Burgess, K.-K. Ho, D. Moye-Sherman, *Synlett* **1994**, 575–583.

95 V. Capriati, S. Florio, R. Luisi, F. M. Perna, J. Barluenga, *J. Org. Chem.* **2005**, *70*, 5852–5858.

96 (a) J. A. Peters, *Synthesis* **1979**, 321–336; (b) J. A. Peters, J. M. A. Baas, B. V. D. Graf, J. M. V. D. Toorn, H. V. Bekkum, *Tetrahedron* **1978**, *34*, 3313–3323; (c) C. Jaime, E. Osawa, Y. Takeuchi, P. Camps, *J. Org. Chem.* **1983**, *48*, 4514–4519; (d) A. Gambacorta, G. Fabrizi, P, Bovicelli, *Tetrahedron* **1992**, *48*, 4459–4464.

97 A. Srikrishna, T. Jagadeeswar Reddy, P. Praveen Kumar, *Synlett* **1997**, 663–664.

98 (a) R. C. Bunce, J. C. Allison, *Synth. Commun.* **1999**, *29*, 2175–2186; (b) R. A. Bunce, C. J. Peeples, P. B. Jones, *J. Org. Chem.* **1992**, *57*, 1727–1733.

99 For a review of the homoerythrina alkaloids, see: R. I. Bick, S. Panichanun, in: *Alkaloids: Chemical and Biological Perspectives, Vol. VII* (Ed.: S. W. Pelletier), Springer-Verlag, New York, **1990**, Chapter 1.

100 M.-A. Le Dréau, D. Desmaële, F. Dumas, J. d'Angelo, *J. Org. Chem.* **1993**, *58*, 2933–2935.

101 (a) J. Aubé, G. L. Milligan, C. J. Mossman, *J. Org. Chem.* **1992**, *57*, 1635–1637; (b) J. Aubé, G. L. Milligan, C. J. Mossman, *J. Am. Chem. Soc.* **1991**, *113*, 8965–8966; (c) W. H. Pearson, J. M. Schkeryantz, *Tetrahedron Lett.* **1992**, *33*, 5291–5294.

102 D. Obrecht, F. Gerber, D. Sprenger, T. Masquelin, *Helv. Chim. Acta* **1997**, *80*, 531–537.

103 O. A. Attanasi, L. De Crescentini, G. Favi, P. Filippone, F. Mantellini, S. Santeusanio, *J. Org. Chem.* **2002**, *67*, 8178–8181.

104 A. Arcadi, F. Marinelli, E. Rossi, *Tetrahedron* **1999**, *55*, 13233–13250.

105 G. Abbiati, A. Arcadi, F. Marinelli, E. Rossi, *Synlett* **2002**, 1912–1916.

106 (a) K. Makino, O. Hara, Y. Takiguchi, T. Katano, Y. Asakawa, K. Hatano, Y. Hamada, *Tetrahedron Lett.* **2003**, *44*, 8925–8929; (b) O. Hara, K. Sugimoto, K. Makino, Y. Hamada, *Synlett* **2004**, 1625–1627.

107 (a) B. Lesch, S. Bräse, *Angew. Chem. Int. Ed.* **2004**, *43*, 115–118; (b) C. F. Nising, U. K. Ohnemüller (née Schmid), S. Bräse, *Angew. Chem. Int. Ed.* **2006**, *45*, 307–309.

108 M. Nakamura, C. Liang, E. Nakamura, *Org. Lett.* **2004**, *6*, 2015–2017.

109 P. Saikia, D. Prajapati, J. S. Sandhu, *Tetrahedron Lett.* **2003**, *44*, 8725–8727.

110 N. Cramer, S. Laschat, A. Baro, H. Schwalbe, C. Richter, *Angew. Chem. Int. Ed.* **2005**, *44*, 820–822.

111 (a) K. Krohn, M. Riaz, *Tetrahedron Lett.* **2004**, *45*, 293–294; (b) K. Krohn, M. Riaz, U. Flörke, *Eur. J. Org. Chem.* **2004**, 1261–1270.

112 A. Alexakis, S. March, *J. Org. Chem.* **2002**, *67*, 8753–8757.

113 (a) For the isolation and absolute configuration determination of clavukerin A, see: M. Kobayashi, B. W. Son, M. Kido, Y. Kyogoku, I. Kitawaga, *Chem. Pharm. Bull.* **1983**, *31*, 2160–2163; (b) For the absolute configuration determination of isoclavukerin, see: T. Kusumi, T. Hamada, M. Hara, M. O. Ishitsuka, H. Ginda, H. Kakisawa, *Tetrahedron Lett.* **1992**, *33*, 2019–2022.

114 (a) T. Fujimoto, Y. Takeuchi, K. Kai, Y. Hotei, K. Ohta, I. Yamamoto, *J. Chem. Soc., Chem. Commun.* **1992**, 1263–1264; (b) T. Fujimoto, Y. Uchiyama, Y. Kodama, K. Ohta, I. Yamamoto, *J. Org. Chem.* **1993**, *58*, 7322–7323; (c) T. Fujimoto, Y. Kodama, I. Yamamoto, *J. Org. Chem.* **1997**, *62*, 6627–6630; (d) T. Nagao, T. Suenaga, T. Ichihashi, T. Fujimoto, I. Yamamoto, A. Kakehi, R. Iriye, *J. Org. Chem.* **2001**, *66*, 890–893; (e) N. Kishimoto, T. Fujimoto, I. Yamamoto, *J. Org. Chem.* **1999**, *64*, 5988–5992.

115 (a) W. S. Browers, T. Ohta, J. S. Cleere, P. A. Marsella, *Science* **1976**, *193*, 542–547; (b) E. E. Schweizer, O. Meeder-Nycz, in: *Chromenes, Chromanes, Chromones* (Ed.: P. Ellis), Wiley-Interscience, New York, **1977**, pp. 11–139; (c) G. P. Ellis, I. M. Lockhart, *Chromans and Tocopherols*, Wiley-Interscience, New York, **1981**; (d) G. Broggini, G. Folcio, N. Sardone, M. Sonzogni, G. Zecchi, *Tetrahedron: Asymm.* **1996**, *7*, 797–806 and references therein.

116 H. L. van Lingen, W. Zhuang, T. Hansen, F. P. J. T. Rutjes, K. A. Jørgensen, *Org. Biomol. Chem.* **2003**, *1*, 1953–1958.

117 G. W. Burton, K. U. Ingold, *Acc. Chem. Res.* **1986**, *19*, 194–201 and references therein.

118 R. Hiessböck, C. Wolf, E. Richter, M. Hitzler, P. Chiba, M. Kratzel, G. Ecker, *J. Med. Chem.* **1999**, *42*, 1921–1926.

119 N. Cohen, G. Weber, B. L. Banner, R. J. Lopresti, B. Schaer, A. Focella, G. B. Zenchoff, A. Chiu, L. Todaro, M. O'Donnell, A. F. Welton, D. Brown, R. Garippa, H. Crowley, D. W. Morgan, *J. Med. Chem.* **1989**, *32*, 1842–1860.

120 C. A. Lipinski, C. E. Aldinger, T. A. Beyer, J. Bordner, D. F. Burdi, D. L. Bussolotti, P. B. Inskeep, T. W. Siegel, *J. Med. Chem.* **1992**, *35*, 2169–2177.

121 J. F. Austin, S.-G. Kim, C. J. Sinz, W.-J. Xiao, D. W. C. MacMillan, *Proc. Natl. Acad. Sci. USA* **2004**, *101*, 5482–5487.

122 (a) J. S. Carle, C. Chrisophersen, *J. Org. Chem.* **1981**, *46*, 3440–3443; (b) J. S. Carle, C. Chrisophersen, *J. Org. Chem.* **1980**, *45*, 1586–1589; (c) J. S. Carle, C. Chrisophersen, *J. Am. Chem. Soc.* **1979**, *101*, 4012–4013.

123 T. Lavoisier-Gallo, E. Charonnet, J. Rodriguez, *J. Org. Chem.* **1998**, *63*, 900–902.

124 A. Zakarian, A. Batch, R. A. Holton, *J. Am. Chem. Soc.* **2003**, *125*, 7822–7824.

125 H. Takikawa, M. Hirooka, M. Sasaki, *Tetrahedron Lett.* **2003**, *44*, 5235–5238.

126 A. F. Barrero, J. E. Oltra, M. Álvarez, *Tetrahedron Lett.* **2000**, *41*, 7639–7643.

127 P. Langer, I. Freifeld, *Chem. Eur. J.* **2001**, *7*, 565–572.

128 C.-W. Cho, M. J. Krische, *Angew. Chem. Int. Ed.* **2004**, *43*, 6689–6691.

129 Y.-L. Lin, H.-S. Kuo, Y.-W. Wang, S.-T. Huang, *Tetrahedron* **2003**, *59*, 1277–1281.

130 M. Engler, T. Anke, O. Sterner, U. Brandt, *J. Antibiot.* **1997**, *50*, 325–329.

131 B. Patterson, S. Marumoto, S. D. Rychnovsky, *Org. Lett.* **2003**, *5*, 3163–3166.

132 D. J. Kopecky, S. D. Rychnovsky, *J. Am. Chem. Soc.* **2001**, *123*, 8420–8421.

133 M. D. D'Ambrosio, A. Guerriero, C. Debitus, F. Pietra, *Helv. Chim. Acta* **1996**, *79*, 51–60.

134 H. Comas, E. Pandolfi, *Synthesis* **2004**, 2493–2498.

135 A. Yungai, F. G. West, *Tetrahedron Lett.* **2004**, *45*, 5445–5448.

136 P. Langer, V. Köhler, *Org. Lett.* **2000**, *2*, 1597–1599.

137 J. P. N. Papillon, R. J. K. Taylor, *Org. Lett.* **2000**, *2*, 1987–1990.

138 C. Wedler, B. Costisella, H. Schick, *J. Org. Chem.* **1999**, *64*, 5301–5303.

139 (a) Y. Wang, C. Zhao, D. Romo, *Org. Lett.* **1999**, *1*, 1197–1199; (b) H. W. Yang, C. Zhao, D. Romo, *Tetrahedron* **1997**, *53*, 16471–16488; (c) H. W. Yang, D. Romo, *J. Org. Chem.* **1997**, *62*, 4–5; (d) H. W. Yang, D. Romo, *J. Org. Chem.* **1998**, *63*, 1344–1347; (e) W. D. Schmitz, N. Messerschmidt, D. Romo, *J. Org. Chem.* **1998**, *63*, 2058–2059.

140 M. Shibasaki, H. Sasai, T. Arai, T. Iida, *Pure & Appl. Chem.* **1998**, *70*, 1027–1034.

141 Y. Yamamoto, N. Momiyama, H. Yamamoto, *J. Am. Chem. Soc.* **2004**, *126*, 5962–5963.

142 A. Ullmann, M. Gruner, H.-U. Reißig, *Chem. Eur. J.* **1999**, *5*, 187–197.

143 S. K. Verma, M. N. Atanes, J. H. Busto, D. L. Thai, H. Rapoport, *J. Org. Chem.* **2002**, *67*, 1314–1318.

144 R. Marshall Werner, L. M. Williams, J. T. Davis, *Tetrahedron* **1998**, *39*, 9135–9138.

145 Justicidin B is widely found in nature. For example, see: (a) J. Asano, K. Chiba, M. Tada, T. Yoshii, *Phytochemistry* **1996**, *42*, 713–717; (b) G. M. Sheria, K. M. A. Amer, *Phytochemistry* **1984**, *23*, 151–153; (c) T. L. Bachmann, F. Ghia, K. B. G. Torsell, *Phytochemistry* **1993**, *33*, 189–191; (d) B. Goezler, G. Arar, T. Goezler, M. Hesse, *Phytochemistry* **1992**, *31*, 2473–2475; (e) G. R. Pettit, G. M. Cragg, M. I. Suffness, D. Gust, F. E. Boettner, M. Williams, J. A. Saenz-Renauld, P. Brown, J. M. Schmidt, P. D. Ellis, *J. Org. Chem.* **1984**, *49*, 4258–4266; (f) M. Okigawa, T. Maeda, N. Kawano, *Tetrahedron* **1970**, *26*, 4301–4305. For previous syntheses, see: (g) T. Ogiku, M. Seki, M. Takahashi, H. Ohmizu, T. Iwasaki, *Tetrahedron Lett.* **1990**, *31*, 5487–5490; (h) J. Banerji, B. Das, A. Chatterjee, J. N. Shoolery, *Phytochemistry* **1984**, *23*, 2323–2327; (i) S. O. de Silva, C. St. Denis, R. Rodrigo, *J. Chem. Soc., Chem. Commun.* **1980**, 995–997; (j) A. Kamal, M. Daneshtalab, R. G. Micetich, *Tetrahedron Lett.* **1994**, *35*, 3879–3882; (k) T. Ogiku, S. Yoshida, H. Ohmizu, T. Iwasaki, *J. Org. Chem.* **1995**, *60*, 4585–4590; (l) K. Kobayashi, J. Tokimatsu, K. Maeda, O. Morikawa, H. Konishi, *J. Chem. Soc., Perkin Trans. 1* **1995**, 3013–3016; (m) J. L. Charlton, C. J. Oleschuk, G.-L. Chee, *J. Org. Chem.* **1996**, *61*, 3452–3457; For biological properties of retro-justicidin B, see: (n) A. Pelter, R. S. Ward, P. Satyanarayana, P. Collins, *J. Chem. Soc., Perkin Trans. 1* **1983**, 643–647; (o) P. Satyanarayana, P. K. Rao, *Indian J. Chem. Sect. B* **1985**, 151–153; (p) C. W. Chang, M. T. Lin, S. S. Lee, K. C. S. C. Liu, F. L. Hsu, J. Y. Lin, *Antiviral Res.* **1995**, *27*, 367–374.

146 D. C. Harrowven, M. Bradley, J. L. Castro, S. R. Flanagan, *Tetrahedron Lett.* **2001**, *42*, 6973–6975.

147 E. J. Corey, G. T. Kwiatkowski, *J. Am. Chem. Soc.* **1966**, *88*, 5654–5656.

148 J. Mulzer, H. J. Martin, B. List, *Tetrahedron Lett.* **1996**, *37*, 9177–9178.

149 (a) R. Schobert, S. Müller, H.-J. Bestmann, *Synlett* **1995**, 425–426; (b) J. Löf-

fler, R. Schobert, *J. Chem. Soc., Perkin Trans 1* **1996**, 2799–2802.

150 P. Molina, P. M. Fresneda, S. Delgado, J. A. Bleda, *Tetrahedron Lett.* **2002**, *43*, 1005–1007.

151 D. Foulkner, *J. Nat. Prod. Rep.* **2001**, *18*, 1–49.

152 (a) N. B. Perry, L. Ettouati, M. Litaudon, J. W. Blunt, M. H. G. Munro, S. Parkin, H. Hope, *Tetrahedron* **1994**, *50*, 3987–3992; (b) G. Trimurtulu, D. J. Faulkner, N. B. Perry, L. Ettouatti, M. Litaudon, J. W. Blunt, M. H. G. Munro, G. B. Jameson, *Tetrahedron* **1994**, *50*, 3993–4000.

153 P. M. Fresneda, P. Molina, S. Delgado, J. A. Bleda, *Tetrahedron Lett.* **2000**, *41*, 4777–4780.

154 Š. Marchalín, K. Cvopová, D.-P. Pham-Huu, M. Chudík, J. Kozisek, I. Svoboda, A. Daïch, *Tetrahedron Lett.* **2001**, *42*, 5663–5567.

155 G. Sartori, R. Maggi, F. Bigi, C. Porta, X. Tao, G. L. Bernardi, S. Ianelli, M. Nardelli, *Tetrahedron* **1995**, *44*, 12179–12192.

156 P. A. Clarke, W. H. C. Martin, *Org. Lett.* **2002**, *4*, 4527–4529.

157 G. Sabitha, G. S. K. Kumar Reddy, M. Rajkumar, J. S. Yadav, K. V. S. Ramakrishna, A. C. Kunwar, *Tetrahedron Lett.* **2003**, *44*, 7455–7457.

158 J. M. Betancort, K. Sakthivel, R. Thayumanavan, C. F. Barbas, III, *Tetrahedron Lett.* **2001**, *42*, 4441–4444.

159 D. F. Fry, C. B. Fowler, R. K. Dieter, *Synlett* **1994**, 836–838.

160 F. F. Fleming, V. Gudipati, O. W. Steward, *J. Org. Chem.* **2003**, *68*, 3943–3946.

161 V. Capriati, S. Florio, R. Luisi, B. Musio, *Org. Lett.* **2005**, *7*, 3749–3752.

162 P. A. Evans, J. Cui, S. J. Gharpure, R. J. Hinkle, *J. Am. Chem. Soc.* **2003**, *125*, 11456–11457.

163 (a) For a recent review, see: P. Langer, M. Döring, *Eur. J. Org. Chem.* **2002**, 221–234; (b) J. T. Anders, H. Görls, P. Langer, *Eur. J. Org. Chem.* **2004**, 1897–1910.

164 R. A. Stockman, A. Sinclair, L. G. Arini, P. Szeto, D. L. Hughes, *J. Org. Chem.* **2004**, *69*, 1598–1602.

165 N. Baricordi, S. Benetti, G. Biondini, C. De Risi, G. P. Pollini, *Tetrahedron Lett.* **2004**, *45*, 1373–1375.

166 D. Enders, V. Braig, M. Boudou, G. Raabe, *Synthesis* **2004**, 2980–2990.

167 (a) M. Shindo, Y. Sato, K. Shishido, *J. Am. Chem. Soc.* **1999**, *121*, 6507–6508; (b)

M. Shindo, K. Matsumoto, Y. Sato, K. Shishido, *Org. Lett.* **2001**, *3*, 2029–2031.

168 S. S. Bahr, M. M. V. Ramana, *J. Org. Chem.* **2004**, *69*, 8935–8937.

169 M. Beller, C. Breindl, T. H. Riermeier, M. Eichberger, H. Trauthwein, *Angew. Chem. Int. Ed.* **1998**, *37*, 3389–3391.

170 G. A. Molander, Y. Le Huérou, G. A. Brown, *J. Org. Chem.* **2001**, *66*, 4511–4516.

171 X.-D. Hu, C.-A. Fan, F.-M. Zhang, Y. Q. Tu, *Angew. Chem. Int. Ed.* **2004**, *43*, 1702–1705.

172 A. Rosales, R. E. Estévez, J. M. Cuerva, J. E. Oltra, *Angew. Chem. Int. Ed.* **2005**, *44*, 319–322.

173 C. Lu, X. Lu, *Org. Lett.* **2002**, *4*, 4677–4679.

174 P. J. Parsons, N. P. Camp, J. M. Underwood, D. M. Harvey, *Tetrahedron* **1996**, *52*, 11637–11642.

175 W. W. Carmichael, D. F. Biggs, P. R. Gorham, *Science* **1975**, *187*, 542–544.

176 (a) A. B. Smith, III, N. Kanoh, H. Ishiyama, N. Minakawa, J. D. Rainier, R. A. Hartz, Y. Shin Cho, H. Cui, W. H. Moser, *J. Am. Chem. Soc.* **2003**, *125*, 8228–8237; (b) A. B. Smith, III, N. Kanoh, H. Ishiyama, R. A. Hartz, *J. Am. Chem. Soc.* **2000**, *122*, 11254–11255; (c) A. B. Smith, III, N. Kanoh, N. Minakawa, J. D. Rainier, F. R. Blase, R. A. Hartz, *Org. Lett.* **1999**, *1*, 1263–1266.

177 R. M. T. Dahlgren, H. T. Clifford, P. F. Yeo, *The Families of the Monocotyledons. Structure Evolution and Taxonomy*, Springer-Verlag, Berlin, **1985**.

178 (a) K. Sakata, K. Aoki, C.-F. Chang, A. Sakurai, S. Tamura, S. Murakoshi, *Agric. Biol. Chem.* **1978**, *42*, 457–463; (b) H. Shinozaki, M. Ishida, *Brain. Res.* **1985**, *334*, 33–40; (c) Y. Ye, G.-W. Qin, R.-S. Xu, *Phytochemistry* **1994**, *37*, 1205–1208.

179 For recent reviews, see: (a) R. A. Pilli, M. Ferriera de Oliviera, *Nat. Prod. Rep.* **2000**, *17*, 117–127; (b) Y. Ye, G.-W. Qin, R.-S. Xu, *J. Nat. Prod.* **1994**, *57*, 665–669.

180 (a) K. Suzuki, *J. Pharm. Soc. Jpn.* **1929**, *49*, 457; (b) K. Suzuki, *J. Pharm. Soc. Jpn.* **1931**, *51*, 419.

181 D. R. Williams, K. Shamim, J. P. Reddy, G. S. Amato, S. M. Shaw, *Org. Lett.* **2003**, *5*, 3361–3364.

182 D. R. Williams, M. H. Osterhout, J. M. McGill, *Tetrahedron Lett.* **1989**, *30*, 1327–1330.

183 D. R. Williams, M. H. Osterhout, J. M. McGill, *Tetrahedron Lett.* **1989**, *30*, 1331–1334.

184 T. Vojkovský, A. Weichsel, M. Pátek, *J. Org. Chem.* **1998**, *63*, 3162–3163.

185 E. W. Tate, S. Z. Zard, *Tetrahedron Lett.* **2002**, *43*, 4683–4686.

186 M. Lounasmaa, L. Miikki, A. Tolvanen, *Tetrahedron* **1996**, *52*, 9925–9930.

187 (a) W. W. Paudler, G. I. Kerley, J. McKay, *J. Org. Chem.* **1963**, *28*, 2194–2197; (b) M. E. Wall, C. R. Eddy, J. J. Willaman, D. S. Cordell, B. G. Schubert, H. S. Gentry, *J. Am. Pharm. Assoc.* **1954**, *43*, 503–507; (c) L. Huang, Z. Xue, in: *The Alkaloids, Vol. 23* (Ed.: A. Brossi), Academic Press, New York, **1984**, pp. 157–226.

188 (a) D. Schinzer, Y. Bo, *Angew. Chem. Int. Ed. Engl.* **1991**, *30*, 687–688; (b) D. Schinzer, E. Langkopf, *Synlett* **1994**, 375–377; (c) D. Schinzer, U. Abel, P. G. Jones, *Synlett* **1997**, 632–634.

189 (a) L. F. Tietze, P. L. Steck, *Eur. J. Org. Chem.* **2001**, 4353–4356; (b) L. F. Tietze, H. Braun, unpublished.

190 T. Cohen, T. Kreethadumrongdat, X. Liu, V. Kulkarni, *J. Am. Chem. Soc.* **2001**, *123*, 3478–3483.

191 (a) G. Fraenkel, J. G. Russell, Y. Chen, *J. Am. Chem. Soc.* **1973**, *95*, 3208–3215; (b) X. Wei, P. Johnson, R. J. K. Taylor, *J. Chem. Soc., Perkin Trans 1* **2000**, 1109–1116.

192 For another total synthesis based on a domino process, see: R. S. Grainger, A. Patel, *Chem. Comm.* **2003**, 1072–1073.

193 X. Wei, R. J. K. Taylor, *Angew. Chem. Int. Ed.* **2000**, *39*, 409–412; see also references cited therein.

194 M. Marigo, T. Schulte, J. Franzén, K. A. Jørgensen, *J. Am. Chem. Soc* **2005**, *127*, 15710–15711.

195 Y. Huang, A. M. Walji, C. H. Larsen, D. W. C. MacMillan, *J. Am. Chem. Soc.* **2005**, *127*, 15051–15053.

196 A. L. de Sousa, R. A. Pilli, *Org. Lett.* **2005**, *7*, 1617–1619.

197 M. Kuramoto, C. Tong, K. Yamada, T. Chiba, Y. Hayashi, D. Uemura, *Tetrahedron Lett.* **1996**, *37*, 3867–3870.

198 (a) M. C. Carreño, C. García Luzón, M. Ribagorda, *Chem. Eur. J.* **2002**, *8*, 208–216; (b) For further applications, see: M. C. Carreño, M. Ribagorda, G. H. Posner, *Angew. Chem. Int. Ed.* **2002**, *41*, 2753–2755.

199 A. Padwa, S. S. Murphree, Z. Ni, S. H. Watterson, *J. Org. Chem.* **1996**, *61*, 3829–3838.

200 (a) H. Sakai, H. Hagiwara, T. Hoshi, T. Suzuki, M. Ando, *Synthetic Commun.* **1999**, *29*, 2035–2042; (b) H. Hagiwara, H. Sakai, M. Kirita, T. Hoshi, T. Suzuki, M. Ando, *Tetrahedron* **2000**, *56*, 1445–1449.

201 (a) H.-J. Gutke, K. Oesterreich, D. Spitzner, N. A. Braun, *Tetrahedron* **2001**, *57*, 997–1003; (b) N. A. Braun, I. Klein, D. Spitzner, B. Vogler, S. Braun, H. Borrmann, A. Simon, *Liebigs Ann.* **1995**, 2165–2169; (c) N. A. Braun, I. Klein, D. Spitzner, B. Vogler, H. Borrmann, A. Simon, *Tetrahedron Lett.* **1995**, *36*, 3675–3678; (d) N. A. Braun, N. Stumpf, D. Spitzner, *Synthesis* **1997**, 917–920.

202 (a) H.-U. Reißig, E. Hirsch, *Angew. Chem. Int. Ed. Engl.* **1980**, *19*, 813–814; (b) E. Wenkert, *Acc. Chem. Res.* **1980**, *13*, 27–31.

203 J.-K. Ergüden, H. W. Moore, *Org. Lett.* **1999**, *1*, 375–377.

204 D. B. Clarke, S. F. R. Hinkley, R. T. Weavers, *Tetrahedron Lett.* **1997**, *38*, 4297–4300.

205 M. Ihara, K. Makita, Y. Tokunaga, K. Fukumoto, *J. Org. Chem.* **1994**, *59*, 6008–6013.

206 H. Paulsen, S. Antos, A. Brandes, M. Lögers, S. N. Müller, P. Naab, C. Schmeck, S. Schneider, J. Stoltefuß, *Angew. Chem. Int. Ed.* **1999**, *38*, 3373–3375.

207 (a) M. Sugano, N. Makino, S. Sawada, S. Otsuka, M. Watanabe, H. Okamoto, M. Kamada, *J. Biol. Chem.* **1998**, *273*, 5033–5036; (b) A. S. Plump, L. Masucci-Magoulas, C. Bruce, C. L. Bisgaier, J. L. Breslow, A. R. Tall, *Arterioscler. Thromb. Vasc. Biol.* **1999**, *19*, 1105–1110; (c) Overview: C. G. Stevenson, *Crit. Rev. Clin. Lab. Sci.* **1998**, *35*, 517–546.

208 K. Kobayashi, M. Uchida, T. Uneda, K. Yoneda, M. Tanmatsu, O. Morikawa, H. Konishi, *Eur. J. Chem. Soc., Perkin Trans. 1* **2001**, 2977–2982.

209 S. Kobayashi, R. Akiyama, M. Moriwaki, *Tetrahedron Lett.* **1997**, *38*, 4819–4822.

210 S. Luo, J. Zhao, H. Zhai, *J. Org. Chem.* **2004**, *69*, 4548–4550.

211 S. Luo, C. A. Zificsak, R. P. Hsung, *Org. Lett.* **2003**, *5*, 4709–4712.

212 C. Palomo, J. M. Aizpurua, S. García-Granda, P. Pertierra, *Eur. J. Org. Chem.* **1998**, 2201–2207.

213 (a) P. D. Edwards, P. R. Bernstein, *Med. Res. Rev.* **1994**, *14*, 127–194; (b) O. A. Mascaretti, C. E. Boschetti, G. O. Danelon, E. G. Mata, O. A. Roveri, *Curr. Med. Chem.* **1995**, *1*, 441–470.

214 F. Denes, F. Chemla, J. F. Normant, *Eur. J. Org. Chem.* **2002**, 3536–3542.

215 J. Jauch, *J. Org. Chem.* **2001**, *66*, 609–611.

216 (a) J. Jauch, *Synlett* **1999**, 1325–1327; (b) 2875; *Angew. Chem., Int. Ed.* **2000**, *39*, 2764–2765.

217 M. Casey, C. M. Keaveney, *Chem. Commun.* **2004**, 184–185.

218 F. D'Onofrio, R. Margarita, L. Parlanti, G. Piancatelli, M. Sbraga, *Chem. Commun.* **1998**, 185–186.

219 S. Kittivarapong, S. Rajviroonkit, T. Siwapinyoyos, C. Thebtaranonth, Y. Thebtaranonth, *Synthesis* **1995**, 136–138.

220 G. Li, H.-X. Wei, B. S. Phelps, D. W. Purkiss, S. H. Kim, *Org. Lett.* **2001**, *3*, 823–826.

221 H.-X. Wei, S. H. Kim, G. Li, *Org. Lett.* **2002**, *21*, 3691–3693.

222 (a) J. S. Yadav, B. V. S. Reddy, G. Parimala, A. Krishnam Raju, *Tetrahedron Lett.* **2004**, *45*, 1543–1546; (b) J. S. Yadav, B. V. S. Reddy, B. Padmavani, *Synthesis* **2004**, 405–408.

223 R. Ballini, L. Barboni, D. Fiorini, A. Palmieri, *Synlett* **2004**, 2618–2620.

224 C. Simon, J.-F. Peyronel, J. Rodriguez, *Org. Lett.* **2001**, *3*, 2145–2148.

225 G. Bose, V. T. H. Nguyen, E. Ullah, S. Lahiri, H. Görls, P. Langer, *J. Org. Chem.* **2004**, *69*, 9128–9134.

226 U. Beifuss, K. Goldenstein, F. Döring, C. Lehmann, M. Noltemeyer, *Angew. Chem. Int. Ed.* **2001**, *40*, 568–570.

227 K. Mori, G. Audran, H. Monti, *Synlett* **1998**, 259–260.

228 (a) J. B. Gloer, H.-J. Wang, J. A. Scott, D. Malloch, *Tetrahedron Lett.* **1995**, *36*, 5847–5850; (b) H. Fujimoto, M. Inagaki, Y. Satoh, E. Yoshida, M. Yamazaki, *Chem. Pharm. Bull.* **1996**, *44*, 1090–1092.

229 (a) K. Takeda, M. Fujisawa, T. Makino, E. Yoshii, K. Yamaguchi, *J. Am. Chem. Soc.* **1993**, *115*, 9351–9352; (b) K. Takeda, I. Nakayama, E. Yoshii, *Synlett* **1994**, 178–178; (c) K. Takeda, K. Kitagawa, I. Nakayama, E. Yoshii, *Synlett* **1997**, 255–256; (d)

For the formation of other carbocycles by using the same methodology, see: K. Takeda, J. Nakatani, H. Nakamura, K. Sako, E. Yoschii, K. Yamaguchi, *Synlett* **1993**, 841–843.

230 For some examples, see: (a) K. Takeda, A. Nakajima, M. Takeda, Y. Okamoto, T. Sato, E. Yoshii, T. Koizumi, M. Shiro, *J. Am. Chem. Soc.* **1998**, *120*, 4947–4959; (b) K. Takeda, Y. Sawada, K. Sumi, *Org. Lett.* **2002**, *4*, 1031–1033; (c) Y. Sawada, M. Sasaki, K. Takeda, *Org. Lett.* **2004**, *6*, 2277–2279; (d) K. Takeda, Y. Ohtani, *J. Org. Chem.* **1999**, *1*, 677–679.

231 (a) K. Takeda, Y. Ohnishi, *Tetrahedron Lett.* **2000**, *41*, 4169–4172; (b) For a related 2.2.2.2 approach, see: K. Tanaka, K. Takeda, *Tetrahedron Lett.* **2004**, *45*, 7859–7861.

232 (a) K. Takeda, E. Kawanishi, M. Sasaki, Y. Takahashi, K. Yamaguchi, *Org. Lett.* **2002**, *4*, 1511–1514; (b) M. Sasaki, E. Kawanishi, Y. Nakai, T. Matsumoto, K. Yamaguchi, K. Takeda, *J. Org. Chem.* **2003**, *68*, 9330–9339; (c) S. Okugawa, K. Takeda, *Org. Lett.* **2004**, *6* 2973–2975.

233 For pertinent reviews, see: (a) T. Hudlicky, G. Butora, S. P. Fearnley, A. G. Gum, M. R. Stabile, *Studies in Natural Products Chemistry*, Vol. 18 (Ed.: Atta-ur-Rahman), Elsevier, Amsterdam, **1996**, pp. 43–154; (b) B. H. Novak, T. Hudlicky, J. W. Reed, J. Mulzer, D. Trauner, *Curr. Org. Chem.* **2000**, *4*, 343–362; (c) K. W. Bentley, *Nat. Prod. Rep.* **2000**, *17*, 247–268.

234 H. Nagata, N. Miyazawa, K. Ogasawara, *Chem. Commun.* **2001**, 1094–1095.

235 (a) J. Mulzer, G. Durner, D. Trauner, *Angew. Chem., Int. Ed. Engl.*, **1996**, *35*, 2830–2832; (b) J. Mulzer, J. W. Bats, B. List, T. Opatz, D. Trauner, *Synlett* **1997**, 441–444; (c) D. Trauner, S. Porth, T. Opatz, J. W. Bats, G. Giester, J. Mulzer, *Synthesis* **1998**, 653–664; (d) D. Trauner, J. W. Bats, A. Werner, J. Mulzer, *J. Org. Chem.* **1998**, *63*, 5908–5918; (e) J. Mulzer, D. Trauner, *Chirality* **1999**, *11*, 475–482.

236 C. W. Brandt, L. G. Neubauer, *J. Chem. Soc.* **1939**, 1031–1037.

237 H. Nagata, N. Miyazawa, K. Ogasawara, *Org. Lett.* **2001**, *3*, 1737–1740.

238 N. Miyazawa, K. Osagawara, *Tetrahedron Lett.* **2002**, *43*, 4773–4776.

239 For pertinent reviews, see: (a) C. Szantay, G. Blasko, K. Honty, G. Dörnyei, in: *The Alkaloids: Coryantheine, Yohimbine, and Related Alkaloids, Vol. 27* (Ed.: A. Brossi), Academic Press, Orlando, **1986**, pp. 131–268; (b) E. W. Baxter, P. S. Mariano, in: *Alkaloids: Chemical and Biological Perspectives, Vol. 8* (Ed.: S. W. Pelletier), Springer-Verlag, New York, **1992**, pp. 197–319; (c) J. Aube, S. Ghosh, in: *Advances in Heterocyclic Natural Product Synthesis Vol. 3* (Ed.: W. H. Pearson), JAI Press, Greenwich, **1996**, pp. 99–150.

240 P. Deslongchamps, in: *Stereoelectronic Effects in Organic Chemistry*, Pergamon Press, Oxford, **1983**, pp. 209–290.

241 A. T. Keene, J. D. Phillipson, D. C. Warhurst, M. Koch, E. Seguin, *Planta Med.* **1987**, 201–206.

242 R. Badorrey, C. Cativiela, M. D. Díaz-de-Villegas, J. A. Gálvez, *Tetrahedron* **1999**, *55*, 7601–7612.

243 A. Guarna, E. G. Occhiato, F. Machetti, A. Trabocchi, D. Scarpi, G. Danza, R. Mancina, A. Comerci, M. Serio, *Bioorg. Med. Chem.* **2001**, *9*, 1385–1393.

244 (a) M. Weymann, M. Schultz-Kukula, H. Kunz, *Tetrahedron Lett.* **1998**, *39*, 7835–7838; (b) S. Knauer, H. Kunz, *Tetrahedron: Asymmetry* **2005**, *16*, 529–539.

245 (a) H. Waldmann, M. Braun, *J. Org. Chem.* **1992**, *57*, 4444–4451; (b) S. Kirschbaum, H. Waldmann, *Tetrahedron Lett.* **1997**, *38*, 2829–2832; (c) R. Lock, H. Waldmann, *Chem. Eur. J.* **1997**, *3*, 143–151; (d) S. Kirschbaum, H. Waldmann, *J. Org. Chem.* **1998**, *63*, 4936–4946.

246 (a) J. Kervin, Jr., S. Danishefsky, *Tetrahedron Lett.* **1982**, *23*, 3739–3742; (b) S. Danishefsky, M. Langer, C. Vogel, *Tetrahedron Lett.* **1985**, *26*, 5983–5986; (c) S. Danishefsky, C. Vogel, *J. Org. Chem.* **1986**, *51*, 3915–3916.

247 H. Kunz, W. Pfrengle, *Angew. Chem. Int. Ed. Engl.* **1989**, *28*, 1067–1068.

248 (a) T. Michel, A. Kirschning, C. Beier, N. Bräuer, E. Schaumann, G. Adiwidjaja, *Liebigs Ann.* **1996**, 1811–1821; (b) Schaumann, *Angew. Chem. Int. Ed. Engl.* **1994**, *33*, 217–218; For related domino processes, see: (c) N. Bräuer, T. Michel, E. Schaumann, *Tetrahedron* **1998**, *54*, 11481–11488; (d) F. Tries, E. Schaumann, *Eur. J. Org. Chem.* **2003**, 1085–1090.

249 L. F. Tietze, H. Geissler, J. A. Gewert, U. Jakobi, *Synlett* **1994**, 511–512.

250 A. B. Smith, III, S. M. Pitram, A. M. Boldi, M. J. Gaunt, C. Sfouggatakis, W. H. Moser, *J. Am. Chem. Soc.* **2003**, *125*, 14435–14445.

251 (a) S. Omura, *Macrolide Antibiotics: Chemistry, Biology, Practice*, Academic Press, New York, **1984**; (b) S. D. Rychnovsky, *Chem. Rev.* **1995**, *95*, 2021–2040.

252 A. B. Smith, III, S. M. Pitram, *Org. Lett.* **1999**, *1*, 2001–2004.

253 C. S. Poss, S. D. Rychnovsky, S. L. Schreiber, *J. Am. Chem. Soc.* **1993**, *115*, 3360–3361.

254 J. E. Audia, J. J. Droste, J. M. Dunigan, J. Bowers, P. C. Heath, D. W. Holme, J. H. Eifert, H. A. Kay, R. D. Miller, J. M. Olivares, T. F. Rainey, L. O. Weigel, *Tetrahedron Lett.* **1996**, *37*, 4121–4124.

255 R. Jansen, J. Velder, S. Blechert, *Tetrahedron* **1995**, *51*, 8997–9004.

256 J. F. Hayes, M. Shipman, H. Twin, *J. Org. Chem.* **2002**, *67*, 935–942.

257 (a) M. Ihara, T. Taniguchi, Y. Tokunaga, K. Fukumoto, *Synthesis* **1995**, 1405–1406; (b) M. Ihara, T. Taniguchi, Y. Tokunaga, K. Fukumoto, *J. Org. Chem.* **1994**, *59*, 8092–8100.

258 I. Kubo, V. Jamalamadaka, T. Kamikawa, K. Takahashi, K. Tabata, T. Kusumi, *Chem. Lett.* **1996**, 441–442.

259 H. Watanabe, T. Onoda, T. Kitahara, *Tetrahedron Lett.* **1999**, *40*, 2545–2548.

260 (a) H. Akiyama, T. Fujimoto, K. Ohshima, K. Hoshino, I. Yamamoto, *Org. Lett.* **1999**, *1*, 427–430; (b) H. Akiyama, T. Fujimoto, K. Oshshima, K. Hoshino, Y. Saito, A. Okamoto, I. Yamamoto, A. Kakehi, R. Iriye, *Eur. J. Org. Chem.* **2001**, 2265–2272.

261 A. Volonterio, M. Zanda, *Tetrahedron Lett.* **2003**, *44*, 8549–8551.

262 D. T. Aragón, G. V. López, A. García-Tellado, J. J. Marrero-Tellado, P. de Armas, D. Terrero, *J. Org. Chem.* **2003**, *68*, 3363–3365. For an interesting review, see: D. Tejedor, D. González-Cruz, A. Santos-Expósito, J. J. Marrero-Tellado, P. de Armas, F. García-Tellado, *Chem. Eur. J.* **2005**, *11*, 3502–3510.

263 S. Lai, C. K. Zercher, J. P. Jasinski, S. N. Reid, R. J. Staples, *Org. Lett.* **2001**, *3*, 4169–4171.

264 J. Blank, M. Kandt, W.-D. Pfeiffer, A. Hetzheim, P. Langer, *Eur. J. Org. Chem.* **2003**, 182–189.

265 K. C. Liu, M. K. Hu, *Arch. Pharm. (Weinheim)* **1986**, *319*, 188–189.

266 K. Kottke, H. Kuehmstedt, I. Graefe, H. Wehlau, D. Knocke, DD 253623, **1988** [*Chem. Abstr.* **1988**, *109*, 17046].

267 C. Cianci, T. D. Y. Chung, N. Manwell, H. Putz, M. Hagen, R. J. Colonno, M. Krystal, *Antiviral Chem. Chemother.* **1996**, *7*, 353–360.

268 X. Linghu, D. A. Nicewicz, J. S. Johnson, *Org. Lett.* **2002**, *4*, 2957–2960.

269 K. Takeda, T. Tanaka, *Synlett* **1999**, 705–708.

270 A. Jung, O. Koch, M. Ries, E. Schaumann, *Synlett* **2000**, 92–94.

271 R. Ballini, G. Bosica, D. Fiorini, M. V. Gil, M. Petrini, *Org. Lett.* **2001**, *3*, 1265–1267.

272 Y. Tanabe, R. Hamasaki, S. Funakoshi, *Chem. Commun.* **2001**, 1674–1675.

273 H. Yoda, Y. Nakaseko, K. Takabe, *Tetrahedron Lett.* **2004**, *45*, 4217–4220.

274 D. Phillips, A. R. Chamberlin, *J. Org. Chem.* **2002**, *67*, 3194–3201.

275 C. Tratrat, S. Giorgi-Renault, H.-P. Husson, *Org. Lett.* **2002**, *4*, 3187–3189.

276 (a) P. Liénard, J.-C. Quirion, H.-P. Husson, *Tetrahedron* **1993**, *49*, 3995–4006; (b) C. Clémencin-Le Guillou, S. Giorgi-Renault, J.-C. Quirion, H.-P. Husson, *Tetrahedron Lett.* **1997**, *38*, 1037–1040; (c) C. Clémencin-Le Guillou, P. Remuzon, D. Bouzard, J.-C. Quirion, S. Giorgi-Renault, H.-P. Husson, *Tetrahedron* **1998**, *54*, 83–96; (d) For a review, see: A. C. Ramos, R. Pelaez-Lamanié de Clairac, M. Medarde, *Heterocycles* **1999**, *51*, 1443–1470.

277 S. A. A. El Bialy, H. Braun, L. F. Tietze, *Angew. Chem. Int. Ed.* **2004**, *43*, 5391–5393.

278 (a) A. Mondon, *Tetrahedron* **1963**, *19*, 911–917; (b) A. Mondon, *Tetrahedron* **1964**, *20*, 1729–1736; (c) A. Mondon, K. F. Hansen, *Tetrahedron Lett.* **1960**, *14*, 5–8; (d) A. Mondon, H. Nestler, *Angew. Chem Int. Ed. Engl.* **1964**, *3*, 588–589; (e) T. Sano, J. Toda, N. Kashiwaba, T. Oshima, Y. Tsuda, *Chem. Pharm. Bull.* **1987**, *35*, 479–500; (f) R. Ahmad-Schofiel, P. S. Mariano, *J. Org. Chem.* **1987**, *52*, 1478–1482; (g) H. Ishibashi, T. Sato, M. Takahashi, M. Hayashi, M. Ikeda, *Heterocycles* **1988**, *27*, 2787–2790; (h) B. Belleau, *Can. J. Chem.* **1957**, *35*, 651–662; (i) V. Prelog, A. Langemann, O. Rodig, M. Ternmah, *Helv. Chim. Acta* **1959**, *42*, 1301–1310; (j) S. Sugasawa, H. Yoshikawa, *Chem. Pharm. Bull.* **1960**, *8*, 290–293; (k) M. Müller, T. T. Grossnickle, V. Boekelheide, *J. Am. Chem. Soc.* **1959**, *81*, 3959–3963; (l) Y. Tsuda, T. Sano, *Studies in Natural Products Chemistry* (Ed.: Atta-ur-Rahman), Elsevier, Amsterdam, **1989**, *3*, 455–493; (m) G. Stork, A. Brizzolara, H. Landesman, J. Szmuszkovicz, R. J. Terrell, *J. Am. Chem. Soc.* **1963**, *85*, 207–222; (n) M. E. Kuehne, W. G. Bornmann, W. H. Parsons, T. D. Spitzer, J. F. Blount, J. Zubieta, *J. Org. Chem.* **1988**, *53*, 3439–3450.

279 (a) A. S. Chawla, V. K. Kapoor, in: *Handbook of Plant and Fungal Toxicants* (Ed.: J. F. D'Mello), CRC-Press, London, **1997**, pp. 37–49; (b) D. S. Bhakuni, *J. Ind. Chem. Soc.* **2002**, *79*, 203–210.

280 (a) V. Boekelheide, in: *Alkaloids* (Ed.: R. H. F. Manske), Academic Press, New York, **1960**, *7*, 201–227; (b) R. K. Hill, in: *Alkaloids* (Ed.: R. H. F. Manske), Academic Press, New York,**1967**, *9*, 483–515; (c) S. F. Dyke, S. N. Quessy, in: *The Alkaloids* (Ed.: R. G. A. Rodrigo), Academic Press, New York, **1981**, *18*, 1–98; (d) A. H. Jackson, *Chem. Biol. Isoquinoline Alkaloids* **1985**, 62–78; (e) A. S. Chawla, A. H. Jackson, *Nat. Prod. Rep.* **1984**, *1*, 371–373; (f) A. S. Chawla, A. H. Jackson, *Nat. Prod. Rep.* **1986**, *3*, 355–364; (g) K. W. Bentley, *Nat. Prod. Rep.* **1991**, *8*, 339–366; (h) K. W. Bentley, *Nat. Prod. Rep.*, **1992**, *9*, 365–391; (i) K. W. Bentley, *Nat. Prod. Rep.* **1993**, *10*, 449–470; (j) K. W. Bentley, *Nat. Prod. Rep.* **1994**, *11*, 555–576; (k) K. W. Bentley, *Nat. Prod. Rep.* **1995**, *12*, 419–441; (l) M. Williams, J. Robinson, *J. Neurosci.* **1984**, *4*, 2906–2911; (m) M. W. Decker, D. J. Anderson, J. D. Brioni, D. L. Donnelly, C. H. Roberts, A. B. Kang, M. O'Neill, S. Piattoni-Kaplan, J. Swanson, J. P. Sullivan, *Eur. J. Pharmacol.* **1995**, *280*, 79–89.

281 J. Wölfling, É. Frank, G. Schneider, L. F. Tietze, *Eur. J. Org. Chem.* **2004**, 90–100.

282 H. Heaney, M. T. Simcox, A. M. Z. Slawin, R. G. Giles, *Synlett* **1998**, 640–642.

283 X. Wei, R. J. K. Taylor, *Angew. Chem. Int. Ed.* **2000**, *39*, 409–412.

284 W. F. Bailey, A. D. Khanolkar, K. V. Gavaskar, *J. Am. Chem. Soc.* **1992**, *114*, 8053–8060.

285 A. Kessler, C. M. Coleman, P. Charoeny-ing, D. F. O'Shea, *J. Org. Chem.* **2004**, *69*, 7836–7846.

286 A. Barco, S. Benetti, G. Spalluto, *J. Org. Chem.* **1992**, *57*, 6279–6286.

287 W. Oppolzer, K. Thirring, *J. Am. Chem. Soc.* **1982**, *104*, 4978–4979.

288 M. K. Gurjar, A. M. S. Murugaiah, P. Rad-hakrishna, C. V. Ramana, M. S. Chor-ghade, *Tetrahedron: Asymm.* **2003**, *14*, 1363–1370.

289 (a) X. Cai, S. Cheah, S. M. Chen, J. Eck-man, J. Ellis, R. Fisher, A. Fura, G. Grewal, S. Hussion, S. Ip, D. B. Killian, L.-L. Garahan, H. Lounsbury, C. G. Qian, R. T. Scannell, D. Yaeger, D. M. Wypij, C. G. Yeh, M. A. Young, S. X. Yu, *Abstracts of Papers of the American Chemical Society* **1997**, *214*, 214-MEDI; (b) Press release from Millennium Pharmaceuticals, Cambridge, USA, 10th October, **2000**; (c) X. Cai, S. Hwang, D. Killan, T. Y. Shen, US Patent 5,648,486, **1997**; (d) X. Cai, G. Grewal, S. Hussion, A. Fura, T. Biftu, US Patent 5,681,966, **1997**.

290 J. D. Yang, M. S. Kim, M. Lee, W. Baik, S. Koo, *Synthesis* **2000**, *6*, 801–804.

291 G. Scheffler, M. Justus, A. Vasella, H. P. Wessel, *Tetrahedron Lett.* **1999**, *40*, 5845–5848.

292 L. Turet, I. E. Markó, B. Tinant, J.-P. De-clercq, R. Touillaux, *Tetrahedron Lett.* **2002**, *43*, 6591–6595.

293 (a) R. Sakai, T. Higa, C. W. Jefford, G. Ber-nardinelli, *J. Am. Chem. Soc.* **1986**, *108*, 6404–6405; (b) R. Sakai, S. Kohmoto, T. Higa, C. W. Jefford, G. Bernardinelli, *Tetrahedron Lett.* **1987**, *28*, 5493–5496.

294 M. W. Nötzel, K. Rauch, T. Labahn, A. de Meijere, *Org. Lett.* **2002**, *4*, 839–841.

295 M. Suginome, Y. Ohmori, Y. Ito, *Chem. Commun.* **2001**, 1090–1091.

296 (a) D. Enders, J. Vázquez, G. Raabe, *Eur. J. Org. Chem.* **2000**, 893–901; (b) D. Enders, J. Vázquez, G. Raabe, *Chem. Commun.* **1999**, 701–702.

297 J.-H. Ye, K.-Q. Ling, Y. Zhang, N. Li, J.-H. Xu, *J. Chem. Soc., Perkin Trans. 1* **1999**, 2017–2023.

298 N. Thasana, S. Ruchirawat, *Tetrahedron Lett.* **2002**, *43*, 4515–4517.

299 L.-J. Lin, G. Topcu, H. Lotter, N. Ruan-grungsi, H. Wagner, J. M. Pezzuto, G. A. Cordell, *Phytochemistry* **1992**, *16*, 537–551.

300 W. F. Bailey, M. W. Carson, *Tetrahedron Lett.* **1997**, *8*, 1329–1332.

301 (a) R. Keuper, N. Risch, U. Flörke, H.-J. Haupt, *Liebigs Ann.* **1996**, 705–715; (b) R. Keuper, N. Risch, *Liebigs Ann.* **1996**, 717–723.

302 (a) E. C. Constable, A. M. W. Cargill Thompson, *J. Chem. Soc., Dalton Trans.* **1992**, 3467–3475; (b) T. W. Bell, J. Liu, *J. Am. Chem. Soc.* **1988**, *110*, 3673–3674; (c) T. W. Bell, H. Jousselin, *Nature* **1994**, *367*, 441–444; (d) W. R. Cannon, J. D. Madura, R. P. Thummel, J. A. McCammon, *J. Am. Chem. Soc.* **1993**, *115*, 879–884; (e) E. C. Constable, M. D. Ward, *J. Chem. Soc., Dalton Trans.* **1990**, 1405–1409; (f) C.-Y. Huang, L. A. Cabell, V. Lynch, E. V. Ans-lyn, *J. Am. Chem. Soc.* **1992**, *114*, 1900–1901; (g) C.-Y. Huang, L. A. Cabell, E. V. Anslyn, *Tetrahedron Lett.* **1990**, *31*, 7411–7414; (h) J.-M. Lehn, *Angew. Chem. Int. Ed. Engl.* **1990**, *29*, 1304–1319; (i) M. Beley, J.-P. Collin, J.-P. Launay, J.-P. Sau-vage, P. Laine, S. Chodorowskikimmes, *Angew. Chem. Int. Ed. Engl.* **1994**, *33*, 1775–1778; (j) L. Mussons, C. Raposo, M. Crego, J. Anaya, C. Caballero, J. R. Moran, *Tetrahedron Lett.* **1994**, *35*, 7061–7064; (k) E. C. Constable, A. M. W. Cargill Thom-pson, P. Harveson, L. Macko, M. Zehn-der, *Chem. Eur. J.* **1995**, *1*, 360–367; (l) D. J. Cram, *Angew. Chem. Int. Ed. Engl.* **1988**, *27*, 1009–1020; (m) J.-M. Lehn, *Su-pramolecular Chemistry – Concepts and Perspectives*, VCH, **1995**; (n) T. W. Bell, Z. Hou, S. C. Zimmerman, P. A. Thiessen, *Angew. Chem. Int. Ed. Engl.* **1995**, *34*, 2163–2165.

303 B. Plietker, D. Seng, R. Fröhlich, P. Metz, *Eur. J. Org. Chem.* **2001**, 3669–3676.

304 D. T. W. Chu, P. B. Fernandes, A. G. Pernet, *J. Med. Chem.* **1986**, *29*, 1531–1534.

305 S. J. Chung, D. H. Kim, *Tetrahedron* **1995**, *51*, 12549–12562.

306 D. Witthaut, R. Fröhlich, H. J. Schäfer, *Angew. Chem. Int. Ed.* **2001**, *40*, 4212–4214.

307 (a) M. Yamashita, K. Okuyama, T. Kawajiri, A. Takada, Y. Inagaki, H. Nakano, M. Tomiyama, A. Ohnaka, I. Terayama, I. Kawasaki, S. Ohta, *Tetrahe-dron* **2002**, *58*, 1497–1505; (b) M. Yama-shita, K. Okuyama, I. Kawasaki, S. Ohta, *Tetrahedron Lett.* **1995**, *36*, 5603–5606.

308 (a) S. J. Selover, P. Crews, *J. Org. Chem.* **1980**, *45*, 69–72; (b) M. Suzuki, E. Kurosawa, *Bull. Chem. Soc. Jpn.* **1979**, *52*, 3352–3354; (c) S. M. Waraszkiewicz, K. L. Erickson, *Tetrahedron Lett.* **1974**, *23*, 2003–2006; (d) T. Irie, M. Suzuki, Y. Hayakawa, *Bull. Chem. Soc. Jpn.* **1969**, *42*, 843–844; (e) A. F. Cameron, G. Ferguson, J. M. Robertson, *Chem. Commun.* **1967**, 271–272.

309 (a) L. W. H. Chaidir, R. Ebel, R. Edrada, V. Wray, M. Nimtz, W. Sumaryono, P. Proksch, *J. Nat. Prod.* **2001**, *64*, 1216–1220; (b) H. Greger, T. Pacher, B. Brem, M. Bacher, O. Hofer, *Phytochemistry* **2001**, *57*, 57–64; (c) M. Dreyer, B. W. Nugroho, F. I. Bohnenstengel, R. Ebel, V. Wray, L. Witte, G. Bringmann, J. Muehlbacher, M. Herold, P. D. Hung, L. C. Kiet, P. Proksch, *J. Nat. Prod.* **2001**, *64*, 415–420; (d) C. Schneider, F. I. Bohnenstengel, B. W. Nugroho, V. Wray, L. Witte, P. D. Hung, L. C. Kiet, P. Proksch, *Phytochemistry* **2000**, *54*, 731–736; (e) D. Engelmeier, F. Hadacek, T. Pacher, S. Vajrodaya, H. Greger, *J. Agric. Food Chem.* **2000**, *48*, 1400–1404; (f) References cited in ref. [2a–e].

310 (a) N. H. Lee, R. C. Larock, *Bull. Korean Chem. Soc.* **2001**, *22*, 857–866; (b) Y. Yoshida, Y. Sato, S. Okamoto, F. Sato, *J. Chem. Soc., Chem. Commun.* **1995**, 811–812; (c) R. C. Larock, N. H. Lee, *J. Org. Chem.* **1991**, *56*, 6253–6254; (d) K. Ohno, H. Nagase, K. Matsumoto, S. Nishio, Eur. Pat. Appl. EP84856, **1983**.

311 M. Yamashita, N. Ohta, T. Shimizu, K. Matsumoto, Y. Matsuura, I. Kawasaki, T. Tanaka, N. Maezaki, S. Ohta, *J. Org. Chem.* **2003**, *68*, 1216–1224.

312 Y. Mimaki, A. Kameyama, Y. Sashida, Y. Miyata, A. Fujii, *Chem. Pharm. Bull.* **1995**, *43*, 893–895.

313 T. J. Beauchamp, J. P. Powers, S. D. Rychnovsky, *J. Am. Chem. Soc.* **1995**, *117*, 12873–12874.

314 (a) R. H. Burnell, D. R. Taylor, *Chem. Ind.* **1960**, 1239–1240; (b) R. H. Burnell, B. S. Mootoo, *Can. J. Chem.* **1961**, *39*, 1090.

315 D. A. Evans, J. R. Scheerer, *Angew. Chem. Int. Ed.* **2005**, *44*, 6038–6042.

316 (a) J. Löffler, R. Schobert, *J. Chem. Soc., Perkin Trans 1* **1996**, 2799–2802; (b) J. Löffler, R. Schobert, *J. Chem. Soc., Perkin Trans 1* **1996**, 2799–2802.

317 H. Al-Badri, N. Collignon, *Synlett* **1999**, 282–285.

318 R. Adams, M. Gianturco, *J. Am. Chem. Soc.* **1957**, *79*, 166–169.

319 C. C. J. Culvenor, T. A. Geissman, *J. Org. Chem.* **1961**, *26*, 3045–3050.

320 (a) E. Wada, G. Kumaran, S. Kanemasa, *Tetrahedron Lett.* **2000**, *41*, 73–76; (b) E. Wada, H. Koga, G. Kumaran, *Tetrahedron Lett.* **2002**, *43*, 9397–9400; (c) H. Koga, E. Wada, *Tetrahedron Lett.* **2003**, *44*, 715–719; (d) E. Wada, M. Yoshinaga, *Tetrahedron Lett.* **2004**, *45*, 2197–2201.

321 E. Wada, M. Yoshinaga, *Tetrahedron Lett.* **2003**, *44*, 7953–7956.

322 J. Zhang, C.-J. Li, *J. Org. Chem.* **2002**, *67*, 3969–3971.

323 L. F. Tietze, G. Kettschau, *Top. Curr. Chem.* **1997**, *189*, 1–120.

324 (a) N. Yamada, S. Kadowaki, K. Takahashi, K. Umezu, *Biochem. Pharmacol.* **1992**, *44*, 1211–1213; (b) K. Faber, H. Stueckler, T. Kappe, *Heterocycl. Chem.* **1984**, *21*, 1177–1181; (c) J. V. Johnson, B. S. Rauckman, D. P. Baccanari, B. Roth, *J. Med. Chem.* **1989**, *32*, 1942–1949.

325 A. Di Salvo, M. Vittoria Spanedda, M. Ourévitch, B. Crousse, D. Bonnet-Delpon, *Synthesis* **2003**, 2231–2235.

326 For related domino strategies employing Lewis acid catalysts, see also: (a) L. S. Povarov, B. M. Mikhailov, *Izv. Akad. Nauk. SSSR, Ser. Khim.* **1964**, *12*, 2221–2222; (b) J. S. Yadav, B. V. S. Reddy, K. Uma Gayathri, A. R. Prasad, *Synthesis* **2002**, 2537–2541; (c) J. S. Yadav, B. V. S. Reddy, R. Srinivasa Rao, S. Kiran Kumar, A. C. Kunwar, *Tetrahedron* **2002**, *58*, 7891–7896; (d) R. A. Batey, D. A. Powell, A. Acton, A. J. Lough, *Tetrahedron Lett.* **2001**, *42*, 7935–7939.

327 K. Paulvannan, R. Hale, R. Mesis, T. Chen, *Tetrahedron Lett.* **2002**, *43*, 203–207.

328 M. S. Novikov, A. F. Khlebnikov, E. S. Sidorina, R. R. Kostikov, *J. Chem., Perkin Trans. 1* **2000**, 231–237.

329 L. G. Arini, P. Szeto, D. L. Hughes, R. A. Stockman, *Tetrahedron Lett.* **2004**, *45*, 8371–8374.

330 (a) M. Kuramoto, C. Tong, K. Yamada, T. Chiba, Y. Hayashi, D. Uemura, *Tetrahedron Lett.* **1996**, *37*, 3867–3870; (b) H. Arimoto, I. Hayakawa, M. Kuramoto, D. Uemura, *Tetrahedron Lett.* **1998**, *39*, 861–862.

331 A. Padwa, T. M. Heidelbaugh, J. T. Kuehne, *J. Org. Chem.* **2000**, *65*, 2368–2378.

332 M. E. L. De Almeida, F. R. Braz, V. von Buelow, O. R. Gottlieb, J. G. S. Maia, *Phytochemistry* **1976**, *15*, 1186–1187.

333 M. O. F. Goulart, A. E. G. Sant'ana, A. B. de Oliveira, G. G. de Oliveira, J. G. S. Maia, *Phytochemistry* **1986**, *25*, 1691–1695.

334 (a) Z. Rehacek, P. Sajdl, *Ergot Alkaloids: Chemistry, Biological Effects, Biotechnology*, Elsevier Science Publishers, Amsterdam, **1990**; (b) P. A. Stadler, K. A. Gigar, in: *Natural Products and Drug Development* (Eds.: P. Krogsgaard-Larsen, S. B. Christensen, H. Koffod), Munksgaard, Copenhagen, **1984**, pp. 463.

335 J. W. Daly, T. F. Spande, in: *Alkaloids: Chemical and Biological Perspectives, Vol. 4* (Ed.: S. W. Pelletier), Wiley-Interscience, New York, **1986**, pp. 1–274.

336 M. E. Kuehne, F. Xu, *J. Org. Chem.* **1997**, *62*, 7950–7960.

337 M. E. Kuehne, F. Xu, *J. Org. Chem.* **1998**, *63*, 9427–9433.

338 M. E. Kuehne, F. Xu, *J. Org. Chem.* **1998**, *63*, 9434–9439.

339 (a) P. Righi, E. Marotta, G. Rosini, *Chem. Eur. J.* **1998**, *4*, 2501–2512; (b) E. Marotta, P. Righi, G. Rosini, *Tetrahedron Lett.* **1998**, *39*, 1041–1044.

340 K. Takeda, H. Haraguchi, Y. Okamoto, *Org. Lett.* **2003**, *5*, 3705–3707.

341 (a) A. D. Allbut, W. A. Ayer, H. J. Brodie, B. N. Johri, H. Taube, *Can. J. Microbiol.* **1971**, *17*, 1401–1407; (b) W. A. Ayer, H. Taube, *Tetrahedron Lett.* **1972**, *19*, 1917–1920; (c) W. A. Ayer, H. Taube, *Can. J. Chem.* **1973**, *51*, 3842–3854; (d) W. A. Ayer, L. L. Carstens, *Can. J. Chem.* **1973**, *51*, 3157–3160; (e) W. A. Ayer, L. M. Browne, J. R. Mercer, D. R. Taylor, D. E. Ward, *Can. J. Chem.* **1978**, *56*, 717–721; (f) W. A. Ayer, S. P. Lee, *Can. J. Chem.* **1979**, *57*, 3332–3337; (g) W. A. Ayer, T. Yoshida, D. M. J. van Schie, *Can. J. Chem.* **1978**, *56*, 2113–2120; (h) W. A. Ayer, T. T. Nakashima, D. E. Ward, *Can. J. Chem.* **1978**, *56*, 2197–2199; (i) W. A. Ayer, S. P. Lee, T. T. Nakashima, *Can. J. Chem.* **1979**, *57*, 3338–3343.

342 K. Takeda, D. Nakane, M. Takeda, *Org. Lett.* **2000**, *2*, 1903–1905.

343 R. Nesi, S. Turchi, D. Giomi, A. Danesi, *Tetrahedron* **1999**, *55*, 13809–13818.

344 K. Mikami, K. Takahashi, T. Nakai, T. Uchimaru, *J. Am. Chem. Soc.* **1994**, *116*, 10948–10954.

345 (a) G. H. Posner, C. Switzer, *J. Am. Chem. Soc.* **1986**, *108*, 1239–1244; (b) F. E. Ziegler, H. Lim, *J. Org. Chem.* **1982**, *47*, 5229–5230; (c) G. H. Posner, J. P. Mallamo, A. Y. Black, *Tetrahedron* **1981**, *37*, 3921–3926; (d) G. Quinckert, W. D. Weber, U. Schwartz, G. Duerner, *Angew. Chem. Int. Ed. Engl.* **1980**, *19*, 1027–1029; (e) G. Quinckert, U. Schwartz, H. Stark, W. D. Weber, H. Baier, F. Adam, G. Duerner, *Angew. Chem. Int. Ed. Engl.* **1980**, *19*, 1029–1030.

346 (a) R. Grigg, *Tetrahedron: Asymmetry* **1995**, 2475–2486; (b) T. Coulter, R. Grigg, J. F. Malone, V. Sridharan, *Tetrahedron Lett.* **1991**, *32*, 5417–5420.

347 H. Li, H.-R. Zhang, J. L. Petersen, K. K. Wang, *J. Org. Chem.* **2001**, *66*, 6662–6668.

348 (a) M. Schmittel, M. Maywald, *Chem. Commun.* **2001**, 155–156; (b) P. G. Wenthold, M. A. Lipton, *J. Am. Chem. Soc.* **2000**, *122*, 9265–9270; (c) S. R. Brunette, M. A. Lipton, *J. Org. Chem.* **2000**, *65*, 5114–5119; (d) P. R. Schreiner, M. Prall, *J. Am. Chem. Soc.* **1999**, *121*, 8615–8627; (e) K. K. Wang, H.-R. Zhang, J. L. Petersen, *J. Org. Chem.* **1999**, *64*, 1650–1656; (f) B. Engels, M. Hanrath, *J. Am. Chem. Soc.* **1998**, *120*, 6356–6361; (g) M. Schmittel, M. Strittmatter, *Tetrahedron* **1998**, *54*, 13751–13760; (h) B. Engels, C. Lennartz, M. Hanrath, M. Schmittel, M. Strittmatter, *Angew. Chem. Int. Ed.* **1998**, *37*, 1960–1963; (i) M. Schmittel, J.-P. Steffen, D. Auer, M. Maywald, *Tetrahedron Lett.* **1997**, *38*, 6177–6180; (j) M. Schmittel, M. Maywald, M. Strittmatter, *Synlett* **1997**, 165–166; (k) M. Schmittel, M. Keller, S. Kiau, M. Strittmatter, *Chem. Eur. J.* **1997**, *3*, 807–816; (l) M. Schmittel, S. Kiau, *Liebigs Ann./Recl.* **1997**, 733–736; (m) M. Schmittel, M. Strittmatter, S. Kiau, *Angew. Chem. Int. Ed. Engl.* **1996**, *35*, 1843–1845; (n) M. Schmittel, S. Kiau, T. Siebert, M. Strittmatter, *Tetrahedron Lett.* **1996**, *37*, 7691–7694; (o) M. Schmittel, M. Strittmatter, K. Vollmann, S. Kiau, *Tetrahedron Lett.* **1996**, *37*, 999–1002; (p) M. Schmittel, M. Strittmatter, S. Kiau, *Tetrahedron Lett.* **1995**, *36*, 4975–4978.

349 V. M. Dong, D. W. C. MacMillan, *J. Am. Chem. Soc.* **2001**, *123*, 2448–2449.

350 For examples of allylic 1,2-strain directed reactions, see: (a) F. Johnson, *Chem. Rev.* **1968**, *68*, 375–413; (b) A. H. Hoveyda, D. A. Evans, G. C. Fu, *Chem. Rev.* **1993**, *93*, 1307–1370.

351 (a) S. Yagi, *Kyoto Igaku Zasshi.* **1909**, *6*, 208–222; For reviews on the *Daphniphyllum* alkaloids, see: (b) S. Yamamura, Y. Hirata, in: *The Alkaloids, Vol. 15* (Ed.: R. H. F. Manske), Academic Press, New York, **1975**, pp. 41; (c) S. Yamamura, Y. Hirata, *Int. Rev. Sci.: Org. Chem., Ser. Two* **1976**, *9*, 161–189; (d) S. Yamamura, in: *The Alkaloids, Vol. 29* (Ed.: A. Brossi), Academic Press, New York, **1986**, pp. 265–286.

352 (a) C. L. Heathcock, *Proc. Natl. Acad. Sci. USA* **1996**, *93*, 14323–14327; For further studies, see: (b) G. A. Wallace, C. H. Heathcock, *J. Org. Chem.* **2001**, *66*, 450–454.

353 (a) C. H. Heathcock, M. M. Hansen, R. B. Ruggeri, J. C. Kath, *J. Org. Chem.* **1992**, *57*, 2544–2553; (b) C. H. Heathcock, J. Stafford, *J. Org. Chem.* **1992**, *57*, 2566–2574; (c) C. H. Heathcock, R. B. Ruggeri, K. F. McClure, *J. Org. Chem.* **1992**, *57*, 2585–2594.

354 For reviews, see: (a) M. Hasegawa, K. Saigo, in: *Photochemistry and Photophysics, Vol. 2* (Ed.: J. F. Rabek), CRC Press, Florida, **1989**, p. 27; (b) K. Saigo, *Prog. Polym. Sci.* **1992**, *17*, 35–86.

355 M. Hasegawa, H. Miura, M. Kuroda, M. Hayashi, Y. Hashimoto, K. Saigo, *Tetrahedron Lett.* **1999**, *40*, 7251–7254.

356 L. A. Paquette, *Eur. J. Org. Chem.* **1998**, 1709–1728.

357 (a) L. A. Paquette, J. Doyon, L. H. Kuo, *Tetrahedron Lett.* **1996**, *37*, 3299–3302; (b) L. A. Paquette, A. T. Hamme, II, L. H. Kuo, J. Doyon, R. Kreuzholz, *J. Am. Chem. Soc.* **1997**, *119*, 1242–1253.

358 L. A. Paquette, J. Tae, *J. Org. Chem.* **1998**, *63*, 2022–2030.

359 L. A. Paquette, T. M. Morwick, *J. Am. Chem. Soc.* **1997**, *119*, 1230–1241.

360 (a) L. A. Paquette, L. H. Kuo, A. T. Hamme, II, R. Kreuzholz, J. Doyon, *J. Org. Chem.* **1997**, *62*, 1730–1736; (b) L. A. Paquette, L. H. Kuo, J. Doyon, *Tetrahedron* **1996**, *52*, 11625–11636; (c) L. A. Paquette, L. H. Kuo, J. Doyon, *J. Am. Chem. Soc.* **1997**, *119*, 3038–3047; (d)

L. A. Paquette, L. H. Kuo, J. Tae, *J. Org. Chem.* **1998**, *63*, 2010–2021.

361 (a) L. A. Paquette, J. Doyon, *J. Am. Chem. Soc.* **1995**, *117*, 6799–6800; (b) L. A. Paquette, J. Doyon, *J. Org. Chem.* **1997**, *62*, 1723–1729.

362 L. A. Paquette, F. Geng, *J. Am. Chem. Soc.* **2002**, *124*, 9199–9203.

363 (a) U. Jahn, *Chem. Comm.* **2001**, 1600–1601; (b) For further applications, see also: U. Jahn, M. Müller, S. Aussieker, *J. Am. Chem. Soc.* **2000**, *122*, 5212–5213.

364 U. Jahn, D. Rudakov, *Synthesis* **2004**, 1207–1210.

365 Y. Guindon, K. Houde, M. Prévost, B. Cardinal-David, S. R. Landry, B. Daoust, M. Bencheqroun, B. Guérin, *J. Am. Chem. Soc.* **2001**, *123*, 8496–8501.

366 (a) C. H. Heathcock, in: *Comprehensive Organic Synthesis* (Ed.: B. M. Trost) Pergamon: Oxford, **1993**; Vol. 2, Chapter 2.4; (b) R. Mahrwald, *Chem. Rev.* **1999**, *99*, 1095–1120.

367 P. Mochirian, B. Cardinal-David, B. Guérin, M. Prévost, Y. Guindon, *Tetrahedron Lett.* **2002**, *43*, 7067–7071.

368 W. R. Bowman, P. T. Stephenson, A. R. Young, *Tetrahedron* **1996**, *52*, 11445–11462.

369 H. M. Walborsky, M. Topolski, *Tetrahedron Lett.* **1993**, *34*, 7681–7684.

370 G. A. Molander, C. R. Harris, *J. Am. Chem. Soc.* **1996**, *118*, 4059–4071.

371 G. A. Molander, C. Kenny, *J. Am. Chem. Soc.* **1989**, *111*, 8236–8246.

372 (a) L. F. Tietze, G. Kettschau, *Top. Curr. Chem.* **1997**, *189*, 1–120; (b) L. F. Tietze, G. Kettschau, J. A. Gewert, A. Schuffenhauer, *Curr. Org. Chem.* **1998**, *2*, 19–62; (c) L. F. Tietze, U. Beifuss, in: *Comprehensive Organic Synthesis, Vol. 2* (Ed.: B. M. Trost), Pergamon Press, Oxford, **1991**, pp. 341–394; (d) L. F. Tietze, N. Rackelmann, *Pure Appl. Chem.* **2004**, *76*, 1967–1983.

373 L. F. Tietze, T. Brumby, S. Brand, M. Bratz, *Chem. Ber.* **1988**, *121*, 499–506.

374 (a) L. F. Tietze, H. Geissler, J. Fennen, T. Brumby, S. Brand, G. Schulz, *J. Org. Chem.* **1994**, *59*, 182–191; (b) L. F. Tietze, C. Ott, H. Geissler, F. Haunert, *Eur. J. Org. Chem.* **2001**, 1625–1630.

375 L. F. Tietze, P. Saling, *Synlett* **1992**, 281–282.

376 L. F. Tietze, P. Saling, *Chirality* **1993**, *5*, 329–333.

377 L.F Tietze, H. Stegelmeier, K. Harms, T. Brumby, *Angew. Chem. Int. Ed. Engl.* **1982**, 21, 863–864.

378 L.F Tietze, P. Steck, in: *High Pressure Chemistry* (Eds.: R. Eldik, F. G. Klärner), Wiley-VCH, Weinheim, **2002**, pp. 239–283.

379 L. F. Tietze, T. Brumby, M. Pretor, G. Remberg, *J. Org. Chem.* **1988**, 53, 810–820.

380 L. F. Tietze, T. Brumby, T. Pfeiffer, *Liebigs Ann. Chem.* **1988**, 9–12.

381 (a) J. Antel, G. M. Sheldrick, T. Pfeiffer, L. F. Tietze, *Acta Crystallogr.* **1990**, C46; 158–159; (b) L. F. Tietze, S. Brand, T. Pfeiffer, J. Antel, K. Harms, G. M. Sheldrick. *J. Am. Chem. Soc.* **1987**, 109, 921–923.

382 M. Sato, S. Sunami, C. Kaneko, S. Satoh, T. Furuya, *Tetrahedron: Asymm.* **1994**, 5, 1665–1668.

383 (a) G. Sabitha, E. V. Reddy, N. Fatima, J. S. Yadav, K. V. S. R. Krishna, A. C. Kunwar, *Synthesis* **2004**, 1150–1154; (b) J. S. Yadav, B. V. S. Reddy, D. Narsimhaswamy, P. N. Lakshmi, K. Narsimulu, G. Srinivasulu, A. C. Kunwar, *Tetrahedron Lett.* **2004**, 45, 3493–3497.

384 J. K. Gallos, A. E. Koumbis, *Arkivoc* **2003**, 6, 135–144.

385 B. B. Snider, Q. Lu, *J. Org. Chem.* **1996**, 61, 2839–2844.

386 M. R. TePaske, J. B. Gloer, D. T. Wicklow, P. F. Dowd, *Tetrahedron Lett.* **1991**, 32, 5687–5690.

387 M. Shanmugasundaram, S. Manikandan, R. Raghunathan, *Tetrahedron* **2002**, 58, 997–1003.

388 S. Manikandan, M. Shanmugasundaram, R. Raghunathan, *Tetrahedron* **2002**, 58, 8957–8962.

389 J. S. Yadav, B. V. S. Reddy, V. Naveenkumar, R. S. Rao, K. Nagaiah, *Synthesis* **2004**, 1783–1788.

390 L. F. Tietze, T. Pfeiffer, A. Schuffenhauer *Eur. J. Org. Chem.* **1998**, 2733–2741.

391 L. F. Tietze, A. Schuffenhauer, P. R. Schreiner *J. Am. Chem. Soc.* **1998**, 120, 7952–7958.

392 G. Cravotto, G. M. Nano, G. Palmisano, S. Tagliapietra, *Tetrahedron: Asymm.* **2001**, 12, 707–709.

393 (a) G. Feuer, in: *Progress in Medicinal Chemistry* (Eds.: G. P. Ellis, G. B. West), North-Holland, New York, **1974**; (b) A. A. Deana, *J. Med. Chem.* **1983**, 26, 580–585;

(c) E. Wenkert, B. L. Buckwalter, *J. Am. Chem. Soc.* **1972**, 94, 4367–4369.

394 (a) D. S. Hewick, J. McEwen, *J. Pharm. Pharmacol.* **1973**, 25, 458–465; (b) L. G. M. Baars, M. T. Schepers, J. J. R. Hermans, H. J. J. Dahlmans, H. H. W. Thijssen, *J. Pharm. Pharmacol.* **1990**, 42, 861–866; (c) I. Manolov, N. D. Danchev, *Eur. J. Med. Chem.* **1995**, 30, 531–535.

395 S. Koser, H. M. R. Hoffmann, *Heterocycles* **1994**, 37, 661–666.

396 L. F. Tietze, H. Evers, unpublished results.

397 (a) K. Watanabe, S. Yano, S. Horie, L. T. Yamamoto, H. Takayama, N. Aimi, S. Sakai, D. Ponglux, P. Tongroach, J. Shan, P. K. T. Pang, *Pharm. Res. Trad. Herb. Med.* **1999**; 163–177; (b) A. Yamane, J. Fujikura, H. Ogawa, J. Mizutani, *J. Chem. Ecol.* **1992**, 18, 1941–1954.

398 H. Takayama, Y. Iimura, M. Kitajima, N. Aimi, K. Konno, H. Inoue, M. Fujiwara, T. Mizuta, T. Yokota, S. Shigeta, K. Tokuhisa, Y. Hanasaki, K. Katsuura, *Bioorg. Med. Chem. Lett.* **1997**, 7; 3145–3148.

399 (a) D. Staerk, E. Lemmich, J. Christensen, A. Kharazmi, C. E. Olsen, J. W. Jaroszewski, *Planta Medica* **2000**, 66, 531–536; (b) M. Barczai-Beke, G. Doernyei, G. Toth, J. Tamas, C. Szantay, *Tetrahedron* **1976**, 32, 1153–1159.

400 (a) T. Kametani, T. Suzuki, E. Sato, M. Nishimura, K. Unno, *J. Chem. Soc. Chem. Com.* **1982**, 20, 1201–1203; (b) G. Massiot, P. Thepenier, M. J. Jacquier, L. Le Men-Olivier, C. Delaude, *Phytochemistry* **1992**, 31, 2873–2876.

401 L. F. Tietze, J. Bachmann, J. Wichmann, Y. Zhou, T. Raschke, *Liebigs Ann/Recueil* **1997**, 881–886.

402 (a) A. Itoh, Y. Ikuta, Y. Baba, N. Tanahashi, N. Nagakura, *Phytochemistry* **1999**, 52, 1169–1176; (b) O. Hesse, *Justus Liebigs Ann. Chem.* **1914**, 405, 1–57.

403 A. Itoh, Y. Ikuta, T. Tanahashi, N. Nagakura, *J. Nat. Prod.* **2000**, 63, 723–725.

404 (a) M. T. Gonzales-Garza, S. A. Martlin, B. D. Mata-Cardena, S. Said-Fernandez, *J. Pharm. Pharmacol.* **1993**, 45, 144–145; (b) Y. F. Liou, I. H. Hall, K. H. Lee, *J. Pharm. Sci.* **1982**, 71, 745–749.

405 (a) I. Marin, J. P. Abad, D. Urena, R. Amils, *Biochemistry* **1995**, 34, 16519–16523; (b) G. T. Tan, J. M. Pezzuto, A. D. Kinghorn, S. H. Hughes, *J. Nat. Prod.* **1991**, 54, 143–154.

406 L. F. Tietze, N. Rackelmann. *Z. Natur-forsch., B: Chem. Sci.* **2004**, 59, 468–477.

407 (a) K. N. Houk, *J. Am. Chem. Soc.* **1973**, 95, 4092–4094; (b) C. P. Dell, *J. Chem. Soc. Perkin Trans. 1* **1998**, 3873–3905; (c) C. Spino, M. Pesant, Y. Dory, *Angew. Chem. Int. Ed. Engl.* **1998**, 37, 3262–3265 (and references cited therein).

408 H. Al-Badri, J. Maddaluno, S. Masson, N. Collignon, *J. Chem. Soc., Perkin Trans. 1* **1999**, 2255–2266.

409 (a) D. B. Ramachary, N. S. Chowdari, C. F. Barbas, III, *Angew. Chem. Int. Ed.* **2003**, 42, 4233–4237; (b) D. B. Rama-chary, K. Anebouselvy, N. S. Chowdari, C. F. Barbas III, *J. Org. Chem.* **2004**, 69, 5838–5849.

410 (a) C. Hongchao, J. Xixian, L. Zhichang, *Huaxi Yaoxue Zazhi* **1995**, 10, 150; (b) D. R. Schroeder, F. R. Stermitz, *Tetrahe-dron* **1985**, 41, 4309–4320; (c) J. N. Xiang, P. Nambi, E. H. Ohlstein, J. D. Elliott, *Bioorg. Med. Chem.* **1998**, 6, 695–700.

411 (a) D. R. Zitsane, I. T. Ravinya, I. A. Riikure, Z. F. Tetere, E. Y. Gudrinietse, U. O. Kalei, *Russ. J. Org. Chem.* **1999**, 35, 1457–1460; (b) D. R. Zitsane, I. T. Ravinya, I. A. Riikure, Z. F. Tetere, E. Y. Gudrinietse, U. O. Kalei, *Russ. J. Org. Chem.* **2000**, 36, 496–501; (c) I. Bonnard, M. Rolland, C. Francisco, B. Banaigs, *Lett. Pept. Sci.* **1997**, 4, 289–292.

412 (a) H. A. Dondas, M. Frederickson, R. Grigg, J. Markandu, M. Thornton-Pett, *Tetrahedron* **1997**, 53, 14339–14354; (b) J. Markandu, H. A. Dondas, M. Frederick-son, R. Grigg, *Tetrahedron* **1997**, 53, 13165–13176.

413 T. Hartmann, L. Witte, in: *Alkaloids: Chemical & Biological Perspectives*, Vol. 9 (Ed.: S. W. Pelletier), Pergamon, Oxford, **1995**, pp. 155–233.

414 A. Goti, M. Cacciarini, F. Cardona, F. M. Cordero, A. Brandi, *Org. Lett.* **2001**, 3, 1367–1369.

415 R. Huisgen, in: *1,3-Dipolar Cycloadditions – Introduction, Survey, Mechanism*, Vol. 1 (Ed.: A. Padwa), Wiley, New York, **1984**, chapter 1, p. 1.

416 O. Tsuge, S. Kanemasa, *Adv. Heterocycl. Chem.* **1989**, 231–249.

417 H. A. Dondas, C. W. G. Fishwick, R. Grigg, C. Kilner, *Tetrahedron* **2004**, 60, 3473–3485.

418 C. Wittland, M. Arend, N. Risch, *Synthesis* **1996**, 367–371.

419 S. Jarosz, S. Skóra, *Tetrahedron: Asym-metry* **2000**, 11, 1425–1432.

420 S. Jarosz, S. Skóra, *Tetrahedron: Asym-metry* **2000**, 11, 1433–1448.

421 (a) C. Herdeis, J. Telser, *Eur. J. Org. Chem.* **1999**, 1407–1414; (b) C. Herdeis, T. Schiffer, *Synthesis* **1997**, 1405–1410; (c) C. Hereis, T. Schiffer, *Tetrahedron* **1996**, 52, 14745–14756; (d) C. Herdeis, T. Schiffer, *Tetrahedron* **1999**, 55, 1043–1056; (e) C. Herdeis, P. Küpper, S. Plé, *Org. Bi-omol. Chem.* **2006**, 4, 524–529.

422 For alkaloids isolated from *Prosopis* or *Microcos philippinensis* species, see: (a) G. Ratle, X. Monseur, B. C. Das, J. Yassi, Q. Khuong-Huu, R. Goutarel, *Bull. Soc. Chim. Fr.* **1996**, 2945–2947; (b) Q. Khuong-Huu, G. Ratle, X. Monseur, R. Goutarel, *Bull. Soc. Chim. Belg.* **1972**, 81, 425–442; (c) Q. Khuong-Huu, G. Ratle, X. Monseur, R. Goutarel, *Bull. Soc. Chim. Belg.* **1972**, 81, 443–458; (d) A. M. Agui-naldo, R. W. Read, *Phytochemistry* **1990**, 29, 2309–2313.

423 (a) T. Kawasaki, R. Terashima, K. Sakagu-chi, H. Sekiguchi, M. Sakamoto, *Tetrahe-dron Lett.* **1996**, 42, 7525–7528; (b) T. Kawasaki, A. Ogawa, Y. Takashima, M. Sakamoto, *Tetrahedron Lett.* **2003**, 44, 1591–1593; (c) T. Kawasaki, M. Shinada, D. Kamimura, M. Ohzono, A. Ogawa, *Chem. Commun.* **2006**, 420–422.

424 (a) S. Takase, M. Iwami, T. Ando, M. Okamoto, *J. Antibiot.* **1984**, 37, 1320–1323; (b) S. Takase, Y. Kawai, I. Uchida, H. Tanaka, H. Aoki, *Tetrahedron Lett.* **1984**, 25, 4673–4676; (c) S. Takase, Y. Kawai, I. Uchida, H. Tanaka, H. Aoki, *Tetrahedron* **1985**, 41, 3037–3048.

425 J. E. Hochlowski, M. M. Mullally, S. G. Spanton, D. N. Whittern, P. Hill, J. B. McAlpine, *J. Antibiot.* **1993**, 46, 380–386.

426 Y. Kimura, T. Hamasaki, H. Nakajima, *Tetrahedron Lett.* **1982**, 23, 225–228.

427 (a) J. S. Carlé, C. Christopherson, *J. Am. Chem. Soc.* **1976**, 101, 4012–4013; (b) J. S. Carlé, C. Christopherson, *J. Org. Chem.* **1980**, 45, 1586–1589; (c) J. S. Carlé, C. Christopherson, *J. Org. Chem.* **1981**, 46, 3440–3443; (d) P. Wulff, J. S. Carlé, C. Christopherson, *Comp. Bio-chem. Physiol.* **1982**, 71B, 523–524; (e) P.

Keil, E. G. Nielsen, U. Anthoni, C. Christopherson, *Acta Chem. Scand. B* **1986**, *40*, 555–558.

428 (a) R. M. Williams, E. M. Stocking, J. F. Sanz-Cervera, *Top. Curr. Chem.* **2000**, *209*, 97–173; (b) P. M. Scott, *Dev. Food Sci.* **1984**, *8*, 463–468.

429 J. W. Dayl, H. M. Garraffo, T. F. Spande, in: *Alkaloids: Chemical and Biological Perspectives, Vol. 13* (Ed. S. W. Pelletier), Pergamon Press, New York, **1999**, Chapter 1, pp. 1–161; (b) T. F. Spande, M. W. Edwards, L. K. Pannell, J. W. Daly, *J. Org. Chem.* **1988**, *53*, 1222–1226; (c) B. P. Smith, M. J. Tyler, T. Kaneko, H. M. Garraffo, T. F. Spande, J. W. Daly, *J. Nat. Prod.* **2002**, *65*, 439–447.

430 F. Palacios, C. Alonso, G. Rubiales, *Tetrahedron* **1995**, *51*, 3683–3690.

431 G. Büchi, A. Rodriguez, K. Yakurshigui, *J. Org. Chem.* **1989**, *54*, 4494–4496.

432 L. Bolger, R. J. Brittain, D. Jack, M. R. Jackson, L. E. Martin, J. Mills, D. Poynter, M. B. Tyers, *Nature* **1972**, *238*, 354–355.

433 S. W. Zito, M. Martínez, *J. Biol. Chem.* **1980**, *255*, 8645–8649.

434 J. N. Jansonius, G. Eichele, G. C. Ford, D. Picot, C. Thaller, M. G. Vincent, in: *Transaminases* (Eds.: P. Christen, D. E. Metzler), Wiley, New York, **1985**, pp. 110–138.

435 S. Arbilla, J. Allen, A. Wick, S. Langer, *Eur. J. Pharmacol.* **1986**, *130*, 257–263.

436 N. F. Ford, L. J. Browne, T. Campbell, C. Gemenden, R. Goldstein, C. Guide, J. N. F. Wasley, *J. Med. Chem.* **1985**, *28*, 164–170.

437 R. Alvarez-Sarandés, C. Peinador, J. M. Quintela, *Tetrahedron* **2001**, *57*, 5413–5420.

438 L. F. Tietze, J. Fennen, H. Geissler, G. Schulz, E. Anders, *Liebigs Ann. Chem.* **1995**, 1681–1687.

439 (a) S. Laschat, J. Lauterwein, *J. Org. Chem.* **1993**, *58*, 2856–2861; (b) O. Temme, S. Laschat, *J. Chem. Soc. Perkin Trans. 1* **1995**, 125–131.

440 U. Beifuss, A. Herde, S. Ledderhose, *Chem. Commun.* **1996**, 1213–1214.

441 R. S. Kumar, R. Nagarajan, P. T. Perumal, *Synthesis* **2004**, 949–959.

442 (a) M. Hiersemann, *Synlett* **2000**, 415–417; (b) M. Hiersemann, *Eur. J. Org. Chem.* **2001**, 483–491.

443 N. Greeves, K. J. Vines, *J. Chem. Soc., Chem. Commun.* **1994**, 1469–1470.

444 H. Nakamura, H. Yamamoto, *Chem. Commun.* **2002**, 1648–1649.

445 Y. Hu, D. Bai, *Tetrahedron Lett.* **2001**, *40*, 545–548.

446 B. Alcaide, P. Almendros, *Tetrahedron Lett.* **1999**, *40*, 1015–1018.

447 (a) A. Hosomi, H. Sakurai, *J. Am. Chem. Soc.* **1977**, *99*, 1673–1675; (b) G. Majetich, J. S. Song, C. Ringold, G. A. Nemeth, M. G. Newton, *J. Org. Chem.* **1991**, *56*, 3973–3988.

448 L. F. Tietze, M. Rischer, *Angew. Chem. Int. Ed. Engl.* **1992**, *31*, 1221–1222.

449 (a) T. V. Ovaska, J. B. Roses, *Org. Lett.* **2000**, *2*, 2361–2364; (b) C. E. McIntosh, I. Martínez, T. V. Ovaska, *Synlett* **2004**, 2579–2581.

450 K. M. Brummond, J. Lu, *Org. Lett.* **2001**, *3*, 1347–1349.

451 K. Sakamoto, E. Tsujii, F. Abe, T. Nakanishi, M. Yamashita, N. Shigematsu, S. Izumi, M. Okuhara, *J. Antibiot.* **1996**, *49*, 37–44.

452 K. M. Brummond, S.-p. Hong, *J. Org. Chem.* **2005**, *70*, 907–916.

453 A. Cooke, J. Bennett, E. McDaid, *Tetrahedron Lett.* **2002**, *43*, 903–905.

454 (a) C. Agami, F. Couty, C. Puchot-Kadouri, *Synlett* **1998**, 449–456; (b) C. Agami, F. Couty, J. Lin, A. Mikaeloff, M. Poursoulis, *Tetrahedron* **1993**, *49*, 7239–7250; (c) C. Agami, M. Cases, F. Couty, *J. Org. Chem.* **1994**, *59*, 7937–7940.

455 S. Baskaran, E. Nagy, M. Braun, *Liebigs Ann./Recueil* **1997**, 311–312.

456 (a) R. C. Pandey, M. W. Toussaint, R. M. Stroshane, C. C. Kalita, A. A. Aszalos, A. L. Garretson, T. T. Wie, K. M. Byrne, F. R. Geoghegan, Jr., R. J. White, *J. Antibiot.* **1981**, *34*, 1389–1401; (b) R. Misra, R. C. Pandey, B. D. Hilton, P. P. Poller, J. V. Silverton, *J. Antibiot.* **1987**, *40*, 786–802.

457 (a) V. J. Santora, H. W. Moore, *J. Org. Chem.* **1996**, *61*, 7976–7977; (b) J. M. MacDougall, V. J. Santora, S. K. Verma, P. Turnbull, C. R. Hernandez, H. W. Moore, *J. Org. Chem.* **1998**, *63*, 6905–6913; (c) S. K. Verma, E. B. Fleischer, H. W. Moore, *J. Org. Chem.* **2000**, *65*, 8564–8573.

458 D. R. Williams, J. T. Reeves, *J. Am. Chem. Soc.* **2004**, *126*, 3434–3435.

459 (a) K. J. Shea, P. D. Davis, *Angew. Chem. Int. Ed. Engl.* **1983**, *22*, 419–420; (b) K. J. Shea, J. W. Gilman, C. D. Haffner, T. K.

Dougherty, *J. Am. Chem. Soc.* **1986**, *108*, 4953–4956; (c) K. J. Shea, C. D. Haffner, *Tetrahedron Lett.* **1988**, *29*, 1367–1370; (d) K. J. Shea, S. T. Sakata, *Tetrahedron Lett.* **1992**, *33*, 4261–4264; (e) R. W. Jackson, R. G. Higby, J. W. Gilman, K. J. Shea, *Tetrahedron* **1992**, *48*, 7013–7032; (f) R. W. Jackson, K. J. Shea, *Tetrahedron Lett.* **1994**, *35*, 1317–1320.

460 S. M. Sheehan, G. Lalic, J. S. Chen, M. D. Shair, *Angew. Chem. Int. Ed.* **2000**, *39*, 2714–2715.

461 C. Chen, M. E. Layton, S. M. Sheehan, M. D. Shair, *J. Am. Chem. Soc.* **2000**, *122*, 7424–7425.

462 (a) A. Padwa, M. D. Danca, K. I. Hardcastle, M. S. McClure, *J. Org. Chem.* **2003**, *68*, 929–941; (b) A. Padwa, T. M. Heidelbaugh, J. T. Kuethe, M. S. McClure, Q. Wang, *J. Org. Chem.* **2002**, *67*, 5928–5937; (c) A. Padwa, T. M. Heidelbaugh, J. T. Kuethe, M. S. McClure, *J. Org. Chem.* **1998**, *63*, 6778–6779; (d) A. Padwa, A. G. Waterson, *Tetrahedron Lett.* **1998**, *39*, 8585–8588.

463 V. U. Ahmad, S. Iqbal, *Phytochemistry* **1993**, *33*, 735–736.

464 (a) D. J. Burton, Z. Y. Yang, *Tetrahedron* **1992**, *48*, 189–275; (b) M. A. McClinton, D. A. McClinton, *Tetrahedron* **1992**, *48*, 6555–6666.

465 (a) R. Filler, Y. Kobayashi, *Biomedical Aspects of Fluorine Chemistry*, Kodansha, Tokyo, **1982**; (b) Rhone Poulenc Conference: *Synthesis of Aromatic and Heteroaromatic Compounds Substituted by a Limited Number of Fluorine Atoms or Short Chain Fluorinated Groups*, Lyon, September, **1986**. Abstracts: *L'Actual. Chim.* **1987**, 168–170; (c) J. T. Welch, *Selective Fluorination in Organic and Bioorganic Chemistry*, ACS Symposium Ser. No. 456, ACS, Washington, DC, **1991**.

466 (a) K. Burger, A. Fuchs, L. Hennig, B. Helmreich, *Tetrahedron Lett.* **2001**, *42*, 1657–1659; (b) K. Burger, A. Fuchs, L. Hennig, B. Helmreich, D. Greif, *Monatsh. Chem.* **2001**, *132*, 929–945.

467 B. Alcaide, P. Almendros, C. Aragoncillo, M. C. Redondo, *Eur. J. Org. Chem.* **2005**, 98–106.

468 (a) C. Kuhn, L. Skaltsounis, C. Monneret, J.-C. Florent, *Eur. J. Org. Chem.* **2003**, 2585–2595; (b) C. Kuhn, E. Roulland, J.-C. Madelmont, C. Monneret, J.-C.

Florent, *Org. Biomol. Chem.* **2004**, *2*, 2028–2039.

469 B. J. Baker, P. J. Scheuer, *J. Nat. Prod.* **1994**, *57*, 1346–1953.

470 (a) M. Kobayashi, T. Yasuzawa, M. Yoshihara, B. W.-Son, Y. Kyogoku, I. Kitagawa, *Chem. Pharm. Bull.* **1983**, *31*, 1440–1443; (b) H. Kikuchi, Y. Tsukitani, K. Iguchi, Y. Yamada, *Tetrahedron Lett.* **1983**, *24*, 1549–1552.

471 K. Iguchi, S. Kaneta, K. Mori, Y. Yamada, A. Honda, Y. Mori, *Tetrahedron Lett.* **1985**, *26*, 5787–5890.

472 S. A. Raw, R. J. K. Taylor, *J. Am. Chem. Soc.* **2004**, *126*, 12260–12261.

473 R. S. Mali, P. Kaur Sandhu, A. Manekar-Tilve, *J. Chem. Soc., Chem. Commun.* **1994**, 251–252.

474 R. D. Murray, J. Mendez, S. A. Brown, *The Natural Coumarins, Occurrence, Chemistry and Biochemistry*, Wiley-Interscience, New York, **1982**.

475 M. L. Bolla, B. Patterson, S. D. Rychnovsky, *J. Am. Chem. Soc.* **2005**, *127*, 16044–16045.

476 R. Grigg, L. Zhang, S. Collard, A. Keep, *Tetrahedron Lett.* **2003**, *44*, 6979–6982.

477 E. Rossi, A. Arcadi, G. Abbiati, O. A. Attanasi, L. De Crescentini, *Angew. Chem. Int. Ed.* **2002**, *41*, 1400–1402.

478 (a) K. Kikukawa, T. Matsuda, *Chem. Lett.* **1977**, *2*, 159–162; (b) K. Kikukawa, K. Ikenaga, K. Kono, K. Toritani, F. Wada, T. Matsuda, *J. Organomet. Chem.* **1984**, *270*, 277–282; (c) K. Kikukawa, K. Nagira, F. Wada, T. Matsuda, *Tetrahedron* **1981**, *37*, 31–36; (d) K. Kikukawa, K. Maemura, Y. Kiseki, F. Wada, T. Matsuda, C. S. Giam, *J. Org. Chem.* **1981**, *46*, 4885–4888; (e) S. Sengupta, S. Bahattacharya, *J. Chem. Soc. Perkin Trans. 1* **1993**, *17*, 1943–1944; (f) Y. Wang, Y. Pang, Z. Y. Zhang, H. W. Hu, *Synthesis* **1991**, 967–969.

479 M. Beller, H. Fischer, K. Kühlein, *Tetrahedron Letters* **1994**, *35*, 8773–8776.

480 X. Marat, N. Monteiro, G. Balme, *Synlett* **1997**, 845–847.

481 B. Clique, N. Monteiro, G. Balme, *Tetrahedron Lett.* **1999**, *40*, 1301–1304.

482 K. Hiroya, S. Itoh, M. Ozawa, Y. Kanamori, T. Sakamoto, *Tetrahedron Lett.* **2003**, *43*, 1277–1280.

483 (a) K. Rück-Braun, C. Möller, *Chem. Eur. J.* **1999**, *5*, 1028–1037; (b) K. Rück-Braun, C. Möller, *Chem. Eur. J.* **1999**, *5*, 1038–

1044; (c) C. Möller, M. Mikulás, F. Wier-
schem, K. Rück-Braun, *Synlett* **2000**, *1*,
182–184.

484 A. F. Abdel-Magid, B. D. Harris, C. A.
Maryanoff, *Synlett* **1994**, 81–83.

485 (a) D. C. Beshore, C. J. Dinsmore, *Tetrahedron Lett.* **2000**, *41*, 8735–8739; (b)
C. J. Dinsmore, J. M. Bergman, *J. Org.
Chem.* **1998**, *63*, 4131–4134; (c) C. J. Dinsmore, C. B. Zartman, *Tetrahedron Lett.*
2000, *41*, 6309–6312.

486 (a) L. F. Tietze, C. Bärtels, *Tetrahedron*
1989, *45*, 681–686; (b) L. F. Tietze, C.
Bärtels, J. Fennen, *Liebigs Ann. Chem.*
1989, 1241–1245.

487 S. P. Chavan, R. Sivappa, *Tetrahedron Lett.*
2004, *45*, 3113–3115.

488 R. Mahrwald, *Curr. Org. Chem.* **2003**, *7*,
1713–1723.

489 For some examples, see: (a) C. Delas, C.
Moïse, *Synthesis* **2000**, *2*, 251–254; (b)
P. M. Bodnar, J. T. Shaw, K. A. Woerpel, *J.
Org. Chem.* **1997**, *62*, 5674–5675; (c) J.
Mlynarski, M. Mitura, *Tetrahedron Lett.*
2004, *45*, 7549–7552.

490 (a) L. Lu, H. Chang, J. Fang, *J. Org.
Chem.* **1999**, *64*, 843–853; (b) J. Hsu, J.
Fang, *J. Org. Chem.* **2001**, *66*, 8573–8584;
(c) T. Kodama, S. Shuto, S. Ichikawa, A.
Matsuda, *J. Org. Chem.* **2002**, *67*, 7706–
7715.

491 C. M. Mascarenhas, S. P. Miller, P. S.
White, J. P. Morken, *Angew. Chem. Int.
Ed.* **2001**, *40*, 601–603.

492 (a) I. Simpura, V. Nevalainen, *Angew.
Chem. Int. Ed.* **2000**, *39*, 3422–3425; (b) V.
Nevalainen, I. Simpura, *Tetrahedron Lett.*
2001, *42*, 3905–3907; (c) I. Simpura, V.
Nevalainen, *Tetrahedron* **2003**, *59*, 7535–
7546; (d) C. Schneider, M. Hansch,
Chem. Commun. **2001**, 1218–1219.

493 C. Schneider, M. Hansch, *Synlett* **2003**,
837–840.

494 J. B. Shotwell, E. S. Krygowski, J. Hines,
B. Koh, E. W. D. Huntsman, H. W. Choi,
J. S. Schneekloth Jr., J. L. Wood, C. M.
Crews, *Org. Lett.* **2002**, *4*, 3087–3089.

495 (a) R. S. Kerbel, *Carcinogenesis* **2000**, *21*,
505–515; (b) M. K. Oehler, R. Bicknell, *Br.
J. Cancer* **2000**, *82*, 749–752.

496 (a) N. Naruse, R. Kageyama-Kawase, Y.
Funahashi, T. Wakabayashi, *J. Antibiot.*
2000, *53*, 579–590; (b) T. Wakabayashi, R.
Kageyama-Kawase, Y. Funahashi, K.

Yoshimatsu, *J. Antibiot.* **2000**, *53*, 591–
596; (c) N. Hata-Sugi, R. Kageyama-
Kawase, T. Wakabayashi, *Biol. Pharm.
Bull.* **2000**, *25*(4), 446–451.

497 Recently, luminacin C2 was isolated in a
screen for Src kinase inhibitors. *In-vitro*
experiments suggest that it elicits some
of its biological effects via disruption of
SH3-mediated association of any number
of intracellular proteins with Src. See: (a)
S. Sharma, C. Oneyama, Y. Yamashita, H.
Nakano, K. Sugawara, M. Hamada, N.
Kosaka, T. Tamaoki, *Oncogene* **2001**, *20*,
2068–2079; (b) C. Oneyama, H. Nakano,
S. Sharma, *Oncogene* **2002**, *21*, 2037–
2050.

498 D. A. Evans, A. H. Hoveyda, *J. Am. Chem.
Soc.* **1990**, *112*, 6447–6449.

499 (a) K. Nishide, M. Ozeki, H. Kunishige,
Y. Shigeta, P. K. Patra, Y. Hagimoto, M.
Node, *Angew. Chem. Int. Ed.* **2003**, *42*,
4515–4517; (b) M. Ozeki, K. Nishide, F.
Teraoka, M. Node, *Tetrahedron: Asym-
metry* **2004**, *15*, 895–907.

500 M. Pal, Y. K. Rao, R. Rajagopalan, P.
Misra, P. M. Kumar, C. S. Rao, World
Patent, WO 01/90097, **2001**; *Chem. Abstr.*
2002, *136*, 5893.

501 P. Prasit, Z. Wang, C. Brideau, C.-C.
Chan, S. Charleson, W. Cromlish, D.
Either, J. F. Evans, A. W. Ford-Hutchin-
son, J. Y. Gauthier, R. Gordon, J. Guay,
M. Gresser, S. Kargman, B. Kennedy, Y.
Leblanc, S. Leger, J. Mancini, G. P. O'
Neil, M. Oullet, M. D. Percival, H. Per-
rier, D. Riendeau, I. Rodger, P. Tagari, M.
Therien, D. Visco, D. Patrick, *Bioorg.
Med. Chem. Lett.* **1999**, *9*, 1773–1778.

502 J. Bosch, T. Roca, J.-L. Catena, O. Llorens,
J.-J. Perez, C. Lagunas, A. G. Fernandez,
I. Miquel, A. Fernandez-Serrat, C. Farrer-
ons, *Bioorg. Med. Chem. Lett.* **2000**, *10*,
1745–1748.

503 V. R. Pattabiraman, S. Padakanti, V. R.
Veeramaneni, M. Pal, K. R. Yeleswarapu,
Synlett **2002**, *6*, 947–951.

504 B. W. Greatrex, D. K. Taylor, E. R. T.
Tiekink, *J. Org. Chem.* **2004**, *69*, 2580–
2583.

505 J. G. Kim, K. M. Waltz, I. F. Garcia, D.
Kwiatkowski, P. J. Walsh, *J. Am. Chem.
Soc.* **2004**, *126*, 12580–12585.

3
Radical Domino Reactions

Besides carbon–carbon-bond formations by aldol reactions, transition metal-cata-lyzed couplings and pericyclic reactions, transformations based on radicals have become an indispensable alternative as a nonpolar method for the connection of carbon atoms. For quite some time, chemists have hesitated to apply transforma-tions involving free radicals for the synthesis of fine chemicals, because they were afraid of the radicals' high reactivity and thus unselectivity, as well as their unpre-dictability in product formation. On the other hand, radical chemistry was always very important for the synthesis of bulk chemicals. During the past few decades, radical reactions have acquired their rehabilitation for the controlled preparation of complex molecules as a versatile mainstream tool in organic synthesis. This was made possible through intense research in physical organic chemistry, bringing light into the rather marginal knowledge of the behavior of radicals. Since radical reactions are ideal for sequencing, due to the very fundamental reason that the product of every radical reaction is a radical, they have opened the door for many efficient and elegant domino processes. Nowadays, domino radical reactions are prized for their capability to build complex, highly substituted ring systems, and for their general tolerance of functionalities in the substrates, allowing transforma-tions with a minimum use of protecting groups. In marked contrast to polar processes, radical transformations can proceed in most cases in the presence of free hydroxyl and amino groups, as well as keto and ester functionalities. Not sur-prisingly, high grades of chemo-, regio- and stereoselectivity can be obtained, a characteristic attributable to the mild reaction conditions, which are applied in radical chemistry. Another advantageous feature to the use of radicals is the fact that they are equally feasible to add to either inactivated double and triple bonds as to those bearing polarizing groups. As a consequence, the development of new domino radical reactions continues at a vigorous pace, their beneficial contribu-tions in the field of organic chemistry having been documented by a multitude of publications.

Usually, free-radical domino processes are characterized by a sequence of in-tramolecular steps, the overall propagation coordinate being unimolecular (exclud-ing initiation and termination steps) (Scheme 3.1) [1].

The most relevant counterpart of these unimolecular reactions is represented by reactions in which one step – in many cases the first – is an intermolecular radical addition to an appropriate functionalized acceptor (Scheme 3.2).

Domino Reactions in Organic Synthesis. Lutz F. Tietze, Gordon Brasche, and Kersten M. Gericke
Copyright © 2006 WILEY-VCH Verlag GmbH & Co. KGaA, Weinheim
ISBN: 3-527-29060-5

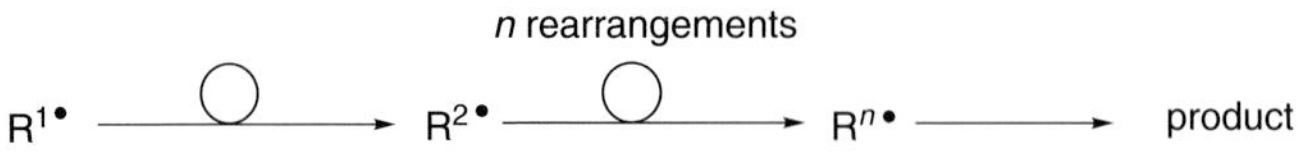

n rearrangements

R^1• ⟶ ⟶ R^2• ⟶ ⟶ R^n• ⟶ product

Scheme 3.1. General scheme for an intramolecular domino radical reaction.

m rearrangements

R• + X=Y $\xrightarrow{\text{intermolecular addition}}$ X–Y•(R) ⟶ ⟶ PR• ⟶ product

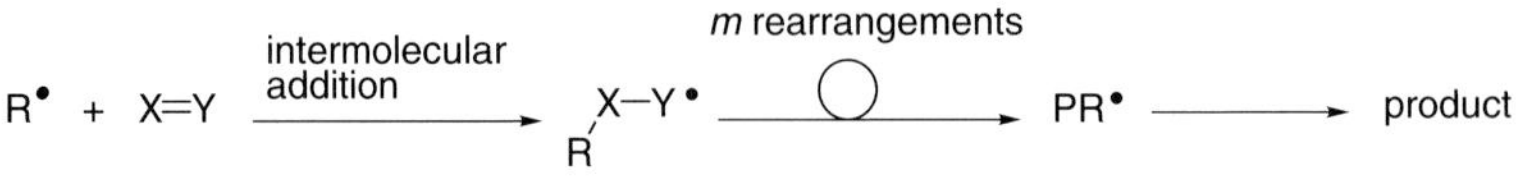

Scheme 3.2. General scheme for an intermolecular domino radical reaction.

Z $\xrightarrow{\text{Initiator}}$ • $\xrightarrow[\text{intermolecular addition}]{X=Y}$ •Y–X

6-*endo*-cycl.   5-*exo*-cycl.

[4+2] annulation   [3+2] annulation

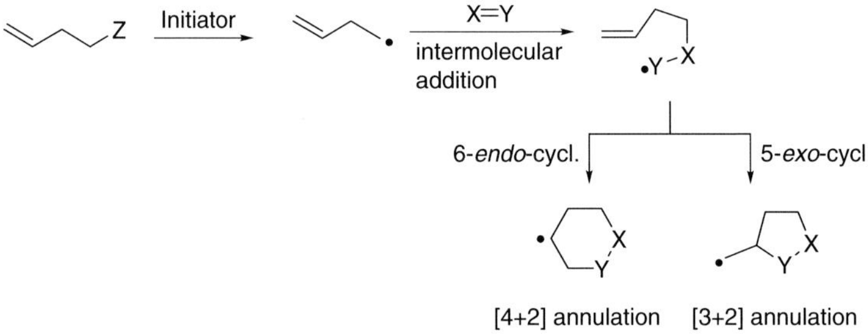

Scheme 3.3. Intermolecular radical addition leading to cyclic products.

If the attacking radical contains an adequately placed radical acceptor functionality, the possibility of a radical cycloaddition is provided, offering a procedure to construct cyclic products from acyclic precursors. For this type of ring-forming process, in which two molecular fragments are united with the formation of two new bonds, the term "annulation" has been adopted (Scheme 3.3).

Oligomerizations and polymerizations in which many radical additions to a limited range of alkenes (or other acceptors) take place will not be discussed in this book, although they are typical domino reactions. However, they usually do not lead to single well-defined products.

Radical domino processes follow a general scheme involving the generation of free radicals as an initiation step. Specifically, the formation of radicals can either proceed by abstraction or substitution utilizing halides, as well as phenylthio or phenylselenium compounds as substrates and stannanes such as nBu$_3$SnH, silanes and germanes as initiators, or by redox processes employing transition metals or lanthanides. Since the often-utilized organo tin compounds bear a toxic potential, there is an ongoing search to use alternative methodologies for the creation of radicals. These should be on the one hand nonpolluting and safe, and on the other hand efficient and reliable. Nevertheless, the provided radical intermediate

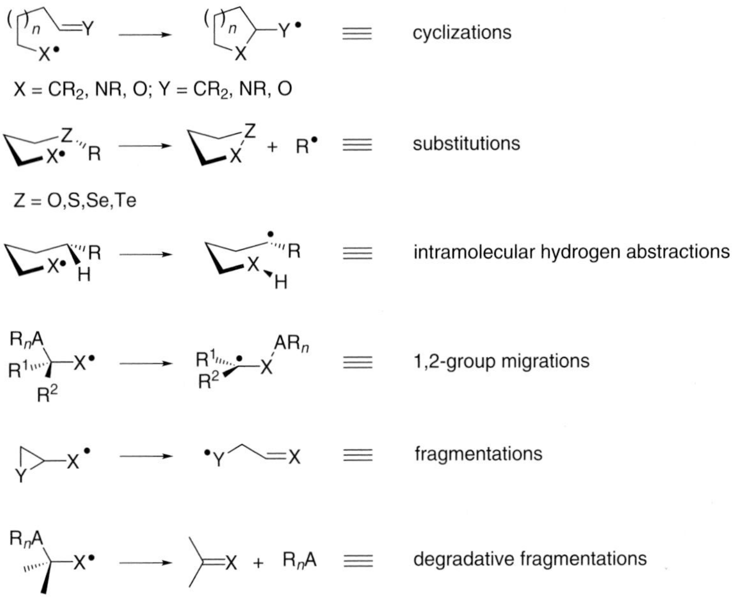

$X = CR_2, NR, O; Y = CR_2, NR, O$ — cyclizations

$Z = O, S, Se, Te$ — substitutions

intramolecular hydrogen abstractions

1,2-group migrations

fragmentations

degradative fragmentations

Scheme 3.4. Walton's and McCarroll's classification of unimolecular free-radical rearrangements.

undergoes either an intermolecular addition to an acceptor molecule or a unimolecular rearrangement. Furthermore, the sequence can be carried on until the final radical is captured by reduction, oxidation or atom transfer, resulting in the formation of the desired product. A fundamental requirement for a radical domino reaction to proceed effectively is that the rates of each individual rearrangement must be rapid compared to the termination reactions (combination, disproportionation, redox) of the radical intermediates and additionally, in comparison with reactions with the solvent, the precursor, or initiator molecules. Another important fact is the selective reaction of the final radical product (and not the remainder of the intermediate radicals) with the designated acceptor. This elementary goal can be achieved by creating a final radical which displays a significant change in the polarity or reactivity, for example an *O*-centered or vinyl-type radical.

An excellent review by Walton and McCarroll has recapitulated the various processes which can be featured during a radical cascade [2]. Moreover, these authors have elaborated a compilation of classes of unimolecular free-radical rearrangements, as illustrated in Scheme 3.4.

The first type of process is characterized by cyclization reactions, which are found in a plethora of examples and hence can be considered as the "flagship" of the different classes being discussed in this section. In spite of the fact that this reaction type distinguishes a broad scope of subsections, the 5-*exo*-trig ring closure can be regarded as the most frequent and productive one. Furthermore, 6-*endo* and 6-

exo processes are also encountered in radical chemistry, though less often. Nevertheless, the well-established Baldwin's rules elucidate the favored and disfavored cyclization modes, providing a helpful guidance for synthetic predictions and mechanistic assumptions [3]. Another important feature that can be imputed to radical cyclization methodologies, is the ability to construct carbon–carbon bonds at centers exhibiting a high sterically demand and in addition congested quaternary stereocenters – one of the most striking tasks in organic chemistry – in an efficient manner.

The second, albeit scarcely applied, class of radical domino reactions includes intramolecular homolytic substitutions, which can also result in the formation of rings. As a consequence, the radical is ejected with the displaced center and does not remain with the main ring; hence, this type of reaction is normally featured as terminating step. The most typical complication occurring within substitutions is an undesired branching of a sequence caused by the released radical.

Another more common group envisages intramolecular hydrogen abstraction processes. Both 1,5- and 1,6-hydrogen migrations should be mentioned at this point as the most prevalent appearing events leading to a translocation of the prior radical center. Moreover, 1,2-group migration typifies a class of radical processes presented as the fourth example in Scheme 3.4. Most of them involve groups that feature some type of unsaturation such as aryl, vinyl, or carbonyl, whereas carbon-centered groups, R_3C, with sp^3 hybridization are usually not able to afford a 1,2-migration.

The final category is represented by fragmentations, whereupon ring-opening motifs, in which a single unsaturated radical is generated, is the most relevant type. Substrates which are well-suited to this type of transformation include cyclopropyl-methyl, oxiranylmethyl and cyclobutylmethyl radicals which exhibit a rapid ring opening. A special case of fragmentation is combined with a coexistent degradation such as decarboxylation.

Furthermore, Walton and McCarroll proposed a very logical and concise system for the classification of free-radical domino reactions, and this is presented in the following [2]. First, a capital letter is assigned to each of the above-described processes: cyclizations (**C**), substitution reactions (**S**), H-abstractions (**H**), 1,2-group migrations (**M**), and fragmentations (**F**). The corresponding symbols are combined with a suffix (*exo* = **x**, *endo* = **n**) clarifying the mechanistic details – for example, $\mathbf{C^{5x}}$ codes a 5-*exo*-cyclization, and $\mathbf{C^{6n}}$ a 6-*endo*-cyclization. By using this abbreviation system the specification of complex radical domino processes becomes much faster and convenient.

It is also worth emphasizing that the initiation and termination steps are not included in the central chain process. For instance, in metal hydride-promoted domino reactions the initial halogen abstraction (or SePh displacement, etc.) and the final hydrogen abstraction from R_nMH are not classified as part of the domino sequence. More precisely, only the propagation steps within the mechanism of this process will be considered as a strict integral part of the domino reaction.

Not surprisingly, the majority of examples presented in this chapter are of pure radical nature, since, as mentioned above, radical reactions are well-suited to sequencing and can be regarded as the prototype of a chain reaction. Nevertheless, a few hetero-radical domino processes, such as combinations with cationic, anionic, pericyclic and oxidative processes, have also been published to date, and insights into these more exotic types of procedure will be described here. As yet, however, no examples in which photochemically induced, transition metal-catalyzed, reductive or enzymatic processes were involved have been identified in the literature.

3.1
Radical/Cationic Domino Processes

We enter into the field of domino radical procedures by presenting a composition of an initiating radical process and a terminating cationic transformation. This methodology, which was described sometime ago by Iwata and coworkers, allows an interesting approach to an enantiopure spiro ether such as **3-6** (Scheme 3.5) [4]. In these studies, the authors made use of the oxidizing properties of ceric ammonium nitrate (CAN) to furnish radical-cation species **3-2** out of optically active 1-arylthiobicyclo[4.1.0]heptane **3-1** by single electron transfer (SET). Radical fission of the cyclopropane moiety and yet another SET process led to the generation of the tertiary carbocation **3-4** with ring expansion. Nucleophilic ring closure gave the spirocyclic ether **3-5** which, on hydrolysis, afforded the desired 1-oxaspiro[4.6]undecane **3-6** in excellent yield (88%) and high optical purity (> 90% ee).

This is the only example of a radical/cationic domino process, which has been found in the literature.

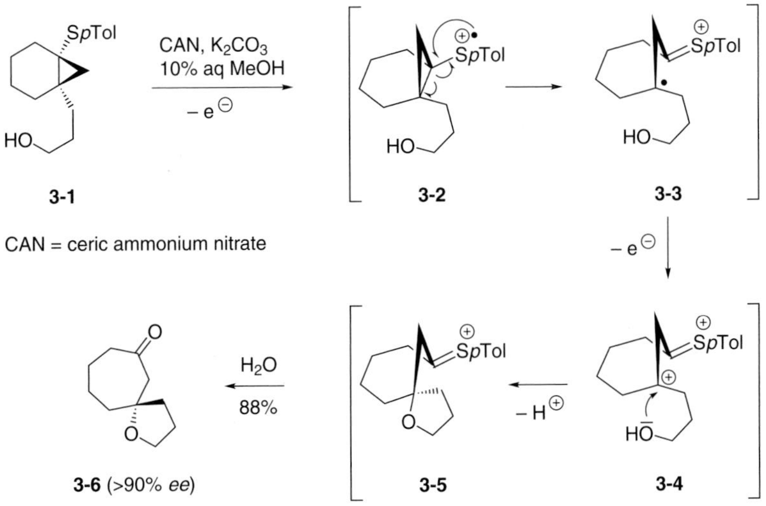

CAN = ceric ammonium nitrate

Scheme 3.5. Rare example of a domino radical ring expansion/cationic cyclization procedure.

3.2

Radical/Anionic Domino Processes

Since its introduction in 1980 by Kagan [5], samarium(II) iodide has been recognized as a versatile and powerful reagent in organic synthesis [6]. In addition to many procedures in which it is used as reducing agent or for the construction of carbon–carbon bonds, only very few examples for the samarium(II) iodide-promoted fragmentation of carbon–carbon bonds have been reported to date. Herein, a new reductive fragmentation/aldol reaction domino sequence by Schwartz and coworkers is highlighted, mediated by samarium(II) iodide [7]. Hence, when 1,4-diketone **3-7** was subjected to a THF solution of samarium(II) iodide and HMPA (this additive was crucial for the success of the process), initially two one-electron transfer processes occurred from samarium(II) iodide to the carbonyl groups of the 1,4-diketone, furnishing intermediate **3-9** bearing two ketyl radicals (Scheme 3.6). Subsequent fragmentation generated a bicyclic samarium enolate which, after tautomerization to **3-10a** and **3-10b**, underwent an aldol reaction to give the two diastereomeric products **3-8a** and **3-8b** in 28% and 20% yield, respectively.

The occurrence of the indole subunit is well established within the class of natural products and pharmaceutically active compounds. Recently, the Reissig group developed an impressive procedure for the assembly of highly functionalized indolizidine derivatives, highlighting again the versatility of domino reactions [8]. The approach is based on a samarium(II) iodide-mediated radical cyclization terminated by a subsequent alkylation which can be carried out in an intermolecular – as well as in an intramolecular – fashion. Reaction of ketone **3-11** with samarium(II) iodide induced a 6-*exo*-trig cyclization, furnishing a samarium enolate intermediate

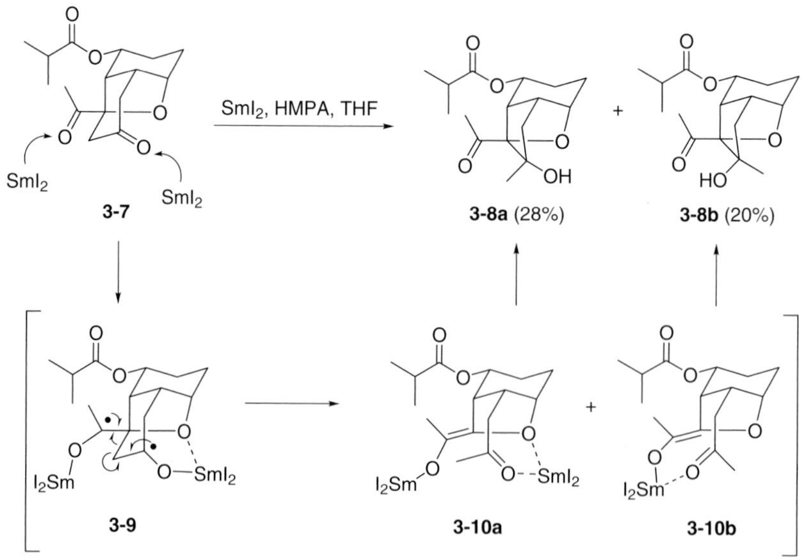

Scheme 3.6. Samarium(II)iodide-promoted domino reductive fragmentation/aldol reaction.

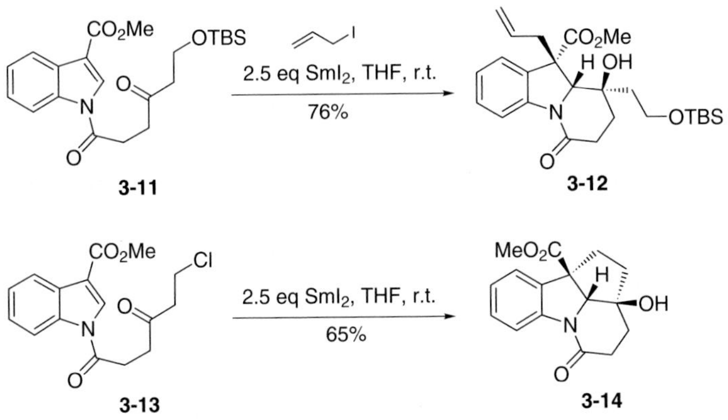

Scheme 3.7. Samarium(II)iodide-induced domino radical cyclization/alkylation procedure

which is consumed by the added allyl iodide; this generates the tricyclic indole derivative **3-12** in good yield and excellent diastereoselectivity (Scheme 3.7). The second example shows an intramolecular alkylation by reaction of **3-13** providing tetracycle **3-14**.

3.3
Radical/Radical Domino Processes

As mentioned above, processes with two or more radical intermediates represent the majority of radical domino reactions. Of special interest is the use of this methodology for the efficient synthesis of natural products.

The naturally occurring alkaloid lysergic acid (**3-15**) has disclosed challenging pharmacological activities, and therefore has attracted the attention of both medicinal [9] and synthetic [10] chemists. Parsons and coworkers developed a radical domino approach for the facile construction of the tetracyclic ring system of **3-17** (Scheme 3.8) [11]. After submitting the aromatic bromide **3-16** to classical radical conditions (tributyltin hydride/AIBN, refluxing benzene), initially a 5-*exo*-trig cyclization took place to afford a *N*-heterocycle, which was followed by a 6-*endo*-trig cyclization. In this way, the desired product **3-17** was obtained as a 3:1 mixture of two diastereoisomers, in 74% yield.

In 2002, Lee and coworkers prepared the unnatural (–)-enantiomer of lasonolide A (**3-21**) using a domino radical cyclization procedure as the key step, and furthermore revised its structure [12]. Lasonolide A possesses a macrolactone structure with two embedded stereochemically demanding tetrahydropyran rings. The compound demonstrates antitumor activity, in particular against leukemia and the highly lethal lung carcinoma. Reaction of β-alkoxyacrylate **3-18** and (bromomethyl)-chloro-dimethylsilane provided **3-19**, which was used for the following radical cyclization (Scheme 3.9). Exposure to tributyltin hydride and AIBN in refluxing benzene led to a 6-*endo*-trig ring closure by the generated α-silyl radical onto the adja-

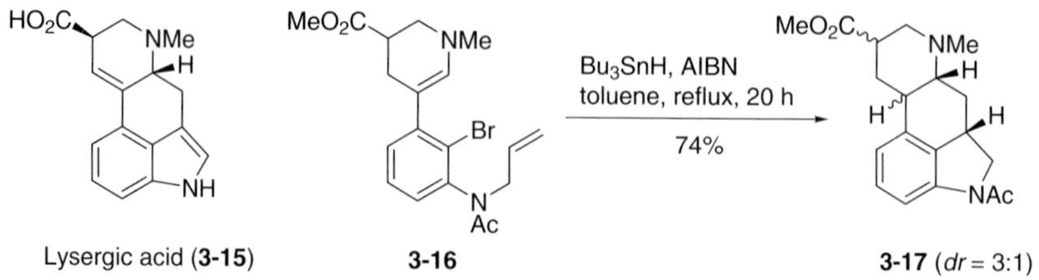

Scheme 3.8. Domino radical cyclization procedure in the synthesis of lysergic acid derivatives.

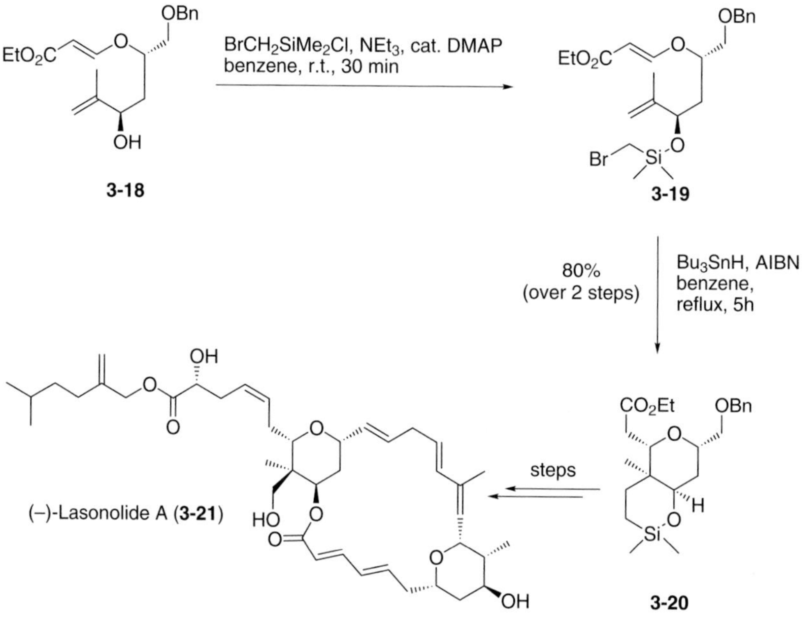

Scheme 3.9. Synthesis of (–)-lasonolide A (**3-21**).

cent double bond. Subsequent 6-*exo*-trig cyclization involving the acrylate moiety delivered the desired bicyclic product **3-20** in 80 % yield (over two steps) as a single diastereoisomer bearing all stereogenic centers with their correct configuration, which was further transformed into **3-21**.

Another effective combination of two radical cyclization steps has been demonstrated by Sha and coworkers during the course of the first total synthesis of (+)-paniculatine (**3-24**), a natural alkaloid belonging to the subclass of *Lycopodium* alkaloids [13]. **3-24** has a unique tetracyclic scaffold with seven stereogenic centers [14]. Although no special features of (+)-paniculatine have so far been documented, other *Lycopodium* alkaloids are reported to be potent acetylcholinesterase inhibitors, or show promising results in the treatment of Alzheimer's disease [15]. When

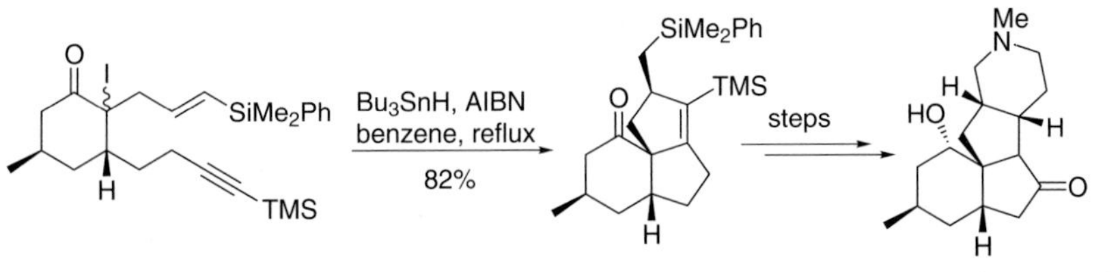

Scheme 3.10. Synthesis of (+)-paniculatine (**3-24**).

standard radical conditions were applied to iodo ketone **3-22**, an α-carbonyl radical is generated, which undergoes a 5-*exo*-dig cyclization forming a bicyclic vinyl radical intermediate (Scheme 3.10). This is followed by a 5-*exo*-trig ring closure onto the adjacent olefin moiety, which then provides the tricyclic core structure **3-23** as a single diastereoisomer, in 82 % yield.

In 1997, a compound termed CP-263,114 (**3-32**) was isolated from an unidentified fungal species by a group at Pfizer (Groton, USA) [16]. The compound exemplified the architecture of unprecedented molecular connectivity, and possessed interesting biological activities such as cholesterol-lowering properties through the inhibition of squalene synthase [16, 17]. Moreover, it was found to inhibit farnesyl transferase, an enzyme implicated in cancer [16, 18]. The first total synthesis was accomplished by the Nicolaou group in 1999 [19]. In another approach to CP-263,114 (**3-32**), the Wood group has established an intramolecular domino radical cyclization to afford the desired isotwistane core (Scheme 3.11) [20]. First, the tertiary alcohol **3-25** was converted into the acetal **3-27** using dibromoethyl ether **3-26** and *N,N*-dimethylaniline. Homolytic abstraction of the bromo atom furnished compounds **3-30** and **3-31** as a mixture of diastereoisomers in 86 % yield over two steps. The process was suggested to proceed *via* initial formation of the primary radical **3-28**, followed by a 5-*exo*-trig ring closure to give **3-29**, by reaction with the maleate moiety and subsequent addition of hydrogen from the less-hindered side.

In attempting to synthesize the natural product azadirachtin (**3-38**) [21], a powerful insect antifeedant and growth regulator, Nicolaou and coworkers created an innovative radical reaction sequence (Scheme 3.12) [22]. Under classical radical conditions (tributyltin hydride/AIBN, refluxing toluene), the bromo acetal **3-33** led to the hexacyclic compound **3-36** in 75 % yield. The reaction is supposed to proceed *via* the radicals **3-34** and **3-35**. The latter is formed by a 5-*exo*-trig cyclization. It follows a 1,5 hydrogen shift with concomitant oxidative rupture of the benzylic bond to give the observed product **3-36**. Further manipulations of **3-36** led to **3-37**, which was again formed in a domino fashion.

Natural products containing a spiropyrrolidinyloxindole nucleus have recently found to exhibit interesting biological activity such as cell-cycle inhibition [23]. This observation encouraged Murphy and coworkers to design a novel domino route to (±)-horsfiline (**3-43**), a natural spiropyrrolidinyloxindole (Scheme 3.13) [24]. Treatment of azide **3-39** with tris(trimethylsilyl)silane (TTMSS) and AIBN as radical starter led to the radical intermediate **3-40** after a first ring closure, which under-

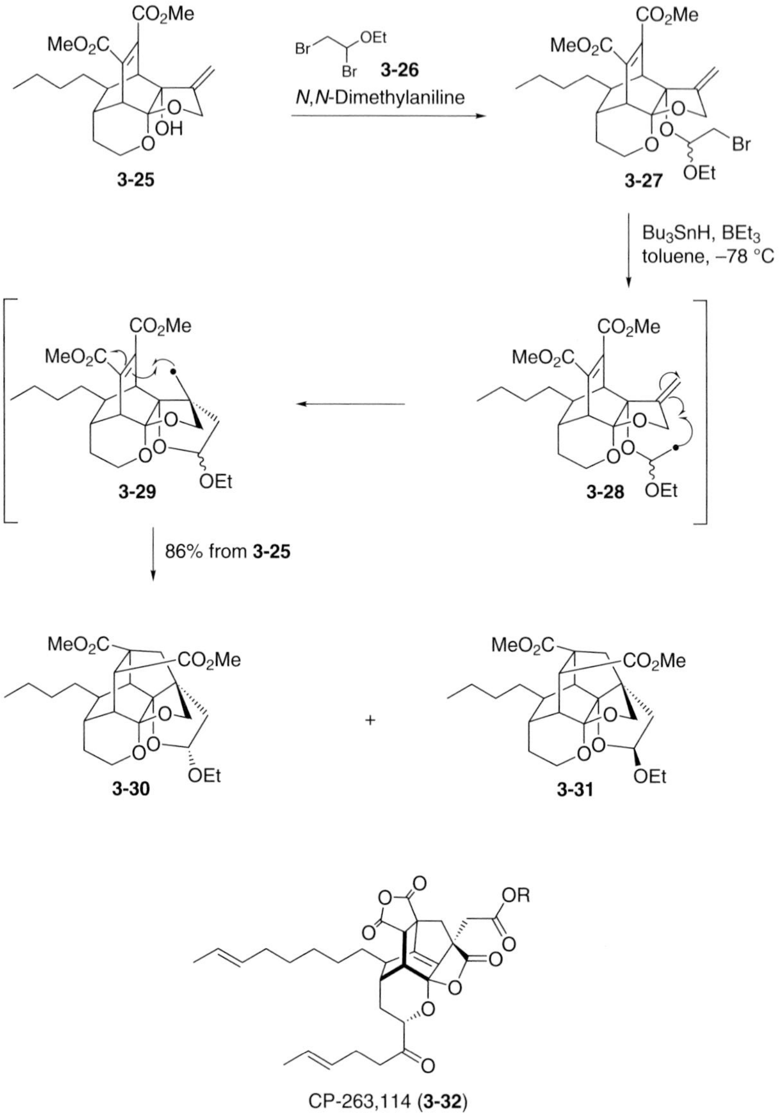

Scheme 3.11. Domino radical cyclization process towards the total synthesis of CP-263,114 (**3-32**).

went a second 5-*exo*-trig cyclization with concomitant loss of nitrogen providing spirocycle **3-42** *via* **3-41**. Methylation and debenzylation completed the synthesis of **3-43**.

Another domino radical cyclization approach, which allows construction of the B- and E-rings of the alkaloid (±)-aspidospermidine (**3-46**), has been described by the same group [25]. Transformation of the iodoazide **3-44** into the tetracycle **3-45** was accomplished in 40 % yield by selective attack at the carbon–iodine bond in the

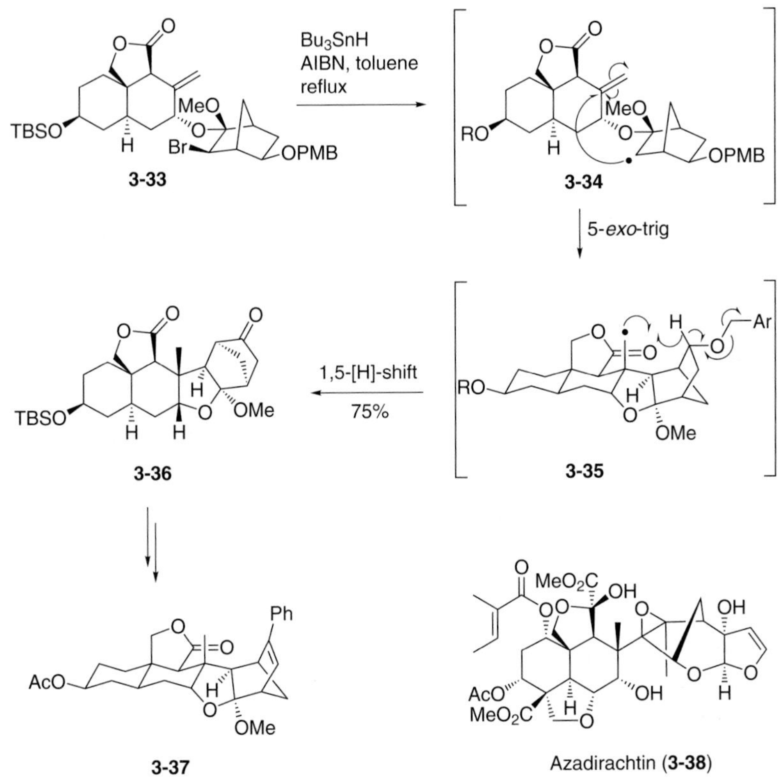

Scheme 3.12. Domino radical cyclization towards the synthesis of azadirachtin (**3-38**).

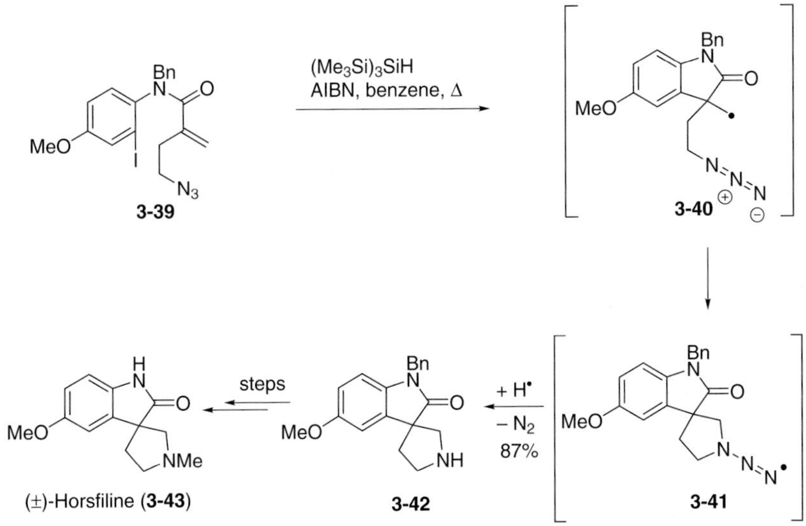

Scheme 3.13. Synthesis of (±)-horsfiline (**3-43**).

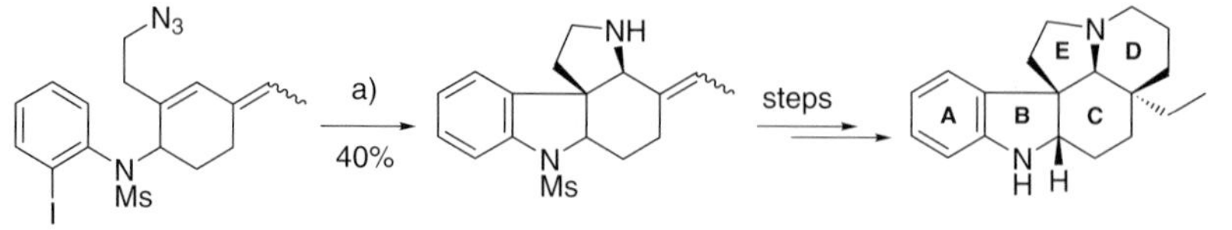

a) TTMSS, AIBN, benzene, reflux.   TTMSS = (Me₃Si)₃SiH

Scheme 3.14. Synthesis of (±)-aspidospermidine (**3-46**).

presence of the azide group, again using TTMSS/AIBN (Scheme 3.14) [26]. The resulting aryl radical initiated a two-fold 5-*exo*-trig cyclization sequence which is accompanied by evolution of nitrogen.

The skeleton of the indole alkaloid (±)-vindoline was prepared in a similar way [27].

Since the discovery that the pyrroloquinoline alkaloid camptothecin (**3-47**) and the related alkaloids mappicine (**3-48a**), nothapodytine B (**3-48b**) and nothapodytine A (**3-48c**) exhibit significant anticancer and antiviral properties [28], many synthetic approaches towards these compounds have been developed. Recently, the Bowman group devised a radical domino reaction protocol for the synthesis of the A–D ring system, featuring a vinyl radical cyclization onto nitriles [29]. Intramolecular cyclization of the vinyl radical **3-50**, generated from vinyl iodide **3-49** and hexamethylditin under photolysis, led to the iminyl radical **3-51** (Scheme 3.15). This can either cyclize in a 6-*endo*-trig fashion to give the π-radical **3-53** directly, or a 5-*exo*-trig cyclization takes place leading to the spirodienyl radical **3-52**, which is then converted into intermediate **3-53** by a neophyl rearrangement. Finally, loss of a hydrogen atom from **3-53** yields the tetracyclic products **3-54**. The mechanism of the final oxidation step is still unclear; however, a H-abstraction caused by methyl radicals, formed from thermal breakdown of trimethyltin radicals, was suggested as an explanation.

Another anti-cancer agent in clinical use is podophyllotoxin (**3-59**); this has an aryl tetrahydronaphthalene lignan lactone skeleton, and demonstrates potent tubulin-binding, anti-mitotic properties (Scheme 3.16) [30]. The Sherburn group [31] prepared this molecule by a tris(trimethylsilyl)silane promoted conversion of thionocarbonate **3-55** into the lactone **3-58**, which proceeded with a yield of 38 %. As intermediates, the radicals **3-56** and **3-57** can be assumed.

Zard and coworkers [32] reported a simple approach to create another group of natural products, namely the lycopodium alkaloids [15]. These authors first investigated the reaction of *O*-benzoyl-*N*-allylhydroxylamide **3-60** with tributyltin hydride and ACCN in refluxing toluene, which led (after formation of the *N*-radical **3-61** in a 5-*exo*-trig/5-*exo*-trig cyclization) to the undesired pyrrolidine **3-62** in 48 % yield. Nevertheless, a small structural modification, namely the placement of a chlorine atom at the allyl moiety as in **3-63**, induced a 5-*exo*-/6-*endo*- instead of the 5-*exo*-/5-

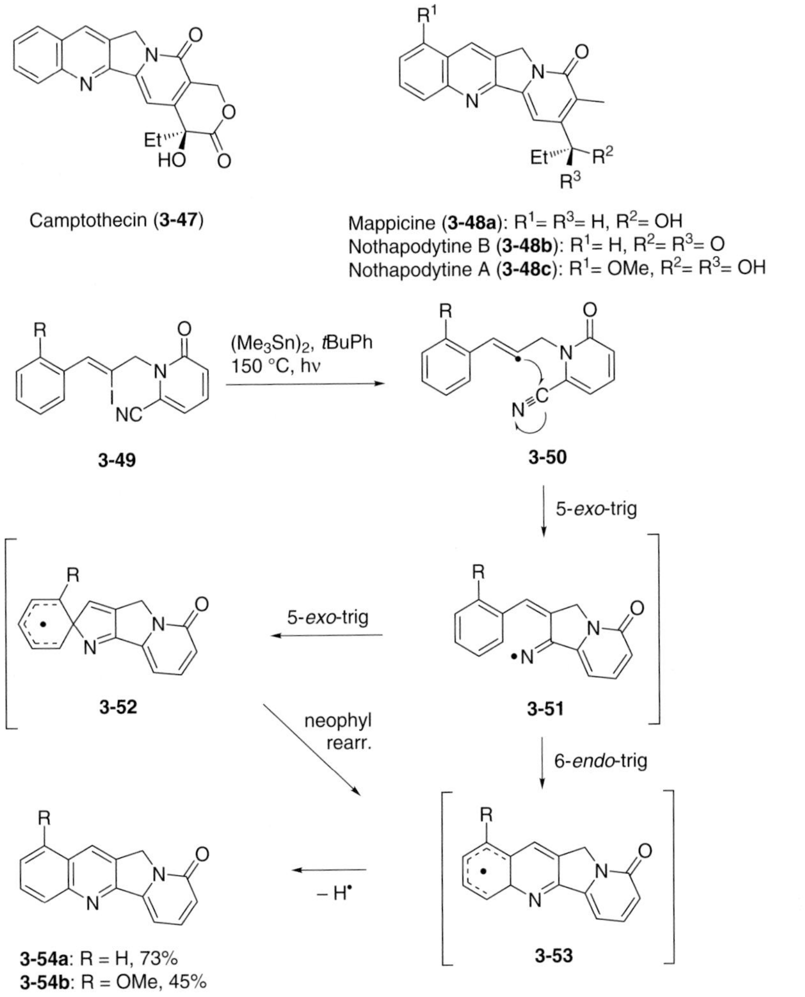
Camptothecin (3-47)
Mappicine (3-48a): R¹= R³= H, R²= OH
Nothapodytine B (3-48b): R¹= H, R²= R³= O
Nothapodytine A (3-48c): R¹= OMe, R²= R³= OH
(Me3Sn)2, tBuPh
150 °C, hν
3-49
3-50
5-exo-trig
3-51
5-exo-trig
3-52
neophyl
rearr.
6-endo-trig
3-53
– H•
3-54a: R = H, 73%
3-54b: R = OMe, 45%

Scheme 3.15. Domino radical procedure in the synthesis of camptothecin (**3-47**) and analogs.

exo-trig cyclization to give the wanted indolizidine **3-64** in 52% yield as a single diastereomer. In this case, a second equivalent of tributyltin hydride was necessary to reductively remove the chlorine atom *in situ* after the twofold cyclization process. **3-64** could be transformed into the lycopodium alkaloid 13-desoxyserratine (**3-65**) (Scheme 3.17).

N-Aziridinylimines are valuable substrates for domino radical cyclizations since they are able to serve simultaneously as radical acceptors and donors. They allow a versatile and general construction of quaternary carbon centers from carbonyl compounds [33]. By employing this methodology, an elegant and stereoselective synthesis of (±)-modhephene (**3-70**), one of the rare naturally occurring [3.3.3]propellanes,

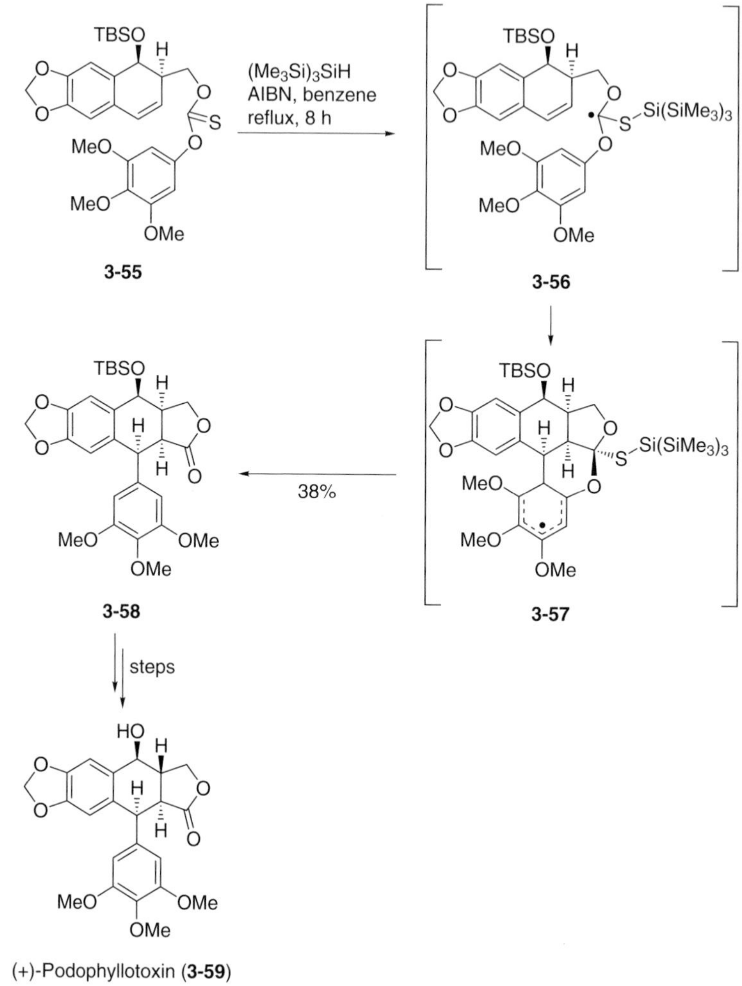

Scheme 3.16. Domino radical cyclization in the total synthesis of (+)-podophyllotoxin (**3-59**).

has been designed by the group of Lee and Kim (Scheme 3.18) [34]. The necessary substrate **3-68** was prepared by treatment of **3-66** with *N*-amino-2-phenylaziridine **3-67** under acid-catalysis. In the presence of tributyltin hydride, **3-68** was transformed almost exclusively into the desired propellane **3-69a**.

In a similar way, α-cedrene has been synthesized by the same group [35]. Moreover, these authors have also developed a further access to modhephene (**3-70**) using a free-radical 6-*endo*-/5-*exo*-domino cyclization of a dieneyne [36].

During the course of a short and efficient total synthesis of (−)-dendrobine (**3-76**), an alkaloid which exhibits antipyretic and hypotensive activities [37], a new domino radical sequence has been exploited by Cassayre and Zard which involves the cycli-

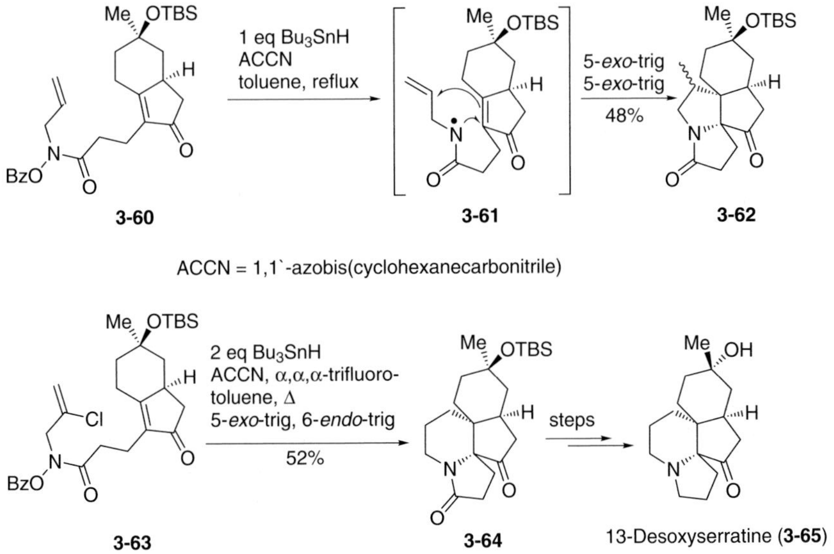

ACCN = 1,1`-azobis(cyclohexanecarbonitrile)

Scheme 3.17. Synthesis of pyrrolidines and indolizidines as well as of the lycopodium alkaloid 13-desoxyserratine (**3-65**).

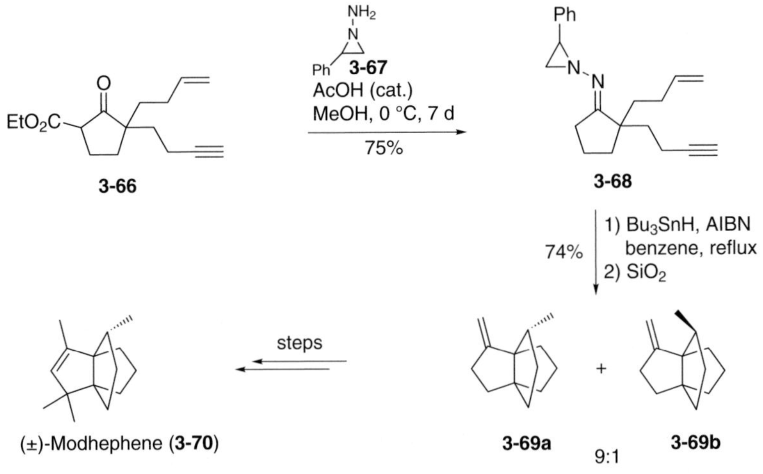

Scheme 3.18. Domino radical cyclization reaction of *N*-aziridinyl imine **3-68** in the total synthesis of (±)-modhephene (**3-70**).

zation of a carbamyl radical [38]. *O*-Benzoyl-*N*-hydroxyurethane **3-71** reacts with tributyltin hydride and ACCN in refluxing toluene to the carbamyl radical **3-72**, which directly undergoes a 5-*exo*-trig cyclization to furnish the oxazolidinone radical **3-73** (Scheme 3.19). The carbon framework of **3-73** then collapses by a radical

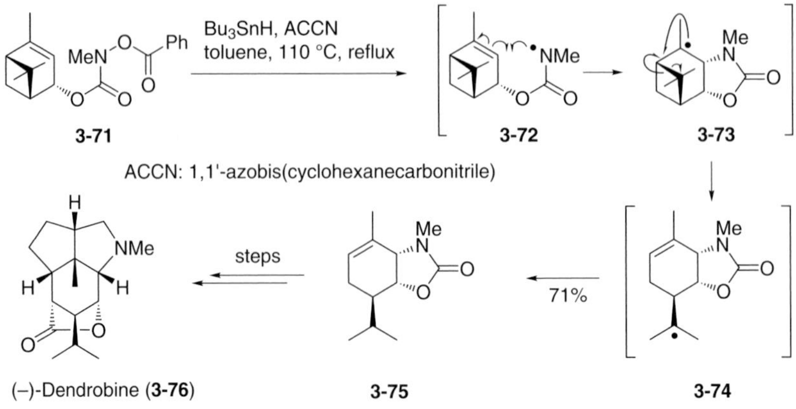

ACCN: 1,1'-azobis(cyclohexanecarbonitrile)

Scheme 3.19. Domino radical cyclization/fragmentation procedure in the synthesis of (–)-dendrobine (**3-76**).

fragmentation, providing the more stable radical **3-74**, which is finally intercepted by tributyltin hydride. This results in formation of the annulated oxazolidinone **3-75** with the desired stereochemistry, in 71 % yield.

In their enantioselective total synthesis of (+)-triptocallol (**3-79**), a naturally occurring terpenoid, Yang and coworkers made use of a concise Mn(OAc)$_3$-mediated and chiral auxiliary-assisted oxidative free-radical cyclization [39]. Reaction of **3-77**, bearing a (*R*)-pulegone-based chiral auxiliary, with Mn(OAc)$_3$ and Yb(OTf)$_3$ yielded tricyclic **3-78** in a twofold ring closure in 60 % yield and a diastereomeric ratio of 9.2:1 (Scheme 3.20). A further two steps led to (+)-triptocallol (**3-79**). For the interpretation of the stereochemical outcome, the authors proposed the hypothetical transition state **TS-3-80**, in which chelation of the β-keto ester moiety with Yb(OTf)$_3$ locks the two carbonyl groups in a *syn* orientation. The attack of the Mn^{III}-oxidation-generated radical onto the proximate double bond is then restricted to the more accessible (*si*)-face, as the (*re*)-face is effectively shielded by the 8-naphthyl moiety.

An exciting example of a rare radical transannular cyclization is the transformation of the iododieneyne **3-81** in the presence of Bu$_3$SnH by Pattenden and coworkers which, *via* **3-82**, led to the two diastereomeric products **3-83** and **3-84** in a 6:1-ratio with 45–60 % yield (Scheme 3.21) [40]. This procedure offers a straightforward approach to the taxane system.

These authors have also used an iodotrienedione in this process, but this led to the desired taxane skeleton in only 25 % yield.

The [6.5.5]-ring fused tricyclic motif is found in many natural products, and has therefore become an important target in synthesis. A convenient access to this structural framework is offered by a radical domino procedure published by the Nagano group [41]. This reaction of optical pure dibromoacetal **3-85** led to the desired tricycle **3-87** *via* **3-86** as a single diastereoisomer in a very respectable yield of 94 % by applying classical radical conditions (excess tributyltin hydride/AIBN, irradia-

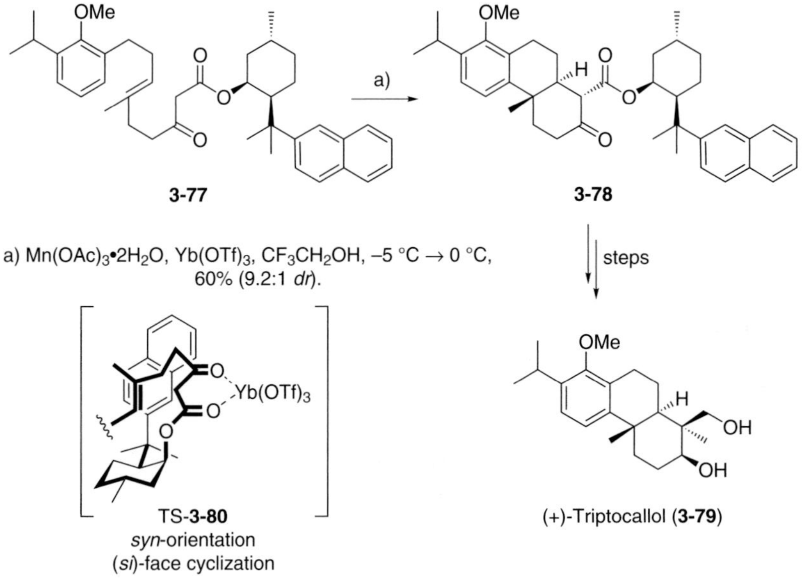

3-77 **3-78**

a) Mn(OAc)₃•2H₂O, Yb(OTf)₃, CF₃CH₂OH, –5 °C → 0 °C,
60% (9.2:1 *dr*).

$$a)\ Mn(OAc)_3\bullet 2H_2O,\ Yb(OTf)_3,\ CF_3CH_2OH,\ -5\ ^\circ C \to 0\ ^\circ C,$$

TS-**3-80**
syn-orientation
(*si*)-face cyclization

(+)-Triptocallol (**3-79**)

Scheme 3.20. Mn(OAc)₃-mediated chiral auxiliary-assisted enantioselective domino radical
cyclization in the total synthesis of (+)-triptocallol (**3-79**).

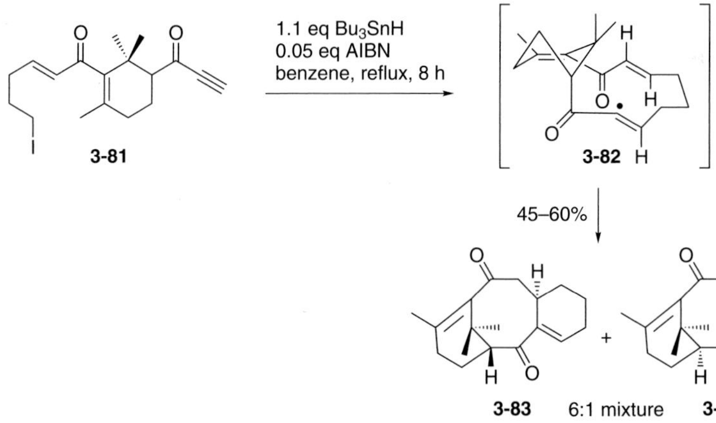

3-81

1.1 eq Bu₃SnH
0.05 eq AIBN
benzene, reflux, 8 h

3-82

45–60%

3-83   6:1 mixture   **3-84**

Scheme 3.21. Domino radical macro-cyclization/transannular-cyclization procedure for the
synthesis of the taxane skeleton.

tion) (Scheme 3.22). Intermediate **3-86** could be isolated when only one equivalent
of tin hydride was used.

The conversion of a glycal to enantiomerically pure [6.5.6]-dioxatricycles containing eight stereogenic centers has been accomplished by Hoffmann and coworkers,

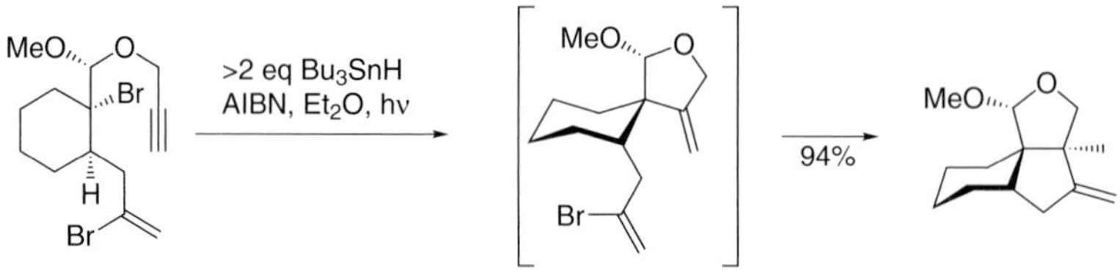

Scheme 3.22. Domino radical cyclization procedure for the synthesis of [6.5.5] fused tricycle **3-87**.

carrying out only three simple steps [42]. As shown in Scheme 3.23, the stereo-chemically pure substrates for the cyclization (*R*)-**3-92a–d** and (*S*)-**3-93a–d** were pre-pared by iodoglycosylation of glycal **3-89** and the racemic silylated ene-ynols **3-88a–d** after chromatographic separation of the formed diastereomers **3-90a–d**, as well as **3-91a–d** and desilylation. The cyclization method of choice appeared to be the air-induced triethylborane protocol, launching the 5-*exo*-trig-/6-*endo*-dig radical cas-cade, which also implies an iodine transfer. The final products (*R*)-**3-94a–d** and (*S*)-**3-95a–d**, respectively, were obtained in low to moderate yield in case of the (*R*)-isomer, and in moderate to good yields in the case of the (*S*)-isomer. Undoubtedly, the difference in outcome of both series is caused by steric factors. Hence, X-ray crystal structure elucidation of the products confirmed that in the (*S*)-series the py-ranoside ring is a well-developed chair, whereas in the (*R*)-series a non-chair confor-mation is adopted.

The first examples of a consecutive radical 5-*exo*-/dig-5-*exo*-dig cyclization of 1,5-diynes have been accomplished by the same researchers [43]. These authors were able to show that their cycloisomerization procedure provides access to strained semicyclic, conjugated dienes with a functionalized dioxatriquinane framework which occurs in the aglycones of steroidal cardiac glycosides, such as isogenine (**3-96**) [44] and C-norcardanolide (**3-97**) (Scheme 3.24) [45].

For example, exposure of the iodotetrahydrofurans **3-98** to triethylborane in re-fluxing benzene in the presence of air and ethyl iodide triggered the formation of the diastereomeric dioxatriquinanes **3-99** and **3-100** (Scheme 3.25).

An efficient methodology for the construction of pyrrolizidines and other polycy-clic nitrogen heterocycles using a radical domino sequence has been revealed by Bowman and coworkers [46]. These authors employed sulfenamides as substrates, which easily form aminyl radicals by treatment with tributyltin hydride and AIBN. For instance, **3-101** smoothly underwent a twofold 5-*exo*-trig cyclization to give the tetracyclic pyrrolizidine product **3-105** in 90% yield (Scheme 3.26). As intermedi-ates, the radicals **3-102** to **3-104** can be assumed.

Free-radical ring-expansion reactions have been established as attractive ap-proaches to standard, medium-sized, and even large rings [47]. The incorporation of an additional, appropriately positioned radical acceptor offers the possibility to extend this methodology to a domino ring expansion/cyclization procedure [48]. A general reaction path is presented in Scheme 3.27, in which readily available sub-strates **3-106** first undergo a radical ring expansion to generate intermediates of type **3-107**. These are then captured by the tethered alkyne moiety, furnishing the

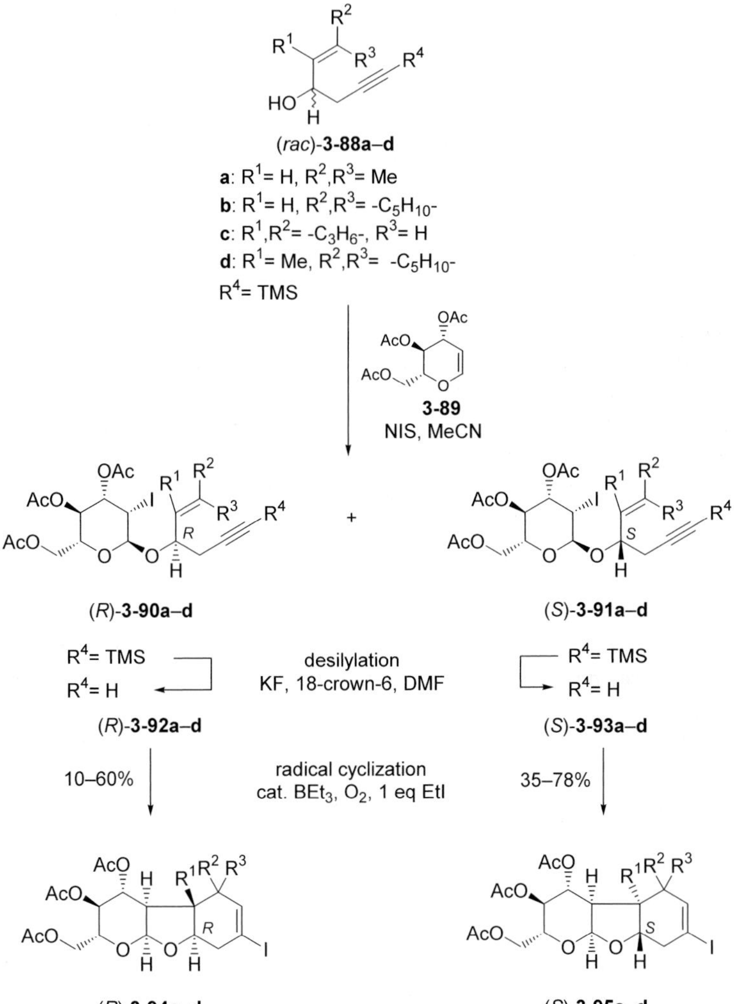

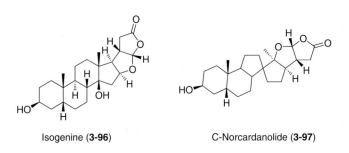

(rac)-**3-88a–d**

a: R^1= H, R^2,R^3= Me
b: R^1= H, R^2,R^3= -C$_5$H$_{10}$-
c: R^1,R^2= -C$_3$H$_6$-, R^3= H
d: R^1= Me, R^2,R^3= -C$_5$H$_{10}$-
R^4= TMS

3-89
NIS, MeCN

(R)-**3-90a–d**   +   (S)-**3-91a–d**

R^4= TMS ⌐   desilylation   ⌐ R^4= TMS
R^4= H   ←   KF, 18-crown-6, DMF   → R^4= H

(R)-**3-92a–d**   (S)-**3-93a–d**

10–60%   radical cyclization   35–78%
                cat. BEt$_3$, O$_2$, 1 eq EtI

(R)-**3-94a–d**   (S)-**3-95a–d**

Scheme 3.23. Radical-mediated domino reaction of glycopyranosides.

Isogenine (**3-96**)   C-Norcardanolide (**3-97**)

Scheme 3.24. Steroids with dioxatriquinane substructure.

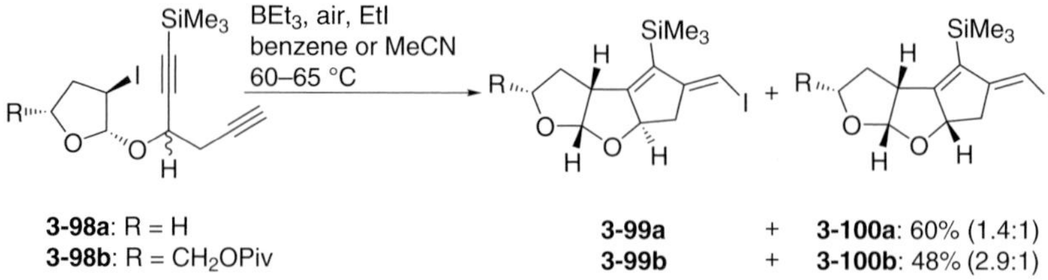

3-98a: R = H
3-98b: R = CH₂OPiv

3-99a + 3-100a: 60% (1.4:1)
3-99b + 3-100b: 48% (2.9:1)

Scheme 3.25. Dioxatriquinanes by triethylborane-induced domino radical atom transfer/cyclization of 1,5-diynes.

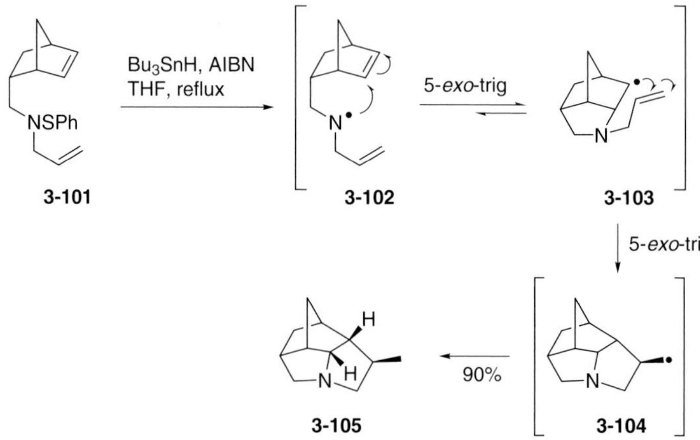

Scheme 3.26. Synthesis of pyrrozidine **3-105** by a domino radical cyclization *via* an aminyl radical.

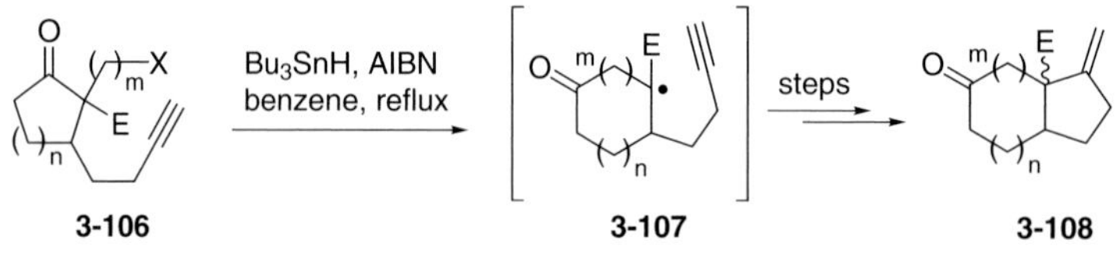

Scheme 3.27. General scheme of domino radical ring expansion/cyclization procedure.

fused bicyclic products **3-108** [49]. As pointed out by Carreira and coworkers, the two reactions may compete with each other. Thus, initially, instead of the ring expansion a 1,5-hydrogen transfer could take place, whereas in the cyclization step with m = 1 a 6-*exo* cyclization might compete.

However, cyclopentanone **3-109** with a *cis* orientation of the iodoalkane group and the alkyne moiety was converted into the fused cyclooctanone **3-110** in 82% yield (Scheme 3.28). In contrast, the corresponding *trans*-isomer only underwent 1,5-hydrogen transfer, leading to a dehalogenated starting material.

Scheme 3.28. Example of a domino radical ring expansion/cyclization procedure.

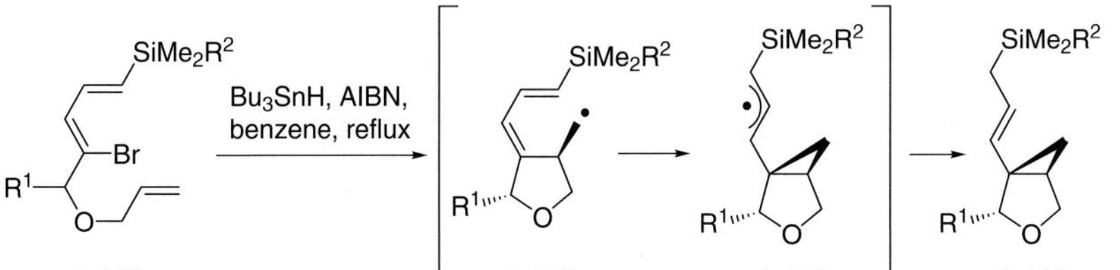

Scheme 3.29. Synthesis of bicyclo[3.1.0]oxahexanes.

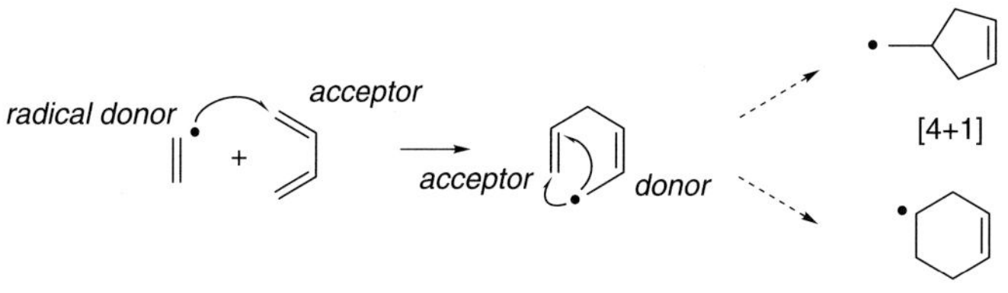

Scheme 3.30. Boomerang-type [4+1] and [4+2] radical sequences.

An early – but mechanistically interesting – construction of a bicyclo[3.1.0]oxa-hexane by a domino radical cyclization was presented by Luh's group [50]. The addition of tributyl tin and AIBN to a solution of bromides **3-111** in refluxing benzene gave **3-114** as single diastereoisomers in acceptable yields *via* the intermediates **3-112** and **3-113** (Scheme 3.29). It is important that the cyclopropyl carbinyl radical intermediate has the correct stability and reactivity, which is achieved by the α-silyl substituent.

In their studies, Takasu, Ihara and coworkers used a new methodology based on a boomerang-type radical domino sequence [51], in which an iodoalkenyl can act as both radical donor and acceptor providing [4+2] or [4+1] cyclization products [52].

Thus, reaction of the 2-iodovinyldienoate **3-115** using different reaction conditions led to the assembly of annulated product **3-116** in moderate to good yields

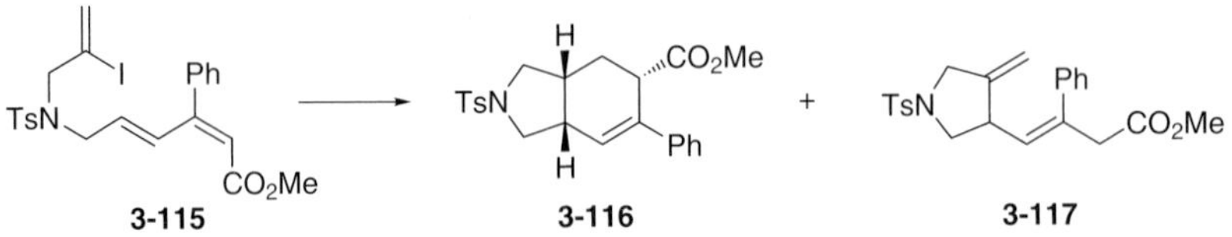

3-115   3-116   3-117

Entry	Conditions	Yield [%]	
		3-116	3-117
1	Bu$_3$SnH, AIBN, benzene, reflux[a]	40	20
2	Bu$_3$SnH, AIBN, benzene, reflux[b]	63	0
3	(TMS)$_3$SiH, AIBN, benzene, reflux[b]	66	0
4	Ni(cyclam)$^{2+}$, DMF, −1.5 V, r.t.	0	85

[a] 1.2 eq Bu$_3$SnH and 0.5 eq AIBN were added dropwise over 1 h.
[b] 1.2 eq hydride and 0.5 eq AIBN were added dropwise over 2 h.

Scheme 3.31. Intramolecular [4+1] and [4+2] annulation reactions employing domino radical cyclizations.

(Scheme 3.31, entries 1–3). The reaction proceeds *via* a 5-*exo*-trig ring closure by attack of the primarily formed vinyl radical onto the adjacent olefin moiety. Further 6-*endo*-trig cyclization involving the initial double bond furnished the desired product **3-116**. Heating of **3-115** in benzene did not lead to the product, which suggests that the [4+2] annulation reactions are not the result of a Diels–Alder process. Ultimately, it should be mentioned that indirect cathodic electrolysis applying Ni(cyclam)$^{2+}$ as mediator afforded only the monocyclic product **3-117** instead of bicycle **3-116**.

Various bicyclo[3.3.0]octanes employing a radical iodine atom transfer reaction have been successfully synthesized by Taguchi and coworkers [53]. This procedure exemplifies the small number of radical domino processes which are initiated by an intermolecular radical addition. The process makes use of the 2-iodomethylcyclopropane derivative **3-118** as precursor of a homoallylic radical **3-120**, which can be captured by 1,4-dienes as well as 1,4-ene-ynes **3-119**. Triethyl borane/air-mediated formation of **3-120** from **3-118** and subsequent Lewis acid-catalyzed formal [3+2] cycloaddition involving a diene or an ene-yne **3-119** led to the radical intermediate **3-121**. Further 5-*exo*-dig(trig) ring closure furnished bicyclic radical **3-122**, which is finally trapped by iodine to give the desired product **3-123**. The products containing an iodalkyl group were converted into the corresponding olefinic compound **3-124** by treatment with DBU, to allow a better separation (Scheme 3.32).

Another Lewis acid-catalyzed atom-transfer domino radical cyclization, to produce various bicyclic and tricyclic ring skeletons, has been developed by Yang and coworkers [54]. Reactions of the α-bromo-β-keto ester **3-125** with Yb(OTf)$_3$ and Et$_3$B/O$_2$ led to the bicycle **3-126** in 85 % yield (Scheme 3.33). The reaction proceeds *via* a 6-*endo*-trig and 5-*exo*-trig cyclization after initial abstraction of the bromine

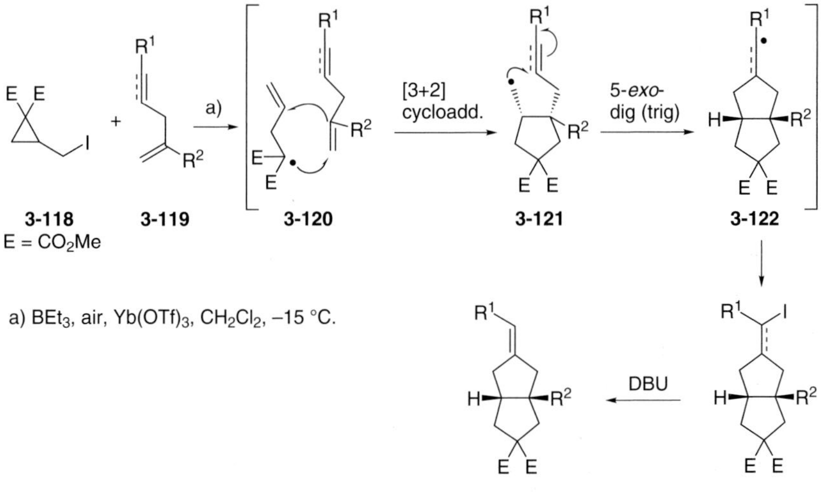

R¹ → R^1

3-118 **3-119** **3-120** **3-121** **3-122**
E = CO₂Me

E = CO_2Me

a) BEt_3, air, $Yb(OTf)_3$, CH_2Cl_2, −15 °C.

[3+2] cycloadd.

5-*exo*-dig (trig)

DBU

3-124 **3-123**

Entry[a]	Diene or Ene-Yne	Product **3-123**	Product **3-124**	Yield [%]
1			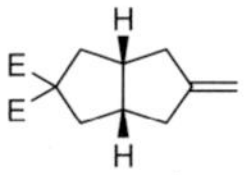	78
2	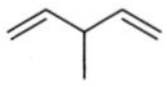		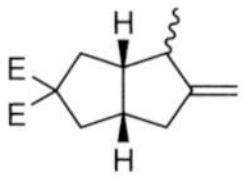	42 (*exo/endo* = 2.8) (**3-123**: 80%)
3	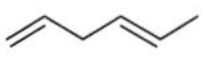		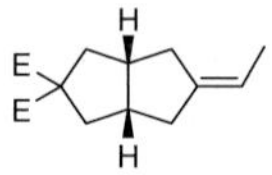	75
4	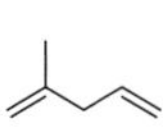		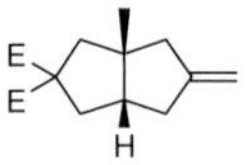	51
5	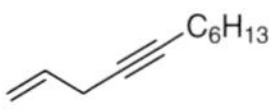 C_6H_{13}	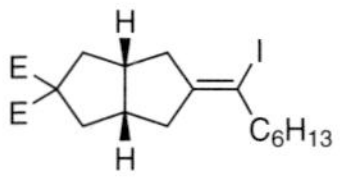		74
6	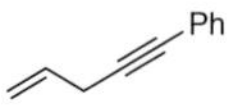 Ph	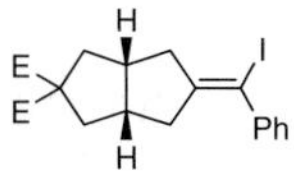		71
7	 OR, R = TPS	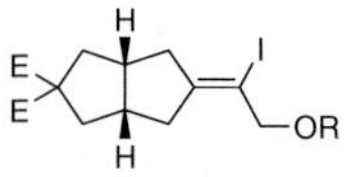		73

[a] Conditions: 0.5 mmol **3-118**, 1 mmol diene or enyne **3-119**, 0.5 mmol BEt_3, 0.5 mmol $Yb(OTf)_3$ in 4 mL CH_2Cl_2 at −15 °C.

Scheme 3.32. Products and mechanism of an iodine atom transfer radical domino reaction.

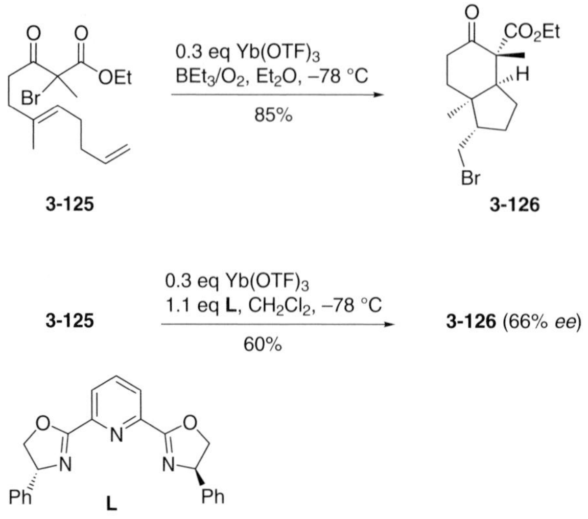

Scheme 3.33. Lewis acid-promoted free radical domino cyclization reaction and enantioselective approach.

atom. The reaction was also carried out in an enantioselective manner using pybox-ligand **L** with 66% *ee* and 60% yield.

The consecutive multiple carbon–carbon bond construction with help of radical-mediated intramolecular domino cyclizations is well established, and has been applied with great success. However, the extension to intermolecular procedures has been severely limited. Nonetheless, a successful example was reported by Yamago, Yoshida and coworkers using a new group-transfer coupling which avoids the problem of selective reaction of transient radicals with coupling partners in radical chain reactions [55]. The reaction was accomplished by heating a neat mixture of benzophenone **3-127**, phenyl acetylene **3-128** and trimethylsilyl phenyl telluride to 100 °C, providing silyl-protected allylic alcohol telluride species **3-129** in excellent 93% yield and 96% (*E*)-selectivity (Scheme 3.34). The transformation is assumed to pass the radical intermediates **3-130** and **3-131** on the way to product **3-129**.

The vinylic carbon–tellurium bond in **3-129** can easily be cleaved by a tributyltin radical to afford vinyl radical **3-131**, which can undergo further transformations as hydrogenation or C–C-bond formation, for example with dimethylfumarate in a (*Z*)-selective mode.

The addition of carbon-centered (alkyl or vinyl) radicals to alkene moieties bearing chiral auxiliaries has been extensively studied [56]. Herein, an approach is highlighted by Malacria's group, relying on the easy introduction and low cost of an enantiopure sulfoxide unit which can be regarded as temporary chiral auxiliary [57]. As can be seen in the general approach [route (1) in Scheme 3.35], the sulfoxide-controlled intramolecular addition of a radical as in **3-132** to a proximate alkene unit furnishes a five-membered ring system **3-133** in a 5-*exo*-trig manner in an diastereoselective way. β-Elimination of the chiral auxiliary then provides the enan-

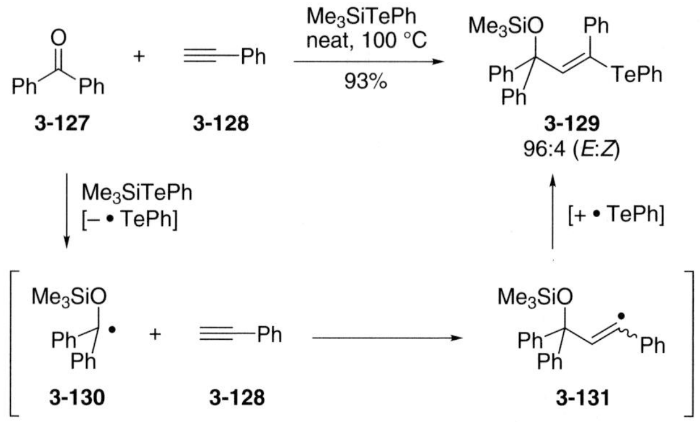

Me₃SiTePh
neat, 100 °C

93%

3-127   3-128

3-129
96:4 (*E:Z*)

Me₃SiTePh
[– • TePh]

[+ • TePh]

3-130   3-128

3-131

Scheme 3.34. Intermolecular domino radical addition procedure for the synthesis of silyl-protected allylic alcohols.

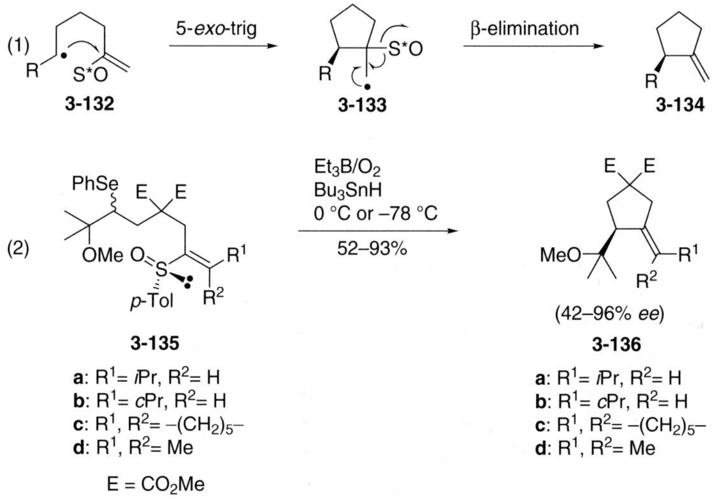

(1)

5-*exo*-trig

β-elimination

3-132   3-133   3-134

(2)

Et₃B/O₂
Bu₃SnH
0 °C or –78 °C

52–93%

3-135

a: R¹= *i*Pr, R²= H
b: R¹= *c*Pr, R²= H
c: R¹, R²= –(CH₂)₅–
d: R¹, R²= Me

E = CO₂Me

(42–96% *ee*)

3-136

a: R¹= *i*Pr, R²= H
b: R¹= *c*Pr, R²= H
c: R¹, R²= –(CH₂)₅–
d: R¹, R²= Me

Scheme 3.35. Domino radical cyclization/β-elimination process involving enantiopure sulfoxides.

tiomerically enriched product **3-134**. Hence, the sulfoxides **3-135** were converted under low-temperature radical conditions (Bu₃SnH, Et₃B/O₂) to the desired products **3-136** in 52–93 % yield and in moderate to high enantiopurity (42–96 % *ee*).

As early as 1986, Stork and coworkers presented an intramolecular radical cyclization/intermolecular trapping methodology within their synthesis of prostaglandin F₂α [58].

The first example of an intermolecular radical addition/intermolecular trapping domino reactions of an acyclic system in a stereocontrolled fashion to build stereogenic centers at the α- and β-carbons was described by Sibi and coworkers [59]. Enantioselective addition of *in-situ*-prepared alkyl radical to crotonate or cinnamate,

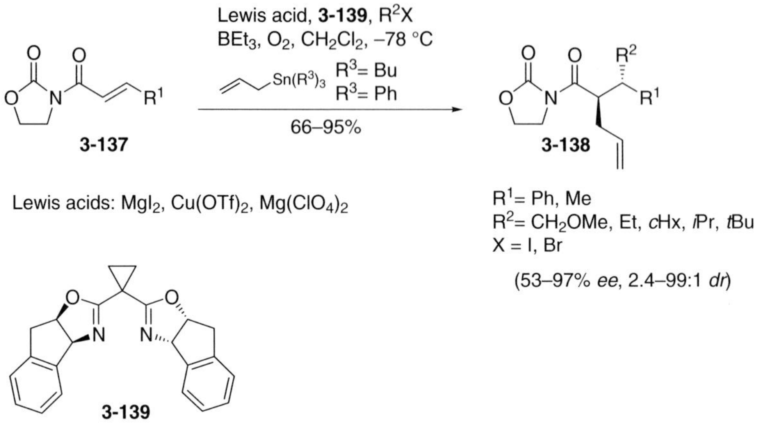

Scheme 3.36. Enantioselective intermolecular domino radical addition procedure.

facilitated by the addition of substoichiometric amounts (30 mol%) of a Lewis acid (MgI_2, $Cu(OTf)_2$ or $Mg(ClO_4)_2$) and chiral bisoxazoline ligand **3-139**, produced a radical in the α-position which could be trapped by a present allylic tin species in an anti-manner (Scheme 3.36). The best results according to stereoselectivity were obtained when the substituents R^1 and R^2 were as large as possible, whereas different Lewis acids gave almost the same results. Thus, utilization of cinnamate **3-137** ($R^1 = Ph$) in combination with tBuI and MgI_2 as Lewis acid gave **3-138** ($R^1 = Ph$, $R^2 = t$Bu) with diastereomeric ratio 99:1 and 97% *ee* in 84% yield.

The classical procedures for the synthesis of β-amino carbonyl compounds are based on a Mannich reaction, which relies on iminium ions. Quite recently, Naito and his group have designed a new type of Mannich reaction involving free-radical chemistry [60]. Thus, submitting substrates **3-140**, containing two electrophilic radical acceptors, to radical conditions (Et_3B, oxygen, refluxing benzene) in the presence of a nucleophilic alkyl radical precursor R^2I, led to a selective addition on the acrylate moiety, furnishing intermediates **3-141** (Scheme 3.37). This radical stays in equilibrium with the aminyl radical **3-142**. Since only the aminyl radical can be intercepted by the highly reactive triethylborane, the equilibrium is pushed to the formation of **3-143**, which gives the β-aminobutyrolactone **3-144** in good yields (64–70%) and high stereoselectivity (12:1).

A rather new concept in the context of domino radical cyclizations has been developed by Gansäuer and coworkers utilizing titanocene-complexes for the radical opening of unsaturated epoxides. The titanocene-catalyzed reactions [61] of **3-145** primarily led to radical **3-146**, which underwent a subsequent intermolecular addition to a present α,β-unsaturated carbonyl compound to form bicyclic carbocycles of type **3-148** *via* the intermediate **3-147** after aqueous work-up (Scheme 3.38) [62]. From a kinetic point of view, the reaction is remarkable since the intermolecular addition of simple radicals to α,β-unsaturated carbonyl compounds is not an easy task, as highlighted above.

BEt3
O2
R2I + Et• ⇌ R2• + EtI
NOBn
R1
3-140
R2
3-141
BEt2
NOBn
BEt3
R2
NHOBn
H2O
64–70%
(dr = 12:1)
3-144
3-143
3-142

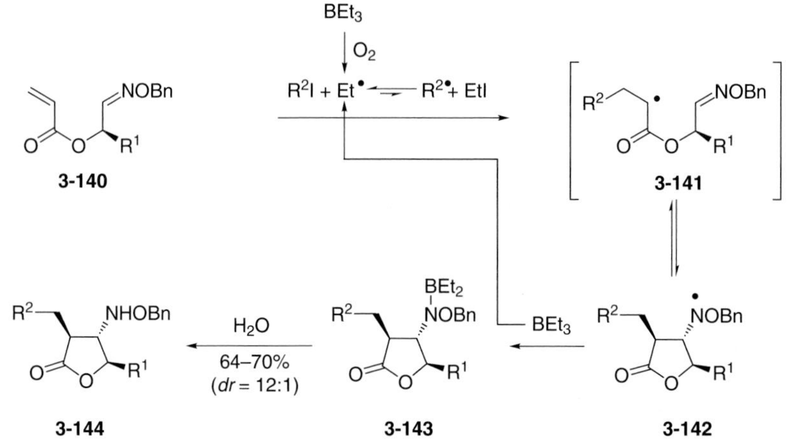

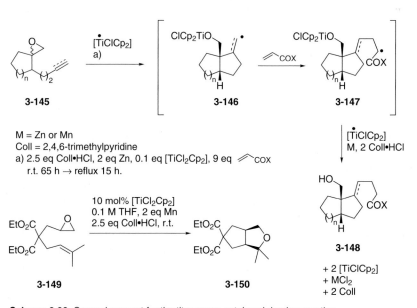

R^1 = TPSOCH$_2$
R^2 = Et, *i*Pr, *c*Hx, *c*Pent
reaction conditions: BEt$_3$, R^2I, benzene, reflux.

Scheme 3.37. Domino radical addition/cyclization-reaction for the asymmetric synthesis of β-aminobutyrolactones.

[TiClCp2]
a)
ClCp2TiO
COX
3-145
3-146
3-147
M = Zn or Mn
Coll = 2,4,6-trimethylpyridine
a) 2.5 eq Coll•HCl, 2 eq Zn, 0.1 eq [TiCl2Cp2], 9 eq COX
r.t. 65 h → reflux 15 h.
[TiClCp2]
M, 2 Coll•HCl
HO
H
3-148
+ 2 [TiClCp2]
+ MCl2
+ 2 Coll
10 mol% [TiCl2Cp2]
0.1 M THF, 2 eq Mn
2.5 eq Coll•HCl, r.t.
EtO2C
EtO2C
3-149
3-150

Scheme 3.38. General concept for the titanocene-catalyzed domino reaction.

Entry	Substrate	Product	dr	Yield [%]
1	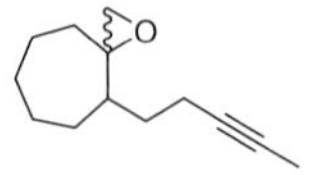	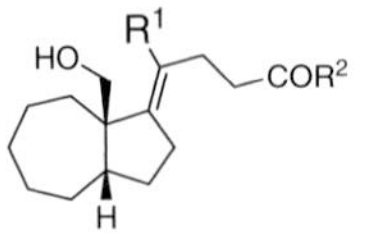 R¹ = H or Me R² = OEt, OtBu or NMe₃	>97:3	55–69
2	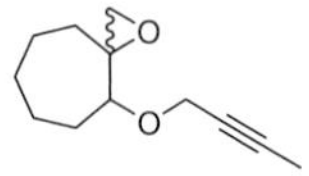	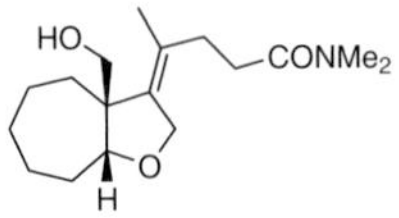	96:4	68
3	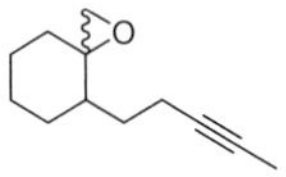	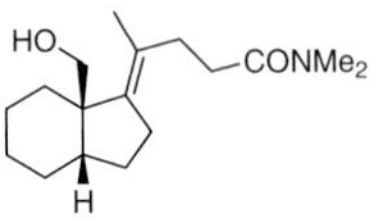	95:5	54
4	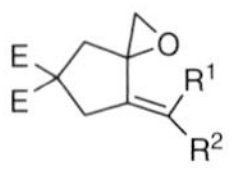	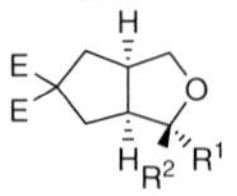 R¹ = R² = Me, Et, –(CH₂)₄–, –(CH₂)₅–, –(CH₂)₆– R¹ = Me, R² = H	80:20	60–66
5	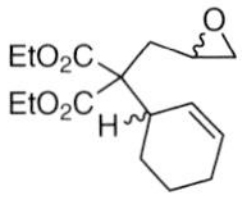	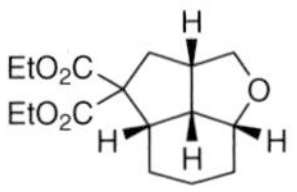	>98:2	63

Scheme 3.39. Examples of titanocene-catalyzed domino reactions.

In a similar manner, the Gansäuer group also used this domino methodology for the constructions of annulated tetrahydrofurans **3-150** using epoxides of type **3-149** [63]. In this way, a broad variety of products could be synthesized in a *cis*-selective fashion in good yields and high diastereoselectivity (Scheme 3.39).

Recently, the Cuerva group also presented a titanocene-catalyzed domino cyclization of an aryl epoxypolyene such as **3-151**, which led to the formation of a *trans/anti/trans*-fused tricyclic compound **3-152**, though in only moderate yield. Nevertheless, six stereogenic centers are formed in this domino process [64]. **3-152** could be transformed into the natural terpenoid stypoldione (**3-153**) (Scheme 3.40) [65].

Another new example using titanocene as catalyst has been revealed by Malacria and coworkers. Here, a previously unknown combination of radical cyclization involving an epoxide-opening of **3-154** and a β-phosphinoyl-elimination takes place to furnish various pyrrolidines **3-155**, bearing a tetrasubstituted *exo*-double bond, in good yields (Scheme 3.41) [66].

Examples of samarium-promoted radical procedures, which were combined with anionic processes, were detailed in Section 3.3. Here, we describe two-fold radical reactions initiated by SmI₂. A combination of two samarium(II) iodide-promoted

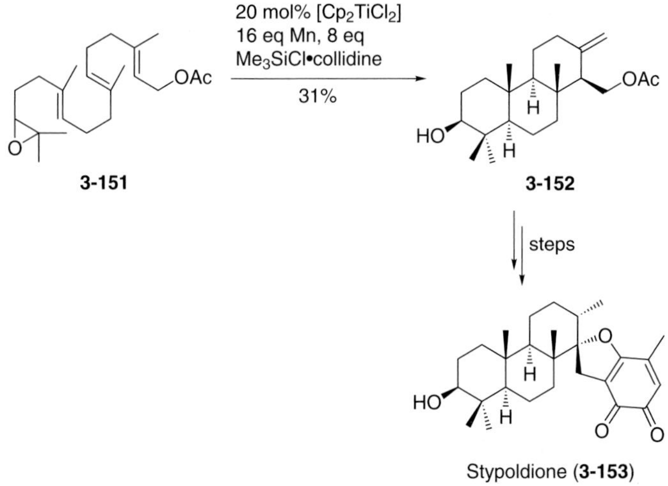

3-151

3-152

20 mol% [Cp₂TiCl₂]
16 eq Mn, 8 eq
Me₃SiCl•collidine
31%

steps

Stypoldione (3-153)

Scheme 3.40. Titanocene-catalyzed synthesis of a tricyclic terpenoid.

radical processes has been used by Molander's group for the synthesis of cyclo-hexanediols [67]. Exposure of α,β-epoxy ketone **3-156** containing an alkene moiety to a THF/HMPA solution of samarium(II) iodide provided *cis*-cyclohexanediols **3-157** and **3-158** as a diastereomeric mixture (12:1), in 88 % yield (Scheme 3.42).

The following mechanism can be assumed. Initially, the ketyl intermediate **3-159** is formed, which leads to **3-160** by a radical epoxide opening. Reaction with the second molecule of SmI₂ gives samarium-chelated hydroxy-ketyl **3-161**, which cyclizes to afford the products **3-157** and **3-158**.

SmI₂ has also been used in the transformation of sugars to give highly oxygenated cyclopentanes, as described by Enholm and coworkers [68]. Treatment of **3-162** with samarium(II) iodide in THF at –78 °C in the presence of a ketone such as **3-163** or an aldehyde **3-165** led to the cyclopentane derivatives **3-164** and **3-166**, respectively, in good yields and excellent to moderate diastereoselectivity (Scheme 3.43). It can be assumed that the reaction proceeds *via* the formation of a ketyl radical from **3-162** and subsequent radical ring closure with the adjacent olefin moiety. The formed radical then undergoes an intermolecular addition to the added carbonyl compound.

Radical ring-opening reactions of cyclopropyl ketones, mediated by samarium(II) iodide and other electron-transfer agents, have been shown to provide a reliable methodology for the construction of carbocycles. Motherwell and coworkers have provided an overview of the great potential of ring-opening reactions of bicyclo[4.1.0]ketones **3-167** to produce radicals of type **3-169** *via* the ketyl radical **3-168**. **3-169** can undergo further domino-type radical cyclization or enolate-trapping (Scheme 3.44) [69].

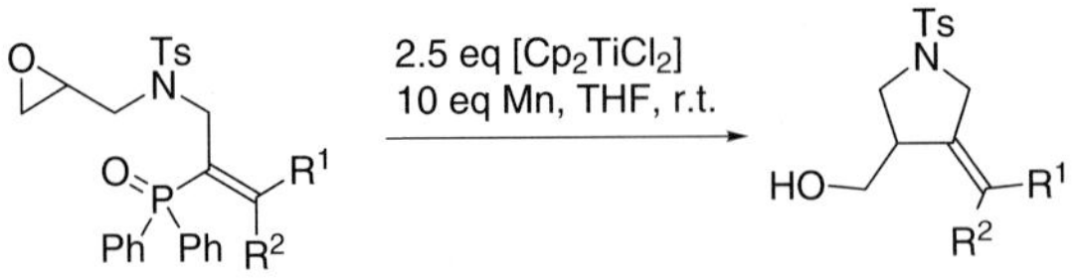

3-154   3-155

Entry	Substrate	Product	Yield [%]
1	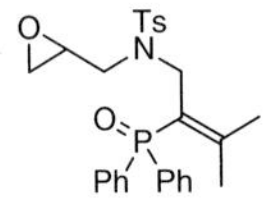	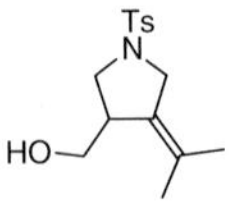	80
2	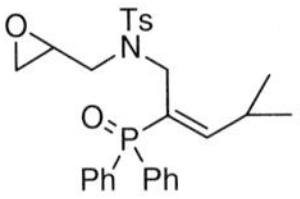	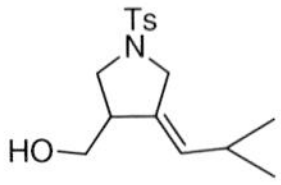	60
3	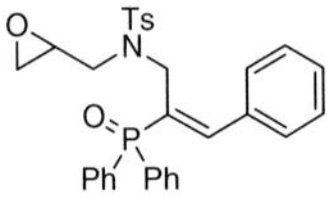	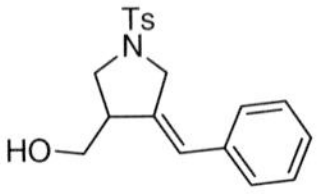	71[a]
4	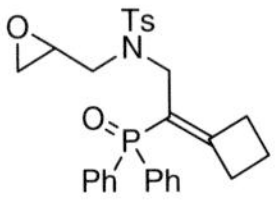	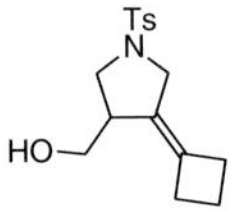	57
5	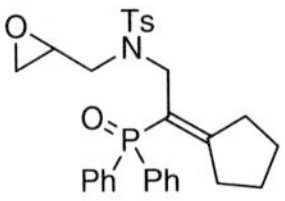	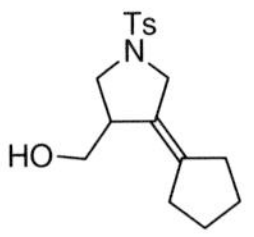	58
6	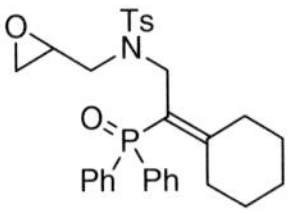	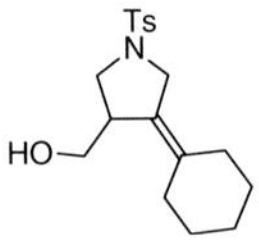	82
7	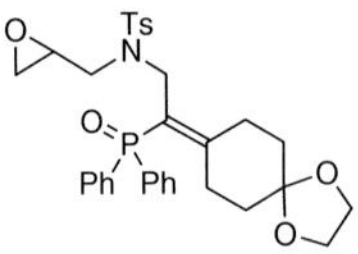	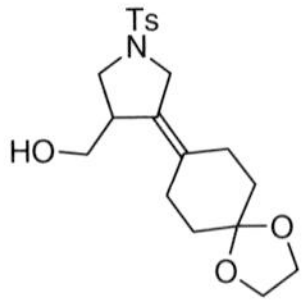	57[b]

[a] *E/Z* ratio 3:1. [b] Hydrolysis with a saturated solution of NH$_4$Cl.

Scheme 3.41. Titanocene-catalyzed synthesis of alkylidenepyrrolidines.

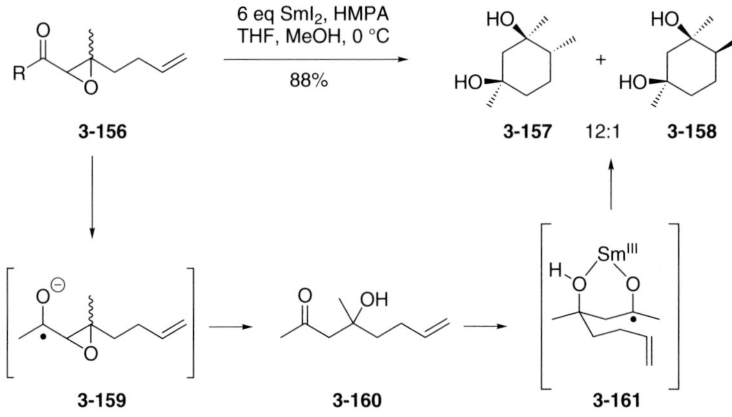

6 eq SmI$_2$, HMPA
THF, MeOH, 0 °C

88%

3-156

3-157   12:1   **3-158**

+

3-159

3-160

3-161

Scheme 3.42. Synthesis of cyclohexan-1,3-diols from epoxides containing a C=C-double bond.

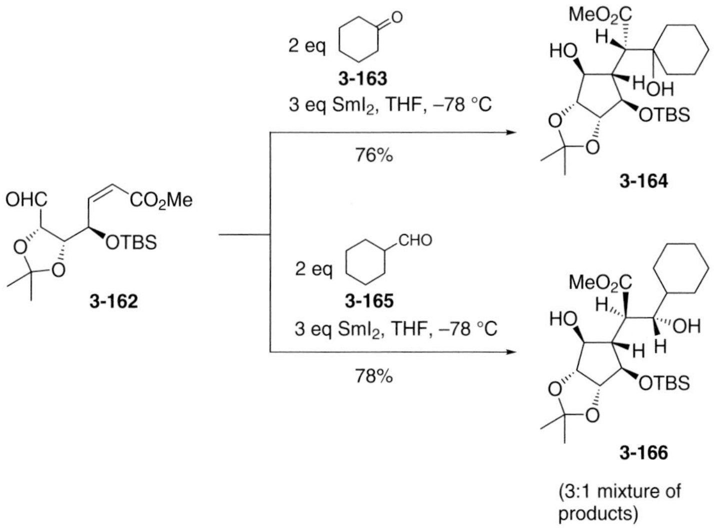

2 eq

3-163

3 eq SmI$_2$, THF, –78 °C

76%

3-164

3-162

2 eq

3-165

3 eq SmI$_2$, THF, –78 °C

78%

3-166

(3:1 mixture of products)

Scheme 3.43. Samarium(II) iodide-promoted domino reaction of carbohydrates.

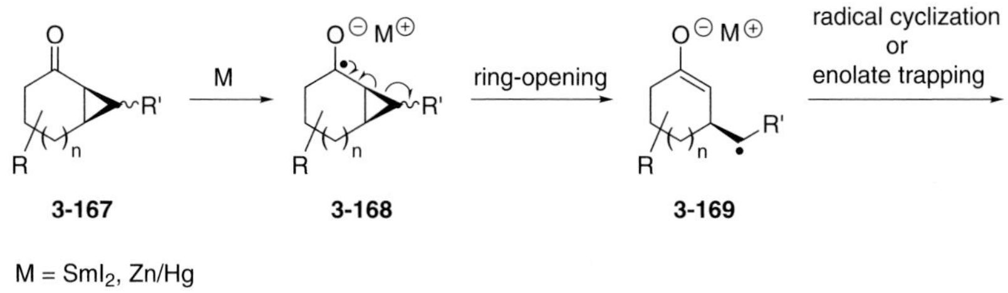

M

ring-opening

radical cyclization
or
enolate trapping

3-167

3-168

3-169

M = SmI$_2$, Zn/Hg

Scheme 3.44. General scheme of domino radical ring-opening reactions of cyclopropyl ketones

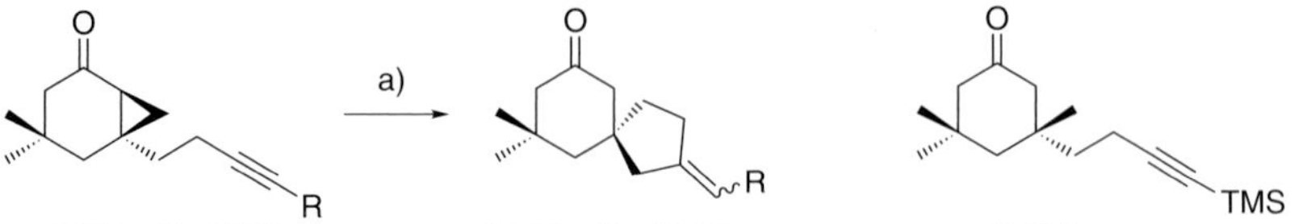

3-170a: R = TMS
3-170b: R = H

3-171a: R = TMS
with Zn/Hg: 52% + 34% of **3-172**
and 9% of starting material
with SmI$_2$: 79%
3-171b: R = H
with SmI$_2$: 57%

3-172

a) Zn/Hg, TMSCl, 2,6-lutidine, THF, reflux
or SmI$_2$, THF/DMPu (9:1), −78 °C → r.t..

Scheme 3.45. Domino radical ring-opening/cyclization procedure for the synthesis
of spiroketone **3-171**.

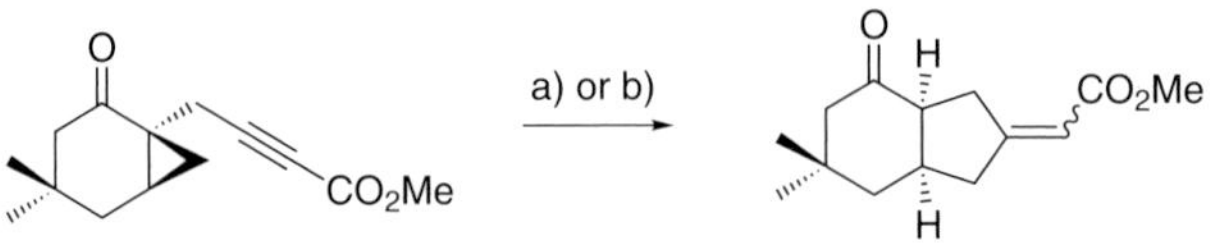

3-173

3-174

a) Zn/Hg, TMSCl, collidine, THF, reflux, 43%.
b) SmI$_2$, DMPU, THF, −78 °C → r.t., 77%.

Scheme 3.46. Domino radical ring-opening/cyclization procedure in the synthesis of
bicyclic ketone **3-174**.

In a first approach, the spiroketone **3-171a** was prepared by treatment of **3-170a**
with a Zn/Hg/TMSCl/2,6-lutidine electron-transfer system (Scheme 3.45). Besides
52% yield of the desired product **3-171a**, 34% of the uncyclized compound **3-172**
and 9% of starting material were isolated.

In contrast, the use of SmI$_2$/DMPU as an electron-transfer system led to the
smooth production of spiroketone **3-171a** in 79% yield, without any side products
(Scheme 3.45). It was also possible to cyclize the unprotected cyclopropyl ketone **3-
170b** to give the spiroketone **3-171b** in 57% yield.

Similar results were observed with **3-173** as substrate. Using zinc amalgam, the
hydridane **3-174** was obtained in 43% yield, whereas samarium(II) iodide furnished **3-174** in 77% yield (Scheme 3.46).

Grignon-Dubois and coworkers have shown that reduction of a quinoline using
zinc and acetic acid in THF gives the dimeric compound **3-177** *via* intermediate **3-
176** (Scheme 3.47) [70]. Usually, a mixture of the *syn-* and *anti*-products is formed;
the substituent has some influence on the regioselectivity of the dimerization and
cyclization step. With R = H and R = 6-Me, only the benzazepine **3-177** were produced, by a head-to-head dimerization.

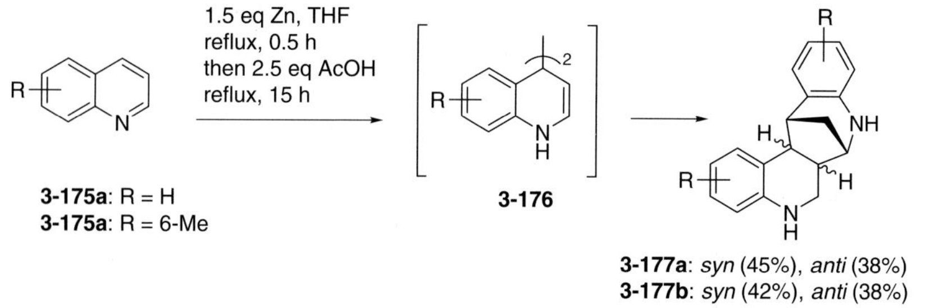
1.5 eq Zn, THF
reflux, 0.5 h
then 2.5 eq AcOH
reflux, 15 h
3-175a: R = H
3-175a: R = 6-Me
3-176
3-177a: syn (45%), anti (38%)
3-177b: syn (42%), anti (38%)

Scheme 3.47. Head-to-head reductive radical dimerization.

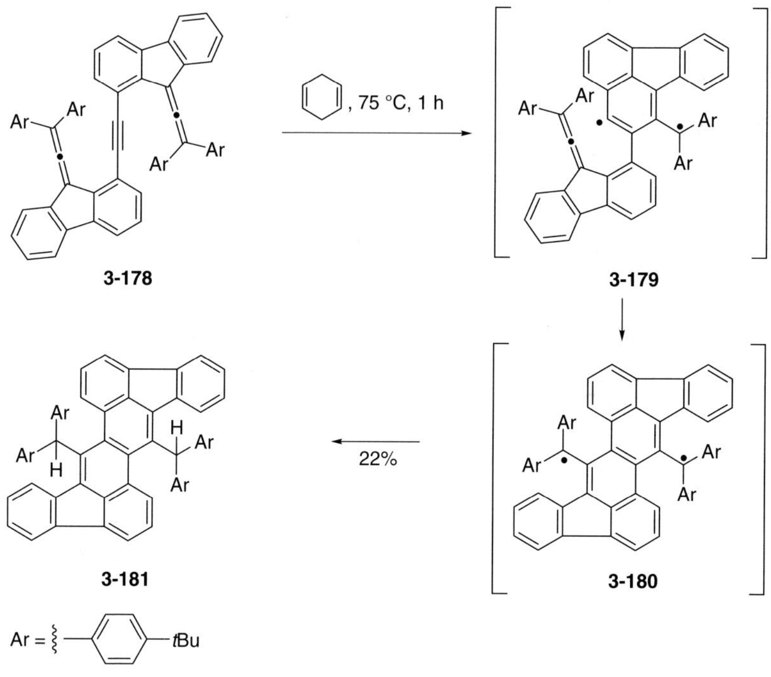
, 75 °C, 1 h
3-178
3-179
3-180
3-181
22%
Ar = tBu

Scheme 3.48. Thermolysis of benzoenynes-allenes initiating a domino radical cyclization.

Finally, thermally induced isomerizations which generate carbon-centered biradical organic molecules have been shown to serve as alternative for conventional chemical and photochemical methods [71]. A straightforward procedure to accomplish such biradicals was described by Myers using a thermal conversion of yne-allenes [72]. According to this scheme, Wang and coworkers [73] heated **3-178** in 1,4-cyclohexadiene to 75 °C and obtained **3-181** in 22 % yield *via* the biradicals **3-179** and **3-180** (Scheme 3.48).

The Pattenden group has recently revealed a new synthetic approach to the pro-pellane sesquiterpen triquinane (±)-modhephene (3-70), introducing an eight-membered ring system with an appropriate thioester. Thus, using standard radical conditions (Bu₃SnH, AIBN) a 5-*exo*-trig/5-*exo*-dig domino cyclization took place to generate the propellane structure in 60% yield [74].

Another very new radical/radical domino procedure was used in the total synthesis of the alkaloid lennoxamine by Ishibashi and coworkers. Here, a 7-*endo* cyclization/homolytic aromatic substitution reaction cascade led to the target compound in 41% yield [75].

3.3.1
Radical/Radical/Anionic Domino Processes

Domino reactions have shown to be useful not only for the composition of molecules, but also for their degradation. This is a concept which is often encountered in Nature. Hence, ribonucleotide reductases (RNRs) represent a type of enzyme that is able to catalyze the formation of DNA monomers from ribonucleotides by radical-mediated 2'-deoxygenation. This process has also been studied *in vitro* utilizing a chemical approach. For example, Robins and coworkers used 6'-O-nitro esters of homonucleosides 3-182 containing a chloro substituent in the 2'-position for their investigations (Scheme 3.49) [76]. Treatment of the latter with nBu₃SnD and AIBN in refluxing benzene resulted in the formation of 6'-oxygen radical 3-183, which allows a 1,5-hydrogen shift resulting in an abstraction of H-3'. The formed C-3' radical loses a chlorine atom, as postulated for the action of ribonucleotide reductase, to furnish enol 3-184 which then undergoes elimination of uracil to give the furanone derivative 3-185 in 75% yield.

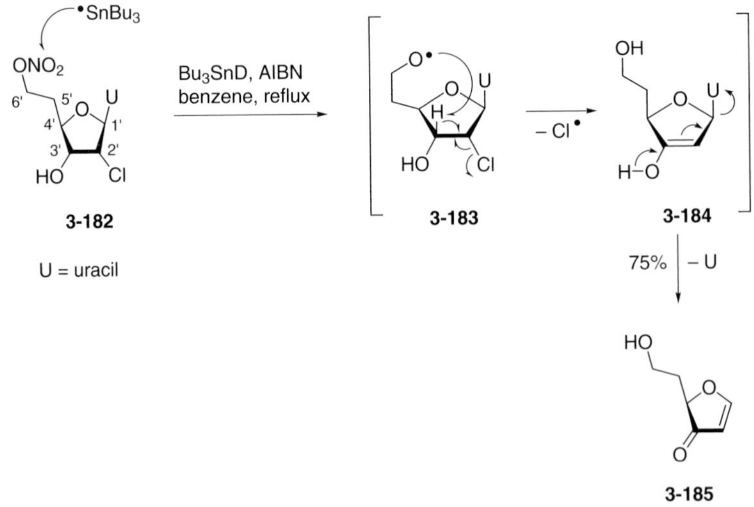

Scheme 3.49. Biomimetic simulation of a free radical-initiated domino reaction.

3.3.2
Radical/Radical/Radical Domino Processes

Instead of simply using two radical reactions in a domino process, the combination of three and more radical C–C- or C–N-bond forming radical transformations is also possible. This makes this methodology one of the most powerful procedures in the synthesis of complex molecules starting from simple substrates [77]. During the years, several strategies have been developed, and these are depicted in Scheme 3.50. The strategies can be classified as three types:

- The so-called "zipper" strategy is characterized by starting the cyclization process in the middle of the chain and working its way back and forth across toward the ends.
- The "macrocyclization/transannular cyclization" process [78], in contrast, is initiated at the ends of the chain and works towards the middle.
- The third, less common, "round trip radical reaction" process starts at one end of the chain and works its way back to the same end [79].

The latter strategy has been employed by Curran and coworkers for their very short total synthesis of the natural products (±)-iso-gymnomitrene **3-191** and (±)-gym-nomitrene **3-194** (Scheme 3.51) [80]. As substrate, the easily accessible iodotriene **3-186** was used; this was converted into a mixture of compounds containing 31% of **3-191** and 3% of **3-194** using a tin hydride derivative. In this process, **3-191** is formed by a 5-*exo*-trig/6-*endo*-trig/5-*exo*-trig domino cyclization *via* **3-187**, **3-188** and

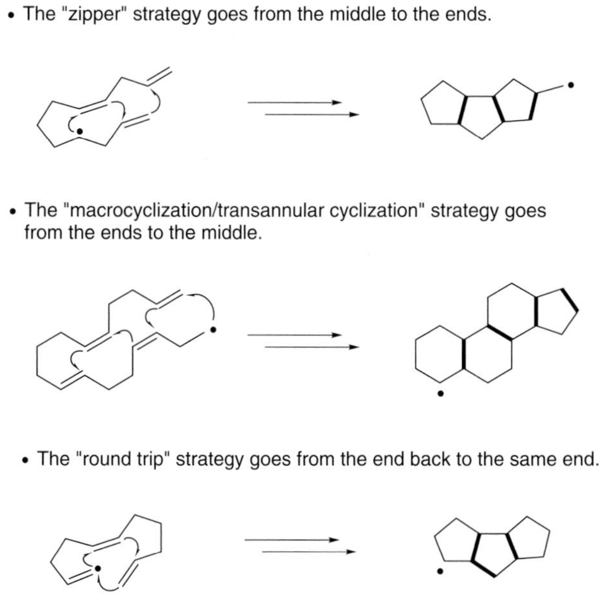

- The "zipper" strategy goes from the middle to the ends.

- The "macrocyclization/transannular cyclization" strategy goes from the ends to the middle.

- The "round trip" strategy goes from the end back to the same end.

Scheme 3.50. Domino radical cyclization strategies from linear acyclic precursors (the formed bonds are highlighted in bold).

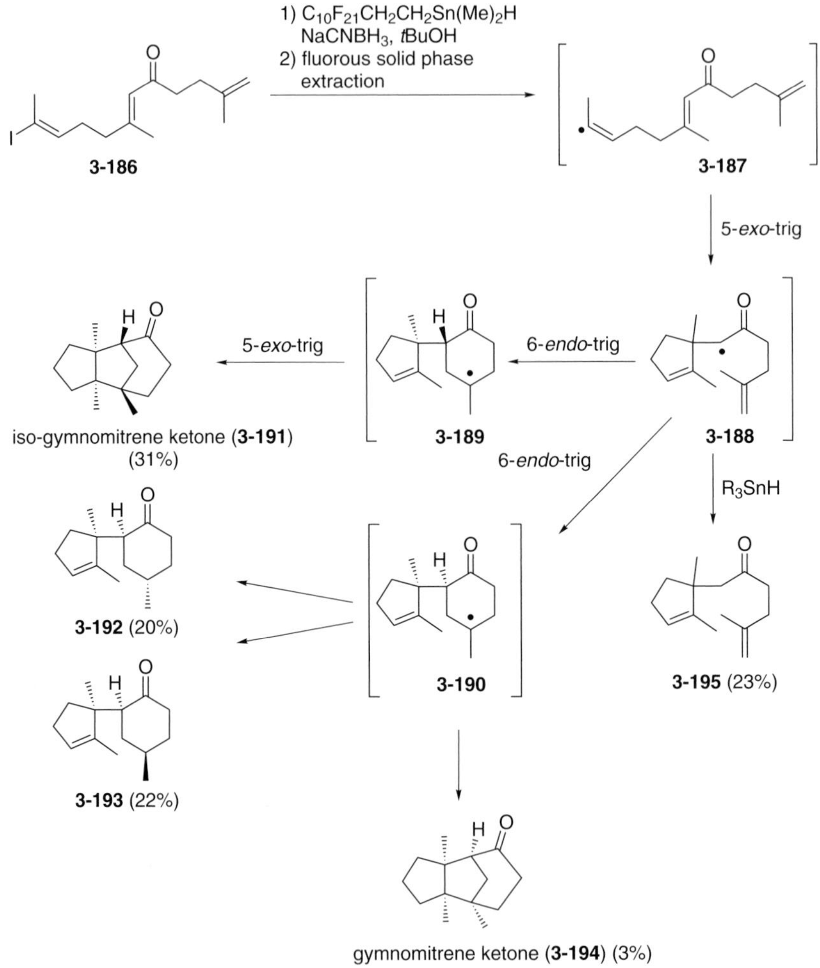

Scheme 3.51. Proposed mechanism for the "round trip" synthesis of iso-gymnomitrene ketone (**3-191**) and gymnomitrene ketone (**3-194**).

3-189. Other compounds obtained are the monocycle **3-195** in 23%, as well as the bicycles **3-192** and **3-193** in 20% and 22% yields, respectively, with **3-190** as a proposed intermediate.

A dramatic improvement in this new "round trip radical domino processes" developed by Curran's group was presented by Takasu, Ihara and coworkers. The new method relies on the introduction of a conjugated ester moiety at the terminal olefin, thereby effecting an acceleration of the domino reaction accompanied with an enhancement of the regio- and stereoselectivity [81]. Thus, reaction of **3-196** with Bu$_3$SnH led to a 4:3 mixture of the two diastereomeric tricycles **3-197** and **3-198** in 83% yield. In this process, the vinyl radical **3-199** is initially formed, but this smoothly cyclizes in 5-*exo*-trig manner to give radical **3-200** (Scheme 3.52). Due to

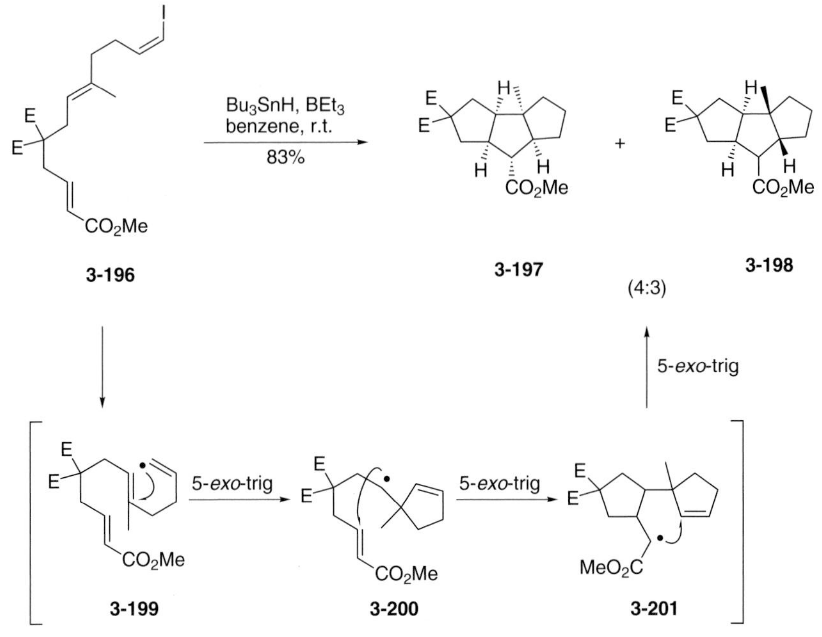

Scheme 3.52. Improved "round trip radical reaction" for the synthesis of linear fused [5.5.5]tricycles.

the high nucleophilicity of this radical, a 5-*exo*-trig addition onto the unsaturated ester moiety providing α-carboxy radical **3-201** is kinetically predominant. The sequence is terminated by a 5-*exo*-trig ring closure onto the electron-rich olefin moiety.

Domino reactions of aryl isocyanides **3-202** with iodopentynes **3-203** for the generation of cyclopentaquinolines **3-204** and **3-205** turned out to be very fruitful for the synthesis of several natural products and analogues belonging to the camptothecin family, as demonstrated by the Curran group [82]. This process can be regarded as [4+1]-radical annulation furnishing geminal carbon–carbon bonds [83]. Interestingly, the second ring can be closed either in a 1,6-fashion which leads, after oxidation, to the tricyclic ring system **3-204**, or in a 1,5-fashion affording a spiro-intermediate which rearranges to give **3-205** (Scheme 3.53).

In a similar way (and as described for the aromatic isocyanides), aliphatic α,β-unsaturated isocyanides can also be used, leading to similar structures with a cyclohexano instead of a benzo moiety [84]. Based on the approach using aromatic isocyanides, a small library of about 20 camptothecin derivatives has been prepared, of which irinotecan® and topotecan® have entered the clinical treatment of cancer [85]. For the synthesis of the camptothecin derivatives, **3-206** was alkylated with the appropriate propargylic bromides **3-207** to give **3-208**, which were irradiated in benzene at 70 °C, together with the respective isocyanide **3-209** and hexamethylditin

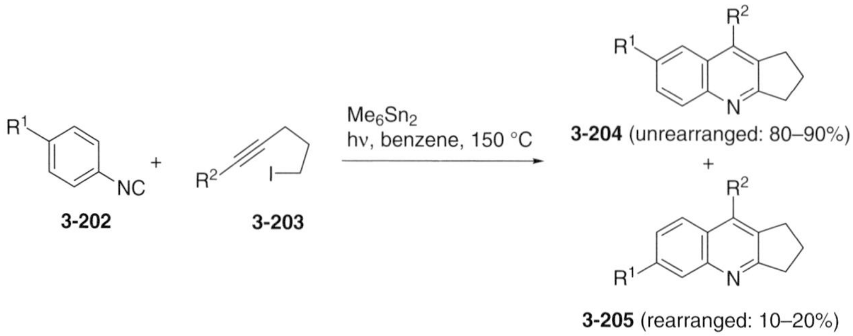
Me6Sn2
hv, benzene, 150 °C
3-202
3-203
3-204 (unrearranged: 80–90%)
3-205 (rearranged: 10–20%)

Scheme 3.53. Synthesis of cyclopentaquinolines from aryl isonitriles and iodopentynes.

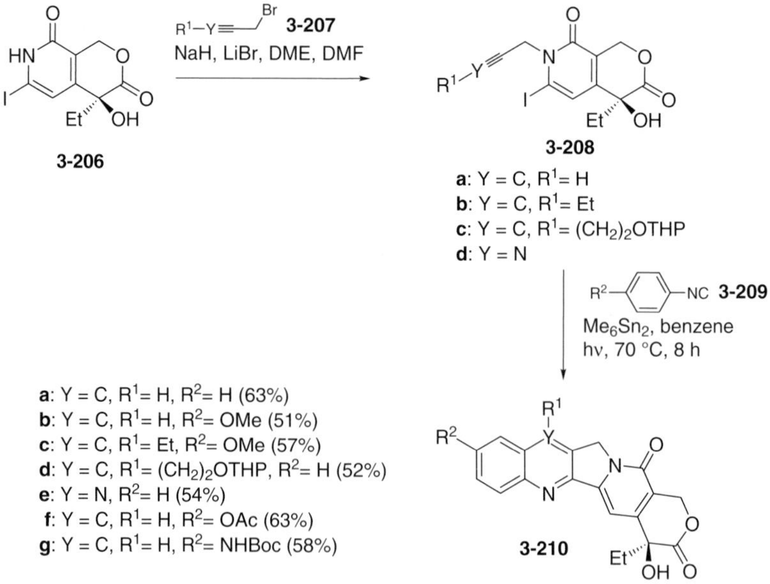
3-207
NaH, LiBr, DME, DMF
3-206
3-208
a: Y = C, R1= H
b: Y = C, R1= Et
c: Y = C, R1= (CH2)2OTHP
d: Y = N
3-209
Me6Sn2, benzene
hv, 70 °C, 8 h
a: Y = C, R1= H, R2= H (63%)
b: Y = C, R1= H, R2= OMe (51%)
c: Y = C, R1= Et, R2= OMe (57%)
d: Y = C, R1= (CH2)2OTHP, R2= H (52%)
e: Y = N, R2= H (54%)
f: Y = C, R1= H, R2= OAc (63%)
g: Y = C, R1= H, R2= NHBoc (58%)
3-210

Scheme 3.54. Domino radical sequence in the synthesis of camptothecin analogues.

[86]. In this way, various pentacycles **3-210** could be obtained in yields of 51 to 63 % (Scheme 3.54). It should be noted that toxic tin organyls could be replaced by silanes such as tris(trimethylsilyl)silane, but under these conditions the reaction rate was significantly reduced.

The same methodology was also used for the synthesis of (*S*)-mappicine (**3-213**), which can be oxidized to produce mappicine ketone, an antiviral lead compound [87]. Reaction of the chiral alcohols **3-211a** and **3-211b**, respectively, with phenylisocyanide **3-212** and hexamethylditin in benzene under irradiation delivered (*S*)-

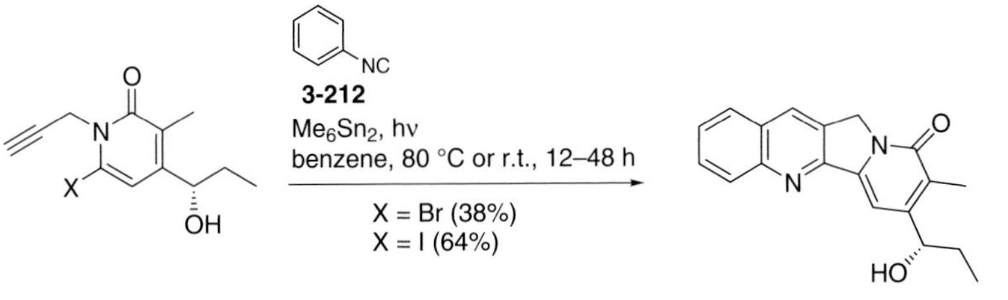

3-211a: X = Br
3-211b: X = I

(*S*)-Mappicine (**3-213**)

Scheme 3.55. Synthesis of (*S*)-mappicine (**3-213**).

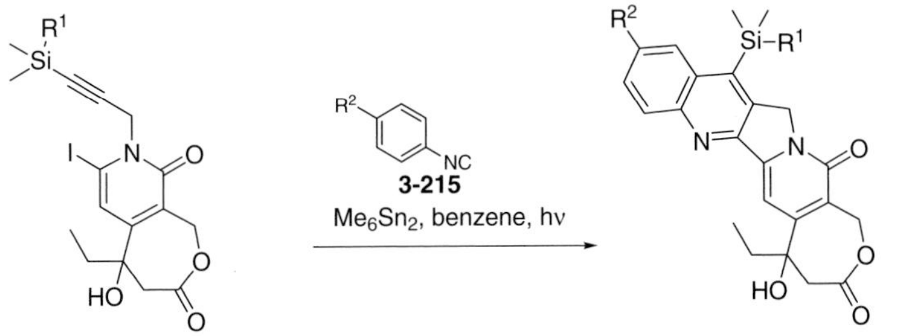

3-214

library of *rac*-homosilatecans (**3-216**)

R^1= e.g. Me, *n*Bu, *n*Oct, $CH_2CH_2CF_3$
CH_2CH_2tBu, Amyl, Ph, tol, Bn
R^2= e.g. H, F, OMe, Me, CF_3, AcO, BocNH

Scheme 3.56. Solution-phase parallel synthesis of *rac*-homosilatecan analogues (**3-216**).

mappicine (**3-213**) in 38 % and 64 % yields, respectively (Scheme 3.55). It is interesting to note that the yield was almost doubled by using the iodide **3-211b** instead of the bromide **3-211a**.

Also in this case, Curran and coworkers produced a library of 64 mappicine analogues by automated solution-phase combinatorial synthesis, as well as a 48-member library of mappicine ketone derivatives [88]. Furthermore, these authors were successful in building up a 115-member library of *rac*-homosilatecans **3-216** using different iodopyridones **3-214** and aryl isocyanides as substrates **3-215** (Scheme 3.56) [89].

It has been shown by Zanardi, Nanni and coworkers that, instead of isocyanides, isothiocyanates can be used in the radical process [90]. Treatment of the mixture of the diazonium tetrafluoroborates **3-217** and the isothiocyanates **3-218** with pyridine or a mixture of 18-crown-6 and potassium acetate in ethyl acetate, furnished the benzothienoquinoline derivatives **3-219**. In addition, in some cases the rearranged products **3-220** were furnished in low to reasonable yields (Scheme 3.57).

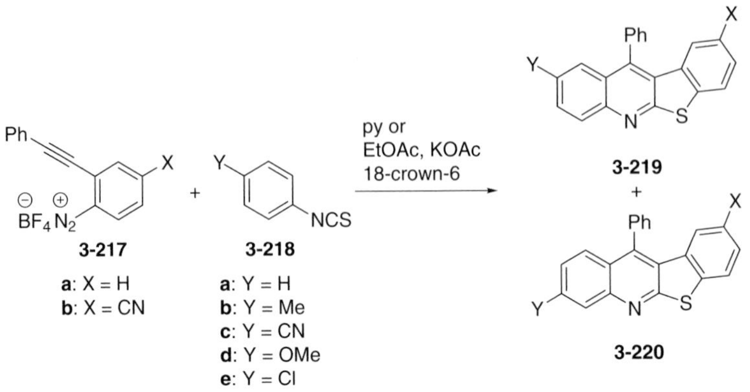

3-217

a: X = H
b: X = CN

3-218

a: Y = H
b: Y = Me
c: Y = CN
d: Y = OMe
e: Y = Cl

Entry	X	Y	3-219 Yield [%]	3-220 Yield [%]
1	H	H	50	
2	H	OMe	39	–
3	H	Me	30	6
4	H	Cl	33	11
5	H	CN	–	–
6	CN	H	60	
7	CN	OMe	52	–
8	CN	Cl	40	16
9	CN	CN	12	21

Scheme 3.57. Domino radical reaction in the synthesis of benzothienoquinoline derivatives.

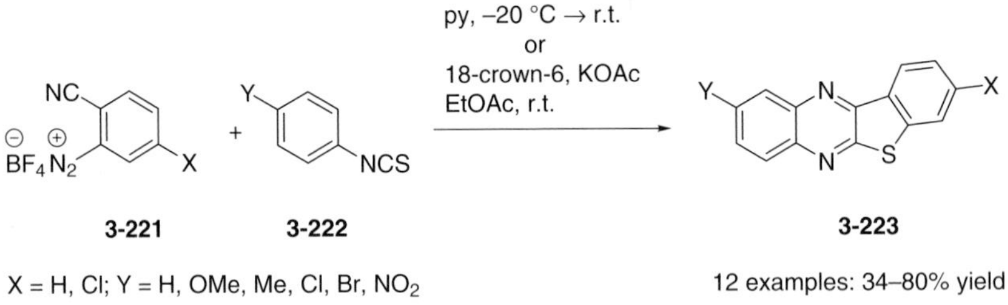

3-221   **3-222**   **3-223**

X = H, Cl; Y = H, OMe, Me, Cl, Br, NO$_2$ 12 examples: 34–80% yield

Scheme 3.58. Domino radical reaction in the synthesis of benzothienoquinoxaline derivatives.

In a similar manner, benzothienoquinoxaline derivatives **3-223** can also be synthesized by employing the diazocyanoaryl tetrafluoroborates **3-221** and the isothiocyanate **3-222** (Scheme 3.58) [91].

Some excellent examples of cationic polycyclizations, especially in the field of steroid synthesis, were described in Chapter 1. However, these polycyclizations can also be performed using a radical as initiator. Such reactions can be divided into those based on serial 6-*endo*-trig cyclizations from polyene acyl precursors [92], radi-

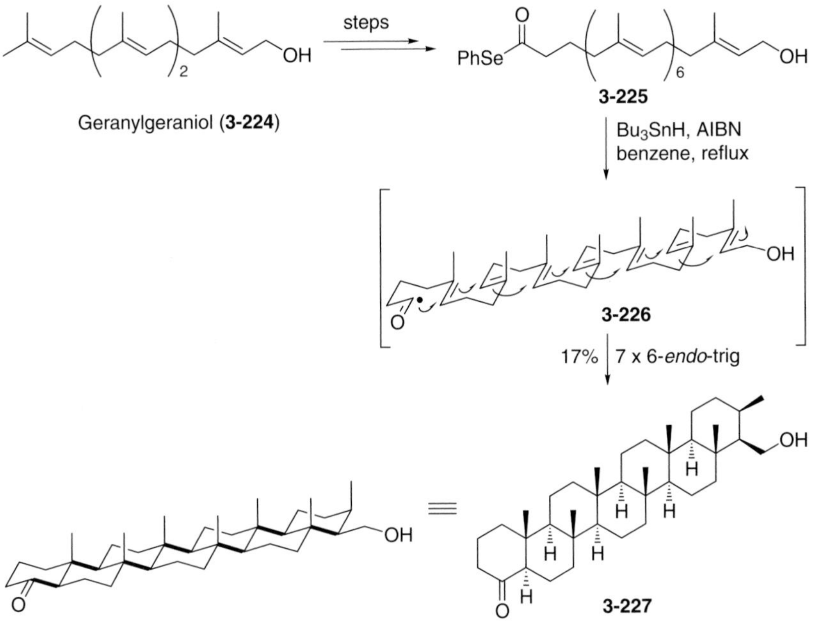

Scheme 3.59. Sevenfold domino radical 6-*endo*-trig cyclization.

cal-mediated macrocyclizations from alkyl radicals followed by radical transannulations [93], consecutive oxy (and aminyl) radical fragmentation/transannulation/cyclization sequences [92], and several combinations of these processes [95].

An impressive example of the power of this approach is the combination of seven 6-*endo*-trig cyclizations in one single event, as devised by Pattenden and his group (Scheme 3.59) [96], The hydroxyheptenselenoate **3-225**, synthesized from all-*E*-geranylgeraniol (**3-224**) involving two successive homologations, was treated with AIBN and tributyltin hydride in refluxing benzene to give the all-*trans-anti* heptacyclic ketone **3-227** in a yield of 17 % *via* the primarily formed radical **3-226**. In order to minimize the formation of reduced byproducts, the Bu$_3$SnH was added over 8 h using a syringe pump – a popular method in the area of radical chemistry.

It could be shown that the stereochemical outcome of such radical polycyclizations is influenced by the nature of the substituents (H, Me, CO$_2$R). For instance, as in the example **3-225**, the all-(*E*)-methyl-substituted polyene **3-228** also gave the corresponding all-*trans-anti* polycycle **3-229** in the presence of Bu$_3$SnH and AIBN. However, the ester-substituted polyene **3-230** led to the *cis-anti-cis-anti-cis* tetracycle **3-231** under similar reaction conditions (Scheme 3.60). A certain degree of preorganization of the precursor is assumed to be the reason for this result [97].

An exact tuning of the reactivities clearly plays an important role in the development of radical domino reactions. Thus, as shown by Pattenden and coworkers, the

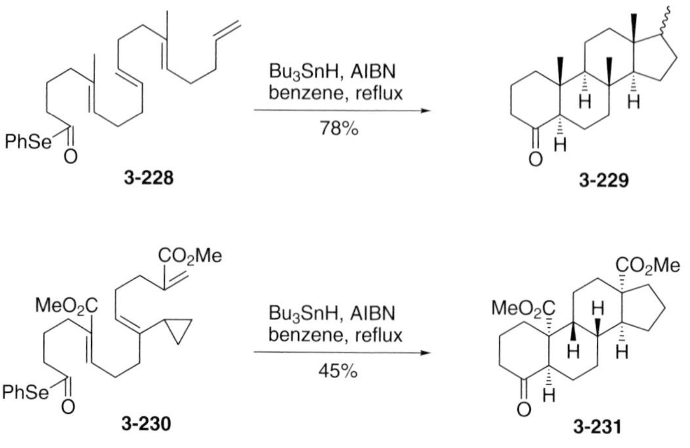

3-228 → 3-229

Bu₃SnH, AIBN
benzene, reflux

78%

3-230 → 3-231

Bu₃SnH, AIBN
benzene, reflux

45%

Scheme 3.60. Influence of substituents on the stereochemical outcome of radical cyclizations.

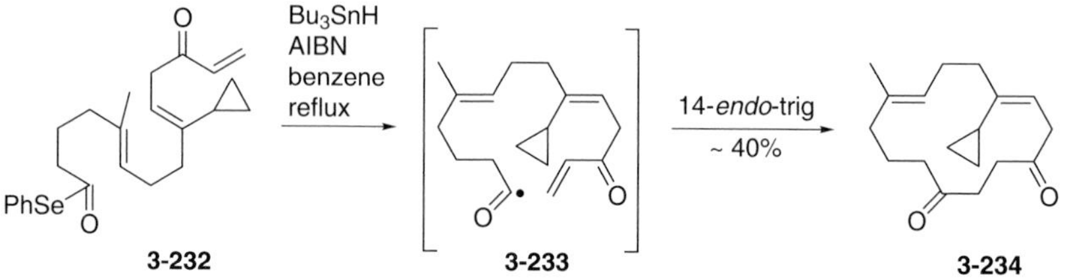

Bu₃SnH
AIBN
benzene
reflux

14-*endo*-trig
~ 40%

3-232   3-233   3-234

Scheme 3.61. Undesired 14-*endo*-trig macrocyclization.

radical **3-233**, obtained from the selenoate **3-232**, does not undergo tetracyclization but rather forms the cyclopropyl-substituted macrocycle **3-234** in about 40 % yield *via* a 14-*endo*-trig-cyclization (Scheme 3.61) [98].

As mentioned above, the class of *Aspidosperma* alkaloids (see Section 3.4) has attracted the attention of the chemical community, due to the compounds' challenging architecture and biological activity [99]. Here, a new approach to the ABCE-tetracycle **3-239** of, for example, aspidospermidine (**3-240**), as developed by the Jones group, is disclosed [100]. Reaction of the substrate **3-235** with tributyltin hydride in refluxing 5-*tert*-butyl-*m*-xylene (ca. 200 °C) resulted in the formation of an aryl radical **3-236** with its well-known ability to abstract a hydrogen atom *via* a six-membered ring transition state, leading to the α-amino radical **3-237**. It follows an attack on the indole systems, which is facilitated by the electron-withdrawing cyano group and a 5-*exo*-trig ring closure of the intermediate **3-238** to give the tetracyclic scaffold **3-239** in 43 % yield as a 8:3:2:1 mixture of diastereoisomers (Scheme 3.62).

The natural spironucleoside hydantocidin (**3-241**), which was discovered in 1991, demonstrates pronounced herbicidal and plant growth-regulatory properties. This

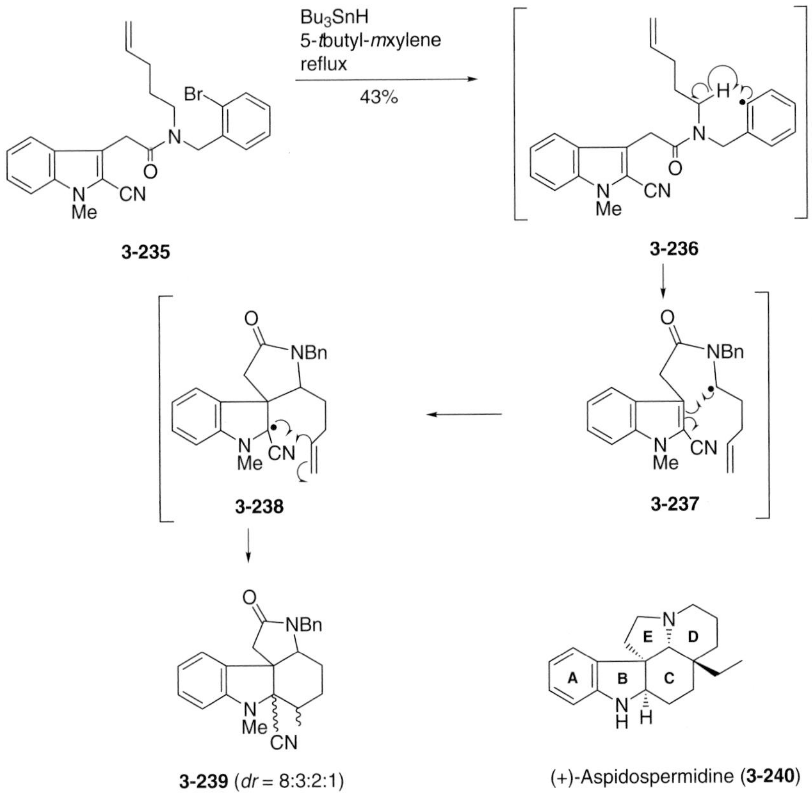

Scheme 3.62. Domino radical hydrogen abstraction-cyclization procedure in the synthesis towards *Aspidosperma* alkaloids.

has in turn led to a stimulation of studies related to the synthesis of anomeric spironucleosides [101].

Herein, two short and efficient synthetic approaches of the group of Chatgilialoglu [102], both based on radical domino reactions, for the preparation of anomeric spironucleosides such as **3-246** and **3-250**, are described. The first method features the conversion of protected 6-hydroxymethylribouridines **3-242a/b** into spironucleosides **3-246a/b** in 36% and 49% yields, respectively, using PhI(OAc) and I_2 (Scheme 3.63). Photolysis of the primarily formed hypoiodite generates an alkoxy radical **3-243**, which directly undergoes a Barton-type hydrogen migration, furnishing the anomeric C-1' radical **3-244**. This forms the corresponding oxenium salts **3-245**. Finally, probably *via* an unstable anomeric iodo intermediate, nucleophilic addition of the hydroxyl group in **3-245** occurs to produce the desired **3-246**.

In the second example, the protected dibromovinyl deoxyuridine **3-247** was transformed into a 2:1-mixture of the desired spirocompounds **3-249** and **3-250** in 57% yield using standard radical conditions with (TMS)$_3$SiH. In addition, 25% of a E/Z-mixture of **3-251** was obtained (Scheme 3.64).

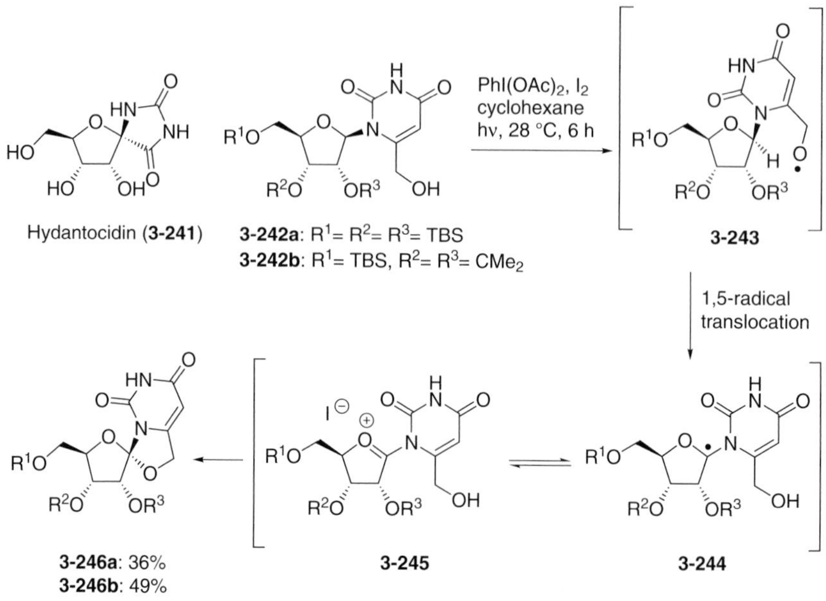

Scheme 3.63. Synthesis of anomeric spironucleosides.

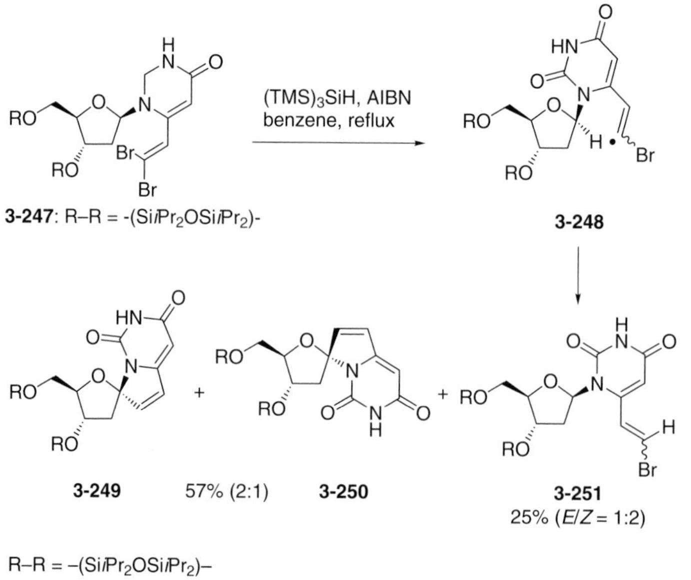

Scheme 3.64. Formation of anomeric spironucleosides of type **3-249** and **3-250**.

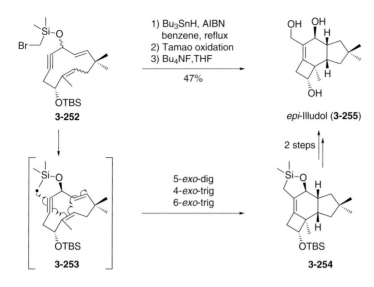

Scheme 3.65. Synthesis of *epi*-illudol.

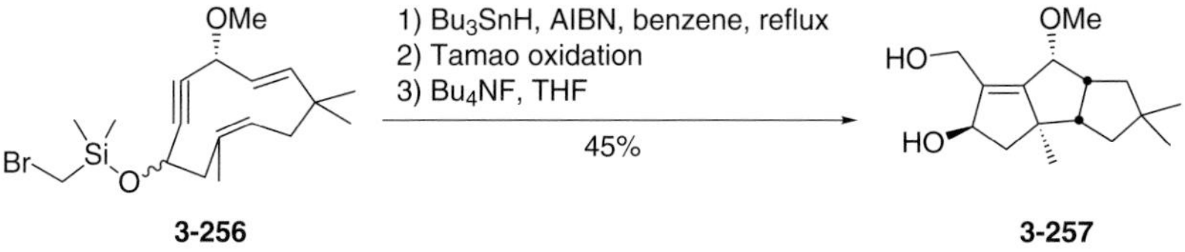

Scheme 3.66. Synthesis of a linear triquinane.

Bromomethyldimethylsilyl (BMDMS) propargyl ethers undergo intramolecular radical 5-*exo*-dig cyclizations using tributyltin hydride; the reaction is initiated by the generation of a stabilized α-silyl radical such as **3-253** [103]. The Malacria group has described the reaction of the monocyclic compound **3-252** which gave the pentacyclic product **3-254** [104]. This was transformed using a Tamao oxidation and subsequent desilylation, to afford the antibiotic *epi*-illudol (**3-255**) in 47 % yield over three synthetic steps (Scheme 3.65).

It should be noted that this strategy could also be utilized to build up the linear triquinane framework **3-257** in 45 % yield, employing the 11-membered ring system **3-256** in which the BMDMS group is simply moved from one to the other side of alkyne moiety as compared to **3-252** (Scheme 3.66) [105].

In another approach, the annulated [5.6.5] ring system **3-259** was obtained from **3-258** in a radical reaction in 61 % yield (Scheme 3.67) [106]. This motif is found as the central core in several important natural products.

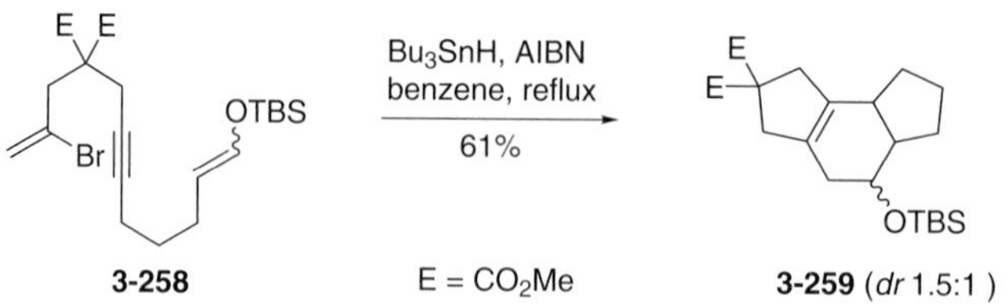

Scheme 3.67. Alternative domino radical procedure for the construction of a [5.6.5] tricycle

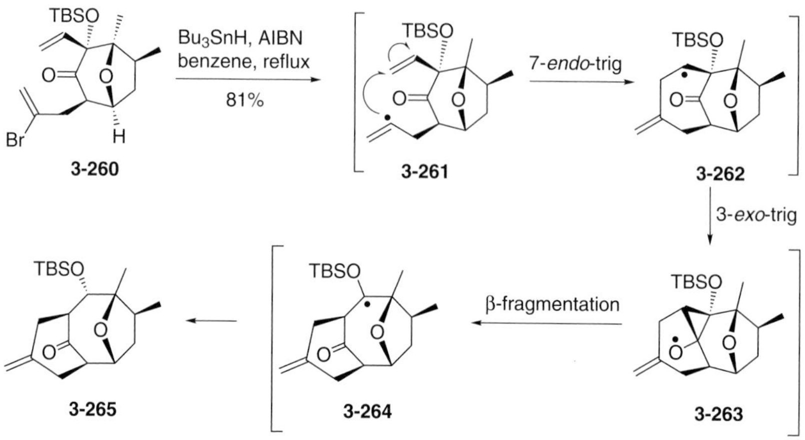

Scheme 3.68. Cyclization/ring expansion for the construction of bridged polycyclic systems.

Mascareñas and coworkers uncovered a rather exceptional procedure featuring a very effective domino radical cyclization/Beckwith–Dowd rearrangement sequence furnishing bicyclo[5.3.1]undecanes as products [107]. On treatment with tributyltin hydride and AIBN, the vinyl bromide **3-260** underwent a smooth conversion to the desired product **3-265** in 81% yield as an approximate 3:1 mixture of inseparable diastereoisomers (Scheme 3.68). The authors proposed a plausible mechanism in which the primarily formed vinyl radical **3-261** cyclizes in a 7-*endo*-trig mode to deliver the carbon-centered secondary radical **3-262**. Afterwards, an internal 1,2-acyl transfer occurs, which in the light of the well-known Beckwith–Dowd ring expansion of α-halomethyl and related cyclic ketones [108], should proceed by a 3-*exo*-trig ring closure between the secondary radical and the carbonyl group to form a strained, short-lived oxycyclopropyl radical intermediate **3-263**. β-Fragmentation of the latter yielded ring-expanded radical **3-264** which, after reaction with Bu₃SnH, provides **3-265**. It should be mentioned that the methodology could be extended to the preparation of the bicyclo[4.3.1]decane scaffolds, though in lower yield.

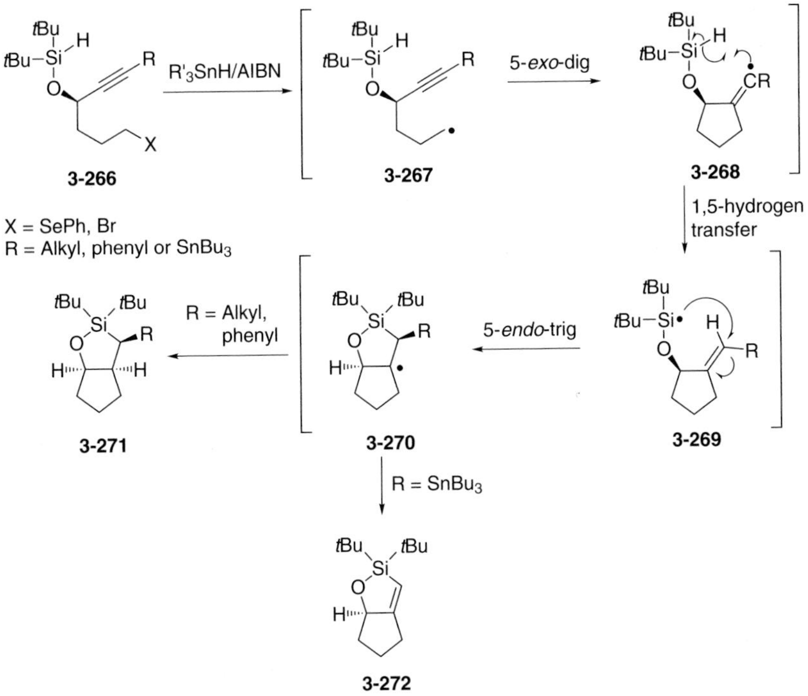

Scheme 3.69. General scheme of a 5-*exo*-dig cyclization/1,5-hydrogen transfer/5-*endo*-trig cyclization.

Clive and coworkers have developed a new domino radical cyclization, by making use of a silicon radical as an intermediate to prepare silicon-containing bicyclic or polycyclic compounds such as **3-271** and **3-272** (Scheme 3.69) [109]. After formation of the first radical **3-267** from **3-266**, a 5-*exo*-dig cyclization takes place followed by an intramolecular 1,5-transfer of hydrogen from silicon to carbon, providing a silicon-centered radical **3-269** *via* **3-268**. Once formed, this has the option to undergo another cyclization to afford the radical **3-270**, which can yield a stable product either by a reductive interception with the present organotin hydride species to obtain compounds of type **3-271**. On the other hand, when the terminal alkyne carries a trimethylstannyl group, expulsion of a trimethylstannyl radical takes place to afford vinyl silanes such as **3-272**.

Scheme 3.70 illustrates three examples in which the highly efficient construction of bi- and polycyclic compounds **3-274**, **3-276** and **3-278** from **3-273**, **3-275** and **3-277**, respectively, is depicted. It should be noted that the carbon–silicon bond in the obtained products can be easily cleaved [110] to achieve valuable synthons for further transformations.

As discussed previously, radical ring-opening reactions of three-membered systems *via* cyclopropylmethyl and oxiranylmethyl radicals represent a fruitful method in organic synthesis [111]. De Kimpe and coworkers have now shown that aziridines can also be used, featuring a radical one-step synthesis of pyrrolizidines **3-280**

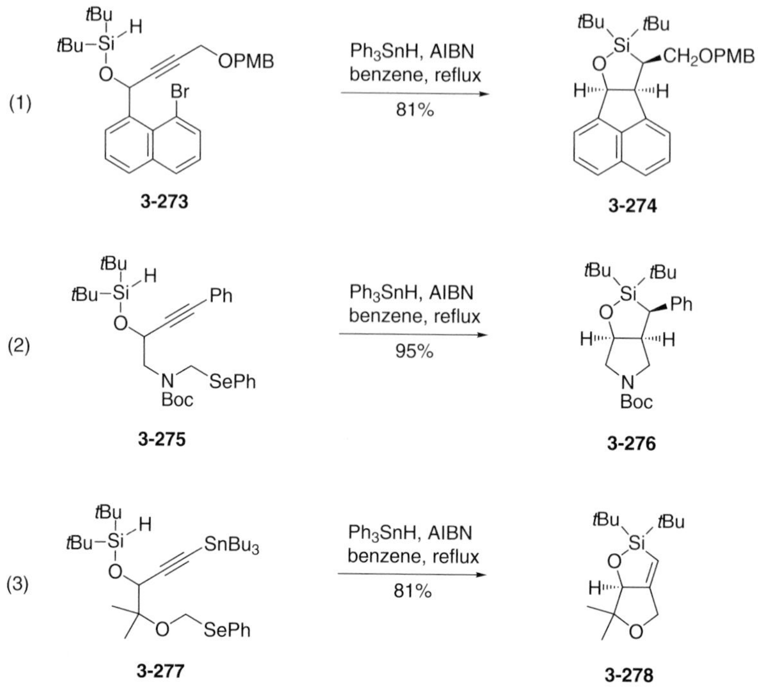

(1)

3-273 → Ph₃SnH, AIBN benzene, reflux 81% → 3-274

(2)

3-275 → Ph₃SnH, AIBN benzene, reflux 95% → 3-276

(3)

3-277 → Ph₃SnH, AIBN benzene, reflux 81% → 3-278

Scheme 3.70. Examples for the 5-*exo*-dig-cyclization/1,5-hydrogen transfer/5-*endo*-trig cyclization domino process.

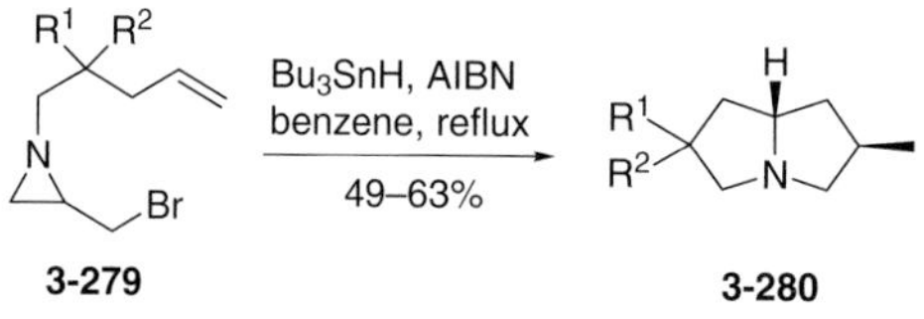

3-279 → Bu₃SnH, AIBN benzene, reflux 49–63% → 3-280

a: R^1= R^2= Me; **b**: R^1= R^2= Ph; **c**: R^1= Me, R^2= H

Scheme 3.71. Domino radical sequence of *N*-alkenylaziridinylmethyl radicals.

from 2-(bromomethyl)aziridines **3-279** by application of standard radical conditions in 49–63% yield (Scheme 3.71) [112].

The reaction sequence is assumed to be launched by the fragmentation of initially formed aziridinylmethyl radicals to give a *N*-allylaminyl radical, which undergoes a twofold 5-*exo*-trig cyclization.

The use of samarium(II) iodide in synthesis permits the assembly of complex molecules as already shown in many examples. They profit from the electron-transfer ability of samarium(II) iodide; thus, if ketones are employed as substrates the furnished ketyl-radical can react in a multitude of different ways.

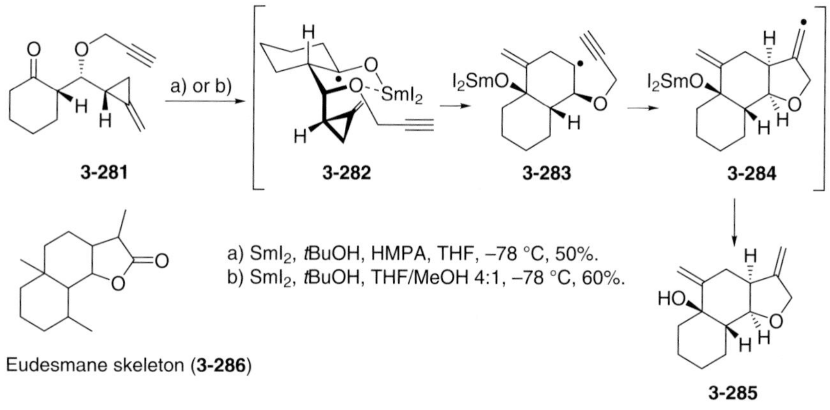

Scheme 3.72. Synthesis of the eudesmane tricyclic framework.

a) SmI$_2$, *t*BuOH, HMPA, THF, –78 °C, 50%.
b) SmI$_2$, *t*BuOH, THF/MeOH 4:1, –78 °C, 60%.

An example, where two C–C-bonds are formed and one C–C-bond is broken is the synthesis of the tricycle **3-285**, which has some similarity with the eudesmane framework **3-286**, developed by Kilburn and coworkers (Scheme 3.72) [113]. Thus, exposure of the easily accessible methylenecyclopropyl–cyclohexanone **3-281** to samarium(II) iodide led to the generation of ketyl radical **3-282**, which builds up a six-membered ring system with simultaneous opening of the cyclopropane moiety. Subsequent capture of the formed radical **3-283** by the adjacent alkyne group afforded the tricycle **3-285** *via* **3-284** as a single diastereoisomer in up to 60 % yield. It should be noted that in this case the usual necessary addition of HMPA could be omitted.

The analgesic properties of paeonilactone B (**3-289**) and its analogues have made these compounds challenging synthetic targets, and again the Kilburn group has presented a samarium(II) iodide-promoted domino radical cyclization for their synthesis [114]. Hence, samarium(II) iodide reaction of the diastereomeric ketones **3-287a** and **3-287b** led to the desired bicycles **3-288** (Scheme 3.73). The stereoselectivity depends heavily on the stereochemistry of the starting material. Thus, the (*S*)-isomer **3-287a** gave a mixture of the diastereomers **3-288a** and **3-288b**, with **3-288a** as the main product. In contrast, **3-287b** exclusively afforded **3-288b**. In this case, the use of an additive such as HMPA or DMPU is crucial for achieving satisfactory yields and high selectivities, with HMPA clearly excelling over DMPU.

In the final section of this chapter, we would like to present some domino processes with a particular high number of reaction steps, and partly unusual transformations.

An unusual formation of alkyl radicals as intermediates was observed in the conversion of the silyl ether **3-290** into the bicyclo[3.1.1]heptanes **3-291** using Bu$_3$SnH, and described by the Malacria group (Scheme 3.74) [115]. The obtained product **3-291** was further transformed either into **3-292a** using methyl lithium, or into **3-292b** by oxidative degradation.

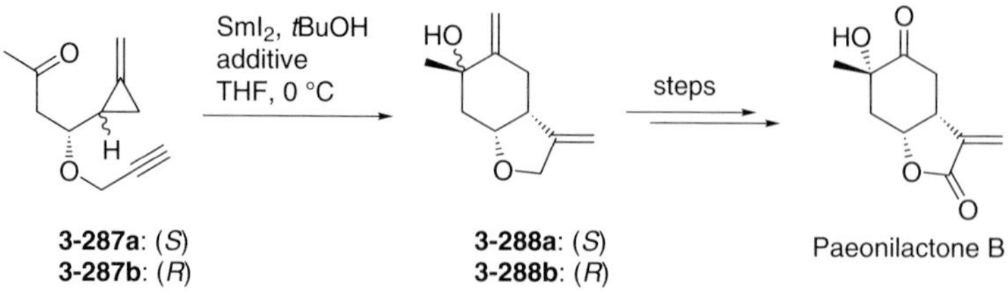

3-287a: (*S*)
3-287b: (*R*)

3-288a: (*S*)
3-288b: (*R*)

Paeonilactone B (**3-289**)

Entry	Substrate **3-287**	Additive	Product **3-288** a:b	Yield [%]
1	a	HMPA	10:1	63
2	a	DMPU	1.5:1	40
3	b	HMPA	<1:30	79
4	b	DMPU	<1:30	62

Scheme 3.73. Samarium(II) iodide-promoted domino radical reaction in the synthesis of paeonilactone B (**3-289**).

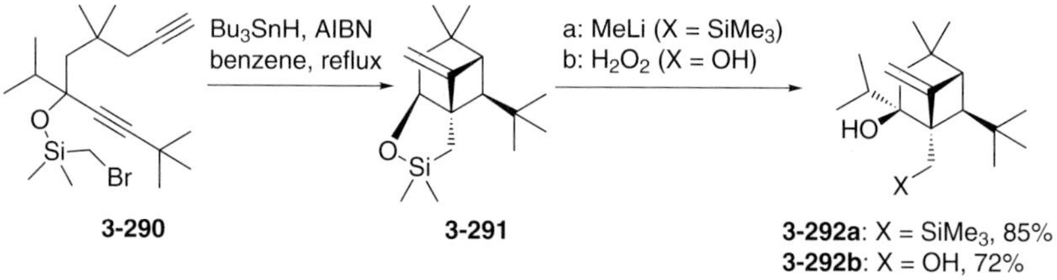

3-290

3-291

3-292a: X = SiMe₃, 85%
3-292b: X = OH, 72%

Scheme 3.74. Synthesis of bicyclo[3.1.1]heptanes.

The transformation is proposed to start with the formation of radical **3-293**, which undergoes a 5-*exo*-dig cyclization providing vinyl radical **3-294** (Scheme 3.75). This does not intercept with the alkyne moiety, but a 1,6-hydrogen transfer occurs to give radical **3-295**, which by 6-*endo*-trig ring closure leads to **3-296** with complete diastereoselectivity. **3-296** is then converted by a previously unknown 4-*exo*-dig ring formation to the vinyl radical **3-297**, followed by a 1,6-hydrogen transfer to give the stabilized α-silyl radical **3-298**, which can be considered as the driving force.

Xanthates serve as a reliable source of electrophilic radicals, and this was exploited by Zard and coworkers for a short synthesis of (±)-matrine (**3-304**), a naturally occurring alkaloid which has been claimed to have anti-ulcerogenic and anti-cancer properties [116]. Heating a mixture of xanthate **3-299** and the radical acceptor **3-300** (3 equiv.) in benzene in the presence of lauroyl peroxide as initiator, gave **3-301** in 30% yield and a 3:1 mixture of the tetracylic products **3-302** and **3-303** in 18% yield (Scheme 3.76) [117]. The three compounds could be converted into the

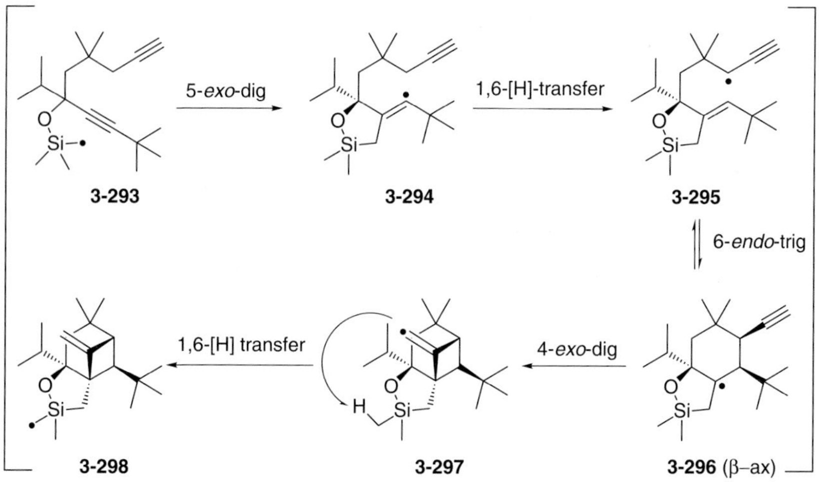

Scheme 3.75. Proposed radical intermediates in the formation of **3-298**.

diastereomeric tetracycles **3-305** and **3-306** utilizing lauroyl peroxide and 2-propanol as solvent. After separation, **3-305** was transformed into (±)-matrine (**3-304**). Although the yield was not very high, during this remarkable process four new bonds (including an intermolecular step) and five contiguous stereogenic centers were created in a single operation, with acceptable stereoselectivity.

Fenestranes [118] possess four rings which share a central carbon atom. As they are conformationally rigid and have a chemically robust structure [119], fenestranes are used as scaffolds for molecular recognition elements [120], catalysts [121], chiral auxiliaries [122], chemical libraries [123], and mechanistic research requiring spatially defined functionalities [124]. The first synthesis of the energetically unfavorable *cis,cis,cis,trans*-[5.5.5.5]-fenestranes **3-311** was discovered by Wender and coworkers, who disposed a remarkably effective domino arene-alkene photocycloaddition-radical cyclization methodology to produces a compound with five rings and eight stereocenters in three overall steps (cf. Chapter 5) [125]. Exposure of **3-307** to an acetonitrile radical generated from benzoyl peroxide in refluxing acetonitrile led, under concomitant opening of the three-membered ring system, to radical **3-309** (Scheme 3.77). After performing two further 5-*exo*-trig cyclizations and a final reductive capture of hydrogen from acetonitrile (as depicted in structure **3-310**), the pentacyclic fenestranes **3-311** were obtained in 35 % and 32 % yields.

Another elegant fenestrane synthesis, discovered by Keese, is described in Chapter 6.

Finally, a rather early (but from a mechanistic viewpoint a very interesting) sequence of radical reactions has been described by Pattenden and coworkers, in which an acetylenic oxime ether **3-312** was converted into the bicyclic oxime **3-319** in 70 % yield (Scheme 3.78) [126]. Hydrolysis of **3-319** led to the bicyclic enone **3-320**, which in fact can also more easily be synthesized by a Robinson annulation.

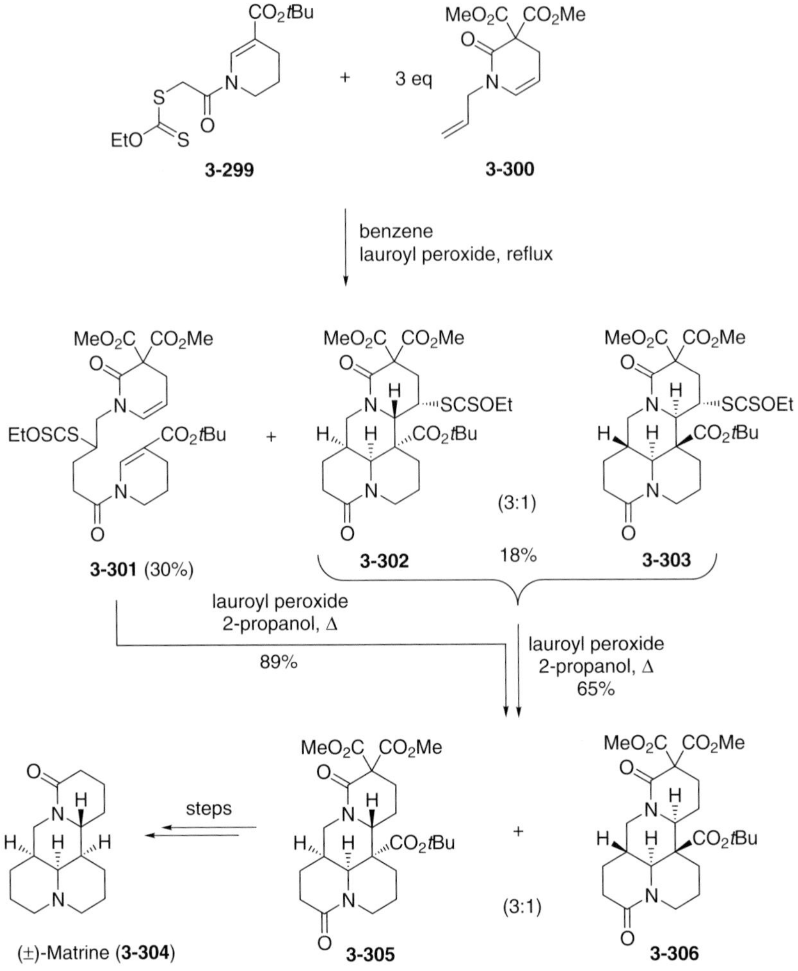

Scheme 3.76. Domino radical sequence in the total synthesis of (±)-matrine (**3-304**).

The proposed mechanism is described in Scheme 3.78, and includes the six radical intermediates **3-313** to **3-318**.

An interesting multicomponent domino free radical reaction in which five new bonds are formed in one operation has recently been applied in the total synthesis of yingzhaosu A by Bachi and coworkers. Thus, a 2,3-dioxabicyclo[3.3.1]nonane system reacts with phenylthiol and 2 equiv. of molecular oxygen in the presence of AIBN, which under irradiation with UV light led to a diastereomeric mixture of endoperoxide-hydroperoxides [127].

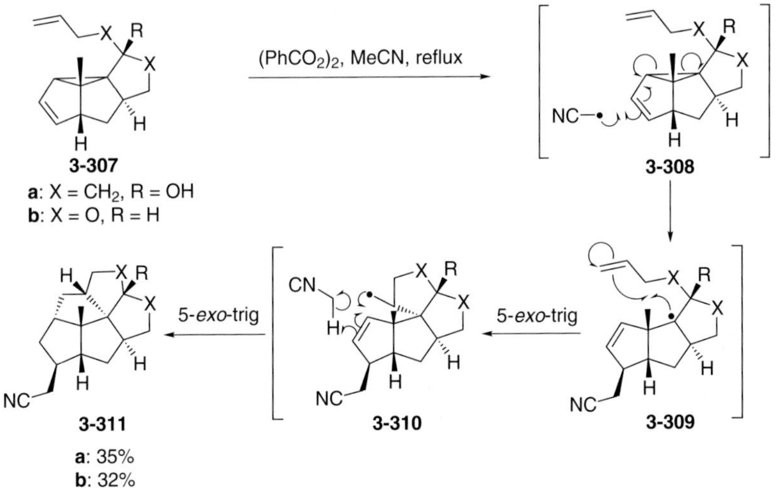

3-307
a: X = CH₂, R = OH
b: X = O, R = H

Scheme 3.77. Domino radical cyclization for the synthesis of fenestranes.

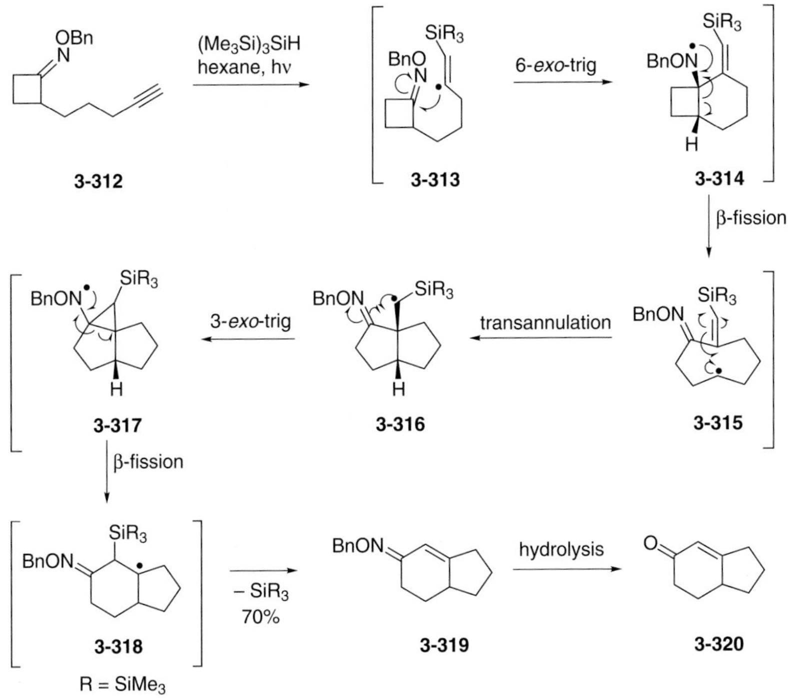

Scheme 3.78. Domino radical double ring expansion/cyclization process with oxime ethers.

3.3.3
Radical/Radical/Pericyclic Domino Processes

The connection of radical and pericyclic transformations in one and the same reaction sequence seems to be "on the fringe" within the field of domino processes. Here, we describe two examples, both of which are highly interesting from a mechanistic viewpoint. The first example addresses the synthesis of dihydroindene **3-326** by Parsons and coworkers, starting from the furan **3-321** (Scheme 3.79) [128]. Reaction of **3-321** with tributyltin hydride and AIBN in refluxing toluene led to the 1,3,5-hexatriene **3-324** *via* the radicals **3-322** and **3-323**. **3-324** then underwent an electrocyclization to yield the hexadiene **3-325** which, under the reaction conditions, aromatized to afford **3-326** in 51% yield.

The second example covers a completely different strategy. Wang and coworkers have exemplified that thermolysis of benzannulated ene-yne-carbodiimides offer an easy access to the corresponding heteroaromatic biradicals [129]. Thus, thermolysis of the ene-yne-carbodiimide **3-327** in chlorobenzene at 132 °C generated the biradical **3-328** which underwent a 1,5-hydrogen shift to give the intermediate **3-329** (Scheme 3.80) [130]. This then lost a molecule of formaldehyde to furnish **3-331**, which could also be regarded as a pyridine analogue of an *o*-quinone methide imine. An intramolecular hetero-Diels–Alder reaction of **3-331** then led to **3-332**. However, the fragmentation process of **3-329** to give **3-331** is relatively slow, and therefore compound **3-330** was produced predominantly in 49% yield.

It should be mentioned that the same group has also developed a related method in which benzoene-yne-allenes were cyclized *via* biradicals in a domino-like manner [131].

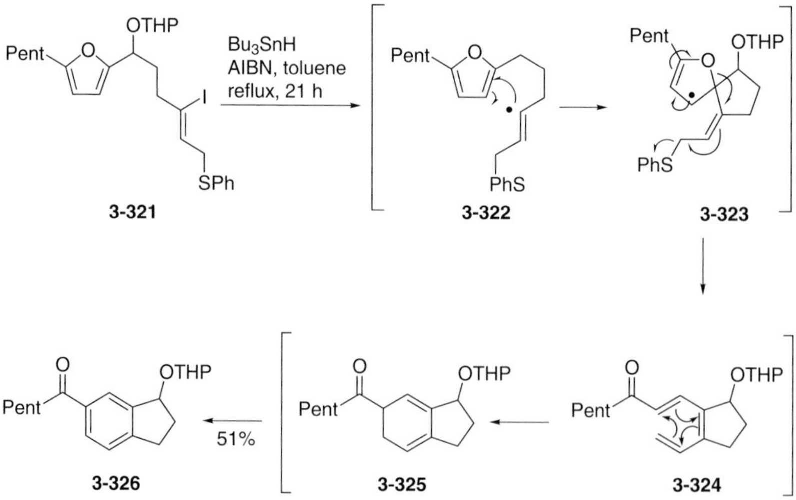

Scheme 3.79. Domino radical cyclization/fragmentation/electrocyclization for the synthesis of a dihydroindene.

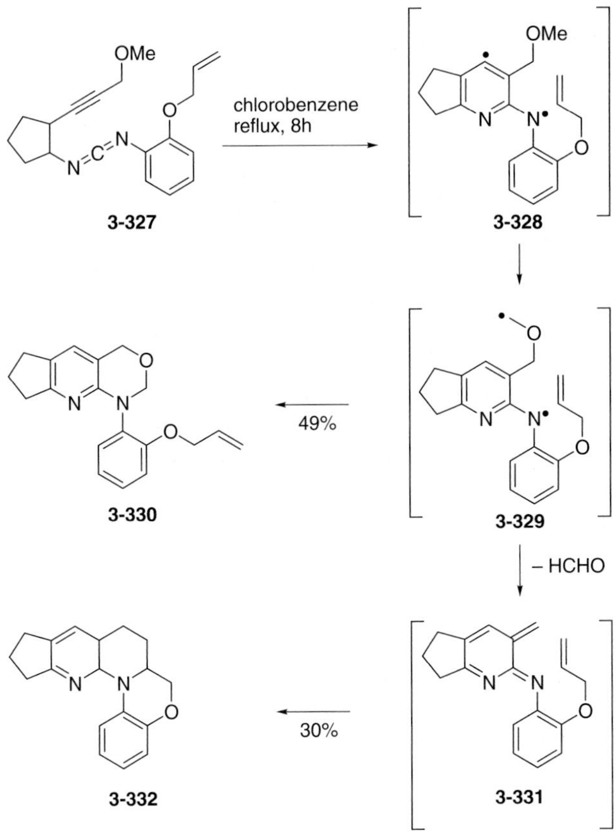

Scheme 3.80. Domino cyclization/1,5-hydrogen shift/hetero-Diels–Alder sequence.

3.3.4
Radical/Radical/Oxidation Domino Processes

The unexpected formation of cyclopenta[*b*]indole **3-339** and cyclohepta[*b*]indole derivatives has been observed by Bennasar and coworkers when a mixture of 2-in-dolylselenoester **3-333** and different alkene acceptors (e. g., **3-335**) was subjected to nonreductive radical conditions (hexabutylditin, benzene, irradiation or TTMSS, AIBN) [132]. The process can be explained by considering the initial formation of acyl radical **3-334**, which carries out an intermolecular radical addition onto the alkene **3-335**, generating intermediate **3-336** (Scheme 3.81). Subsequent 5-*endo*-trig cyclization leads to the formation of indoline radical **3-337**, which finally is oxidized *via* an unknown mechanism (the involvement of AIBN with **3-338** as intermediate is proposed) to give the indole derivative **3-339**.

As seen in Scheme 3.82 the yields of the cyclopenta[*b*]indoles obtained by this method are moderate to good; in some cases (entries 3 and 4), significant amounts of the uncyclized addition products were formed as byproducts. Interestingly, when methyl acrylate was applied, a bis-addition took place, leading to a cyclohepta[*b*]in-dole derivative in 61 % yield (entry 5).

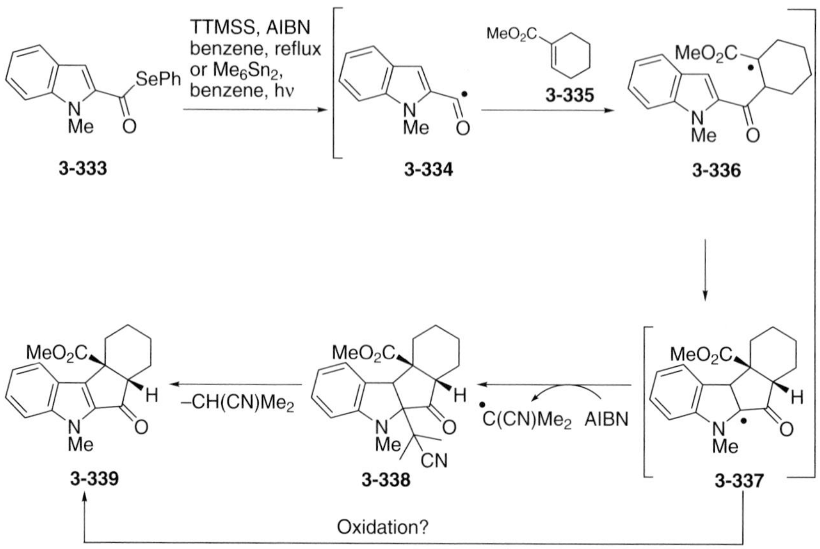

Scheme 3.81. Domino 2-indolylacyl radical addition/cyclization reaction.

Entry	Alkene Acceptor	Addition-cyclization Product	Yield [%]	Addition Product	Yield [%]
1	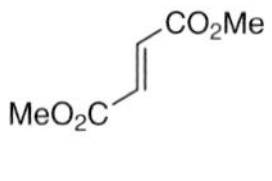	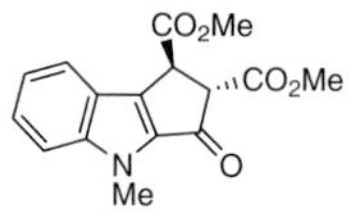	45	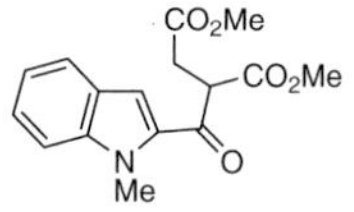	0
2	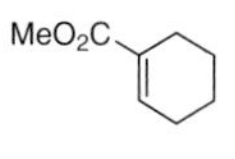	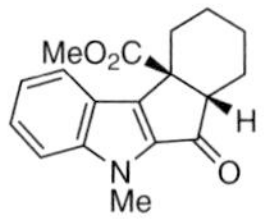	53	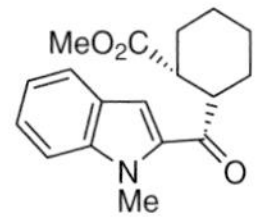	0
3	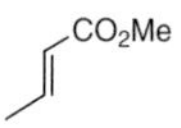	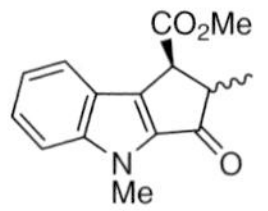	71[a]	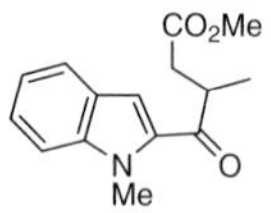	8
4	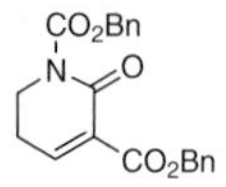	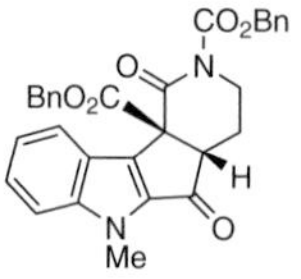	41	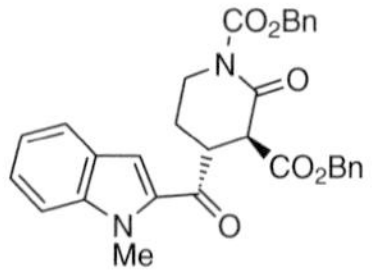	27
5		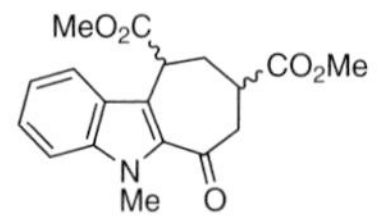	61[b]	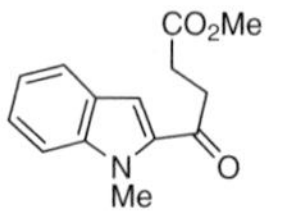	0

[a] 4:3 mixture of *cis*/*trans* stereoisomers.
[b] 3:1 mixture of *cis*/*trans* stereoisomers.

Scheme 3.82. Examples for the domino 2-indolylacyl radical addition/cyclization reaction.

3.4
Radical/Pericyclic Domino Processes

One of the very rare examples of a combination of a radical with a pericyclic reaction – in this case a [4+2] Diels–Alder cycloaddition – is depicted in Scheme 3.83 [133]. The sequence, elaborated by Malacria and coworkers, is based on the premise that the vinyl radical **3-341** formed from the substrate **3-340** using tributyltin hydride exists mainly in the "*Z*"-form. This is reduced by a hydrogen atom to form a 1,3-diene, which can undergo an intramolecular Diels–Alder reaction *via* an *exo*-transition state reaction (the chain lies away from diene).

In addition to **3-343**, triene **3-346** was also formed in a ratio of 1:4 in relation to **3-343**, presumably *via* the intermediates **3-344** and **3-345**. The overall yield of **3-343** and **3-346** was quoted as a respectable 72%.

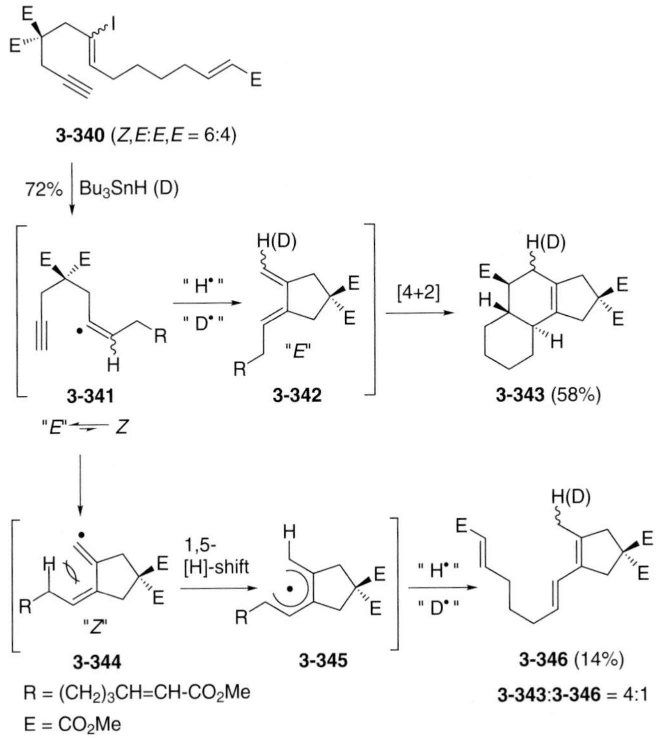

Scheme 3.83. First domino radical cyclization/intramolecular Diels–Alder reaction process.

References

1 A. J. McCarroll, J. C. Walton, *J. Chem. Soc. Perkin Trans. 1* **2001**, 3215–3229.

2 A. J. McCarroll, J. C. Walton, *Angew. Chem. Int. Ed.* **2001**, *40*, 2224–2248.

3 (a) A. L. J. Beckwith, K. U. Ingold, in: *Rearrangements in Ground and Excited States, Vol. 1* (Ed.: P. de Mayo), Academic Press, New York, **1980**, pp. 161–310; (b) J. E. Baldwin, *J. Chem. Soc. Chem. Commun.* **1976**, 734–736.

4 Y. Takemoto, T. Ohra, H. Koike, S.-i Furuse, C. Iwata, *J. Org. Chem.* **1994**, *59*, 4727–4729.

5 P. Girard, J. L. Namy, H. B. Kagan, *J. Am. Chem. Soc.* **1980**, *102*, 2693–2698.

6 (a) G. A. Molander, *Chem. Rev.* **1992**, *92*, 29–68; (b) G. A. Molander, C. R. Harris, *Chem. Rev.* **1996**, *96*, 307–338.

7 A. Schwartz, C. Seger, *Monatsh. Chem.* **2001**, *132*, 855–858.

8 S. Gross, H.-U. Reissig, *Org. Lett.* **2003**, 4305–4307.

9 P. A. Stadler, K. A. Giger, in: *Natural Products and Drug Development*, (Eds.: P. K. Larsen, S. B. Christensen, H. Kofod), Munksgaard, Copenhagen, **1984**, pp. 463.

10 I. Ninomiya, T. Kiguchi, in: *The Alkaloids, Vol. 38*, (Ed.: A. Brossi), Academic Press, New York, p. 1.

11 Y. Ozlu, D. E. Cladingboel, P. J. Parsons, *Synlett* **1993**, 357–358.

12 E. Lee. H. Y. Song, J. W. Kang, D.-S. Kim, C.-K. Jung, J. M. Joo, *J. Am. Chem. Soc.* **2002**, *124*, 384–385.

13 (a) M. Castillo, G. Morales, L. A. Loyola, I. Singh, C. Calvo, H. L. Holland, D. B. MacLean, *Can. J. Chem.* **1975**, *53*, 2513–2514; (b) M. Castillo, G. Morales, L. A. Loyola, I. Singh, C. Calvo, H. L. Holland, D. B. MacLean, *Can. J. Chem.* **1976**, *54*, 2900–2908.

14 C.-K. Sha, F.-K. Lee, C.-J. Chang, *J. Am. Chem. Soc* **1999**, *121*, 9875–9876.

15 (a) W. A. Ayer, *Nat. Prod. Rep.* **1990**, *8*, 455–463; (b) W. A. Ayer, *The Alkaloids, Vol. 45*, Academic Press, New York, **1994**, pp. 233–266.

16 (a) T. T. Dabrah, T. Kaneko, W. Massefski, Jr., E. B. Whipple, *J. Am. Chem. Soc.* **1997**, *119*, 1594–1598; (b) T. T. Dabrah, H. J. Harwood, Jr., L. H. Huang, N. D. Jankovich, T. Kaneko, J.-C. Li, S. Lindsey,

P. M. Moshier, T. A. Subashi, M. Therrien, P. C. Watts, *J. Antibiot.* **1997**, *50*, 1–7.

17 S. A. Biller, K. Neuenschwander, M. M. Ponpipom, C. D. Poulter, *Curr. Pharm. Design* **1996**, *2*, 1–40.

18 D. M. Leonard, *J. Med. Chem.* **1997**, *40*, 2971–2990.

19 K. C. Nicolaou, P. S. Baran, Y.-L. Zhong, H.-S. Choi, W. H. Yoon, Y. He, K. C. Fong, *Angew. Chem.* **1999**, *111*, 1781–1784; *Angew. Chem. Int. Ed.* **1999**, *38*, 1669–1675.

20 J. T. Njardarson, I. M. McDonald, D. A. Spiegel, M. Inoue, J. L. Wood, *Org. Lett.* **2001**, *3*, 2435–2438.

21 J. H. Butterworth, E. D. Morgan, *Chem. Commun.* **1968**, 23–24.

22 K. C. Nicolaou, A. J. Roecker, H. Monenschein, P. Guntupalli, M. Follmann, *Angew. Chem. Int. Ed.* **2003**, *42*, 3637–3642.

23 (a) A. H. Osada, C.-B. Cui, R. Onose, F. Hanaoka, *Bioorg. Med. Chem.* **1997**, *5*, 193–203; (b) S. Edmondson, S. J. Danishefsky, L. Sepp-Lorenzino, N. Rosen, *J. Am. Chem. Soc.* **1999**, *121*, 2147–2155.

24 D. Lizos, R. Tripoli, J. A. Murphy, *Chem. Commun.* **2001**, 2732–2733.

25 B. Patro, J. A. Murphy, *Org. Lett.* **2000**, *2*, 3599–3601.

26 (a) S. Kim, G. H. Joe, J. Do, *J. Am. Chem. Soc.* **1994**, *116*, 5521–5522; (b) O. Callaghan, M. Kizil, J. A. Murphy, B. Patro, *J. Org. Chem.* **1999**, *64*, 7856–7862; (c) J. A. Murphy, M. Kizil, *J. Chem. Soc., Chem. Commun.* **1995**, 1409–1410.

27 S. Zhou, S. Bommezijn, J. A. Murphy, *Org. Lett.* **2002**, *4*, 443–445.

28 M. Potmesil, H. Pinedo, *Camptothecins: New Anticancer Agents*, CRC Press, Boca Raton, Florida, **1995**.

29 W. R. Bowman, C. F. Bridge, P. Brookes, M. O. Cloonan, D. C. Leach, *J. Chem. Soc. Perkin Trans. 1* **2002**, 58–68.

30 Y. Damayanthi, J. W. Lown, *Curr. Med. Chem.* **1998**, *5*, 205–252.

31 A. J. Reynolds, A. J. Scott, C. I. Turner, M. S. Sherburn, *J. Am. Chem. Soc.* **2003**, *125*, 12108–12109.

32 J. Cassayre, F. Gagosz, S. Z. Zard, *Angew. Chem. Int. Ed.* **2002**, *41*, 1783–1785.

33 S. Kim, I. S. Kee, S. Lee, *J. Am. Chem. Soc.* **1991**, *113*, 9882–9883.

34 H.-Y. Lee, D.-I. Kim, S. Kim, *Chem. Commun.* **1996**, 1539–1540.

35 H.-Y. Lee, S. Lee, D. Kim, B. K. Kim, J. S. Bahn, S. Kim, *Tetrahedron Lett.* **1998**, *39*, 7713–7716.

36 H.-Y. Lee, D. K. Moon, J. S. Bahn, *Tetrahedron Lett.* **2005**, *46*, 1455–1458.

37 L. Porter, *Chem. Rev.* **1967**, *67*, 441–464.

38 J. Cassayre, S. Z. Zard, *J. Organomet. Chem.* **2001**, *624*, 316–326.

39 D. Yang, M. Xu, M.-Y. Bian, *Org. Lett.* **2001**, *3*, 111–114.

40 S. A. Hitchcock, S. J. Houldsworth, G. Pattenden, D. C. Pryde, N. M. Thomson, A. J. Blake, *J. Chem. Soc. Perkin Trans. 1* **1998**, 3181–3206.

41 H. Nagano, Y. Ohtani, E. Odake, J. Nakagawa, Y. Mori, T. Yajima, *J. Chem. Res. Synopses* **1999**, 338–339.

42 H. M. R. Hoffmann, U. Herden, M. Breithor, O. Rhode, *Tetrahedron* **1997**, *53*, 8383–8400.

43 T. J. Woltering, H. M. R. Hoffmann, *Tetrahedron* **1995**, *51*, 7389–7402.

44 A. F. Krasso, M. Binder, C. Tamm, *Helv. Chim. Acta* **1972**, *55*, 1352–1371.

45 G. R. Pettit, T. R. Kasturi, J. C. Knight, J. Occolowitz, *J. Org. Chem.* **1970**, *35*, 1404–1410.

46 W. R. Bowman, D. N. Clark, R. J. Marmon, *Tetrahedron* **1994**, *50*, 1295–1310.

47 P. Dowd, W. Zhang, *Chem. Rev.* **1993**, *93*, 2091–2115.

48 D. P. Curran, in: *Comprehensive Organic Synthesis, Vol. 4* (Ed.: B. M. Trost, I. Fleming), Pergamon, Oxford, **1991**, p. 779.

49 C. Wang, X. Gu, M. S. Yu, D. P. Curran, *Tetrahedron* **1998**, *54*, 8355–8370.

50 W.-W. Weng, T.-Y. Luh, *J. Org. Chem.* **1993**, *58*, 5574–5575.

51 (a) A. L. J. Beckwith, D. M. O'Shea, *Tetrahedron Lett.* **1986**, *27*, 4525–4528; (b) G. Stork, R. Mook, Jr., *Tetrahedron Lett.* **1986**, *27*, 4529–4532; (c) D. P. Curran, S. Sun, *Aust. J. Chem.* **1995**, *48*, 261–267; (d) B. P. Haney, D. P. Curran, *J. Org. Chem.* **2000**, *65*, 2007–2013; (e) O. Kitagawa, Y. Yamada, H. Fujiwara, T. Taguchi, *J. Org. Chem.* **2002**, *67*, 922–927; (f) K. Takasu, J. Kuroyanagi, A. Katsumata, M. Ihara, *Tetrahedron Lett.* **1999**, *40*, 6277–6280; (g) K. Takasu, S. Maiti, A. Katsumata, M. Ihara, *Tetrahedron Lett.* **2001**, *42*, 2157–2160.

52 K. Takasu, H. Ohsato, J. Kuroyanagi, M. Ihara, *J. Org. Chem.* **2002**, *67*, 6001–6007.

53 O. Kitagawa, Y. Yamada, A. Sugawara, T. Taguchi, *Org. Lett.* **2002**, *4*, 1011–1013.

54 D. Yang, S. Gu, Y.-L. Yan, H.-W. Zhao, N.-Y. Zhu, *Angew. Chem. Int. Ed.* **2002**, *41*, 3014–3017.

55 S. Yamago, H. Miyoshi, H. Miyazoe, J. Yoshida, *Angew. Chem. Int. Ed.* **2002**, *41*, 1407–1409.

56 (a) N. A. Porter, B. Lacher, V. H.-T. Chang, D. R. Magnin, *J. Am. Chem. Soc.* **1989**, *111*, 8309–8310; (b) N. A. Porter, D. M. Scott, B. Lacher, B. Giese, H. G. Zeitz, H. J. Lindner, *J. Am. Chem. Soc.* **1989**, *111*, 8311–8312; (c) D. M. Scott, A. T. McPhail, N. A. Porter, *Tetrahedron Lett.* **1990**, *31*, 1679–1682; (d) N. A. Porter, D. M. Scott, I. J. Rosenstein, B. Giese, A. Veit, H. G. Zeitz, *J. Am. Chem. Soc.* **1991**, *113*, 1791–1799.

57 E. Lacôte, B. Delouvrié, L. Fensterbank, M. Malacria, *Angew. Chem. Int. Ed.* **1998**, *37*, 2116–2118.

58 (a) G. Stork, P. M. Sher, H.-L. Chen, *J. Am. Chem. Soc.* **1986**, *108*, 6384–6385; (b) G. Stork, P. M. Sher, *J. Am. Chem. Soc.* **1986**, *108*, 303–304.

59 M. P. Sibi, J. Chen, *J. Am. Chem. Soc.* **2001**, *123*, 9472–9473. For recent applications, see: (a) M. P. Sibi, H. Hasegawa, *Org. Lett.* **2002**, *4*, 3347–3349; (b) M. P. Sibi, H. Miyabe, *Org. Lett.* **2002**, *4*, 3435–3438; (c) M. P. Sibi, M. Aasmul, H. Hasegawa, T. Subramanian, *Org. Lett.* **2003**, *5*, 2883–2886.

60 H. Miyabe, K. Fujii, T. Goto, T. Naito, *Org. Lett.* **2000**, *2*, 4071–4074.

61 (a) A. Gansäuer, M. Pierobon, H. Bluhm, *Angew. Chem. Int. Ed.* **1998**, *37*, 101–103; (b) A. Gansäuer, H. Bluhm, M. Pierobon, *J. Am. Chem. Soc.* **1998**, *120*, 12849–12859; (c) A. Gansäuer, T. Lauterbach, H. Bluhm, M. Noltemeyer, *Angew. Chem. Int. Ed.* **1999**, *38*, 2909–2910; (d) A. Gansäuer, H. Bluhm, *Chem. Rev.* **2000**, *100*, 2771–2788; (e) A. Gansäuer, M. Pierobon, H. Bluhm, *Synthesis* **2001**, 2500–2520.

62 A. Gansäuer, M. Pierobon, H. Bluhm, *Angew. Chem. Int. Ed.* **2002**, *41*, 3206–3208.

63 A. Gansäuer, B. Rinker, M. Pierobon, S. Grimme, M. Gerenkamp, C. Mück-Lichtenfeld, *Angew. Chem. Int. Ed.* **2003**, *42*, 3687–3690.

64 J. Justicia, J. E. Oltra, J. M. Cuerva, *J. Org. Chem.* **2004**, *69*, 5803–5806.

65 J. Justicia, A. Rosales, E. Buñuel, J. L. Oller-López, M. Valdivia, A. Haïdour, J. E. Oltra, A. F. Barrero, D. J. Cárdenas, J. M. Cuerva, *Chem. Eur. J.* **2004**, *10*, 1778–1788.

66 D. Leca, L. Fensterbank, E. Lacôte, M. Malacria, *Angew. Chem. Int. Ed.* **2004**, *43*, 4220–4222.

67 G. A. Molander, C. del Pozo Losada, *J. Org. Chem.* **1997**, *62*, 2935–2943.

68 E. J. Enholm, A. Trivellas, *Tetrahedron Lett.* **1994**, *35*, 1627–1628.

69 R. A. Batey, J. D. Harling, W. B. Motherwell, *Tetrahedron* **1996**, *52*, 11421–11444.

70 J.-C. Gauffre, M. Grignon-Dubois, B. Rezzonico, J.-M. Léger, *J. Org. Chem.* **2002**, *67*, 4696–4701.

71 W. T. Borden, *Diradicals*, Wiley-Interscience, New York, **1982**.

72 (a) A. G. Myers, E. Y. Kuo, N. S. Finney, *J. Am. Chem. Soc.* **1989**, *111*, 8057–8059; (b) A. G. Myers, P. S. Dragovich, *J. Am. Chem. Soc.* **1989**, *111*, 9130–9132.

73 K. K. Wang, H.-R. Zhang, J. L. Petersen, *J. Org. Chem.* **1999**, *64*, 1650–1656.

74 B. De Boeck, N. M. Harrington-Frost, G. Pattenden, *Org. Biomol. Chem.* **2005**, *3*, 340–347.

75 T. Taniguchi, K. Iwasaki, M. Uchiyama, O. Tamura, H. Ishibasi, *Org. Lett.* **2005**, *7*, 4389–4390.

76 M. J. Robins, Z. Guo, M. C. Samano, S. F. Wnuk, *J. Am. Chem. Soc.* **1999**, *121*, 1425–1433.

77 (a) C. P. Jasperse, D. P. Curran, T. L. Fevig, *Chem. Rev.* **1991**, *91*, 1237–1286; (b) D. P. Curran, *Synlett* **1991**, 63–72; (c) D. P. Curran, in: *Comprehensive Organic Synthesis, Vol. 4* (Ed.: B. M. Trost, I. Fleming), Pergamon, Oxford, **1991**, p. 779; (d) P. J. Parsons, C. S. Penkett, A. J. Shell, *Chem. Rev.* **1996**, *96*, 195–206; (e) M. Malacria, *Chem. Rev.* **1996**, *96*, 289–306; (f) B. B. Snider, *Chem. Rev.* **1996**, *96*, 339–363.

78 (a) N. A. Porter, V. H. T. Chang, D. R. Magnin, B. T. Wright, *J. Am. Chem. Soc.* **1988**, *110*, 3554–3560; (b) S. Handa, G. Pattenden, *J. Chem. Soc. Perkin Trans 1* **1999**, 843–845; (c) S. A. Hitchcock, S. J. Houldsworth, G. Pattenden, D. C. Pryde, N. M. Thomson, A. J. Blake, *J. Chem. Soc. Perkin Trans. 1* **1998**, 3181–3206; (d) U. Jahn, D. P. Curran, *Tetrahedron Lett.* **1995**, *36*, 8921–8924.

79 (a) I. Ryu, N. Sonoda, D. P. Curran, *Chem. Rev.* **1996**, *96*, 177–194; (b) D. P. Curran, S. Sun, *Aust. J. Chem.* **1995**, *48*, 261–267.

80 B. P. Haney, D. P. Curran, *J. Org. Chem.* **2000**, *65*, 2007–2013.

81 K. Takasu, S. Maiti, A. Katsumata, M. Ihara, *Tetrahedron Lett.* **2001**, *42*, 2157–2160.

82 D. P. Curran, H. Liu, *J. Am. Chem. Soc.* **1991**, *113*, 2127–2132.

83 D. P. Curran, H. Liu, H. Josien, S.-B. Ko, *Tetrahedron* **1996**, *52*, 11385–11404.

84 I. Lenoir, M. L. Smith, *J. Chem. Soc. Perkin Trans. 1* **2000**, 641–643.

85 (a) D. P. Curran, H. Liu, *J. Am. Chem. Soc.* **1992**, *114*, 5863–5864; (b) D. P. Curran, *J. Chin. Chem. Soc. (Taipei)* **1993**, *40*, 1–6; (c) D. P. Curran, J. Sisko, P. E. Yeske, H. Liu, *Pure Appl. Chem.* **1993**, *65*, 1153–1159; (d) D. P. Curran, S.-B. Ko, H. Josien, *Angew. Chem. Int. Ed.* **1996**, *34*, 2683–2684.

86 H. Josien, S.-B. Ko, D. Bom, D. P. Curran, *Chem. Eur. J.* **1998**, *4*, 67–83.

87 H. Josien, D. P. Curran, *Tetrahedron* **1997**, *53*, 8881–8886.

88 O. de Frutos, D. P. Curran, *J. Comb. Chem.* **2000**, *2*, 639–649.

89 A. E. Garbada, D. P. Curran, *J. Comb. Chem.* **2003**, *5*, 617–624.

90 L. Benati, R. Leardini, M. Minozzi, D. Nanni, P. Spagnolo, G. Zanardi, *J. Org. Chem.* **2000**, *65*, 8669–8674.

91 R. Leardini, D. Nanni, P. Pareschi, A. Tundo, G. Zanardi, *J. Org. Chem.* **1997**, *62*, 8394–8399.

92 L. Chen, G. B. Gill, G. Pattenden, H. Simonian, *J. Chem. Soc. Perkin Trans. 1* **1996**, 31–43.

93 S. A. Hitchcock, S. J. Houldsworth, G. Pattenden, D. C. Pryde, N. M. Thomson, A. J. Blake, *J. Chem. Soc. Perkin Trans. 1* **1998**, 3181–3206.

94 G. J. Hollingworth, G. Pattenden, D. J. Schulz. *Aust. J. Chem.* **1995**, *48*, 381–399.

95 S. Handa, G. Pattenden, *Contemp. Org. Synth.* **1997**, *4*, 196–215.

96 S. Handa, G. Pattenden, *J. Chem. Soc. Perkin Trans. 1* **1999**, 843–845.

97 (a) H. M. Boehm, S. Handa, G. Pattenden, L. Roberts, A. J. Blake, W.-S. Li, *J. Chem. Soc. Perkin Trans. 1* **2000**, 3522–3538; (b) S. Handa, P. S. Nair, G. Pattenden, *Helv. Chim. Acta* **2000**, *83*, 2629–2643.

98 S. Handa, G. Pattenden, W.-S. Li, *Chem. Commun.* **1998**, 311–312.

99 J. E. Saxton, *Nat. Prod. Rep.* **1996**, *14*, 559–590.

100 S. T. Hilton, T. C. T. Ho, G. Pljevaljcic, M. Schulte. K. Jones, *Chem. Commun.* **2001**, 209–210.

101 (a) M. Nakajima, K. Itoi, Y. Tamamatsu, T. Kinoshita, T. Okasaki, K. Kawakubo, M. Shindou, T. Honma, M. Tohjigamori, T. Haneishi, *J. Antibiot.* **1991**, *44*, 293–300; (b) H. Haruyama, T. Takayama, T. Kinoshita, M. Kondo, M. Nakajima, T. Haneishi, *J. Chem. Soc. Perkin Trans. 1* **1991**, 1637–1640; (c) H. Sano, S. Mio, M. Shindou, T. Honma, S. Sugai, *Tetrahedron* **1995**, *46*, 12563–12572, and references therein.

102 C. Chatgilialoglu, T. Gimisis, G. P. Spada, *Chem. Eur. J.* **1999**, *5*, 2866–2876.

103 L. Fensterbank, M. Malacria, S. M. Sieburth, *Synthesis* **1997**, 813–854.

104 C. Aïssa, B. Delouvrié, A.-L. Dhimane, L. Fensterbank, M. Malacria, *Pure Appl. Chem.* **2000**, *72*, 1605–1613.

105 A.-L. Dhimane, C. Aïssa, M. Malacria, *Angew. Chem. Int. Ed.* **2002**, *41*, 3284–3287.

106 L. Fensterbank, E. Mainetti, P. Devin, M. Malacria, *Synlett* **2000**, 1342–1344.

107 J. R. Rodríguez, L. Castedo, J. L. Mascareñas, *Org. Lett.* **2001**, *3*, 1181–1183.

108 P. Dowd, W. Zhang, *Chem. Rev.* **1993**, *93*, 2091–2115.

109 D. L. J. Clive, W. Yang, A. C. MacDonald, Z. Wang, M. Cantin, *J. Org. Chem.* **2001**, *66*, 1966–1983.

110 M. Sannigrahi, D. L. Mayhew, D. L. J. Clive, *J. Org. Chem.* **1999**, *64*, 2776–2788.

111 (a) D. C. Nonhebel, *Chem. Soc. Rev.* **1993**, 347–359; (b) D. J. Pasto, *J. Org. Chem.* **1996**, *61*, 252–256.

112 D. De Smaele, P. Bogaert, N. De Kimpe, *Tetrahedron Lett.* **1998**, *39*, 9797–9800.

113 F. C. Watson, J. D. Kilburn, *Tetrahedron Lett.* **2000**, *41*, 10341–10345.

114 (a) R. J. Boffey, M. Santagostino, W. G. Whittingham, J. D. Kilburn, *Chem. Commun.* **1998**, 1875–1876; (b) R. J. Boffey, W. G. Whittingham, J. D. Kilburn, *J. Chem. Soc. Perkin Trans. 1* **2001**, 487–496.

115 S. Bogen, L. Fensterbank, M. Malacria, *J. Am. Chem. Soc.* **1997**, *119*, 5037–5038.

116 K. A. Aslanov, Y. K. Kushmuradov, S. Sadykov, *Alkaloids* **1987**, *31*, 117–192.

117 L. Boiteau, J. Boivin, A. Liard, B. Quiclet-Sire, S. Z. Zard, *Angew. Chem. Int. Ed.* **1998**, *37*, 1128–1131.

118 For reviews, see: (a) A. K. Gupta, X. Fu, J. P. Snyder, J. M. Cook, *Tetrahedron*, **1991**, *47*, 3665–3710; (b) B. R. Venepalli, W. C. Agosta, *Chem. Rev.* **1987**, *87*, 399–410.

119 (a) M. Trachsel, R. Keese, *Helv. Chim. Acta* **1988**, *71*, 363–368; (b) A. P. Davis, *Chem. Soc. Rev.* **1993**, 243–253.

120 (a) R. P. Bonar-Law, J. K. M. Sanders, *J. Am. Chem. Soc.* **1995**, *117*, 259–271 and references cited therein; (b) U. Maitra, B. G. Bag, *J. Org. Chem.* **1994**, *59*, 6114–6115 and references cited therein.

121 D. Roy, D. M. Birney, *Synlett* **1994**, 798–800 and references cited therein.

122 P. Mathivanan, U. Maitra, *J. Org. Chem.* **1995**, *60*, 364–369 and references cited therein.

123 R. Boyce, G. Li, H. P. Nestler, T. Suenaga, W. C. Still, *J. Am. Chem. Soc.* **1994**, *116*, 7955–7956 and references cited therein.

124 K. Kumar, Z. Lin, D. H. Waldeck, M. B. Zimmt, *J. Am. Chem. Soc.* **1996**, *118*, 243–244.

125 P. A. Wender, T. M. Dore, M. A. deLong, *Tetrahedron Lett.* **1996**, *37*, 7687–7690.

126 G. Pattenden, D. J. Schulz, *Tetrahedron Lett.* **1993**, *34*, 6787–6790.

127 (a) A. M. Szpilman, E. E. Korshin, H. Rozenberg, M. D. Bachi, *J. Org. Chem.* **2005**, *70*, 3618–3632; (b) E. E. Korshin, R. Hoos, A. M. Szpilman, L. Kostantinovski, G. H. Posner, M. D. Bachi, *Tetrahedron* **2002**, *58*, 2449–2469.

128 A. Demircan, P. J. Parsons, *Synlett* **1998**, 1215–1216.

129 (a) C. Shi, Q. Zhang, K. K. Wang, *J. Org. Chem.* **1999**, *64*, 925–932; (b) M. Schmittel, J.-P. Steffen, B. Engels, C. Lennartz, M. Hanrath, *Angew. Chem. Int. Ed.* **1998**, *37*, 2371–2373.

130 H. Li, J. L. Petersen, K. K. Wang, *J. Org. Chem.* **2003**, *68*, 5512–5518.

131 H.-R. Zhang, K. K. Wang, *J. Org. Chem.* **1999**, *64*, 7996–7999.

132 M.-L. Bennasar, T. Roca, R. Griera, J. Bosch, *J. Org. Chem.* **2001**, *66*, 7547–7551.

133 M. Journet, M. Malacria, *J. Org. Chem.* **1994**, *59*, 6885–6886.

4
Pericyclic Domino Reactions

The combination of pericyclic transformations as cycloadditions, sigmatropic rearrangements, electrocyclic reactions and ene reactions with each other, and also with non-pericyclic transformations, allows a very rapid increase in the complexity of products. As most of the pericyclic reactions run quite well under neutral or mild Lewis acid acidic conditions, many different set-ups are possible. The majority of the published pericyclic domino reactions deals with two successive cycloadditions, mostly as [4+2]/[4+2] combinations, but there are also [2+2], [2+5], [4+3] (Nazarov), [5+2], and [6+2] cycloadditions. Although there are many examples of the combination of hetero-Diels–Alder reactions with 1,3-dipolar cycloadditions (see Section 4.1), no examples could be found of a domino all-carbon-[4+2]/[3+2] cycloaddition. Co-catalyzed [2+2+2] cycloadditions will be discussed in Chapter 6.

The combination of two successive [4+2] cycloadditions has already been described by Diels and Alder [1a] for the reaction of dimethyl acetylenedicarboxylate with an excess of furan. A beautiful, more modern, example is the synthesis of pagodane (4-5) by Prinzbach [2], in which an intermolecular Diels–Alder reaction of 4-1 and 4-2 to give 4-3 is followed by an intramolecular cycloaddition. The obtained 4-4 is then transformed into 4-5 (Scheme 4.1).

Another impressive example is the synthesis of paracyclophanes as 4-9 by Hopf [3], starting from a 1,2,4,5-hexatetraene 4-6 and an electron-deficient alkyne 4-7 to give 4-9 *via* the intermediate 4-8 (Scheme 4.2).

An impressive combination of two Diels–Alder reactions is also described by Winkler [4] for the synthesis of the taxane skeleton, though two different Lewis acids must be used for the two cycloadditions. Thus, it does not strictly match the definition of a domino reaction.

The second largest group of pericyclic domino reactions starts with a sigmatropic rearrangement, which is most often a Claisen or an oxa- and aza-Cope rearrangement; however, some processes also exist with a 2,3-sigmatropic rearrangement as the second step.

A beautiful example of a domino [3+3]-sigmatropic rearrangement is the synthesis of the enantiopure antifungal antibiotic (–)-preussin (4-14) by Overman [5], which starts from the amine 4-10 and decanal to give the iminium ion 4-11 (Scheme 4.3). This undergoes a [3+3]-sigmatropic rearrangement to provide 4-12, followed by a Mannich reaction with the formation of 4-13.

Domino Reactions in Organic Synthesis. Lutz F. Tietze, Gordon Brasche, and Kersten M. Gericke
Copyright © 2006 WILEY-VCH Verlag GmbH & Co. KGaA, Weinheim
ISBN: 3-527-29060-5

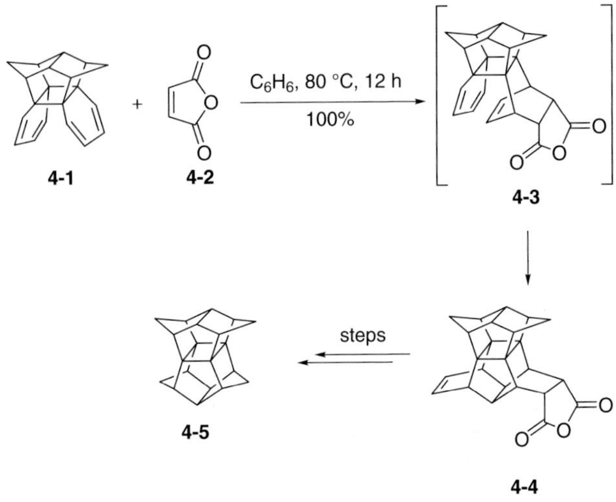

Scheme 4.1. Synthesis of pagodane (**4-5**).

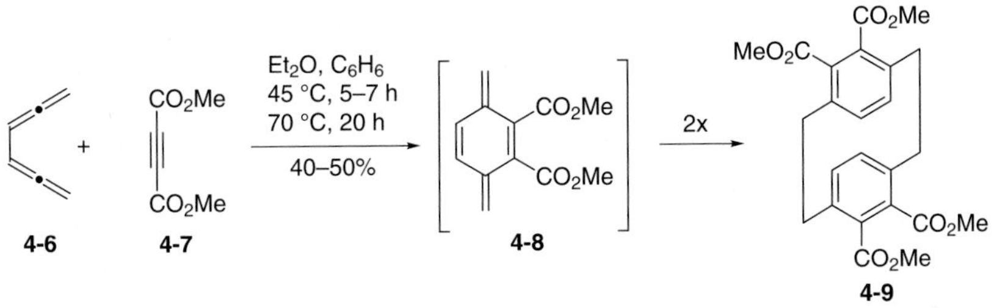

Scheme 4.2. Synthesis of paracyclophanes.

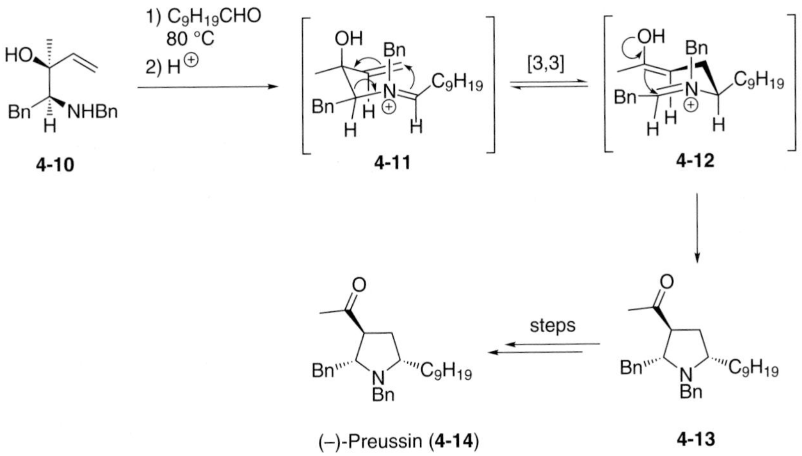

Scheme 4.3. Synthesis of (−)-preussin (**4-14**).

To date, only a few examples have been identified of electrocyclic transformations and ene reactions as the initiating event in a domino process.

4.1
Diels–Alder Reactions

Since the number of domino processes which start with a Diels–Alder reaction is rather large, we have subdivided this section of the chapter according to the second step, which might be a second Diels–Alder reaction, a 1,3-dipolar cycloaddition, or a sigmatropic rearrangement. However, there are also several examples where the following reaction is not a pericyclic but rather is an aldol reaction; these examples will be discussed under the term "Mixed Transformations".

4.1.1
Diels–Alder/Diels–Alder Reactions

Dailey and coworkers [6] extended the studies of Prinzbach using **4-1** as substrate. These authors found that, by employing dicyanoacetylene **4-15** in the reaction with **4-1** the domino adduct **4-16** but not **4-17**, as expected, is the main product (Scheme 4.4). In the formation of **4-16** one of the two 1,3-butadiene moieties in **4-1** has reacted with the dienophile **4-14** from the inside, followed by a second [4+2] cycloaddition of the formed dicyanoethene moiety. **4-17** is observed as a side product; here, in the first step, the dienophile reacts from the outside, while in the second step the other formed dienophile moiety undergoes a cycloaddition with the second 1,3-butadiene moiety. This mode of action is actually favored in the reaction of all other dienophiles employed, due to their larger size when compared to **4-15**.

In further studies, the group of Dailey [6] also investigated the domino Diels–Alder reaction of tetraenes of type **4-18** with dicyanoacetylene **4-15**, which led exclusively to product **4-19** in good yield (Scheme 4.4).

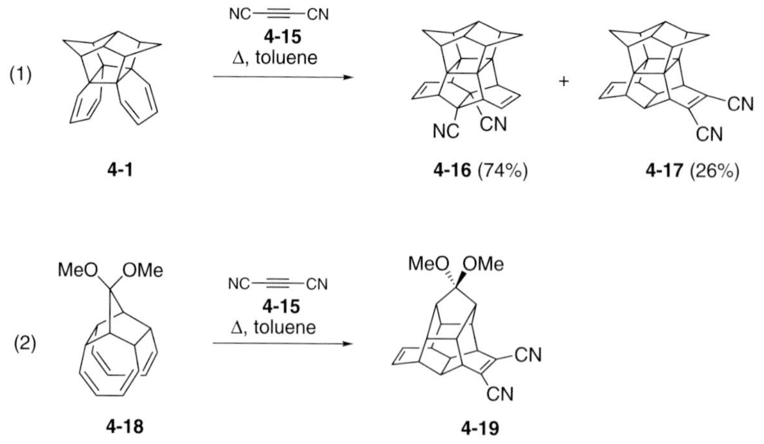

Scheme 4.4. Domino Diels–Alder reactions of **4-1** and **4-18**.

Slee and LeGoff performed further investigations on the reaction of dimethyl acetylenedicarboxylate **4-20** with an excess of furan **4-21**, as first described by Diels and Alder (Scheme 4.5) [1a]. At 100 °C, **4-24** and **4-25** were not produced (as proposed), but rather **4-22** and **4-23**, since at elevated temperature an equilibrium takes place and the primarily formed **4-24** and **4-25** isomerize to give a 6:1-mixture of the *exo-endo* and the *exo-exo* products **4-22** and **4-23**, respectively. However, at lower temperature, in the primarily formed [4+2] cycloadduct the double bond substituted with the two carbomethoxy group acts as the dienophile to give the two products **4-24** and **4-25** in a 3:1 ratio with 96 % yield within five weeks, as has been shown by Diels and Olsen [1a,1c]. For a differentiation of these two types of adducts, Paquette and coworkers [7] used a "domino and pincer product". The Cram group [8] described one of the first examples of a reaction of a tethered bisfuran **4-26** with dimethyl acetylenedicarboxylate **4-20a** to give **4-27**.

Lautens and coworkers [9] ultimately used this approach to prepare a multitude of different polyheterocyclic ring systems **4-29**, using **4-20** and **4-28** as substrates (Table 4.1). Unsymmetrical tethered bisfurans and acetylene dicarboxyclic acid derivatives have also been used in this domino process to allow the formation of three new rings with up to six stereogenic centers.

The development of efficient syntheses of steroids remains an important target. For example, Sherburn and coworkers [10a] developed a new strategy for the pre-

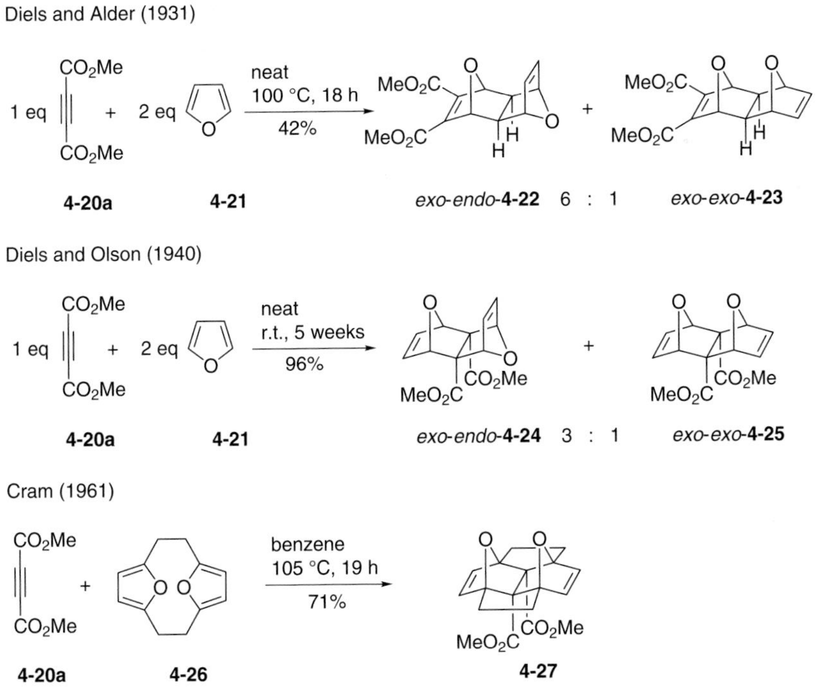

Diels and Alder (1931)

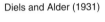

1 eq **4-20a** + 2 eq **4-21** neat 100 °C, 18 h 42% → *exo-endo*-**4-22** 6 : 1 *exo-exo*-**4-23**

Diels and Olson (1940)

1 eq **4-20a** + 2 eq **4-21** neat r.t., 5 weeks 96% → *exo-endo*-**4-24** 3 : 1 *exo-exo*-**4-25**

Cram (1961)

4-20a + **4-26** benzene 105 °C, 19 h 71% → **4-27**

Scheme 4.5. Reaction of acetylenedicarboxylate and furan.

Entry	Dienophile	Bis-diene		Product	Yield [%]
	CO₂R ⫼ CO₂R	O O X		O O X RO₂C CO₂R	
	4-20a: R = Me	**4-28**		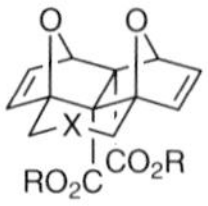 **4-29**	
	4-20b: R = H				
1	a	a	X = CH₂	aa	74
2	b	a	X = CH₂	ba	71
3	a	b	X = O	ab	76
4	a	c	X = NPMP	ac	63
5	a	d	X = NBn	ad	72
6	a	e	X = NPNB	ae	65
7	a	f	X = S	af	71

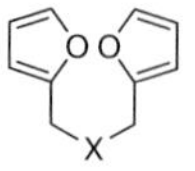

Table 4.1. Reaction of acetylenedicarboxylate and tethered bisfurans.

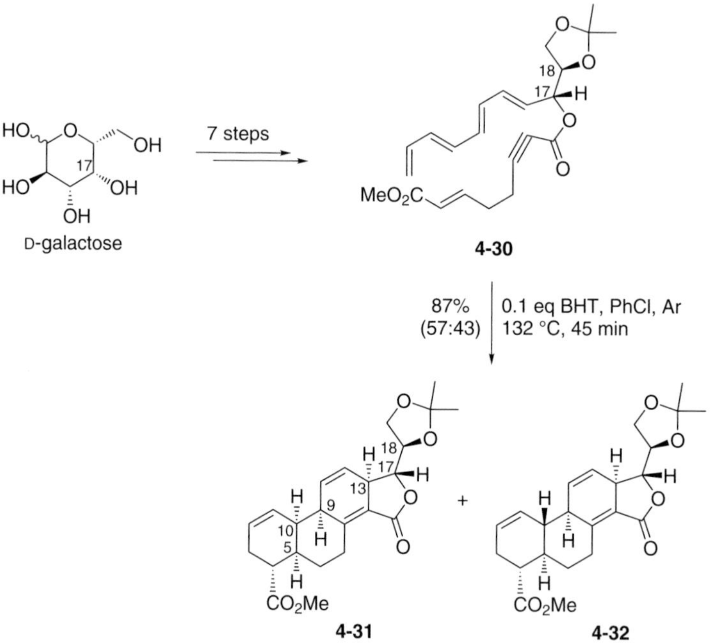

Scheme 4.6. Synthesis of steroid analogues.

paration of tetracarbocyclic products of D-homosteroids using a novel Lewis acid-promoted domino process of two intramolecular Diels–Alder reactions. This approach can also be used for the construction of the steroid skeleton. Thus, heating enantiopure **4-30** obtained from D-galactose gave the tetracycles **4-31** and **4-32** in 87% yield as a 57:43 mixture (Scheme 4.6) [10b].

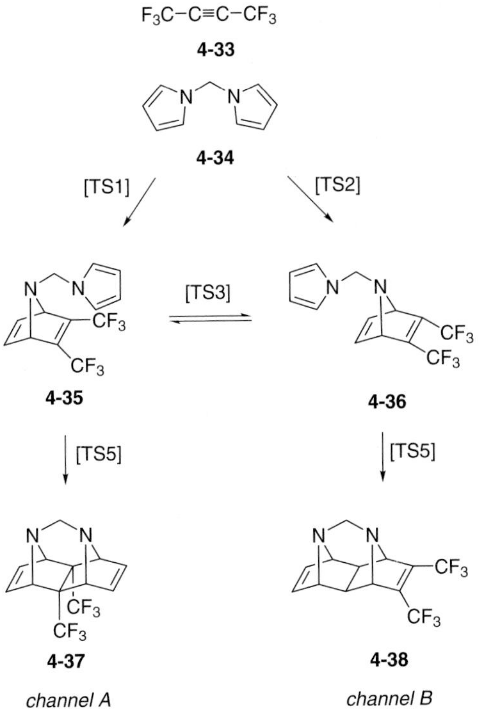

F₃C–C≡C–CF₃

4-33

4-34

[TS1]   [TS2]

[TS3]

4-35   **4-36**

[TS5]   [TS5]

4-37   **4-38**

channel A   *channel B*

Scheme 4.7. Theoretical investigations on the reaction of **4-33** and **4-34**.

Domingo and coworkers [11] have contributed an important theoretical input for the understanding of domino reactions. An interesting example is the domino Diels–Alder reaction of **4-33** and **4-34**, in which the products **4-37** and **4-38** could be formed *via* **4-35** and **4-36**, respectively (Scheme 4.7). Visnick and Battiste [12] had shown that, at room temperature, only cycloadduct **4-37** is formed, whereas with heat **4-38** is obtained quantitatively. This is in line with the calculations showing that TS5 is higher in energy than TS4 (74.5 and 55.3 kJ mol^{-1}, respectively); on the other hand, cycloadduct **4-38** is more stable (−92.9 kJ mol^{-1}) than cycloadduct **4-37** (−78.7 kJ mol^{-1}), which explains the formation of **4-38** under thermodynamic control. Calculations have also been performed for the bisfuran system **4-28a** [13].

4.1.2
Diels–Alder Reactions/Sigmatropic Rearrangements

In contrast to the lack of examples of the domino Diels–Alder reaction/1,3-dipolar cycloaddition, the combination of a Diels–Alder reaction with a sigmatropic rearrangement has been used intensively.

Serrano's group described asymmetric domino Diels–Alder/Cope reactions using 1,3-nitrocyclohexadienes containing a sugar moiety such as **4-39** and cy-

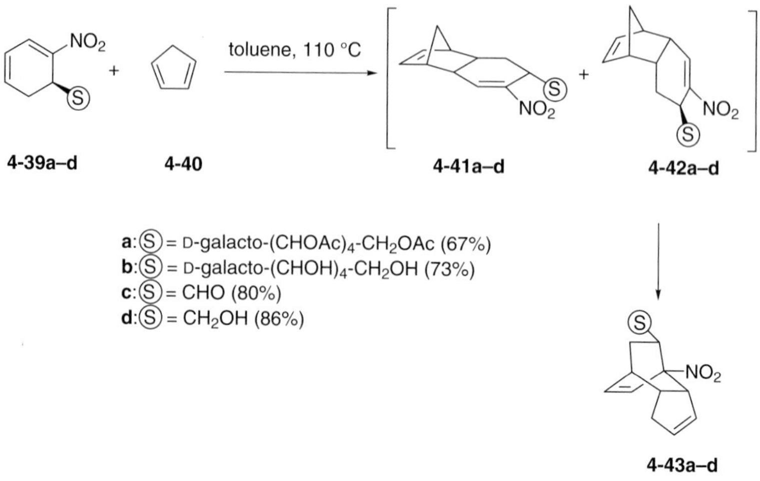

Scheme 4.8. Reaction of nitrocyclohexadienes with cyclopentodiene.

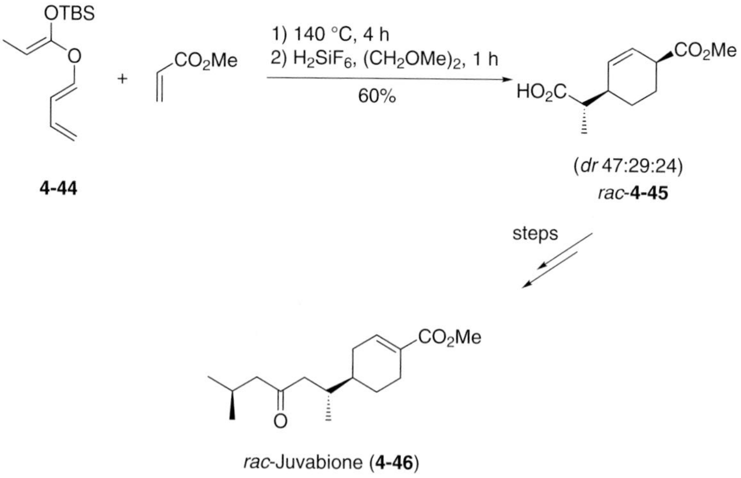

Scheme 4.9. Synthesis of *rac*-juvabione (**4-46**).

clopentadiene **4-40** (Scheme 4.8) [14]. These authors obtained **4-43** as a more or less single product upon heating. The proposed cycloadducts **4-41** and **4-42** could also be isolated in very small quantities, and it could be shown that only the *endo* adduct **4-42** undergoes Cope rearrangement to give **4-43**.

Neier and coworkers have used a domino Diels–Alder/Ireland–Claisen process for the synthesis of (*rac*)-juvabione **4-46** and (*rac*)-epijuvabione [15]. Since neither the Diels–Alder reaction of the acetal **4-44** and methyl acrylate nor the sigmatropic rearrangement seemed to be stereoselective, these authors obtained the cyclohexene derivative **4-45** as a mixture of three diastereomers (Scheme 4.9).

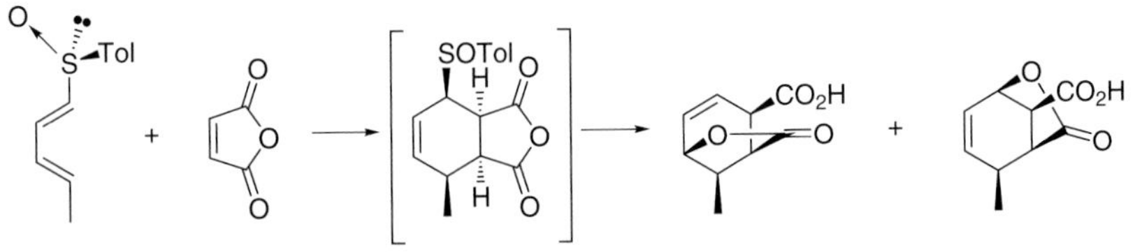

Entry	Solvent	4-2 [eq]	T	t [d]	Pressure	4-48/4-49a/4-49b	Yield [%][a]	4-49b [% ee]
1	MeCN	3.0	70 °C	2	-	0:70:30	60	50
2	CH₂Cl₂	1.2	r.t.	1	13 kbar	70:30:0	90	82

[a] Calculated from the crude reaction mixture.

Scheme 4.10. Reaction of chiral butadienes and maleic anhydride.

Carreño, Ruano and coworkers [16] designed a very useful combination of a Diels–Alder reaction and a 2,3-sigmatropic rearrangement of a sulfoxide which allowed the synthesis of enantioenriched or even almost enantiopure products. The approach had been used for the synthesis of many structurally diversified compounds. In an earlier report, these authors had synthesized bridged lactones **4-49a** and **4-49b**, starting from the chiral butadiene **4-47** and maleic anhydride **4-2** (Scheme 4.10). Interestingly, high pressure accelerates the formation of the cycloadduct **4-48**, but not rearrangement of the sulfoxide moiety in **4-48**.

The same authors also used this approach for an enantioselective synthesis of the natural product (+)-royleanone (**4-54**), a member of the abietane diterpenoid family [17]. The enantiopure sulfoxide **4-50** was oxidized using DDQ to give crude 1,4-benzoquinone **4-51**, which by reaction with the diene **4-52** in CH₂Cl₂ under high pressure led to the tricyclic compound **4-53** with 97% *ee* and 60% yield based on **4-50** (Scheme 4.11). Hydrogenation of the unconjugated double bond in **4-53** afforded 35% of the desired compound **4-54** after crystallization to separate it from the unwanted *cis*-isomer.

The procedure can also be initiated by a hetero-Diels–Alder reaction [18].

Recently, Carreño, Urbano and coworkers were also able to synthesize almost enantiopure [7]helicene bisquinones **4-58** and **4-59** (96% *ee*) by reaction of the sulfoxide (*S,S*)-**4-55** with the diene **4-56** in dichloromethane at –20 °C (Scheme 4.12) [19]. This six-step domino process includes a double Diels–Alder reaction, sulfoxide elimination, and aromatization of the rings B and F of the intermediate **4-57** to give **4-58**, which could be oxidized to the fully aromatized **4-59**.

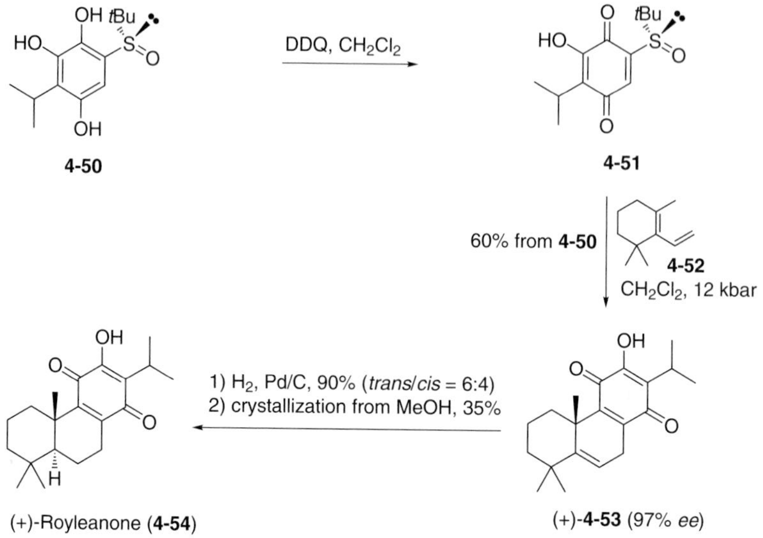

Scheme 4.11. Synthesis of (+)-royleanone (**4-54**).

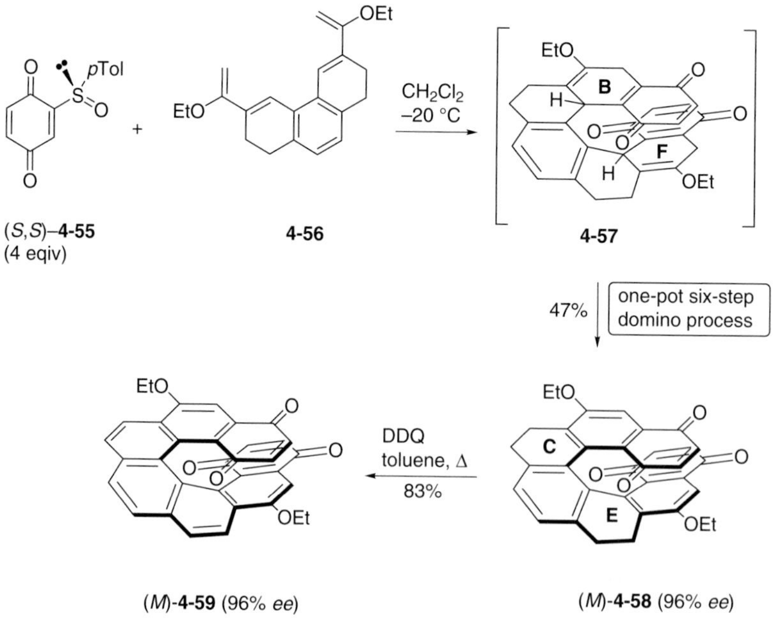

Scheme 4.12. Asymmetric synthesis of [7]helicene bisquinones.

4.1.3
Diels–Alder/Retro-Diels–Alder Reactions

Diels–Alder reactions can also be coupled with retro-Diels–Alder reactions to form interesting sequences. An illustrative example is the reaction of 4-60 to give 4-62 *via* the nonisolatable Diels–Alder adduct 4-61 used by Jacobi and coworkers in a synthesis of paniculide-A (4-63) (Scheme 4.13). The elimination of acetonitrile from the primary cycloadduct 4-61 in this sequence is worthy of note [20]. The synthesis of ansa compounds from steroids by Winterfeldt and coworkers [21], in which the steroid framework is almost completely dismantled, is an attractive example of a Diels–Alder/retro-Diels–Alder process. Thus, reaction of ergosterolacetate 4-64 with propargylaldehyde in refluxing toluene led directly to the ansa compound 4-66. On the other hand, the primary cycloadduct 4-65 can be isolated if Lewis acids are used at room temperature. Heating leads again to 4-66 (Scheme 4.14).

In their approach towards novel antitumor agents, Danishefsky and coworkers [22] recently published another interesting example of a domino Diels–Alder/retro-Diels–Alder reaction. These authors were attracted by the 14-membered resorcinylic macrolide radicol (4-67) which inhibits the Hsp90 molecular chaperone [23, 24]. However, in order to avoid the epoxide moiety in 4-67, which might serve as a locus of nondiscrimimating cell toxicity, these authors focused their activity on the cyclopropane-analogue 4-70 by heating a mixture of 4-68 and 4-69, which led to 4-70 with extrusion of isobutene in 75% yield (Scheme 4.15).

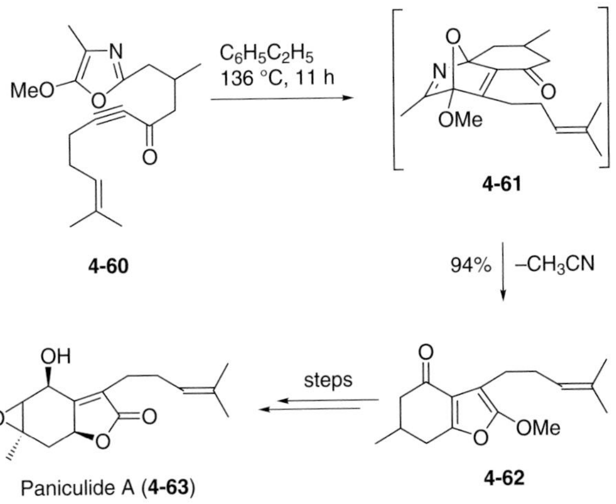

Scheme 4.13. Synthesis of paniculide A (4-63).

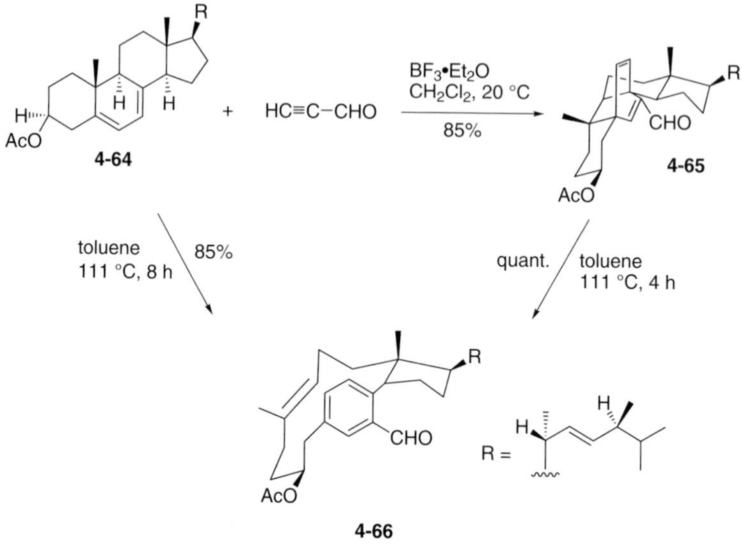

Scheme 4.14. Synthesis of macrocycles from steroids.

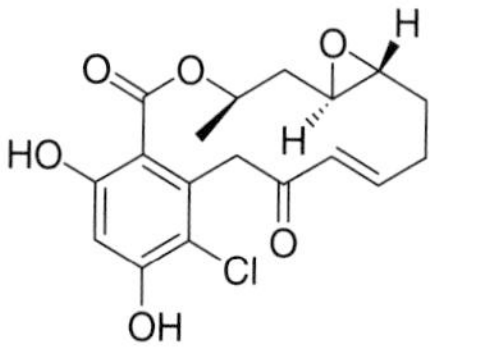

Radicicol (**4-67**)

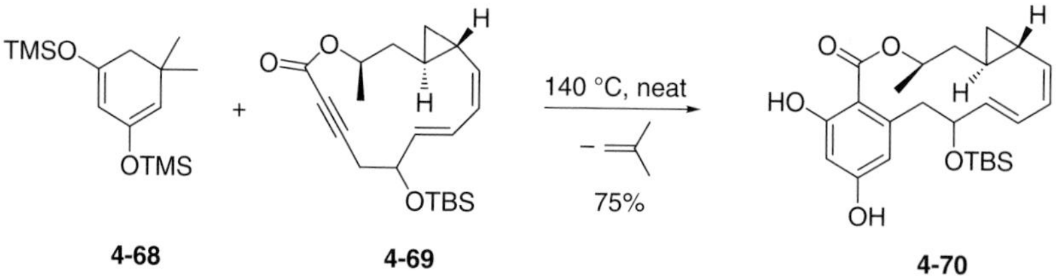

Scheme 4.15. Synthesis of a radicicol analogue.

4.1.4
Diels–Alder Reactions/Mixed Transformations

All examples which do not belong to the above-described sections have been collected under the heading of Diels–Alder/mixed transformations. It is quite astounding which combinations have been put together. The following transformations have been used as a second step after the Diels–Alder reaction: [2+2] cycload-

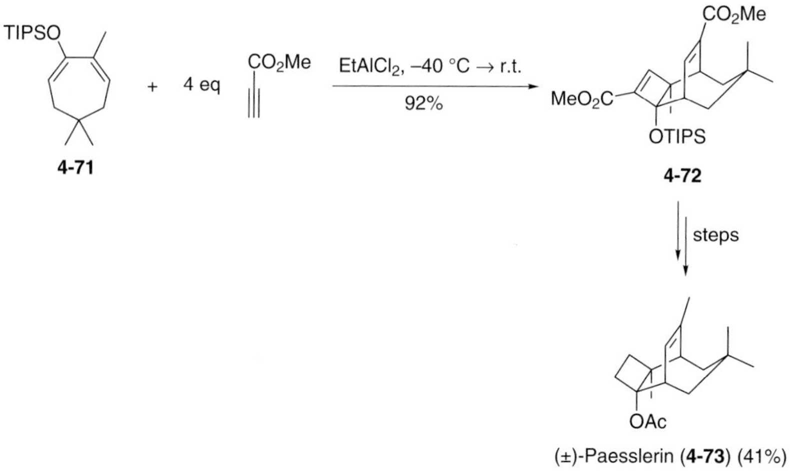

Scheme 4.16. Synthesis of (±)-paesslerin (**4-73**).

ditions, aldol reactions, allylations of aldehydes, a Schmidt rearrangement, a Morita–Baylis–Hillman cyclization, a retro-Prins reaction, aromatizations, and metal-organic transformations.

Takasu, Ihara and coworkers described an efficient synthesis of (±)-paesslerin A (**4-73**) using a combination of a [4+2] and a [2+2] cycloaddition (Scheme 4.16) [25]. Reaction of **4-71** and propargylic acid methyl ester in the presence of the Lewis acid EtAlCl$_2$ led to **4-72** in 92 % yield, which was converted in six steps into the desired natural product **4-73** by transformation of one of the ester moieties into a methyl group, hydrogenation of one double bond, removal of the other ester moiety, and exchange of the TIPS group for an acetate.

Deslongchamps and coworkers [26] used a combination of a transannular Diels–Alder cycloaddition and an intramolecular aldol reaction in the synthesis of the unnatural enantiomer of a derivative of the (+)-aphidicolin (**4-74**), which is a diterpenoic tetraol isolated from the fungus *Cephalosporium aphidicolia*. This compound is an inhibitor of DNA polymerase, and is also known to act against the herpes simplex type I virus. In addition, it slows down eukaryotic cell proliferation, which makes it an interesting target as an anticancer agent.

The γ-epiaphidicoline skeleton as in **4-75** was synthesized from the macrocyclic precursor **4-76** containing an electron-poor dienophile and a butadiene moiety, and which undergoes a transannular Diels–Alder reaction to give **4-78** *via* the transition state **4-77** (Scheme 4.17). Due to the vicinity of the formyl moiety and the methylene group in **4-78**, an intramolecular aldol reaction can take place to form **4-79**. The overall domino reaction is highly diastereoselective to give **4-79** as a single diastereomer in 81 % yield, being controlled by the two stereogenic centers in the substrate **4-76**. In addition to **4-79**, 8 % of the intermediate cycloadduct is obtained. In the reaction, three new rings and six new stereogenic centers are formed. Further transformations of **4-79** then led to **4-75**.

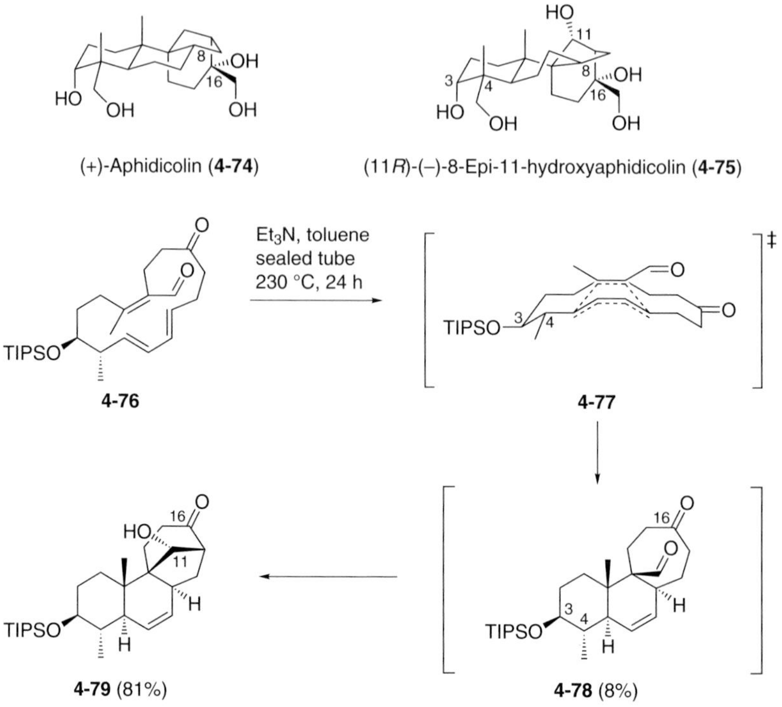

(+)-Aphidicolin (**4-74**)

(11*R*)-(−)-8-Epi-11-hydroxyaphidicolin (**4-75**)

4-76 → Et₃N, toluene sealed tube 230 °C, 24 h → **4-77**

4-79 (81%) ← **4-78** (8%)

Scheme 4.17. Synthesis of aphidicolin analogue.

In 1987, Vaultier and coworkers [27] developed a combination of a [4+2] cycloaddition of a bora-1,3-diene to provide an allylborane, which then reacts with an aldehyde to give a highly functionalized alcohol. The Lallemand group, as well as Hall and colleagues, has recently used this procedure. In an approach for the synthesis of the antifeedant natural product clerodin (**4-83**), Lallemand and coworkers performed a three-component domino reaction of **4-80**, **4-81** and methyl acrylate to give **4-82** (Scheme 4.18) [28].

Hall also prepared condensed cyclohexenes **4-89** by reaction of **4-84**, **4-85** and **4-87** in excellent yield and with high selectivity *via* the proposed intermediate and transition state **4-86** and **4-88**, respectively (Scheme 4.19) [29].

Partially hydrogenated indoles are useful heterocycles. They can easily be obtained by a domino Diels–Alder/Schmidt process, as described by Aubé and coworkers [30]. An example is the reaction of the enone **4-90** with a butadiene **4-91** in the presence of the Lewis acid MeAlCl₂, which led to tricyclic compounds as **4-93** *via* **4-92** in over 80 % yield (Scheme 4.20). The procedure has also been used for the synthesis of pyrroloisoquinolones, azepinoindolones, and perhydroindoles.

Roush and coworkers developed a new one-pot sequence consisting of an intramolecular Diels–Alder- and an intramolecular vinylogous Morita–Baylis–Hillman-cyclization for the synthesis of spinosyns [31]. These compounds are polyketide natural products possessing extraordinary insecticidal activity.

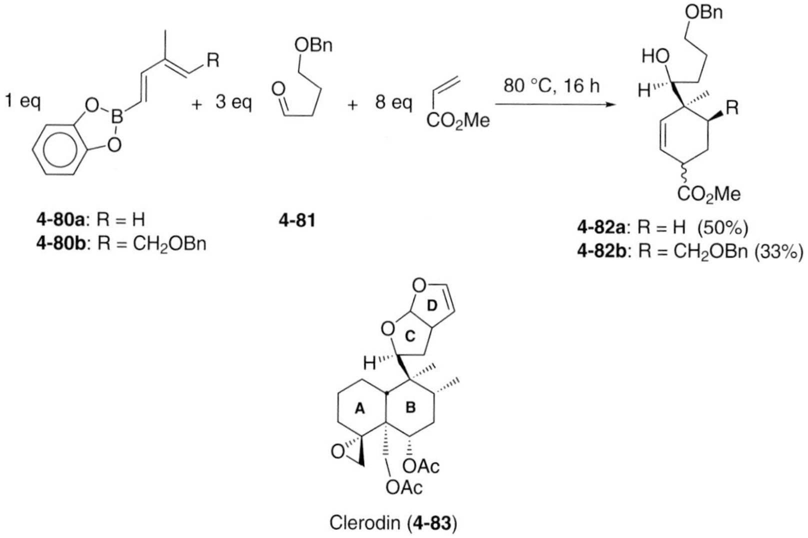

Scheme 4.18. Approach towards clerodin (**4-83**) by combination of a Diels–Alder reaction and an allylation.

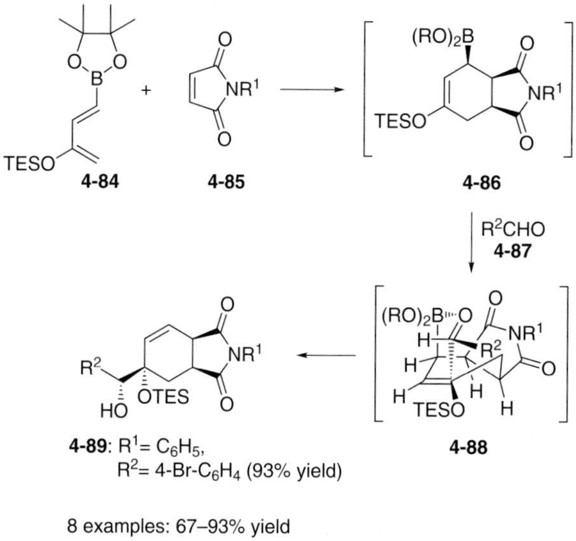

8 examples: 67–93% yield

Scheme 4.19. Synthesis of condensed cyclohexenes **4-89**.

The reaction of **4-95** to give a 96:4-mixture of **4-96** and **4-97** containing the central core of **4-94** was performed by heating **4-95** for 67 h at 40 °C and then adding PMe₃ to induce the Morita–Baylis–Hillman reaction (Scheme 4.21).

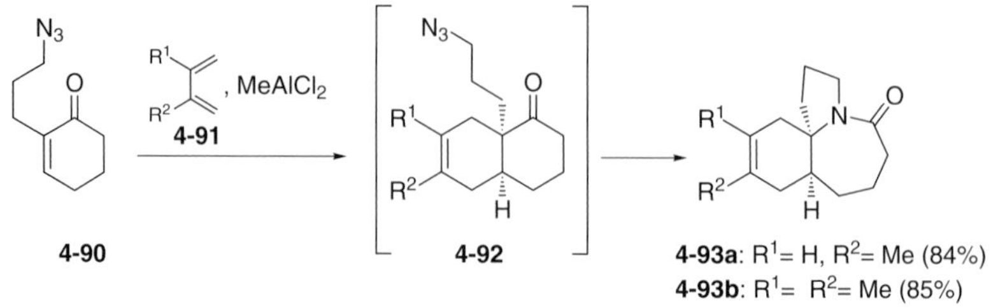

Scheme 4.20. Synthesis of indole derivatives.

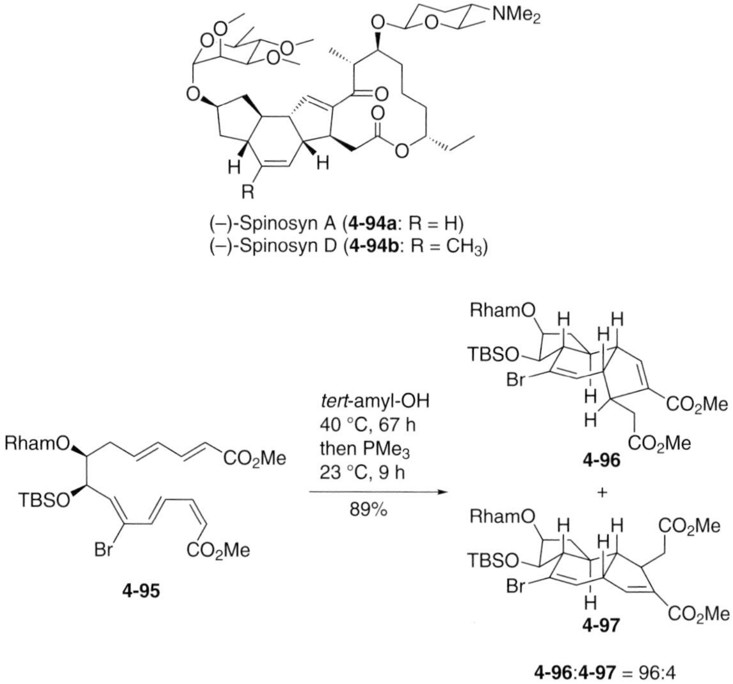

(−)-Spinosyn A (**4-94a**: R = H)
(−)-Spinosyn D (**4-94b**: R = CH$_3$)

4-96:4-97 = 96:4

Scheme 4.21. Synthesis of spinosyns.

Padwa's group has not only developed highly efficient domino reactions using transition metal catalysis, but they are also well known for their unique combinations of a cycloaddition and a *N*-acyliminium ion cyclization. An example of this strategy, which is very suitable for the synthesis of heterocycles and alkaloids, is the reaction of **4-98** to give **4-101** *via* the intermediates **4-99** and **4-100** (Scheme 4.22). Furthermore, **4-101** was transformed into the alkaloid (+)-γ-lycorane **4-102** [32].

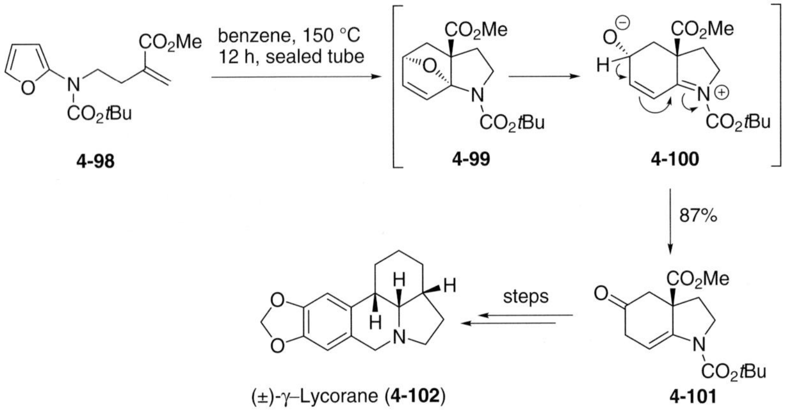

Scheme 4.22. Synthesis of (±)-γ-lycorane (**4-102**).

The sequence could even be prolonged by including a Pummerer reaction. Thus, treatment of **4-103** with trifluoroacetic acid (TFA) gave the furan **4-104**, which underwent a cycloaddition to furnish **4-105**; the erythryna skeleton **4-109** was obtained after subsequent addition of a Lewis acid such as $BF_3 \cdot Et_2O$ (Scheme 4.23) [33]. It can be assumed that **4-106**, **4-107** and **4-108** act as intermediates. In a more recent example, these authors also used the procedure for the synthesis of indole alkaloids of the Aspidosperma type [34].

A combination of a [4+2] cycloaddition of an electron-deficient 1,3-diene with an enamine followed by an elimination was described by the Bodwell group [35].

Reaction of **4-110** with the enamines **4-111** at room temperature gave the 2-hydroxybenzophenones **4-114** *via* **4-112** and **4-113** in usually good yield, especially using cyclic enamines with a ring size of five to seven (Scheme 4.24); in contrast, the reaction rate fell markedly with eight-membered enamines.

A combination of a Diels–Alder and a Fisher carbene-cyclopentannulation is described as the last example in this subgroup. Thus, Barluenga and coworkers used a [4+2] cycloaddition of 2-amino-1,3-butadienes **4-115** with a Fischer alkoxy-arylalkynylcarbene complex **4-116**; this is followed by a cyclopenta-annulation reaction with the aromatic ring in **4-116** to give **4-117** (Scheme 4.25) [36]. An extension of this domino process is the reaction of **4-118** with 2 equiv. of the alkynyl carbene **4-119** containing an additional C–C-double bond (Table 4.2) [37]. The final product **4-120**, which was obtained in high yield, is formed by a second [4+2] cycloaddition of the primarily obtained cyclopenta-annulated intermediate.

Quite recently, a domino Diels–Alder/Prins/pinacol reaction was reported by Barriault's group [38]. This novel method is very reliable and efficient for the synthesis of highly functionalized bicyclo[*m.n.*1]alkanones. In addition, Aubé and coworker [39] used a combination of a Diels–Alder and a Schmidt reaction within the total synthesis of the *Stemona* alkaloid stenine [40].

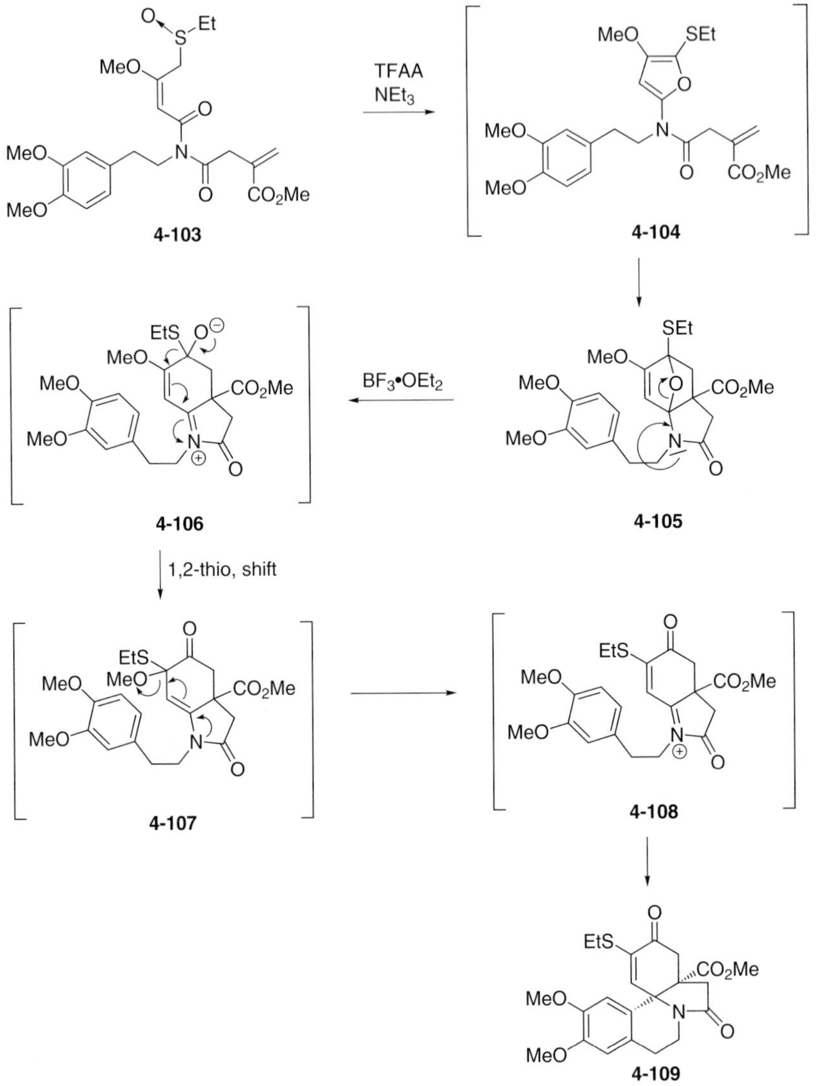

Scheme 4.23. Synthesis of erythryna alkaloids.

4.1.5
Hetero-Diels–Alder Reactions

As with all-carbon Diels–Alder reactions, the hetero-Diels–Alder reaction [41] can also be used as the first step in many combinations with other transformations. In contrast to the normal Diels–Alder reaction, several examples are known where the first step is followed by a 1,3-dipolar cycloaddition. This type of domino reaction has been especially investigated by Denmark and coworkers, and used for the synthesis of several complex natural products. Since Denmark has reviewed his studies in

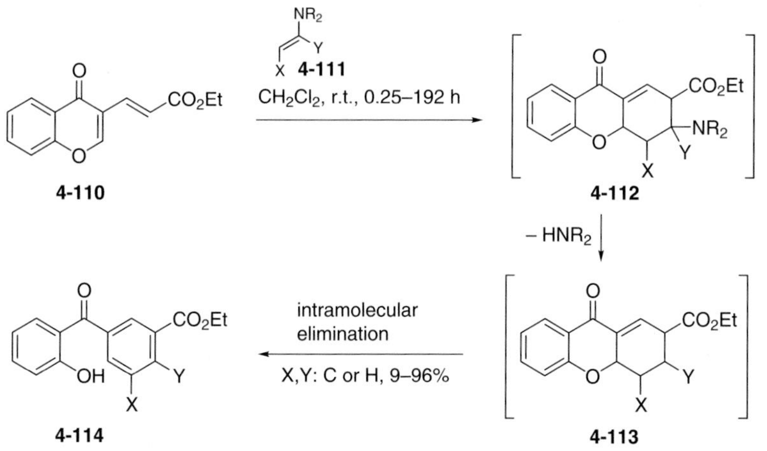

NR$_2$

X **4-111**

CH$_2$Cl$_2$, r.t., 0.25–192 h

CO$_2$Et

4-110

CO$_2$Et

NR$_2$

Y

X

4-112

– HNR$_2$

CO$_2$Et

OH

X

Y

4-114

intramolecular
elimination

X,Y: C or H, 9–96%

CO$_2$Et

Y

X

4-113

Scheme 4.24. Combination of a cycloaddition with an elimination.

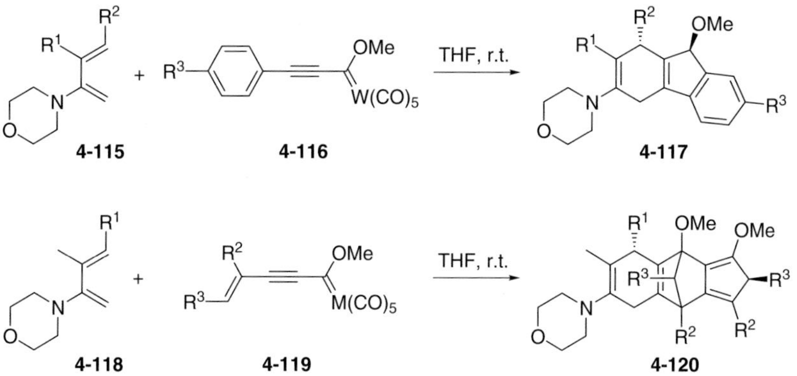

R^2

R^1

N

O

4-115

+

R^3

OMe

W(CO)$_5$

4-116

THF, r.t.

R^2 OMe

R^1

N

O

R^3

4-117

R^1

N

O

4-118

+

R^2

R^3

OMe

M(CO)$_5$

4-119

THF, r.t.

R^1 OMe OMe

R^3

R^3

R^2 R^2

N

O

4-120

Scheme 4.25. Combination of a Diels–Alder reaction with a carbene-cyclopentannulation.

Entry	Substrates **4-118** and **4-119**				Product **4-120**	
	R^1	R^2	R^3	M	Time	Yield [%]
1	H	H	Ph	W	20 min	95
2	H	H	Ph	Cr	30 min	95
3	CH$_2$OMe	H	Ph	W	25 min	–
4	H	CH$_2$(CH$_2$)$_2$CH$_2$		Cr	2 days	86
5	CH$_2$OMe	CH$_2$(CH$_2$)$_2$CH$_2$		Cr	6 days	85

Table 4.2. Results of the domino reaction of **4-118** and **4-119**.

detail, only a few newer examples will be discussed at this point [42]. For the synthesis of (+)-1-epiaustraline (**4-127**), these authors used an inter/intramolecular [4+2]/[3+2] domino reaction [40]. In the first step, the nitroalkene **4-121** undergoes a

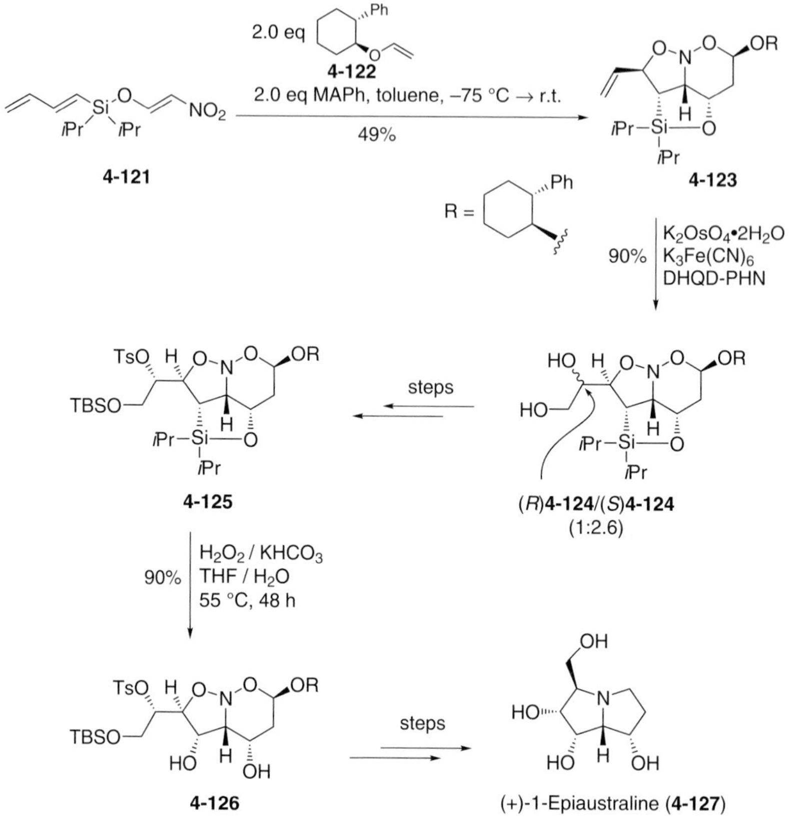

Scheme 4.26. Synthesis of (+)-1-epiaustraline (**4-127**).

cycloaddition with the chiral enol ether **4-122** to give a nitrone; this is intercepted by the vinylsilane moiety in **4-122** to furnish **4-123** as a single diastereomer (Scheme 4.26). In this process, four of the five stereogenic centers in **4-127** are already established. The final stereogenic center was installed by a diastereoselective dihydroxylation of **4-123** to give **4-124**, which was then further transformed to furnish (+)-1-epiaustraline (**4-127**) *via* **4-125** and **4-126**.

Examples for [4+2]/[3+2] domino processes, in which both cycloadditions proceed in an intramolecular mode, are the SnCl$_4$-catalyzed transformation of **4-128** and **4-131**, respectively to give **4-130** and **4-133** *via* the corresponding nitronates **4-129** and **4-132** (Scheme 4.27) [44].

This approach allows linear polyenes to be converted to functionalized polycyclic systems bearing up to six stereogenic centers. Another interesting use of the method deals with the synthesis of azapropellanes [45].

Another combination of hetero-Diels–Alder reactions and a [3+2] cycloaddition of a nitroalkene was described by Avalos and coworkers [46]. Using the chiral substrate **4-134** derived from a sugar, the domino process can be performed as a three-component transformation using an electron-rich dienophile and an electron-poor

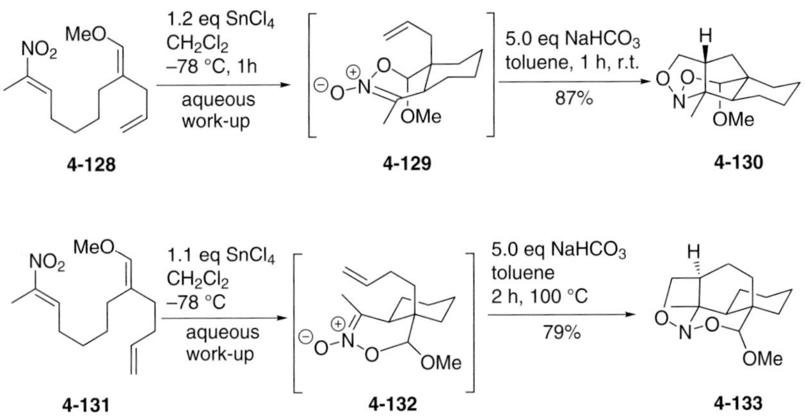

Scheme 4.27. Intramolecular domino hetero-Diels–Alder/[3+2] cycloaddition.

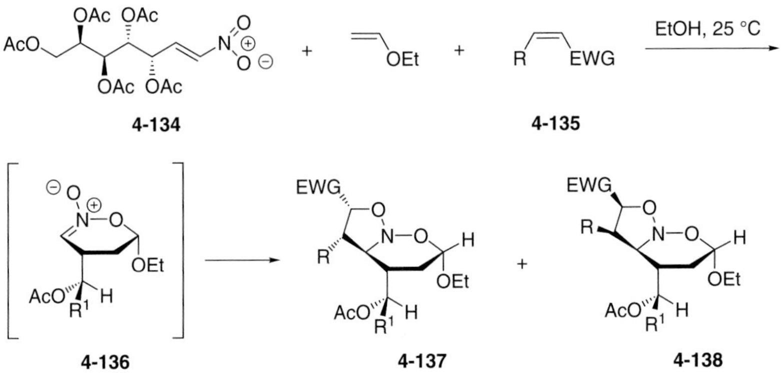

R^1 = D-lyxo-$(CHOAc)_3CH_2OAc$

Entry	EWG	Yield [%][a]	
		4-137	**4-138**
1	COMe	70	0[b]
2	CO$_2$Me	75	4
3	CN	60	10
4	CO$_2$Me	50	0[b]

[a] Isolated yields after crystallization, [b] Not isolated

Scheme 4.28. Reaction of **4-134** with ethyl vinyl ether and electron-deficient alkenes.

1,3-dipolarophile. Thus, reaction of a mixture of **4-134**, **4-135** and ethyl vinyl ether led to **4-137** as the major diastereomer, together with **4-138** *via* the primarily formed [4+2]-cycloadduct **4-136** (Scheme 4.28).

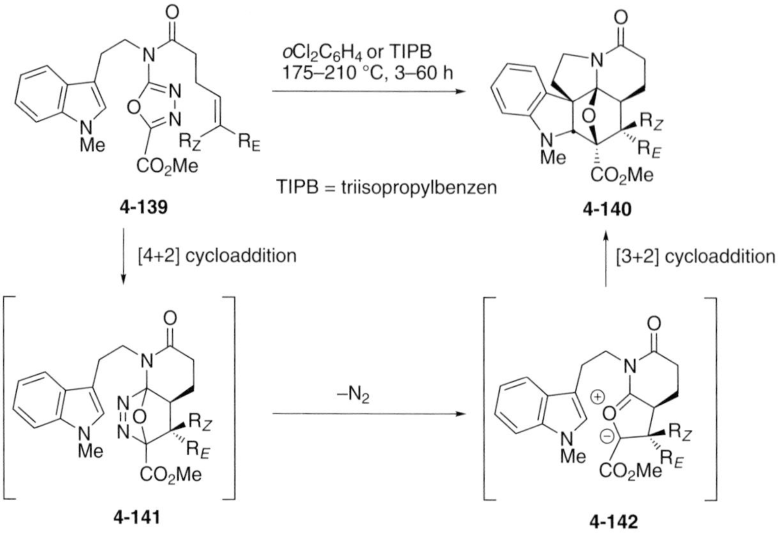

4-139 → **4-140**

oCl$_2$C$_6$H$_4$ or TIPB
175–210 °C, 3–60 h

TIPB = triisopropylbenzen

[4+2] cycloaddition

[3+2] cycloaddition

4-141   –N$_2$ **4-142**

Entry	4-139	R	Yield [%] of 4-140
1	a	R = H	87
2	b	R_E = Me	65
3	c	R_E = CH$_2$OTBS	86
4	d	R_E = Ph	61
5	e	R_E = OBn	88
6	f	R_Z = OBn	41
7	g	R_E = CO$_2$Me	71
8	h	R_Z = CO$_2$Me	62

Scheme 4.29. Synthesis of analogues of the indole alkaloid vindoline.

Having an efficient total synthesis of the indole alkaloid vindoline in mind, the Boger group [47] developed a facile entry to its core structure using a domino [4+2]/[3+2] cycloaddition. Reaction of the 1,3,4-oxadiazoles **4-139** led to **4-140** in high yield and excellent stereoselectivity *via* the intermediates **4-141** and **4-142** (Scheme 4.29).

As already described for the all-carbon-Diels–Alder reaction, a hetero-Diels–Alder reaction can also be followed by a retro-hetero-Diels–Alder reaction. This type of process, which has long been known, is especially useful for the synthesis of heterocyclic compounds. Sánchez and coworkers described the synthesis of 2-aminopyridines [48] and 2-glycosylaminopyridines **4-144** [49] by a hetero-Diels–Alder reaction of pyrimidines as **4-143** with dimethyl acetylenedicarboxylate followed by extrusion of methyl isocyanate to give the desired compounds (Scheme 4.30). This approach represents a new method for the synthesis of 2-aminopyridine nucleoside analogues. In addition to the pyridines **4-144**, small amounts of pyrimidine derivatives are formed by a Michael-type addition.

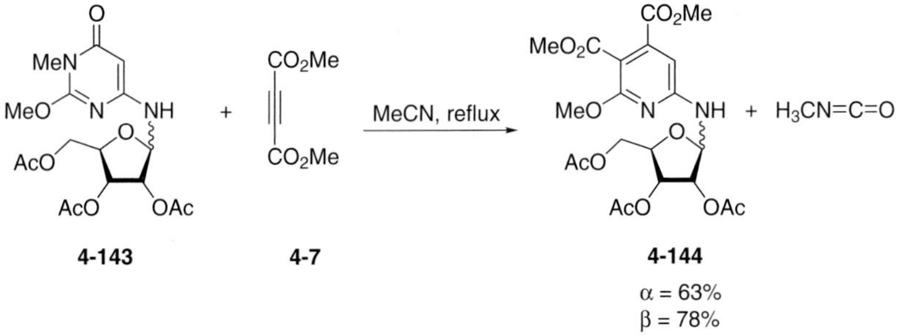

MeN

O

MeO N NH

AcO O

AcO OAc

4-143

+

CO₂Me

CO₂Me

4-7

MeCN, reflux

CO₂Me

MeO₂C

MeO N NH

AcO O

AcO OAc

4-144

α = 63%
β = 78%

+ H₃CN=C=O

Scheme 4.30. Synthesis of 2-aminopyridine nucleoside analogues.

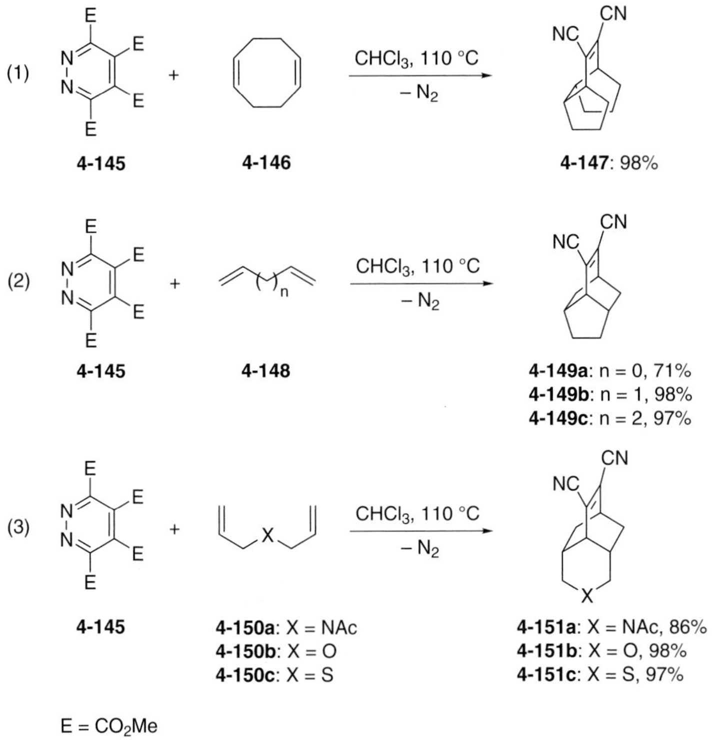

(1)

E
N
N E
E
E

4-145

+

4-146

CHCl₃, 110 °C
– N₂

NC CN

4-147: 98%

(2)

E
N
N E
E
E

4-145

+

4-148

CHCl₃, 110 °C
– N₂

NC CN

4-149a: n = 0, 71%
4-149b: n = 1, 98%
4-149c: n = 2, 97%

(3)

E
N
N E
E
E

4-145

+

X

4-150a: X = NAc
4-150b: X = O
4-150c: X = S

CHCl₃, 110 °C
– N₂

NC CN

X

4-151a: X = NAc, 86%
4-151b: X = O, 98%
4-151c: X = S, 97%

E = CO₂Me

Scheme 4.31. Synthesis of carbo- and hetero-cage systems.

4,5-Dicyanopyridazine **4-145** has been used for the synthesis of carbo- and hetero-cage systems employing nonconjugated dienes such as cyclooctadiene **4-146**, or **4-148** or **4-150** to give **4-147**, **4-149** and **4-151**, respectively (Scheme 4.31) [50]. In a similar way, dihydrofurans, dihydropyrans, pyrrolines, and enol ethers have also been used [51].

The Hall group [52] has developed a new three-component domino reaction of 1-aza-4-borono-1,3-butadiene **4-152**, a dienophile and an aldehyde to give α-hydroxy-methylpiperidine derivatives. In the first step, a hetero-Diels–Alder reaction takes place, which is followed by allylboration. As an example, reaction of **4-152** with the maleimide **4-153** in the presence of benzaldehyde furnished **4-154** in yields of up to 80% using the three substrates in a 1:2:1 ratio (Scheme 4.32).

In a similar way, Carreaux and coworkers [53] used 1-oxa-1,3-butadienes **4-155** carrying a boronic acid ester moiety as heterodienes [54], enol ethers and saturated as well as aromatic aldehydes. Thus, reaction of **4-155** and ethyl vinyl ether was carried out for 24 h in the presence of catalytic amounts of the Lewis acid $Yb(fod)_3$ (Scheme 4.33). Without work-up, the mixture was treated with an excess of an aldehyde **4-156** to give the desired α-hydroxyalkyl dihydropyran **4-157**. Although this is not a domino reaction, it is nonetheless a simple and useful one-pot procedure.

Vassilikogiannakis and coworkers described a simple sequential process for the biomimetic synthesis of litseaverticillol B (**4-159**) which includes a cycloaddition of **4-158** and singlet oxygen to give **4-160**, followed by ring opening to afford the hydrogenperoxide **4-161** (Scheme 4.34) [55]. Reduction of **4-161** led to the hemiacetal **4-162**, which underwent an aldol reaction to afford **4-159**.

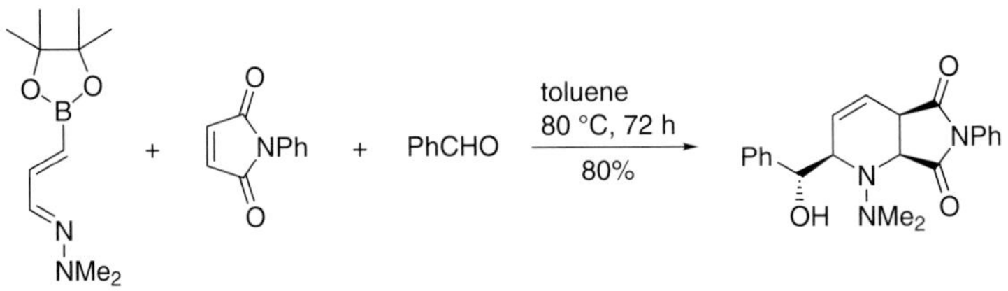

4-152 **4-153** **4-154**

Scheme 4.32. Combination of a hetero-Diels–Alder reaction with an allylation.

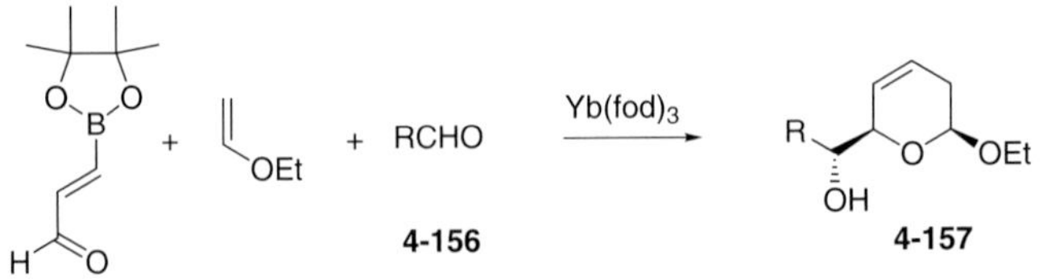

4-155 **4-156** **4-157**

9 examples 45–95% yield
(R = 4-$NO_2C_6H_4$: 95%,
R = C_6H_{13}: 45%)

Scheme 4.33. Synthesis of dihydropyrans **4-157**.

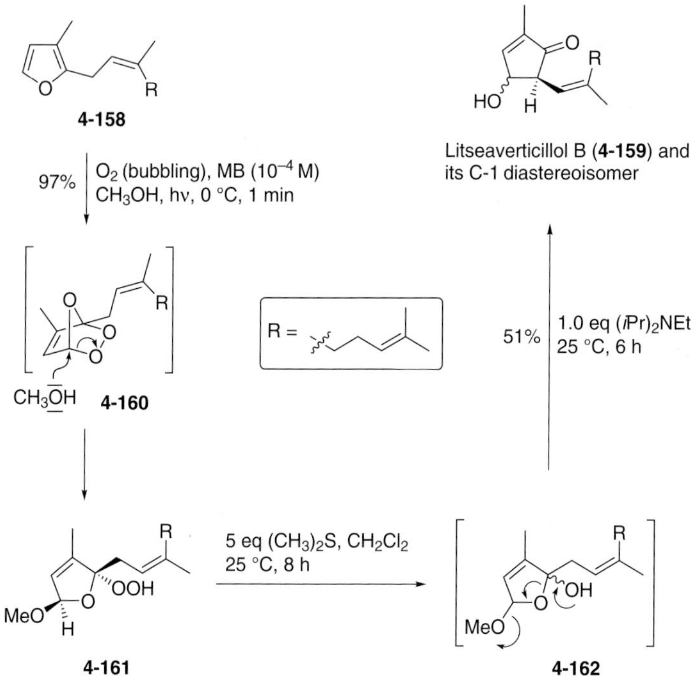

4-158

97% | O$_2$ (bubbling), MB (10^{-4} M)
CH$_3$OH, hv, 0 °C, 1 min

Litseaverticillol B (**4-159**) and
its C-1 diastereoisomer

4-160

R =

51% | 1.0 eq (*i*Pr)$_2$NEt
25 °C, 6 h

4-161

5 eq (CH$_3$)$_2$S, CH$_2$Cl$_2$
25 °C, 8 h

4-162

Scheme 4.34. Singlet oxygen-initiated transformation of furan **4-158** into litseaverticillol B (**4-159**).

4.2
1,3-Dipolar Cycloadditions

As with the Diels–Alder reaction, 1,3-dipolar cycloadditions can be used to start a domino process. In some examples, a second 1,3-dipolar cycloaddition, a rearrangement, a cleavage of the formed heterocycle, or an elimination may follow.

Giomi's group developed a domino process for the synthesis of spiro tricyclic nitroso acetals using α,β-unsaturated nitro compounds **4-163** and ethyl vinyl ether to give the nitrone **4-164**, which underwent a second 1,3-dipolar cycloaddition with the enol ether (Scheme 4.35) [56]. The diastereomeric cycloadducts formed, **4-165** and **4-166** can be isolated in high yield. However, if R is hydrogen, an elimination process follows to give the acetals **4-167** in 56% yield.

Knölker and coworkers also used a domino [3+2] cycloaddition for the clever formation of a bridged tetracyclic compound **4-172**, starting from a cyclopentanone **4-168** and containing two exocyclic double bonds in the α-positions (Scheme 4.36) [57]. The reaction of **4-168** with an excess of allylsilane **4-169** in the presence of the Lewis acid TiCl$_4$ led to the spiro compound **4-170** in a *syn* fashion. It follows a Wagner–Meerwein rearrangement to give a tertiary carbocation **4-171**, which acts as an electrophile in an electrophilic aromatic substitution process. The final step is the

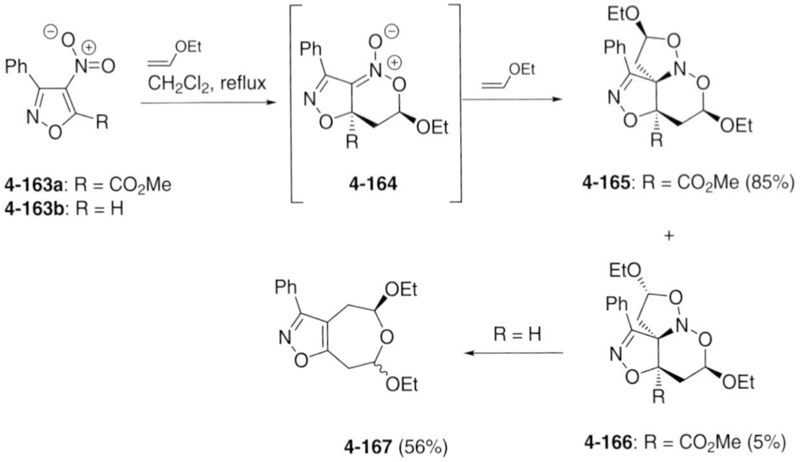

4-163a: R = CO$_2$Me
4-163b: R = H

4-164

4-165: R = CO$_2$Me (85%)

+

4-167 (56%)

R = H

4-166: R = CO$_2$Me (5%)

Scheme 4.35. Synthesis of spiro tricyclic compounds by a domino 1,3-dipolar cycloaddition.

4-168

11.0 eq ⌒SiPr$_3$
4-169

1.1 eq TiCl$_4$, CH$_2$Cl$_2$
reflux, 7 d

4-170

4-172

47%

4-171

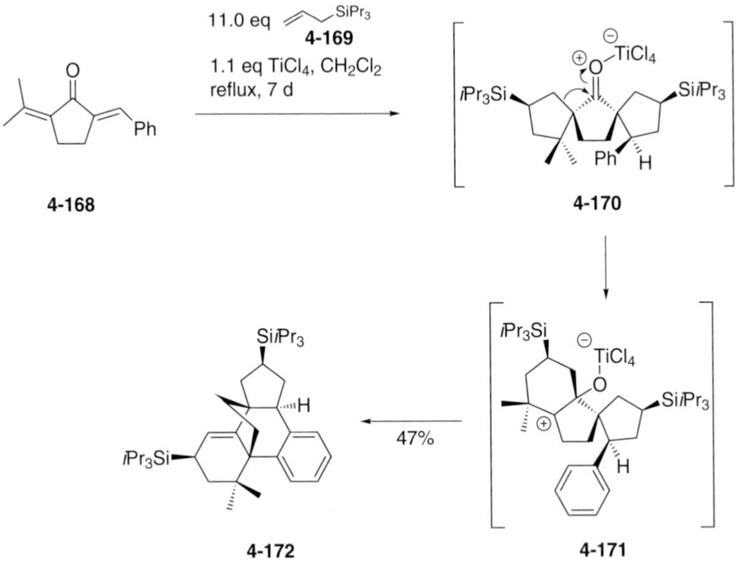

Scheme 4.36. Reaction of **4-168** with an allylsilane **4-169**.

elimination of the tertiary alcohol during hydrolytic work-up to give **4-172**, in 47 % overall yield.

Another domino process which is initiated by a 1,3-dipolar cycloaddition, and which allows the synthesis of either β-lactams or thiiranes, was recently reinvesti-

gated by Babiono's group [58]. For the synthesis of β-lactams **4-176**, a 1,3-oxa-zolium-5-olate (münchnone) **4-173** is treated with an imine to give a cycloadduct **4-174**, which then undergoes two rearrangements with **4-175** as a proposed intermediate (Scheme 4.37) [59].

Fu and coworkers described a copper/phospha-ferrocene-oxazoline-catalyzed procedure for the enantioselective coupling of alkynes with nitrones also to give β-lactams with excellent *ee*-values (up to 93%) [60]. The reaction can be performed in an intermolecular mode, but is even more interesting in an intramolecular mode and also tolerates many functional groups. An example, which in addition includes an α-alkylation after cyclization in the presence of the chiral ligand, is the reaction of the heterocyclic nitrone **4-177**. This substrate, in the presence of allyl iodide (**4-178**) and the chiral ligand **4-179**, gave the β-lactone **4-180** in 70% yield and 90% *ee* (Scheme 4.38). The first example of this type of transformation using stoichiometric amounts of a copper acetylide with a nitrone was described by Kinugasa [61].

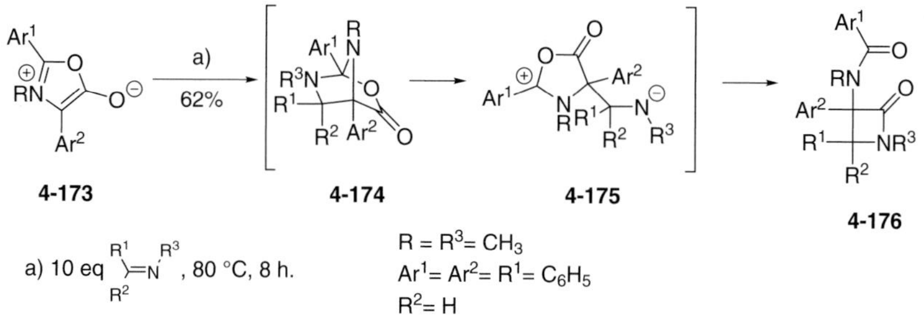

a) 10 eq $\overset{R^1}{\underset{R^2}{>}}=N{\cdot}R^3$, 80 °C, 8 h.

$R = R^3 = CH_3$
$Ar^1 = Ar^2 = R^1 = C_6H_5$
$R^2 = H$

Scheme 4.37. Synthesis of β-lactams.

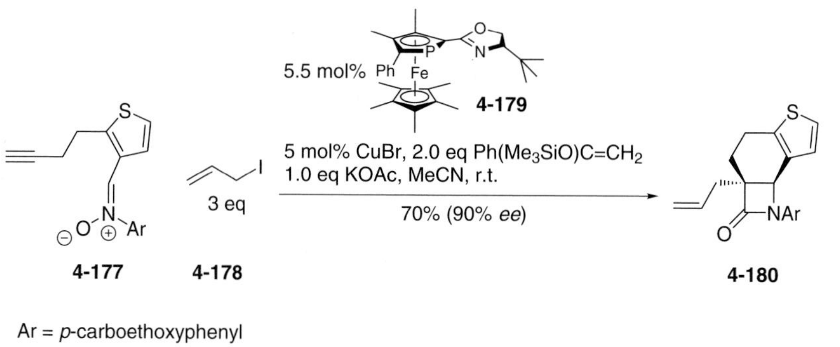

Ar = *p*-carboethoxyphenyl

Scheme 4.38. Enantioselective formation of β-lactams from nitrones and alkynes.

Pagenkopf's group developed a novel domino process for the synthesis of pyrroles **4-183**, which allows for the control over the installation of substituents at three positions and seems to be very suitable for combinatorial chemistry [62]. The process consists of a 1,3-dipolar cycloaddition of an intermediate 1,3-dipole formed from the cyclopropane derivative **4-181** with a nitrile to give **4-182** followed by dehydration and isomerization (Scheme 4.39). The yield ranges from 25 to 93 %, and the procedure also works well with condensed cyclopropanes.

Reaction of the nitrone **4-184** with allenic esters **4-185** as described by Ishar and coworkers led to the benzo[*b*]indolizines **4-186**, together with small quantities of **4-187** (<5 %) (Scheme 4.40) [63]. The first transformation is a 1,3-dipolar cycloaddition; this is followed by four further steps, including a [4+2] cycloaddition of an intermediate 1-aza-1,3-butadiene.

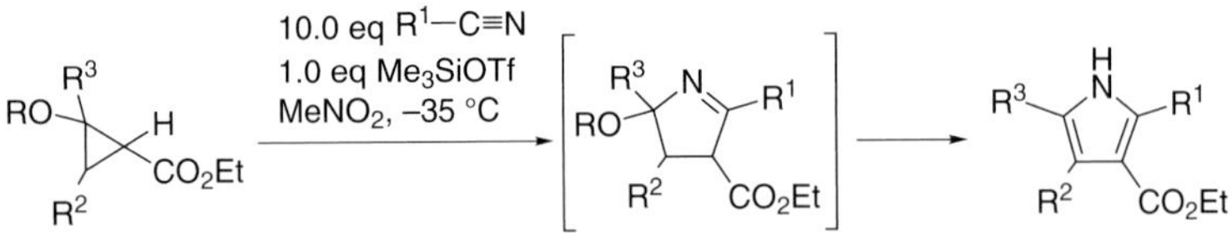

Selected examples: **4-183a**: R¹= Me, R²= R³= H (80%)
4-183b: R¹= CH=CH₂Ph, R²= R³= H (39%)

Scheme 4.39. Synthesis of pyrroles.

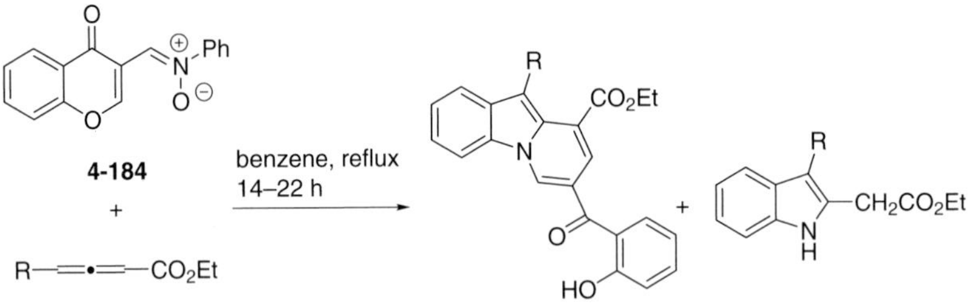

4-186a: R = H (35%)
4-186b: R = Me (55%)
4-186c: R = Et (60%)

4-187 (minor)

Scheme 4.40. Synthesis of benzo[*b*]indolizines.

4.3
[2+2] and Higher Cycloadditions

In this section, those domino reactions are described which start with a cycloaddition not of the [4+2] and [3+2] types. The variety referred to is quite high; consequently, in addition to the well known [2+2] scheme, [4+3] cycloaddition and the more exotic [5+2] cycloaddition are also outlined.

One of the best methods to synthesize cyclopentenone derivatives is the Pauson–Khand procedure. However, Shindo's group have recently developed a domino process consisting of a [2+2] cycloaddition of a ketone with an ynolate, followed by a Dieckmann condensation to give a β-lactone as **4-190** which is decarboxylated under reflux in toluene in the presence of silica gel to afford cyclopentenones [64a]. Thus, the reaction of **4-188** and **4-189** led to **4-190**, which on heating furnished the linear cucumin **4-191** (Scheme 4.41). This natural product has been isolated from the mycelial cultures of the agaric *Macrocystidia cucumis* [65, 66]. The domino procedure described was also used to synthesize dihydrojasmone and α-cuparenone. Moreover, the [2+2] cycloaddition can be combined with a Michael reaction [64b].

3,6-Dihydro-2*H*-pyran-2-ones (e. g., **4-195**) are valuable intermediates in the synthesis of several natural products [67]. Hattori, Miyano and coworkers [68] have recently shown that these compounds can be easily obtained in high yield by a Pd^{2+}-catalyzed [2+2] cycloaddition of α,β-unsaturated aldehydes **4-192** with ketene **4-193**, followed by an allylic rearrangement of the intermediate **4-194** (Scheme 4.42). In this reaction the Pd^{2+}-compound acts as a mild Lewis acid. α,β-unsaturated ketones can also be used, but the yields are below 20%.

An unusual domino process was observed by Biehl and coworkers [69] in the reaction of 2-bromo-1-naphthol **4-196** with arylacetonitriles in the presence of LDA or LiTMP; by employing 3-thienylacetonitrile **4-197**, the tetracyclic compound **4-200** was obtained in 57% yield (Scheme 4.43). The reaction probably includes the formation of an aryne and a ketenimine which undergo [2+2] cycloaddition to give **4-198**, followed by rearrangement and allylic addition to the intermediately formed aryl cyano compound **4-199**.

Kanematsu's group used a combination of an intramolecular [2+2] cycloaddition of an allenyl ether **4-202** followed by a [3+3] sigmatropic rearrangement (Scheme 4.44) [70]. The substrate for the domino reaction can be obtained *in situ* by treat-

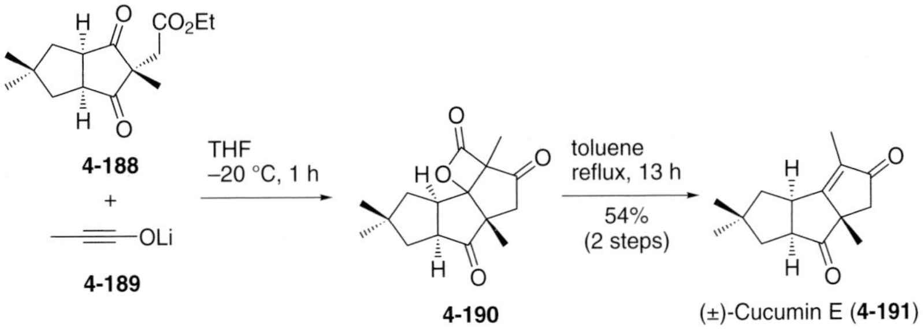

Scheme 4.41. Synthesis of (±)-cucumin E (**4-191**).

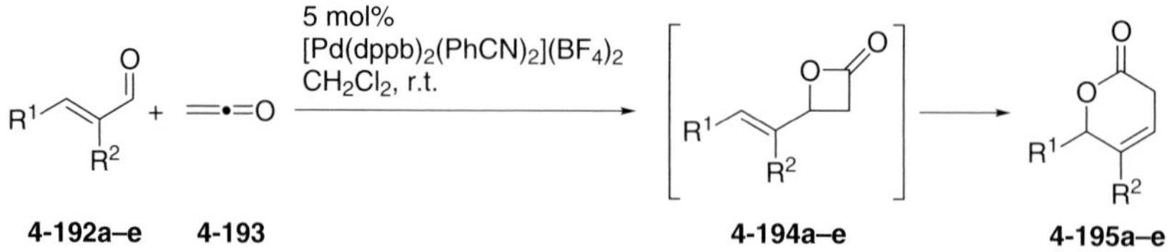

4-192	R¹	R²	Yield [%] of 4-195
a	Me	H	81
b	H	H	0[a]
c	Ph	H	77
d	pentyl	H	58
e	Me	Et	66

[a] 96% of **4-194b**.

Scheme 4.42. Synthesis of dihydropyranones.

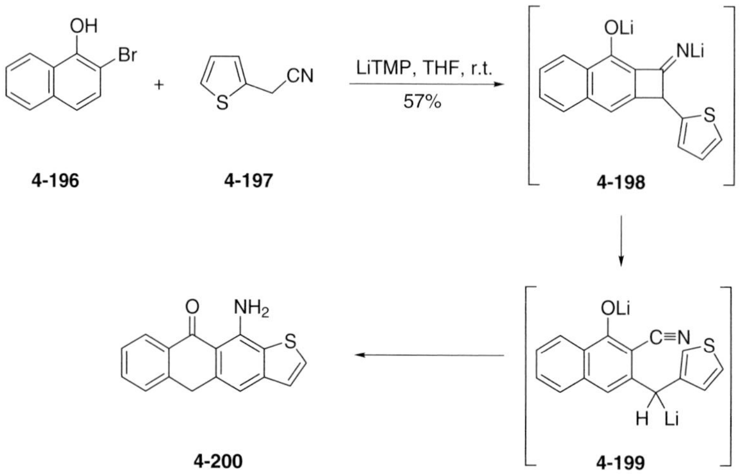

Scheme 4.43. Unusual domino process of **4-196** and **4-197**.

ment of the propargyl ether **4-201** with *t*BuOK to give the enantiopure oxataxane derivative **4-203** in quantitative yield *via* **4-202**.

Quinoline derivatives are of major interest as bioactive compounds. A new method for the synthesis of polycyclic fused quinolines **4-207** was developed by Rossi and coworkers [71], who used the [2+2] cycloaddition of an alkyne **4-204** with enamines **4-205** to give **4-206** (Scheme 4.45). Thereby follows an annulation, with the formation of **4-207a–c**.

West and coworkers developed two new domino processes in which a [4+3] cycloaddition (Nazarov electrocyclization) of 1,4-dien-3-ones is succeeded by either an

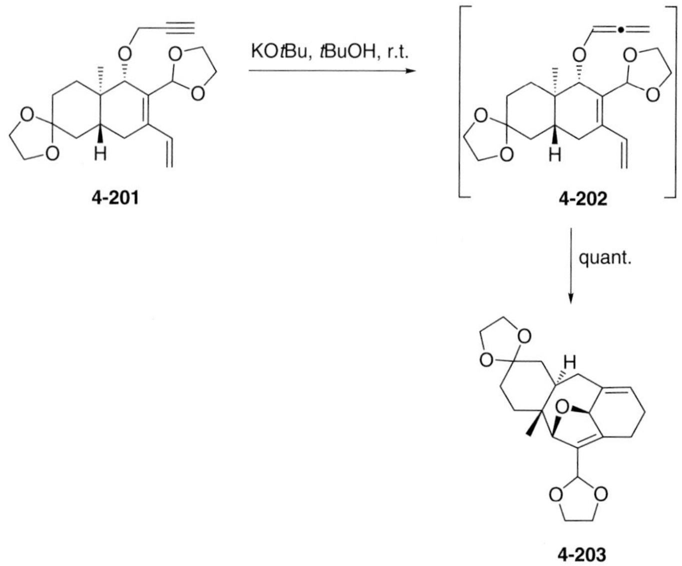

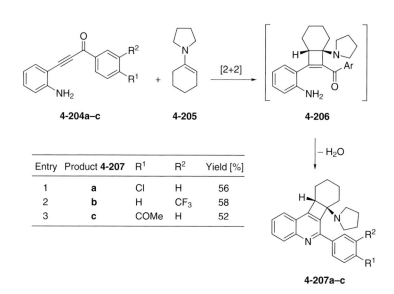

Scheme 4.44. Synthesis of an oxataxane skeleton.

Entry	Product **4-207**	R¹	R²	Yield [%]
1	**a**	Cl	H	56
2	**b**	H	CF₃	58
3	**c**	COMe	H	52

Scheme 4.45. Synthesis of polycyclic fused quinolines.

addition of allylsilanes [69] or a [2+3] cycloaddition [73]. The Lewis acid-catalyzed re-action of **4-208** and allylsilane **4-209** first led stereoselectivity to the cyclopentenyl carbocation **4-210**, which is attacked by the allylsilane present in the reaction mix-ture to give the two diastereomeric carbocations **4-211a** and **4-211b**, probably as a 2:1 mixture (Scheme 4.46). Further reaction of the two cations involves either a nu-

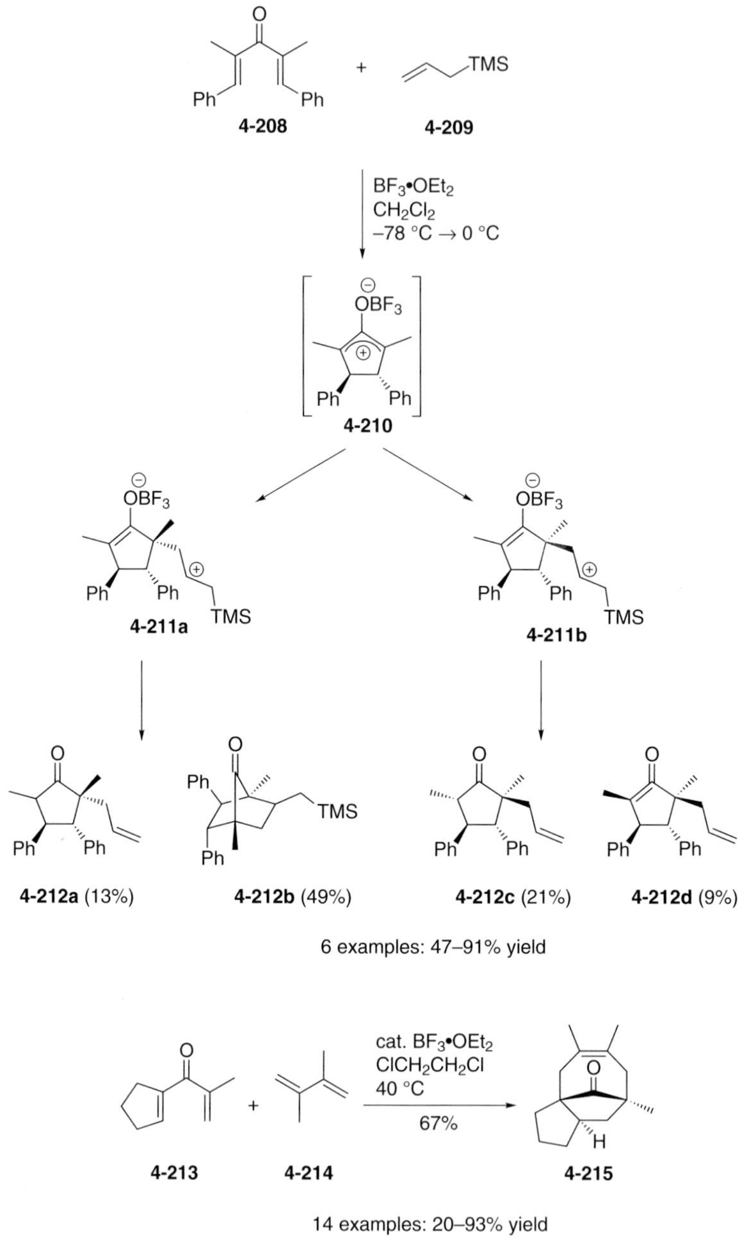

Scheme 4.46. Domino [4+3] cycloaddition.

cleophilic attack at the silicon atom to give **4-212a**, **4-212c** and **4-212d**, respectively, or addition of the enolate functionality present in **4-211** to the cationic center. The latter reaction occurs only in the diastereomer **4-211a** to give **4-212b**. In a similar fashion, treatment of a mixture of the 1,4-dienone **4-213** and the 1,3-butadiene **4-214** with BF$_3$·Et$_2$O yielded the bridged compound **4-215** as a single isomer.

A remarkable domino process which leads to products of high complexity with good stereoselectivity was observed by Engler and coworkers by treating a mixture of **4-216** and 2.2 equiv. of **4-217** in the presence of BF$_3$·Et$_2$O at −78 °C to −20 °C (Scheme 4.47) [74].

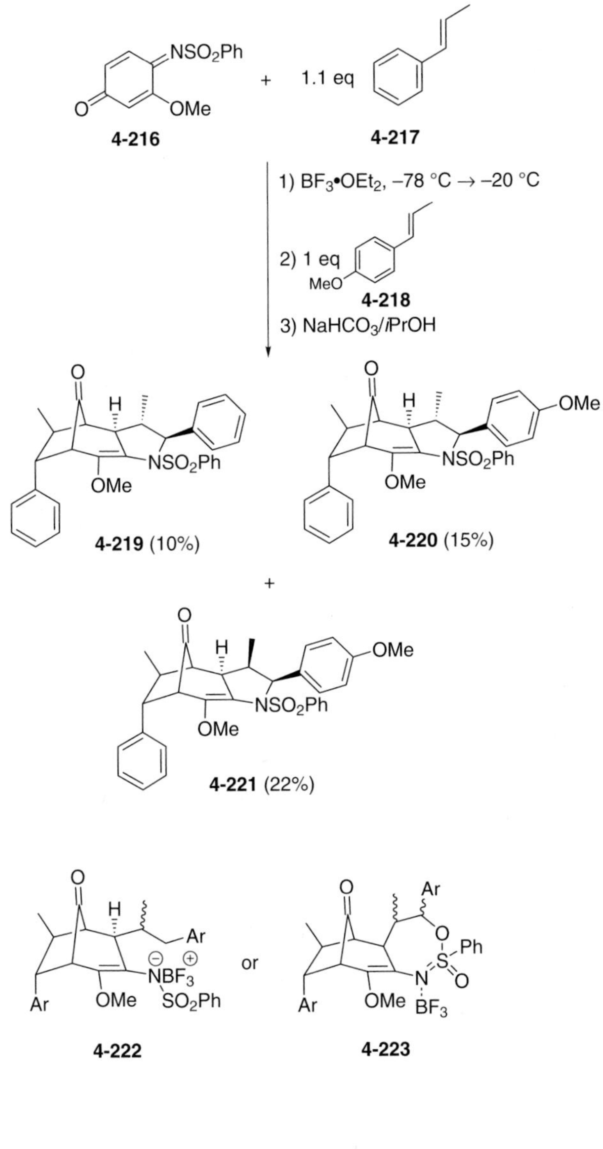

Scheme 4.47. Three-component reaction of **4-216**, **4-217** and **4-218**.

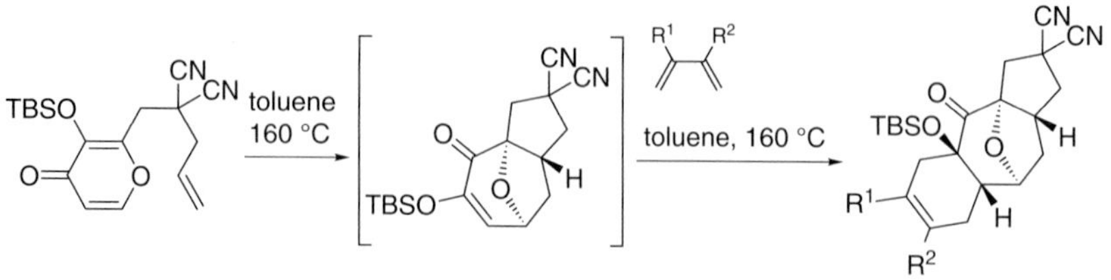

4-224   **4-225**   **4-226a**: R^1= R^2= Me, 81%
                                      4-226b: R^1= H, R^2= OTBS, 83%

Scheme 4.48. Synthesis of tetracyclic compounds from pyrone **4-224**.

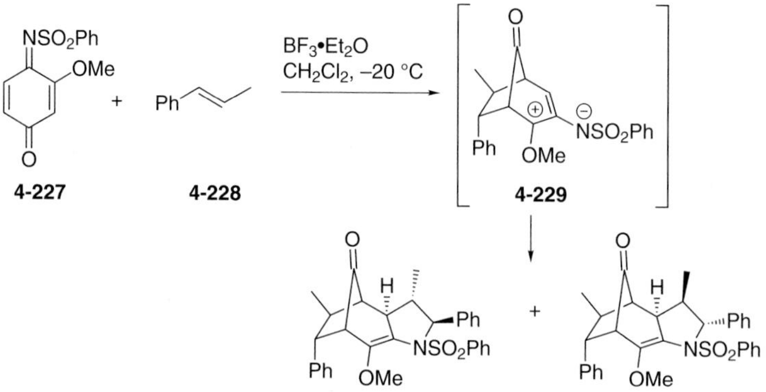

4-227   **4-228**   **4-229**

4-230 (58%)   **4-231** (16%)

Scheme 4.49. Domino [5+3]/[3+2] cycloaddition.

As main product, **4-219** (56 %) was obtained together with 9 % of a stereoisomer. The reaction can also be performed as a three-component reaction using **4-216**, **4-217** and **4-218**, which gave a mixture of **4-219**, **4-220** and **4-221**. The latter transformation indicates that intermediates such as **4-222** or **4-223** might play a role in this domino process.

A interesting combination of a [5+2] and a [4+2] cycloaddition was used by Mascareñas and coworkers [75] for the synthesis of 6,7,5-tricarbocyclic compounds bearing a 1,4-oxa-bridge. In this process, the entropic acceleration of an intramolecular process compared to an intermolecular cycloaddition was used to allow differentiation. Simple heating of a 1:5 mixture of the pyrone **4-224** and 2,3-dimethylbutadiene at 160 °C in toluene in a sealed tube for 30 h led to **4-226a** in 81 % yield (Scheme 4.48); in this reaction, **4-225** could be identified as intermediate. Similarly, reaction of **4-224** and 2-TBSO-1,3-butadiene gave **4-226b** in 83 % yield. The process allows the straightforward formation of four carbon–carbon bonds, and creates three new cycles and five stereogenic centers.

Engler and coworkers [76] developed a new domino process which consists of a [5+3] cycloaddition of a *p*-quinone monoimide with a styrene derivative followed by a [3+2] or [3+3] cycloaddition. The reaction allows the formation of two additional rings and up to eight stereogenic centers, with high selectivity. The best results, with 58 % yield of **4-230**, were obtained in the transformation of **4-227** and **4-228** in the presence of $BF_3 \cdot Et_2O$ at –20 °C (Scheme 4.49). In addition, the diastereomer **4-231** was obtained in 16 % yield. It can be assumed that the cation **4-229** functions as an intermediate. The process also functions with quinones, though much less efficiently.

4.4
Sigmatropic Rearrangements

Domino reactions starting with a sigmatropic rearrangement have long been known. Most of these processes use a Cope or Claisen rearrangement [77], followed by another Cope or Claisen rearrangement.

This sequence was successfully employed by Thomas [78] for the synthesis of sesquiterpenes such as β-sinesal (**4-232**), and by Raucher and coworkers [79] for the synthesis of the germanocrolid (+)-dihydrocostunolide (**4-233**) (Scheme 4.50).

Jacobi and coworkers used an oxy-Cope/Diels–Alder sequence to synthesize the tumor inhibitor gnididione (**4-234**) [80]. A similar sequence was also used by Kraus and coworkers for the synthesis of 11-deoxydaunomycinone (**4-235**) [81].

A domino Claisen/ene strategy was employed for the synthesis of (+)-9(11)-dehydroestrone [82] methyl ether, while an example of a domino aza-Cope rearrangement/Mannich reaction is the above-mentioned synthesis of (–)-preussin (**4-14**) [5].

Clearly, during the past few years the development of domino reactions with a sigmatropic rearrangement as the first step has increased steadily.

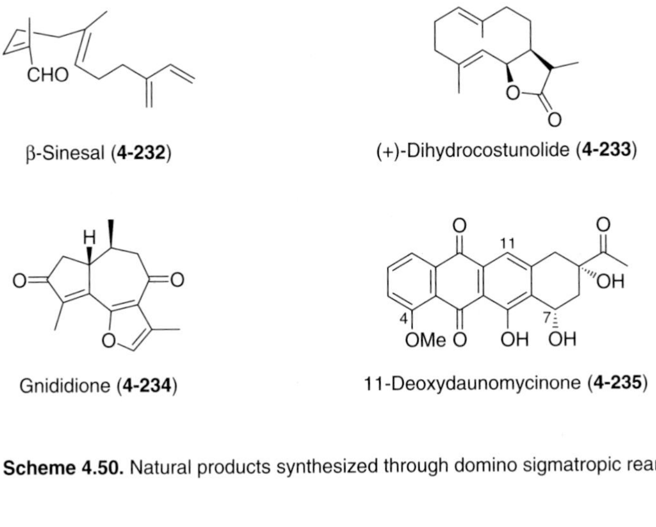

β-Sinesal (**4-232**)

(+)-Dihydrocostunolide (**4-233**)

Gnididione (**4-234**)

11-Deoxydaunomycinone (**4-235**)

Scheme 4.50. Natural products synthesized through domino sigmatropic rearrangements.

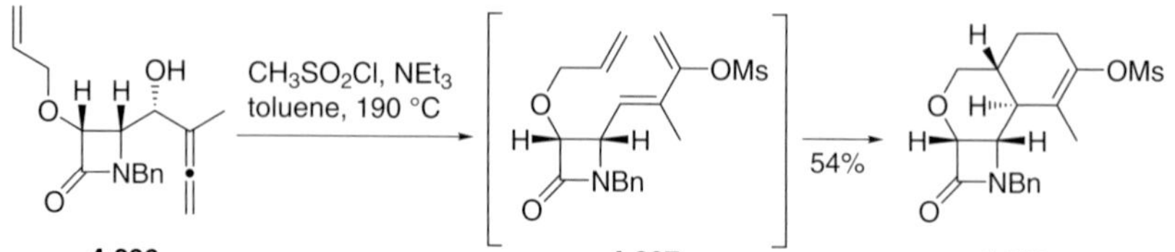

Scheme 4.51. Synthesis of the cycloadduct **4-238**.

Alcaide, Almendros and coworkers developed a combination of a 3,3-sigmatropic rearrangement of the methanesulfonate of an α-allenic alcohol to give a 1,3-butadiene which is intercepted by a dienophile present in the molecule to undergo an intramolecular Diels–Alder reaction [83]. Thus, on treatment of **4-236** with CH_3SO_2Cl, the methanesulfonate was first formed as intermediate, and at higher temperature this underwent a transposition to give **4-237** (Scheme 4.51). This then led directly to the cycloadduct **4-238** *via* an *exo* transition state.

Several natural products such as forbesione (**4-242a**), which is isolated from the genus *Garcinia* of the *Guttiferane* family of plants, contain a 4-oxatricyclo[4.3.1.0]decan-2-one ring system. It has been proposed that it is formed in Nature from a prenylated xanthone by a Claisen rearrangement, followed by a Diels–Alder reaction [84]. Accordingly, Nicolaou and coworkers heated the substrate **4-339** for 20 min at 120 °C in dimethylformamide (DMF) and obtained two major products – the desired 1-*O*-methylforbesione **4-242b** in 63 % yield, and compound **4-243** in 26 % yield (Scheme 4.52) [85]. It can be proposed that **4-242** is formed *via* pathway A and **4-243** *via* pathway B, with **4-240** and **4-241**, respectively, as intermediates.

Using a similar approach, Theodorakis and coworkers aimed at synthesizing lateriflorone (**4-244**) from **4-246** [86]. The synthesis was unsuccessful, but the domino Claisen/Diels–Alder reaction allowed an efficient access to **4-245**, which is a proposed precursor in the biosynthesis of **4-244**. With **4-246** in hand, its treatment in toluene at 110 °C gave **4-248** *via* **4-247** as a single compound. Interestingly, **4-248** was also obtained from **4-246** in excellent yield, simply upon standing at room temperature for several days (Scheme 4.53). Further transformations of **4-248** led to **4-249** and, furthermore, to **4-250**, which was then converted into *seco*-lateriflorone (**4-245**). The process could be even prolonged by another Claisen rearrangement, which allowed the total synthesis of racemic forbesione acetate (**4-252**) starting from **4-251**; **4-252** was then converted into forbesione (**4-242a**) [87]. Studies on the timing of this domino reaction suggest that the initial step of the reaction is the C-ring Claisen rearrangement [88].

Baldwin and coworkers described an interesting and high-yielding pericyclic domino process, consisting of a Cope and a Diels–Alder reaction, which on thermal treatment of the tetraene **4-253** led to tricyclic compound **4-254** (Scheme 4.54) [89].

An unpredicted rearrangement was found by Mal and coworkers when heating the anthraquinone **4-255** in toluene at reflux [90]. These authors obtained the rear-

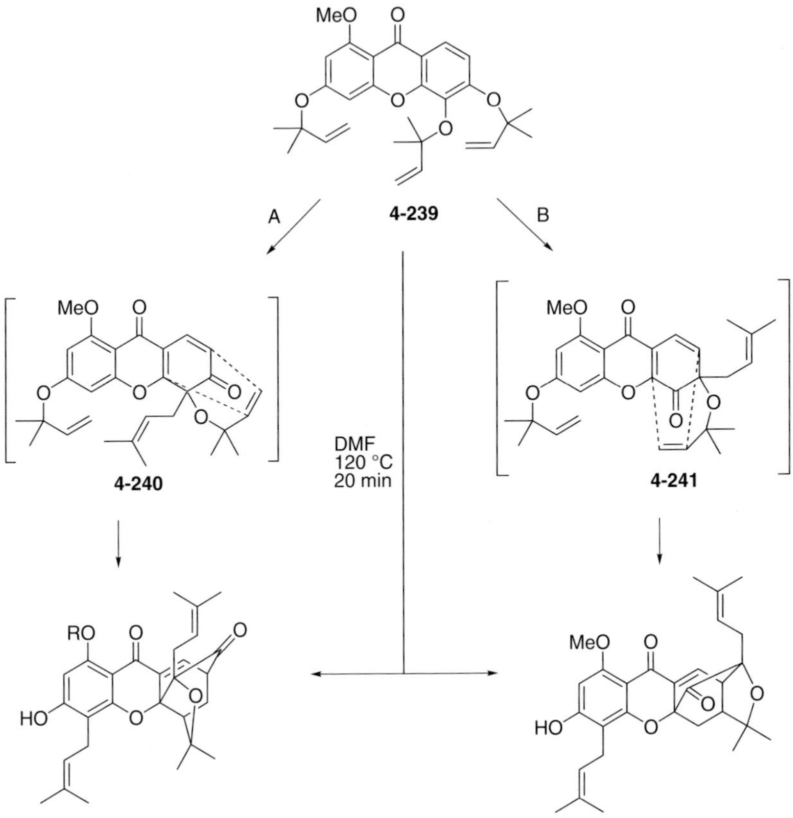

4-240

4-239

4-241

DMF
120 °C
20 min

4-242a: R = H: Forbesione
4-242b: R = Me (63%): 1-*O*-Methylforbesione

4-243b (26%)

Scheme 4.52. Synthesis of 1-*O*-methylforbesione (**4-242b**).

ranged anthraquinone **4-256** in high yield, which might be formed from **4-255** by a Cope rearrangement to give **4-257** followed by a cheletropic CO elimination and an aromatization *via* a 1,5-sigmatropic hydrogen shift (Scheme 4.55).

Benzoxazoles are of interest due to their optical applications. Hiratani and co-workers prepared novel bis(benzoxazole) derivatives by two consecutive Claisen rearrangements of **4-258** [91]. Heating **4-285a** to 180 °C without solvent gave a mixture of **4-259a** to **4–261a** in 71%, 13%, and 5% yields, respectively (Scheme 4.56). In a similar reaction, **4-258b** led to 15% of **4-259b** and 82% of **4-260b** with a trace of **4-261b**. **4-259b** was seen to emit blue fluorescence.

Barriault developed a new pericyclic domino process for the synthesis of the bioactive diterpenoid vinigrol (**4-262**), which was isolated from *Virgaria nigra* [92]. The natural product possesses antihypertensive and anti-platelet-aggregating properties. **4-262** contains a unique tricyclo [4.4.4.0^{4a,8a}]tetradecane framework, which could be obtained by a combination of an oxy-Cope, a Claisen and an ene reaction

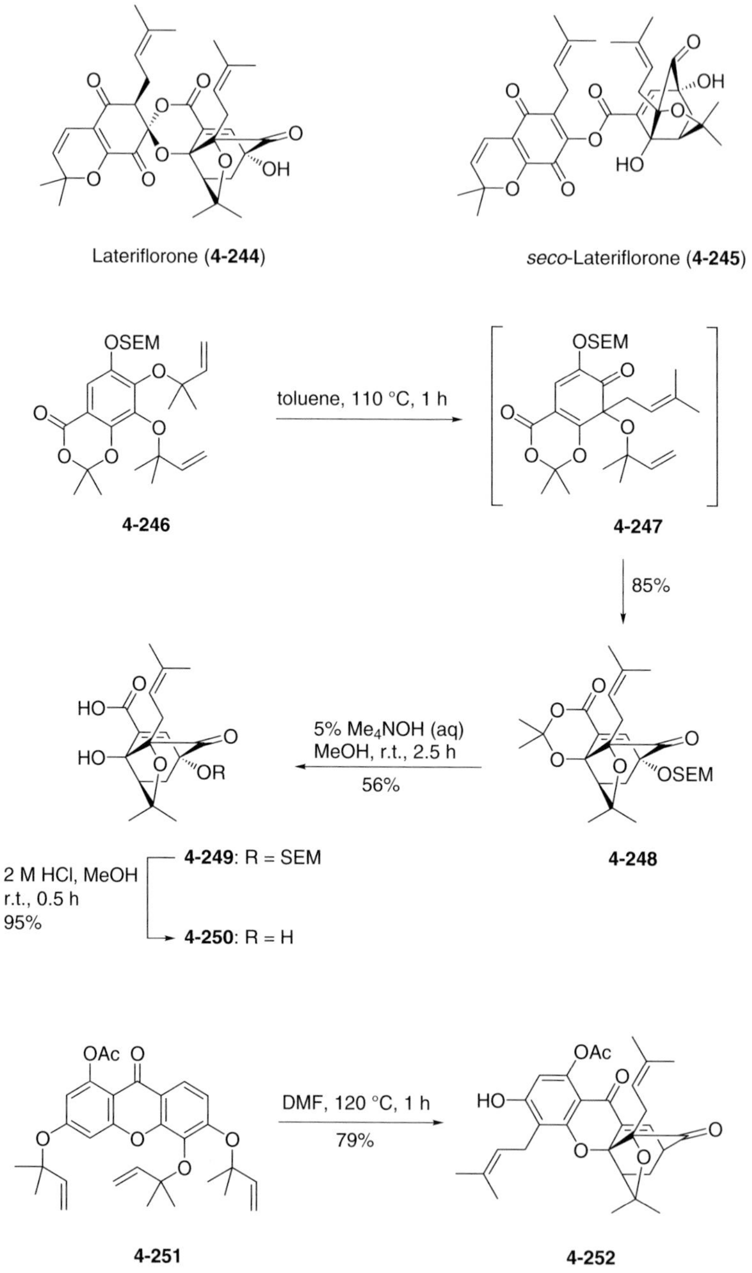

Lateriflorone (**4-244**)

seco-Lateriflorone (**4-245**)

Scheme 4.53. Synthesis of *seco*-lateriflorone and 1-acetylforbesione.

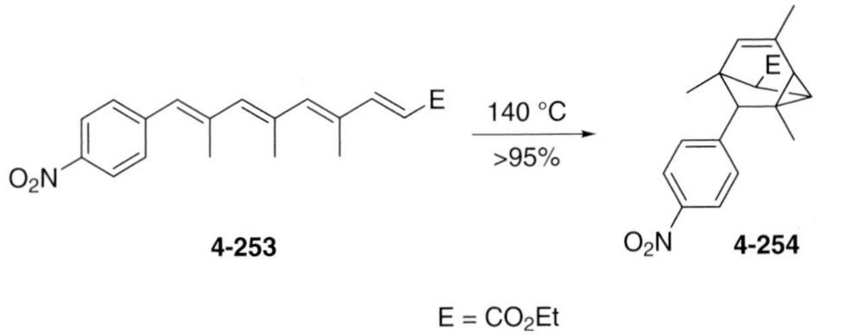

4-253

140 °C
>95%

O₂N   **4-254**

E = CO₂Et

Scheme 4.54. Domino Cope/Diels–Alder reaction of **4-253**.

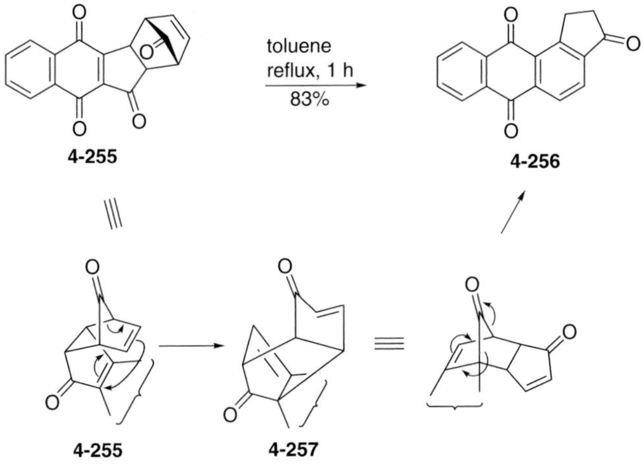

4-255

toluene
reflux, 1 h
83%

4-256

4-255   **4-257**

Scheme 4.55. Rearrangement in the synthesis of anthraquinone **4-256**.

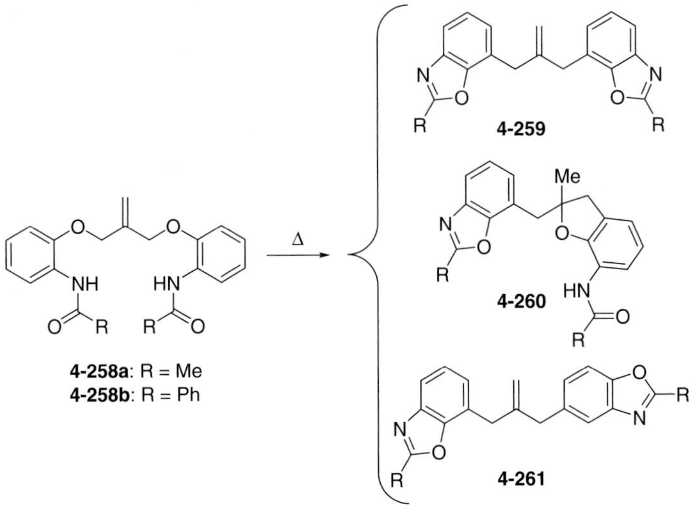

4-259

Me

4-260

Δ

4-258a: R = Me
4-258b: R = Ph

4-261

Scheme 4.56. Synthesis of novel benzoxazoles.

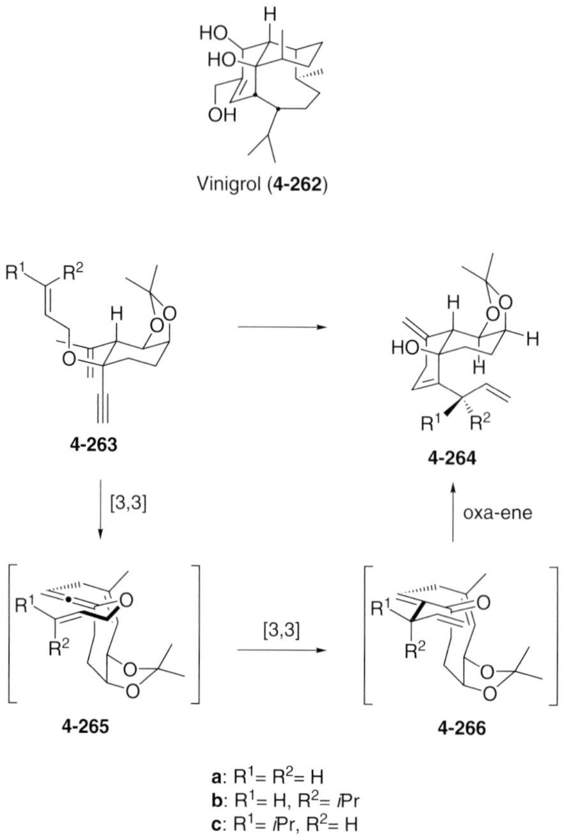

Vinigrol (**4-262**)

4-263

[3,3]

4-264

oxa-ene

4-265

[3,3]

4-266

a: R^1= R^2= H
b: R^1= H, R^2= *i*Pr
c: R^1= *i*Pr, R^2= H

Scheme 4.57. Investigations towards the synthesis of vinigrol (**4-262**).

starting from **4-263** using NaH and microwave irradiation (Scheme 4.57). In this domino process, **4-265** and **4-266** are the proposed intermediates. Whereas **4-263a** gave **4-264a** in 86% yield, neither **4-263a** nor **4-263c** containing the necessary *i*Pr-group, as in **4-262**, gave the desired products **4-264b** or **4-264c**.

Barriault and coworkers also examined the synthesis of wiedemannic acid (**4-267**), using a similar approach. This natural product is a diterpene of the abietane group (**4-268**), which has been isolated from *Salvia wiedemannii*, a native plant of central Turkey. A major challenge in the synthesis was the efficient formation of the five contiguous stereogenic centers C-4, C-5, C-8, C-9, and C-10, of which three carbons are quaternary centers [93]. The goal was accomplished by the combination of an oxy-Cope, a Claisen and an ene reaction, starting from **4-269** under microwave irradiation to give **4-270** in 90% yield with an excellent diastereoselectivity (>25:1) (Scheme 4.58). It can be assumed that **4-271** and **4-272** are intermediates in this domino process. **4-270** could be transformed into the desired **4-267** in 13 steps, in 23% yield.

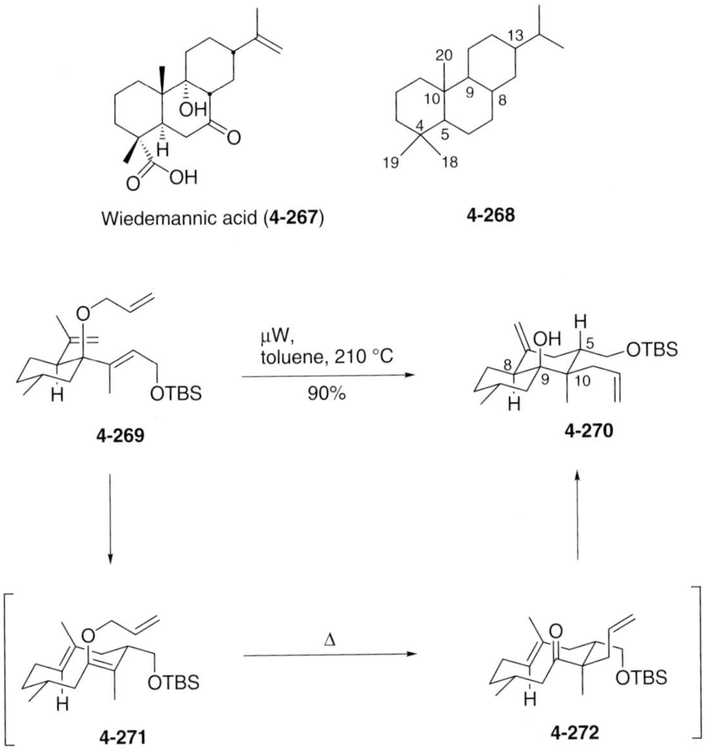

Wiedemannic acid (**4-267**)　　　　　**4-268**

Scheme 4.58. Total synthesis of wiedemannic acid (**4-267**).

In another approach aimed at the synthesis of the sesquiterpene (+)-arteannuin M **4-274**, the same group used a domino oxy-Cope/ene process [94]. **4-274** had been isolated by Brown and coworkers [95] in 1998, from *Artemisia annua* L., a plant found in the mountains of Sichuan province in Southern China. The absolute configuration of **4-274** was not known at that time, and neither was the relative configuration at C-4. As substrate for the domino process the 1,5-diene **4-276** was used, having been prepared from enantiopure **4-273** by reaction with the Li-compound **4-275**. Heating of **4-276** in the presence of DBU gave **4-280**, which was further transformed into (+)-arteannuin M (*ent*-**4-274**), thus clarifying the structure of the natural product as the (–)-enantiomer (Scheme 4.59). The domino Cope/ene reaction has some interesting stereochemical features. The relative configuration in **4-280** is determined by the equatorially oriented side chain in the transition structure **4-279**; the other possible diastereomer was not observed. Also of worthwhile note is the chirality transfer, since the intermediate **4-277** does not contain a stereogenic center. However, the compound is chiral due to a planar chirality, and racemization of **4-277** is slow but not zero; thus, **4-280** is obtained only with 78 % *ee*.

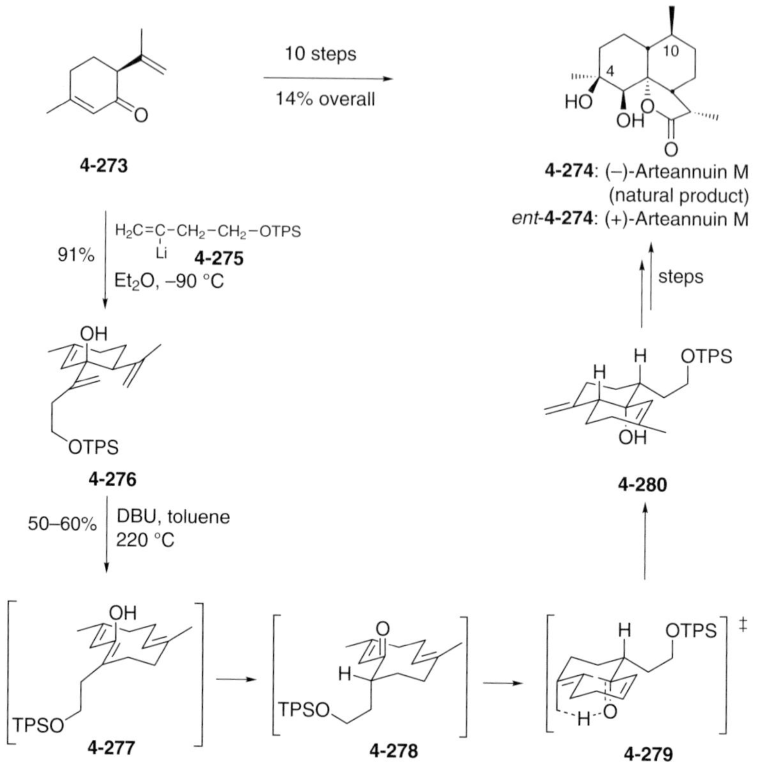

Scheme 4.59. Synthesis of (+)-arteannuin M (*ent*-**4-274**).

A domino process based on the twofold addition of alkenyl anions to a squarate ester was used by Paquette and coworkers [96] for the total synthesis of the triquinane sesquiterpene hypnophilin (**4-284**). The three-component reaction of **4-281**,**4-282** and vinyl lithium gave primarily the *trans*- and *cis*-adducts A and B, which furnished D either by an electrocyclic ring opening/ring closure *via* C or a dianionic oxy-Cope rearrangement (Scheme 4.60). Further transformations led to E and F, which resulted in the formation of **4-283** on treatment with acid.

Jasmonates are important odorant compounds. For the synthesis of new substances of this type, Giersch and Forris developed a domino Claisen/ene/retro-ene process which allows the acid-catalyzed transformation of sorbyl alcohol **4-285** and the cyclic acetals **4-286** into the cycloalkenone **4-292** with the proposed intermediates **4-287** to **4–291** (Scheme 4.61) [97]. A similar domino process had been described by Srikrishna and coworkers [98]. Acyclic acetals gave only the Claisen rearrangement products.

Several examples of domino oxy-Cope carbonyl-ene (Prins) reactions have been described by Hiersemann and coworkers, leading to mono- and bicyclic compounds [99]. Examples are the transformations of **4-293a–c** to give **4-294a–c** as single diastereomers (Scheme 4.62).

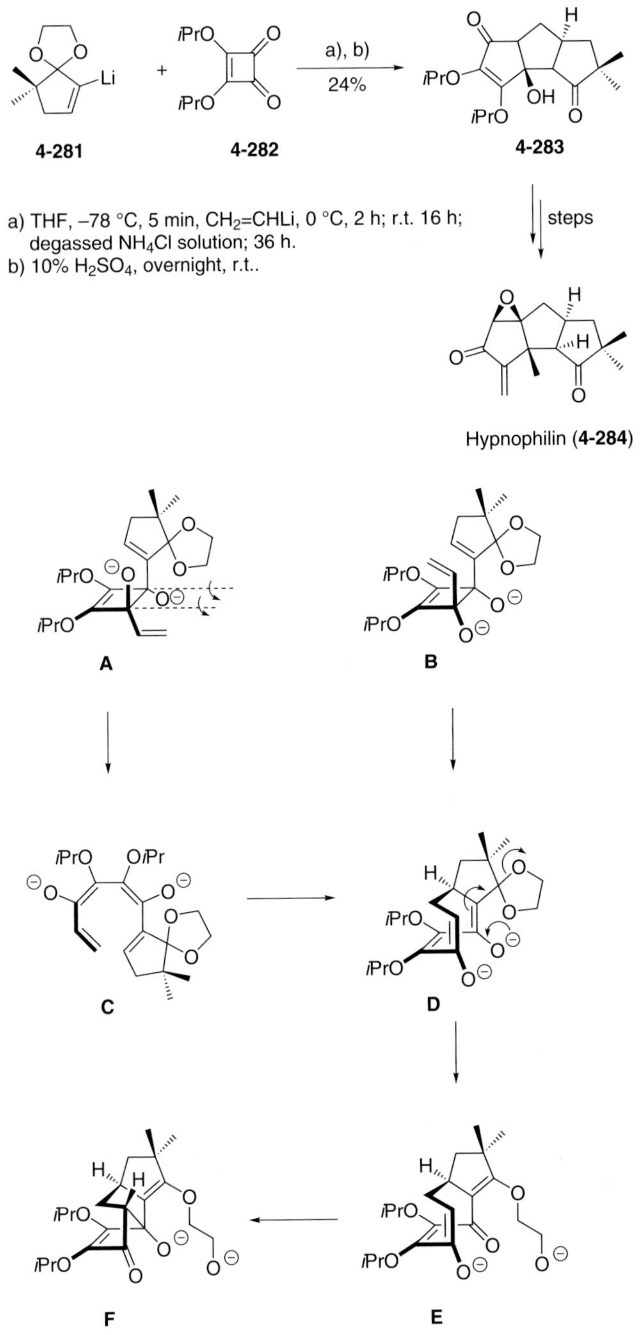

4-281
4-282
a), b)
24%
4-283
a) THF, –78 °C, 5 min, CH2=CHLi, 0 °C, 2 h; r.t. 16 h; degassed NH4Cl solution; 36 h.
b) 10% H2SO4, overnight, r.t..
steps
Hypnophilin (4-284)
A
B
C
D
E
F

Scheme 4.60. Synthesis of hypnophilin (**4-284**).

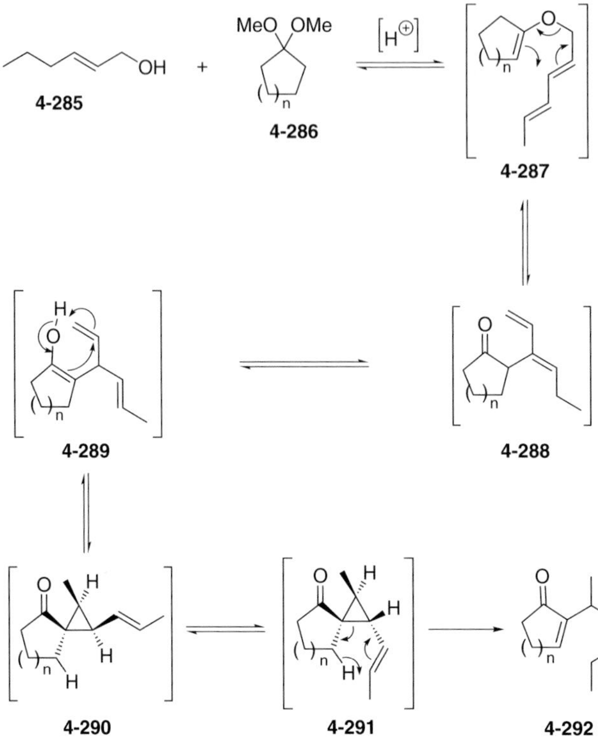
MeO OMe
[H⊕]
OH
4-285
4-286
4-287
4-288
4-289
4-290
4-291
4-292

n	Yield [%]
1	73
2	68
3	83
4	68

Scheme 4.61. Synthesis of jasmonate analogoues.

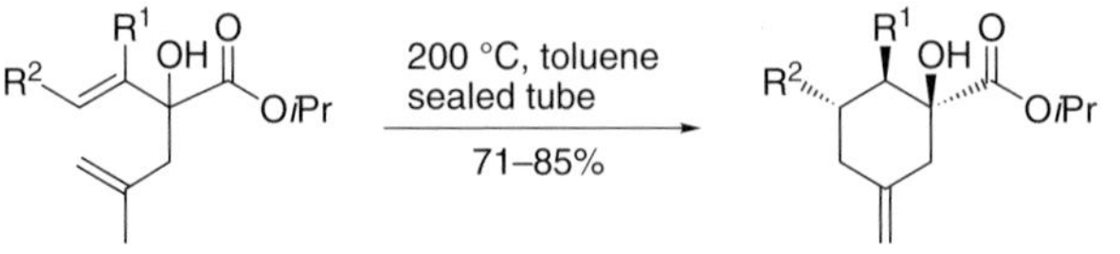
200 °C, toluene
sealed tube
71–85%

4-293a–c

4-294a: $R^1 = R^2 = H$
4-294b: $R^1 = CH_3$, $R^2 = H$
4-294c: $R^1 = H$, $R^2 = Ph$

Scheme 4.62. Domino oxy-Cope carbonyl ene reactions.

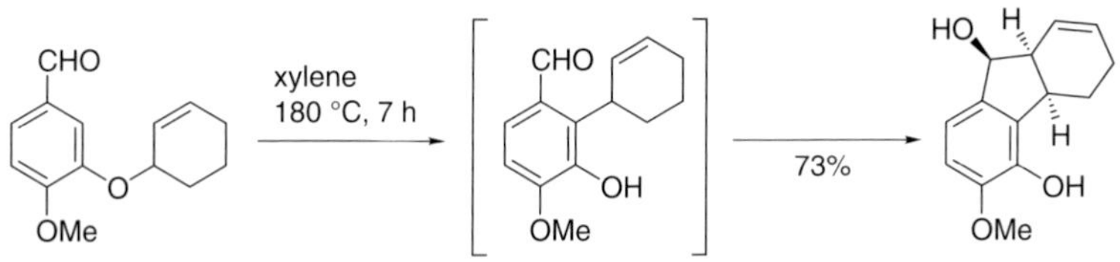

4-295 **4-296** **4-297**

Scheme 4.63. Synthesis of fluorenes.

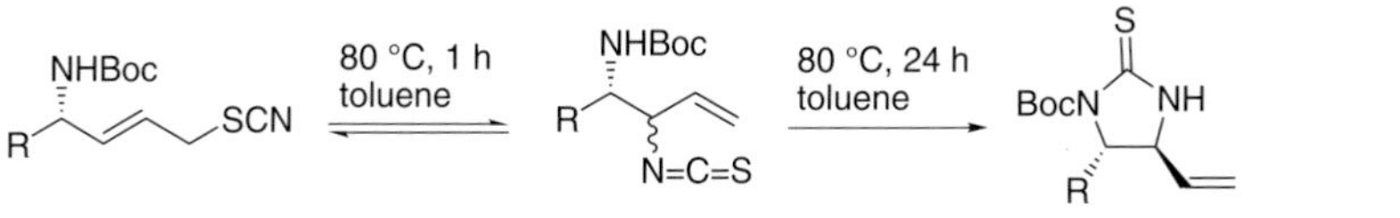

4-298 **4-299** **4-300**

R = Me, Et, *i*Pr, Bn, *i*Bu (*ds:* 98:2 to 99.5:0.5, 85–90% yield)

Scheme 4.64. Synthesis of imidazolethiones.

A simple synthesis of fluorenes as **4-297** was developed by Schäfer and coworkers, also using a combination of a Claisen rearrangement and a carbonyl ene reaction (Scheme 4.63) [100]. Heating **4-295** in xylene at 180 °C led to **4-297** as a single diastereomer in 73 % yield; the phenol **4-296** can be assumed as an intermediate, but this could not be detected in the reaction mixture.

Gonda and coworkers developed a new domino process to give substituted diastereomerically pure imidazolethiones from thiocyanates, which includes a 3,3-sigmatropic rearrangement and an intramolecular amine addition [101]. Heating **4-298** at 80 °C in toluene for 1 h led to the two diastereomers **4-299** in a 1:1-ratio, whereas prolonged heating of the mixture for 26 h gave **4-300** as an almost single product (Scheme 4.64).

A dianionic dioxy-Cope rearrangement followed by an intramolecular aldol reaction was described by Saito and coworkers for the formation of enantiopure 5,6-disubstituted cyclopentene carbaldehydes [102]. Thus, treatment of **4-301** containing two (*E*)-double bonds with $NaN(TMS)_2$ in the presence of crown ether afforded enantiopure (–)-**4-302** in 70 % yield after aqueous work-up as a single diastereomer (Scheme 4.65). It can be assumed that **4-301** is transformed into the dianion **4-303**, which undergoes a Cope rearrangement to give **4-304**. Aldol addition and elimination of water then led to (–)-**4-302** *via* **4-305**. Due to the stereospecific nature of the Cope rearrangement, the all-(*Z*)-compound **4-306** exclusively afforded (+)-**4-302** in 63 % yield. Moreover, the *E*,*Z*-substrate gave (±)-**4-302**.

Majumdar and coworkers used a combination of a 3,3-sigmatropic rearrangement followed by an intramolecular [1,6]-Michael addition for the synthesis of pyrimidine-annulated heterocycles as **4-308** from **4-307** (Scheme 4.66) [103].

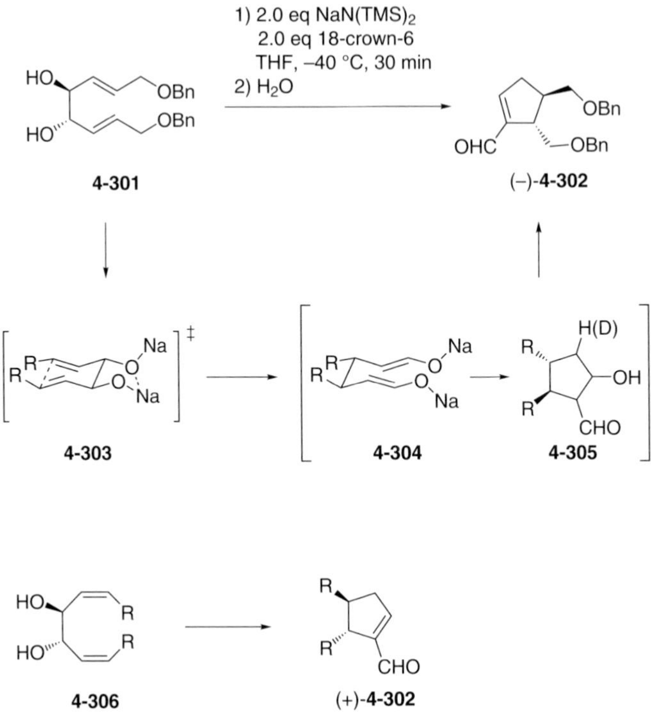

Scheme 4.65. Domino Cope/aldol reaction.

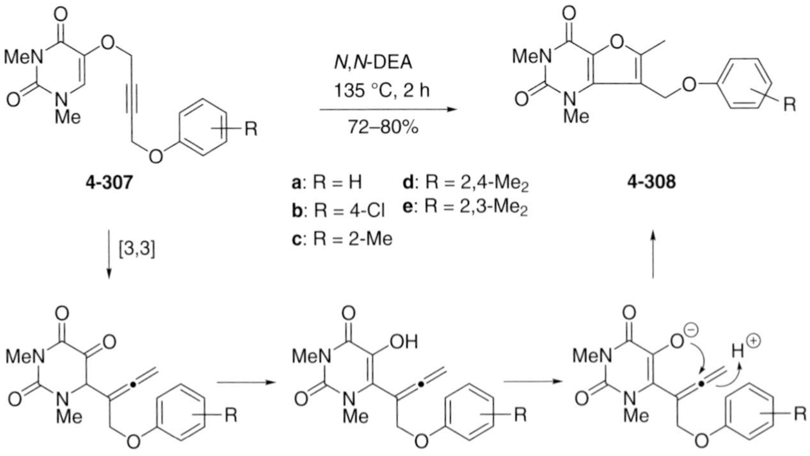

a: R = H   d: R = 2,4-Me$_2$
b: R = 4-Cl   e: R = 2,3-Me$_2$
c: R = 2-Me

Scheme 4.66. Synthesis of pyrimidine-annelated heterocycles.

Wipf and coworkers used a Claisen rearrangement of allyl phenyl ethers **4-309** followed by an enantioselective carboalumination using the chiral Zr-complex **4-310** and trimethyl aluminum (Scheme 4.67) [104]. After an oxidative work-up of the intermediate trialkylalane, the corresponding alcohols **4-311** were obtained with up to 80 % *ee* and 78 % yield. One can also transfer an ethyl group using triethyl aluminum with even better *ee*-values (up to 92 %), but the yields were rather low (42 %) due to a more sluggish oxidative cleavage of the Al–C bond.

A combination of a Claisen rearrangement, a Wittig rearrangement and a Wittig reaction was described by Mali and coworkers for the synthesis of 6-prenylcoumarins. In these transformations, 2-prenyloxybenzaldehydes was employed as substrate [105].

A combination of 2,3 sigmatropic rearrangement (Pummerer-type reaction) followed by an electrophilic aromatic substitution of the intermediate sulfenium ion, the formation of an iminium ion and, finally, a second electrophilic aromatic substitution, was used by Daïch and coworkers for the synthesis of *iso*-indolo-isoquinolinones as **4-314** (Scheme 4.68) [106]. Thus, reaction of the two diastereomeric sulfoxides **4-313**, easily obtainable from **4-312** by a Grignard reaction and oxidation, led to **4-314** as a single product after crystallization in 42 % yield.

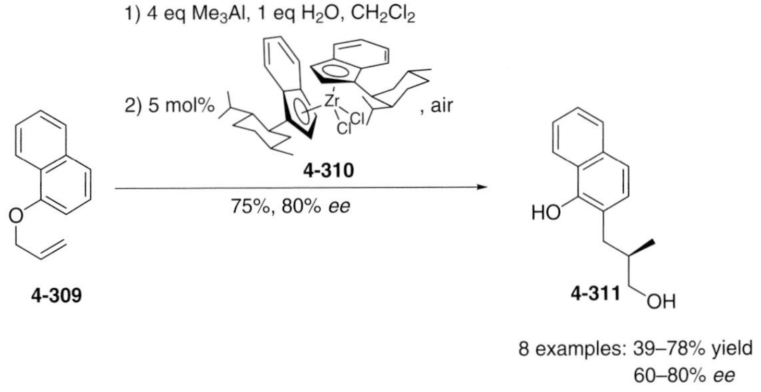

1) 4 eq Me$_3$Al, 1 eq H$_2$O, CH$_2$Cl$_2$

2) 5 mol%   , air

4-310

75%, 80% *ee*

4-309

4-311

8 examples: 39–78% yield
60–80% *ee*

Scheme 4.67. Enatioselective formation of alcohols of type **4-311**.

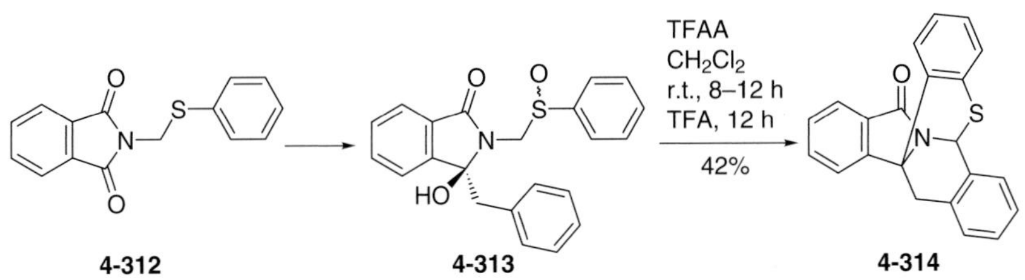

TFAA
CH$_2$Cl$_2$
r.t., 8–12 h
TFA, 12 h

42%

4-312   **4-313**   **4-314**

Scheme 4.68. Combination of a 2,3-sigmatropic rearrangement with electrophilic aromatic substitutions.

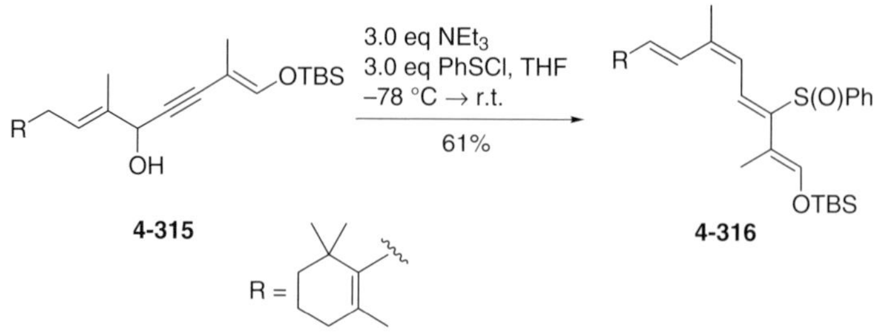

Scheme 4.69. Synthesis of conjugated tetraenes.

Another domino process starting with a [2,3] sigmatropic rearrangement allows transformation of the propargylic alcohol **4-315** into the conjugated tetraenes **4-316** on treatment with phenylsulfenyl chloride, as described by Lera and coworkers (Scheme 4.69) [107].

Quite recently, Wipf and coworkers developed a combination of a Claisen rearrangement with the addition of an organo aluminum species to the newly formed carbonyl moiety [108]. The new process is based on an already-discussed (see above) transformation by the same author [104].

4.5
Electrocyclic Reactions

To date, only a few examples are known where a domino reaction starts with an electrocyclic reaction, although the value of this approach is clearly demonstrated by the beautiful synthesis of estradiol methyl ether **4-319** through a domino electrocyclic/cycloaddition process. There is also an impressive example of a double thermal electrocyclization being used; however, the starting material for this domino reaction was prepared *in situ* by a transition metal-catalyzed transformation, and is therefore discussed in Chapter 6.

For the synthesis of estradiol methyl ether **4-319**, the cyclobutene derivative **4-317** was heated to give the orthoquinonedimethane **4-318** which cyclized in an intramolecular Diels–Alder reaction [109]. The thermally permitted, conrotatory electrocyclic ring-opening of benzocyclobutenes [110] with subsequent intramolecular cycloaddition also allowed the formation of numerous complex frameworks (Scheme 4.70).

Fukumoto and coworkers used this approach for the synthesis of (±)-genesine (**4-324**). The key step is the thermal electrocyclic ring opening of **4-320** with the formation of **4-321**, followed by an electrocyclic ring closure to give **4-322**. This then undergoes a [3,3] sigmatropic rearrangement to give **4-323** (Scheme 4.71) [111].

Sorensen and coworkers used a domino conrotatory electrocyclic ring-opening/6π-disrotatory electrocyclization for the formation of ring C in the total synthesis of (±)-viridin (**4-327**) (Scheme 4.72) [112]. Heating **4-325** in the presence of a base followed by *in-situ* oxidation with DDQ afforded the tetracycle **4-326** in 83% yield.

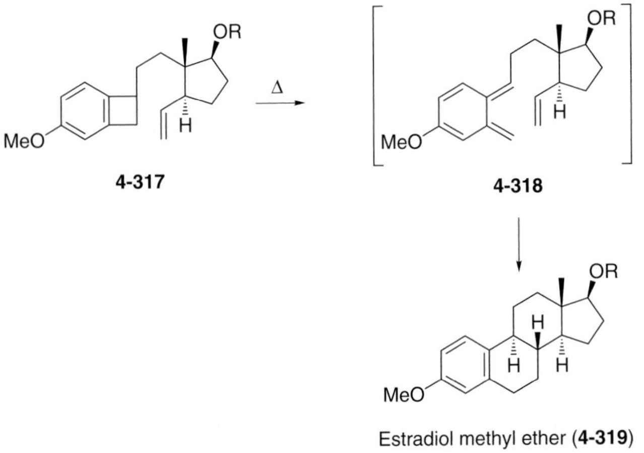

Estradiol methyl ether (**4-319**)

Scheme 4.70. Synthesis of estradiol methyl ether (**4-319**).

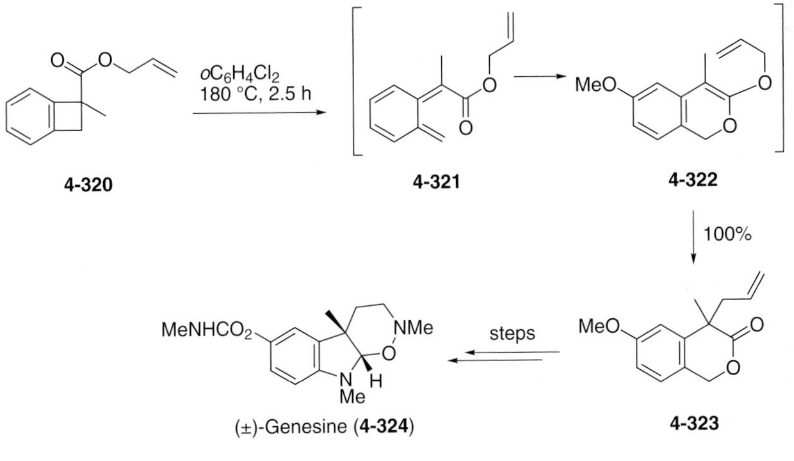

Scheme 4.71. Synthesis (±)-genesine (**4-324**).

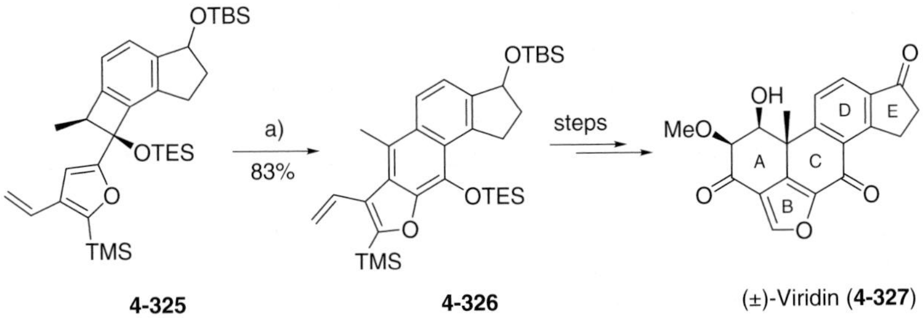

a) 2 eq *i*Pr$_2$EtN, xylene (degassed), 140 °C, 3.5 h; then 1 eq DDQ, r.t., 15 min.

Scheme 4.72. Synthesis of (±)-viridine (**4-327**).

TAN-1085 **4-331** is a tetracyclic antibiotic that is produced by *Streptomyces* sp. S-11106 and inhibits not only angiogenesis but also the enzyme aromatase. Suzuki and coworkers have recently published the first total synthesis using a domino electrocyclic process as the key step [113]. For this purpose, the benzocyclobutene **4-328** containing an allylic alcohol was oxidized to the corresponding α,β-unsaturated aldehyde, which underwent two electrocyclic reactions, namely the ring opening of the cyclobutene moiety and the subsequent 6π-closure to give the biaryl dialdehyde **4-329** (Scheme 4.73). Formation of the tetracycle **4-330** was then achieved by treatment with SmI$_2$. Interestingly, the corresponding TBS ether of the allylic alcohol moiety in **4-328** did not undergo the desired process, whereas the α,β-unsaturated aldehyde of **4-328** – probably due to the cooperative effect of the electron-donating (2 × OMe) and the electron-withdrawing aldehyde group – underwent the reaction at only room temperature.

Very recently, de Meijere and coworkers reported on a 6π electrocyclization/ Diels–Alder reaction [114]. The domino transformation provides access to highly substituted tri- and tetracyclic compounds.

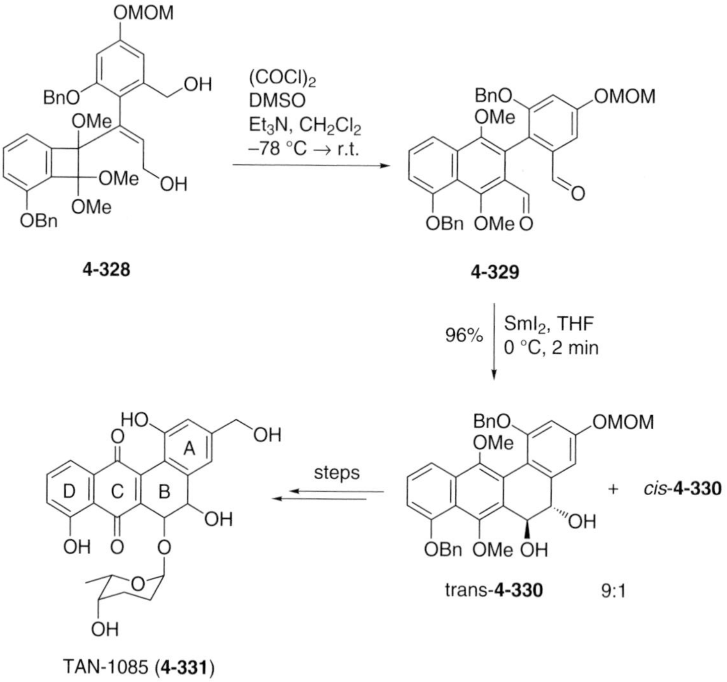

Scheme 4.73. Synthesis of TAN-1085 (**4-331**).

4.6
Ene Reactions

Ene reactions offer an excellent means of commencing a domino process, and typical carbon-ene reactions are described in this section (for details of the reaction mechanisms, see Chapter 1).

A three-component domino process consisting of an ene reaction followed by the addition of an allylsilane to afford polysubstituted tetrahydropyrans in generally good yield was described by Markó and coworkers [115]. However, the nature of the products formed in this process depends heavily on the Lewis acid employed as a catalyst. Thus, reaction of the allylsilane 4-333 with an aldehyde in the presence of $BF_3 \cdot Et_2O$ led to the domino products 4-334, whereas in the presence of $TiCl_4$ the diol 4-332, and in the presence of Et_2AlCl the alcohol 4-335, were obtained (Scheme 4.74). The latter compound can then be transformed in stepwise manner to 4-334, using the same aldehyde as previously. However, it is also possible to use another aldehyde to prepare tetrahydropyrans of type 4-336.

Cohen and coworkers also used a metallo-ene reaction starting from allylic phenyl thioethers such as 4-337 (Scheme 4.75) [116]. With $tBuOK/nBuLi$, 4-338 is

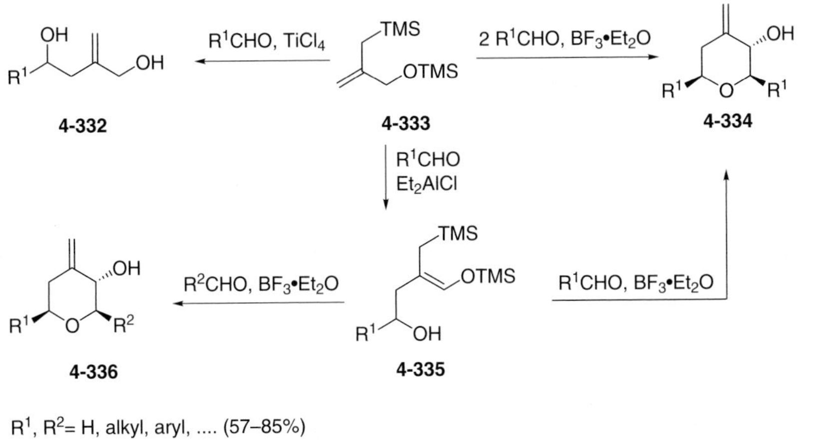

R^1, R^2= H, alkyl, aryl, ... (57–85%)

Scheme 4.74. Synthesis of tetrahydropyrans.

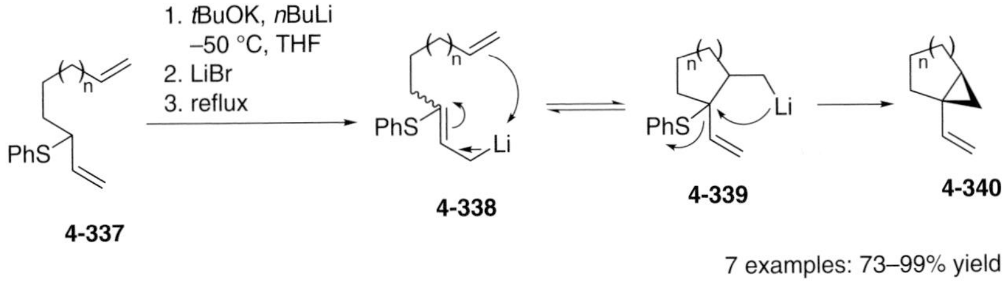

7 examples: 73–99% yield

Scheme 4.75. Domino metalle-ene/substitution process.

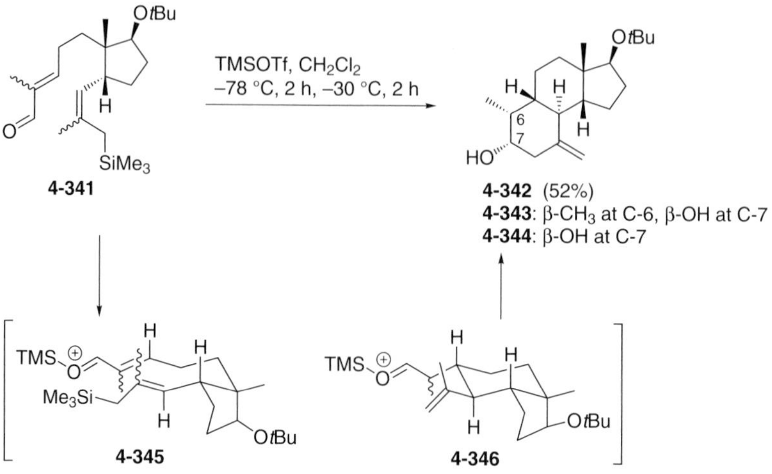

Scheme 4.76. Synthesis of the BCD-ring portion of steroids.

formed and undergoes an intramolecular ene reaction to give a cycloalkane **4-339**, followed by the formation of a cyclopropane moiety as in **4-340**.

Tietze and coworkers [117] also used a metallo-ene reaction as the initiating step for the synthesis of the BCD-ring portion of steroids, which is then followed by a carbonyl-ene reaction.

Treatment of the enantiopure substrate **4-341** containing a sila-ene moiety with trimethylsilyl trifluoromethanesulfonate (TMS OTf) gave **4-342** in a highly stereoselective fashion and 52% yield. It can be assumed that the reaction passes through the two chairlike transition structures **4-345** and **4-346** (Scheme 4.76). It is of interest that the stereochemistry of the two C–C-double bonds in **4-341** did not influence the configuration of the product. Moreover, when using EtAlCl$_2$ as mediator, a 3:1-mixture of **4-343** and **4-344** is obtained in 40% yield. This can be explained by an axial orientation of the carbonyl moiety in the transition state.

Some older examples of this type of process include the studies of Markó on the synthesis of pseudomonic acid C analogues [118], the preparation of indanones by Snider [119], and synthesis of the skeleton of the sesquiterpenes khusiman and zizaen by Wenkert and Giguere and their coworkers [120, 121].

4.7
Retro-Pericyclic Reactions

In this section are described those domino reactions which start with a retro-pericyclic reaction. This may be a retro-Diels–Alder reaction, a retro-1,3-dipolar cycloaddition, or a retro-ene reaction, which is then usually followed by a pericyclic reaction as the second step. However, a combination is also possible with another type of transformation as, for example, an aldol reaction.

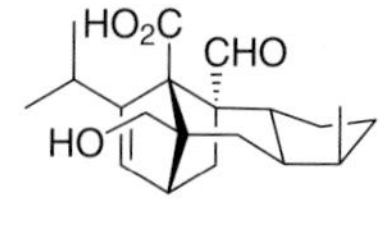

Sordaricin (**4-347**)

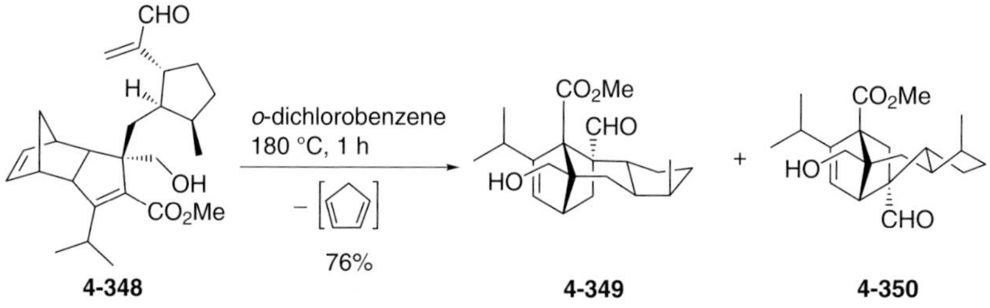

Scheme 4.77. Synthesis of sordaricin (**4-347**).

Recently, Mander and coworkers [122] reported the total synthesis of sordaricin (**4-347**), the aglycone of the potent antifungal diterpene sordarin which was first isolated in 1971 from the ascomycete *Sordaria araneosa*. Two approaches were explored: the first method utilized a possible biogenetic Diels–Alder reaction; the second was based on a domino retro-Diels–Alder/intramolecular Diels–Alder process. Thus, heating of **4-348** led, with extrusion of cyclopentadiene, to a 1,3-butadiene as intermediate which underwent an intramolecular Diels–Alder reaction to give the desired **4-349** as the main product, together with a small amount of **4-350** (Scheme 4.77).

A retro-1,3-dipolar cycloaddition followed by an 1,3-dipolar cycloaddition was used for a highly efficient total synthesis of (–)-histrionicotoxin (**4-354**) (HTX) by Holmes and coworkers [123]. HTX is a spiropiperidine-containing alkaloid which was isolated by Doly, Witkop and coworkers [124] from the brightly colored poison-arrow frog *Dendrobates histrionicus*. It is of great pharmacological interest as a non-competitive inhibitor of acetylcholine receptors.

The key step in the synthesis of **4-354** is the retro-1,3-dipolar cycloaddition of the isoxazolidine **4-351** to give the nitronate **4-352**, which underwent an intramolecular 1,3-dipolar cycloaddition. The obtained cycloadduct **4-353** can be transformed in a few steps into the desired target **4-354** (Scheme 4.78).

Goti, Brandi and coworkers developed an effective synthesis of (–)-rosmarinecine (**4-357**) *via* a domino cycloreversion-intramolecular nitrone cycloaddition of **4-355**, which led to **4-356** (Scheme 4.79) [125].

A combination of an anionic oxy retro-ene and an aldol reaction to give annulated cyclopentenones **4-361** from **4-358** was described by Jung and coworkers (Scheme 4.80) [126]. It can be assumed that, in the presence of KH, the potassium alkoxide **4-359** is first formed; this leads to **4-360** and finally to **4-361** in an intramolecular aldol reaction.

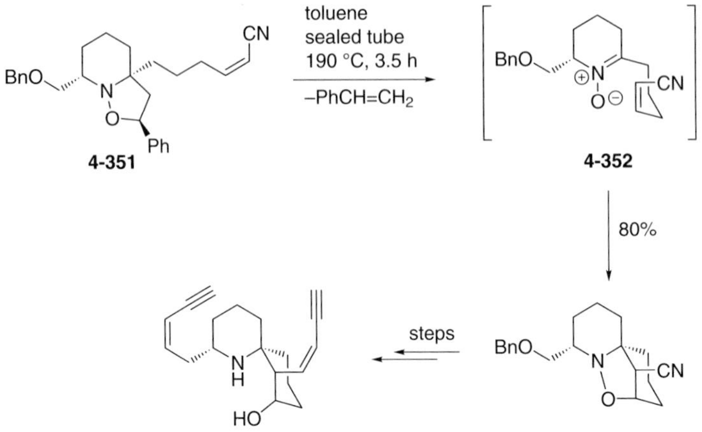

4-351 4-352

(–)-Histrionicotoxin (**4-354**) **4-353**

Scheme 4.78. Synthesis of (–)-histrionicotoxin (**4-354**).

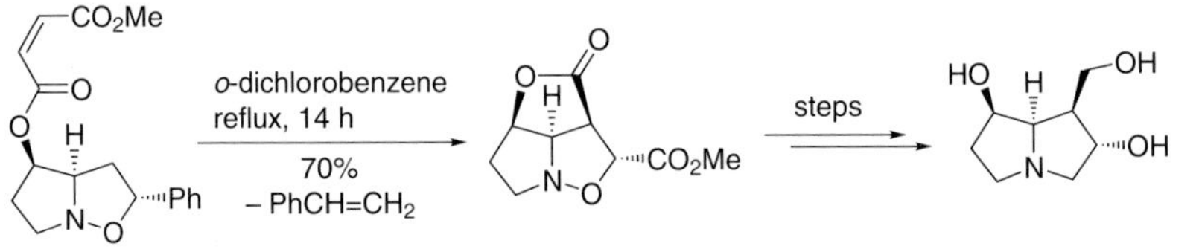

4-355 **4-356** (–)-Rosmarinecine (**4-357**)

Scheme 4.79. Synthesis of (–)-rosmarinecine (**4-357**).

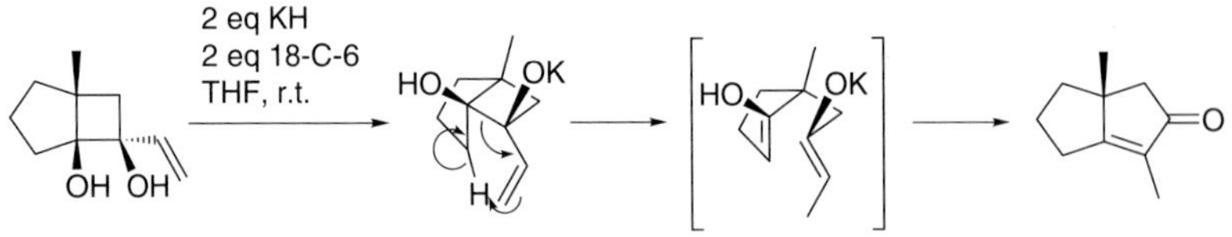

4-358 **4-359** **4-360** **4-361**

Scheme 4.80. Domino oxy-retro-ene/aldol reaction.

References

1 (a) O. Diels, K. Alder, *Justus Liebigs Ann. Chem.* **1931**, *490*, 243; (b) O. Diels, S. Olsen, *J. Prakt. Chem.* **1940**, *156*, 285–314; (c) J. D. Slee, E. LeGoff, *J. Org. Chem.* **1970**, *35*, 3897–3901.

2 (a) F.-G. Klärner, U. Artschwager-Perl, W.-D. Fessner, C. Grund, R. Pinkos, J.-P. Melder, H. Prinzbach, *Tetrahedron Lett.* **1989**, *30*, 3137–3140; (b) W.-D. Fessner, C. Grund, H. Prinzbach, *Tetrahedron Lett.* **1989**, *30*, 3133–3136; (c) W.-D. Fessner, G. Sedelmeier, P. R. Spurr, G. Rihs, H. Prinzbach, *J. Am. Chem. Soc.* **1987**, *109*, 4626–4642.

3 (a) H. Hopf, I. Böhm, J. Kleinschroth, *Org. Synth.* **1981**, *60*, 41–48; (b) H. Hopf, F. Th. Lenich, *Chem. Ber.* **1974**, *107*, 1891–1902.

4 J. D. Winkler, H. S. Kim, S. Kim, *Tetrahedron Lett.* **1995**, *36*, 687–690.

5 W. Deng, L. E. Overman, *J. Am. Chem. Soc.* **1994**, *116*, 11241–11250.

6 T. D. Golobish, J. K. Burke, A. H. Kim, S. W. Chong, E. L. Probst, P. J. Carroll, W. P. Dailey, *Tetrahedron* **1998**, *54*, 7013–7024

7 L. A. Paquette, M. J. Wyvratt, H. C. Berk, R. E. Moerck, *J. Am. Chem. Soc.* **1978**, *100*, 5845–5855.

8 (a) D. J. Cram, G. R. Knox, *J. Am. Chem. Soc.* **1961**, *83*, 2204–2209; (b) D. J. Cram, C. S. Montgomery, G. R. Knox, *J. Am. Chem. Soc.* **1966**, *88*, 515–525.

9 M. Lautens, E. Fillion, *J. Org. Chem.* **1997**, *62*, 4418–4427.

10 (a) M. Nörret, M. S. Sherburn, *Angew. Chem. Int. Ed.* **2001**, *40*, 4074–4076; (b) C. I. Turner, R. M. Williamson, P. Turner, M. S. Sherburn, *Chem. Comm.* **2003**, 1610–1611.

11 L. R. Domingo, M. Arnó, J. Andrés, *J. Am. Chem. Soc.* **1998**, *120*, 1617–1618.

12 M. Visnick, M. A. Battiste, *J. Chem. Soc., Chem. Commun.* **1985**, 1621–1622.

13 L. R. Domingo, M. T. Picher, J. Andrés, *J. Org. Chem.* **2000**, *65*, 3473–3477.

14 E. Román, M. Baños, F. J. Higes, J. A. Serrano, *Tetrahedron: Asymm.* **1998**, *9*, 449–458.

15 N. Soldermann, J. Velker, O. Vallat, H. Stöckli-Evans, R. Neier, *Helv. Chim. Acta* **2000**, *83*, 2266–2276.

16 M. C. Carreño, M. Belén Cid, J. L. García Ruano, *Tetrahedron: Asymm.* **1996**, *7*, 2151–2158.

17 M. C. Carreño, J. L. García Ruano, M. A. Tuledo, *Chem. Eur. J.* **2000**, *6*, 288–291.

18 Y. Arroyo, J. F. Rodríguez, M. Santos, M. A. Sanz Tejedor, I. Vaca, J. L. García Ruano, *Tetrahedron Asymm.* **2004**, 1059–1063.

19 M. C. Carreño, M. González-López, A. Urbano, *Chem. Commun.* **2005**, 611–613.

20 P. A. Jacobi, C. S. R. Kaczmarek, U. E. Udodong, *Tetrahedron Lett.* **1984**, *25*, 4859–4862.

21 P. K. Chowdhury, A. Prelle, D. Schomburg, M. Thiemann, E. Winterfeldt, *Liebigs Ann. Chem.* **1987**, 1095–1099.

22 Z.-Q. Yang, S. J. Danishefsky, *J. Am. Chem. Soc.* **2003**, *125*, 9602–9603.

23 (a) P. Delmontte, J. Delmontee-Plaquée, *Nature* **1953**, *171*, 344–347; (b) W. A. Ayer, S. P. Lee, A. Tsuneda, Y. Hiratsuka, *Can. J. Microbiol.* **1980**, *26*, 766–773.

24 S. M. Roe, C. Prodromou, P. O'Brien, J. E. Ladbury, P. W. Piper, L. H. Pearl, *J. Med. Chem.* **1999**, *42*, 260–266.

25 K. Inanaga, K. Takasu, M. Ihara, *J. Am. Chem. Soc.* **2004**, *126*, 1352–1353.

26 (a) F. Bilodeau, L. Dubé, P. Deslongchamps, *Tetrahedron* **2003**, *59*, 2781–2791; (b) G. Bélanger, P. Deslongchamps, *Org. Lett.* **2000**, *2*, 285–287; (c) D. G. Hall, P. Deslongchamps, *J. Org. Chem.* **1995**, *60*, 7796–7814.

27 M. Vaultier, F. Truchet, B. Carboni, R. W. Hoffmann, I. Denne, *Tetrahedron Lett.* **1987**, *28*, 4169–4172.

28 J.-Y. Lallemand, Y. Six, L. Ricard, *Eur. J. Org. Chem.* **2002**, 503–513.

29 X. Gao, D. G. Hall, *Tetrahedron Lett.* **2003**, *44*, 2231–2235.

30 Y. Zeng, D. S. Reddy, E. Hirt, J. Aubé, *Org. Lett.* **2004**, *6*, 4993–4995.

31 D. J. Mergott, S. A. Frank, W. R. Roush, *Org. Lett.* **2002**, *4*, 3157–3160.

32 A. Padwa, M. A. Brodney, S. M. Lynch, *J. Org. Chem.* **2001**, *66*, 1716–1724.

33 A. Padwa, *Chem. Commun.* **1998**, *14*, 1417–1424.

34 A. Padwa, M. B. Brodney, S. M. Lynch, P. Rashatasakhon, Q. Wang, H. Zhang, *J. Org. Chem.* **2004**, *69*, 3735–3745.

35 G. J. Bodwell, K. M. Hawco, R. P. da Silva, *Synlett* **2003**, *2*, 179–182.

36 J. Barluenga, F. Aznar, S. Barluenga, *J. Chem. Soc., Chem. Commun.* **1995**, 1973–1974.

37 J. Barluenga, F. Aznar, S. Barluenga, A. Martín, S. García-Granda, E. Martín, *Synlett* **1998**, 473–474.

38 R. M. A. Lavigne, M. Riou, M. Girardin, L. Morency, L. Barriault, *Org. Lett.* **2005**, *7*, 5921–5923.

39 Y. Zeng, J. Aubé, *J. Am. Chem. Soc.* **2005**, *127*, 15712–15713.

40 (a) R.-S. Xu, *Stud. Nat. Prod. Chem.* **2000**, *21*, 729–772; (b) R. A. Pilli, M.d.C. Ferreira de Oliveira, *Nat. Prod. Rep.* **2000**, *17*, 117–127.

41 L. F. Tietze, G. Kettschau, *Top. Curr. Chem.* **1997**, *189*, 1–120.

42 S. E. Denmark, A. Thorarensen, D. S. Middleton, *J. Am. Chem. Soc.* **1996**, *118*, 8266–8277.

43 S. E. Denmark, J. J. Cottell, *J. Org. Chem.* **2001**, *66*, 4276–4284.

44 S. E. Denmark, L. Gomez, *Org. Lett.* **2001**, *3*, 2907–2910.

45 S. E. Denmark, D. S. Middleton, *J. Org. Chem.* **1998**, *63*, 1604–1618.

46 M. Avalos, R. Babiano, P. Cintas, J. L. Jiménez, J. C. Palacios, M. A. Silva, *Chem. Commun.* **1998**, 459–460.

47 (a) G. D. Wilkie, G. I. Elliott, B. S. J. Blagg, S. E. Wolkenberg, D. R. Soenen, M. M. Miller, S. Pollack, D. L. Boger, *J. Am. Chem. Soc.* **2002**, *124*, 11292–11294; (b) Z. Q. Yuan, H. Ishikawa, D. L. Boger, *Org. Lett.* **2005**, *7*, 741–744; (c) Y. Choi, H. Ishikawa, J. Velcicky, G. I. Elliott, M. M. Miller, D. L. Boger, *Org. Lett.* **2005**, *7*, 4539–4542; (d) G. I. Elliott, J. Velcicky, H. Ishikawa, Y. K. Li, D. L. Boger, *Angew. Chem. Int. Ed.* **2006**, *45*, 620–622.

48 J. Cobo, C. García, M. Melguizo, A. Sánchez, M. Nogueras, *Tetrahedron* **1994**, *50*, 10345–10358.

49 J. Cobo, M. Melguizo, A. Sánchez, M. Nogueras, E. De Clercq, *Tetrahedron* **1996**, *52*, 5845–5856.

50 D. Giomi, R. Nesi, S. Turchi, R. Coppini, *J. Org. Chem.* **1996**, *61*, 6028–6030.

51 D. Giomi, M. Cecchi, *J. Org. Chem.* **2003**, *68*, 3340–3343.

52 (a) B. B. Touré, H. R. Hoveyda, J. Tailor, A. Ulaczyk-Lesanko, D. G. Hall, *Chem. Eur. J.* **2003**, *9*, 466–474; (b) J. Tailor, D. G. Hall, *Org. Lett.* **2000**, *2*, 3715–3718.

53 M. Deligny, F. Carreaux, B. Carboni, L. Toupet, G. Dujardin, *Chem. Commun.* **2003**, 276–277.

54 L. F. Tietze, G. Kettschau, J. A. Gewert, A. Schuffenhauer, *Curr. Org. Chem.* **1998**, *2*, 19–62.

55 G. Vassilikogiannakis, I. Margaros, T. Montagnon, *Org. Lett.* **2004**, *6*, 2039–2042.

56 D. Giomi, S. Turchi, A. Danesi, C. Faggi, *Tetrahedron* **2001**, *57*, 4237–4242.

57 H. J. Knölker, E. Baum, R. Graf, P. G. Jones, O. Spieß, *Angew. Chem. Int. Ed.* **1999**, *38*, 2583–2585.

58 M. Avalos, R. Babiano, P. Cintas, M. B. Hursthouse, J. L. Jiménez, M. E. Light, I. Lopéz, J. C. Palacios, G. Silvero, *Chem. Eur. J.* **2001**, *7*, 3033–3042.

59 E. Funke, R. Huisgen, *Chem. Ber.* **1971**, *104*, 3222–3228.

60 R. Shintani, G. C. Fu, *Angew. Chem. Int. Ed.* **2003**, *42*, 4082–4085.

61 M. Kinugasa, S. Hashimoto, *J. Chem. Soc. Chem. Commun.* **1972**, 466–467.

62 M. Yu, B. L. Pagenkopf, *Org. Lett.* **2003**, *5*, 5099–5101.

63 M. P. S. Ishar, K. Kumar, *Tetrahedron Lett.* **1999**, *40*, 175–176.

64 (a) M. Shindo, Y. Sato, K. Shishido, *Tetrahedron Lett.* **2002**, *43*, 5039–5041; (b) M. Shindo, *Synthesis* **2003**, 2275–2288.

65 V. Hellwig, J. Dasenbrock, S. Schumann, W. Steglich, K. Leonhardt, T. Anke, *Eur. J. Org. Chem.* **1998**, 73–79.

66 The first total synthesis of cucumin E: G. Mehta, J. D. Umarye, *Tetrahedron Lett.* **2001**, *42*, 1991–1993.

67 Z.-C. Yang, W.-S. Zhou, *J. Chem. Soc. Chem. Commun.* **1995**, 743–744.

68 (a) T. Hattori, Y. Suzuki, O. Uesugi, S. Oi, S. Miyano, *Chem. Commun.* **2000**, 73–74; (b) T. Hattori, Y. Suzuki, Y. Ito, D. Hotta, S. Miyano, *Tetrahedron* **2002**, *58*, 5215–5223.

69 S. Tandel, A. Wang, H. Zhang, P. Yousuf, E. R. Biehl, *J. Chem. Soc. Perkin Trans. 1* **2000**, 3149–3153.

70 S.-K. Yeo, N. Hatae, M. Seki, K. Kanematsu, *Tetrahedron* **1995**, *51*, 3499–3506.

71 E. Rossi, G. Abbiati, A. Arcadi, F. Marinelli, *Tetrahedron Lett.* **2001**, *42*, 3705–3708.

72 S. Giese, L. Kastrup, D. Stiens, F. G. West, *Angew. Chem. Int. Ed.* **2000**, *39*, 1970–1973.

73 Y. Wang, B. D. Schill, A. M. Arif, F. G. West, *Org. Lett.* **2003**, *5*, 2747–2750.

74 T. A. Engler, C. M. Scheibe, *J. Org. Chem.* **1998**, *63*, 6247–6253.

75 J. R. Rodríguez, A. Rumbo, L. Castedo, J. L. Mascareńas, *J. Org. Chem.* **1999**, *64*, 966–970.

76 T. A. Engler, C. M. Scheibe, R. Iyengar, *J. Org. Chem.* **1997**, *62*, 8274–8275.

77 (a) J. D. Winkler, *Chem. Rev.* **1996**, *96*, 167–176; (b) F. E. Ziegler, *Chem. Rev.* **1988**, *88*, 1423–1452.

78 A. F. Thomas, *J. Am. Chem. Soc.* **1969**, *91*, 3281–3289.

79 S. Raucher, K.-W. Chi, K.-J. Hwang, J. F. Burks, Jr., *J. Org. Chem.* **1986**, *51*, 5503–5505.

80 P. A. Jacobi, H. G. Selnick, *J. Am. Chem. Soc.* **1984**, *106*, 3041–3043.

81 G. A. Kraus, S. H. Woo, *J. Org. Chem.* **1987**, *52*, 4841–4846.

82 K. Mikami, K. Takahashi, T. Nakai, T. Uchimaru, *J. Am. Chem. Soc.* **1994**, *116*, 10948–10954.

83 B. Alcaide, P. Almendros, C. Aragoncillo, M. Redondo, *Chem. Commun.* **2002**, 1472–1473.

84 A. J. Quillinan, F. Scheinmann, *Chem. Commun.* **1971**, 966–967.

85 (a) K. C. Nicolaou, J. Li, *Angew. Chem. Int. Ed.* **2001**, *40*, 4264–4268; (b) K. C. Nicolaou, H. Xu, M. Wartmann, *Angew. Chem. Int. Ed.* **2005**, *44*, 756–761.

86 E. J. Tisdale, B. G. Vong, H. Li, S. H. Kim, C. Chowdhury, E. A. Theodorakis, *Tetrahedron* **2003**, *59*, 6873–6887.

87 E. J. Tisdale, I. Slobodov, E. A. Theodorakis, *Org. Biomol. Chem.* **2003**, *1*, 4418–4422.

88 E. J. Tisdale, I. Slobodov, E. A. Theodorakis, *Proc. Natl. Acad. Sci. USA* **2004**, *101*, 12030–12035.

89 J. E. Moses, J. E. Baldwin, R. M. Adlington, A. R. Cowley, R. Marquez, *Tetrahedron Lett.* **2003**, *44*, 6625–6627.

90 D. Mal, N. K. Hazra, *Chem. Commun.* **1996**, 1181–1182.

91 E. Koyama, G. Yang, K. Hiratani, *Tetrahedron Lett.* **2000**, *41*, 8111–8116.

92 L. Morency, L. Barriault, *Tetrahedron Lett.* **2004**, *45*, 6105–6107.

93 E. L. O. Sauer, L. Barriault, *Org. Lett.* **2004**, *6*, 3329–3332.

94 (a) L. Barriault, J. M. Warrington, G. P. A. Yap, *Org. Lett.* **2000**, *2*, 663–665; (b) L.

Barriault, D. H. Deon, *Org. Lett.* **2001**, *3*, 1925–1927.

95 G. D. Brown, L.-K. Sy, R. Haynes, *Tetrahedron* **1998**, *54*, 4345–4356.

96 (a) F. Geng, J. Liu, L. A. Paquette, *Org. Lett.* **2002**, *4*, 71–73; (b) Review: L. A. Paquette, *Eur. J. Org. Chem.* **1998**, 1709–1728.

97 W. Giersch, I. Farris, *Helv. Chim. Acta* **2004**, *87*, 1601–1606.

98 A. Srikrishna, S. Venkateswarlu, S. Nagaraju, K. Krishuan, *Tetrahedron* **1994**, *50*, 8765–8772.

99 M. Hiersemann, L. Abraham, A. Pollex, *Synlett* **2003**, 1088–1095.

100 S. Lambrecht, H. J. Schäfer, R. Fröhlich, M. Grehl, *Synlett* **1996**, 283–284.

101 J. Gonda, M. Martinková, J. Imrich, *Tetrahedron* **2002**, *58*, 1611–1616.

102 S. Saito, T. Yamamoto, M. Matsuoka, T. Moriwake, *Synlett* **1992**, 239–240.

103 K. C. Majumdar, U. Das, *J. Org. Chem.* **1998**, *63*, 9997–10000.

104 P. Wipf, S. Ribe, *Org. Lett.* **2001**, *3*, 1503–1505.

105 R. S. Mali, P. P. Joshi, P. K. Sandhu, A. Manekar-Tilve, *J. Chem. Soc. Perkin Trans. 1*, **2002**, 371–376.

106 N. Hucher, A. Daïch, B. Decroix, *Org. Lett.* **2000**, *2*, 1201–1204.

107 B. Iglesias, A. Torrado, A. R. de Lera, S. López, *J. Org. Chem.* **2000**, *65*, 2696–2705.

108. P. Wipf, D. L. Waller, J. T. Reeves, *J. Org. Chem.* **2005**, *70*, 8096–8102.

109 (a) M. B. Groen, F. J. Zeelen, *Recl. Trav. Chim. Pays-Bas* **1986**, *105*, 465–487, and references therein; (b) R. L. Funk, K. P. C. Vollhardt, *Chem. Soc. Rev.* **1980**, *9*, 41–69; (c) T. Kametani, H. Nemeto, *Tetrahedron* **1981**, *37*, 3–16.

110 (a) W. Oppolzer, *Synthesis* **1978**, 793–802; (b) J. L. Charlton, M. M. Alauddin, *Tetrahedron* **1987**, *43*, 2873–2889.

111 (a) K. Shishido, K. Hiroya, H. Komatsu, K. Fukumoto, T. Kametani, *J. Chem. Soc. Perkin Trans. 1* **1987**, 2491–2495; (b) K. Shishido, E. Shitara, H. Komatsu, K. Hiroya, K. Fukumoto, T. Kametani, *J. Org. Chem.* **1986**, *51*, 3007–3011.

112 E. A. Anderson, E. J. Alexanian, E. J. Sorensen, *Angew. Chem. Int. Ed.* **2004**, *43*, 1998–2001.

113 K. Ohmori, K. Mori, Y. Ishikawa, H. Tsuruta, S. Kuwahara, N. Harada, K.

Suzuki, *Angew. Chem. Int. Ed.* **2004**, *43*, 3167–3171.

114 R. von Essen, D. Frank, H. W. Sünnemann, D. Vidović, J. Magull, A. de Meijere, *Chem. Eur. J.* **2005**, *11*, 6583–6592.

115 I. E. Markó, R. Dumeunier, C. Leclercq, B. Leroy, J.-M. Plancher, A. Mekhalfia, D. J. Bayston, *Synthesis* **2002**, *7*, 958–972.

116 D. Cheng, K. R. Knox, T. Cohen, *J. Am. Chem. Soc.* **2000**, *122*, 412–413.

117 L. F. Tietze, M. Rischer, *Angew. Chem. Int. Ed.* **1992**, *31*, 1221–1222.

118 I. E. Markó, J.-M. Plancher, *Tetrahedron Lett.* **1999**, *40*, 5259–5262.

119 (a) B. B. Snider, B. E. Goldman, *Tetrahedron* **1986**, *42*, 2951–2956; (b) B. B. Snider, E. A. Deutsch, *J. Org. Chem.* **1983**, *48*, 1822–1829.

120 C. F. Huebner, E. Donoghue, L. Dorfman, F. A. Stuber, N. Danieli, E. Wenkert, *Tetrahedron Lett.* **1966**, *7*, 1185–1191.

121 (a) R. J. Giguere, A. M. Namen, B. O. Lopez, A. Arepally, D. E. Ramos, G. Majetich, J. Defauw, *Tetrahedron Lett.* **1987**, *28*, 6553–6556; (b) R. J. Giguere, P. G. Harran, B. O. Lopez, *Synth. Commun.* **1990**, *20*, 1453–1462.

122 L. N. Mander, R. J. Thomson, *Org. Lett.* **2003**, *5*, 1321–1324.

123 G. M. Williams, S. D. Roughley, J. E. Davies, A. B. Holmes, *J. Am. Chem. Soc.* **1999**, *121*, 4900–4901.

124 J. W. Daly, I. Karle, C. W. Myers, T. Tokuyama, J. A. Waters, B. Witkop, *Proc. Natl. Acad. Sci. USA* **1971**, *68*, 1870–1875.

125 A. Goti, S. Cicchi, M. Cacciarini, F. Cardona, V. Fedi, A. Brandi, *Eur. J. Org. Chem.* **2000**, 3633–3645.

126 M. E. Jung, P. Davidov, *Org. Lett.* **2001**, *3*, 3025–3027.

5
Photochemically Induced Domino Processes

Photochemical reactions usually proceed *via* an excited electronic state. Unique products can therefore be obtained, which are not accessible by thermally induced transformations. Furthermore, selectivity in these reactions has recently been improved lately [1]. On the other hand, processes involving excited states are somehow limited in their substrates. That may be the reason why, until now, the combination of photochemical and thermal processes in a domino fashion has received such scarce attention. Thus, there are respectively only a few examples – or even no examples – for the different categories, which we have used for classification. However, since light is an environmentally safe "reagent" and allows many interesting transformations, we believe that there is ample space to conduct research in the field of photo-induced domino processes. The usual photochemical transformations which could be used in this field are photolytic cleavages, cycloadditions, isomerizations (including rearrangements), oxidations, and reductions – including photo-induced electron transfer (PET) reactions.

5.1
Photochemical/Cationic Domino Processes

PET reactions [2] can be considered as versatile methods for generating radical cations from electron-rich olefins and aromatic compounds [3], which then can undergo an intramolecular cationic cyclization. Niwa and coworkers [4] reported on a photochemical reaction of 1,1-diphenyl-1,*n*-alkadienes in the presence of phenanthrene (Phen) and 1,4-dicyanobenzene (DCNB) as sensitizer and electron acceptor to construct 5/6/6- and 6/6/6-fused ring systems with high stereoselectivity.

Hence, irradiation of 1,1-diphenyl-8-methyl-1,7-nonadiene (**5-1a**) in the presence of DCNB and Phen through an aqueous $CuSO_4$ filter solution ($\lambda > 334$ nm) afforded octahydro-phenylanthracene **5-2a** in 45 % yield as the sole isomer (Scheme 5.1). In a similar way, **5-1b** and **5-3** led to **5-2b** and **5-4** in 57 % and 71 % yields, respectively.

It can be assumed that in the domino process of, for example **5-3**, a reactive radical cation intermediate **5-5** is initially formed [5]. The intramolecular cyclization then proceeds almost exclusively through a stable, chair-like, six-membered transition state **5-8** to give a distonic radical cation **5-9**, which is trapped by the aromatic

Domino Reactions in Organic Synthesis. Lutz F. Tietze, Gordon Brasche, and Kersten M. Gericke
Copyright © 2006 WILEY-VCH Verlag GmbH & Co. KGaA, Weinheim
ISBN: 3-527-29060-5

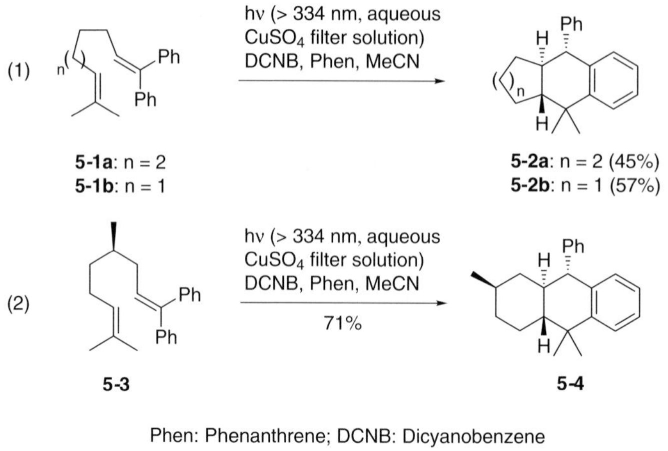

(1)

hv (> 334 nm, aqueous
CuSO₄ filter solution)
DCNB, Phen, MeCN

5-1a: n = 2
5-1b: n = 1

5-2a: n = 2 (45%)
5-2b: n = 1 (57%)

(2)

hv (> 334 nm, aqueous
CuSO₄ filter solution)
DCNB, Phen, MeCN

71%

5-3

5-4

Phen: Phenanthrene; DCNB: Dicyanobenzene

Scheme 5.1. Stereoselective domino cyclization *via* photoinduced electron-transfer reaction.

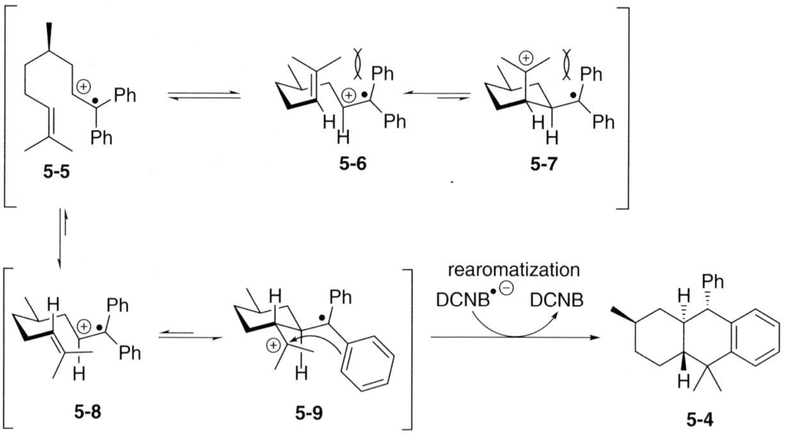

5-5

5-6

5-7

5-8

5-9

rearomatization
DCNB•⊖   DCNB

5-4

Scheme 5.2. Proposed mechanism of the domino cyclization of photochemically induced reaction of **5-4**.

ring to give **5-4** (Scheme 5.2). The process continues under electron transfer from the DCNB⁻-radical and rearomatization [6]. In theory, **5-5** could also exist in conformation **5-6**, which would give **5-7** in a first cyclization. However, the latter radical cations seems not be able to cyclize due to steric interactions.

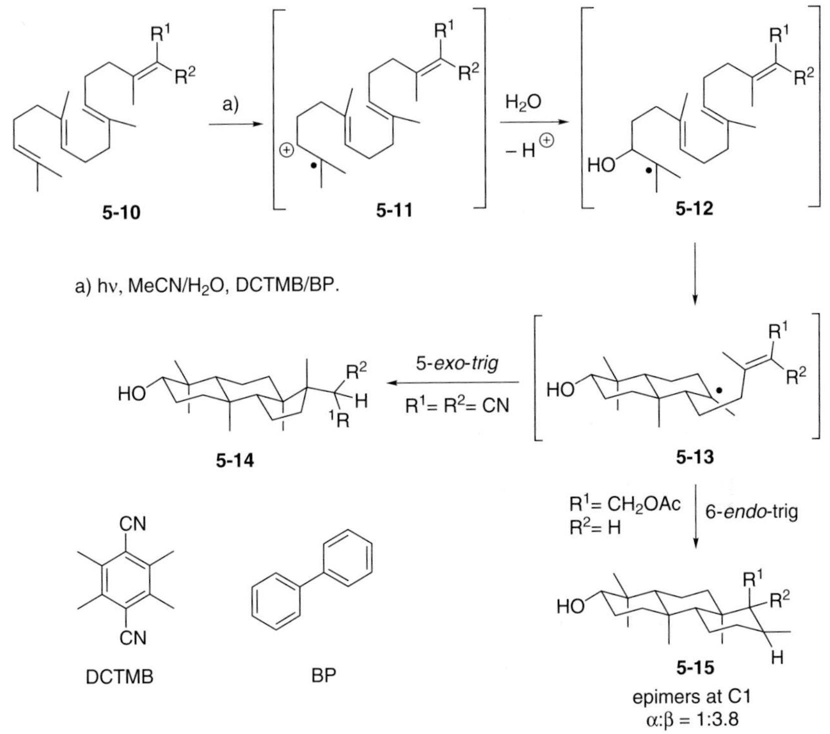

Scheme 5.3. Cyclization of polyalkene by photoinduced electron transfer.

5.2
Photochemical/Anionic Domino Processes

In a similar approach, Demuth and coworkers used PET and employed 1,4-dicyano-tetramethylbenzene (DCTMB) and 1,1'-biphenyl (BP) to form radical cations as **5-11** from tetraenes as **5-10**.

Irradiation of **5-10** in MeCN/H_2O in the presence of DCTMB and BP gives rise to a DCTMB$^-$/BP$^+$-radical pair. Subsequent electron transfer from the tetraene **5-10** to the BP$^+$-radical oxidizes regioselectively the ω-alkene moiety of **5-10** to give **5-11**, and in addition restores the uncharged cosensitizer. After nucleophilic attack of water to **5-11**, deprotonation takes place, delivering the neutral β-hydroxy radical **5-12**, which then undergoes the desired cyclizations. Depending on the substitution pattern, the cyclization proceeds either in a 5-*exo*- or 6-*endo*-trig fashion to give the products **5-14** and **5-15**, respectively.

The procedure has been used for the total synthesis of the natural product stypoldione (**5-18**) starting from **5-16** to give the 6/6/6-membered tricyclic compound **5-17** which was further manipulated (Scheme 5.4) [7]. **5-17** was obtained as a mixture of two diastereomers in low yield ranging from 20 to 30% [8, 9]. However, three

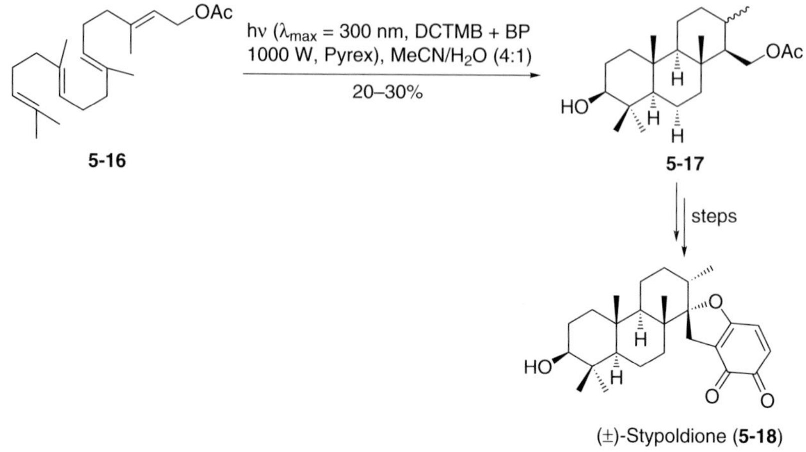

Scheme 5.4. Synthesis of a stypoldione precursor.

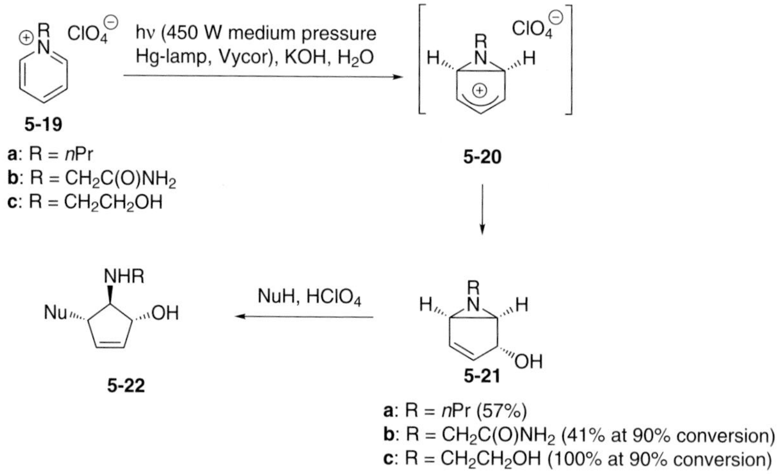

Scheme 5.5. Photochemical reaction of **5-19**.

new C–C bonds, three six-membered rings, and seven stereogenic centers were formed in this process.

The research groups of Mariano and West developed a photoinduced electrocyclization/nucleophilic addition sequence. Thus, irradiation of N-alkylpyridinium perchlorates as **5-19** in an aqueous solution led to the aziridine cations **5-20**, which react in a nucleophilic addition with OH⁻ to give the isolable azabicyclo[3.1.0]hex-2-enols **5-21**. These can be further transformed by a nucleophilic ring-opening of the aziridine moiety under acidic conditions to lead to useful unsymmetrically *trans,trans*-trisubstituted cyclopentenes **5-22** (Scheme 5.5) [10].

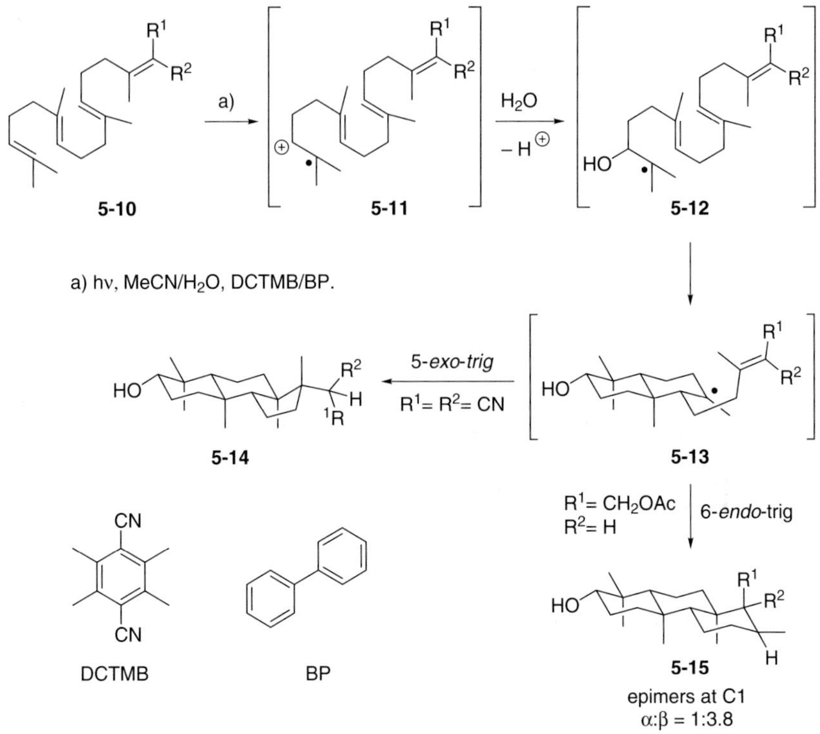

a) hν, MeCN/H₂O, DCTMB/BP.

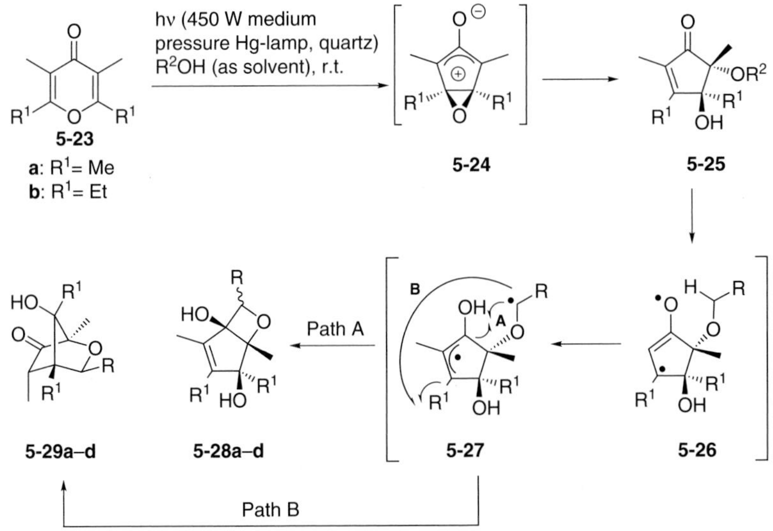

Entry	1	R¹	R²OH	R	Products (yield [%])	
1	a	Me	MeOH	H	5-28a (30)	5-29a (29)
2	a	Me	MeOH	H	5-28b (23)	5-29b (29)
3	b	Et	EtOH	Me	5-28c (45)[a]	5-29c (19)
4	b	Et	EtOH	Me	5-28d (35)[b]	5-29d (14)

[a] Isolated as 2:1 mixture of β-methyl/α-methyl diastereomers.
[b] Isolated as 1.5:1 mixture of β-methyl/α-methyl diastereomers.

Scheme 5.6. Synthesis of fused oxetanes **5-28** and bicyclic ethers **5-29**.

A direct formation of **5-22** [e. g., of 2-amino-4-cyclopentene-1,3-diol (**5-22**, Nu = OH)], is observed if the photochemical reaction of **5-19** is performed under acidic conditions using $HClO_4$. After acetylation, the formed diacetate can be isolated in about 25 % yield.

In a comparable manner, irradiation of an alcoholic solution of tetraalkyl pyran-4-ones **5-23** led to 2-alkoxycyclopentenones **5-25** by nucleophilic trapping with the solvent of the intermediate oxyallylic zwitterion **5-24** (Scheme 5.6) [11]. However, in contrast to the aziridines **5-20**, further exposure to light induces a subsequent reaction in which hydrogen abstraction from the excited enone chromophore in **5-25** furnishes a 1,4-biradical **5-26**. This can undergo a radical coupling in two ring-closure modes (Scheme 5.6; Paths A and B) to give 6-oxabicyclo[3.2.0]hept-2-ene-1,4-diols **5-28** and 2-oxabicyclo[2.2.1]heptane-6-ones **5-29** as a mixture of compounds. Despite the moderate yields and selectivities, the increase of complexity in this domino process is high.

A remarkable combination of a photochemical [2+2] cycloaddition with a retro-aldol reaction and cyclization is the so-called de Mayo reaction. Büchi and coworkers have used this method for the total synthesis of loganin (**5-35**). Thus, reac-

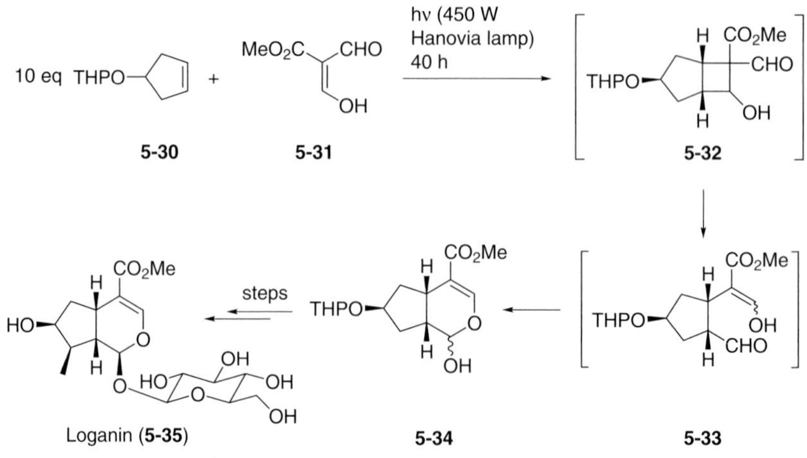

Scheme 5.7. Synthesis of loganin (**5-35**).

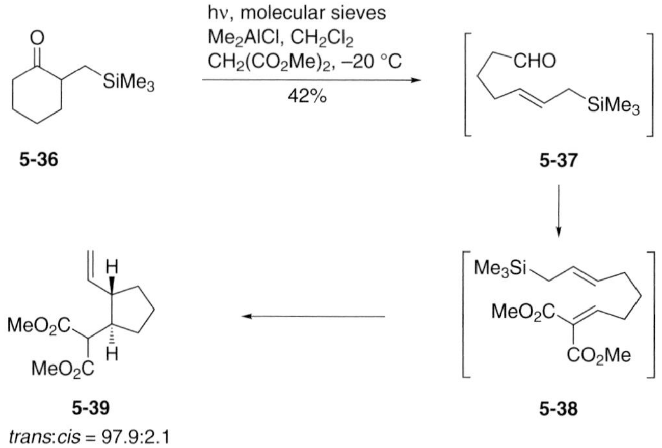

Scheme 5.8. Synthesis of 1,2-disubstituted cyclopentanes.

tion of the alkene **5-30** with methyl diformylacetate (**5-31**) under irradiation with a high-pressure mercury lamp led to the cyclobutane derivative **5-32**, which opens up to give **5-33** and, after recyclization, the cyclopentadihydropyran **5-34**. Tietze and co-workers used a similar approach for the first total synthesis of secologanin, which is a key intermediate in the biosynthesis of the monoterpenoic indole alkaloids, the cinchona, pyrroloquinoline and ipecacuanha alkaloids [12].

Another photochemically induced domino process consisting of three steps was employed for the formation of 1,2-disubstituted cyclopentanes **5-39**, as described by Tietze and coworkers. Irradiation of a mixture of **5-36**, dimethyl malonate and catalytic amounts of the Lewis acid Me$_2$AlCl in a Pyrex flask caused a Norish Type I cleavage of **5-36**, followed by an intramolecular hydrogen shift to give the acyclic

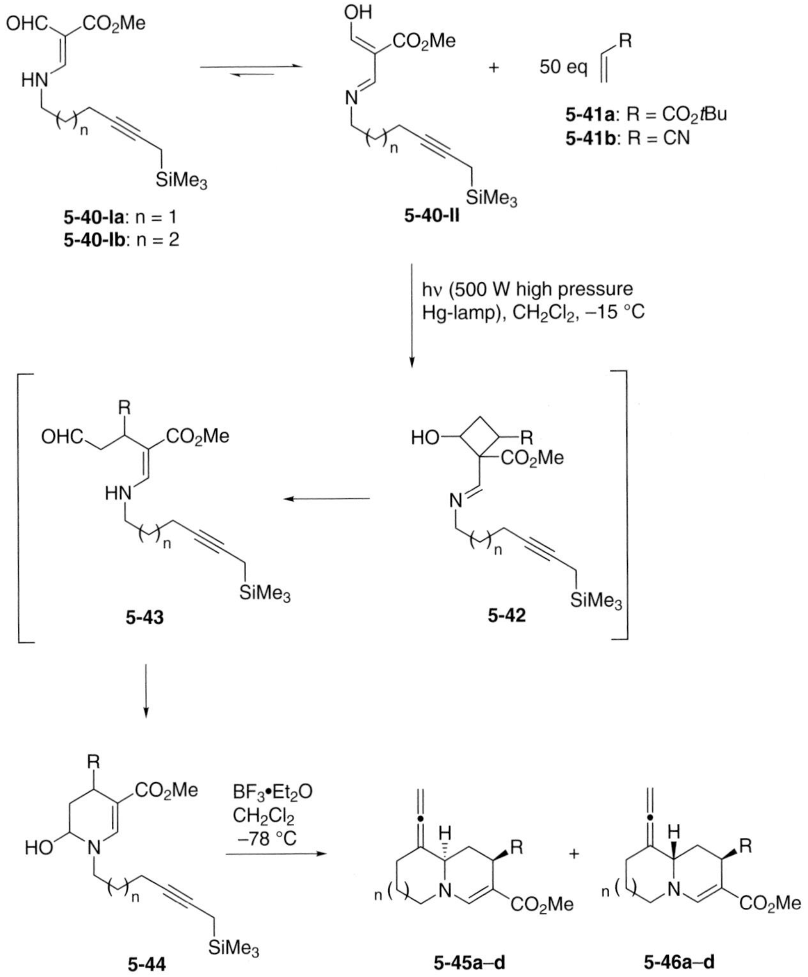

Scheme 5.9. Photochemical cycloaddition/iminium ion cyclization affording quinolizidines and higher analogues.

Entry	5-40	n	5-41	R	5-45/5-46	dr	Yield [%]
1	a	1	a	CO_2tBu	a	63.8:1	60
2	a	1	b	CN	b	86.3:1	48
3	b	2	a	CO_2tBu	c	32.2:1	68
4	b	2	b	CN	d	69.3:1	41

aldehyde **5-37** with an ω-allylsilane moiety. **5-37** underwent a Knoevenagel condensation with dimethyl malonate to give **5-38**, followed by a Lewis acid-catalyzed sila-ene reaction. The final product is the cyclopentane **5-39** with a *trans*-1,2-disubstitution with very good diastereoselectivity [13].

The same group has developed another synthetically useful photochemically induced domino transformation. Irradiation of the enaminecarbaldehydes **5-40a** or **5-40b** in the presence of acrylic acid ester **5-41a** or acrylonitrile **5-41b** afforded the quinolizidines **5-45a** and **5-45b** as well as the pyrido[1,2-*a*]azepines **5-45c** and **5-45d**, respectively, with high stereoselectivity [14]. Only very small amounts of the corresponding diastereomers **5-46a–d** were detected.

It can be assumed that, upon irradiation, tautomer **5-40-II** reacts with the alkene **5-41** in a highly regioselective [2+2] cycloaddition to give the cyclobutane **5-42** as an intermediate. Subsequent retro-aldol-type reaction and hemiacetal formation produces **5-44** *via* **5-43**. After addition of the Lewis acid (BF$_3 \cdot$Et$_2$O), cyclization takes place to give the desired products. It should be noted that the excess of alkene must be removed under reduced pressure before addition of the Lewis acid in order to avoid polymerization.

5.3
Photochemical/Radical Domino Processes

Photochemically formed radicals can undergo normal radical reactions. In these transformations, the radical cation being formed by PET gives a neutral radical spe-

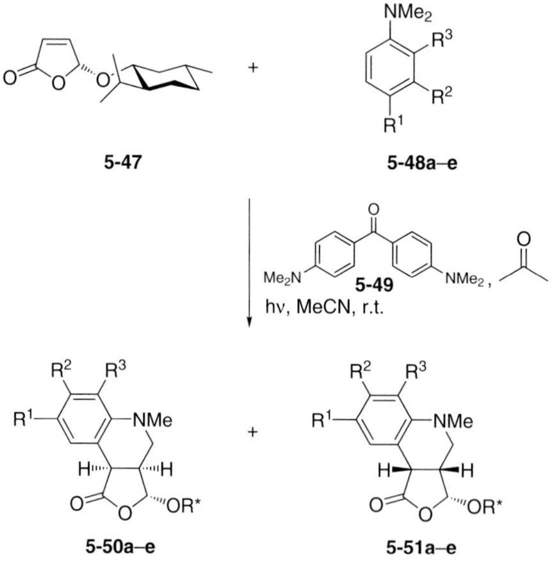

Entry	5-48	R^1	R^2	R^3	5-50/5-51	Yield [%] 5-50	Yield [%] 5-51
1	a	H	H	H	a	74	3
2	b	H	H	Me	b	60	2
3	c	Me	H	H	c	78	3
4	d	H	Me	Me	d	64	4
5	e	H	–(CH$_2$)$_2$–		e	53	3

Scheme 5.10. Synthesis of quinoline derivatives.

cies through a simple proton abstraction (Scheme 5.10). Thus, irradiation of a mixture of enantiopure (5R)-5-menthyloxy-2,5-dihydrofuran-2-one **5-47**, the aniline **5-48** and the benzophenone derivative **5-49** as sensitizer in the presence of acetone (as reported by Hoffmann and coworkers) led to enantiopure 1,2,3,4-tetrahydroquinolines of type **5-50** and **5-51** in reasonable yields (Scheme 5.10) [15].

According to the proposed mechanism, the radical cation **5-52** is first formed by the PET process sensitized by excited Michler's ketone **5-49**. The aminoalkyl radical **5-53** is then formed by proton abstraction, which adds to **5-47** to give the oxoallyl radical **5-54** (Scheme 5.11). There follows an intramolecular addition process in the *ortho* position of the electron-rich aromatic ring to give **5-55**, and a rearomatization using acetone as oxidant. The formed radical **5-56** can then abstract a hydrogen from **5-48** to give again the aminoalkyl radical **5-53**.

Quite recently, Mattay and coworkers reported on several related photochemically induced domino processes for the synthesis of steroids [16]. In these cases, the initially formed radical cation is stabilized by formation of an oxenium ion. For in-

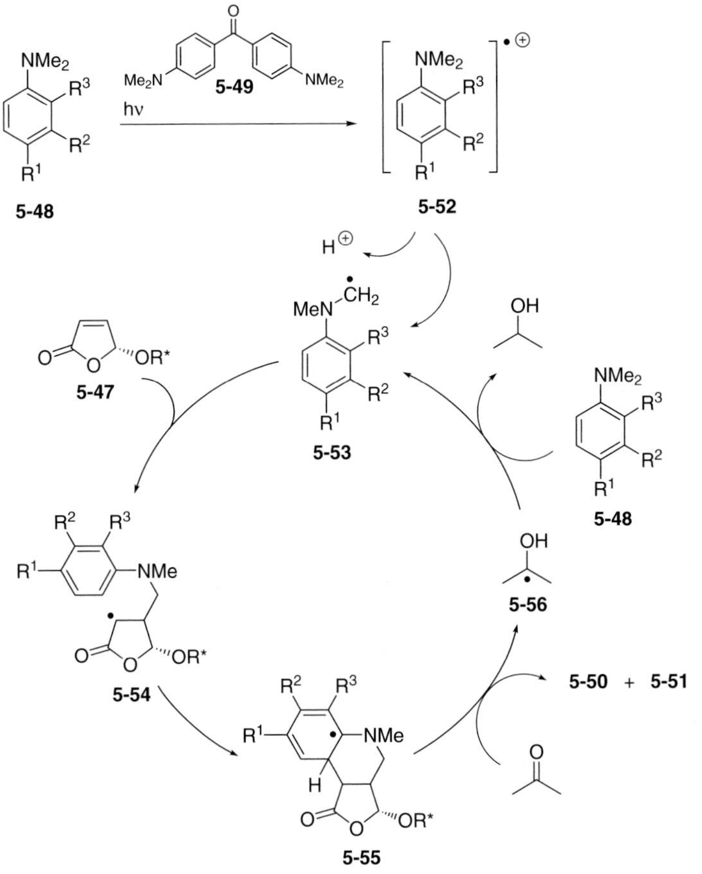

Scheme 5.11. Proposed mechanism for the photochemical reaction of **5-47** and **5-53**.

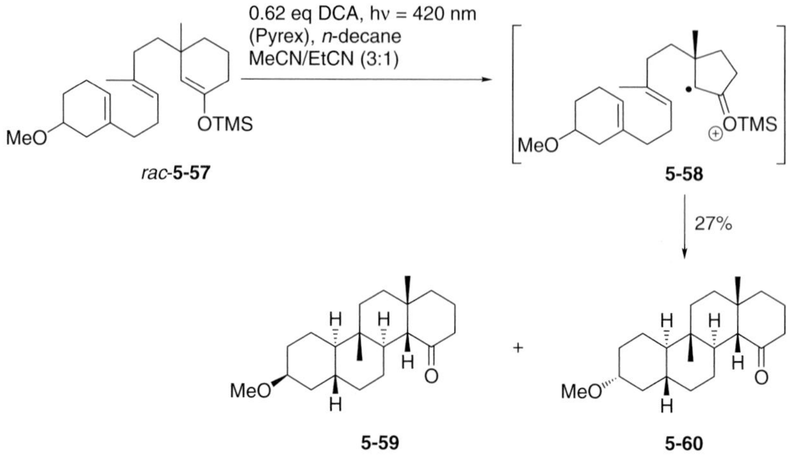

Scheme 5.12. Synthesis of steroid derivatives.

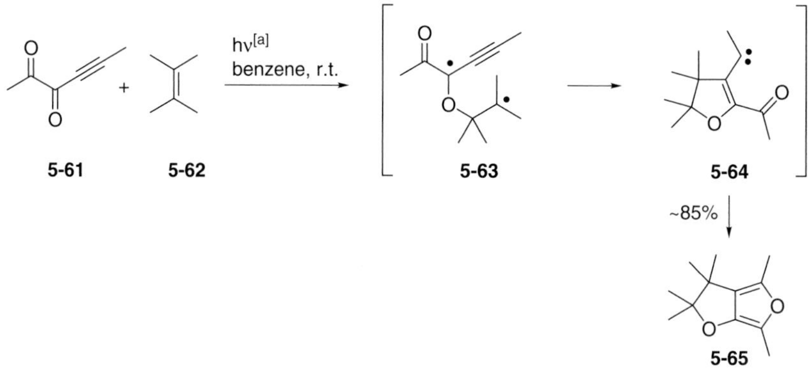

[a] >340 nm, 450 W medium pressure Hg-lamp, Pyrex, uranium-glass filter.

Scheme 5.13. Photochemically induced formation of a tetrasubstituted furan.

stance, irradiation of a diastereomeric mixture of racemic silylenol ether **5-57** in the presence of 9,10-dicyanoanthracene (DCA) as sensitizer led to the two compounds **5-59** and **5-60** in 27% combined yield (Scheme 5.12).

One of the best-known and highly useful photochemical synthetic procedures is the Paterno–Büchi reaction [17]. This transformation has also been adapted as basic principle for domino processes by different research groups. Agosta and coworkers published a procedure by which tetrasubstituted furans such as **5-65** can be built up from **5-61** and **5-62** (Scheme 5.13) [18].

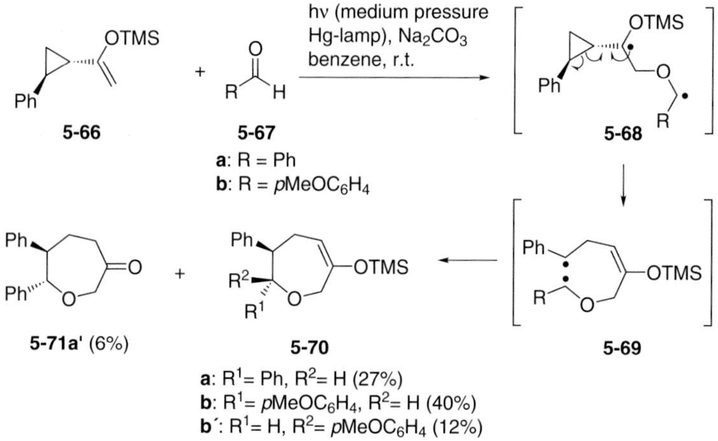

Scheme 5.14. Synthesis of tetrahydrooxepins.

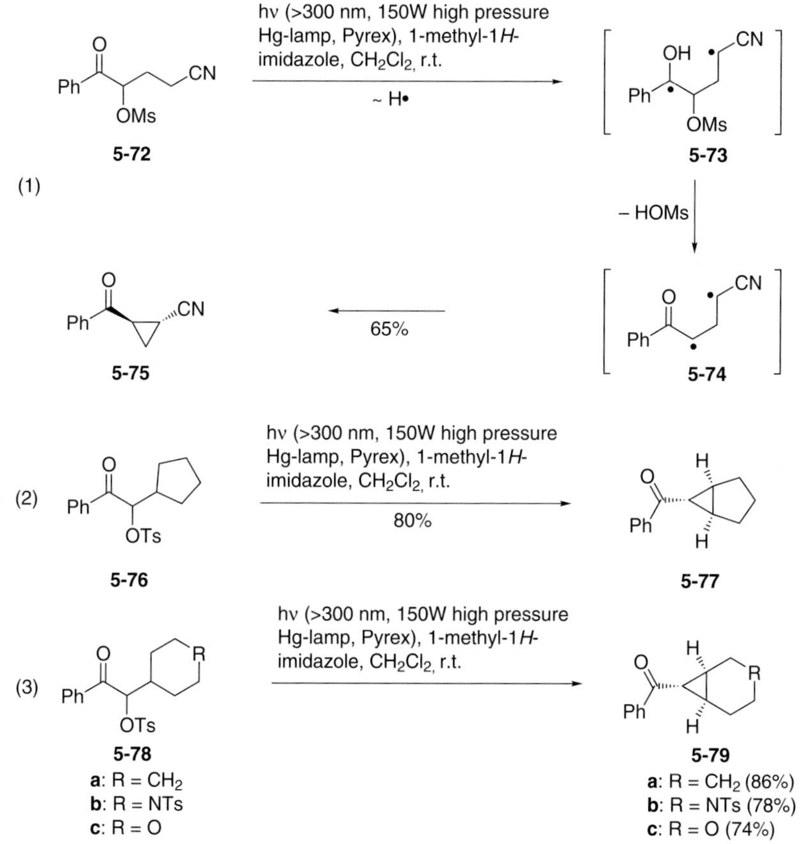

Scheme 5.15. Photochemical synthesis of functionalized cyclopropyl ketones.

It is assumed that first biradical **5-63** is formed on addition of **5-62** to photochemically excited **5-61**. There follows a ring closure involving the triple bond to yield α-acyl-α, β-unsaturated carbene **5-64**. The carbene moiety then undergoes a final 6π electrocyclic ring closure on the carbonyl group to give the desired furan **5-65**.

A more recent approach by the research group of Lambert gives access to substituted tetrahydrooxepins of type **5-70** (Scheme 5.14) [19]. Here, irradiation of the starting materials **5-66** and **5-67** again did not lead to the usually observed oxetane, but rather to a biradical **5-68**. There follows a cyclopropyl ring fragmentation to give a new biradical **5-69**, which finally undergoes a radical recombination generating the tetrahydrooxepins **5-70**, though in relatively low yields and poor selectivities.

The Norrish–Yang reaction [20] is based on the photochemical excitation of ketones followed by an intramolecular hydrogen transfer with the formation of biradicals. Wessig and coworkers used this procedure to prepare functionalized cyclopropyl ketones as **5-75** from **5-72** (Scheme 5.15) [21]. The substrate employed con-

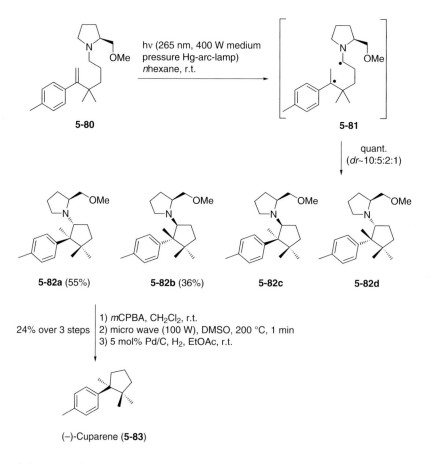

Scheme 5.16. Synthesis of (−)-cuparene (**5-83**) *via* a 5.3 sequence.

tains a leaving group adjacent to the C=O group. In that way, the primarily formed biradical **5-73** undergoes a rapid elimination to give a new biradical **5-74**, which then cyclizes to afford the product. In a similar way, **5-76** was transformed into **5-77**, whereas **5-78** led to **5-79**. All reactions proceed with high yield.

Grainger's group has developed an asymmetric route to (–)-cuparene (**5-83**) [22] using another photoinduced generation of a biradical (Scheme 5.16) [23]. Thus, irradiation of (*S*)-proline-derived **5-80** resulted initially in the formation of **5-81**, which subsequently cyclizes in almost quantitative yield to afford a mixture of the four possible diastereomers **5-82a–d** in an approximate 10:5:2:1 ratio. The two major isomers could be separated by column chromatography to provide **5-82a** in 36 % yield and the desired **5-82b** in 55 %, which was converted into the natural product **5-83** in 24 % yield over three steps.

The final example to be discussed in this chapter is a domino β-fragmentation/ hydrogen abstraction process developed by Suárez and coworkers [24]. Besides other related examples reported by the same group [25], the formation of **5-91** from **5-84** is probably the most striking due to its similarity to the natural product limonin (**5-92**) [26].

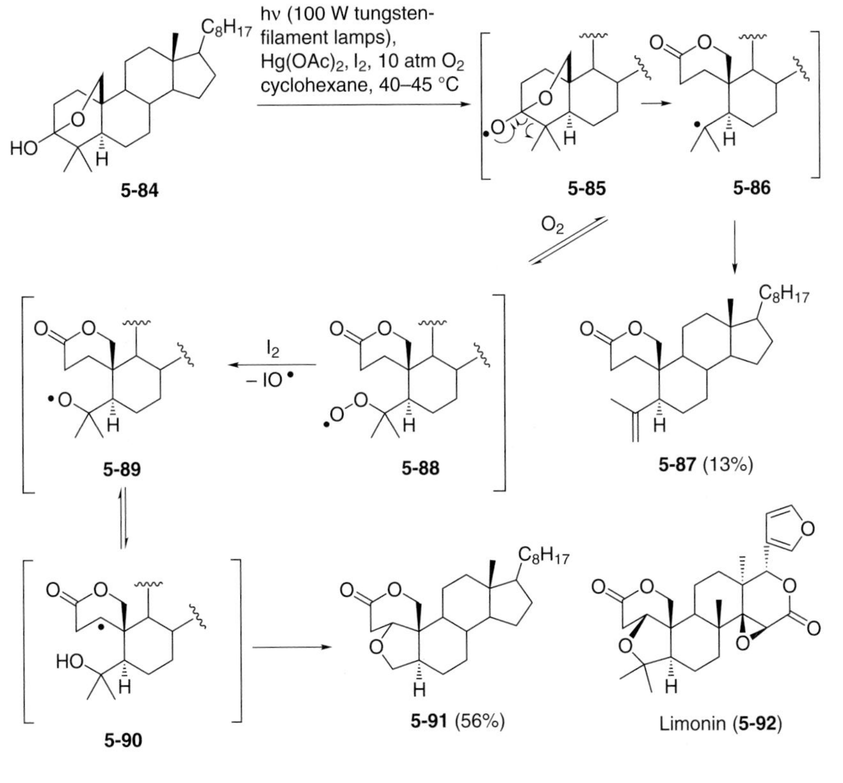

Scheme 5.17. Photochemical transformation of **5-84**.

Irradiation of the hemiacetal **5-84** in the presence of Hg(OAc)$_2$, iodine, and molecular oxygen with visible light affords the alkoxy radical **5-85**. Subsequently, a β-fragmentation/peroxidation sequence gives **5-88** via **5-86**, which then reacts with iodine leading to the alkoxy radical **5-89** and, furthermore, to the carbon radical **5-90** by hydrogen transfer. Trapping of **5-90** in a ring closure yields product **5-91** in satisfactory 56% yield, along with a byproduct **5-87**.

5.4
Photochemical/Pericyclic Domino Processes

Among photochemical/pericyclic domino processes, sequences that involve an initial photoenolization to afford reactive hydroxy-*o*-quinodimethane species which can be subsequently trapped by dienophiles in either an inter- or an intramolecular Diels–Alder fashion belong to the most elegant and efficient domino reactions in this class. Their potency for the construction of complex molecules has been demonstrated by Barton and Quinkert or, more recently, by Kraus [27], Moorthy and Venugopalan [28], as well as by Bach using an enantioselective approach [29]. In addition, Nicolaou's group has developed a straightforward approach to highly substituted tetralines **5-96** and **5-99** employing aromatic aldehydes of type **5-93** or **5-97** as substrates (Scheme 5.18) [29]. In the reaction of **5-93**, an intermolecular cycloaddition of the formed *o*-quinodimethane **5-94** with an alkene **5-95** took place, whereas with **5-97** an intramolecular reaction of the intermediate **5-98** occurred.

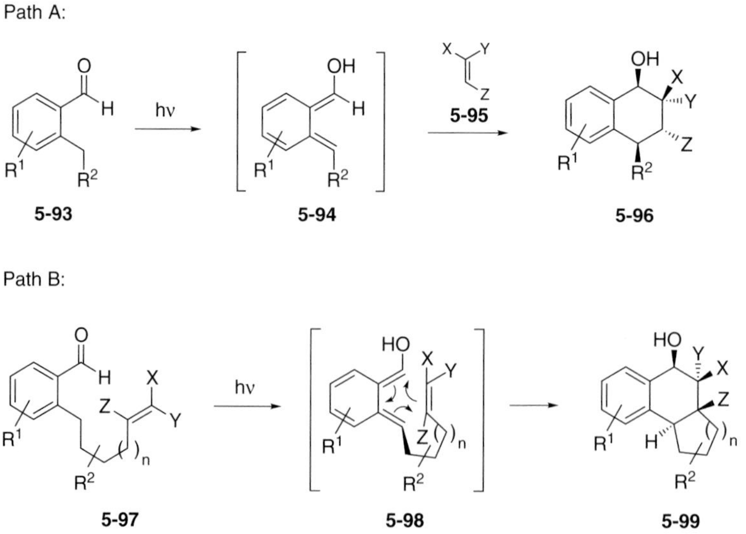

Path A:

Path B:

Scheme 5.18. General strategy for the synthesis of benzannulated compounds from benzaldehydes by the photogeneration and trapping of hydroxy-*o*-quinodimethanes.

Entry	Aldehyde	Dienophile	Product	dr[b]	Yield [%][a]
1				ca. 8:1	71
2				ca. 3:1	83
3				ca. 9:3:1	84
4					95
5					83
6					75

[a] Reaction conditions: hν (450 W medium pressure Hg-lamp, Pyrex), toluene, r.t..
[b] Major diastereomers shown.

Table 5.1. Synthesis of tetralines by intermolecular trapping of hydroxy-*o*-quinodimethanes.

In Table 5.1 are listed six selected examples, which show that many functional groups are tolerated, and that the intermolecular (Path A) as well as the intramolecular Diels–Alder mode (Path B) provide good yields and reasonable to very high selectivities.

Nicolaou and coworkers used this approach also for the synthesis of hamigerans A and B [30], as well as of several of their epimers [29, 31]. The group of Kraus succeeded in a formal total synthesis of the anticancer agent podophyllotoxin **5-103** from **5-100** [32] (Scheme 5.19) [33]. The method allows a rapid access to the central core **5-102** via **5-101**.

Another photoinduced 1,5-hydrogen shift/twofold Diels–Alder process to give substituted aromatic compounds has been reported by Nair and coworkers [34].

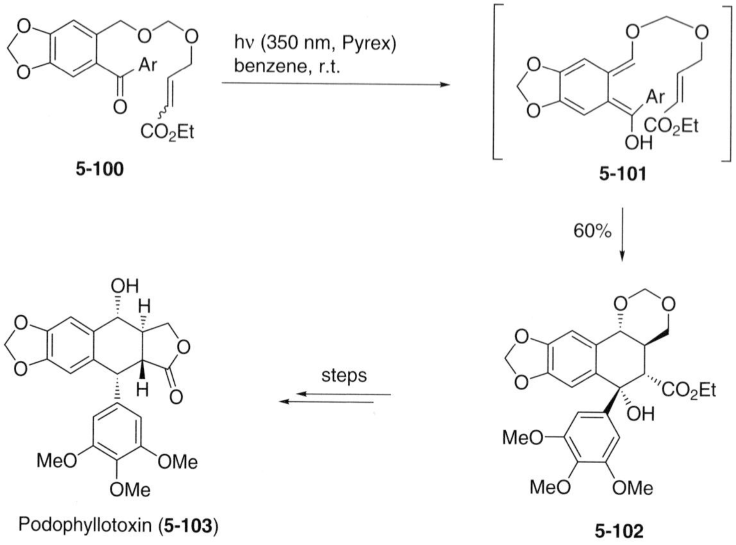

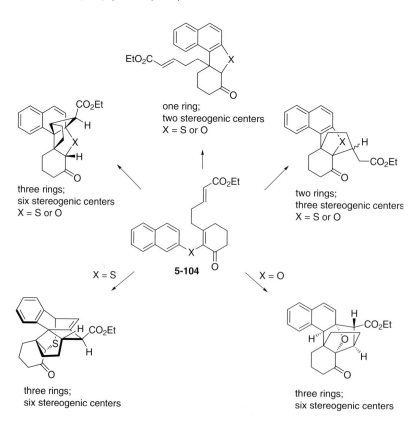

Scheme 5.19. Photoenolization/Diels–Alder reaction sequence is the synthesis of podophyllotoxin (**5-103**).

Scheme 5.20. Photoinduced cyclizations of **5-104**.

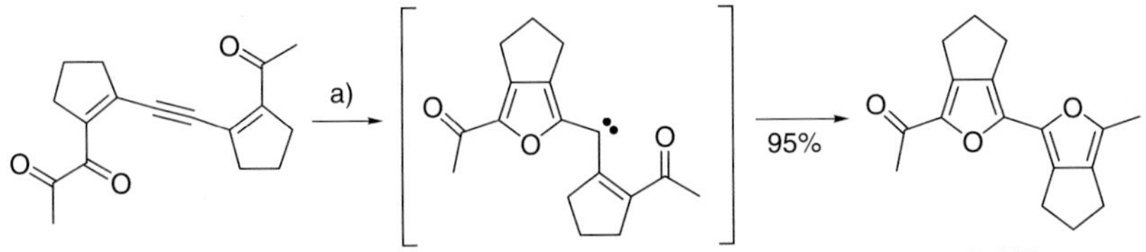

a) hν (365 nm, Pyrex), benzene,
r.t..

Scheme 5.21. Photocyclization/electrocyclization sequence for the synthesis of **5-107**.

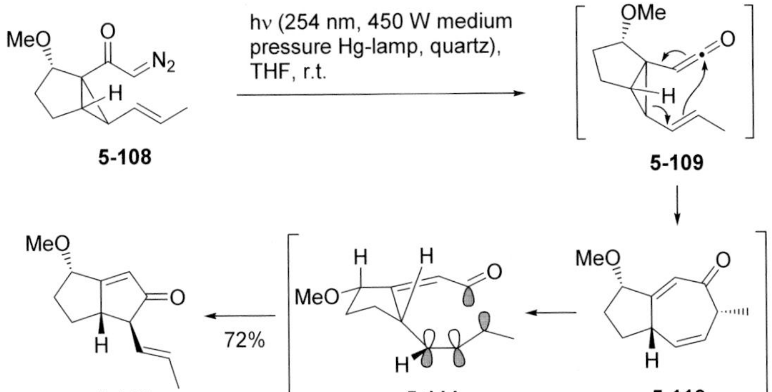

Four further examples:
69–80% yield

Scheme 5.22. Domino Wolff/Cope rearrangement/Norrish type I fragmentation/recombination
process.

Likewise, Sheridan's group developed a photoassisted twofold [6π + 2π] cyclization generating tetracycloundecadienes [35], while Dittami's group has reported on several impressive domino photocyclization/pericyclic processes of substrates of type **5-104**, which are represented in Scheme 5.20 [36].

The group of Nakatani introduced photochemically induced cyclization that takes advantage of a system in which a carbene generator and a carbene trap are combined in the same molecule [37]. Thus, irradiation of compound **5-105** induced a cyclization to give an intermediate carbene **5-106**, which underwent an intramolecular trapping by a pericyclic 6π electrocyclization to afford **5-107** in a very good yield of 95% (Scheme 5.21).

Finally, a nice combination of a light-induced Wolff reaction of a diazoketone (e. g., **5-108**) with a thermal Cope rearrangement, a light-induced Norrish type I cleavage and a recombination was developed by Stoltz and coworkers (Scheme 5.22) [38]. Here, irradiation of **5-108** at 254 nm in a photoreactor afforded the bicyclic compound **5-112** in 72% yield. As intermediate, the cyclopentanoheptadienone **5-110** is formed *via* **5-109**, which on prolonged exposure to light leads to a Norrish type I fragmentation to give the transient allylic radical **5-111**. Immediate recombination of that biradical then leads to **5-112**. It is worth noting that the high diastereoselectivity of the formal 1,3-acyl migration affords the fused bicyclo[3.3.0]octane **5-112** as a single isomer.

5.5
Photochemical/Photochemical Domino Processes

The photochemical *meta* cycloaddition of arenes and alkenes, in which three C–C bonds are formed, is a powerful sequence for the construction of complex molecules [39]. The photoaddition, starting from simple substrates, proceeds with almost quantitative yield. The process has been employed by Wender and co-workers in various ways for the synthesis of natural products such as silphenes, subergorgic acids, grayanotoxin, and retigeranic acid (**5-116**) [40]. The *meta* photoaddition of enantiomerically pure **5-113** led to the angular annulated tricyclopentane system **5-114** which, after addition of formamide, gave triquinane **5-115** in a second photochemical step with opening of the vinylcyclopropane. The complex pentacyclic retigeranic acid (**5-116**) could then be formed from **5-115** in a few steps [40b]. This process includes an intramolecular Diels–Alder reaction for the formation of the 5,6-ring system.

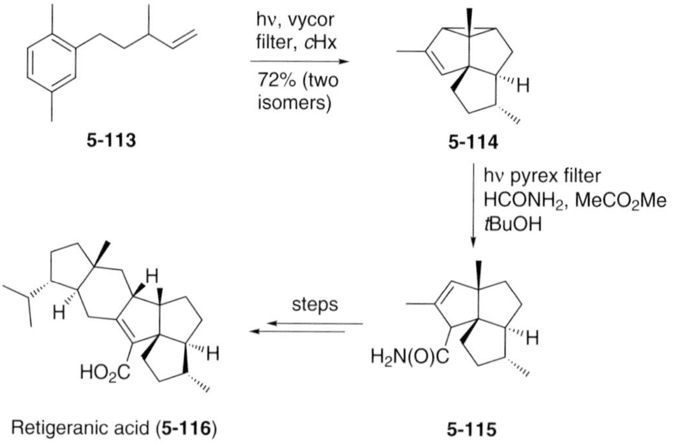

Scheme 5.23. Synthesis of retigeranic acid (**5-116**) by photochemical *meta*-cycloaddition.

5.6
Photochemical/Transition Metal-Catalyzed Domino Processes

To date, only one example of a combination of a photochemically induced transformation with a transition metal-catalyzed reaction has been found in the literature. This hv/Pd⁰-promoted process allows the synthesis of five-membered cyclic γ-keto esters **5-119** from 5-iodoalkenes **5-117** in the presence of CO and an alcohol **5-118** as a nucleophile (Scheme 5.24) [41]. The yields are high, and differently substituted iodoalkenes can be employed.

Komatsu and coworkers also carried out mechanistic studies on their new domino process, showing there is an interplay of two reactive intermediates: radicals and organopalladium complexes (Scheme 5.25). It is proposed that under the influence of both, the Pd⁰ catalyst and light, radical **5-120** is formed first. Subsequent CO addition furnishes the acyl radical **5-121**, which is trapped in a 5-*exo*-trig cycliza-

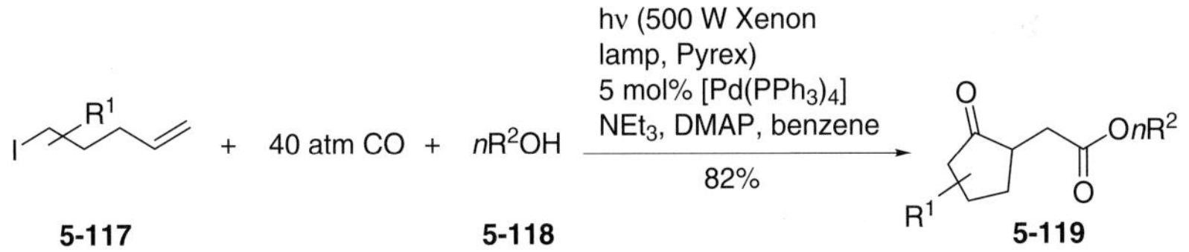

Entry	Substrate 5-117	Nucleophile 5-118	Product 5-119	Yield [%]
1	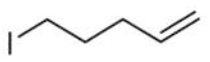	nBuOH	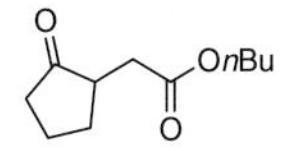	82
2		BnOH	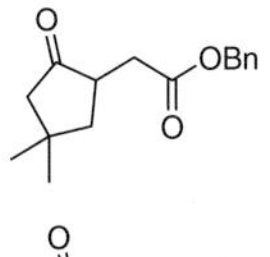	83
3	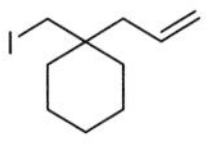	EtOH	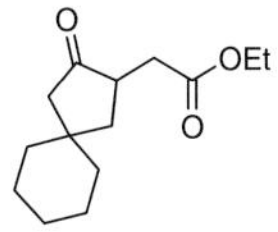	76
4	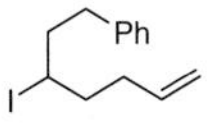	MeOH	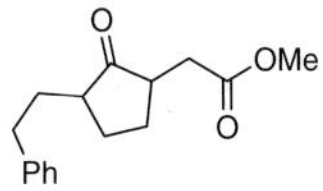 cis:trans = 44:56	72
5	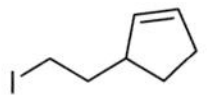	MeOH	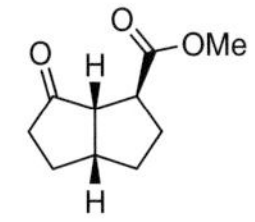	78

Scheme 5.24. Light-Pd⁰-induced cyclization/carbonylation sequence.

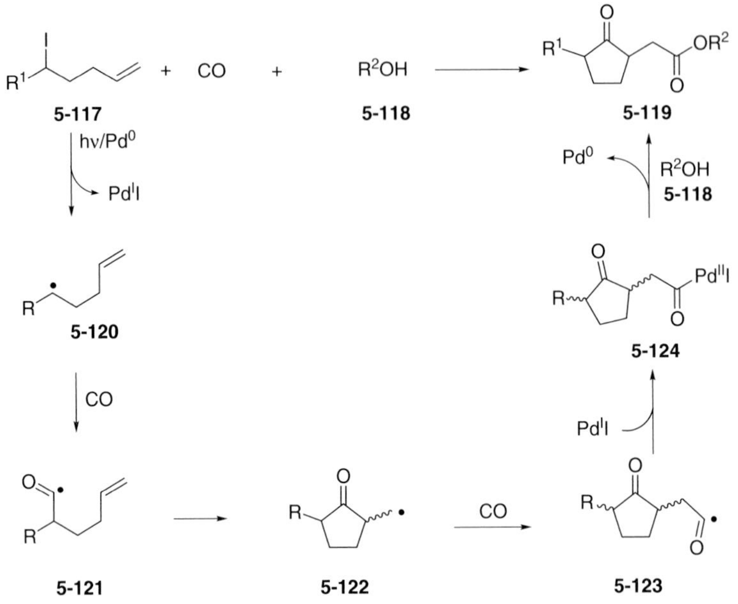

Scheme 5.25. Proposed mechanism of the Pd/light-induced domino sequence.

tion, after which a second carbonylation gives **5-123** via **5-122**. This radical couples with a Pd^I species to form an acyl-palladium intermediate **5-124**, which is attacked by the nucleophile R^{2}OH **5-118** to afford the final products **5-119** by a reductive elimination of Pd^0.

To date, no examples have been found in the literature for the combination of photochemically induced transformations with reductive/oxidative or enzymatic processes.

References

1 (a) A. Bauer, F. Westkämper, S. Grimme, T. Bach, *Nature* **2005**, *436*, 1139–1140; (b) B. Basler, O. Schuster, T. Bach, *J. Org. Chem.* **2005**, *70*, 9798–9808; c) M. Kemmler, E. Herdtweck, T. Bach, *Eur. J. Org. Chem.* **2004**, 4582–4595.

2 (a) J. Mattay, *Synthesis* **1989**, 233–252; (b) G. J. Kavarnos, *Fundamentals of Photoinduced Electron Transfer*, VCH, New York, **1993**.

3 (a) R. A. Neunteufel, D. R. Arnold, *J. Am. Chem. Soc.* **1973**, *95*, 4080–4081; (b) T. Majima, C. Pac, A. Nakasone, H. Sakurai, *J. Am. Chem. Soc.* **1981**, *103*, 4499–4508;

(c) S. L. Mattes, S. Farid, *J. Am. Chem. Soc.* **1986**, *108*, 7356–7361; (d) U. Hoffmann, Y. Gao, B. Pandey, S. Klinge, K.-D. Warzecha, C. Krüger, H. D. Roth, M. Demuth, *J. Am. Chem. Soc.* **1993**, *115*, 10358–10359; (e) C. Heinemann, M. Demuth, *J. Am. Chem. Soc.* **1997**, *119*, 1129–1130.

4 H. Ishii, R. Yamaoka, T. Imai, T. Hirano, S. Maki, H. Niwa, D. Hashizume, F. Iwasaki, M. Ohashi, *Tetrahedron Lett.* **1998**, *39*, 9501–9504.

5 (a) C. Pac, A. Nakasone, H. Sakurai, *J. Am. Chem. Soc.* **1977**, *99*, 5806–5808; (b)

T. Asanuma, T. Gotoh, Y. Tsuchida, M. Yamamoto, Y. Nishijima, *J. Chem. Soc., Chem. Commun.* **1977**, 485–486.

6 S. L. Mattes, S. Farid, *J. Am. Chem. Soc.* **1986**, *108*, 7356–7361.

7 (a) W. H. Gerwick, W. Fenical, N. Fritsh, J. Clardy, *Tetrahedron Lett.* **1979**, *20*, 145–148; (b) W. H. Gerwick, W. Fenical, *J. Org. Chem.* **1981**, *46*, 22–27.

8 (a) C. Heinemann, X. Xing, K.-D. Warzecha, P. Ritterskamp, H. Görner, M. Demuth, *Pure Appl. Chem.* **1998**, *70*, 2167–2176; (b) K.-D. Warzecha, X. Xing, M. Demuth, *Pure Appl. Chem.* **1997**, *69*, 109–112; (c) K.-D. Warzecha, X. Xing, M. Demuth, *Helv. Chim. Acta* **1995**, *78*, 2065–2076; (d) C. Heinemann, M. Demuth, *J. Am. Chem. Soc.* **1997**, *119*, 1129–1130; (e) V. Rosales, J. Zambrano, M. Demuth, *Eur. J. Org. Chem.* **2004**, 1798–1802.

9 (a) X. Xing, M. Demuth, *Eur. J. Org. Chem.* **2001**, 537–544; (b) X. Xing, M. Demuth, *Synlett* **1999**, 987–990.

10 R. Ling, M. Yoshida, P. S. Mariano, *J. Org. Chem.* **1996**, *61*, 4439–4449.

11 (a) M. Fleming, R. Basta, P. V. Fisher, S. Mitchell, F. G. West, *J. Org. Chem.* **1999**, *64*, 1626–1629; For related sequences, see: (b) F. G. West, P. V. Fisher, A. M. Arif, *J. Am. Chem. Soc.* **1993**, *115*, 1595–1597; (c) F. G. West, D. W. Willoughby, *J. Org. Chem.* **1993**, *58*, 3796–3797.

12 (a) G. Büchi, J. A. Carlson, J. E. Powell, Jr., L. F. Tietze, *J. Am. Chem. Soc.* **1970**, *92*, 2165–2167; (b) G. Büchi, J. A. Carlson, J. E. Powell, Jr., L. F. Tietze, *J. Am. Chem. Soc.* **1973**, *95*, 540–545; (c) L. F. Tietze, *Angew. Chem. Int. Ed. Engl.* **1983**, *22*, 828–841.

13 L. F. Tietze, J. R. Wünsch, *Synthesis* **1990**, *11*, 985–990.

14 L. F. Tietze, J. R. Wünsch, M. Noltemeyer, *Tetrahedron* **1992**, *48*, 2081–2099.

15 (a) S. Bertrand, N. Hoffmann, J.-P. Pete, V. Bulach, *Chem. Commun.* **1999**, 2291–2292; (b) S. Bertrand, N. Hoffmann, S. Humbel, J. P. Pete, *J. Org. Chem.* **2000**, *65*, 8690–8703.

16 (a) J. O. Bunte, S. Rinne, J. Mattay, *Synthesis* **2004**, 619–633; (b) H. Rinderhagen, J. Mattay, *Chem. Eur. J.* **2004**, *10*, 851–874.

17 (a) E. Paterno, G. Chieffi, *Gazz. Chim. Ital.* **1909**, *39*, 341–361; (b) G. Büchi, C. G. Inman, E. S. Lipinsky, *J. Am. Chem. Soc.* **1954**, *76*, 4327–4331.

18 A. K. Mukherjee, P. Margaretha, W. C. Agosta, *J. Org. Chem.* **1996**, *61*, 3388–3391.

19 C. Yong Gan, J. N. Lambert, *J. Chem. Soc., Perkin Trans. 1* **1998**, 2363–2372.

20 (a) P. J. Wagner, *Acc. Chem. Res.* **1971**, *4*, 168–177; (b) P. J. Wagner, *Top. Curr. Chem.* **1976**, *66*, 1–52; (c) P. J. Wagner, *Acc. Chem. Res.* **1989**, *22*, 83–91; (d) P. J. Wagner, B.-S. Park, in: *Organic Photochemistry* (Ed.: A. Padwa), Marcel Dekker Inc., New York-Basel, **1991**, p. 227; (e) P. J. Wagner, in: *CRC Handbook of Organic Photochemistry and Photobiology* (Eds.: W. M. Horspool, P. S. Song), CRC Press, Boca Raton, New York, London, Tokyo, **1995**, p. 449.

21 P. Wessig, O. Mühling, *Helv. Chim. Acta* **2003**, *86*, 865–893.

22 For another total synthesis based on a domino process, see: T. Cohen, T. Kreethadumrongdat, X. Liu, V. Kulkarni, *J. Am. Chem. Soc.* **2001**, *123*, 3478–3483.

23 (a) R. S. Grainger, A. Patel, *Chem. Comm.* **2003**, 1072–1073; for related examples, see: (b) F. D. Lewis, G. D. Reddy, D. M. Bassani, *J. Am. Chem. Soc.* **1993**, *115*, 6468–6469; (c) F. D. Lewis, G. D. Reddy, D. M. Bassani, S. Schneider, M. Gahr, *J. Am. Chem. Soc.* **1994**, *116*, 597–605.

24 A. Boto, R. Freire, R. Hernández, E. Suárez, M. S. Rodríguez, *J. Org. Chem.* **1997**, *62*, 2975–2981.

25 A. Boto, R. Hernández, E. Suárez, C. Betancor, M. S. Rodríguez, *J. Org. Chem.* **1995**, *60*, 8209–8217.

26 For reviews, see: (a) D. L. Dreyer, in: *Fortschritte der Chemie Organischer Naturstoffe, Vol. 26* (Ed.: L. Zechmeister), Springer-Verlag, Wien, **1968**, pp. 190–244; (b) D. A. H. Taylor, in: *Fortschritte der Chemie Organischer Naturstoffe, Vol. 45* (Ed.: W. Herz, H. Grisebach, G. W. Kirby), Springer-Verlag, Wien, **1984**, pp. 1–102; (c) J. D. Connolly, K. C. Overton, J. Polonsky, in: *Progress in Phytochemistry, Vol. 2* (Eds.: L. Reinhold, Y. Liwschitz), Interscience, London, **1970**, pp. 385–484.

27 (a) G. A. Kraus, L. Chen, *Synth. Commun.* **1993**, *14*, 2041–2049; (b) G. A. Kraus, G. Zhao, *J. Org. Chem.* **1996**, *61*, 2770–2773.

28 J. N. Moorthy, P. Mal, N. Singhal, P. Venkatakrishnan, R. Malik, P. Venugopalan, *J. Org. Chem.* **2004**, *69*, 8459–8466.

29 (a) B. Grosch, C. N. Orlebar, E. Herdtweck, M. Kaneda, T. Wada, Y. Inoue, T.

Bach, *Chem. Eur. J.* **2004**, *10*, 2179–2189; (b) K. C. Nicolaou, D. Gray, J. Tae, *Angew. Chem. Int. Ed.* **2001**, *40*, 3675–3678.

30 K. D. Wellington, R. C. Cambie, P. S. Rutledge, P. R. Bergquist, *J. Nat. Prod.* **2000**, *63*, 79–85.

31 K. C. Nicolaou, D. Gray, J. Tae, *Angew. Chem. Int. Ed.* **2001**, *40*, 3679–3683.

32 T. W. Doyle, in: *Etoposide (VP-16) Current Status and New Developments*, Academic Press, New York, **1984**.

33 G. A. Kraus, Y. Wu, *J. Org. Chem.* **1992**, *57*, 2922–2925.

34 V. Nair, G. Anilkumar, C. N. Jayan, N. P. Rath, *Tetrahedron Lett.* **1998**, *39*, 2437–2440.

35 W. Chen, K. Chaffee, H.-J. Chung, J. B. Sheridan, *J. Am. Chem. Soc.* **1996**, *118*, 9980–9981.

36 J. P. Dittami, X. Y. Nie, H. Nie, H. Ramanathan, C. Buntel, S. Rigatti, J. Bordner, D. L. Decosta, P. Williard, *J. Org. Chem.* **1992**, *57*, 1151–1158.

37 K. Nakatani, K. Adachi, K. Tanabe, I. Saito, *J. Am. Chem. Soc.* **1999**, *121*, 8221–8228.

38 R. Sarpong, J. T. Su, B. M. Stoltz, *J. Am. Chem. Soc.* **2003**, *125*, 13624–13625.

39 (a) A. Gilbert, *Pure Appl. Chem.* **1980**, *52*, 2669–2682; (b) D. Bryce-Smith, *Pure Appl. Chem.* **1973**, *34*, 193–212; (c) J. Mattay, *Angew. Chem. Int. Ed. Engl.* **1987**, *26*, 825–849; (d) P. A. Wender, L. Siggel, J. M. Nuss, *Org. Photochem.* **1989**, *10*, 357–473.

40 (a) P. A. Wender, R. Ternansky, M. deLong, S. Singh, A. Olivero, K. Rice, *Pure Appl. Chem.* **1990**, *62*, 1597–1602; (b) P. A. Wender, S. K. Singh, *Tetrahedron Lett.* **1990**, *31*, 2517–2520.

41 I. Ryu, S. Kreimerman, F. Araki, S. Nishitani, Y. Oderaotoshi, S. Minakata, M. Komatsu, *J. Am. Chem. Soc.* **2002**, *124*, 3812–3813.

6

Transition Metal-Catalyzed Domino Reactions

Transition metal-catalyzed transformations are of major importance in synthetic organic chemistry [1]. This reflects also the increasing number of domino processes starting with such a reaction. In particular, Pd-catalyzed domino transformations have seen an astounding development over the past years with the Heck reaction [2] – the Pd-catalyzed transformation of aryl halides or triflates as well as of alkenyl halides or triflates with alkenes or alkynes – being used most often. This has been combined with another Heck reaction or a cross-coupling reaction [3] such as Suzuki, Stille, and Sonogashira reactions. Moreover, several examples have been published with a Tsuji–Trost reaction [1b, 4], a carbonylation, a pericyclic or an aldol reaction as the second step.

In a similar manner, cross-coupling reactions have been used as the first step, followed by a second Pd-catalyzed transformation or other reactions.

Several Pd-catalyzed domino processes start with a Tsuji–Trost reaction, a palladation of alkynes or allenes [5], a carbonylation [6], an amination [7] or a Pd(II)-catalyzed Wacker-type reaction [8]. A novel illustrious example of this procedure is the efficient enantioselective synthesis of vitamin E [9].

Within this chapter, two sections are devoted to rhodium and ruthenium. The two main procedures using rhodium are: first, the formation of 1,3-dipoles from diazocompounds followed by a 1,3-dipolar cycloaddition [10]; and second, hydroformylation [11]. The ruthenium-catalyzed domino reactions are mostly based on metathesis [12], with the overwhelming use of Grubbs I and Grubbs II catalysts.

In the final section of the chapter, the use of the remaining transition metals in domino processes, including cobalt, nickel, copper, titanium, and iron, will be discussed.

In transition metal-catalyzed domino reactions, more than one catalyst is often employed. In Tietze's definition and the classification of domino reactions, no distinction has been made between transformations where only one or more transition metal catalyst is used for the different steps, provided that they take place in a chronologically distinct order. Poli and coworkers [13] differentiated between these processes by calling them "pure-domino reactions" (which consisted of a single catalytic cycle driven by a single catalytic system) or "pseudo-domino reactions". The latter type was subdivided into:

- Type I reactions, which consisted of two or more independent catalytic cycles employing one catalyst; and
- Type II reactions, which employed two or more different catalysts.

Domino Reactions in Organic Synthesis. Lutz F. Tietze, Gordon Brasche, and Kersten M. Gericke
Copyright © 2006 WILEY-VCH Verlag GmbH & Co. KGaA, Weinheim
ISBN: 3-527-29060-5

Such an inter-type difference will not be utilized in this book, mainly because it complicates the classification and is not necessary as the focus is placed on the substrates and the products. The argument is also valid for enzymatic transformations [12d, 14], where one enzymatic system with one enzyme or different independent enzymatic systems with one or more enzymes may be used. In Nature, as well as in several artificial enzymatic domino reactions, a mixture of different enzymes catalyzing independent cycles is employed.

6.1
Palladium-Catalyzed Transformations

Palladium has the advantage of being compatible with many functional groups, and it is therefore an ideal catalyst for domino reactions. As in all processes of this type, an adjustment of the reactivity of the functionalities involved in the different steps is necessary. This may be done by taking advantage, for example, of the different reactivity of aryl iodides compared to aryl bromides, or of vinyl bromides compared to aryl bromides. An illustrious example of this is the synthesis of estradiol **6/1-5** by Tietze and coworkers, who used two Heck reactions starting with the substituted anisol **6/1-1**, which undergoes a Heck reaction with the indene derivative **6/1-2** to give **6/1-3** [15]. The transformation takes place exclusively at the vinyl bromide moiety in **6/1-1**. In a second Heck reaction, **6/1-3** was transformed into **6/1-4**, which led to the enantiopure estradiol (**6/1-5**) in three steps (Scheme 6/1.1).

Another possibility of allowing differentiation is preference in the formation of different ring sizes in the different steps, as in the reaction of **6/1-6**, where a four- or a six-membered ring could be formed in the first step.

Thus, an early example of two successive Heck reactions is the formation of the condensed bicyclic compound **6/1-7** from the acyclic precursor **6/1-6** by Overman

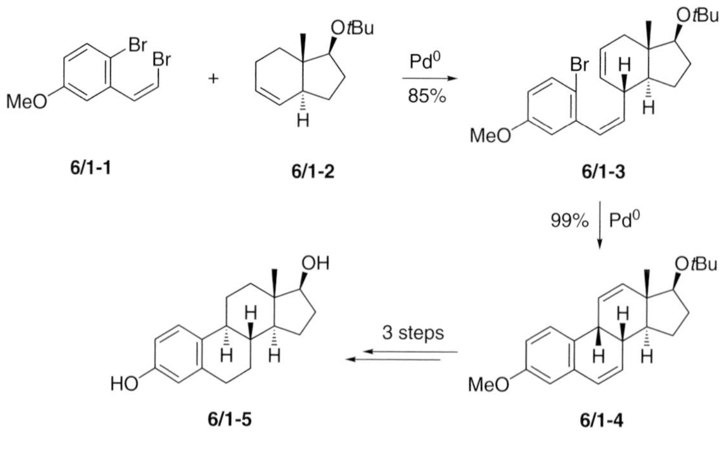

Scheme 6/1.1. Synthesis of enantiopure estradiol (**6/1-5**).

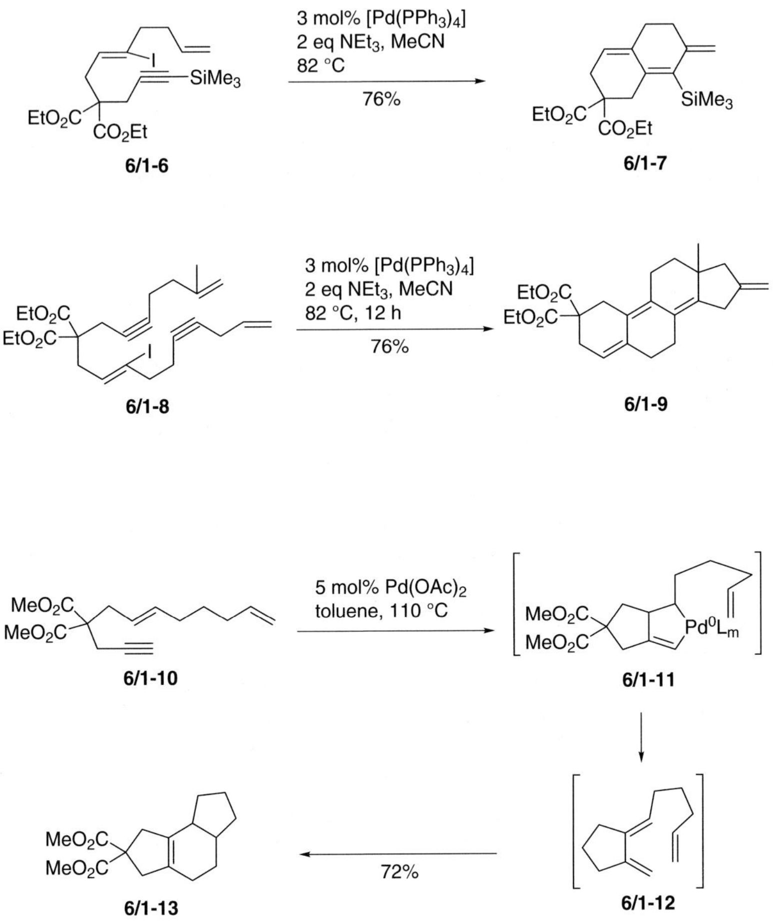

Scheme 6/1.2. Formation of polycyclic compounds.

and coworkers [16]. It was later shown by Negishi that a multitude of rings, as in 6/1-9, can be prepared by this process, starting from 6/1-8 [17].

The variety of products can be increased by an anion-capturing process of the intermediate Pd-compounds, and this has been intensively explored by Grigg and coworkers [18].

One productive facet of Pd-catalyzed domino reactions is the cycloisomerization of enynes and allenes, as shown by Trost and coworkers [19]. Thus, transformation of the dienyne 6/1-10 using Pd(OAc)$_2$ led to 6/1-13 in 72 % yield, in which the last step is a Diels–Alder reaction of the intermediate 6/1-12 (Scheme 6/1.2).

6.1.1
The Heck Reaction

6.1.1.1 Domino Heck Reactions

Overcrowded tetrasubstituted alkenes can possess remarkable switching properties under ultraviolet (UV) light [20]. They are therefore an interesting class of compounds for the design of organic switches and the development of reversible optical data storage. Tietze and coworkers [21] have prepared a wide variety of this type of compounds **6/1-16** by double Heck processes in high yield, and with complete control of the configuration of the formed double bond using aryl bromides **6/1-14a–f** as substrates which contain a triple bond and an allylsilane moiety. The best results were obtained using the Hermann-Beller catalyst **6/1-15** [22]. It can be assumed that the Pd species **6/1-18**, **6/1-19** and **6/1-20** are intermediates, whereby the allylsilane moiety in **6/1-20** is responsible for selective formation of the tertiary stereogenic center (Scheme 6/1.3). Thus, Tietze and coworkers have shown that the inherent disadvantage of Heck reactions using alkenes with α- and α'-hydrogens to produce mixtures of double bond isomers can be avoided by employing allylsilanes [23]. In such cases, the Pd–H elimination of the intermediate Pd-complex is highly regioselective. When irradiated with a high-pressure mercury lamp, (*E*)-**6/1-16** gives a 1:1-mixture of the (*Z*)- and (*E*)-isomers.

The double-Heck-approach can also be employed for the preparation of novel heterocyclic compounds as **6/1-25** and **6/1-26** (Scheme 6/1.4) [24]. Thus, the palladium-catalyzed reaction of **6/1-21** and the cyclic enamide **6/1-22** gave a 1.2:1-mixture of **6/1-23** and **6/1-24**, which in a second Heck reaction using the palladacene **6/1-15** led to **6/1-25** and **6/1-26** in an overall yield of 44–49 %. The synthesis can also be performed as a domino process using a mixture of Pd(OAc)$_2$ and the palladacene **6/1-15**.

A clever synthesis of (+)- and (–)-scopadulcic acid A **6/1-27** was described by Overman and coworkers [25] employing a Pd-catalyzed domino process of the methylene cycloheptene iodide **6/1-33**. This generates the B, C, and D rings of the scopadulan skeleton as a single stereoisomer in 90 % yield following the retrosynthetic analysis of **6/1-27** to give **6/1-28**. The synthesis started with **6/1-29**, which allowed an enantioselective reduction of the carbonyl moiety using (*R*)-*B*-isopinocampheyl-9-borabicyclo[3.3.1]nonane[(*R*)-Alpine-Borane] following the procedure by Brown and Midland [26] to generate the corresponding secondary alcohol in 94 % yield and 88 % *ee*. Transformation of the CH$_2$OTBS group into the organolithium-derivative followed by coupling with the amide **6/1-30** gave **6/1-31**. Cope rearrangement of the corresponding enoxysilane of **6/1-31** then led to **6/1-32** which was transformed into **6/1-33** using classical reactions. The domino Heck reaction of **6/1-33** proceeded cleanly to afford **6/1-34** after desilation as a single tricyclic product. The addition of Ag$_2$CO$_3$ is necessary to suppress migration of the double bond in the initially formed intermediate. In a seven-step synthesis, **6/1-34** was then transformed into **6/1-35** which, in eight steps, afforded the desired product **6/1-27** (Scheme 6/1.5).

The Shibasaki group [27] developed an enantioselective total synthesis of (+)-xestoquinone (**6/1-38**) using an asymmetric double Heck reaction with BINAP as chi-

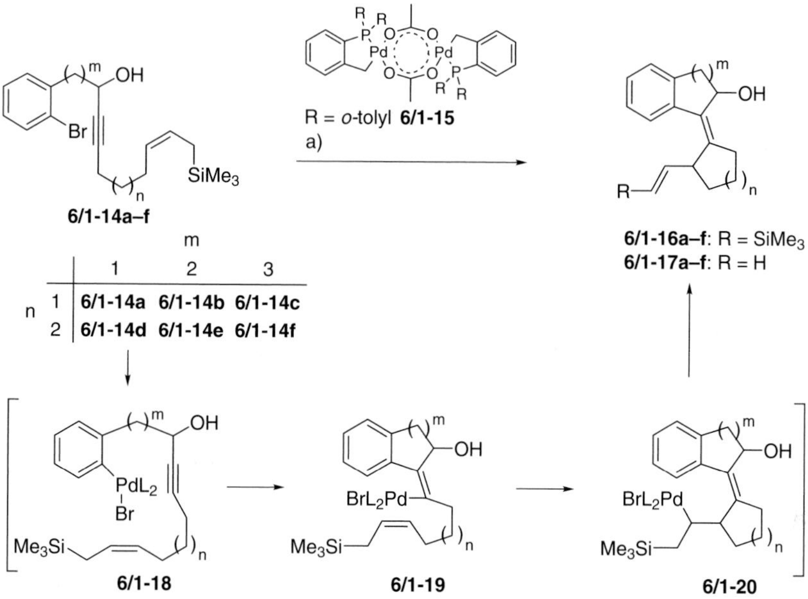

	m		
	1	2	3
n 1	6/1-14a	6/1-14b	6/1-14c
2	6/1-14d	6/1-14e	6/1-14f

6/1-16a–f: R = SiMe₃
6/1-17a–f: R = H

Entry	Substrate	m/n	Conditions	Temp [°C]	Time [h]	dr of 6/1-16	Yield [%] 6/1-16 (6/1-17)
1	6/1-14a	1/1	a	80	4.5	1.2:1	71 (3)
2	6/1-14b	2/1	a	95	22	1.3:1	66 (5)
3	6/1-14c	3/1	b	130	21	2.6:1	43 (3)
4	6/1-14d	1/2	a	80	15	4.5:1	62 (8)
5	6/1-14e	2/2	a	100	17	9.4:1	61 (18)
6	6/1-14f	3/2	b	130	21	20:1	43 (7)

conditions: a) Pd(OAc)₂, PPh₃, KOAc, Pr₄NBr, DMF. b) Hermann–Beller catalyst 6/1-15, KOAc, Pr₄NBr, DMF.

Scheme 6/1.3. Allylsilane-terminated domino Heck reactions of 6/1-14a–f.

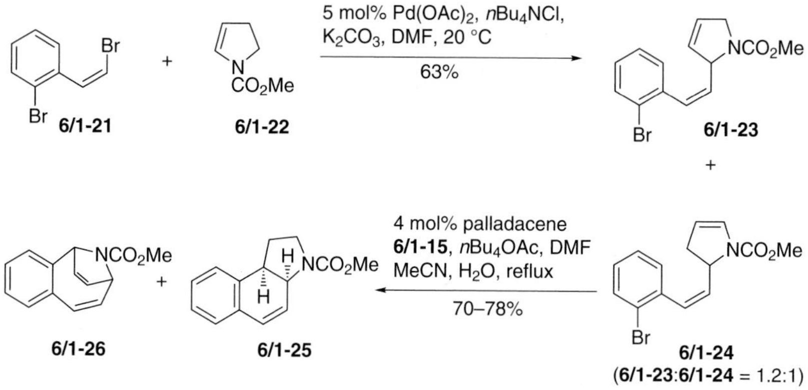

Scheme 6/1.4. Double Heck-reaction for the synthesis of heterocycles.

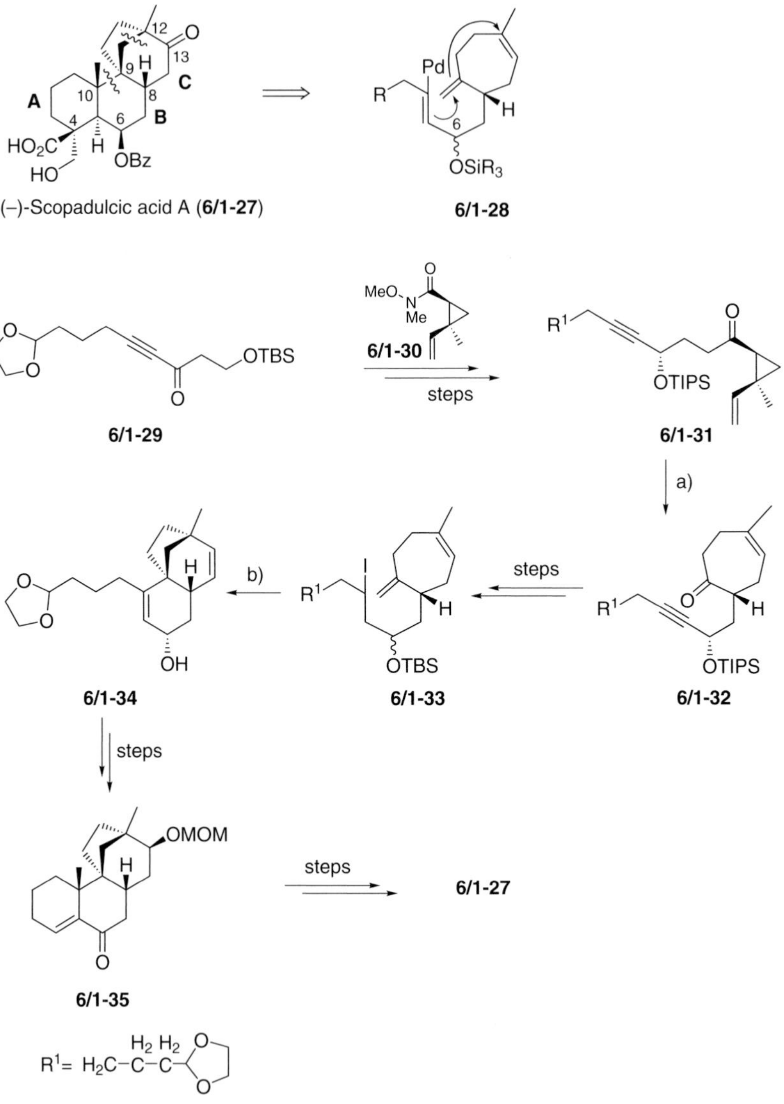

(−)-Scopadulcic acid A (**6/1-27**)

6/1-28

6/1-29

6/1-30

6/1-31

6/1-32

6/1-33

6/1-34

6/1-35

6/1-27

R¹ = H₂C-C-C-

Reaction conditions: a) LDA, 20% HMPA/THF, −78 °C, TMSCl, −78 °C → 0 °C, HCl/H₂O.
b) 30% Pd(OAc)₂, 60% PPh₃, Ag₂CO₃, THF, reflux, TBAF, THF, r.t..

Scheme 6/1.5. Synthesis of (−)-scopadulcic acid A (**6/1-27**).

ral ligand. Reaction of **6/1-36a** with Pd₂(dba)₃·CHCl₃ in the presence of (*S*)-BINAP gave **6/1-37** in 39% yield and 63% *ee*, which was transformed into xestoquinone **6/1-38** (Scheme 6/1.6). It is interesting that in the second Heck reaction an *endo*-trig approach takes place to give a six-membered ring instead of the seemingly more favored *exo*-trig reaction with the formation of a five-membered ring. Keay

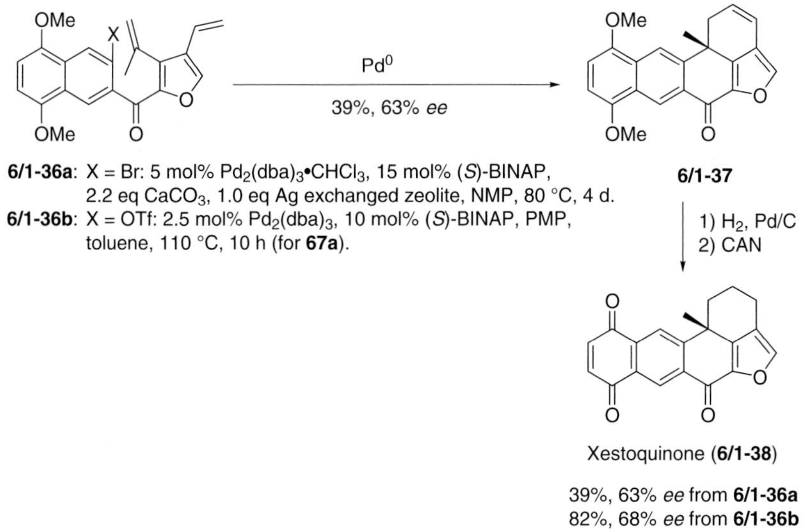
OMe
X
OMe
O
O
Pd^0
39%, 63% *ee*
OMe
OMe
O
O
6/1-36a: X = Br: 5 mol% $Pd_2(dba)_3$•$CHCl_3$, 15 mol% (*S*)-BINAP,
2.2 eq $CaCO_3$, 1.0 eq Ag exchanged zeolite, NMP, 80 °C, 4 d.
6/1-36b: X = OTf: 2.5 mol% $Pd_2(dba)_3$, 10 mol% (*S*)-BINAP, PMP,
toluene, 110 °C, 10 h (for **67a**).
6/1-37
1) H_2, Pd/C
2) CAN
O
O
O
O
Xestoquinone (**6/1-38**)
39%, 63% *ee* from **6/1-36a**
82%, 68% *ee* from **6/1-36b**

Scheme 6/1.6. Synthesis of Xestoquinone (**6/1-38**).

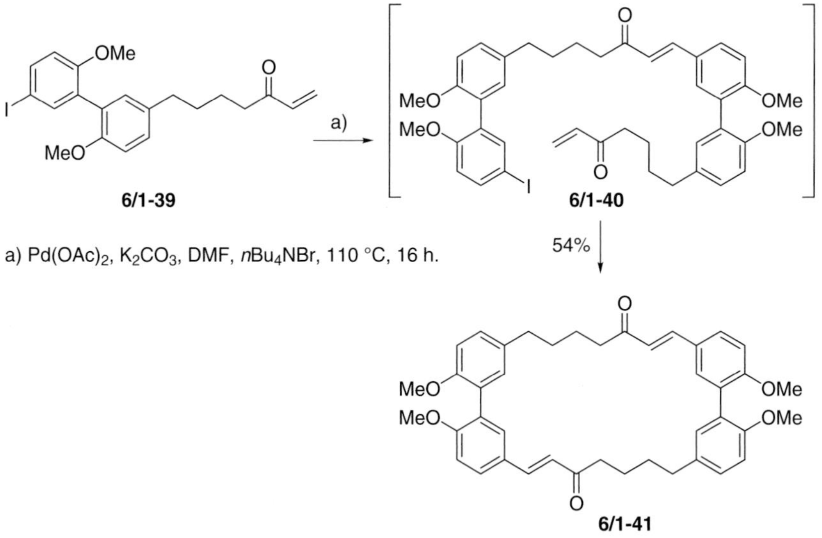
OMe
O
I
MeO
6/1-39
a)
O
MeO
MeO
OMe
OMe
I
O
6/1-40
54%
a) $Pd(OAc)_2$, K_2CO_3, DMF, nBu_4NBr, 110 °C, 16 h.
O
MeO
MeO
OMe
OMe
O
6/1-41

Scheme 6/1.7. Synthesis of macrocyclic compounds.

and coworkers [28] recently reported on an improvement of this process using the triflate **6/1-36b** as substrate. They obtained **6/1-37** with a much higher yield and slightly improved enantioselectivity.

The intramolecular Heck reaction is an appropriate method for the synthesis of all types of cyclic compounds such as normal, medium, and large rings [29]. Using

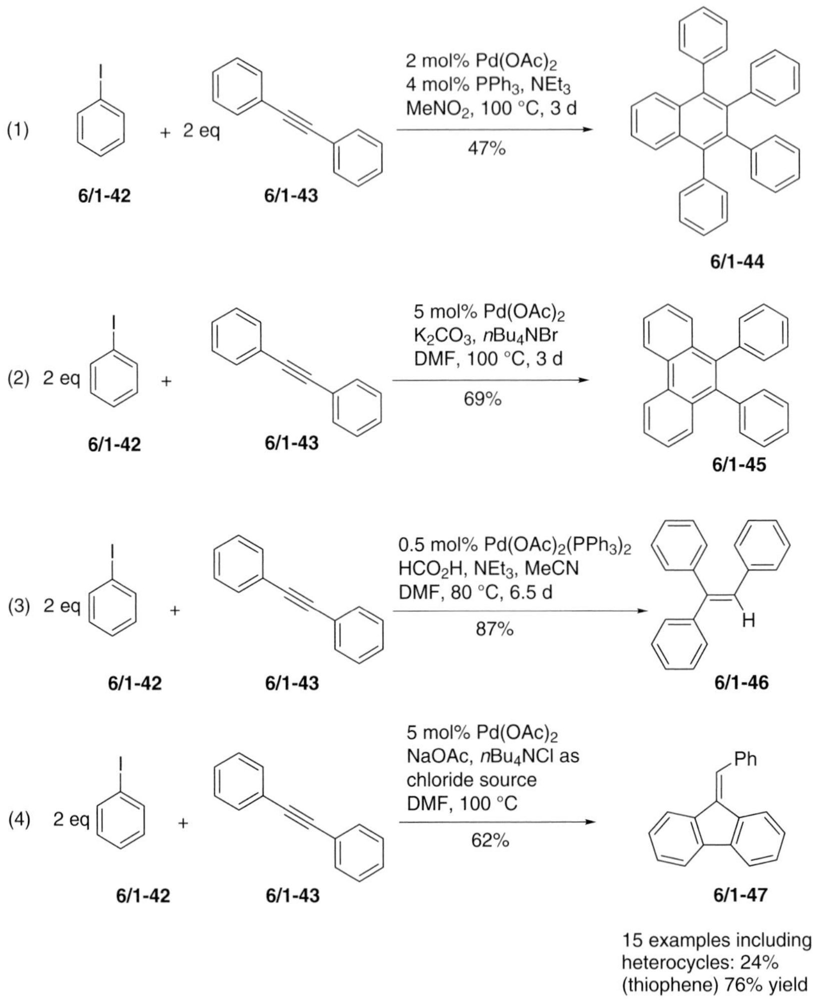

(1) 6/1-42 + 2 eq 6/1-43

2 mol% Pd(OAc)₂
4 mol% PPh₃, NEt₃
MeNO₂, 100 °C, 3 d

47%

6/1-44

(2) 2 eq 6/1-42 + 6/1-43

5 mol% Pd(OAc)₂
K₂CO₃, nBu₄NBr
DMF, 100 °C, 3 d

69%

6/1-45

(3) 2 eq 6/1-42 + 6/1-43

0.5 mol% Pd(OAc)₂(PPh₃)₂
HCO₂H, NEt₃, MeCN
DMF, 80 °C, 6.5 d

87%

6/1-46

(4) 2 eq 6/1-42 + 6/1-43

5 mol% Pd(OAc)₂
NaOAc, nBu₄NCl as
chloride source
DMF, 100 °C

62%

6/1-47

15 examples including
heterocycles: 24%
(thiophene) 76% yield

Scheme 6/1.8. Reaction of **6/1-42** and **6/1-43** under different conditions.

ω-haloallenes, 20-membered carbocycles can be obtained in over 85 % yield; however, by using the same reaction conditions, 13-membered rings were only formed in 18 % yield. In agreement with these basics is the observation that the Pd⁰-catalyzed transformation of substrate **6/1-39** does not give an intramolecular cyclization, but first by an intermolecular C–C-bond formation to afford the pseudo-dimer **6/1-40** which then cyclizes to yield the 26-membered carbocycle **6/1-41** in 54 % yield (Scheme 6/1.7) [30].

It is well known that minor changes in conditions can have dramatic effects on the products obtained. For example, Heck's group [31] described the palladium-catalyzed reaction of iodobenzene **6/1-42** and 2 equiv. of diphenylacetylene **6/1-43** in

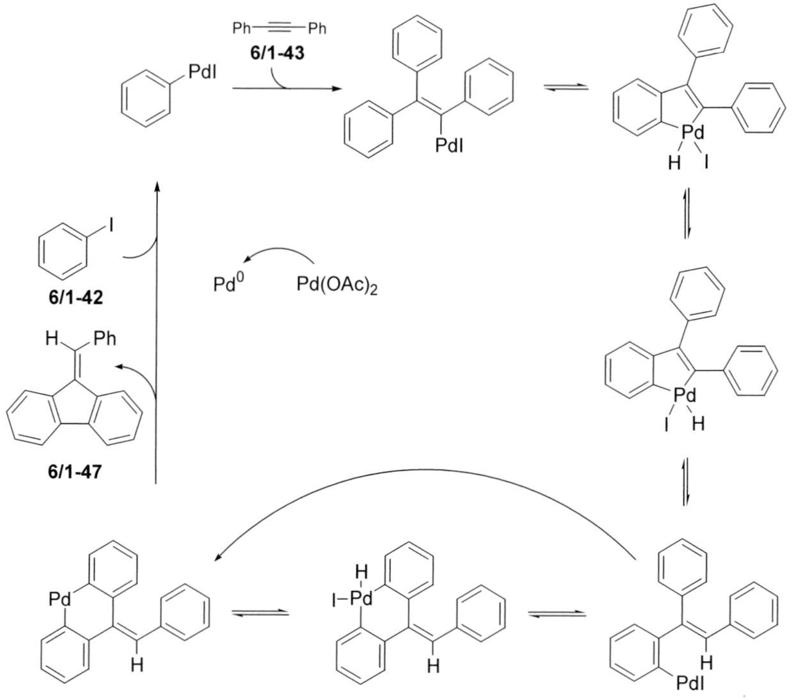

Scheme 6/1.9. Proposed mechanism for the formation of 6/1-47 from 6/1-42 and 6/1-43.

the presence of triphenylphosphane and triethylamine as base to give the substituted naphthalene 6/1-44. Dyker and coworkers [32] used a 2:1-mixture of 6/1-42 and 6/1-43 in the presence of nBu$_4$NBr and K$_2$CO$_3$ as base, and obtained the phenanthrene 6/1-45 (Scheme 6/1.8). Moreover, Cacchi's group [33] used the same conditions but added a formate and obtained the triphenylethene 6/1-46.

Finally, Larock and coworkers [34] recently reported on an efficient synthesis of 9-alkylidene- and 9-benzylidene-9H-fluorenes 6/1-47, again using 6/1-42 and 6/1-43 as substrates. The best results were obtained with sodium acetate and nBu$_4$NCl, which allowed 6/1-47 to be obtained in 62 % yield. A proposed mechanism is given in Scheme 6/1.9, suggesting a migration of palladium from a vinylic to an arylic position [35].

De Meijere and coworkers developed a nice example where three C–C-bonds are formed in one domino process [36]. Thus, reaction of 6/1-48 with the Hermann–Beller catalyst 6/1-15 led to 6/1-50 as the only product (Scheme 6/1.10). It can be assumed that the Pd-compound 6/1-49 is an intermediate.

Another cyclotrimerization was observed in the Pd^0-catalyzed transformation of substrates of type 6/1-51 which led to annulated naphthalenes 6/1-52, as described by Grigg and coworkers [37] (Scheme 6/1.11). The reaction can also be performed as a two-component transformation involving a combination of an intra- and an intermolecular process.

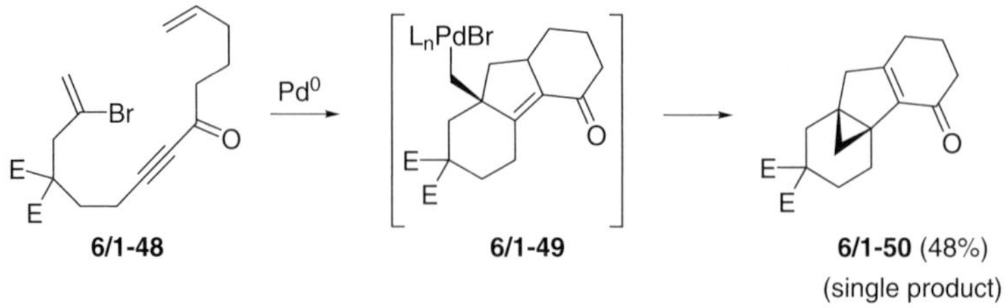

6/1-48 **6/1-49** **6/1-50** (48%)
(single product)

Scheme 6/1.10. Synthesis of a tetracyclic cyclopropane derivative.

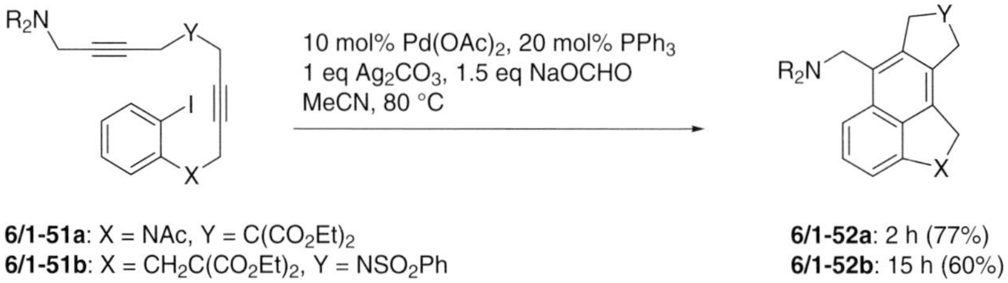

10 mol% Pd(OAc)₂, 20 mol% PPh₃
1 eq Ag₂CO₃, 1.5 eq NaOCHO
MeCN, 80 °C

6/1-51a: X = NAc, Y = C(CO₂Et)₂
6/1-51b: X = CH₂C(CO₂Et)₂, Y = NSO₂Ph

6/1-52a: 2 h (77%)
6/1-52b: 15 h (60%)

Scheme 6/1.11. Cyclotrimerisation of **6/1-51**.

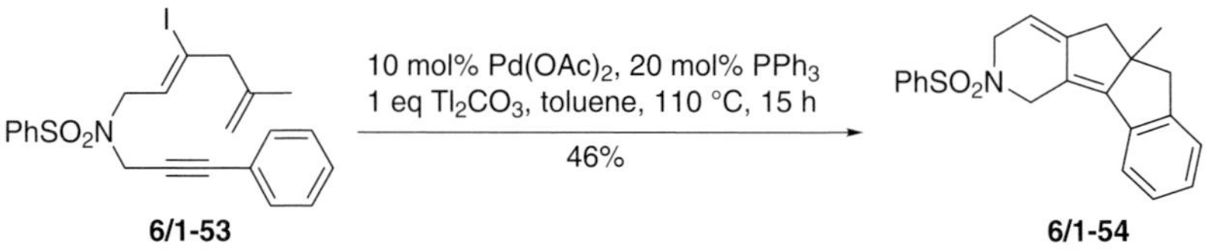

10 mol% Pd(OAc)₂, 20 mol% PPh₃
1 eq Tl₂CO₃, toluene, 110 °C, 15 h

46%

6/1-53 **6/1-54**

Scheme 6/1.12. Double Heck/Friedel–Crafts reaction.

The same group [38] also developed a double Heck reaction which was then terminated by a Friedel–Crafts alkylation to give **6/1-54** from **6/1-53** (Scheme 6/1.12); this involved an attack of an alkylpalladium(II) intermediate on an aryl or heteroaryl moiety. Noteworthy is the finding that the formal Friedel–Crafts alkylation occurs on both electron-rich and electron-poor heteroaromatic rings, as well as on substituted phenyl rings. Single Heck/Friedel–Crafts alkylation combinations have also been performed.

To date, only double Heck reactions have been described, but triple Heck reactions are also possible [39]. Reaction of the alkynyl aryl iodide **6/1-55** with norbornene in the presence of Pd(OAc)₂, triphenylphosphane and triethylamine as base led to the cyclopropanated norbornene derivate **6/1-59** as a single diastereomer in 40% yield (Scheme 6/1.13). It can be assumed that the alkenyl Pd-species **6/1-56** is first formed stereoselectively, and this undergoes a Heck reaction with norbornene

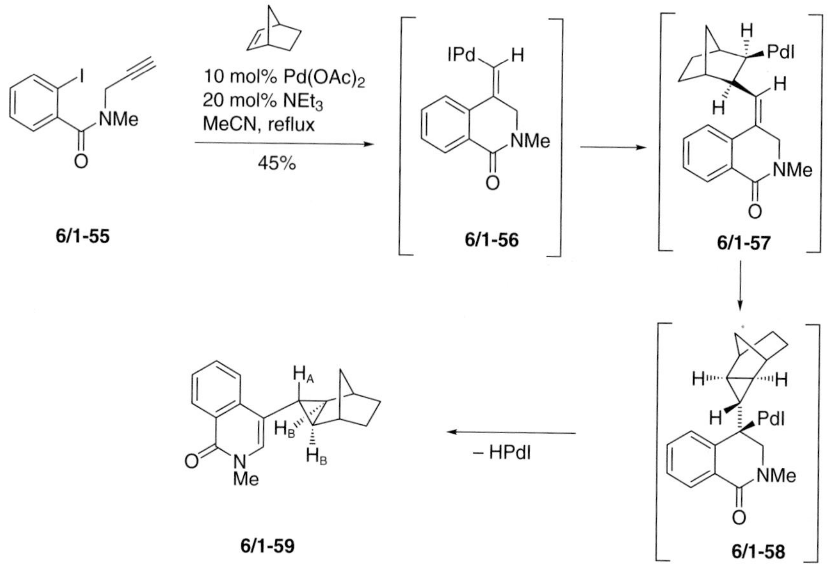

Scheme 6/1.13. Triple Heck reaction.

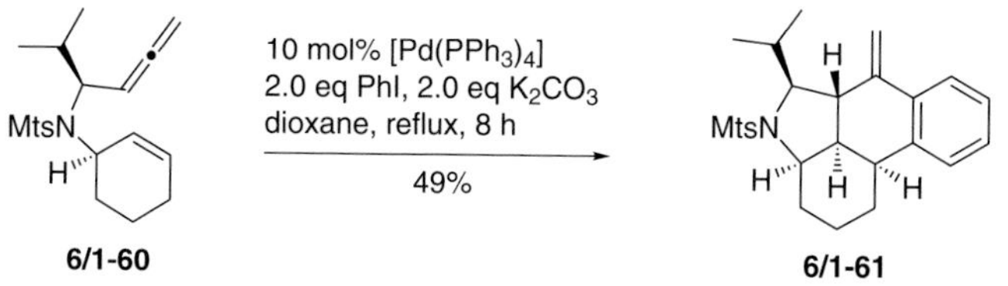

Scheme 6/1.14. Intermolecular polycyclisation.

to give **6/1-57** in a *syn*-fashion, followed by formation of the cyclopropanated intermediate **6/1-58**, which loses HPdI.

As shown in the preceding examples, although intramolecular Pd-catalyzed polycyclization is a well-established procedure, some few examples exist of polycyclizations where the first step is an intermolecular process. In this respect, the Pd⁰-catalyzed domino reaction of allenes in the presences of iodobenzene reported by Tanaka and coworkers [40] is an intriguing transformation. As an example the Pd-catalyzed reaction of **6/1-60** in the presence of iodobenzene led to **6/1-61** in 49 % yield, allowing the formation of three rings in one sequence (Scheme 6/1.14).

Domínguez and coworkers [41] used a twofold Heck reaction for the construction of annulated *N*-heterocycles such as **6/1-64a–c**, starting from the enamides **6/1-63** which can be easily obtained from the corresponding amines **6/1-62a–c** and *o*-iodo-benzoic acid chloride (Scheme 6/1.15).

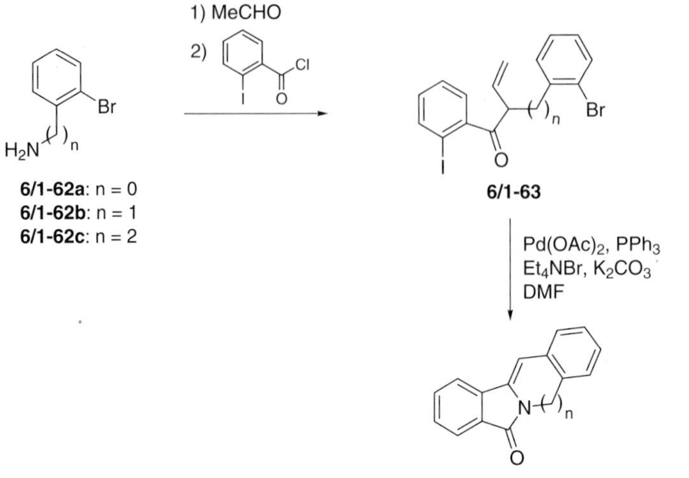
1) MeCHO
2)
Cl
I
O
Br
H_2N
n
Br
n
I
O
6/1-63
$Pd(OAc)_2$, PPh_3
Et_4NBr, K_2CO_3
DMF
N
n
O

6/1-62a: n = 0
6/1-62b: n = 1
6/1-62c: n = 2

6/1-64a: 2 h, 130 °C, (81%)
6/1-64b: 6 h, 100 °C, (87%)
6/1-64c: 24 h, 130 °C, (47%)

Scheme 6/1.15. Synthesis of annulated heterocycles.

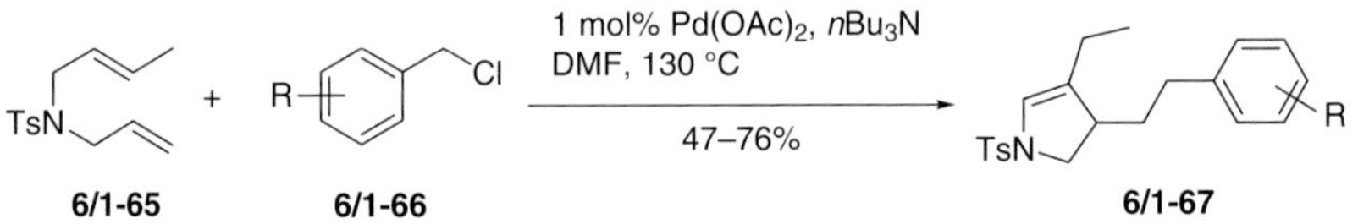
TsN
+
R
Cl
1 mol% $Pd(OAc)_2$, nBu_3N
DMF, 130 °C
47–76%
TsN
R

6/1-65 **6/1-66** **6/1-67**

Scheme 6/1.16. Synthesis of pyrrolines.

Another double Heck reaction was reported by Pan and coworkers [42] to prepare substituted pyrrolines **6/1-67** using the bisallylamine **6/1-67** and benzylchlorides as substrates (Scheme 6/1.16).

Keay and coworkers [43] observed an unusual remote substituent effect on the enantioselectivity in intramolecular domino Heck processes to give the tetracyclic products **6/1-69** to **6/1-72** from **6/1-68a–d** (Scheme 6/1.17). The enantioselectivity varies strongly with the substitution on pattern; thus, the transformation of **6/1-68a** and **6/1-68c** containing one methyl group gave the corresponding products **6/1-69** and **6/1-71** with excellent enantioselectivity up to 96 %, whereas the substrates with no or two methyl groups gave much lower *ee*-values.

6.1.1.2 Heck/Cross-Coupling Reactions

The combination of a Heck and a cross-coupling reaction has not been widely exploited. However, there are some reactions where, following oxidative addition, a

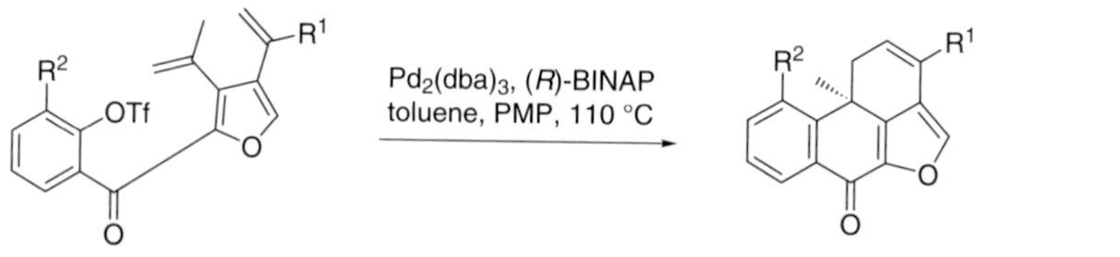

6/1-68a: R^1= Me, R^2= H
6/1-68b: R^1= R^2= H
6/1-68c: R^1= H, R^2= Me
6/1-68d: R^1= R^2= Me

6/1-69: R^1= Me, R^2= H, 78%, 90% *ee* (*R*)
6/1-70: R^1= R^2= H, 83%, 71% *ee* (*R*)
6/1-71: R^1= H, R^2= Me, 71%, 96% *ee* (*R*)
6/1-72: R^1= R^2= Me, 68%, 71% *ee* (*R*)

Scheme 6/1.17. Influence of the substitution pattern on the enantioselectivity in intramolecular double Heck reaction.

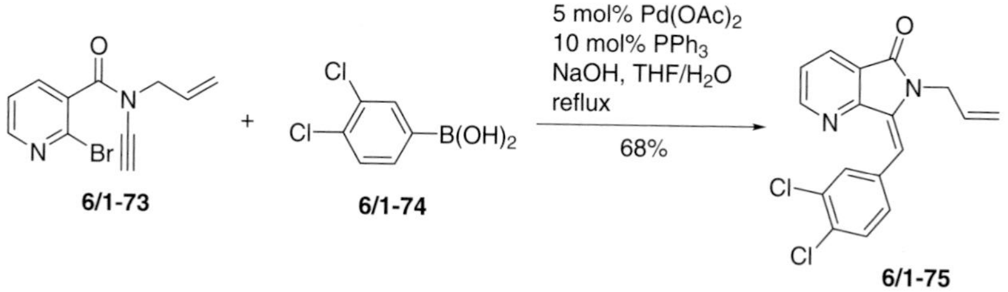

Scheme 6/1.18. Synthesis of azaindoles.

cis-palladation of an alkyne takes place, and this is followed by a Suzuki, Stille, or Sonogashira reaction. This approach utilizes the comparatively high reaction rate of the *cis*-palladation of alkynes. Most often, the first step is an intramolecular transformation which utilizes the entropic effect, though this is not a necessary requirement.

Ternary Pd-catalyzed coupling reactions of bicyclic olefins (most often norbornadiene is used) with aryl and vinyl halides and various nucleophiles have been investigated intensively over the past few years [44]. A new approach in this field is to combine Heck and Suzuki reactions using a mixture of phenyliodide, phenylboronic acid and the norbornadiene dicarboxylate. Optimizing the conditions led to 84% of the desired biphenylnorbornene dicarboxylate [45]. Substituted phenyliodides and phenylboronic acids can also be used, though the variation at the norbornadiene moiety is highly limited.

Cossy and coworkers described a precise combination of a Heck and a Suzuki–Miyama reaction using ynamides and boronic acids to give indole and 7-azaindole derivatives [46]. Thus, reaction of **6/1-73** with **6/1-74** using Pd(OAc)$_2$ as catalyst led to **6/1-75** in 68% yield (Scheme 6/1.18).

Ahn, Kim and coworkers, in the preparation of 4-alkylidene-3-arylmethylpyrrolidines **6/1-77**, used a Heck reaction of the vinyl bromide **6/1-76** in the presence of an arylboronic acid [47] (Scheme 6/1.19). It has been assumed that the interme-

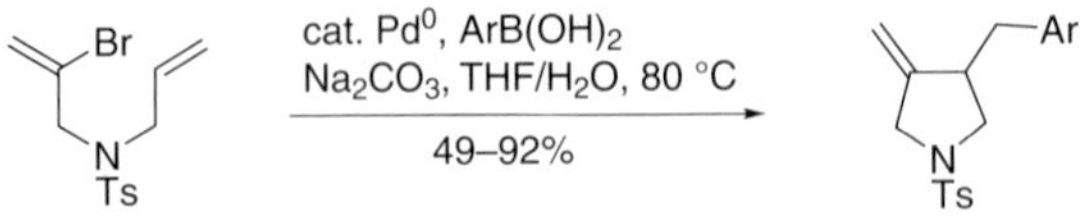

Scheme 6/1.19. Synthesis of pyrrolidines.

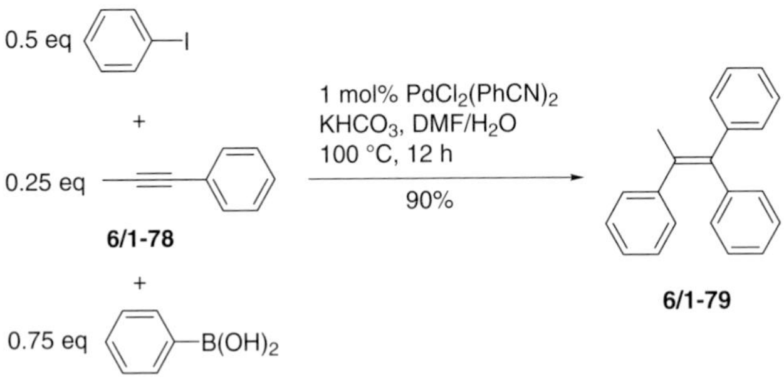

Scheme 6/1.20. Synthesis of tamoxifen analogues.

diately formed alkylpalladium species is stabilized by a coordination with one of the N-sulfonyl oxygens which might suppress the Pd-β-hydride elimination to allow the Suzuki reaction to proceed.

Recently, Larock and coworkers used a domino Heck/Suzuki process for the synthesis of a multitude of tamoxifen analogues [48] (Scheme 6/1.20). In their approach, these authors used a three-component coupling reaction of readily available aryl iodides, internal alkynes and aryl boronic acids to give the expected tetrasubstituted olefins in good yields. As an example, treatment of a mixture of phenyliodide, the alkyne **6/1-78** and phenylboronic acid with catalytic amounts of PdCl$_2$(PhCN)$_2$ gave **6/1-79** in 90% yield. In this process, substituted aryl iodides and heteroaromatic boronic acids may also be employed. It can be assumed that, after Pd^0-catalyzed oxidative addition of the aryl iodide, a *cis*-carbopalladation of the internal alkyne takes place to form a vinylic palladium intermediate. This then reacts with the ate complex of the aryl boronic acid in a transmetalation, followed by a reductive elimination.

A combination of an intramolecular Pd^0-catalyzed alkenylation of an alkyne and a Stille reaction followed by a 8π-electrocyclization was developed by Suffert and coworkers [49]. Thus, treatment of the diol **6/1-80** with the stannyldiene **6/1-81** in the presence of [Pd(PPh$_3$)$_4$] led to **6/1-83** in 16% yield *via* the intermediate **6/1-82**. At a higher temperature, **6/1-83** can undergo a 4π conrotatory ring opening to afford a mixture of **6/1-85** and **6/1-86** *via* **6/1-84** (Scheme 6/1.21).

Müller and coworkers developed a concise domino process in which termination of an intramolecular arylation of an alkyne was achieved by a Sonogashira alkynyla-

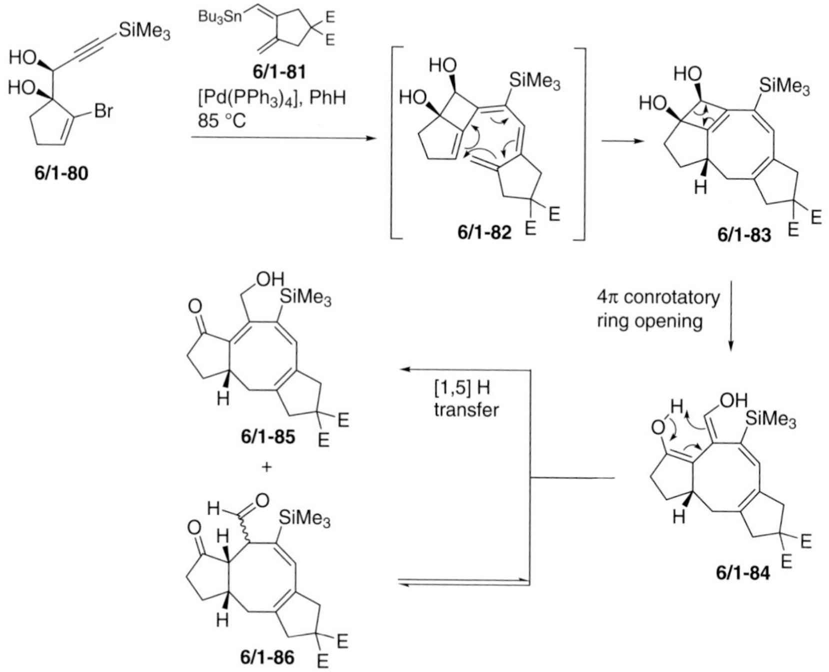

Scheme 6/1.21. Combination of a domino Heck/Stille reaction with an electrocyclization.

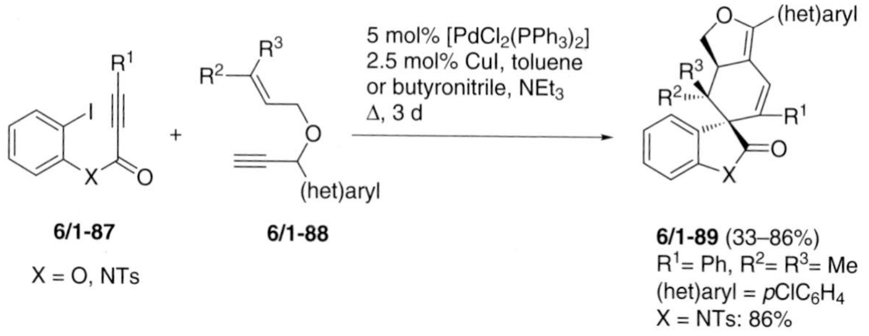

Scheme 6/1.22. Synthesis of spiro compounds **6/1-89**.

tion [50]. The transformation could even be carried on by an intramolecular cycloaddition. Treatment of a mixture of **6/1-87** and **6/1-88** in the presence of catalytic amounts of [PdCl$_2$(PPh$_3$)$_2$] and CuI in toluene/butyronitrile/triethylamine under reflux led to hitherto unknown spiro compounds **6/1-89** in up to 86 % yield (Scheme 6/1.22).

6.1.1.3 Heck/Tsuji–Trost Reactions

To date, only a few examples have been described where a Heck reaction has been combined with a Pd⁰-catalyzed nucleophilic substitution and the yields are less satisfying. However, the opposite variation – namely the combination of a Tsuji–Trost and a Heck reaction – has been used more often (see Section 6.1.3).

Helmchen and coworkers employed α,ω-amino-1,3-dienes as substrates [51]. By using palladium complexes with chiral phosphino-oxazolines L* as catalysts, an enantiomeric excess of up to 80% was achieved. In a typical experiment, a suspension of Pd(OAc)$_2$, the chiral ligand L*, the aminodiene **6/1-90** and an aryltriflate in dimethylformamide (DMF) was heated at 100 °C for 10 days. *Via* the chiral palladium complex **6/1-91**, the resulting cyclic amine derivative **6/1-92** was obtained in 47% yield and 80% *ee* (Scheme 6/1.23). Using aryliodides the reaction time is shorter, and the yield higher (61%), but the enantiomeric excess is lower (67% *ee*). With BINAP as a chiral ligand for the Pd⁰-catalyzed transformation of **6/1-90** and aryliodide, an *ee*-value of only 12% was obtained.

Within a total synthesis of the neurotoxin (–)-pumiliotoxin C [52], Minnaard, Feringa and coworkers used a domino Heck/Tsuji–Trost reaction of **6/1-93** and **6/1-94** to give the perhydroquinoline **6/1-95** in 26% yield after hydrogenation [53] (Scheme 6/1.24).

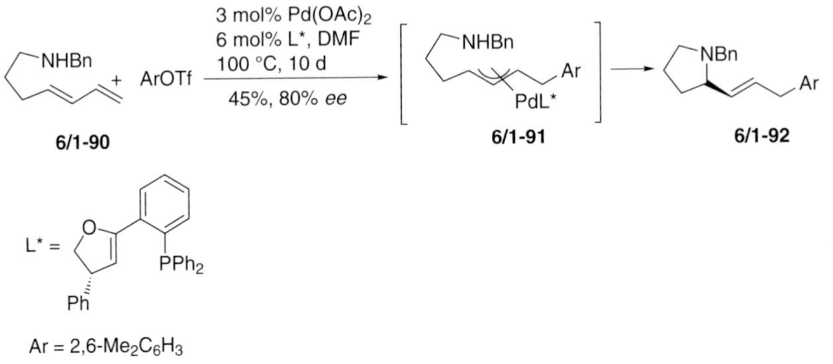

Scheme 6/1.23. Enantioselective domino Heck/allylation process.

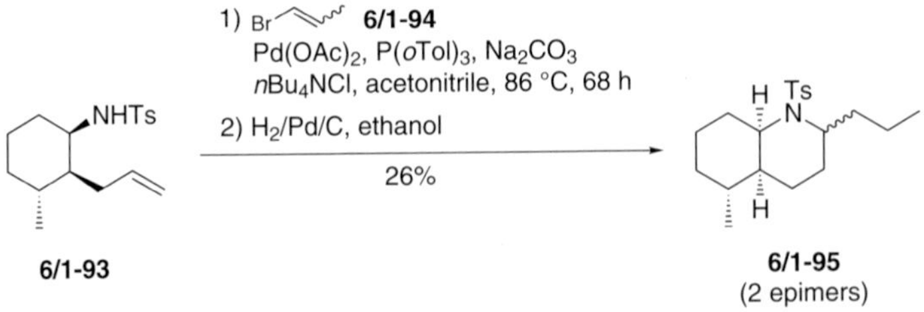

Scheme 6/1.24. Towards the synthesis of pumiliotoxin C.

Similarly, acyclic ω-olefinic *N*-tosyl amides with vinyl bromides have also been used to give pyrrolidones and piperidones in 49 to 82% yield (eight examples) [54].

6.1.1.4 Heck Reactions/CO-Insertions

The insertion of CO into an organic Pd species is a very common procedure, and may also form part of a domino process, for example, after a Heck reaction.

In an approach towards a total synthesis of the marine ascidian metabolite perophoramidine (**6/1-96**) [55], Weinreb and coworkers developed a domino Heck/carbonylation process [56]. This allowed construction of the C,E,F-ring system of **6/1-96**, together with the C-20 quaternary center and the introduction of a functionality at C-4 (Scheme 6/1.25). Thus, reaction of **6/1-97** in the presence of catalytic amounts of Pd(OAc)$_2$ and P(oTol)$_3$ under a CO atmosphere in DMA/MeOH led to **6/1-98** in 77% yield.

Bicyclic lactones such as **6/1-101** were synthesized by Negishi and coworkers [57] using a domino Heck carbopalladation as the key step of vinyl halides as **6/1-99** to give **6/1-100**. The product can be transformed into the desired lactone **6/1-101** in a few steps (Scheme 6/1.26).

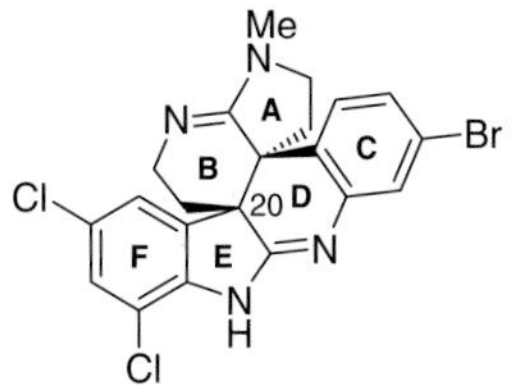

Perophoramidine (**6/1-96**)

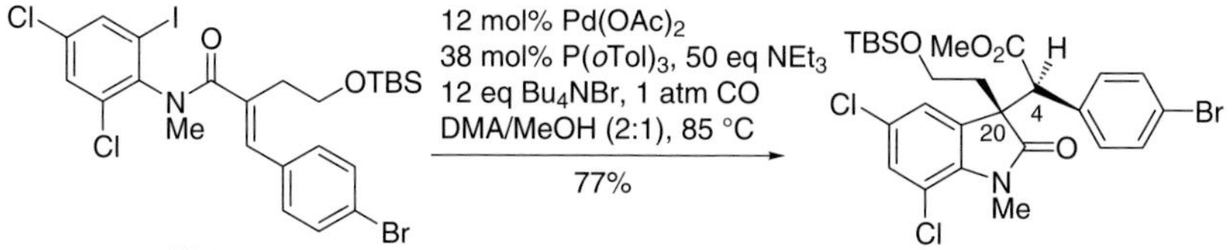

Scheme 6/1.25. Synthesis of the C,E,F-ring system of perophoramidine (**6/1-96**).

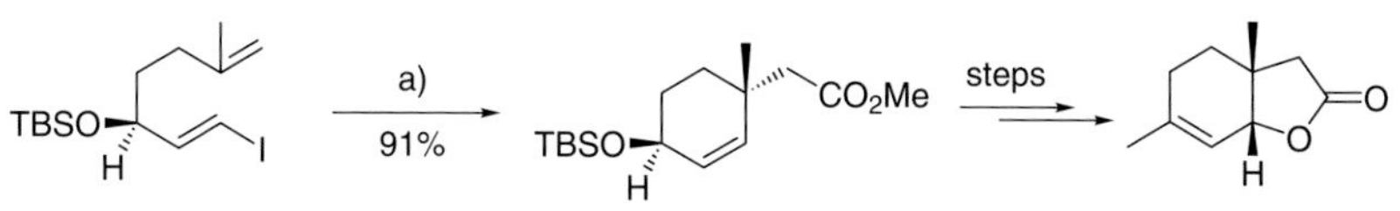

a) 5 mol% Cl$_2$Pd(PPh$_3$)$_2$, 1 atm CO, NEt$_3$, MeOH, H$_2$O, DMF, O$_2$, 85 °C, 0.5–1 h.

Scheme 6/1.26. Synthesis of bicyclic lactones.

6.1.1.5 Heck Reactions/C–H-Activations

A domino process, which has been encountered with increasing frequency during the past few years, is the combination of a Heck reaction with a C–H-activation. This type of transformation takes place if the intermediate organopalladium compound cannot react in the usual manner, and an aryl-H-bond is in close vicinity. C–H-activation can also occur if the arylhalide is used in excess.

Carretero and coworkers [58] encountered three C–H-activations after a first Heck reaction using α,β-unsaturated sulfones **6/1-102** and iodobenzene. Under normal conditions, the expected Heck product **6/1-103** is formed; however, if an excess of phenyliodide is used, then **6/1-104** is obtained in high yield. In this transformation three molecules of phenyliodide are incorporated into the final product (Scheme 6/1.27).

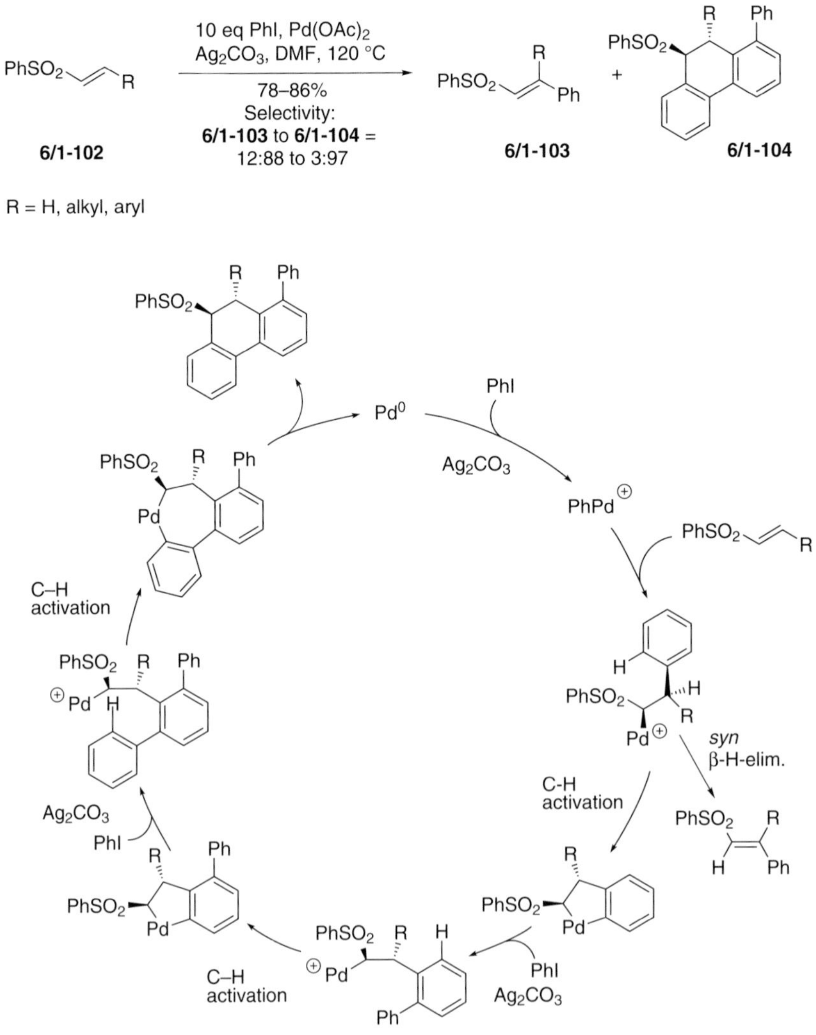

Scheme 6/1.27. Domino Heck reaction/CH-activation of α,β-unsaturated sulfones and PhI.

Larock and coworkers identified a Pd-catalyzed double C–H-activation by reaction of N-(3-iodophenyl)anilines as **6/1-105** with alkynes as **6/1-106** [59]. Thus, reaction of **6/1-105** with **6/1-106** in the presence of Pd⁰ led to **6/1-107** in reasonable yield. It can be assumed that the two palladacycles **6/1-108** and **6/1-109** act as intermediates. The procedure allows the efficient synthesis of substituted carbazoles (Scheme 6/1.28).

Tietze and coworkers [60] observed a combination of a Heck reaction and a C–H-activation by treatment of the alkyne **6/1-111** with Pd⁰. These authors aimed at compound **6/1-112**, but **6/1-110** was obtained as a single product in high yield (Scheme 6/1.29). It can again be assumed that after oxidative addition a *cis*-carbopalladation of the triple bond takes place to give an alkenyl Pd intermediate which undergoes the C–H-insertion into the neighboring naphthalene and not into the aryl ether moiety.

Based on a transformation described by Catellani and coworkers [61], the Lautens group [62] developed a three-component domino reaction catalyzed by palladium for the synthesis of benzo annulated oxacycles **6/1-114** (Scheme 6/1.30). As substrates, these authors used a *m*-iodoaryl iodoalkyl ether **6/1-113**, an alkene substi-

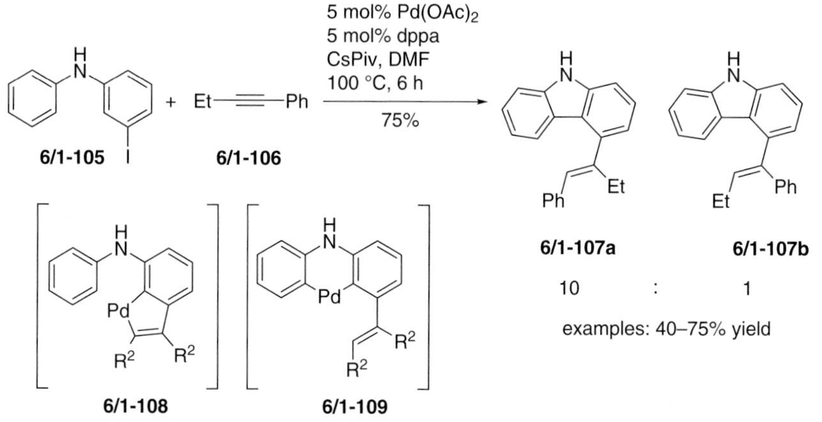

Scheme 6/1.28. Synthesis of carbazoles.

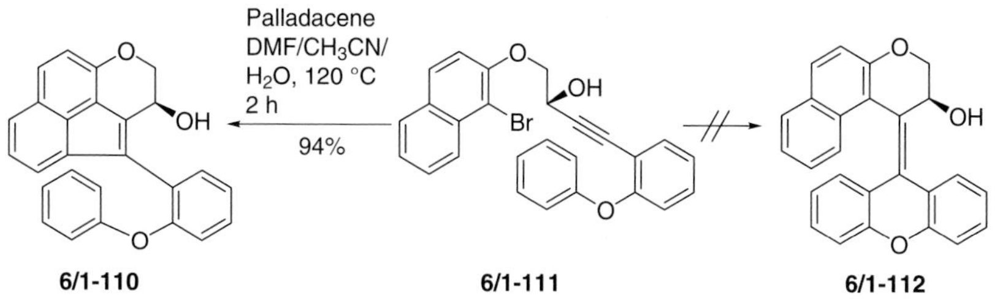

Scheme 6/1.29. Synthesis of acenaphtylenes.

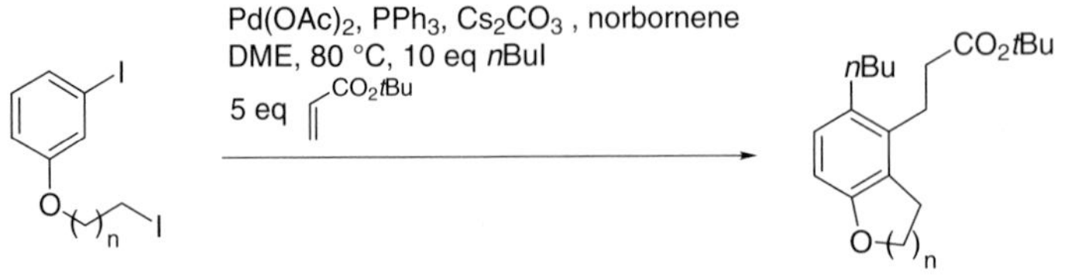

6/1-113a: n = 1
6/1-113b: n = 2
6/1-113c: n = 3

6/1-114a: n = 1, 85%
6/1-114b: n = 2, 62%
6/1-114c: n = 3, 69%

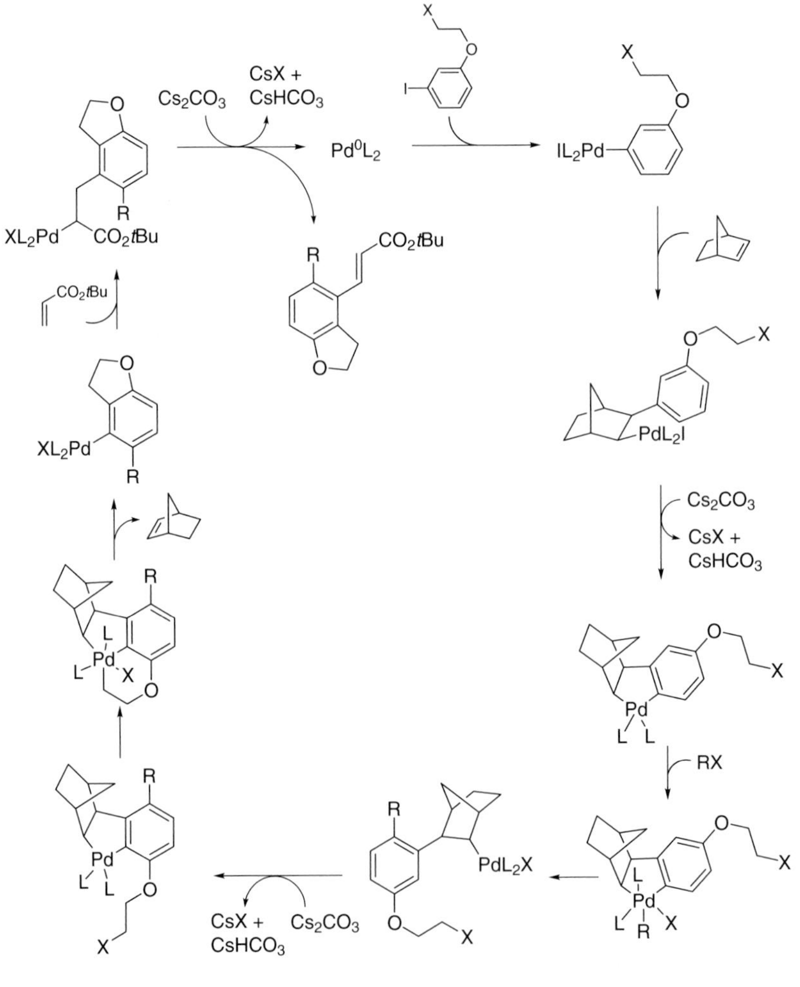

Scheme 6/1.30. Postulated mechanism for the reaction of **6/1-113** to give annulated oxacycles **6/1-113**.

tuted with an electron-withdrawing group (such as *t*-butyl acrylate), and an iodoalkane (such as *n*BuI) in the presence of norbornene. The yields are high in most cases. It is proposed that, after the oxidative addition of the aryliodide, a Heck-type reaction with norbornene first takes place to form a palladacycle, which is then alkylated with the iodoalkane. Finally, norbornene is eliminated under formation of the oxacycle and the obtained Pd-aryl species reacts with the acrylate.

6.1.1.6 Heck Reactions: Pericyclic Transformations

The most common combination of a Heck reaction in a domino process with a non-Pd-catalyzed transformation is that with a pericyclic reaction, especially a Diels–Alder cycloaddition. This is reasonable, since a Heck reaction allows an easy formation of a 1,3-butadiene.

The de Meijere group [63] prepared interesting spiro-compounds containing a cyclopropyl moiety using a combination of a Heck and a Diels–Alder reaction, with bicyclopropylidene **6/1-115** as the starting material. The transformation can be performed as a three-component process. Thus, reaction of **6/1-115**, iodobenzene and acrylate gave **6/1-116** in excellent yield. With vinyliodide, the tricyclic compound **6/1-117** was obtained (Scheme 6/1.31). Several other examples were also described.

Another domino Heck/Diels–Alder process described by the same group [64] implies the Pd⁰-catalyzed reaction of **6/1-118** in the presence of acrylate or methyl vinyl ketone to give the corresponding bicyclic compounds **6/1-120** and **6/1-121** *via* the transient **6/1-119** (Scheme 6/1.32). Good yields were obtained only if potassium carbonate is used as base.

Similarly, the Pd-catalyzed arylation of 1,3-dicyclopropyl-1,2-propadiene **6/1-122** with iodobenzene in the presence of dimethyl maleate led to the diastereomeric cyclopropane derivatives **6/1-124** and **6/1-125** *via* **6/1-123** in 86 % yield as a 4:1-mixture [65] (Scheme 6/1.33). Several other aryl halides and dienophiles have been used in this reaction.

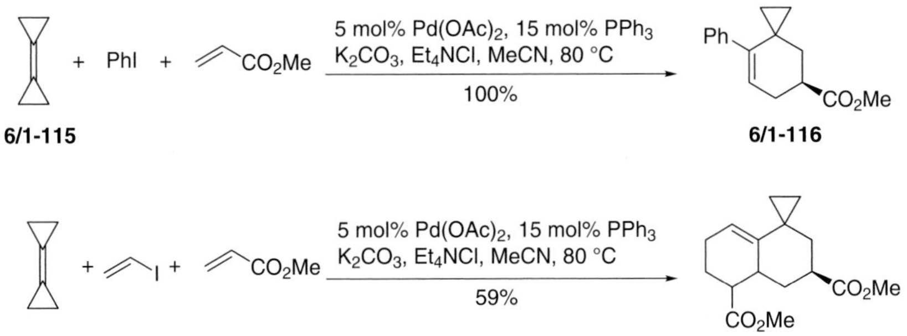

Scheme 6/1.31. Domino Heck reaction/cycloaddition.

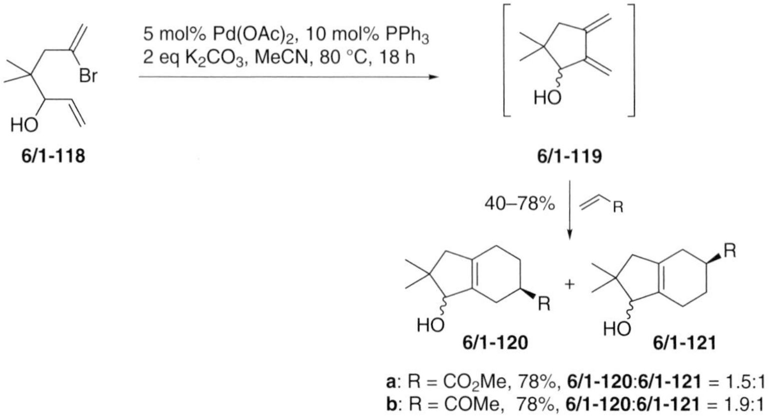

Scheme 6/1.32. Synthesis of indene derivatives.

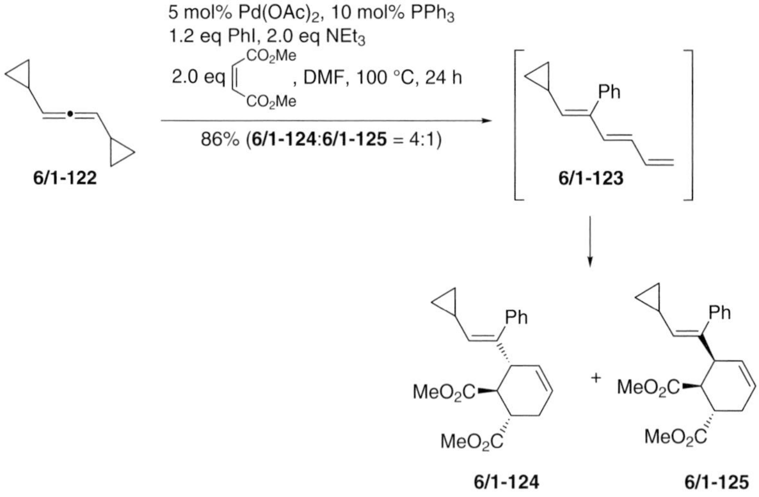

Scheme 6/1.33. Synthesis of cyclopropane derivatives.

Suffert and coworkers [66] described several examples of a carbopalladative cyclization followed by an electrocyclic reaction. Treatment of a mixture of the *trans*-bis(tributylstannyl)ethylene and the alkyne **6/1-128c**, which can easily be obtained from 2-bromocyclohexenone **6/1-126a** and the lithium compound **6/1-127**, with catalytic amounts of Pd(Ph₃P)₄ at 90 °C led to **6/1-130a**, presumably *via* the cyclobutane **6/1-129** in 62 % yield. The ring size of the starting material seems to have a pronounced influence on the reaction, since **6/1-128d** obtained from **6/1-126b** gave **6/1-130b** only in 24 % yield. The developed methodology could eventually be applied for the synthesis of ascosalipyrrolidinone (**6/1-131**) [67] (Scheme 6/1.34).

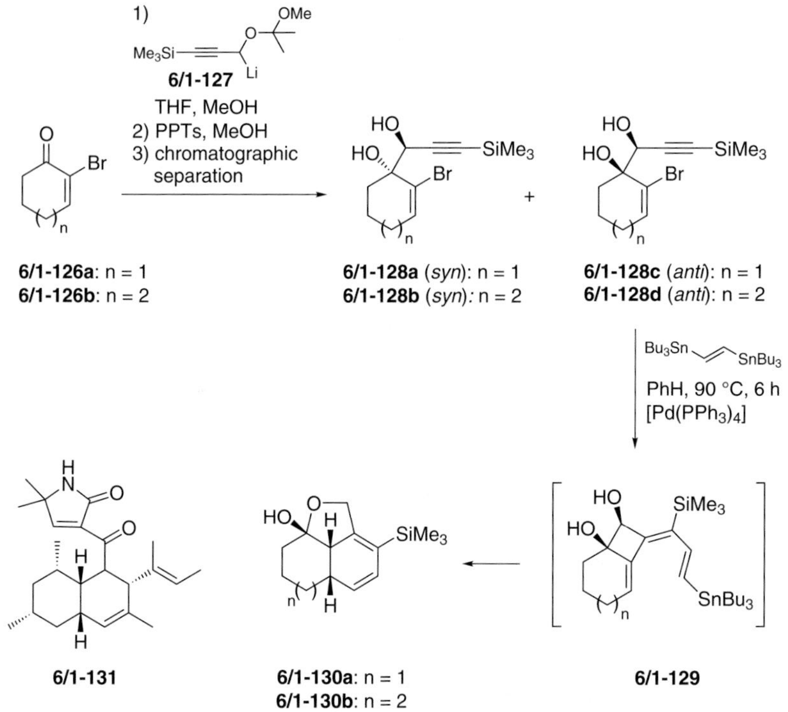

Scheme 6/1.34. Synthesis of ascosalipyrrolidinone (**6/1-131**).

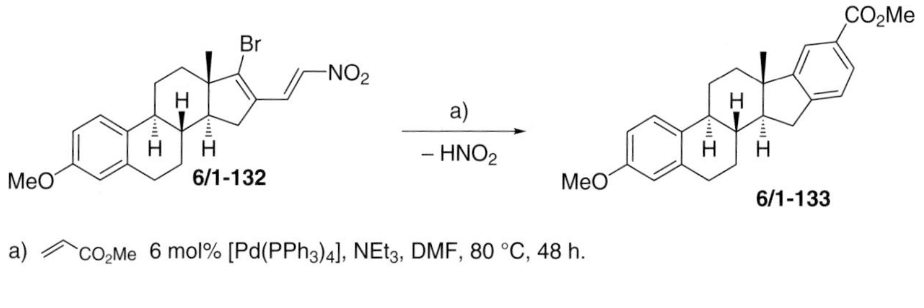

a) ⟋⟍CO$_2$Me 6 mol% [Pd(PPh$_3$)$_4$], NEt$_3$, DMF, 80 °C, 48 h.

Scheme 6/1.35. Synthesis of areno-annulated steroids.

Thiemann and coworkers [68] sought novel types of steroids with different bio-logical activity, and in doing so prepared areno-annulated compounds such as **6/1-133** (Scheme 6/1.35). This is achieved with a Heck reaction of **6/1-132** with an acry-late, followed by an electrocyclic ring closure of the formed hexatriene. The reaction is then terminated by removal of the nitro group, with formation of the aromatic ring system.

6.1.1.7 Heck Reactions/Mixed Transformations

Heck reactions can also be combined with anion capture processes, aminations, metatheses, aldol and Michael reactions, and isomerizations. The anion capture process has also been widely used with other Pd-catalyzed transformations. Outstanding examples of many different combinations have been developed by Grigg and coworkers, though not all of them match the requirements of a domino process. All of these reactions will be detailed here, despite the fact the nature of these intermediate transformations would also have permitted their discussion in Chapter 2.

Grigg and coworkers developed bimetallic domino reactions such as the electrochemically driven Pd/Cr Nozaki–Hiyama–Kishi reaction [69], the Pd/In Barbier-type allylation [70], Heck/Tsuji–Trost reaction/1,3 dipolar cycloaddition [71], the Heck reaction/metathesis [72], and several other processes [73–75]. A first example for an anion capture approach, which was performed on solid phase, is the reaction of 6/1-134 and 6/1-135 in the presence of CO and piperidine to give 6/1-136. Liberation from solid phase was achieved with HF, leading to 6/1-137 (Scheme 6/1.30) [76].

Another example, which includes a metathesis and a 1,3-dipolar cycloaddition, is the reaction of a mixture of iodothiophene, the allylamine 6/1-138 and allene with 10 mol% Pd(OAc)$_2$, 20 mol% PPh$_3$ and K$_2$CO$_3$ in toluene at 80 °C for 36 h under 1 atm CO to give the enone 6/1-139. Subsequent metathesis reactions using Grubbs' second-generation catalyst 6/1-140 led to the Δ^3-pyroline 6/1-141, which can undergo a 1,3-dipolar cycloaddition [77]. Thus, in the presence of AgOAc and the imine 6/1-142, the cycloadduct 6/1-144 was obtained with intermediate formation of the 1,3-dipole 6/1-143 (Scheme 6/1.37).

Yamazaki, Kondo and coworkers [78] reported on a combination of a Heck reaction and an amination on solid phase for the synthesis of indole carboxylate 6/1-148, employing an acetylated immobilized enamide 6/1-145 and a bifunctionalized

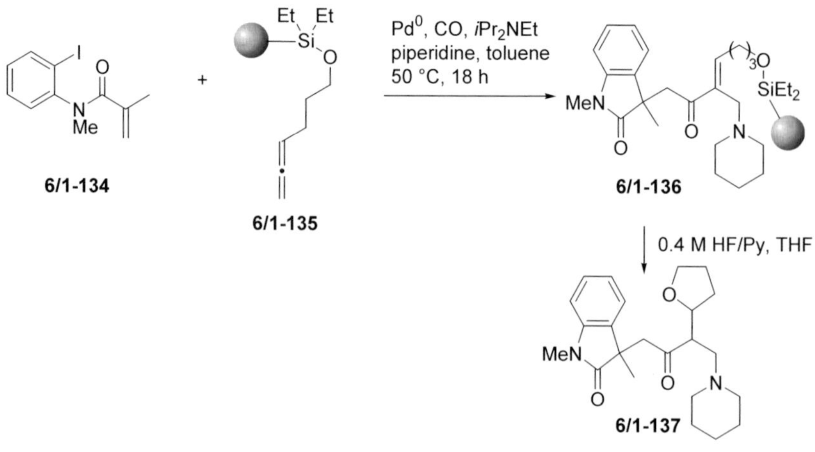

Scheme 6/1.36. Domino Heck/anion-capture reaction.

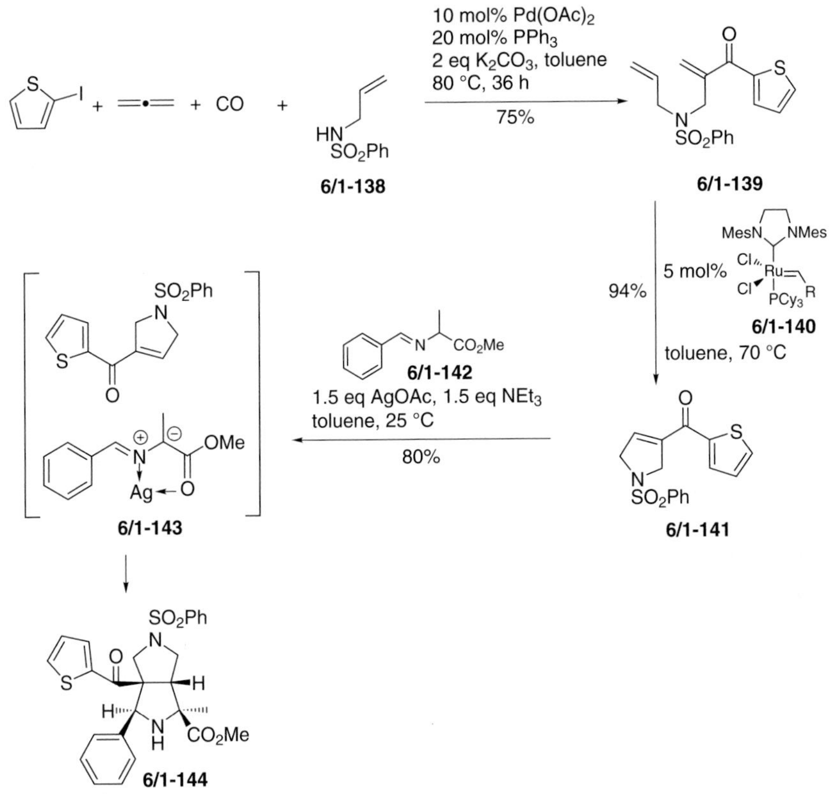

10 mol% Pd(OAc)$_2$
20 mol% PPh$_3$
2 eq K$_2$CO$_3$, toluene
80 °C, 36 h

75%

HN–SO$_2$Ph

6/1-138

6/1-139

94%

5 mol%

MesN NMes

6/1-140

toluene, 70 °C

6/1-142

1.5 eq AgOAc, 1.5 eq NEt$_3$
toluene, 25 °C

80%

6/1-143

6/1-141

6/1-144

Scheme 6/1.37. Synthesis of **6/1-144**.

arene **6/1-146**; in these transformations, **6/1-147** can be assumed as the interme-
diate (Scheme 6/1.38). The best results were obtained with o-dibromoarenes as **6/1-146b** and **6/1-146d** using the Pd$_2$(dba)$_3$/tBu$_3$P/Cy$_2$NMe catalytic system developed by Fu and coworkers [79]. The protocol was also extended to the synthesis of isoquinolines [80]. Indoles were additionally synthesized employing an opposite protocol, with the initial step being the amination reaction [81].

The Balme group [82] also used the anion capture approach to develop a short entry to triquinanes as **6/1-150** forming a transient Pd^{II}-complex in a Heck reaction of the vinyl iodide **6/1-149** (Scheme 6/1.39). The latter reacts with the carbanion of a malonate moiety in the substrate; however, the products of the normal Heck reaction are also formed.

A combination of a Heck reaction with an aldol condensation is observed on treatment of aromatic aldehydes or ketones as **6/1-151** with allylic alcohols as **6/1-152**, as described by Dyker and coworkers [83]. The Pd-catalyzed reaction led to **6/1-154** *via* **6/1-153**, in 55 % yield (Scheme 6/1.40).

Hallberg and coworkers [84] developed a synthesis of monoprotected 3-hydroxy-indan-1-ones **6/1-156** in moderate to good yields using salicylaldehyde triflates and

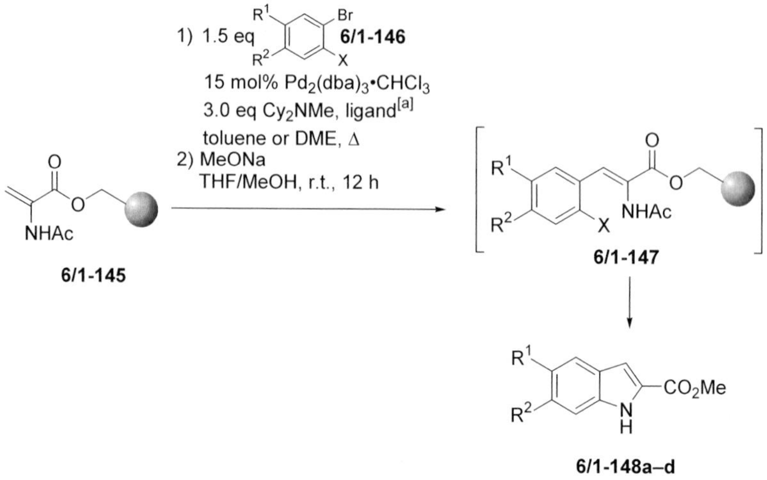

6/1-146	R^1	R^2	X	6/1-148	R^1	R^2	Yield [%][a,b]
a	H	H	I	**a**	H	H	46
b	H	H	Br	**a**	H	H	78
c	H	H	OTf	**a**	H	H	41
d	Me	Me	Br	**b**	Me	Me	82
e	OMe	H	Br	**c**	OMe	H	39
				d	H	OMe	31

[a] A 0.5 M stock solution of P*t*Bu$_3$ in toluene was used for **6/1-146a,b**, and *t*Bu$_3$PHBF$_4$ was used for **6/1-146c–e**.
[b] Isolated yield of **6/1-148** after SiO$_2$ column chromatography based on the loading of **6/1-145**.

Scheme 6/1.38. Synthesis of indoles.

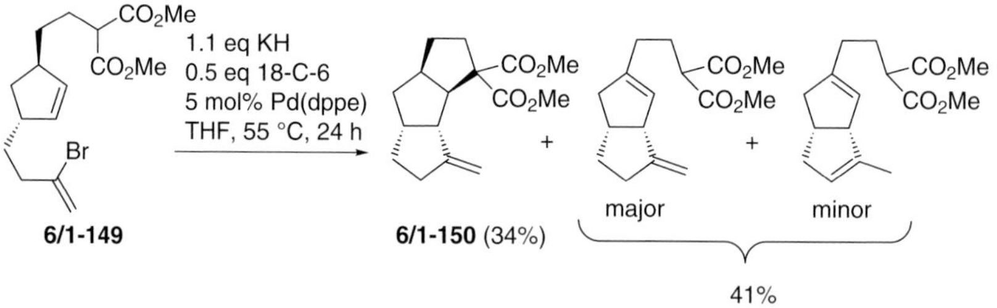

Scheme 6/1.39. Synthesis of triquinanes.

2-hydroxyethyl vinyl ether, presumably *via* a transient **6/1-155** (Scheme 6/1.41). The best yields were obtained using PMP as base; however, salicylcarbaldehydes with electron-withdrawing groups give always lower yield, and in the reaction of the nitro compound a product was not observed.

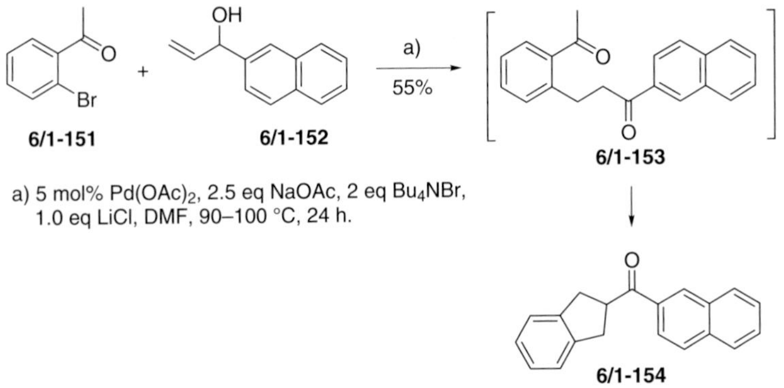

Scheme 6/1.40. Domino Heck/aldol reaction.

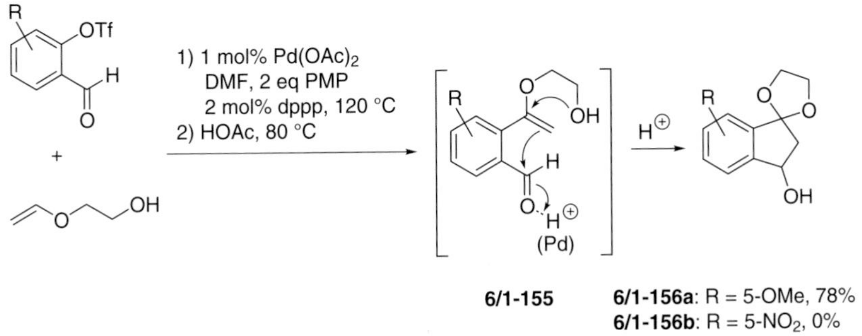

6/1-156a: R = 5-OMe, 78%
6/1-156b: R = 5-NO₂, 0%

Scheme 6/1.41. Synthesis of hydroxyindanones.

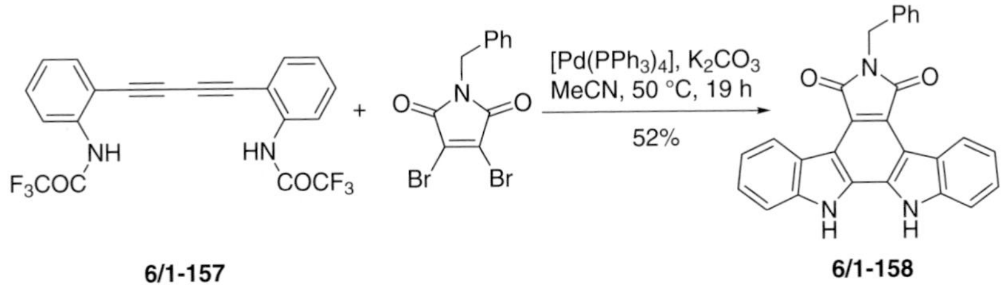

Scheme 6/1.42. Synthesis of the indolocarbazole skeleton.

The Cacchi group [85] developed a Pd-catalyzed domino process between *o*-alky-nyltrifluoroacetanilides as **6-157** and aryl or alkenyl halides, which leads to substituted pyrroles within an indole system. This scheme was successfully applied to the preparation of indolo[2,3-*a*]carbazoles as **6-158** using *N*-benzyl-3,4-dibro-momaleimide (Scheme 6/1.42). The indolocarbazole is found in several bioactive natural products as arcyriaflavin A and the cytotoxic rebeccamycin.

Very recently, Dongol and coworker have developed a one-pot synthesis of isoxazolidinones starting from *O*-homoallyl hydroxylamines and aryl halides. After a Heck reaction of the substrates, a subsequent C–N bond formation took place to furnish the target compounds in up to 79 % yield [86].

6.1.2
Cross-Coupling Reactions

Domino transition metal-catalyzed processes can also start with a cross-coupling reaction; most often, Suzuki, Stille and Sonogashira reactions are used in this context. They can be combined with another Pd-catalyzed transformation, and a number of examples have also been reported where a pericyclic reaction, usually a Diels–Alder reaction, follows. An interesting combination is also a Pd-catalyzed borination followed by a Suzuki reaction.

6.1.2.1 Suzuki Reactions

Interestingly, not only the combination of a Heck with a Suzuki reaction (as described above) but also a Suzuki with a Heck reaction is possible. However, a fine tuning of the reactivity of the different functionalities is necessary.

Shibasaki and coworkers [87] described the first enantioselective combination of this type in their synthesis of halenaquinone (**6/1-162**) (Scheme 6/1.43). The key step is an intermolecular Suzuki reaction of **6/1-159** and **6/1-160**, followed by an enantioselective Heck reaction in the presence of (*S*)-BINAP to give **6/1-161**. The *ee*-value was good, but the yield was low.

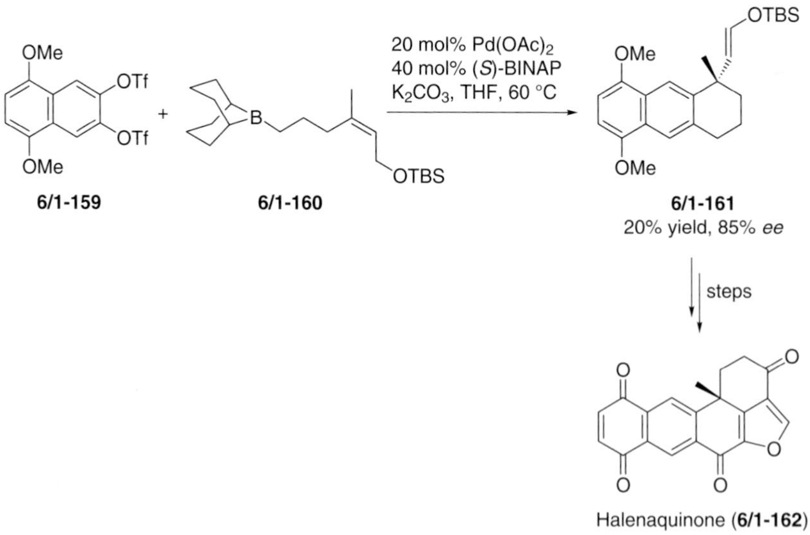

Scheme 6/1.43. Synthesis of halenaquinone (**6/1-162**).

Similarly, in the reaction of **6/1-163** and **6/1-164**, the corresponding boron compound is first formed *in situ* by addition of 9-BBN **6/1-163**; this then undergoes cross-coupling with **6/1-164**, followed by a Heck reaction to give the tricyclic carbon skeleton **6/1-165**. The use of triphenylarsine as co-ligand has a pronounced positive effect. Thus, without any additional ligand lower yields (53 %), several side products were obtained. Somewhat surprisingly, by using triphenylphosphane **6/1-165** is obtained in only 4 % yield (Scheme 6/1.44) [88].

Polycyclic aromatic hydrocarbons such as fluoranthrene or C_{60}-fullerene are structures of great interest. A straightforward entrance to analogues and partial structure, respectively, has now been developed by de Meijere and coworkers [89], using a combination of a Suzuki and a Heck-type coupling. Thus, reaction of 1,8-dibromophenanthrene **6/1-166** and *o*bromphenylboronic acid **6/1-167** employing 20 mol% of the Pd^0 catalyst led to **6/1-168** and **6/1-169** in 54 % yield as a 1:1-mixture (Scheme 6/1.45).

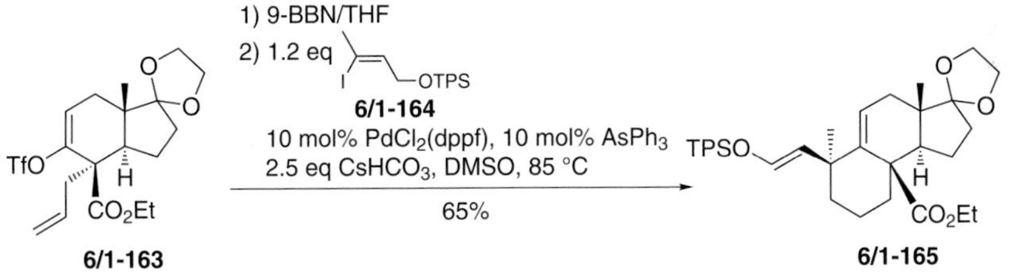

Scheme 6/1.44. Synthesis of tricyclic **6/1-165**.

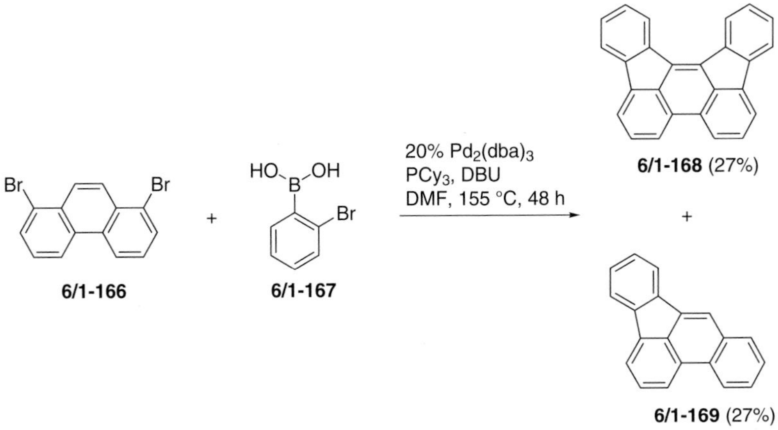

Scheme 6/1.45. Synthesis of fluoranthrene.

6.1.2.2 Stille Reactions

In those domino processes which start with a Stille reaction, the second step is usually a Diels–Alder reaction. However, there are two examples where an electrocyclization follows.

The natural product panepophenanthrin (**6/1-170**), isolated in 2002 from the fermented broth of the mushroom strain *Panus radus* IFO 8994 [90], is the first example of an inhibitor of the ubiquitin-activating enzyme [91]. Retrosynthetic analysis based on a biomimetic analysis led to the conjugated diene **6/1-172** by a retro-Diels–Alder reaction *via* the hemiacetal **6/1-171**. Further disconnections of **6/1-172** produces the vinyl stannane **6/1-173** and the vinyl bromide **6/1-174** [92].

Indeed, a Stille reaction of the TES-protected (±)-bromoxone (**6/1-175**) with the stannane **6/1-173** led to **6/1-176** upon standing overnight, *via* the diene **6/1-172** (Scheme 6/1.46). This synthesis is a classic example of a transition metal-catalyzed formation of a 1,3-diene, followed by a cycloaddition.

Martin and coworkers [93] described a highly efficient enantioselective total synthesis of manzamine A (**6/1-177**) with a concise domino Stille/Diels–Alder reaction to construct the tricyclic ABC ring core in **6/1-177** as the key step. Reaction of **6/1-178** with vinyl tributylstannane in the presence of $(Ph_3P)_4Pd$ afforded the triene

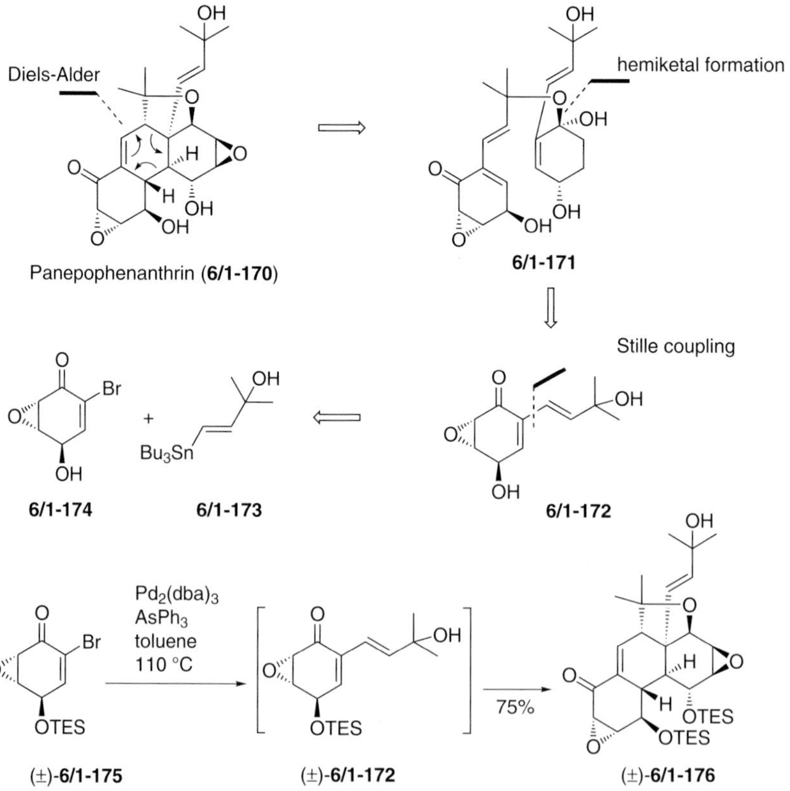

Scheme 6/1.46. Synthesis of panepophenanthrin (**6/1-170**).

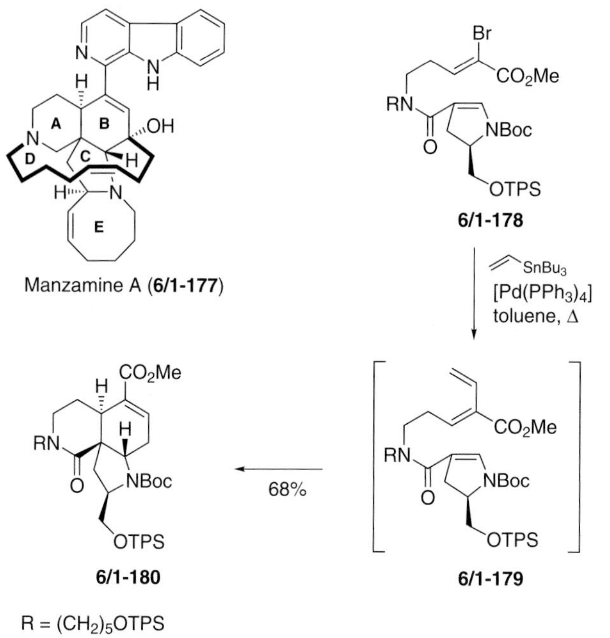

Scheme 6/1.47. Synthesis of manzamine A (**6/1-177**).

6/1-179 as intermediate, and this underwent an intramolecular Diels–Alder reaction upon heating to yield **6/1-180** as the only product, in 68% yield (Scheme 6/1.47). In this process, three new carbon–carbon bonds and three new stereogenic centers with excellent induced diastereoselectivity under the existing stereogenic center in **6/1-178** are formed.

Suffert and coworkers [94] developed a useful procedure for the synthesis of polycyclic ring systems by employing an intramolecular Stille reaction of a tributylstannyl diene as **6/1-182** or **6/1-183** containing a vinyl triflate moiety, followed by a transannular Diels–Alder reaction. Thus, the intramolecular Stille reaction of **6/1-183a** at room temperature did not lead to the expected macrocycle **6/1-184**, but rather to the annulated polycyclic compound **6/1-185**. This is slowly converted by oxidation to give the pentacycle **6/1-186** in an overall yield of 48% (Scheme 6/1.48). Better yields were obtained with the monomethoxy-compound **6-183b** (61%) and the dimethoxy-compound **6/1-183c** (70%). In addition, similar products containing a six- or seven-membered ring E **6/1-181a** and **6/1-181b** were also prepared, starting from **6/1-182**. A puzzling result is the fact that the transannular Diels–Alder reaction [95] takes place at only room temperature, perhaps due to Pd-catalysis [96]; however, final proof of this assumption has not yet been provided.

The combination of a Stille and a Diels–Alder reaction has also been used for a derivatization of steroids, for example, in the synthesis of pentacyclic compounds such as **6/1-189**, as described by Kollár and coworkers [97]. Reaction of 17-iodoandrosta-16-ene **6/1-187** with vinyltributyltin and various dienophiles such as diethyl

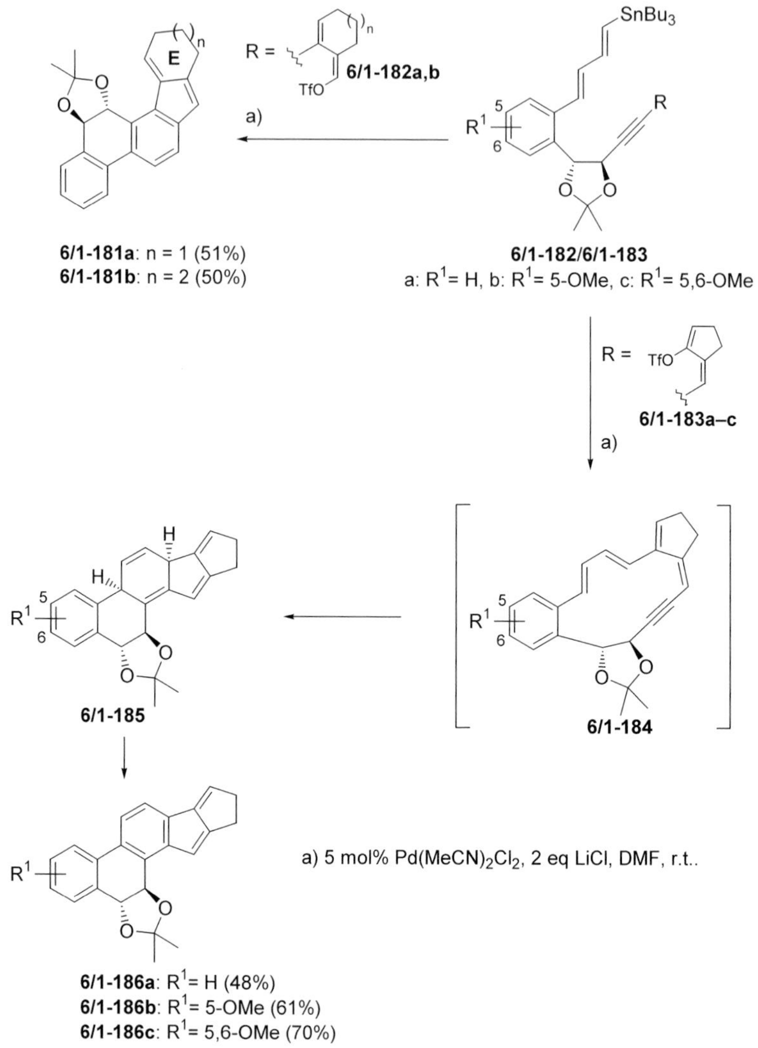
R =
TfO
6/1-182a,b
a)
SnBu3
R
6/1-181a: n = 1 (51%)
6/1-181b: n = 2 (50%)
6/1-182/6/1-183
a: R1= H, b: R1= 5-OMe, c: R1= 5,6-OMe
R =
TfO
6/1-183a–c
a)
6/1-185
6/1-184
a) 5 mol% Pd(MeCN)2Cl2, 2 eq LiCl, DMF, r.t..
6/1-186a: R1= H (48%)
6/1-186b: R1= 5-OMe (61%)
6/1-186c: R1= 5,6-OMe (70%)

Scheme 6/1.48. Domino Stille/transannular-Diels–Alder reaction.

maleate in the presence of catalytic amounts of Pd^0 led to **6/1-189** *via* the 1,3-butadiene **6/1-188** (Scheme 6/1.49).

The same group has also performed a combination of a Stille reaction and 1,3-dipolar cycloaddition using substrate **6/1-187** [98].

As mentioned previously, the Stille reaction can also be combined with an electrocyclization. Trauner and coworkers [99] used this approach for the synthesis of a part of SNF4435C (**6/1-190**) and its natural diastereomer. SNF4435C, which was isolated from the culture broth of an Okinawan strain of *Streptomyces spectabilis*, acts as an immunosuppressant and multidrug resistance reversal agent [100]. In order to form the annulated cyclobutane skeleton in **6/1-190**, the vinyl iodide

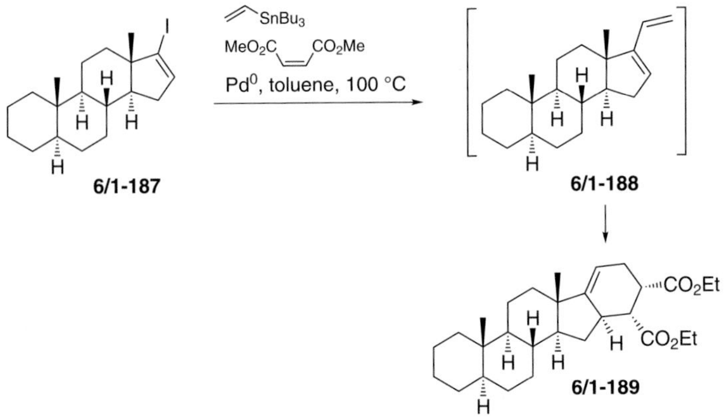

Scheme 6/1.49. Derivatisation of steroids.

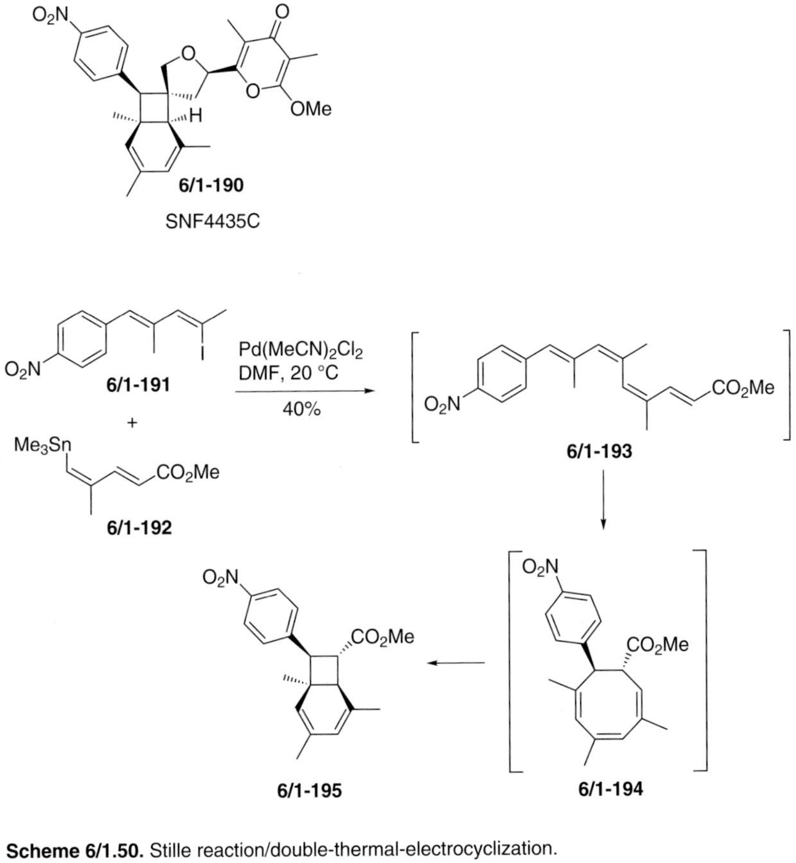

6/1-190

SNF4435C

Scheme 6/1.50. Stille reaction/double-thermal-electrocyclization.

Synthesis 6/1.51. Synthesis of the immunosuppressant SNF4435 C (**6/1-190**).

6/1-191 and the tin compound **6/1-192** were treated with Pd⁰ to give the intermediate *E,E,Z,E*-phenyloctatetraene **6/1-193**, which cyclizes to the cyclooctatriene **6/1-194**. There followed a ring contraction, leading to the desired cyclobutane derivative **6/1-195** in 40 % overall yield (Scheme 6/1.50).

Finally, Parker and coworkers [101] were able to use this approach for the total synthesis of SNF4435C (6/1-190) (Scheme 6/1.51). The Pd-catalyzed reaction of 6/1-191 and 6/1-195 gave a 4:1-mixture of 6/1-190 and its *endo*-diastereomer 6/1-199 in 53 % yield. In this transformation, the tetraene 6/1-196 can be assumed as intermediate, which theoretically could undergo an $8\pi/6\pi$ electrocyclization to give the *endo*-products *via* conformation 6/1-197 and the *exo*-products *via* conformation 6/1-198. However, only the two *endo*-products 6/1-190 and its diastereomer 6/1-199 are found, and not 6/1-200, which is consistent with the most likely nonenzymatic formation of 6/1-190 and its diastereomer in Nature from their co-metabolite spectabilin.

6.1.2.3 Sonogashira Reactions

In domino Sonogashira processes, the second step is usually an amination or a hydroxylation to give γ-lactones, furans, or indoles; however, there is also the possibility of performing a Heck reaction as a second step.

Thus, Alami and coworkers [102] have shown that benzylhalides as 6/1-201 can react with 1-alkynes as 6/1-202 in the presence of Pd^0 and CuI in a Sonogashira reaction which is followed by a Heck and a second Sonogashira reaction to give tetra-substituted alkenes 6/1-203 in yields of 22 to 90 % (Scheme 6/1.52).

Fiandanese and coworkers [103] described a new approach for the synthesis of the butenolides xerulin (6/1-207) and dihydroxerulin (6/1-208), which are of interest as potent noncytotoxic inhibitors of the biosynthesis of cholesterol (Scheme 6/1.53). The key transformation is a Pd^0-catalyzed Sonogashira/addition process of 6/1-204 or 6/1-206 with (Z)-3-iodo-2-propenoic acid 6/1-205, which is followed by the formation of a lactone to give 6/1-207 and 6/1-208, respectively.

Snieckus and his group members [104] used the known domino Sonogashira/Castro–Stephens reaction [105, 106] for the synthesis of the natural product plicadin (6/1-209), this having been isolated from *Psorelia plicata* in 1991 [107]. In this synthesis, Pd^0-catalyzed reaction of the alkyne 6/1-210 and the iodobenzene derivative 6/1-211 in the presence of CuI led to the furan 6/1-212, which was transformed into 6/1-209 *via* 6/1-213 (Scheme 6/1.54). There are some discrepancies of the physical data of the natural and the synthetic product; thus, it might be possible that the natural product has a different structure. It should also be mentioned that the

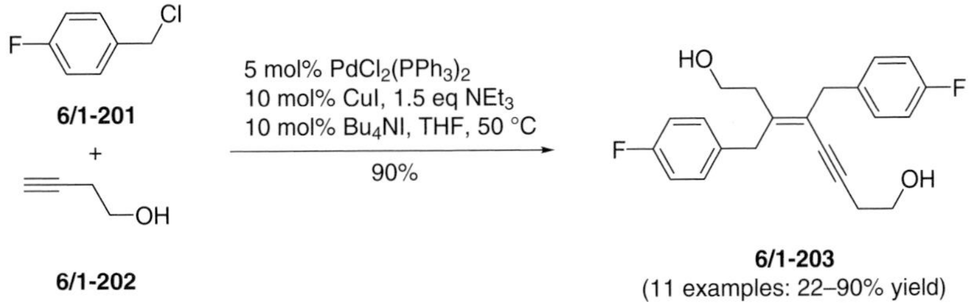

Scheme 6/1.52. Synthesis of tetra-substituted alkenes.

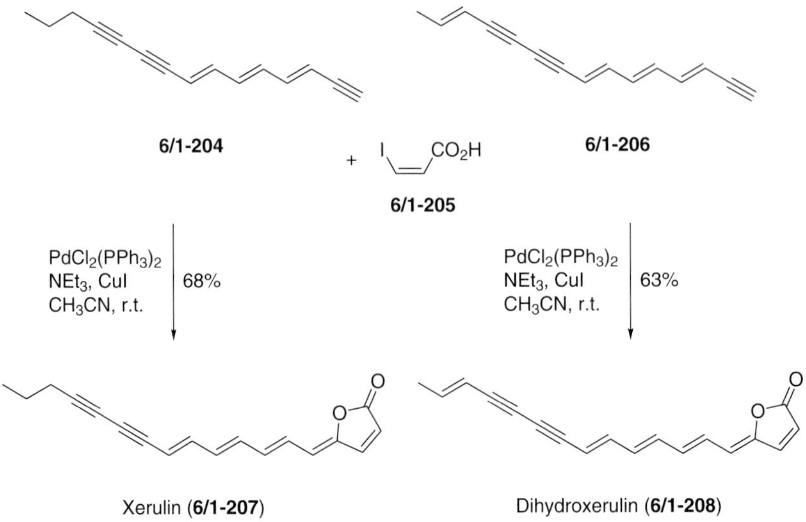
6/1-204
+
I
CO2H
6/1-205
6/1-206
PdCl2(PPh3)2
NEt3, CuI
CH3CN, r.t.
68%
PdCl2(PPh3)2
NEt3, CuI
CH3CN, r.t.
63%
O
O
O
O
Xerulin (6/1-207)
Dihydroxerulin (6/1-208)

Scheme 6/1.53. Synthesis of butenolides.

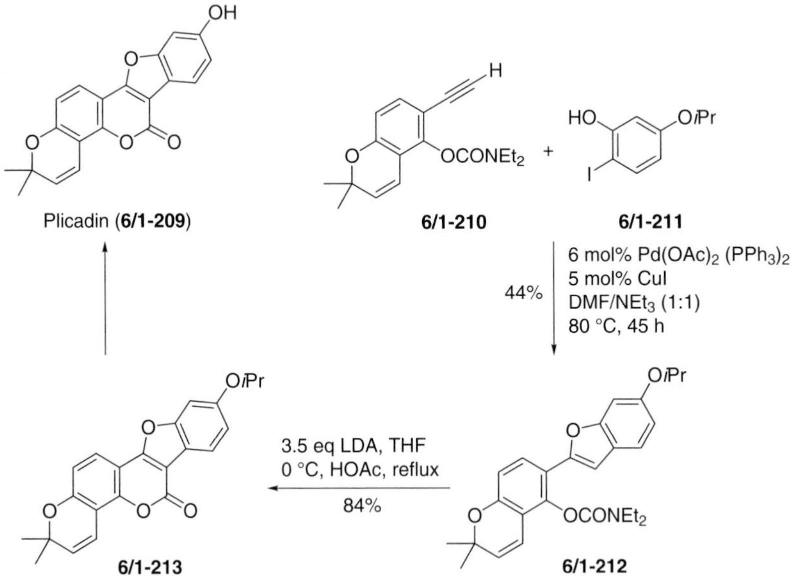
OH
O
O
O
O
Plicadin (6/1-209)
H
O
OCONEt2
+
HO
OiPr
I
6/1-210
6/1-211
44%
6 mol% Pd(OAc)2 (PPh3)2
5 mol% CuI
DMF/NEt3 (1:1)
80 °C, 45 h
OiPr
O
O
O
O
6/1-213
3.5 eq LDA, THF
0 °C, HOAc, reflux
84%
OiPr
O
O
OCONEt2
6/1-212

Scheme 6/1.54. Synthesis of plicadin (**6/1-209**).

Snieckus group also prepared **6/1-209** using a different route, which gave **6/1-209** in 20.5 % overall yield in five steps, whereas the described synthesis led to **6/1-209** in 6.8 % in six steps.

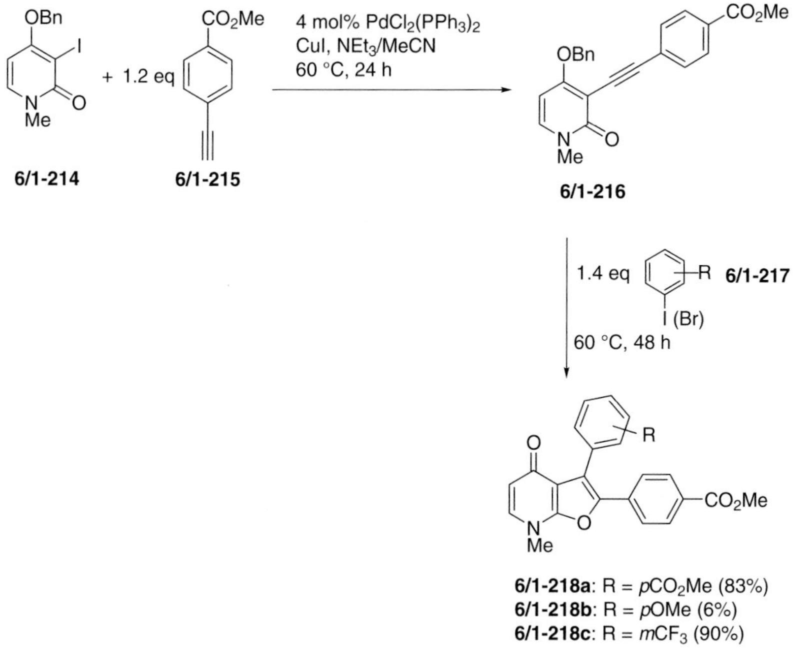

Scheme 6/1.55. Synthesis of furopyridones.

Useful sequential one-pot transformations for the synthesis of heterocycles again employing two Pd-catalyzed steps were developed by Balme and coworkers [108]. A recent example is the synthesis of furo[2,3-*b*]pyridones **6/1-218** [109] by reaction of iodopyridone **6/1-214** [110] and the alkyne **6/1-215** in the presence of catalytic amounts of PdCl$_2$(PPh$_3$)$_2$ and CuI in MeCN/Et$_3$N, followed by addition of the aryl iodide **6/1-217** to give **6/1-218** (Scheme 6/1.55). A wide variety of other pyridones, alkynes and arylhalides have been used as substrates, with yields ranging from 6% (**6/1-218**, R = *p*OMe) to 90% (**6/1-218**, R = *m*CF$_3$).

Rubrolide A (**6/1-221b**) is a marine tunicate metabolite with some *in-vitro* antibiotic activity; it contains a (Z)-γ-alkylidene butenolide moiety which was synthesized by Negishi and coworkers using a domino cross-coupling/lactonization process [111]. Reaction of alkyne **6/1-219** and (Z)-3-iodocinnamic acid **6/1-220** gave the diacetate of rubrolide A in 38% yield using Pd(PPh$_3$)$_4$ as catalyst and CuI in acetonitrile (Scheme 6/1.56).

The diacetates of rubrolide C, D, and E were prepared in 50 to 54% yield, in similar manner.

Pal and coworkers described a Pd^0-catalyzed water-based synthesis of indoles **6/1-224** using iodoanilines as **6/1-222** and alkynes **6/1-223** in up to 89% yield (Scheme 6/1.57) [112].

In a similar way, Alami and coworkers described a three-component Pd-catalyzed domino process of *o*-iodophenols or *o*-iodoanilides in the presence of a secondary

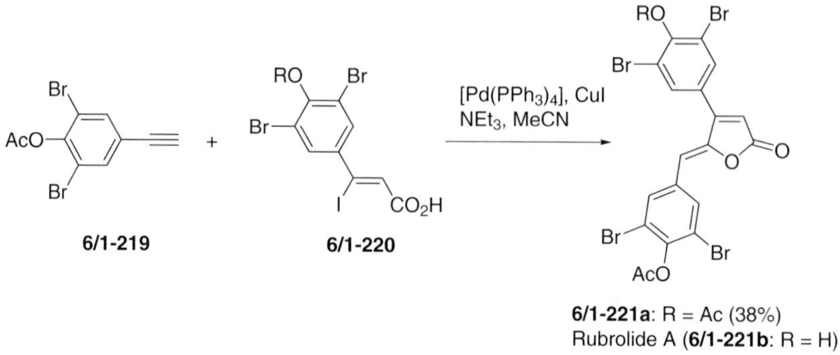

6/1-221a: R = Ac (38%)
Rubrolide A (**6/1-221b**: R = H)

Scheme 6/1.56. Synthesis of rubrolide A (**6/1-221b**).

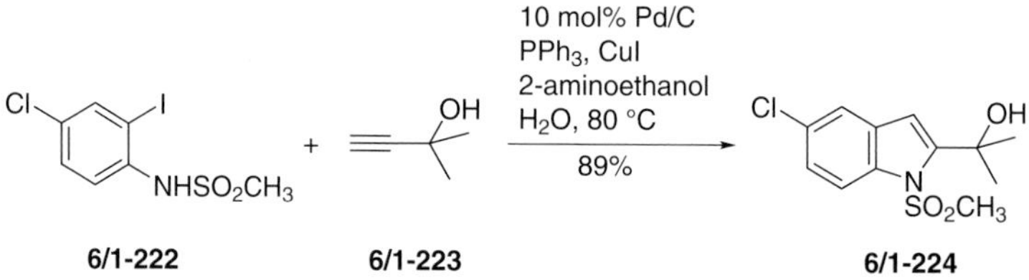

Scheme 6/1.57. Synthesis of indoles.

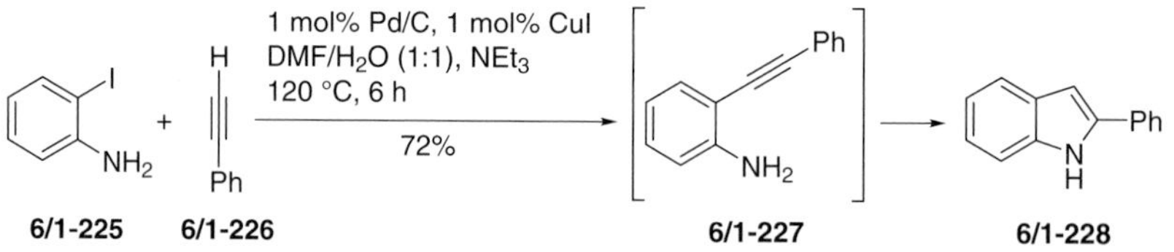

Scheme 6/1.58. Domino reaction with palladium on charcoal as catalyst.

amine and propargyl bromide to give benzofurans and indoles, respectively, in good to excellent yield [113].

Djakovitch and coworkers [114] have finally shown that Pd on activated carbon can also be used for these multistep transformations. Reaction of 2-iodoaniline **6/1-225** and phenylacetylene **6/1-226** in the presence of 1 mol% Pd/C and 1 mol% CuI at 120 °C led to the indole **6/1-228**, probably *via* **6/1-227**, though the latter compound was not detected in the reaction mixture (Scheme 6/1.58).

6.1.2.4 Other Cross-Coupling Reactions

A variety of less common cross-coupling reactions are discussed in this section, including borylations, silastannylations, and even cross-coupling with diindium compounds.

For the domino transition metal-catalyzed synthesis of macrocycles, conditions must be found for two distinct cross-coupling reactions, of which one is inter- and the other intramolecular. For this purpose, Zhu's group [115] has developed a process of a Miyura arylboronic ester formation followed by an intramolecular Suzuki reaction to give model compounds of the biphenomycin structure **6/1-232** containing an endo-aryl-aryl bond.

Thus, reaction of **6/1-229** with the diborane **6/1-230** in the presence of catalytic amounts of Pd(dppf)$_2$Cl$_2$ in high dilution afforded **6/1-231** in 45% yield (Scheme 6/1.59).

A similar reaction (which is not a domino process in its strict definition but a useful one-pot transformation) was described by Queiroz and coworkers [116]. Reaction of bromobenzothiophene **6/1-233** with pinacolborane in the presence of Pd(OAc)$_2$ and a phosphine followed by addition of 2-bromonitrostyrene **6/1-234** led to **6/1-235**, normally in good yields (Scheme 6/1.60).

A regio- and diastereoselective Pd-catalyzed domino silastannylation/allyl addition of allenes **6/1-236** containing a carbonyl moiety with Bu$_3$Sn-SiMe$_3$ **6/1-237** is described by Kang and coworkers [117]. The reaction allows the synthesis of hetero- and carbocyclic compounds with a ring size of five and six. It can be assumed that

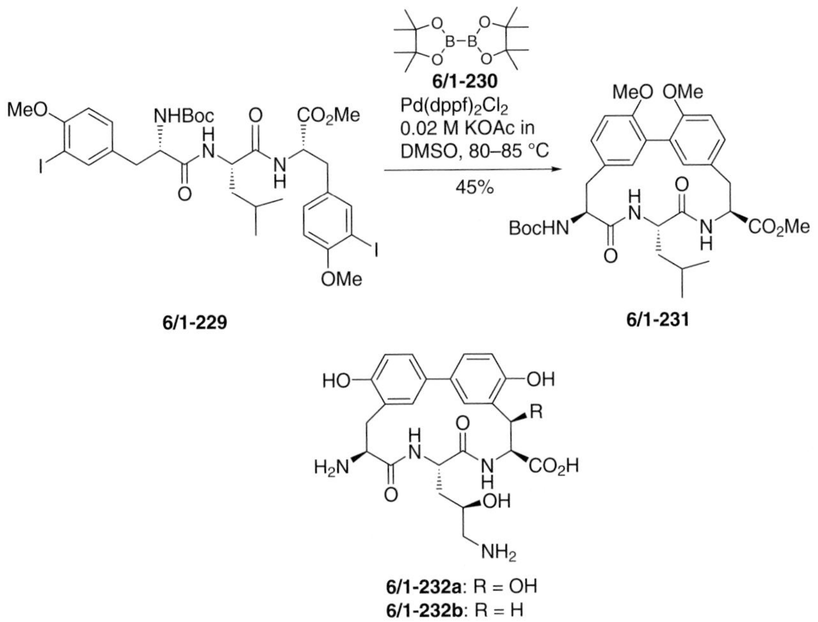

Scheme 6/1.59. Synthesis of biphenomycin analogue.

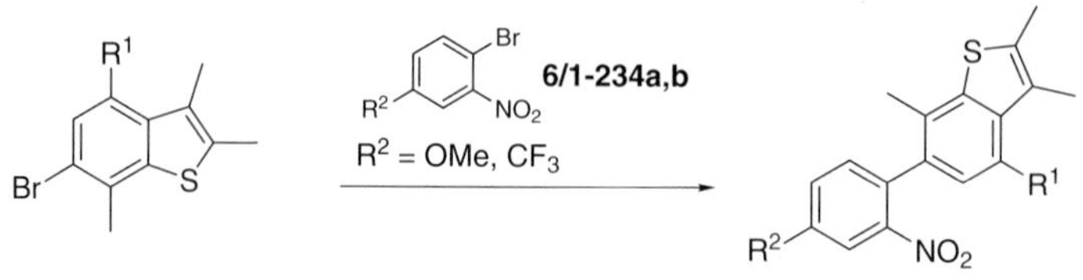

6/1-234a,b

R^2 = OMe, CF$_3$

6/1-233a: R^1= H
6/1-233b: R^1= Me

6/1-235a: R^1= H, R^2= OMe (80%)
6/1-235b: R^1= H, R^2= CF$_3$ (70%)
6/1-235c: R^1= Me, R^2= CF$_3$ (50%)

Reagents and conditions:
3.0 eq pinacolboran, 5 mol% Pd(OAc)$_2$, 20 mol% 2-(dicyclohexylphosphino)biphenyl,
4.0 eq NEt$_3$, dioxane, 1 h, 80 °C
then 0.7 eq **6/1-234**, 3.0 eq Ba(OH)$_2$•8H$_2$O, 100 °C, 1.5 h.

Scheme 6/1.60. Synthesis of biphenyls.

Me$_3$SiPdSnBu$_3$ is formed primarily from **6/1-237**, which then adds to the allene moiety in **6/1-236** to give a σ- or π-allyl palladium complex. This undergoes an intramolecular carbonyl allyl addition to afford the *cis*-cycloalkanols **6/1-238** (Scheme 6/1.61).

An unusual Pd-catalyzed cross-coupling reaction of a diindium reagent obtained from 3-bromo-1-iodopropene **6/1-239** was recently described by Hirashita and coworkers [118] to afford homoallylic alcohols **6/1-240** (Scheme 6/1.62).

6.1.3
Nucleophilic Substitution (Tsuji–Trost Reaction)

The Pd^0-catalyzed nucleophilic substitution of allylic acetates, carbonates or halides (also known as the Tsuji–Trost reaction) is a powerful procedure for the formation of C–C, C–O, and C–N bonds. One of the early impressive examples, where this transformation had been combined with a pallada-ene reaction, was developed by Oppolzer and coworkers [119]. In general, the Tsuji–Trost reaction can be combined with other Pd-catalyzed transformations as a Heck or a second Tsuji–Trost reaction; however, Michael reactions are also known as the second step. In their enantioselective total synthesis of the alkaloid cephalotaxin **6/1-244**, Tietze and coworkers [120] used a combination of a Tsuji–Trost and a Heck reaction (Scheme 6/1.63). Again, it was necessary to adjust the reactivity of the two Pd-catalyzed transformations to allow a controlled process. Reaction of **6/1-241a** using Pd(PPh$_3$)$_4$ as catalyst led to **6/1-242**, which in a second Pd-catalyzed reaction led to **6/1-243**. In this process, nucleophilic substitution of the allylicacetate is faster than the oxidative addition of the arylbromide moiety in **6/1-241a**; however, if one uses the iodide **6/1-241b**, the yields fall dramatically due to an increased rate of the oxidative addition.

The formation of **6/1-243** from **6/1-241a** is not a domino reaction, however, as a change of the catalyst is necessary for the second step. The Pd-catalyzed transfor-

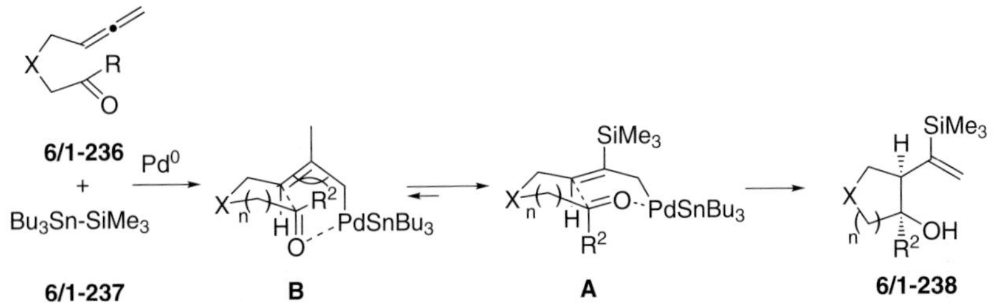

Scheme 6/1.61. Synthesis of cycloalkanols.

Entry	Substrate **6/1-236**	Product **6/1-238**	Yield [%]
1	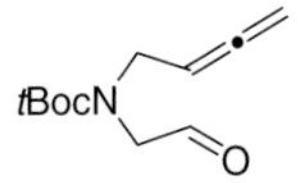	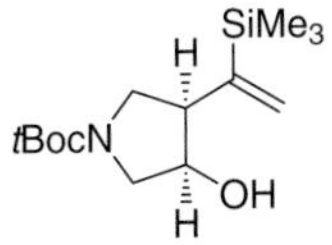	66
2	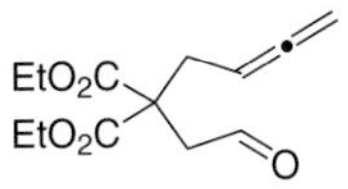	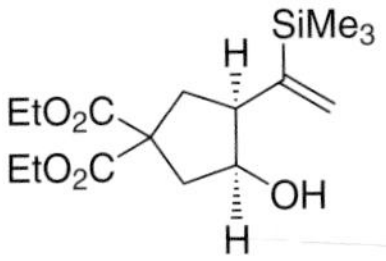	63
3	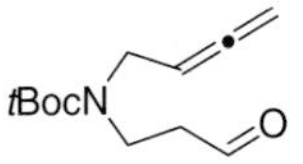	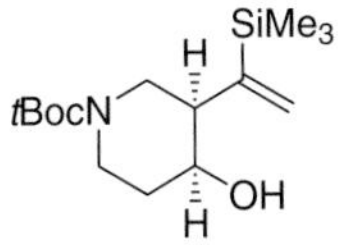	64
4	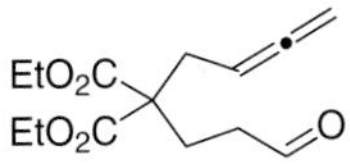	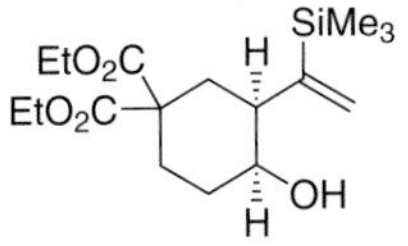	63
5	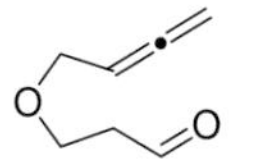	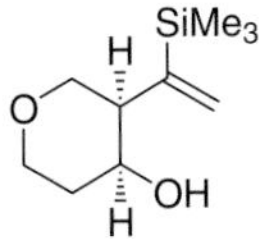	68

mation of **6/1-245** to give **6/1-246**, which is also described by Tietze and coworkers [121], is a "real" domino process, however (Scheme 6/1.64). Here, substrate **6/1-245d** gave the best result, as the reaction rate of the oxidative addition in comparison to the nucleophilic substitution is reduced due to the electron-donating group at the arylhalide.

A combination of a Tsuji–Trost and a Heck reaction was also used by Poli and Giambastiani [122] for the synthesis of the aza analogues **6/1-252** of the natural products podophyllotoxin **6/1-247a** and etoposide **6/1-247b**, which show pronounced

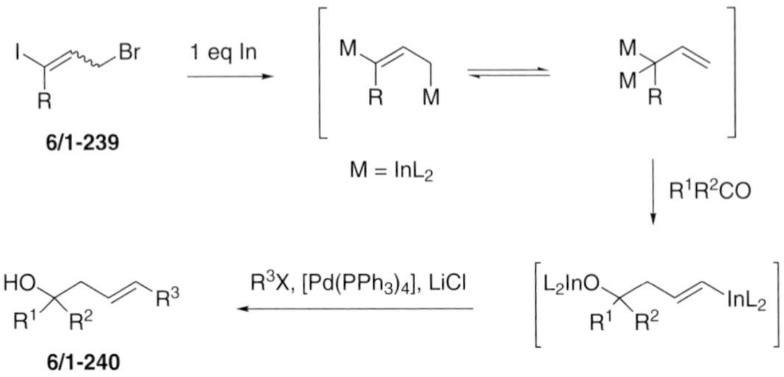

6/1-239

$M = InL_2$

R^1R^2CO

6/1-240

$R^3X, [Pd(PPh_3)_4], LiCl$

Scheme 6/1.62. Cross-coupling reactions with indium compounds.

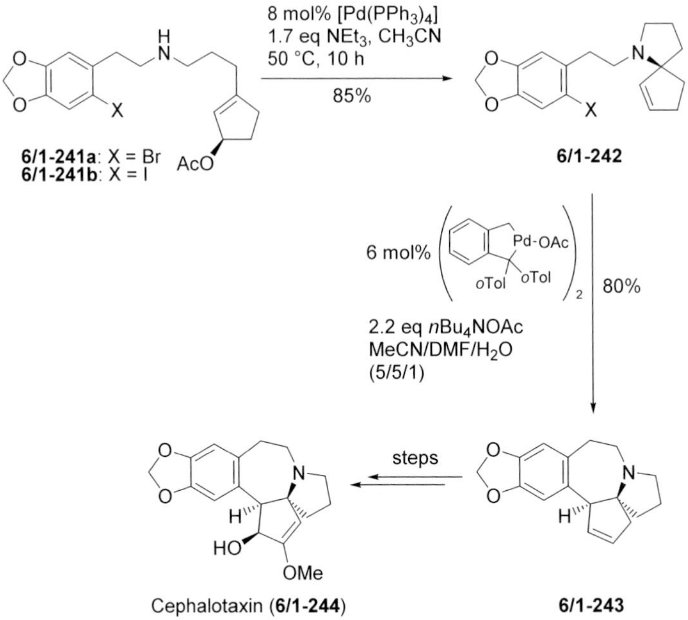

8 mol% [Pd(PPh$_3$)$_4$]
1.7 eq NEt$_3$, CH$_3$CN
50 °C, 10 h
85%

6/1-241a: X = Br
6/1-241b: X = I   AcO

6/1-242

6 mol%   Pd-OAc $oTol$ $oTol$ $_2$ 80%

2.2 eq nBu$_4$NOAc
MeCN/DMF/H$_2$O
(5/5/1)

Cephalotaxin (**6/1-244**)

steps

6/1-243

Scheme 6/1.63. Enantioselective synthesis of cephalotaxin (**6/1-244**).

antimitotic and antitumor activities, respectively [123]. For the synthesis of **6/1-252**, compound **6/1-248**, obtained from piperonal, and the amide **6/1-249** containing an allylacetate moiety, were transformed into **6/1-250**, which served as the starting material for the domino reaction. Treatment of **6/1-250** with catalytic amounts of Pd(OAc)$_2$ in the presence of dppe at 145 °C led to the tetracyclic structure **6/1-251** in 55 % yield as a 75:25 mixture of two (out of four possible) diastereomers (Scheme

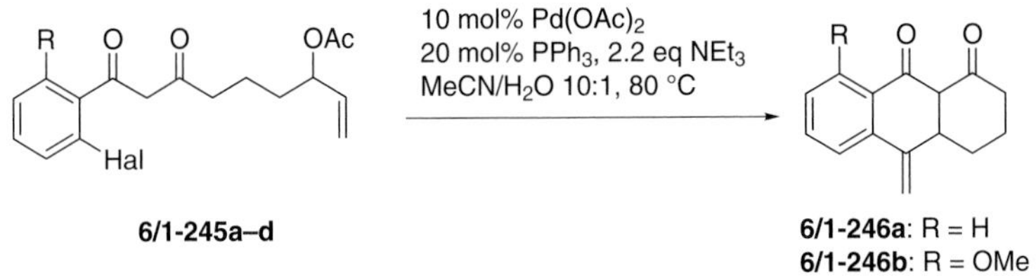

6/1-245a–d

6/1-246a: R = H
6/1-246b: R = OMe

6/1-245	R	Hal	6/1-246	Yield [%]
a	H	Br	a	49
b	H	I	a	23
c	OMe	Br	b	77
d	OMe	I	b	89

Scheme 6/1.64. Domino Tsuji–Trost/Heck reaction.

6/1.65). It has been shown that the nucleophilic substitution of the allylacetate moiety takes place at only 85 °C, whereas the following Heck reaction requires a higher reaction temperature.

Lamaty and coworkers described a straightforward combination of three Pd-catalyzed transformations: first, an intermolecular nucleophilic substitution of an allylic bromide to form an aryl ether; second, an intramolecular Heck-type transformation in which as the third reaction the intermediate palladium species is intercepted by a phenylboronic acid [124]. Thus, the reaction of a mixture of 2-iodophenol (**6/1-253**), methyl 2-bromomethylacrylate **6/1-254** and phenylboronic acid in the presence of catalytic amounts of Pd(OAc)$_2$ led to 3,3-disubstituted 2,3-dihydrobenzofuran **6/1-255** (Scheme 6/1.66). In addition to phenylboronic acid, several substituted boronic acids have also been used in this process.

An intramolecular Pd-catalyzed ring closure of an allylic halide, acetate or carbonate containing an allene moiety as **6/1-256**, followed by a Suzuki reaction, was used by Zhang and coworkers for the synthesis of five-membered carbo- and heterocycles **6/1-257** and **6/1-258** (Scheme 6/1.67) [125].

In their elegant synthesis of (+)-hirsutene **6/1-261**, Oppolzer and coworkers [126] combined an intramolecular Pd-catalyzed nucleophilic substitution of the allylcarbonate **6/1-259** with carbonylation; this is followed by an insertion of the formed carbonyl Pd-species into the double bond, obtained in the former reaction. The last step is a second carbonylation to give **6/1-260** after methylation of the formed acid moiety (Scheme 6/1.68).

A combination of a Pd-catalyzed nucleophilic substitution by a phenol and a ring expansion was described by Ihara and coworkers [127] using *cis*- or *trans*-substituted propynylcyclobutanols **6/1-262a** or **6/1-262b**. The product ratio depends on the stereochemistry of the cyclobutanols and the acidity of the phenol **6/1-263**. Thus, reaction of **6/1-262b** with *p*-methoxyphenol **6/1-263** (X = *p*OMe) led exclu-

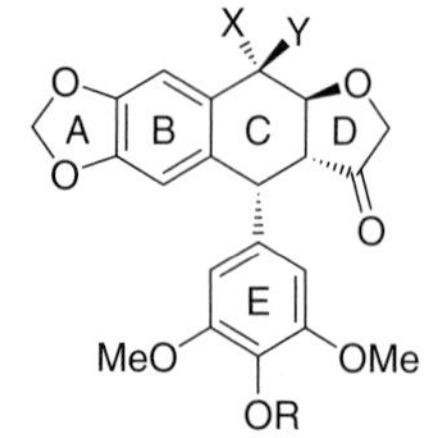

Podophyllotoxin (**6/1-247a**): X = OH, R = Me, Y = H
Etoposide (**6/1-247b**): X = H, R = H, Y = Me

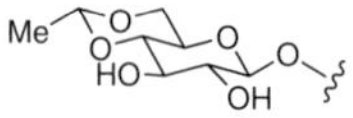

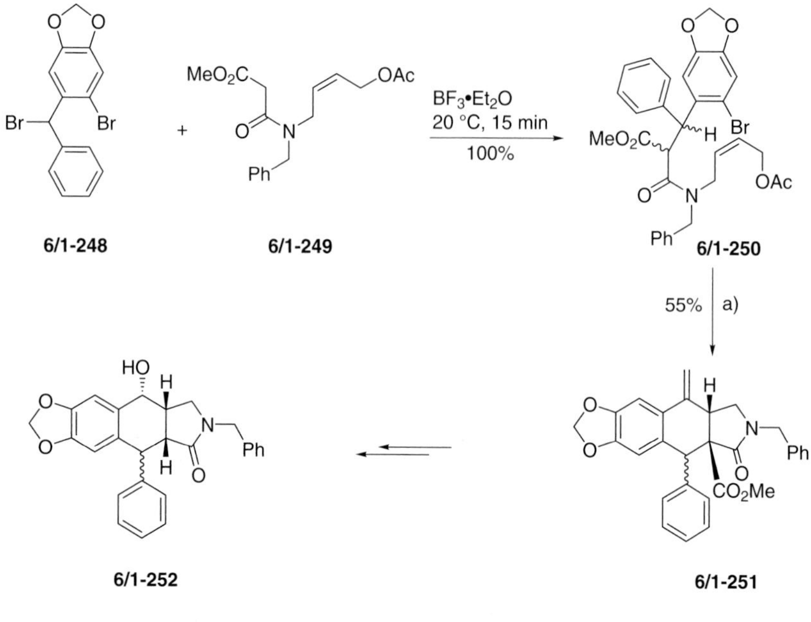

a) 1.1 eq NaH then 10 mol% Pd(OAc)₂, 20 mol% dppe
2.0 eq KOAc, DMF, 145 °C, 25 min.

Scheme 6/1.65. Synthesis of aza analogues of podophyllotoxin.

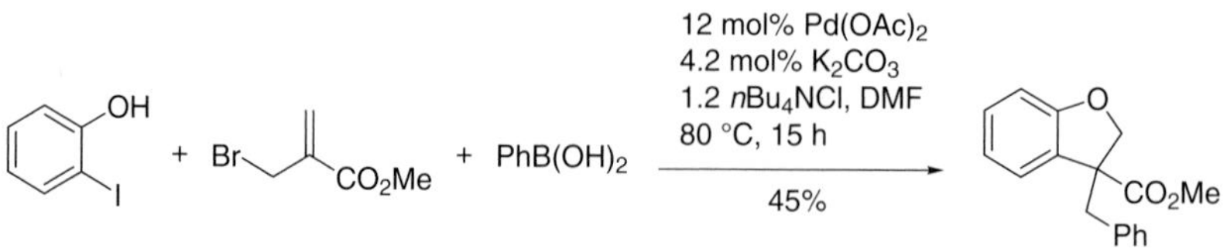

12 mol% Pd(OAc)₂
4.2 mol% K₂CO₃
1.2 nBu₄NCl, DMF
80 °C, 15 h

45%

Scheme 6/1.66. Synthesis of benzofurans.

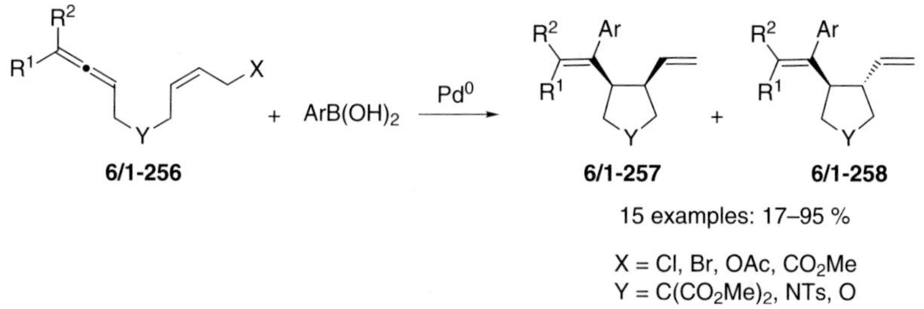

Scheme 6/1.67. Synthesis of five-membered hetero- and carbocycles.

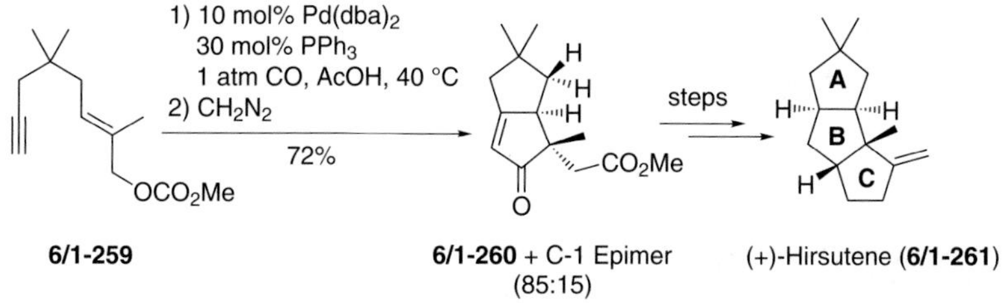

15 examples: 17–95 %

X = Cl, Br, OAc, CO₂Me
Y = C(CO₂Me)₂, NTs, O

Scheme 6/1.68. Synthesis of (+)-hirsutene (6/1-261).

sively to **6/1-264a** in 98 % yield, whereas with *p*-nitrophenol **6/1-263** (X = *p*NO₂) only **6/1-265** was obtained, though in modest yield. In contrast, the Pd-catalyzed reaction of **6/1-262a** with *p*-methoxyphenol **6/1-263** (X = *p*MeOH) gave a mixture of **6/1-264b** and **6/1-265** in a 36:64 ratio in 98 % yield, whereas with *p*-nitrophenol **6/1-263** (X = *p*NO₂) again only **6/1-265** was obtained.

It can be assumed that, according to the proposed mechanism, the β,γ-unsaturated ketone **6/1-264a** or **6/1-264b** is first formed, and this then isomerizes to give the α,β-unsaturated ketone **6/1-265** (Scheme 6/1.69).

A combination of a Tsuji–Trost and a Michael addition was used for the synthesis of (+)-dihydroerythramine **6/1-269**, as reported by Desmaële and coworkers [128]. The Pd-catalyzed reaction of the allylic acetate **6/1-267** with the nitromethylarene **6/1-266** in the presence of Cs₂CO₃ as base led to the domino product **6/1-268** as a 4:1 mixture of two diastereomers in 79 % yield. Further manipulation of **6/1-268a** yielded the desired dihydroerythramine **6/1-269** (Scheme 6/1.70). Interestingly, using the corresponding allylic carbonate without additional base gave the monoalkylated product only.

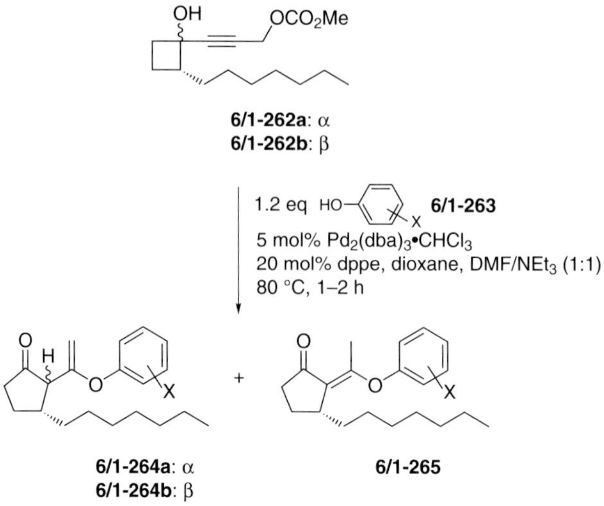

6/1-262a: α
6/1-262b: β

1.2 eq HO—⬡—X **6/1-263**

5 mol% Pd$_2$(dba)$_3$•CHCl$_3$
20 mol% dppe, dioxane, DMF/NEt$_3$ (1:1)
80 °C, 1–2 h

6/1-264a: α   **6/1-265**
6/1-264b: β

Entry	Substrate	X	Product	Yield [%]
1	6/1-262a	4-OMe	6/1-264b:6/1-265 = 36:64[a]	98
2	6/1-262a	2,4,6-Trimethyl	6/1-264b:6/1-265 = 23:77[a]	98
3	6/1-262a	4-Me	6/1-265	92
4	6/1-262a	2-OMe	6/1-265	93
5	6/1-262a	H	6/1-265	97
6	6/1-262a	4-Cl	6/1-265	96
7	6/1-262a	4-NO$_2$	6/1-265	70
8	6/1-262b	4-OMe	6/1-264a	98
9	6/1-262b	2,4,6-Trimethyl	6/1-264a	90
10	6/1-262b	4-Me	6/1-264a	80
11	6/1-262b	2-OMe	6/1-264a	93
12	6/1-262b	H	6/1-264a:6/1-265 = 65:35[a]	94
13	6/1-262b	4-Cl	6/1-264a:6/1-265 = 68:32[a]	67
14	6/1-262b	4-NO$_2$	6/1-265	23

[a] The product ratio was determined by ^{1}H-NMR.

Scheme 6/1.69. Domino Tsuji–Trost reaction/rearrangement.

6.1.4
Reactions of Alkynes and Allenes

The Pd[0]-catalyzed transformation of enediynes represents a highly efficient and effective approach for the synthesis of polycyclic compounds, with different ring sizes being obtained by a variation of the tether [129]. In this respect, reaction of **6/1-270** led to the tricyclic product **6/1-271** as a single diastereomer. The initial step is a chemoselective hydropalladation of the propargylic ester moiety in **6/1-270** to give an alkenyl-Pd-species, according to the mechanism depicted in Scheme 6.71. A hexatriene is formed as a byproduct.

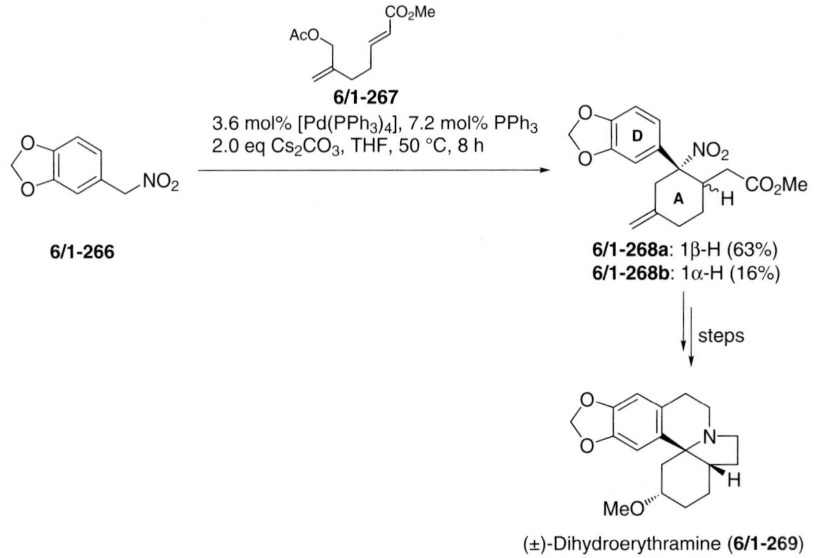

Scheme 6/1.70. Synthesis of (±)-dihydroerythramine (**6/1-269**).

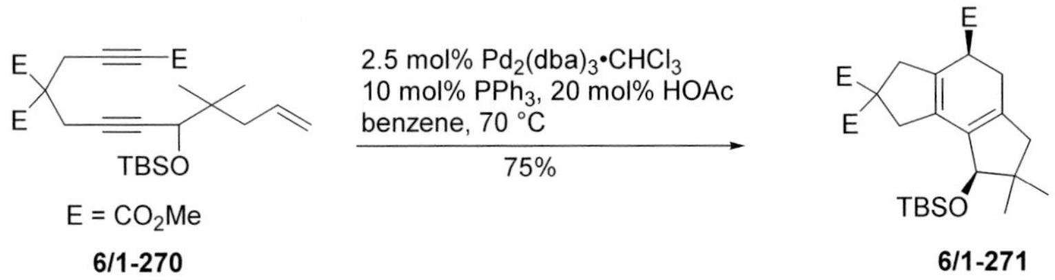

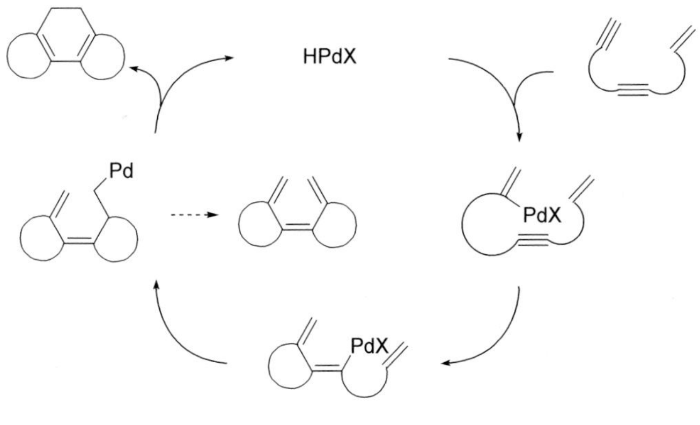

Scheme 6/1.71. Mechanism of the hydropalladation of enediynes.

In a similar way, enetetraynes can undergo a pentacyclization as observed for compound **6/1-272**, which gave **6/1-273** in 66 % yield (Scheme 6/1.72). In this case the initial alkenyl-Pd-species is formed by an oxidative addition of Pd^0 to the alkenyliodide moiety in **6/1-272** [130]. In comparison to the reaction of **6/1-270**, this procedure has the advantage of complete control of the chemoselective formation of the alkenyl-Pd-species, but the disadvantage that the starting material is less easily accessible.

Trost and coworkers [131] synthesized oxaheterocycles by a Pd-catalyzed addition of terminal alkynes onto hydroxyalkynoates, followed by an intramolecular addition of the hydroxyl functionality on the triple bond. Simple lactonization may take place as a side reaction.

The best results were obtained using 10 mol% of $Pd(OAc)_2$ and TDMPP at room temperature; under these conditions, **6/1-277** could be obtained from **6/1-274** and **6/1-275**, probably *via* **6/1-276** in 61 % yield; the amount of the lactone **6/1-278** was below 5 % (Scheme 6/1.73). Several other terminal alkynes and hydroxyalkynoates have been used to produce dihydropyrans in yields of 41 to 71 %.

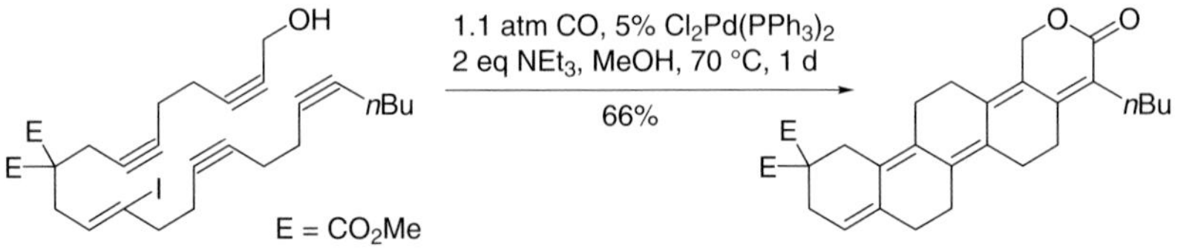

6/1-272 **6/1-273**

Scheme 6/1.72. Polycyclisation of enetetraynes.

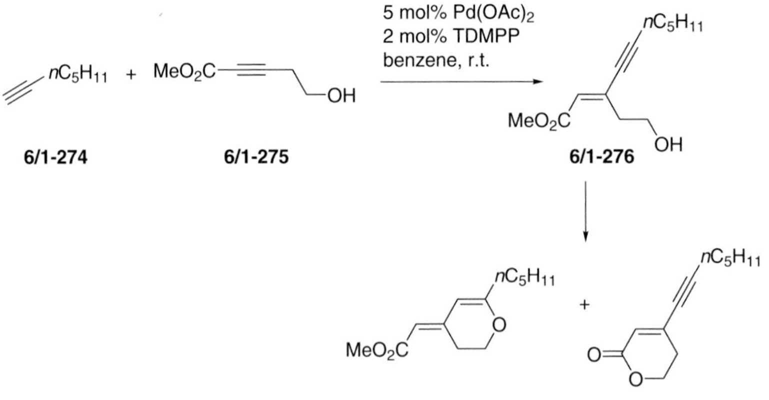

6/1-277 (61%) **6/1-278** (traces)

Scheme 6/1.73. Synthesis of dihydropyrans.

Holzapfel and coworkers [132] used the carbopalladation of alkynes followed by a cyclization for the synthesis of tricyclic compounds as **6/1-280**, derived from the sugar derivative **6/1-279** (Scheme 6/1.74).

Another versatile domino process for the synthesis of carbocycles as well as heterocycles is the Pd-catalyzed reaction of organic halides or triflates with alkynes or allenes, which contain a carbo- or heteronucleophile in close vicinity to these functionalities (see Scheme 6/1.75) [133].

Some single examples of this type of reaction were discussed in earlier sections, but due to the importance of these transformations, an additional overview will be provided here. One of the first transformations based on this strategy was published by Inoue and coworkers [134] using propargylic alkoxide, an arylhalide or vinyl bromide and CO_2 to give cyclic carbonates. The Balme group used this ap-

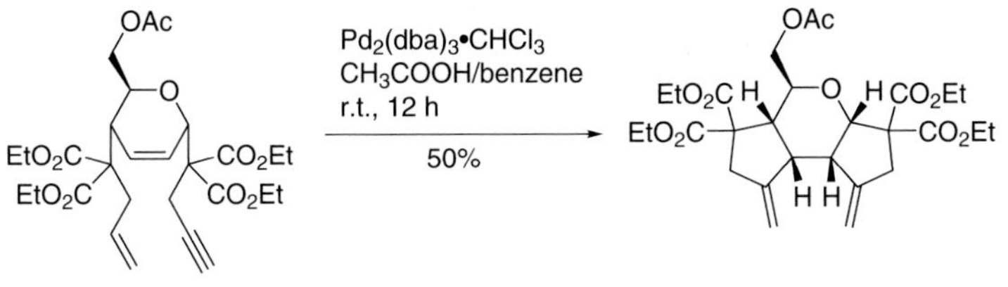

Scheme 6/1.74. Synthesis of tricyclic compounds from sugars.

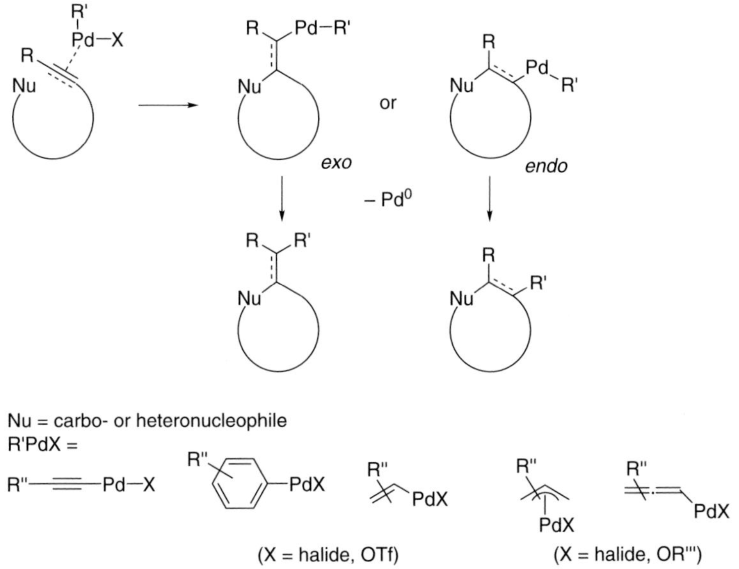

Nu = carbo- or heteronucleophile

R'PdX =

(X = halide, OTf)   (X = halide, OR''')

Scheme 6/1.75. Pd-catalyzed transformation of organic halides or triflates with alkynes or allenes containing a nucleophilic functionality.

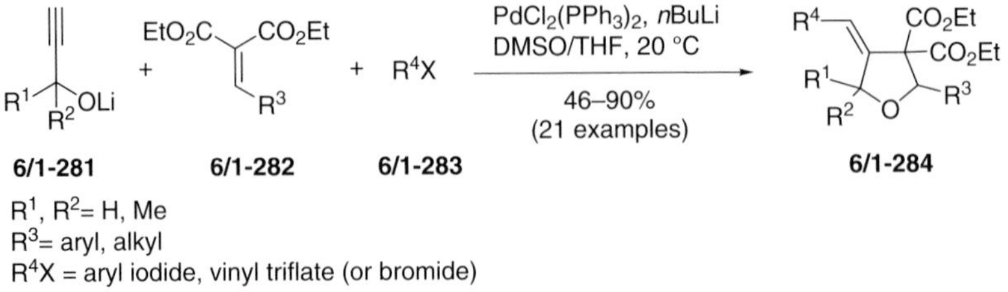
PdCl2(PPh3)2, nBuLi
DMSO/THF, 20 °C
46–90%
(21 examples)
6/1-281
6/1-282
6/1-283
6/1-284
R1, R2= H, Me
R3= aryl, alkyl
R4X = aryl iodide, vinyl triflate (or bromide)

Scheme 6/1.76. Synthesis of furan derivatives.

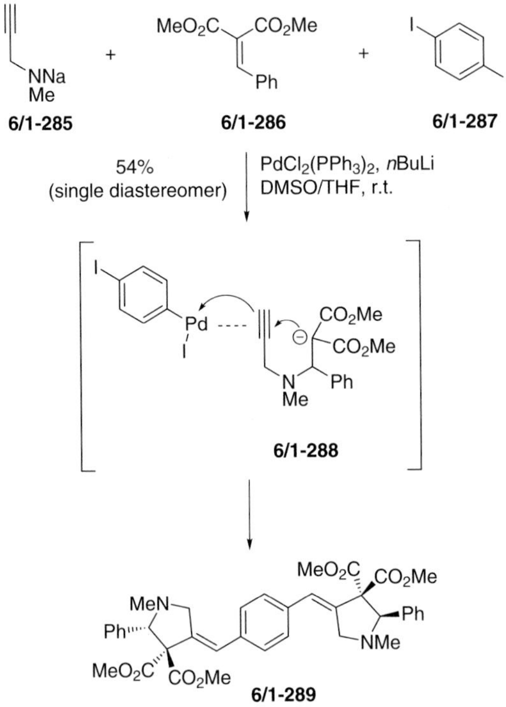
6/1-285
6/1-286
6/1-287
54%
(single diastereomer)
PdCl2(PPh3)2, nBuLi
DMSO/THF, r.t.
6/1-288
6/1-289

Scheme 6/1-77. Synthesis of pyrrolidines.

proach for the synthesis of tetrahydrofurans [135], as well as for benzylidene and alkylidene tetrahydrofurans **6/1-284** [136], starting from an alkylidene or benzylidene malonate **6/1-282**, an aryl halide or vinyl triflate (bromide) **6/1-283** and an allylic alkoxide or propargylic alkoxides as **6/1-281**, the latter being more reactive with the malonate **6/1-282** (Scheme 6/1.76).

Allylic chlorides can also be used as coupling partners, as shown by Lu and Liu [137]. In addition, the domino process could be used for the synthesis of benzylidene pyrrolidines, as shown for the reaction of **6/1-285** with benzylidene malonate

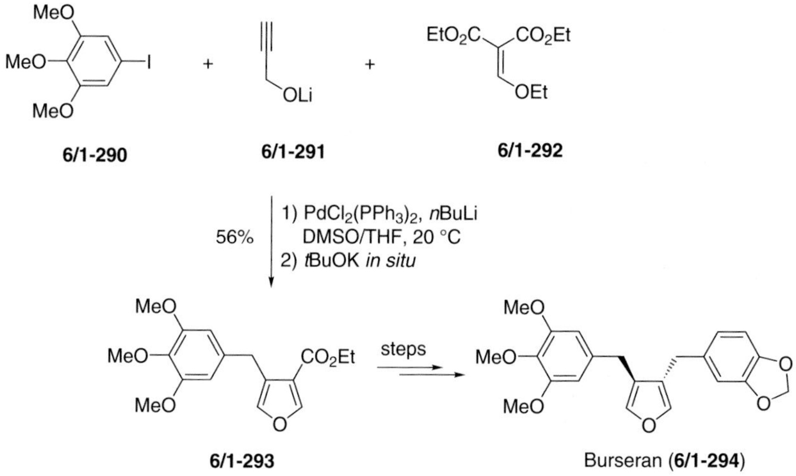

Scheme 6/1.78. Synthesis of burseran (**6/1-294**).

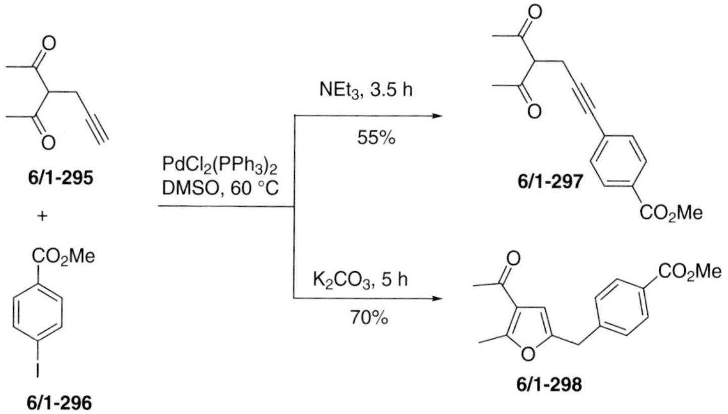

Scheme 6/1.79. Reaction of propynyl-1,3-dicarbonyls.

6/1-286 and diiodobenzene **6/1-287** to give **6/1-289** as a single diastereomer in 54 % yield with **6/1-288** as a proposed intermediate (Scheme 6/1.77) [138].

The approach also allows the synthesis of furans by employing ethoxymethylene malonate, followed by an eliminative decarboxylation. This method was used by Balme for a formal synthesis of the antitumor lignan burseran (**6/1-294**), starting from **6/1-290,6/1-291** and **6/1-292** *via* the furan **6/1-293** (Scheme 6/1.78) [139].

Furans as **6/1-298** can also be obtained by Pd-catalyzed reaction of 2-propynyl-1,3-dicarbonyls **6/1-295** with aryl halides **6/1-296** in DMF, using potassium carbonate as base, as shown by Arcadi, Cacchi and coworkers (Scheme 6/1.79) [140].

This method has also been applied to the synthesis of steroidal derivatives as **6/1-300** and **6/1-302** from **6/1-299** and **6/1-301**, respectively (Scheme 6/1.80).

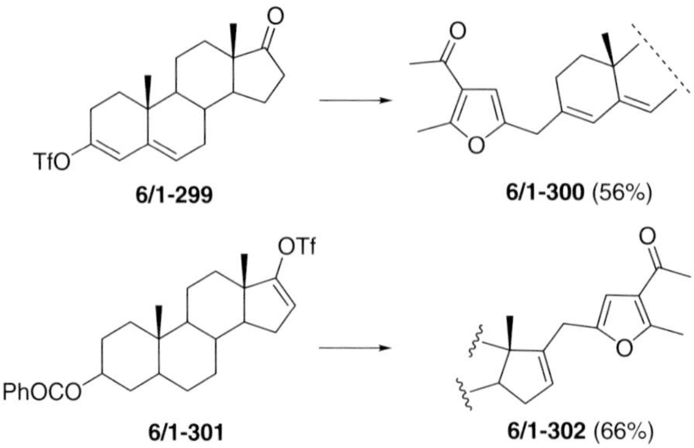

Scheme 6/1.80. Introduction of a furan moiety to steroids.

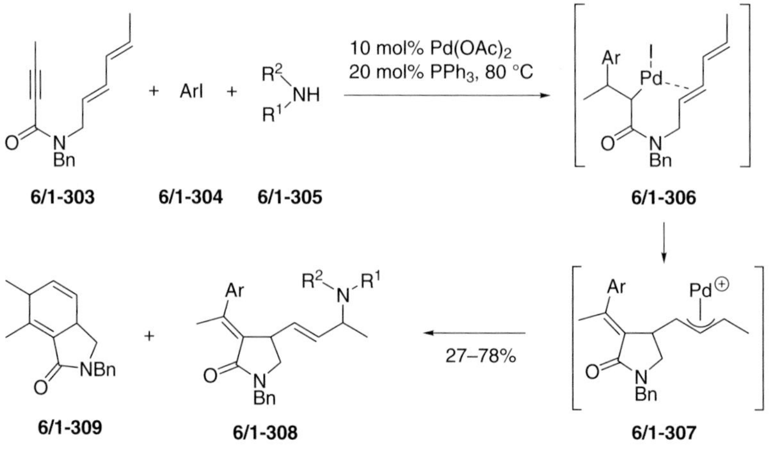

Scheme 6/1.81. Synthesis of pyrrolidones.

Furthermore, the Wu group [141] has prepared isoquinolines and iso-indoles starting from 2-(2-phenylethynyl)-benzonitrile, while Yamamoto and coworkers [142] prepared benzopyrans from alkynylbenzaldehydes, the Cacchi group [143] synthesized benzofurans, and Balme and coworkers [144] furo[2,3-*b*]pyridones.

Instead of alkynes, allenes can also be used as substrates in this type of approach. Finally, one can also apply carbon-nucleophiles such as butadienes in this domino process. Thus, Lu and Xie [145] have treated the alkyne **6/1-303** with an aryl halide **6/1-304** and an amine **6/1-305** to give the substituted pyrrolidinone **6/1-308** *via* the proposed intermediates **6/1-306** and **6/1-307**. As a side product, **6/1-309** is found to have been formed by a cycloaddition of **6/1-303** (Scheme 6/1.81).

Mouriño and coworkers have shown, in a latest publication, an impressive combination of a palladium-catalyzed 6-*exo*-cyclocarbopalladation and a Negishi coupling. Using this strategy, the authors were able to synthesize bioactive metabolites of vitamin D_3 [146].

Another recent development in the field of palladium-catalyzed reactions with alkynes is a novel multicomponent approach devised by the Lee group. Starting from α-bromovinyl arenes and propargyl bromides, the assembly of eight-membered carbocycles can be realized via a cross-coupling/[4+4] cycloaddition reaction. The authors also presented the combination of a cross-coupling and homo [4+2], hetero [4+2], hetero [4+4] or [4+4+1] annulation leading to various cyclic products [147].

6.1.5
Other Pd⁰-Catalyzed Transformations

A very common combination in Pd-catalyzed domino reactions is the insertion of CO as the last step (this was discussed previously). However, there is also the possibility that CO is inserted as the first step after oxidative addition. This process, as well as the amidocarbonylation, the amination, the arylation of ketones, the isomerization of epoxides, and the reaction with isonitriles will be discussed in this section.

A typical second step after the insertion of CO into aryl or alkenyl-Pd(II) compounds is the addition to alkenes [148]. However, allenes can also be used (as shown in the following examples) where a π-allyl-η³-Pd-complex is formed as an intermediate which undergoes a nucleophilic substitution. Thus, Alper and coworkers [148], as well as Grigg and coworkers [149], described a Pd-catalyzed transformation of o-iodophenols and o-iodoanilines with allenes in the presence of CO. Reaction of **6/1-310** or **6/1-311** with **6/1-312** in the presence of Pd⁰ under a CO atmosphere (1 atm) led to the chromanones **6/1-314** and quinolones **6/1-315**, respectively, via the π-allyl-η³-Pd-complex **6/1-313** (Scheme 6/1.82). The enones obtained can be transformed by a Michael addition with amines, followed by reduction to give γ-amino alcohols. Quinolones and chromanones are of interest due to their pronounced biological activity as antibacterials [150], antifungals [151] and neurotrophic factors [152].

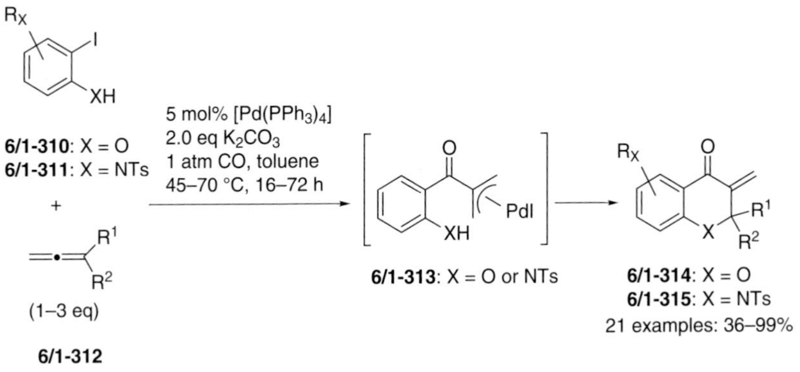

Scheme 6/1.82. Synthesis of chromanones and quinolones.

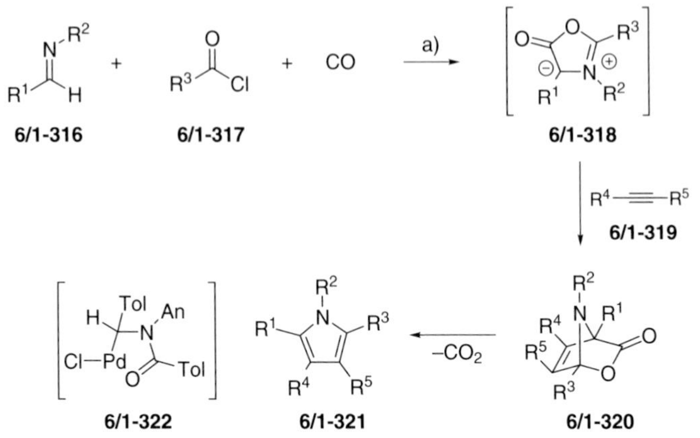

a) Preformed 5 mol% Pd-catalyst, 15 mol% P(*o*tolyl)$_3$, *i*Pr$_2$NEt, CH$_3$CN/THF.
Tol = *p*tolyl; An = *p*C$_6$H$_4$OMe; R^1= R^2= R^3= aryl, alkyl; R^4= R^5= H, aryl, electron-withdrawing groups.

Scheme 6/1.83. The four-component assembly of pyrroles.

Another option is the *in situ* reaction of the obtained enones in a 1,3-dipolar cycloaddition using nitrones or azomethineylides formed from the corresponding imines with DBU in the reaction mixture [153].

Arndtsen and coworkers [154] described the first Pd-catalyzed synthesis of münchnones **6/1-318** from an imine **6/1-316**, a carboxylic acid chloride **6/1-317** and CO. The formed 1,3-dipol **6/1-318** can react with an alkyne **6/1-319** present in the reaction mixture to give pyrroles **6/1-321** *via* **6/1-320**, in good yields. The best results in this four-component domino process were obtained with the preformed catalyst **6/1-322** (Scheme 6/1.83).

As described above, Pd-catalyzed transformations can also be combined with a Cope rearrangement. The scope of the Cope rearrangement was greatly increased by the introduction of the anionic oxy-Cope approach, which allows a rate acceleration of about 10^{10}- to 10^{17}-fold [155]. However, in some cases even this is insufficient (as shown in the synthesis of phomoidrides [156]) due to an insufficient overlap of the involved orbitals [157]. Thus, Leighton and coworkers proposed the use of a bicyclic 1,5-diene **6/1-326** with a lactone moiety having a significant strain and twisting, which would be released in the Cope rearrangement to give **6/1-327**. This was indeed successful, and allowed the rearrangement to take place at 110 °C, thereby avoiding the formation of side products [158]. The lactone moiety in **6/1-326** was introduced by a Pd-catalyzed carbonylation of the vinyl triflate **6/1-325** at 75 °C, followed by the rearrangement at 110 °C (Scheme 6/1.84).

N-Acyl-α-amino acids are important compounds in both chemistry and biology. They are easily obtained in a transition metal-catalyzed, three-component domino reaction of an aldehyde, an amide, and CO. Whereas cobalt was mainly used for this process, Beller and coworkers [159] have recently shown that palladium has a

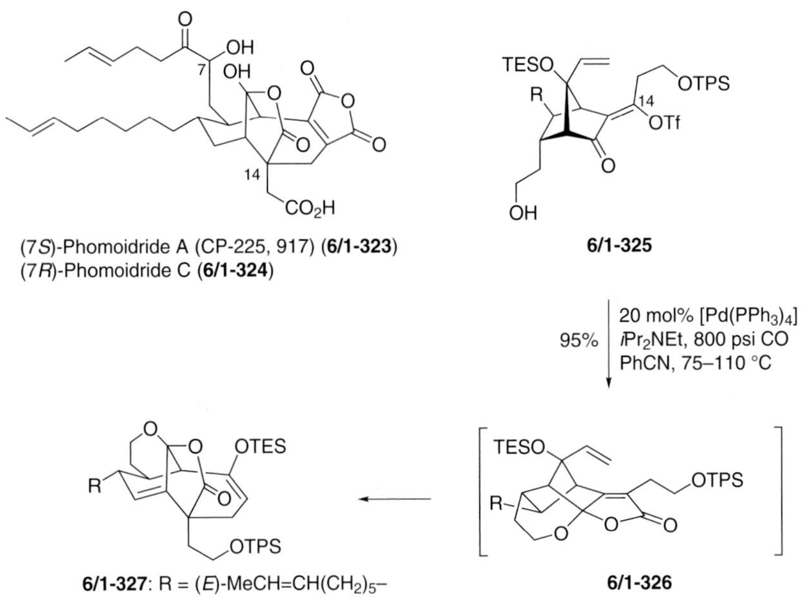

(7S)-Phomoidride A (CP-225, 917) **(6/1-323)**
(7R)-Phomoidride C **(6/1-324)**

6/1-325

$$95\% \quad \begin{array}{|l} \text{20 mol\% [Pd(PPh}_3)_4] \\ i\text{Pr}_2\text{NEt, 800 psi CO} \\ \text{PhCN, 75–110 °C} \end{array}$$

6/1-327: R = (E)-MeCH=CH(CH$_2$)$_5$–

6/1-326

Scheme 6/1.84. Synthesis of the phomoidride skeleton.

higher catalyst activity and can be used under milder conditions. Among the many reactions developed by Beller and coworkers, the amidocarbonylation of acetamides **6/1-328** and aromatic aldehydes **6/1-329** to give **6/1-330** is described (Scheme 6/1.85). Electron-donating substituents at the aromatic aldehyde have been shown to increase the reaction rate.

A very useful Pd-catalyzed domino reaction consisting of a formal Buchwald–Hartwig amidation and a C–H activation/aryl-aryl bond-forming process was recently developed by Zhu and coworkers [160]. The products are polyheterocycles such as **6/1-332**, obtained from linear diamides as **6/1-331**. If great intrigue is the possibility also of preparing azaphenanthrenes **6/1-333** to **6/1-336**, fused with medium-sized and macrocyclic ring systems and using a longer tether between the amide functionalities (Scheme 6/1.86). Also of interest is the observation that replacement of one of the iodo atoms in **6/1-331** did not lead to the expected 1,4-benzodiazepine-2,5-dione. Thus, it can be assumed that there is some kind of cooperative effect in the two distinct Pd-catalyzed bond-forming processes.

Zhu and coworkers have also discovered an interesting switch in the synthesis of different heterocyclic scaffolds from the same starting material, simply by changing the metal [161].

Again, using a Pd-catalyzed amidation as the first step, Edmondson and coworkers [162] developed a synthesis of 2,3-disubstituted indoles **6/1-340** by reaction of **6/1-337** and **6/1-338** with catalytic amounts of Pd$_2$(dba)$_3$ and the ligand **6/1-341**. On order to achieve good results, a second charge of Pd had to be added after 12 h. In the first step the enaminone **6/1-339** is formed, which then cyclizes in a Heck-

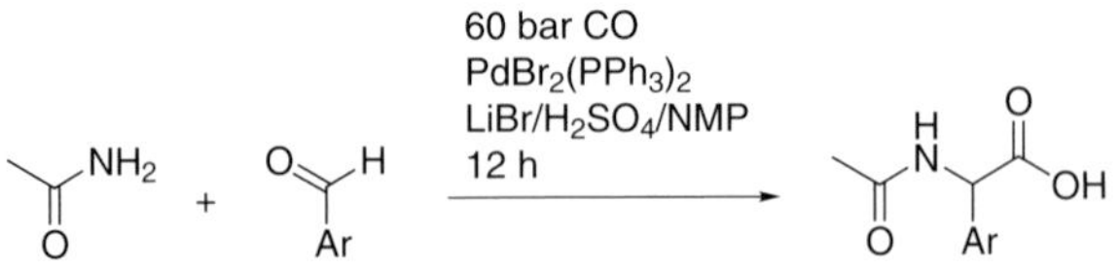

60 bar CO
PdBr$_2$(PPh$_3$)$_2$
LiBr/H$_2$SO$_4$/NMP
12 h

6/1-328   6/1-329   →   6/1-330

Entry	Ar	σp[a]	(100 °C, 12 h)		(120 °C, 15 h)	
			Yield [%][b]	TON[c]	Yield [%][b]	TON[c]
1	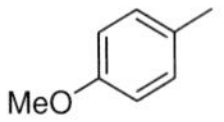 MeO–	−0.27	75	300	−	−
2	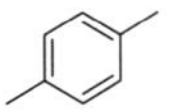	−0.17	86	344	95	380
3	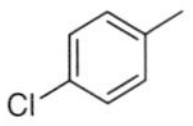 Cl–	0.30	65	260	−	−
4	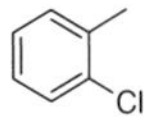 (2-Cl)	−	56	224	−	−
5	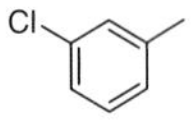 (3-Cl)	−	63	252	−	−
6	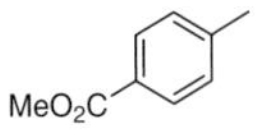 MeO$_2$C–	0.39	52	208	89	356
7	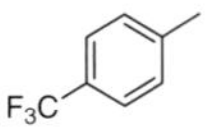 F$_3$C–	0.54	42	168	82	328
8	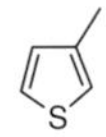 (thiophene)	−	−	−	42[d]	168

[a] Hammett constants for *para*-substituents. [b] Yield of isolated product.
[c] TON [(mol product)/(mol cat.)]. [d] 60 h, 125 °C.

Scheme 6/1.85. Palladium-catalyzed amidocarbonylation of aromatic aldehydes.

type reaction to give **6/1-340**. In a similar way, reaction of **6/1-337** and **6/1-342** led to **6/1-343** (Scheme 6/1.87). The driving force for the aryl migration is, presumably, the formation of a stable indole moiety.

The Alper group [163] reported on a highly efficient double carbohydroamination for the preparation of α-amino carboxylic acid amides **6/1-345**, starting from aryl iodides and a primary amine **6/1-344**, in usually high yield (Scheme 6/1.88); both, aryl iodides with electron-donating and electron-withdrawing groups can be used.

A combination of a Pd-catalyzed arylation of a ketone followed by intramolecular cyclization of the formed enolate with an allylic silyl ether moiety in one of the substrates led to the direct formation of a 1-vinyl-1*H*-isochromene, as described by Wills and coworkers [164].

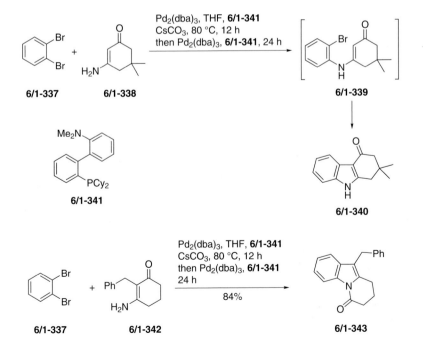

6/1-331a–e

Pd(dppf)Cl$_2$, DMSO
KOAc, 120 °C

6/1-332a: R^1= H (51%)
6/1-332b: R^1= Me (98%)
6/1-332c: R^1= Bn (91%)
6/1-332d: R^1= iPr (83%)
6/1-332e: R^1= Me$_2$CHCH$_2$ (94%)

6/1-333 (18%)

6/1-334 (57%)

6/1-335 (48%)

6/1-336 (47%)

Scheme 6/1.86. Synthesis of substituted 5,6-dihydro-8*H*-[5,7-*a*]diazacyclohepta[*jk*]
phenanthrene-4,7-diones.

6/1-337

6/1-338

Pd$_2$(dba)$_3$, THF, **6/1-341**
CsCO$_3$, 80 °C, 12 h
then Pd$_2$(dba)$_3$, **6/1-341**, 24 h

6/1-339

Me$_2$N

PCy$_2$

6/1-341

6/1-340

6/1-337

Ph

H$_2$N

6/1-342

Pd$_2$(dba)$_3$, THF, **6/1-341**
CsCO$_3$, 80 °C, 12 h
then Pd$_2$(dba)$_3$, **6/1-341**
24 h

84%

6/1-343

Scheme 6/1.87. Synthesis of indoles.

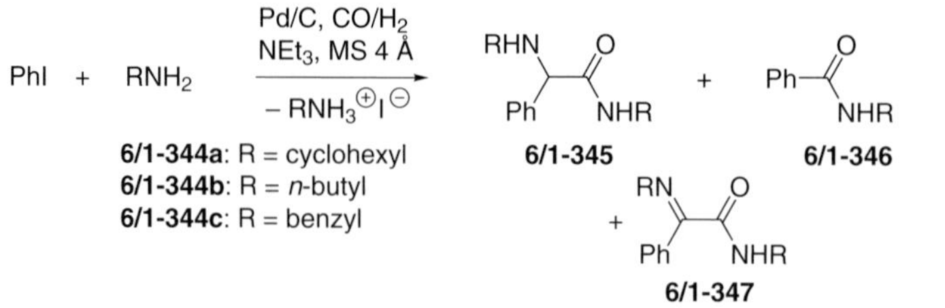

6/1-344a: R = cyclohexyl
6/1-344b: R = n-butyl
6/1-344c: R = benzyl

Entry	Amine	CO [psi]	H$_2$ [psi]	T [°C]	Products [%][a]		
					6/1-345	6/1-346	6/1-347
1	6/1-344a	800	100	120	87	5	0
2	6/1-344a	800	100	120	59	41	trace
3	6/1-344a	800	100	120	66	18	10
4	6/1-344a	800	200	105	46	18	18
5	6/1-344a	700	300	120	77	9	trace
6	6/1-344a	800	400	120	73	8	0
7	6/1-344b	700	300	120	67	29	trace
8	6/1-344c	800	100	120	24	34	5

[a] Reaction conditions: 1 mmol ArI, 10 mmol amine, 3 mL NEt$_3$, 1 g MS 4 Å , 10% of 0.02 mmol Pd/C and CO/H$_2$ were employed in the reactions for 24 h.

Scheme 6/1.88. Palladium-catalyzed double carbohydroamination of iodobenzene.

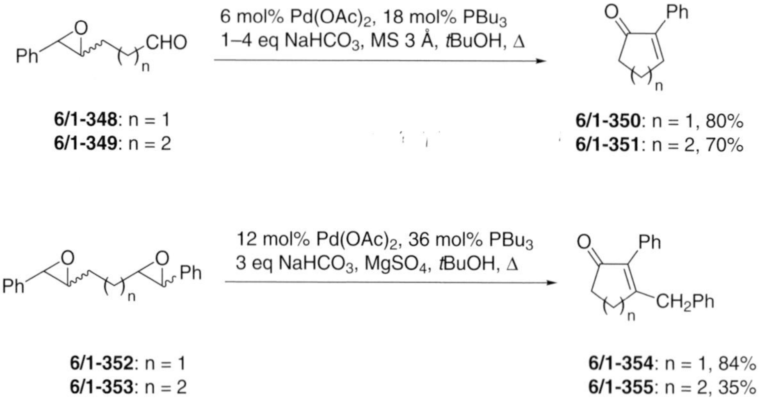

6/1-348: n = 1
6/1-349: n = 2

6/1-350: n = 1, 80%
6/1-351: n = 2, 70%

6/1-352: n = 1
6/1-353: n = 2

6/1-354: n = 1, 84%
6/1-355: n = 2, 35%

Scheme 6/1.89. Synthesis of cyclopentenones and cyclohexenones.

Epoxides are reactive substrates, which can easily be isomerized to give aldehydes or ketones. Kulawiec and coworkers have combined a Pd-catalyzed isomerization of mono and diepoxide **6/1-348** or **6/1-349** and **6/1-352** or **6/1-353**, followed by an aldol condensation to give either cyclopentenones or cyclohexenones **6/1-350, 6/1-351, 6/1-354** and **6/1-355**, respectively (Scheme 6/1.89) [165].

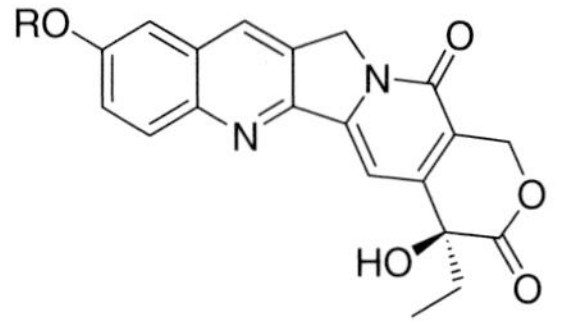

(S)-camptothecin (**6/1-356**)

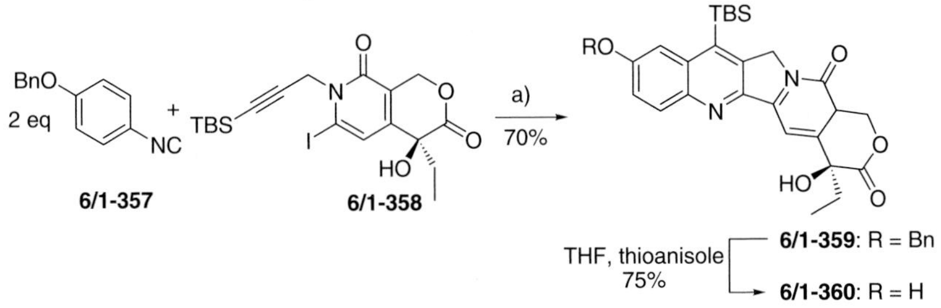

a) 20 mol% Pd(OAc)$_2$, 1.5 eq Ag$_2$CO$_3$, toluene, r.t., 20 h;
   repeat with 10 mol Pd(OAc)$_2$, 0.7 eq Ag$_2$CO$_3$.

Scheme 6/1.90. Synthesis of the camptothecin analogue DB-67 (**6/1-360**).

Finally, Curran and coworkers [166] developed a Pd^0-catalyzed domino process for the synthesis of camptothecin (**6/1-356**) and analogues. This natural product is a very potent anticancer agent [167]; it contains a 11*H*-indolizino[1,2-*b*]quinolin-9-one skeleton, which is also found in mappicine [168] and the promising new analogue DB-67 (**6/1-360**) [169], which is currently in preclinical development for the treatment of cancer. The skeleton can be provided simply by a domino radical reaction of an aryl isonitrile and a 6-iodo-*N*-propargylpyridone in 40–60 % yield [170]. However, the product is also accessible by a Pd-catalyzed domino process. Reaction of the isonitrile **6/1-357** and the iodopyridone **6/1-358** occurred in the presence of 1.5 equiv. Ag$_2$CO$_3$ and 20 mol% Pd(OAc)$_2$ in toluene, though the transformation of **6/1-359** was not complete. This problem was solved by filtration and evaporation of the product mixture and addition of a new charge of Pd(OAc)$_2$ and Ag$_2$CO$_3$, employing the same reaction conditions. Using this procedure the anticancer agent DB-67 **6/1-360** was synthesized in 53 % yield from **6/1-357** and **6/1-358** after deprotection (Scheme 6/1.90).

6.1.6
Pd(II)-Catalyzed Transformations

The most important reaction based on Pd^{II}-catalysis is the Wacker oxidation [171], which is used industrially for the synthesis of acetaldehyde, starting from ethane. This process can be combined with a Heck reaction and has been used by Tietze and coworkers [172] for an efficient enantioselective synthesis of vitamin E (**6/1-**

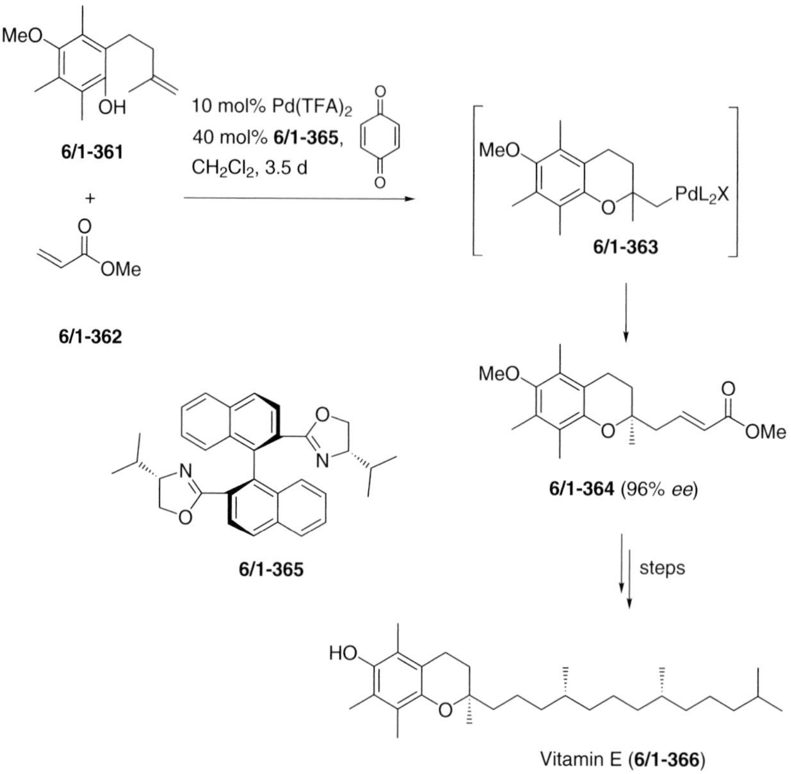

Scheme 6/1.91. Enantioselective synthesis of vitamin E (**6/1-366**).

366) [173] using a BOXAX ligand (Scheme 6/1.91) [174]. In this way, the chromane ring and parts of the side chain of vitamin E can be introduced in one process. Thus, reaction of **6/1-361** with acrylate **6/1-362** in the presence **6/1-365**, benzoquinone and Pd(TFA)$_2$ led to **6/1-364** in 84 % yield and 96 % *ee*. It can be assumed that **6/1-363**, which reacts with acrylate, is an intermediate in this process. Methylvinylketone has also been used with similar results. The product of the domino process had been transformed into vitamin E (**6/1-366**) within a few steps [175].

Another example of a Pd^{II}-driven domino process is the reaction of the alkyne **6/1-367** with CO on a solid support to give the benzo[*b*]furan **6/1-368** (Scheme 6/1.92). Ninety different benzo[*b*]furans were prepared using this procedure, with > 90 % purity in most cases [176].

Nemoto, Ihara and coworkers [177] used a Pd^{II}-catalyzed domino process for the synthesis of the steroid equilenin (**6/1-373**) (Scheme 6/1.93); this includes a ring enlargement followed by a Heck-type reaction. One interesting point in this process is the near-complete reversion of the facial selectivity using different solvents. Thus, reaction of compound **6/1-369** with Pd(OAc)$_2$ at room temperature in HMPA-THF (1:4) gave the two diastereomers **6/1-372a** and **6/1-372b** in a 73:27 ratio in 60 % yield, whereas in dichloroethane only **6/1-372b** was formed in 63 % yield. As intermedi-

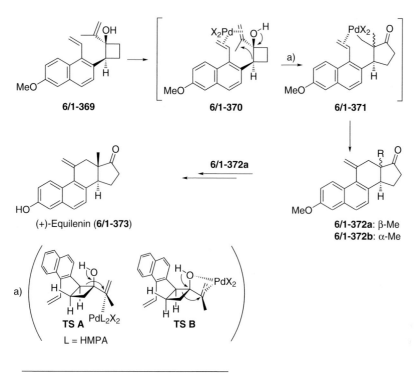

Scheme 6/1.92. Synthesis of benzofurans.

Scheme 6/1.93. Synthesis of (+)-equilenin (**6/1-373**).

ates, the Pd-complexes **6/1-370** and **6/1-371** can be assumed as the two transition structures **TS A** and **TS B**, thereby explaining the different facial selectivity.

Toyota, Ihara and coworkers [178] used a combination of a Wacker- and a Heck-type transformation to construct the cedrane skeleton. Thus, reaction of **6/1-374** using 10 mol% Pd(OAc)₂ under an atmosphere of O₂ led to the domino product **6/1-375** in 30% yield. In addition, 58% of the mono-cyclized compound **6/1-376** was obtained (Scheme 6/1.94).

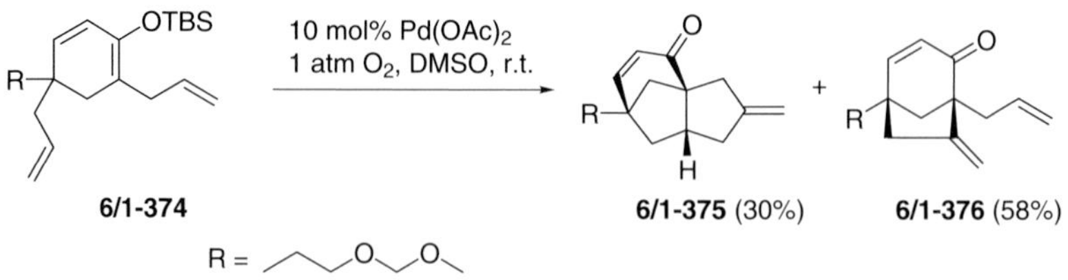

6/1-374

R =

6/1-375 (30%)

6/1-376 (58%)

Scheme 6/1.94. Palladium-catalyzed domino cycloalkenylation.

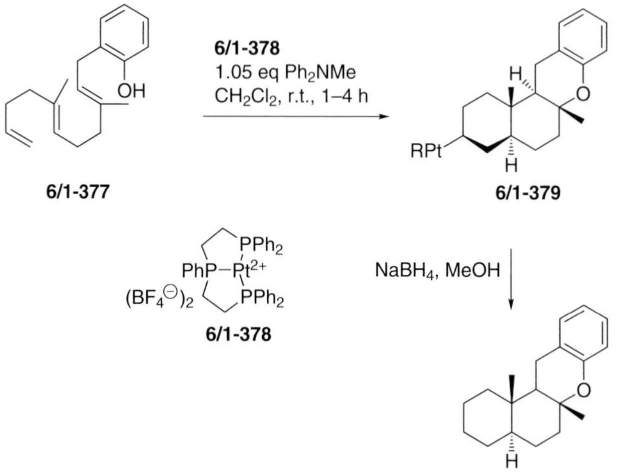

6/1-377

6/1-378
1.05 eq Ph₂NMe
CH₂Cl₂, r.t., 1–4 h

6/1-379

6/1-378

NaBH₄, MeOH

6/1-380 (*dr* 98:2)

Scheme 6/1.95. Pt^{II}-mediated polycyclisation.

A Pd^{II}- or Pt^{II}-induced polycyclization of compounds containing two or even three C=C-double bonds and a hydroxyl group was described by Gagné and coworkers [179]. Thus, reaction of **6/1-377** with equimolar amounts of the Pt^{II} complex **6/1-378** led to the Pt–alkyl complex **6/1-379**. Treatment of **6/1-379** with NaBH₄ provided the tetracyclic **6/1-380** as a 96:3:1 mixture in 86 % yield (Scheme 6/1.95). It was proposed that, in the domino process, carbocations are formed as intermediates.

Holzapfel and coworkers [180] transformed the pseudoglycal **6/1-381** into the di-hydropyran **6/1-382** using PdCl₂(MeCN)₂ in the presence of CuCl₂ in acetic acid/acetonitrile (Scheme 6/1.96). The process contains an opening of the acetal moiety followed by a Wacker-like reaction, elimination of HPdL₂Cl, and isomerization of the formed double bond.

Lloyd-Jones, Booker-Milburn and coworkers described a novel Pd^{II}-catalyzed di-amination of 1,3-butadienes as **6/1-384** with ureas **6/1-383** to give **6/1-385** (Scheme 6/1.97) [181]. The best results were obtained in the presence of *p*-benzoquinone, whereas with Cu-salts the turnover was inhibited.

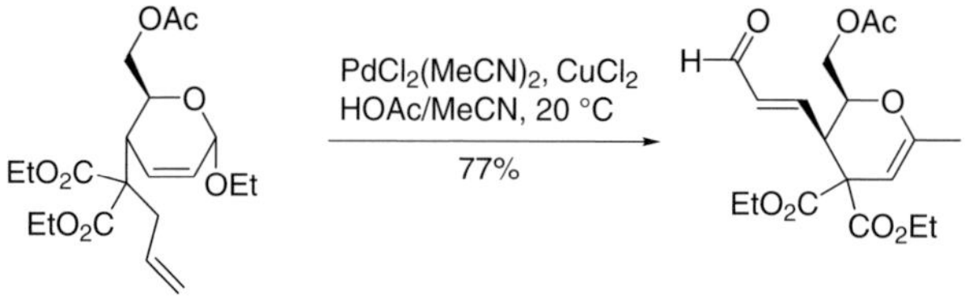

Scheme 6/1.96. Pd^{II}-catalysed formation of dihydropyrans.

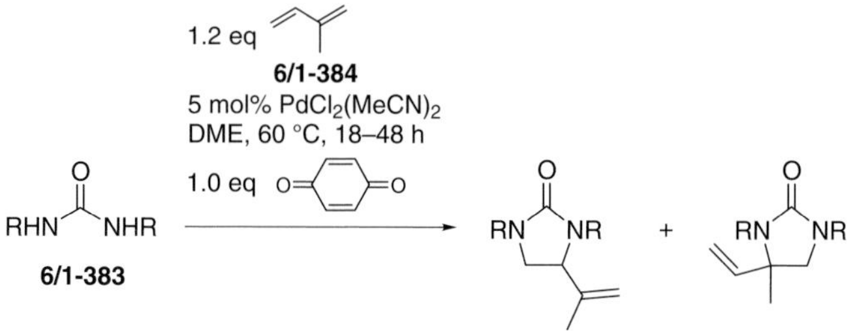

Entry	R	Ratio 6/1-385a/6/1-385b	Yield [%] 6/1-385a/b
1	Me	67/33	29
2	Et	77/23	81
3	Bu	90/10	82

Scheme 6/1.97. Pd^{II}-catalyzed 1,2-diamination of isoprene with *N,N*-dialkyl ureas.

As described in the preceding sections, many domino reactions start with the formation of vinyl palladium species, these being formed by an oxidative addition of vinylic halides or triflates to Pd^0. On the other hand, such an intermediate can also be obtained from the addition of a nucleophile to a divalent palladium-coordinated allene. Usually, some oxidant must be added to regenerate Pd^{II} from Pd^0 in order to achieve a catalytic cycle. Lu and coworkers [182] have used a protonolysis reaction of the formed carbon–palladium bond in the presence of excess halide ions to regenerate Pd^{2+} species. Thus, reaction of **6/1-386** and acrolein in the presence of Pd^{2+} and LiBr gave mainly **6/1-388**. In some reactions **6/1-389** was formed as a side product (Scheme 6/1.98).

A similar approach was used by the Alcaide group [183] in the synthesis of tricyclic β-lactams **6/1-391** from **6/1-390** (Scheme 6/1.99). In this domino process the primarily obtained π-allyl-Pd-complex reacts with the N-nucleophile of the urethane moiety to form a C–N-bond and a vinyl halide. The final step is then an intramolecular Heck-type reaction of the vinyl halide with the alkyne moiety and re-

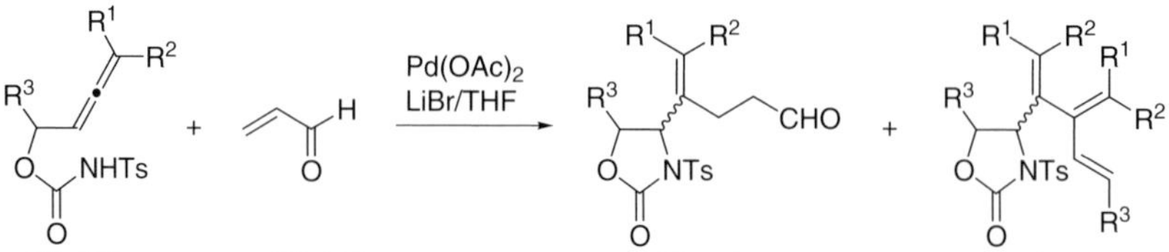

6/1-386 6/1-387 6/1-388 6/1-389

8 examples
e.g. $R^1 = R^2 =$ Me, $R^3 =$ Et
6/1-388: 71% (1.7:1), **6/1-389**: trace

Scheme 6/1.98. Synthesis of urethanes.

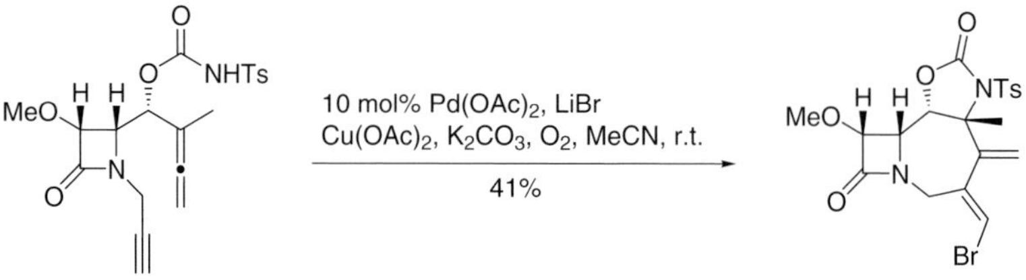

6/1-390 6/1-391

Scheme 6/1-99. Synthesis of tricyclic β-lactams.

placement of Pd by bromide. In this transformation, $O_2/Cu(OAc)_2$ was used to reoxidize the formed Pd^0 to Pd^{II}.

The latest example of a Pd^{II}-catalyzed Wacker/Heck methodology was published by Rawal and coworkers. During the total synthesis of mycalamide A, an intermolecular Wacker oxidation with methanol acting as nucleophile and a subsequent ring closure via Heck reaction led to a tetrahydropyran moiety in a 5.7:1 diastereomeric mixture [184].

6.2
Rhodium-Catalyzed Transformations

There are two important rhodium-catalyzed transformations that are broadly used in domino processes as the primary step. The first route is the formation of keto carbenoids by treatment of diazo keto compounds with Rh^{II} salts. This is then followed by the generation of a 1,3-dipole by an intramolecular cyclization of the keto carbenoid onto an oxygen atom of a neighboring keto group and an inter- or intramolecular 1,3-dipolar cycloaddition. A noteworthy point here is that the insertion can also take place onto carbonyl groups of aldehydes, esters, and amides. Moreover, cycloadditions of Rh–carbenes and ring chain isomerizations will also be discussed in this section.

The second rhodium-catalyzed route which is widely used in connection with domino processes is that of hydroformylation. This by itself is a very important industrial process for the formation of aldehydes using an alkene and carbon monoxide. Finally, rhodium catalysts have also been used in this respect.

6.2.1
Formation of Carbenes

Many examples of the Rh^{II}-catalyzed formation of carbenes from diazo keto compounds (especially in the synthesis of natural products) have been published by Padwa and his group [185]. Thus, the pentacyclic skeleton **6/2-3** of aspidosperma alkaloids could be constructed by reaction of the diazo compound **6/2-1** with Rh^{II} in 95 % yield *via* the carbonylylide **6/2-2** [186]. In a similar approach, the core skeleton **6/2-9** of ribasine (**6/2-4**) and related natural products **6/2-5**–**6/2-7** has been obtained from **6/2-8** with Rh^{II} in the presence of an imine which undergoes a 1,3-dipolar cycloaddition with the intermediate formed carbonylylide [187]. Intensive studies have been performed with different types of carbonyl compounds [188]. In another approach, diazo ketones containing an alkyne group have been used; these undergo an insertion of the primarily formed keto carbenoid to form a new carbenoid, which then reacts with a carbonyl group to give a furan. If the compounds contain an additional C–C-double bond, an intramolecular [4+2] cycloaddition can follow as in **6/2-10** to give **6/2-11** [189] (Scheme 6/2.1).

Several other groups have employed a similar approach. Muthusamy and coworkers prepared cyclooctanoid ring systems such as **6/2-14** and **6/2-15**, respectively, starting from **6/2-12** either with *N*-phenylmaleimide or dimethyl ethynedicarboxylate as 1,3-dipolarophiles *via* the dipole **6/2-13** (Scheme 6/2.2) [190].

These authors also prepared novel epoxy-bridged cyclooxaalkanones; in this process, the carbonyl group always acts as 1,3-dipolarophile, even if one employs α,β-unsaturated aldehydes. Thus, reaction of **6/2-16** with aliphatic or aromatic aldehydes **6/2-17** in the presence of catalytic amounts of rhodium acetate gave **6/2-18**, regioselectively. With the α,β-unsaturated aldehydes **6/2-20**, only cycloadducts **6/2-21** were obtained using the diazo compound **6/2-19** as substrate (Scheme 6/2.3) [191].

Schmalz and coworkers [192] developed an efficient entry to the skeleton of colchicin (**6/2-22**) by reaction of **6/2-23** with catalytic amounts of rhodium acetate to give almost exclusively *rac*-**6/2-25** in 62 % yield *via* the 1,3-dipole **6/2-24**. Small amounts of diastereomer **6/2-26** were also found as a side product (Scheme 6/2.4).

Chiu's group [193] used this domino process for an entry to pseudolaric acid **6/2-27**, starting from **6/2-28**, to yield **6/2-29** and **6/2–30** as an almost 1:1-mixture of diastereomers (Scheme 6/2.5). Attempts to improve the stereoselectivity by using chiral rhodium complexes did not change the picture very much. The pseudolaric acids A, B and C are diterpenoids, which were isolated from the root bark of *Pseudolarix kaempferi* Gordon (Pinaceae), and are components of the traditional Chinese medicine called *tujinpi*. They reveal antifungal activity and cytotoxicities at submicromolar levels [194].

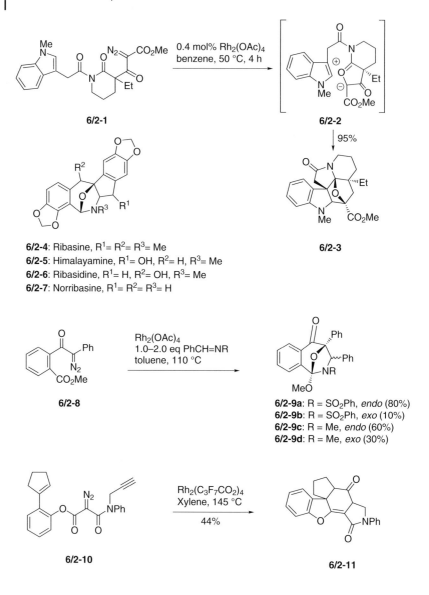

Scheme 6/2.1. Rhodium(II)catalyzed formation of carbenes from diazo compounds.

Recently, it has been shown by Hodgson and coworkers [195] that this domino reaction can indeed be performed in an enantioselective manner. Thus, treatment of **6/2-31** using the BINOL-phosphate Rh$_2$-complex **6/2-34** at −15 °C gave (+)-**6/2-33**, probably *via* **6/2-32**, in 66% yield and 90% *ee* (Scheme 6/2.6). Several other substrates and chiral catalysts have also been employed, though with lower selectivity.

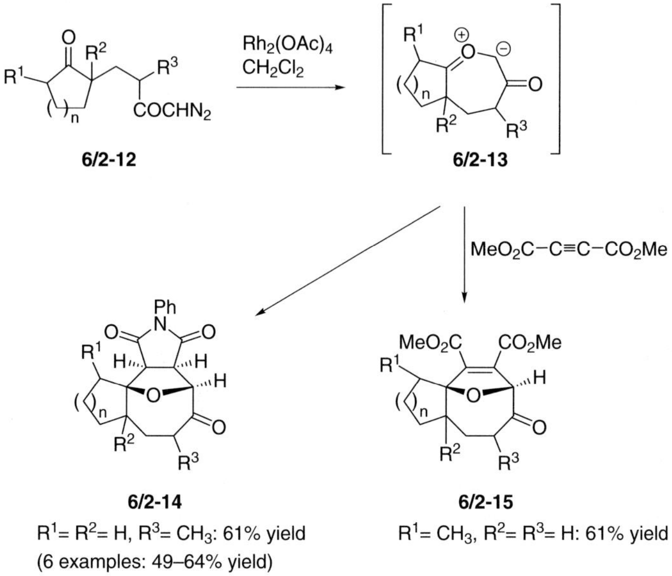

6/2-12

Rh$_2$(OAc)$_4$
CH$_2$Cl$_2$

6/2-13

MeO$_2$C−C≡C−CO$_2$Me

6/2-14

R^1= R^2= H, R^3= CH$_3$: 61% yield
(6 examples: 49–64% yield)

6/2-15

R^1= CH$_3$, R^2= R^3= H: 61% yield

Scheme 6/2.2. Synthesis of annulated cyclooctanoids.

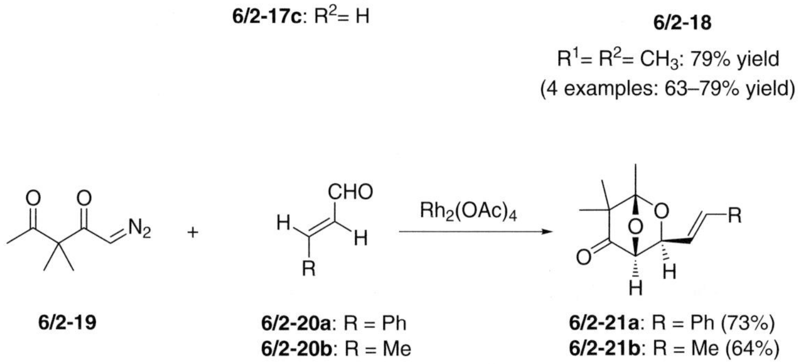

6/2-16

6/2-17a: R^2= Me
6/2-17b: R^2= OMe
6/2-17c: R^2= H

Rh$_2$(OAc)$_4$

6/2-18

R^1= R^2= CH$_3$: 79% yield
(4 examples: 63–79% yield)

6/2-19

6/2-20a: R = Ph
6/2-20b: R = Me

Rh$_2$(OAc)$_4$

6/2-21a: R = Ph (73%)
6/2-21b: R = Me (64%)

Scheme 6/2.3. Synthesis of epoxy-bridged cyclooxaalkanones.

The domino reaction of a carbonylylide from a diazoketone followed by a 1,3-dipolar cycloaddition has also been investigated using ruthenium(II)porphyrins **6/2-39** as catalyst [196]. Moreover, Che and coworkers [197] have used the Ru-cata-

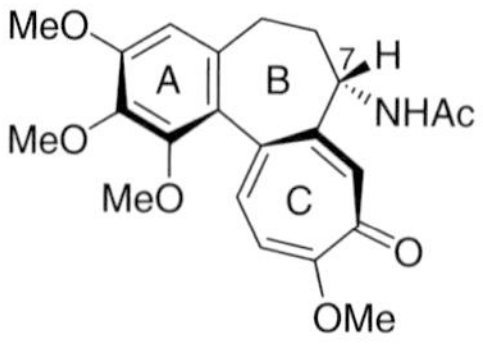

(a*R*,7*S*)-Colchicin (**6/2-22**)

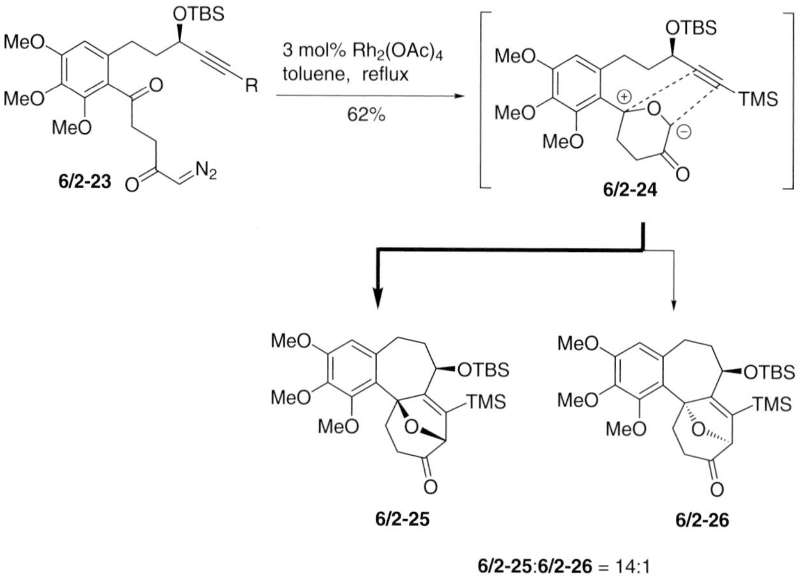

Scheme 6/2.4. Synthesis of the core of colchicin.

lyst **6/2-39** for a three-component transformation of an imine **6/2-35**, an α-diazoester **6/2-36** and an alkene **6/2-37** to provide pyrrolidines **6/2-38** with an azomethine ylide instead of a carbonyl ylide as intermediate (Scheme 6/2.7). The domino process also works with alkynes instead of alkenes to give 2,3-pyrolines, and providing even higher yields.

Another Rh^{II}-catalyzed decomposition of a α-diazoester as described by Sabe and coworkers [198] was used for the synthesis of indolizidine alkaloids (Scheme 6/2.8). It can be assumed that, first, an ammonium ylide is formed which then undergoes a 1,2-shift with ring-expansion. Thus, reaction of **6/2-40** with Rh$_2$(OAc)$_4$ led to a 72:28 mixture of **6/2-41** and **6/2-42** in 85 % yield. Cu(acac)$_2$ can also be used with even better yields, but lower selectivity (65:35).

The examples described so far clearly show the value of the rhodium-catalyzed carbene transfer obtained from diazo compounds onto carbonyl and imino groups. However, the scope is even broader, as the formed carbene can also undergo an ad-

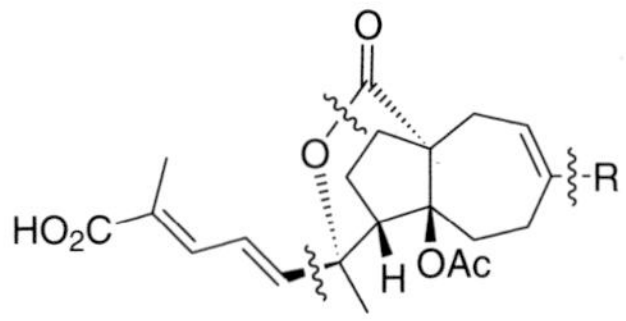

6/2-27a: R = Me Pseudolaric acid A
6/2-27b: R = CO_2Me Pseudolaric acid B

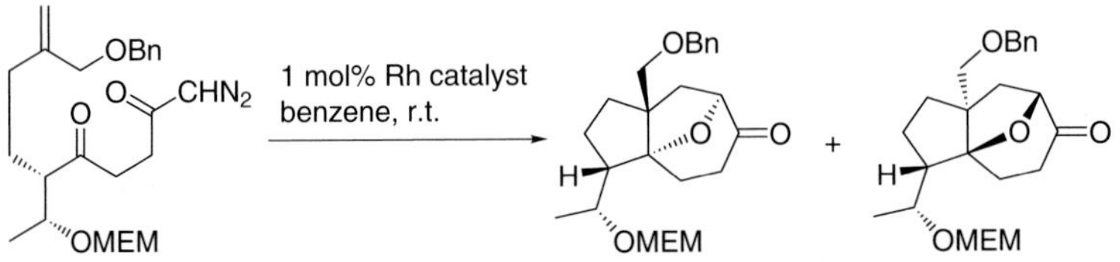

6/2-28

Catalyst	Yield [%]	6/2-29		6/2-30
$Rh_2(OAc)_4$	66	1.25	:	1
$Rh_2(cap)_4$	69	1.1	:	1

Scheme 6/2.5. Entry to pseudolaric acid.

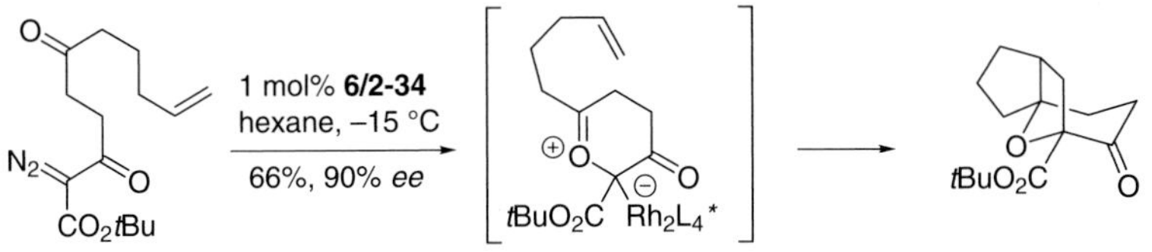

6/2-31 **6/2-32** (+)-**6/2-33**

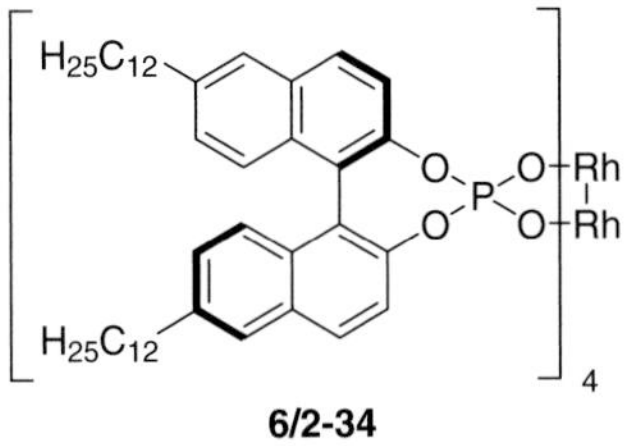

6/2-34

Scheme 6/2.6. Enantioselective domino carbene-formation/1,3-dipolar cycloaddition.

dition onto C=C-double bonds, followed by a ring-chain isomerization or a cycloaddition to form a cyclopropane followed by a sigmatropic rearrangement.

Maas and coworkers [199] showed that the rhodium(II)-catalyzed decomposition of vinyldiazoacetate **6/2-44** in the presence of semicyclic enaminocarbonyl compounds **6/2-43** gives betaines as **6/2-45**, with formation of a spiro compound as in-

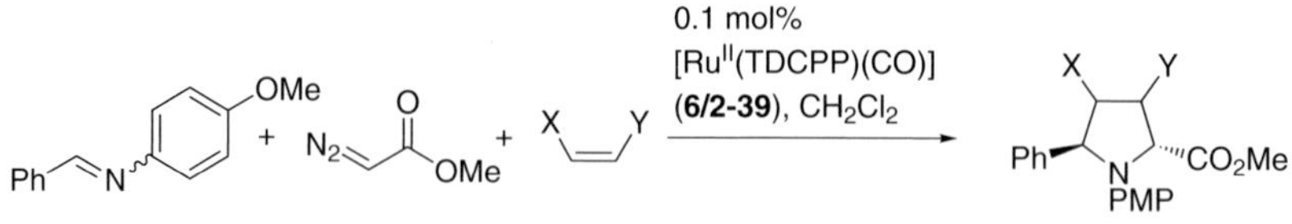

6/2-35 6/2-36 6/2-37 6/2-38

Entry	Dipolarphile	Product	T [°C]	Yield [%]
1	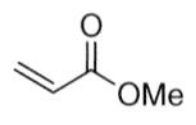	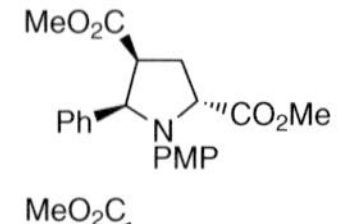	r.t.	63
2	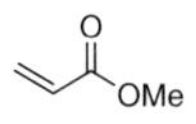	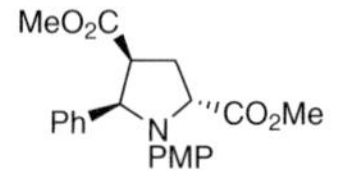	r.t.	57
3	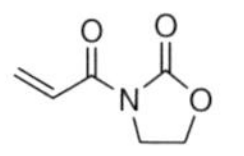	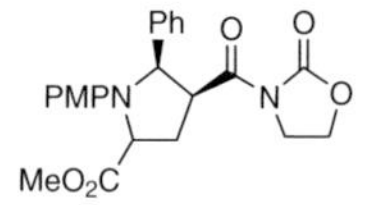	r.t.	61
4	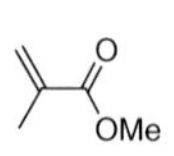	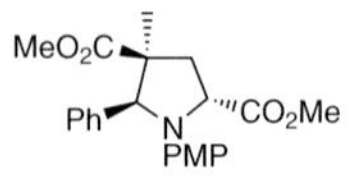	50	45
5	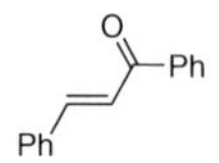	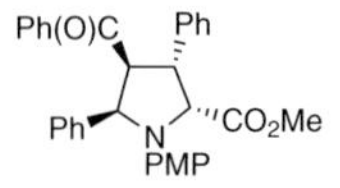	50	51
6	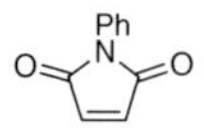	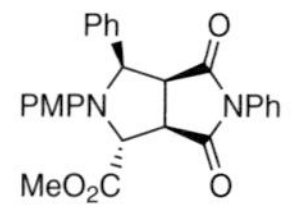	50	52

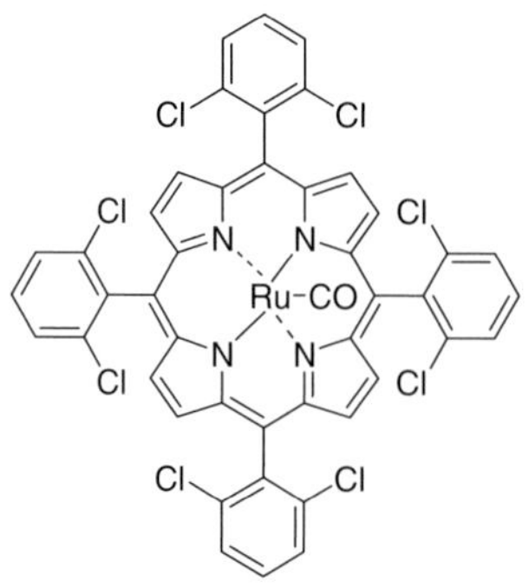

6/2-39

Scheme 6/2.7. Three-component reaction of a α-diazoester, an imine and an alkene.

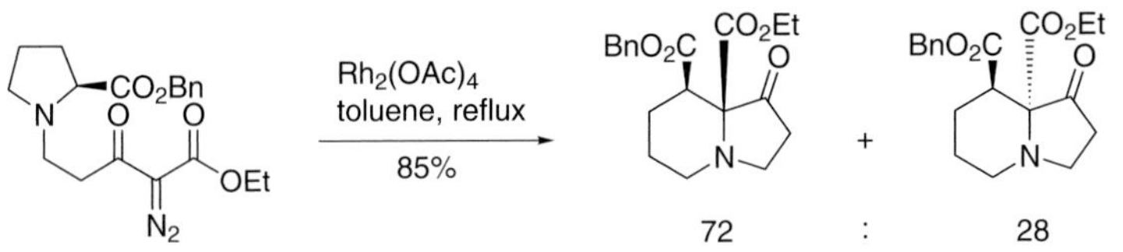

6/2-40 **6/2-41** **6/2-42**

Scheme 6/2.8. Synthesis of indolizidine alkaloids.

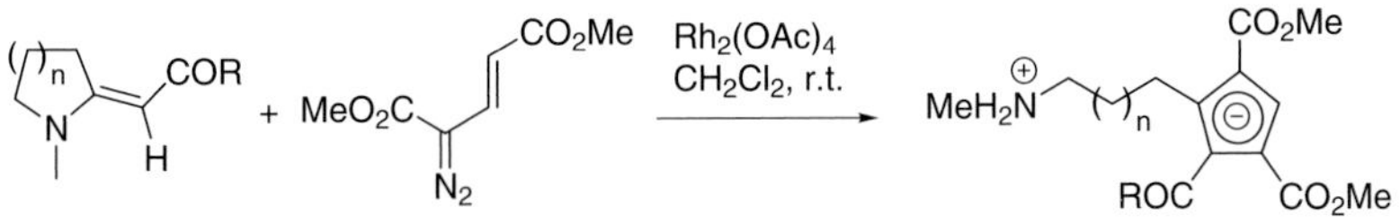

6/2-43 **6/2-44** **6/2-45**

R = aryl, hetaryl, OMe, O*t*Bu
n = 1,2,3

R = thienyl, n = 1: 94% yield
12 examples: 23–94% yield

Scheme 6/2.9. Rhodium-catalyzed decomposition of vinyldiazoacetate with ring-chain isomerisation.

termediate (Scheme 6/2.9); this is followed by the above-mentioned ring-chain isomerization.

An example involving a Bamford–Stevens and a Claisen rearrangement was published by the Stoltz group [200], in which a rhodium-catalyzed stereoselective hybrid migration of a diazoalkane **6/2-47** led to a 1,5-diene **6/2-49** *via* the carbene **6/2-48** (Scheme 6/2.10). The 1,5-diene then underwent a Claisen rearrangement to give the aldehydes **6/2-50** in good yields and, usually, high diastereoselectivity. The necessary diazoalkanes **6/2-47** can be formed *in situ* by a thermal decomposition of the hydrazones **6/2-46**. The reaction is quite general, tolerating aromatic, alkyl, alkenyl substitution adjacent to the hydrazone functionality. The reaction can be extended by performing a Prins reaction using a suitable starting material through treatment of the reaction mixture with Me_2AlCl.

The Davies group has described several examples of a rhodium-catalyzed decomposition of a diazo-compound followed by a [2+1] cycloaddition to give divinyl cyclopropanes, which then can undergo a Cope rearrangement. Reaction of the pyrrol derivative **6/2-51** and the diazo compound **6/2-52** led to the tropane nucleus **6/2-54** *via* the cyclopropane derivative **6/2-53** (Scheme 6/2.11) [201]. Using (*S*)-lactate and (*R*)-pantolactone as chiral auxiliaries at the diazo compound, a diastereoselectivity of around 90:10 could be achieved in both cases.

These authors were also able to perform this domino process in an enantioselective fashion using the Rh–proline derivative **6/2-57** (Rh_2(S-DOSP)$_4$) as chiral catalyst for the cyclopropanation [202]. Reaction of 2-diazobutenoate **6/2-56** and alkenes **6/2-55** in the presence of the catalyst **6/2-57** led primarily to the cyclopropane derivative

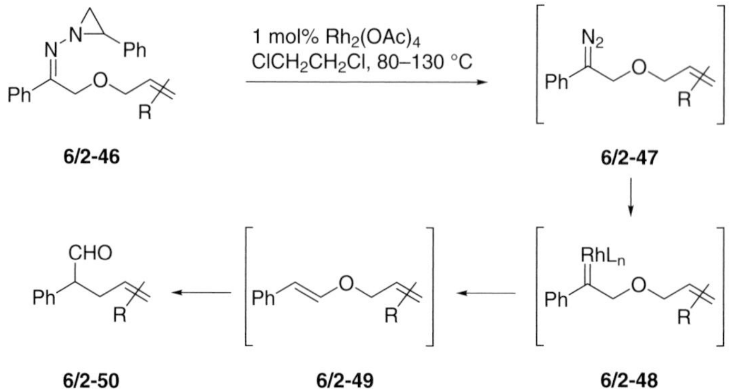

Entry	Substrate	Product	Yield
1	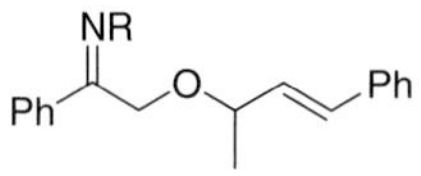	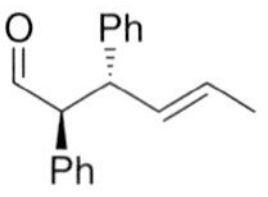	79% (*dr* > 20:1)
2	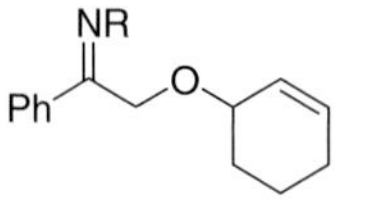	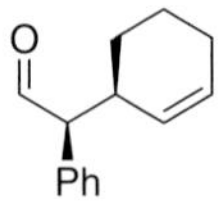	72% (*dr* > 6:1)[a]
3	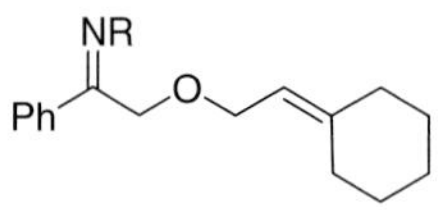	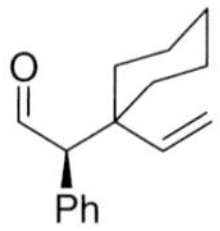	86%
4	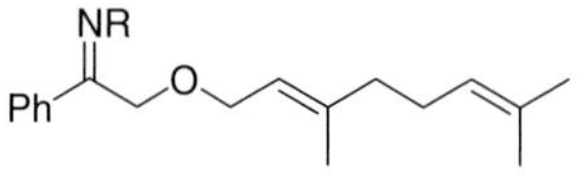	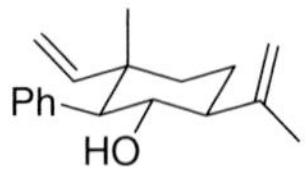	68% (*dr* > 20:1)[a]
5	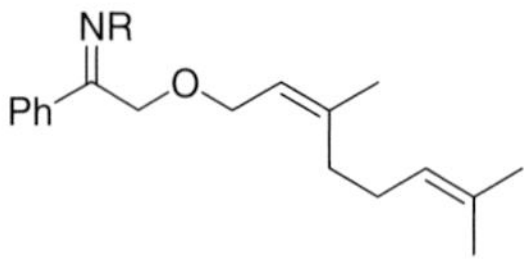	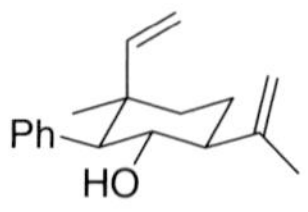	63% (*dr* > 20:1)[a]

[a] Subsequent treatment with Me₂AlCl at –40 °C.

Scheme 6/2.10. Synthesis of aldehydes and "Prins-products" from hydrazones.

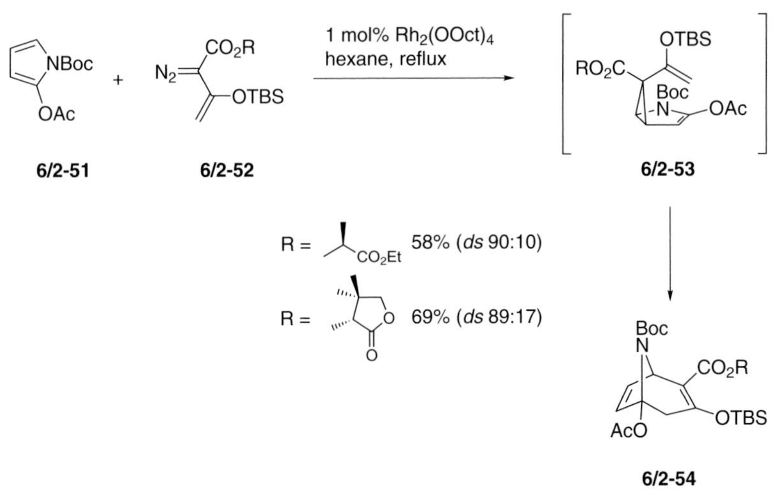

Scheme 6/2.11. Synthesis of the tropane nucleus.

6/2-58, which was rearranged to give the cycloheptadienes **6/2-59** *via* a boat-like transition structure. Enantioselectivities up to 98 % *ee* and good to excellent yields were obtained. The transformations can also be performed in an intramolecular mode, with up to 93 % *ee* [203]. However, the intermolecular reaction usually gives better enantioselectivities. The method could also be used for the formation of annulated cycloheptadienes as **6/2-62** from **6/2-60** and **6/2-61** (Scheme 6/2.12).

6.2.2
Hydroformylations

The synthesis of aldehydes from alkenes known as hydroformylation using CO and hydrogen and a homogeneous catalyst is a very important industrial process [204]. Today, over seven million tons of oxoproducts are formed each year using this procedure, with the majority of butanal and butanol from propene. To further increase the efficiency of this process it can be combined with other transformations in a domino fashion. Eilbracht and coworkers [205] used a Mukaiyama aldol reaction as a second step, as shown for the substrate **6/2-63** which, after 3 days led to **6/2-65** in 91 % yield *via* the primarily formed adduct **6/2-64** (Scheme 6/2.13). However, employing a reaction time of 20 h gave **6/2-64** as the main product.

In a similar way, carbocycles having a quaternary center could be obtained from acyclic unsaturated 1,3-dicarbonyl compounds [206]. Other combinations are the domino hydroformylation/Wittig olefination/hydrogenation described by Breit and coworkers [207]. The same group also developed the useful domino hydroformylation/Knoevenagel/hydrogenation/decarboxylation process (Scheme 6/2.14) [208]; a typical example is the reaction of **6/2-66** in the presence of a monoester of malonic acid to give **6/2-67** in 41 % yield in a *syn:anti*-ratio of 96:4. Compounds **6/2-68** and **6/2-69** can be assumed as intermediates.

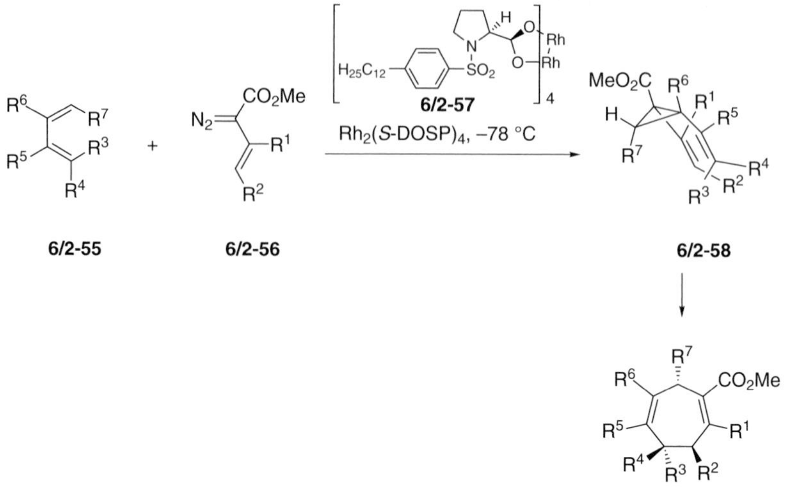

6/2-55 6/2-56 6/2-58

6/2-59 (60–98% *ee*)

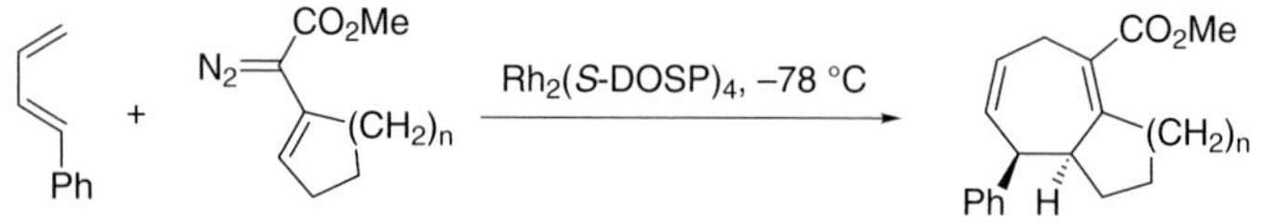

6/2-60a **6/2-61a**: n = 1 **6/2-61b**: n = 2 **6/2-62**

6/2-62	n	*ee* [%]	Yield [%]
a	2	94	62
b	1	81	60

Scheme 6/2.12. Enantioselective formation of cycloheptadiene derivatives.

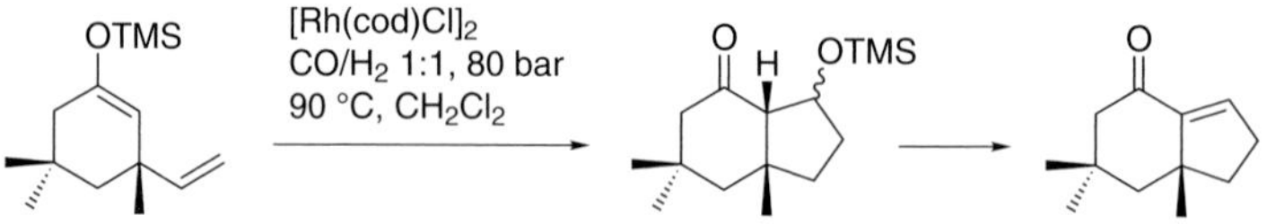

6/2-63 **6/2-64** **6/2-65**

	6/2-64	6/2-65
20 h	50% (*dr* 6:1)	minor amounts
3 d	–	91%

Scheme 6/2.13. Domino hydroformylation/Mukaiyama reaction.

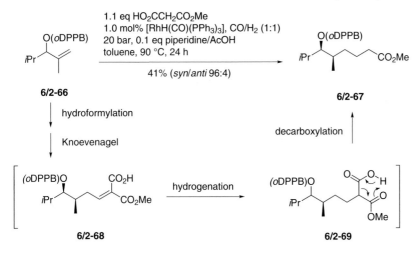

1.1 eq HO$_2$CCH$_2$CO$_2$Me
1.0 mol% [RhH(CO)(PPh$_3$)$_3$], CO/H$_2$ (1:1)
20 bar, 0.1 eq piperidine/AcOH
toluene, 90 °C, 24 h

41% (syn/anti 96:4)

6/2-66

6/2-67

hydroformylation

Knoevenagel

decarboxylation

hydrogenation

6/2-68

6/2-69

Scheme 6/2.14. Domino hydroformylation/Knoevenagel/hydrogenation/decarboxylation process.

Aflatoxin B$_2$ (**6/2-70**)

CO/H$_2$, [Ru], 120 °C

6/2-71

6/2-72a: R^1= R^2= H (69%)
6/2-72b: R^1= Me, R^2= H (47%)
6/2-72c: R^1= H, R^2= Me (62%)

Scheme 6/2.15. Synthesis of internal acetals.

Internal acetals are found in several biological active natural products such as di-hydroclerodin [209] and aflatoxins as **6/2-70** [210]. An efficient formation of this functionality was described by Eilbracht and coworkers [211], using a hydroformylation of an enediol as **6/2-71** to give the tetrahydrofurobenzofurans **6/2-72** (Scheme 6/2.15).

The same group also developed an efficient entry to indoles using a combination of a hydroformylation and a Fischer indole synthesis [212]. Under optimized conditions employing 0.5 mol% of RhCl(COD)$_2$ or Rh(CO)$_2$(acac), 50 bar CO and 20 bar H$_2$ at 100–120 °C, a mixture of 1,1-disubstituted alkenes **6/2-73**, phenylhydrazine **6/2-74** and PTSA gave the indoles **6/2-75** in around 50–70% yield (Scheme 6/2.16).

As demonstrated by Hoffmann and coworkers, hydroformylation can also be combined with an allylboration and a second hydroformylation, which allows the formation of carbocycles and also heterocycles [213]. A good regioselectivity in favor of the linear aldehyde was obtained by use of the biphephos ligand [214]. Reaction of the allylboronate **6/2-76** having an *E*-configuration with CO/H$_2$ in the presence of catalytic amounts of Rh(CO)$_2$(acac) and biphephos led to the lactol **6/2-80** *via* **6/2-77–79** (Scheme 6/2.17). In a separate operation, **6/2-80** was oxidized to give the lactone **6/2-81** using tetrabutyl ammonium perruthenate/*N*-methylmorpholine *N*-oxide.

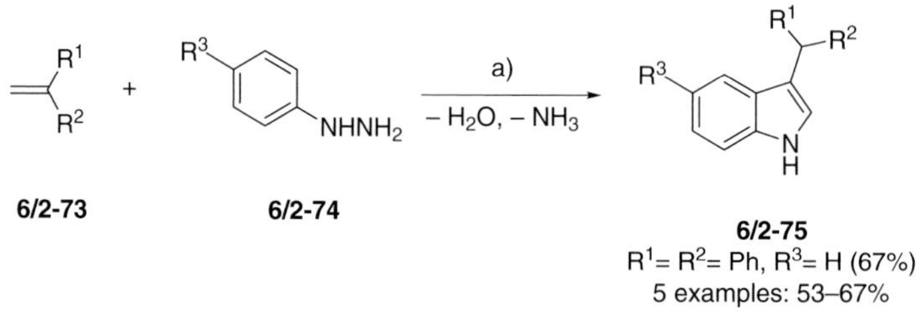

6/2-73   6/2-74

6/2-75
R^1= R^2= Ph, R^3= H (67%)
5 examples: 53–67%

a) 0.5 mol% [RhCl(COD)]$_2$ or Rh(CO)$_2$(acac), 50 bar CO, 20 bar H$_2$, 1.0 eq *p*TsOH, dioxane or toluene, 100–120 °C.

Scheme 6/2.16. Synthesis of indoles.

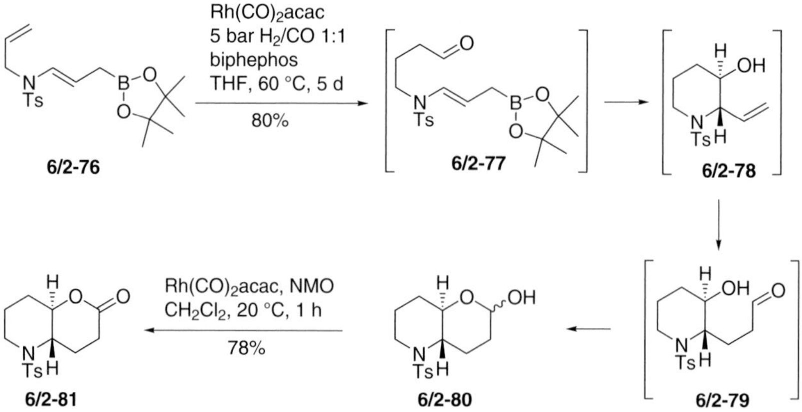

6/2-76   6/2-77   6/2-78

6/2-81   6/2-80   6/2-79

Scheme 6/2.17. Domino hydroformylation/allylboration/hydroformylation process.

Lazzaroni and coworkers described a short synthesis of compounds of type **6/2-86** with an indolizidine skeleton using a domino process consisting of a hydroformylation of **6/2-82** to give **6/2-83**, followed by an aldol-type reaction and an elimination of water from the formed amino alcohol **6/2-85** [215]. The amount of aldehyde **6/2-84** formed as a side product could be reduced by carrying out the hydroformylation at 125 °C and lower pressure (30 atm; COH$_2$ = 1:1). Under these conditions, **6/2-86** and **6/2-84** were obtained in an 85:15 ratio, without affecting the enantiopurity of the product (Scheme 6/2.18). The indolizidine skeleton is a common motive in nature, and the indolizidine alkaloids usually express a pronounced bioactivity, especially as cytotoxic agents. However, due to their high general toxicity, anticancer drugs based on this heterocyclic system have not yet been approved.

In a similar way as described for the hydroformylation, the rhodium-catalyzed silaformylation can also be used in a domino process. The elementary step is the formation of an alkenyl-rhodium species by insertion of an alkyne into a Rh–Si bond (silylrhodation), which provides the trigger for a carbocyclization, followed by an insertion of CO. Thus, when Matsuda and coworkers [216] treated a solution of the 1,6-enyne **6/2-87** in benzene with the dimethylphenylsilane under CO pressure (36 kg cm^{-2}) in the presence of catalytic amounts of Rh$_4$(CO)$_{12}$, the cyclopentane derivative **6/2-88** was obtained in 85 % yield. The procedure is not restricted to the formation of carbocycles; rather, heterocycles can also be synthesized using 1,6-enynes as **6/2-89** and **6/2-90** with a heteroatom in the tether (Scheme 6/2.19). Interestingly, **6/2-91** did not lead to the domino product; neither could 1,7-enynes be used as substrates, while the Thorpe–Ingold effect (geminal substitution) seems important in achieving good yields.

Leighton and coworkers [217] have also used this approach to develop efficient strategies for the synthesis of polyketide-derived natural products [218]. A main motif of these compounds is a skipped polyol structure, as in **6/2-94**; this can easily be prepared by a novel Rh-catalyzed domino reaction of a diallylsilyl ether in the presence of CO, followed by a Tamao oxidation [219]. Thus, reaction of, for example, the silane **6/2-93**, which is readily prepared from the corresponding ho-

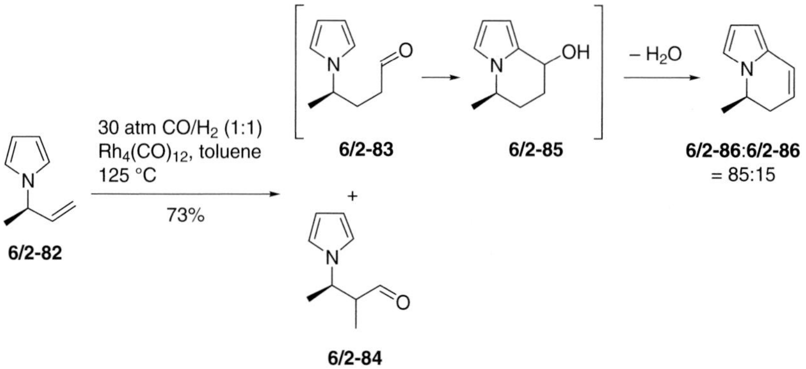

Scheme 6/2.18. Synthesis of indolizidines.

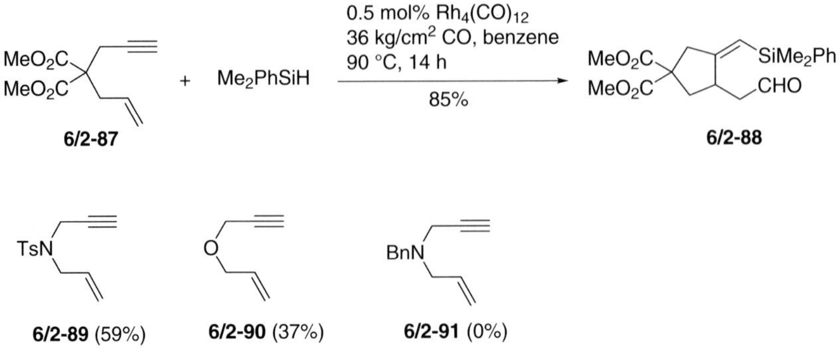

6/2-87 + Me$_2$PhSiH

0.5 mol% Rh$_4$(CO)$_{12}$
36 kg/cm^2 CO, benzene
90 °C, 14 h

85%

6/2-88

6/2-89 (59%)   **6/2-90** (37%)   **6/2-91** (0%)

Scheme 6/2.19. Domino silaformylation.

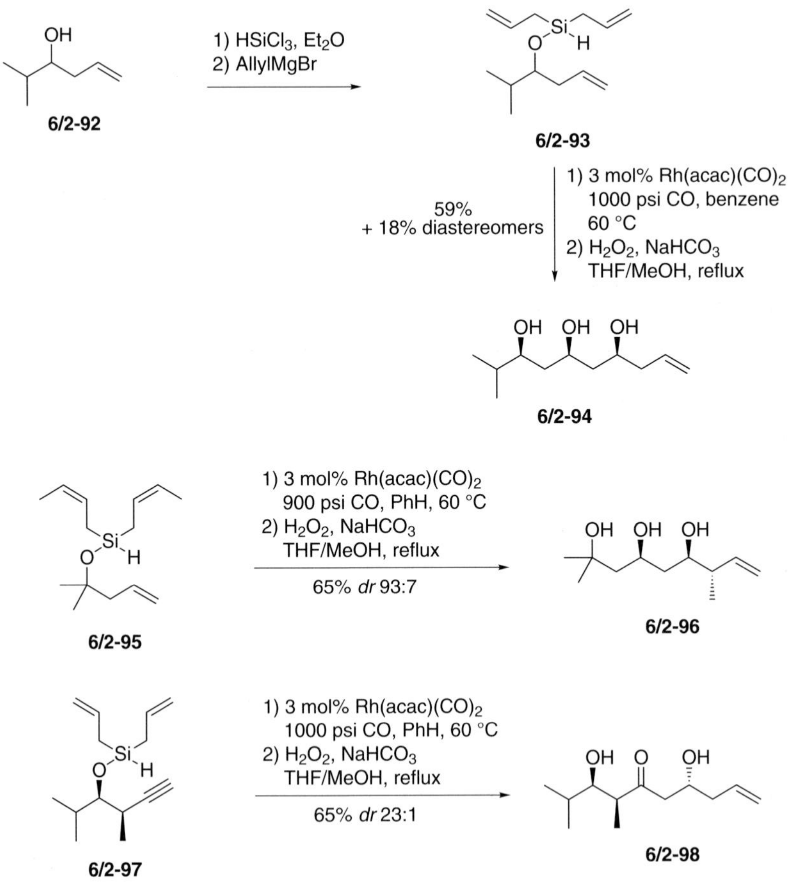

6/2-92

1) HSiCl$_3$, Et$_2$O
2) AllylMgBr

6/2-93

59%
+ 18% diastereomers

1) 3 mol% Rh(acac)(CO)$_2$
   1000 psi CO, benzene
   60 °C
2) H$_2$O$_2$, NaHCO$_3$
   THF/MeOH, reflux

6/2-94

6/2-95

1) 3 mol% Rh(acac)(CO)$_2$
   900 psi CO, PhH, 60 °C
2) H$_2$O$_2$, NaHCO$_3$
   THF/MeOH, reflux

65% *dr* 93:7

6/2-96

6/2-97

1) 3 mol% Rh(acac)(CO)$_2$
   1000 psi CO, PhH, 60 °C
2) H$_2$O$_2$, NaHCO$_3$
   THF/MeOH, reflux

65% *dr* 23:1

6/2-98

Scheme 6/2.20. Synthesis of polyols.

Lazzaroni and coworkers described a short synthesis of compounds of type **6/2-86** with an indolizidine skeleton using a domino process consisting of a hydroformylation of **6/2-82** to give **6/2-83**, followed by an aldol-type reaction and an elimination of water from the formed amino alcohol **6/2-85** [215]. The amount of aldehyde **6/2-84** formed as a side product could be reduced by carrying out the hydroformylation at 125 °C and lower pressure (30 atm; $COH_2 = 1:1$). Under these conditions, **6/2-86** and **6/2-84** were obtained in an 85:15 ratio, without affecting the enantiopurity of the product (Scheme 6/2.18). The indolizidine skeleton is a common motive in nature, and the indolizidine alkaloids usually express a pronounced bioactivity, especially as cytotoxic agents. However, due to their high general toxicity, anticancer drugs based on this heterocyclic system have not yet been approved.

In a similar way as described for the hydroformylation, the rhodium-catalyzed silaformylation can also be used in a domino process. The elementary step is the formation of an alkenyl-rhodium species by insertion of an alkyne into a Rh–Si bond (silylrhodation), which provides the trigger for a carbocyclization, followed by an insertion of CO. Thus, when Matsuda and coworkers [216] treated a solution of the 1,6-enyne **6/2-87** in benzene with the dimethylphenylsilane under CO pressure (36 kg cm^{-2}) in the presence of catalytic amounts of Rh$_4$(CO)$_{12}$, the cyclopentane derivative **6/2-88** was obtained in 85 % yield. The procedure is not restricted to the formation of carbocycles; rather, heterocycles can also be synthesized using 1,6-enynes as **6/2-89** and **6/2-90** with a heteroatom in the tether (Scheme 6/2.19). Interestingly, **6/2-91** did not lead to the domino product; neither could 1,7-enynes be used as substrates, while the Thorpe–Ingold effect (geminal substitution) seems important in achieving good yields.

Leighton and coworkers [217] have also used this approach to develop efficient strategies for the synthesis of polyketide-derived natural products [218]. A main motif of these compounds is a skipped polyol structure, as in **6/2-94**; this can easily be prepared by a novel Rh-catalyzed domino reaction of a diallylsilyl ether in the presence of CO, followed by a Tamao oxidation [219]. Thus, reaction of, for example, the silane **6/2-93**, which is readily prepared from the corresponding ho-

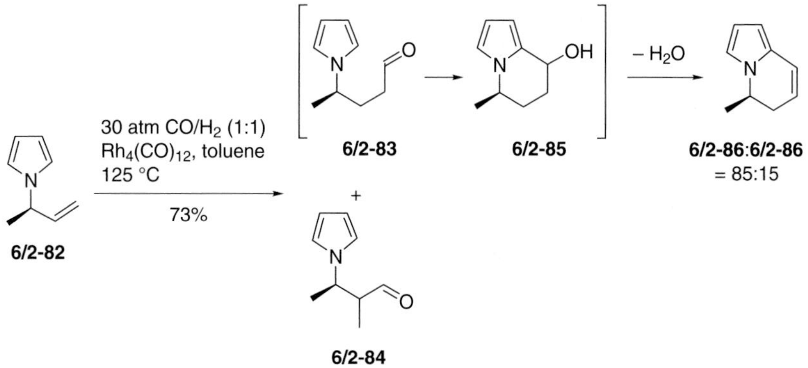

Scheme 6/2.18. Synthesis of indolizidines.

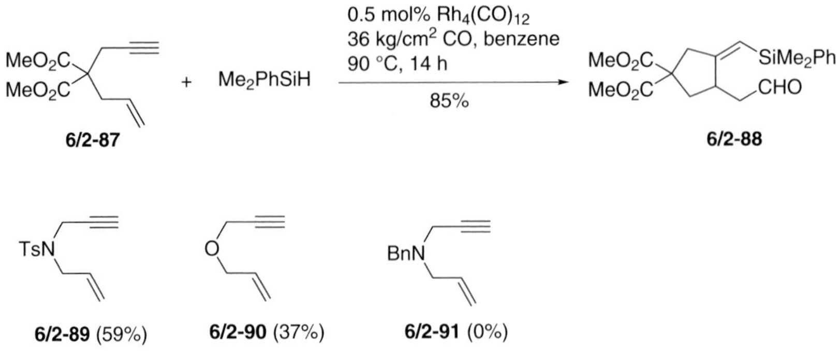

Scheme 6/2.19. Domino silaformylation.

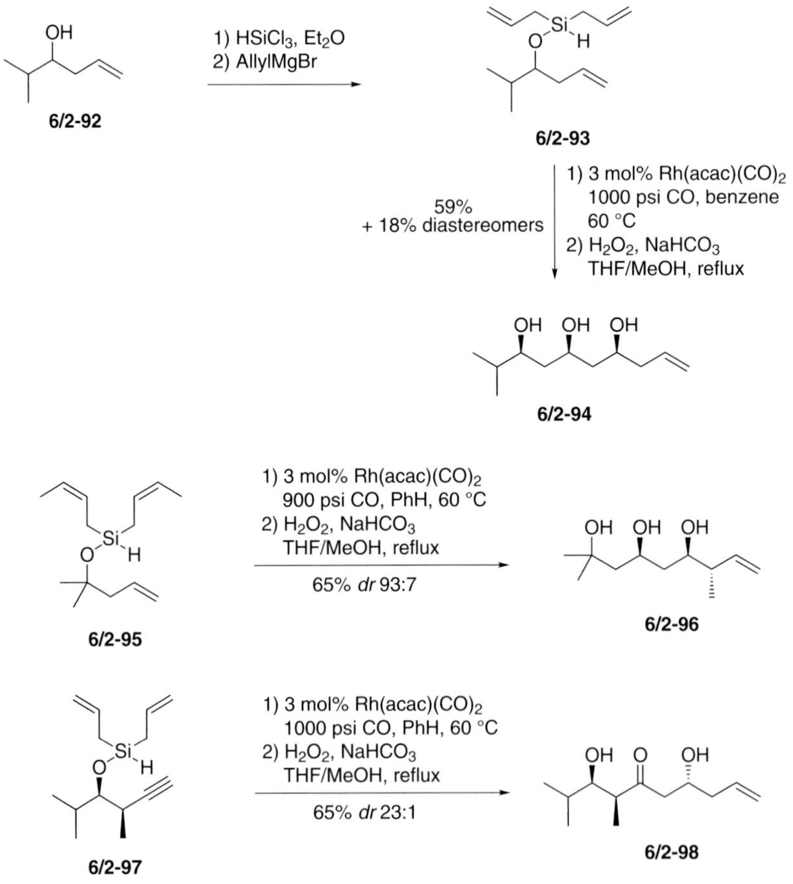

Scheme 6/2.20. Synthesis of polyols.

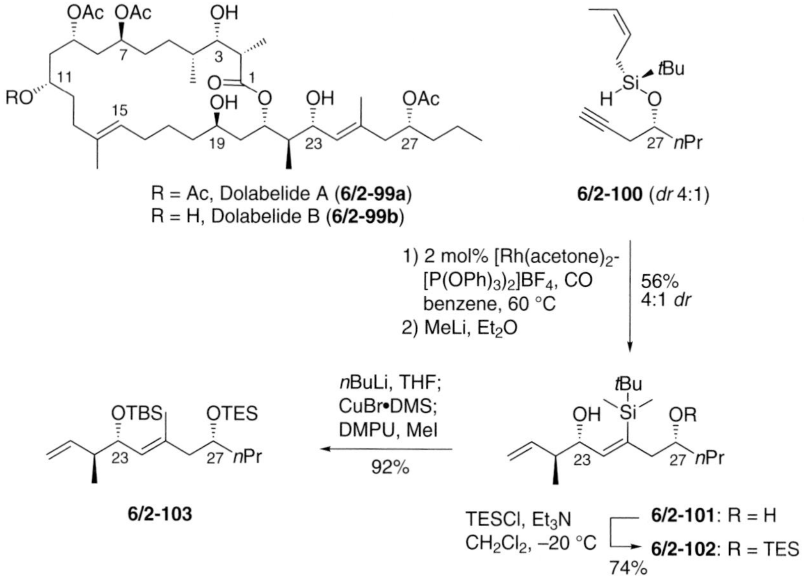

Scheme 6/2.21. Synthesis of dolabelides fragments.

moallylic alcohol **6/2-92** under 1000 psi CO in the presence of a catalytic amount of Rh(acac)(CO)$_2$ followed by oxidation with H$_2$O$_2$, led to the *syn,syn*-triol **6/2-94** in 59% yield, together with some diastereomers (18%). The process also works with substituted allylsilanes, as shown for the di-*cis* crotylsilane **6/2-95** which yielded the *syn* polyol **6/2-96** in 65% yield and a diastereoselectivity of 93:7. Moreover, reaction of the alkyne **6/2-97** using an oxidative work-up led to the dihydroxyketone **6/2-98** in 65% yield and excellent diastereoselectivity of 23:1, though with a 1,5-*anti* orientation (Scheme 6/2.20).

Leighton and coworkers have used their approach for the synthesis of the C(15)–C(30) fragment of the dolabelides A and B (**6/2-99a,b**) [220]. Employing the silyl ether **6/2-100** as substrate, these authors obtained **6/2-101** in 56% yield. Protection gave **6/2-102** which, by replacement of the silicon moiety by methyl, led to **6/2-103** (Scheme 6/2.21).

6.2.3
Other Rhodium-Catalyzed Transformations

Besides the formation of carbenes from diazo compounds and the hydroformylation, rhodium (as described previously for palladium) has also been used as catalyst in domino processes involving cycloadditions. Thus, Evans and coworkers developed a new Rh(I)-catalyzed [4+2+2] cycloaddition for the synthesis of eight-membered rings as **6/2-105** using a lithium salt of N-tosylpropargylamines as **6/2-104**, allyl carbonates and 1,3-butadiene (Scheme 6/2.22) [221]. The first step is an al-

cat. Rh(PPh₃)₃Cl

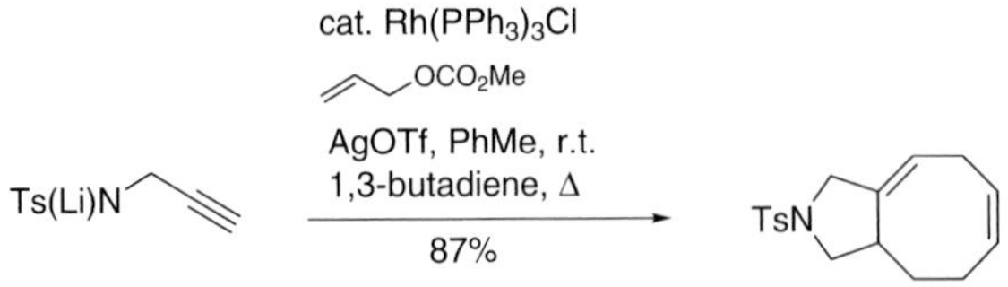

OCO₂Me

AgOTf, PhMe, r.t.
1,3-butadiene, Δ

87%

Ts(Li)N

6/2-104

TsN

6/2-105

Scheme 6/2.22. Synthesis of cyclooctadiene derivatives.

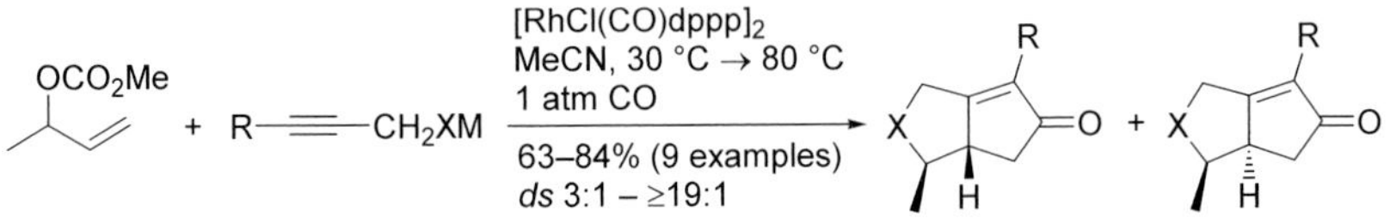

OCO₂Me

+ R——CH₂XM

[RhCl(CO)dppp]₂
MeCN, 30 °C → 80 °C
1 atm CO

63–84% (9 examples)
ds 3:1 – ≥19:1

6/2-106   6/2-107

R

X   =O   +   X   =O

H   H

6/2-108   6/2-109
main product

X = C(CO₂Me), TsN, O
M = Li or Na
R = H, Me, Ph

Scheme 6/2.23. Rhodium-catalyzed domino allylation/Pause–Khand process.

lylic amination to give an ene-yne, which undergoes Rh-catalyzed cyclization followed by a cycloaddition with 1,3-butadiene.

A first example of a combination of a Rh-catalyzed allylic substitution and a Pauson–Khand annulation reaction has also been developed by the same group [222]. Thus, [RhCl(CO)dppp]₂ is able to catalyze both transformations at different reaction temperatures. Treatment of the allylic carbonate **6-106** with the alkyne derivative **6-107** led to a diastereomeric mixture of **6-108** and **6-109** in 63–84 % yield, with **6-108** as the main product (Scheme 6/2.23).

Itoh and coworkers [223] have shown that fullerene derivatives as **6/2-113**, which to date have been prepared in a stepwise procedure, can be obtained in a three-component domino process by treatment of diynes **6/2-109**, dimethylphenylsilane **6/2-110** and fullerene (C₆₀) in the presence of a Rh-catalyst [223]. Interestingly, using maleic anhydride as dienophile failed to give the desired cycloadduct, whereas C₆₀ – in spite of its strong tendency to form complexes with various transition metals [224] – never suppressed the catalytic silylative cyclization step to give the diene **6/2-112** (Scheme 6/2.24).

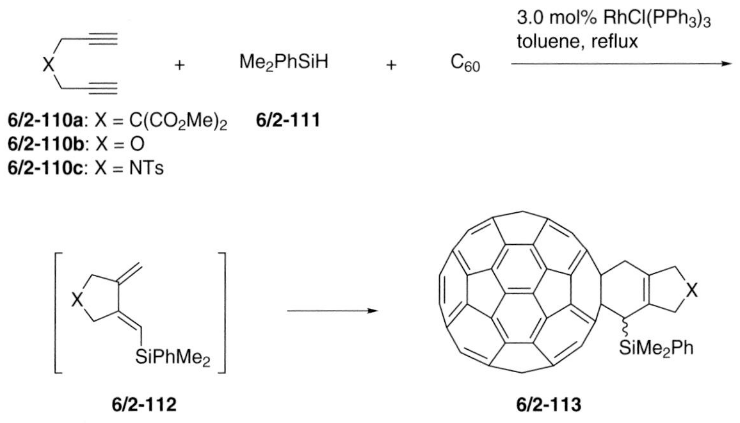

3.0 mol% RhCl(PPh$_3$)$_3$
toluene, reflux

X + Me$_2$PhSiH + C$_{60}$

6/2-110a: X = C(CO$_2$Me)$_2$ **6/2-111**
6/2-110b: X = O
6/2-110c: X = NTs

SiPhMe$_2$

6/2-112

SiMe$_2$Ph

6/2-113

Entry	Substrate 6/2-110	Product 6/2-113	t [min]	Yield [%][a]
1	a	a	40	53 (36)
2	a	a	60	71 (58)
3	a	a	110	63 (34)
4	b	b	200	41 (34)
5	c	c	45	32 (28)

[a] The yield is based on comsumed C$_{60}$ with isolated yield shown in parenthesis.

Scheme 6/2.24. Rhodium-catalyzed three-component coupling of C$_{60}$ with diynes and a silane.

6.3
Ruthenium-Catalyzed Transformations

The most important ruthenium-catalyzed domino process is based on a metathesis reaction. Nonetheless, a few other ruthenium-catalyzed processes have been employed for the synthesis of substituted β,γ-unsaturated ketones, as well as unsaturated γ-lactams and allylic amines.

6.3.1
Metathesis Reactions

Today, the metathesis reaction is one of the most useful transformations for C–C-couplings. Moreover, the combination of two or more metathesis reactions has increased its efficiency immensely; in particular, the domino ring-opening/ring-closure metathesis has been widely used [225]. One of the first examples is a route to capuellone **6/3-6** reported by Grubbs and coworkers, starting from the norbornene derivate **6/3-1**. This compound, on treatment with the Tebbe reagent, led to **6/3-4**

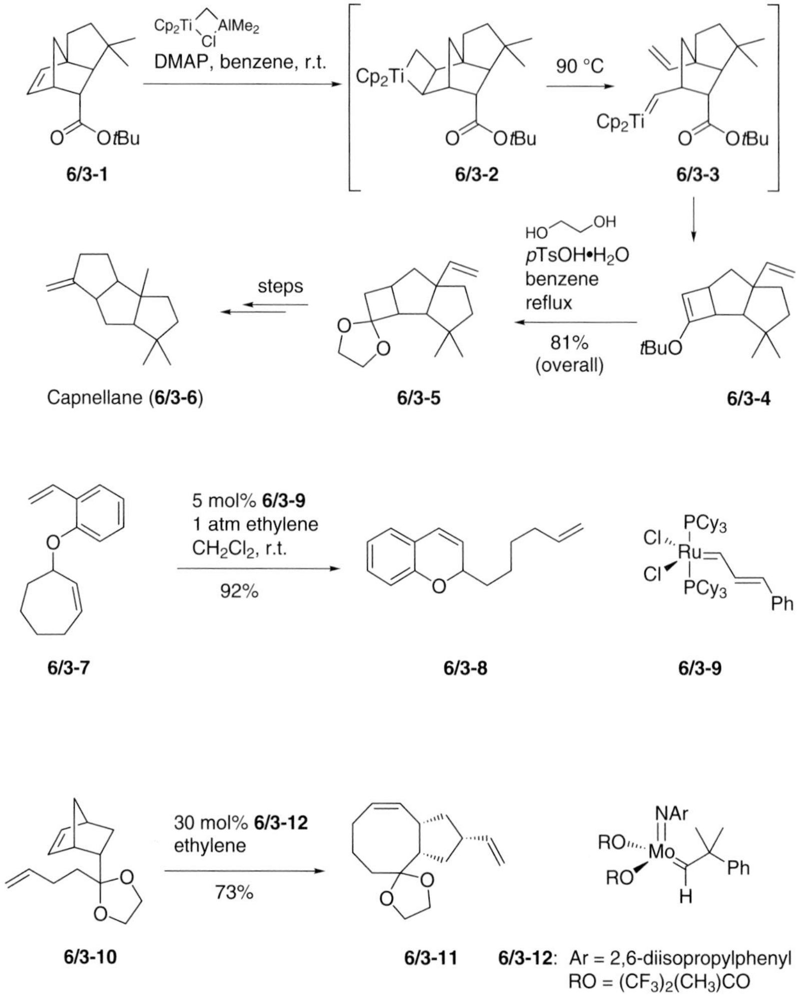

Scheme 6/3.1. Early examples of dominometatheses.

via **6/3-2** and **6/3-3** [226]. **6/3-4** was then transformed into the acetal **6/3-5** and further on to capnellane (**6/3-6**) (Scheme 6/3.1).

Hoveyda and coworkers [227] used a domino process to give chromanes **6/3-8** by treatment of **6/3-7** in the presence of ethylene. One of the first-generation Grubbs' catalyst **6/3-9** and one of Blechert's [228] early examples allowed the synthesis of bicyclic compounds of different sizes, depending on the length of the tether; thus, the reaction of **6/3-10** led to **6/3-11** using 30 mol% of the Schrock Mo complex **6/3-12**.

In recent years, in addition to the ring-opening metathesis (ROM) and ring-closing metathesis (RCM), the enyne metathesis and the cross-metathesis (CM) have

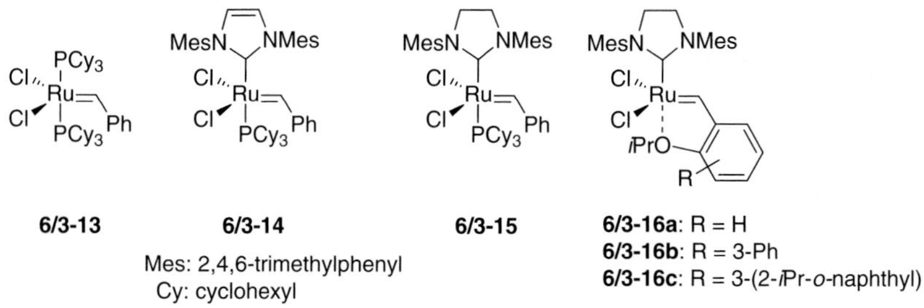

Scheme 6/3.2. Ru-catalyst for metathesis reactions.

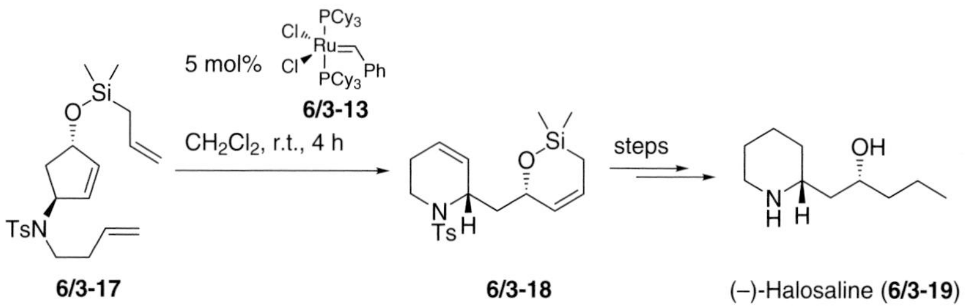

Scheme 6/3.3. Synthesis of (–)-halosaline (6/3-19).

obtained increasing importance. Thus, the CM has found several industrial applications such as the Shell Higher Olefin Process (SHOP) [229] and the Phillips Triolefin Process [230].

The main reason for the rapid development of metathesis reactions on a laboratory scale (the reaction itself had been known for quite a long time) has been the development of active and robust second-generation ruthenium catalysts (6/3-14 to 6/3-16), which usually provide better yields than the first-generation Grubbs' catalysts (6/3-9 or 6/3-13) (Scheme 6/3.2). This also reflects the huge number of domino processes based on ruthenium-catalyzed metathesis, which is usually followed by a second or even a third metathesis reaction. However, examples also exist where, after a metathesis, a second transition metal-catalyzed transformation or a pericyclic reaction takes place.

6.3.1.1 Metathesis-Metathesis Processes

An excellent example of a RCM/ROM domino process is shown in the total synthesis of the piperidine alkaloid (–)-halosaline (6/3-19) by Blechert and coworkers (Scheme 6/3.3) [231]. The key step is the reaction of the enantiopure cyclopentene derivative 6/3-17 to give 6/3-18 with 5 mol% of the catalyst 6/3-13. Further transformations of 6/3-18 led to the natural product 6/3-19.

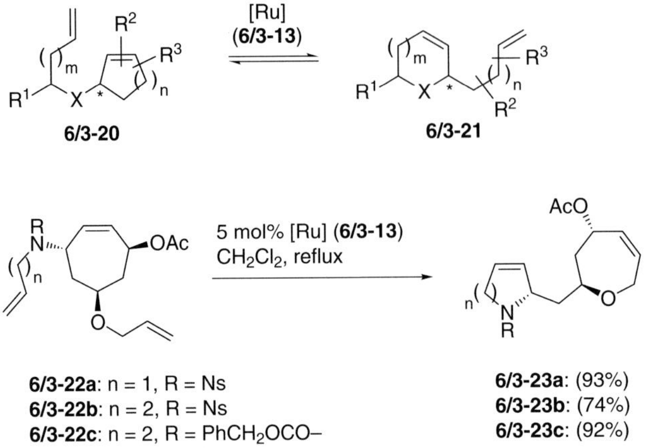
[Ru]
(6/3-13)
6/3-20
6/3-21
5 mol% [Ru] (6/3-13)
CH_2Cl_2, reflux
6/3-22a: n = 1, R = Ns
6/3-22b: n = 2, R = Ns
6/3-22c: n = 2, R = $PhCH_2OCO$–
6/3-23a: (93%)
6/3-23b: (74%)
6/3-23c: (92%)

Scheme 6/3.4. Synthesis of heterocycles.

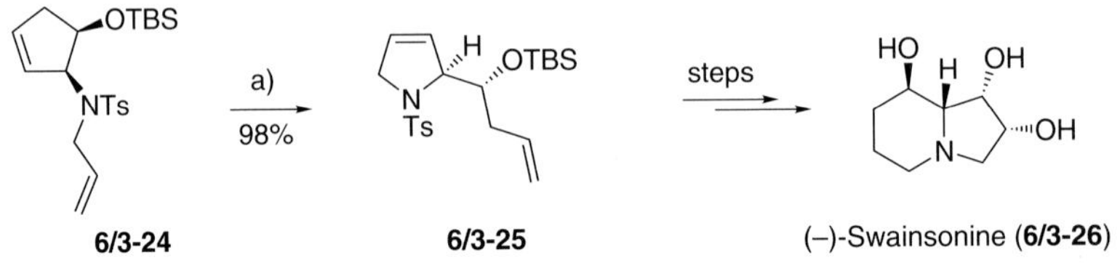
a)
98%
steps
6/3-24
6/3-25
(–)-Swainsonine (6/3-26)

a) 5 mol% [Ru] (**6/3-13**), $CH_2{=}CH_2$, CH_2Cl_2, 40 °C, 1 h.

Scheme 6/3.5. Synthesis of (–)-swainsonine (**6/3-26**).

In a similar process, the same group prepared a variety of substituted aza- and oxacycles **6/3-21** from **6/3-20**. Using this procedure, dihydrofurans, tetrahydropyrans, dihydropyroles, and tetrahydropyridines have been prepared in excellent yield. A further goal was the synthesis of tetrahydrooxepines **6/3-23**; this was accomplished using **6/3-22** as substrates, again in excellent yield (Scheme 6/3.4). Here, the ring rearrangement is accompanied by a RCM as an additional thermodynamic driving force [232].

In a related transformation, the indolizidine alkaloid (–)-swainsonine (**6/3-26**) was synthesized from **6/3-24** *via* **6/3-25** (Scheme 6/3.5) [233].

Another example of this useful domino process is the enantioselective synthesis of the quinozilidine alkaloid (–)-lasubine II [234]. Condensed tricyclic compounds as **6/3-28** can also be prepared from norbornene derivatives **6/3-27** in excellent yield, as shown by Funel and coworkers (Scheme 6/3.6) [235].

Piva and coworkers [236] prepared functionalized butenolides and pyrones using a combination of a ring-closing and a cross-coupling metathesis. Thus, reaction of a

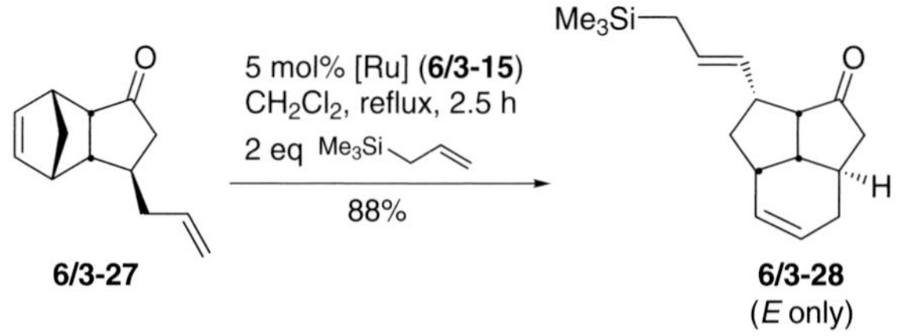

Scheme 6/3.6. Synthesis of condensed tricyclic compounds.

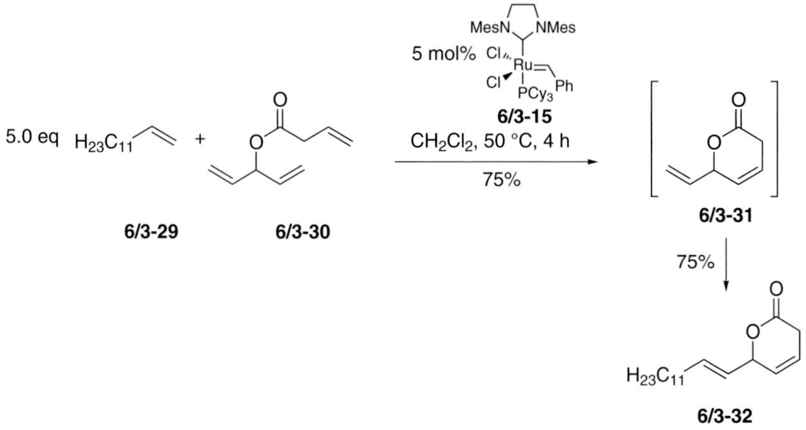

Scheme 6/3.7. Synthesis of pyrones.

mixture of the alkene **6/3-29** and the dienol ester **6/3-30** in the presence of the Grubbs II catalyst (**6/3-15**) led to **6/3-32** with an (*E*)-configuration in 75 % yield, together with 17 % of **6/3-31**, after a reaction time of 4 h. Interestingly, the sequential process with addition of the alkene after 2 h gave only traces of the desired compound **6/3-32** (Scheme 6/3.7).

Another combination of a ROM/RCM was described by Aubé and coworkers [237], in their synthesis towards dendrobatid alkaloids using **6/3-33** under an atmosphere of ethylene to give **6/3-34** (Scheme 6/3.8).

Grubbs and coworkers [238] used the ROM/RCM to prepare novel oxa- and aza-heterocyclic compounds, using their catalyst **6/3-15** (Scheme 6/3.9; see also Table 6/3.1). As an example, **6/3-35** gave **6/3-36**, by which the more reactive terminal alkene moiety reacts first and the resulting alkylidene opens the five-membered ring. In a similar reaction, namely a domino enyne process, fused bicyclic ring systems were formed. In this case the catalyst also reacts preferentially with the terminal alkene moiety.

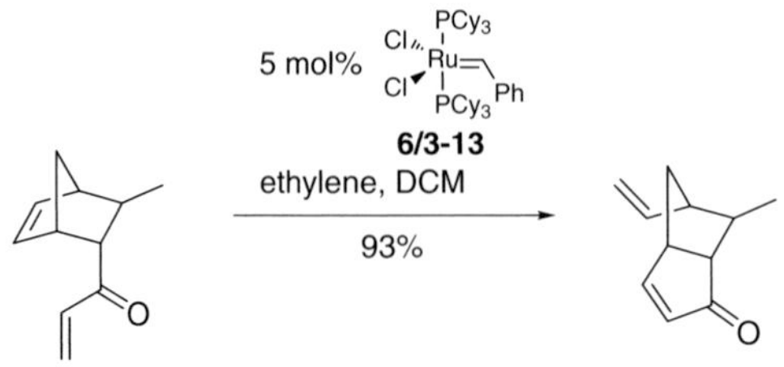

5 mol%

6/3-13

ethylene, DCM

93%

6/3-33

6/3-34

Scheme 6/3.8. Approach to dendrobatid alkaloids.

Substrate[a]	Concentration [M] of **6/3-15**	Product	Yield [%]
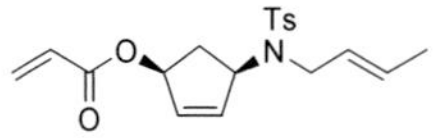	0.05	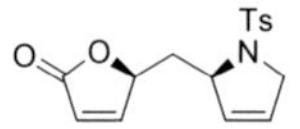	89
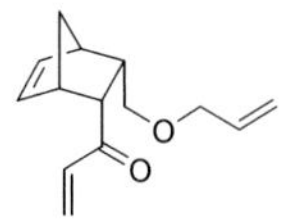	0.005	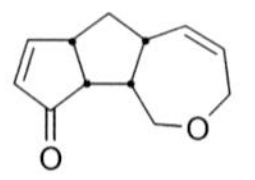	45
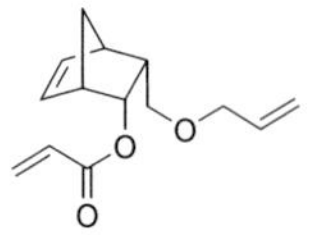	0.005	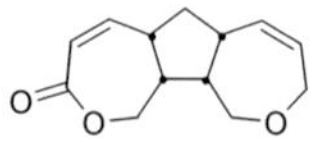	47
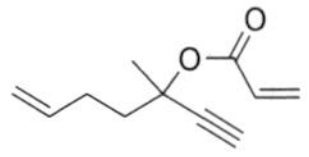	0.03	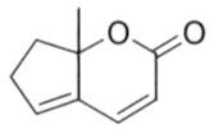	95
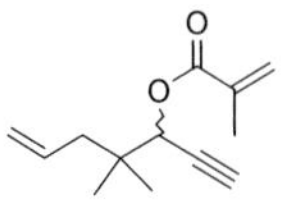	0.03	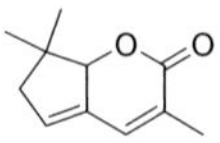	86
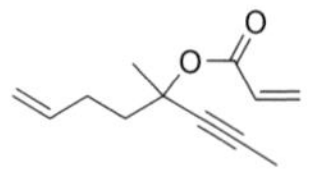	0.03	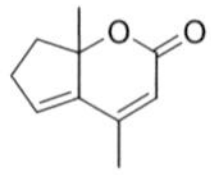	72
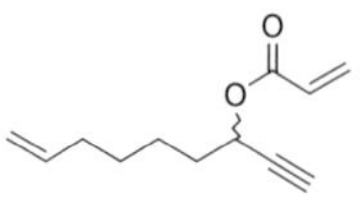	0.015	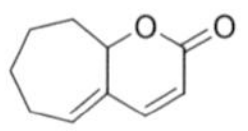	68
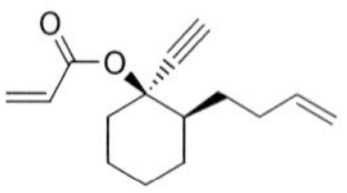	0.03	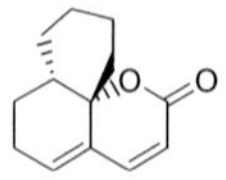	100
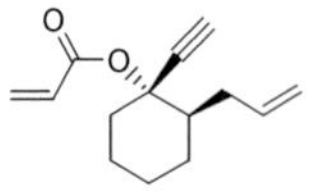	0.06	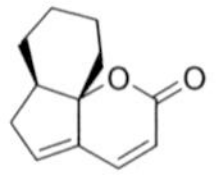	74

[a] 40 °C in CH_2Cl_2 for 6–12 h.

Table 6/3.1. Examples for domino ROM/RCM reactions.

Polyether-type structures such as **6/3-38** are frequently found in bioactive compounds (e. g., maitansine). Nicolaou and coworkers [239] have developed a new, efficient approach to these compounds, which is based on a domino ROM/RCM using the second-generation Grubbs' catalyst **6/3-15**. Thus, the cyclobutene derivative **6/3-37** could be transformed into **6/3-38** in 80% yield (Scheme 6/3.10).

A threefold domino RCM was observed by Harrity and coworkers [240] using substrates of type **6/3-39** which, on treatment with the Ru-catalyst **6/3-13**, led to the tricyclic compound **6/3-40** in 72% yield. Surprisingly, the diastereomer **6/3-41** gave **6/3-42** in only 38% yield, even using 20 mol% of the catalyst (Scheme 6/3.11).

Natural products are valuable substrates for further transformations, especially if they contain one or more stereogenic centers of defined absolute configuration. One of the most frequently used natural products in this respect is the monoterpene α-pinene (**6/3-43**), which has also been restructured using domino processes. Mehta and coworkers [241] have used the alcohols **6/3-44**, easily accessible from α-pinene (**6/3-43**), for a ROM/RCM reaction to give **6/3-45** and **6/3-46**, which were

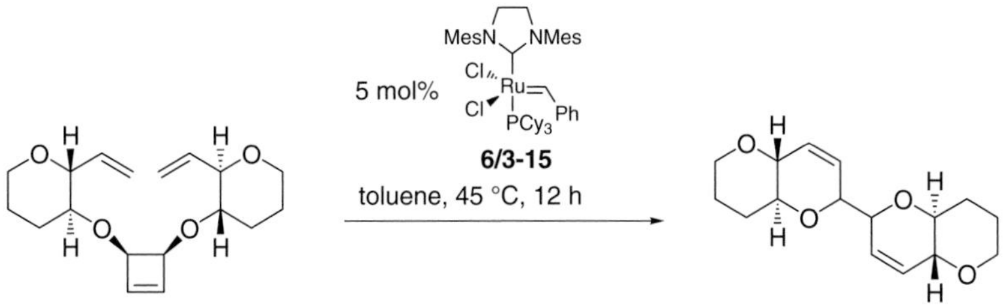

Scheme 6/3.10. Synthesis of polyethers.

Entry	Substrate	Product	Yield [%]
1	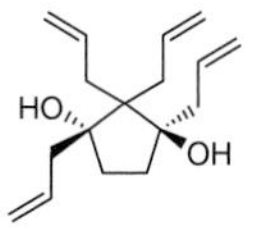 6/3-39	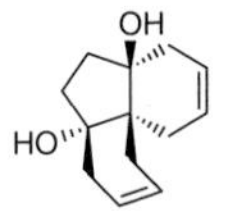 6/3-40	72[a]
2	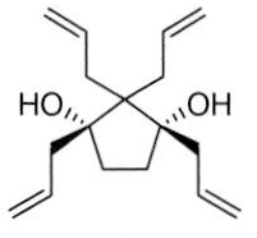 6/3-41	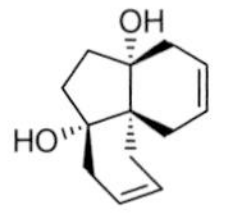 6/3-42	38[b]

[a] 10 mol% Grubbs I (**6/3-13**), dichloroethane, 60 °C, 48 h.
[b] 20 mol% Grubbs I (**6/3-13**), dichloroethane, 60 °C, 48 h.

Scheme 6/3.11. Ru-catalyzed preparation of angular fused tricycles.

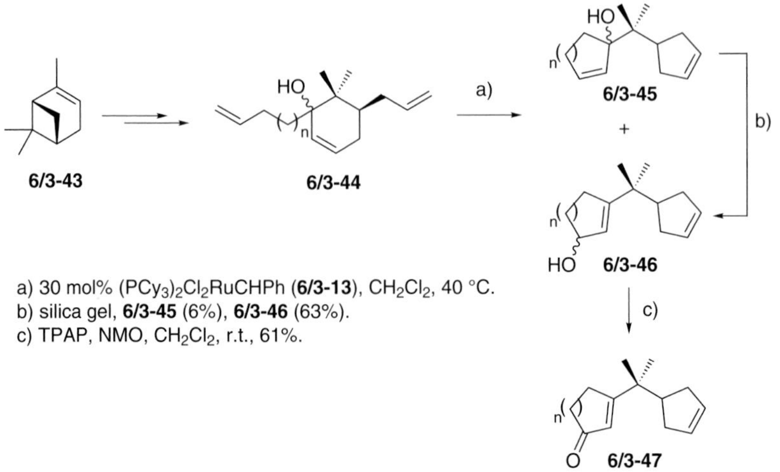
HO
a)
b)
c)
6/3-43
6/3-44
6/3-45
6/3-46
6/3-47
a) 30 mol% (PCy3)2Cl2RuCHPh (6/3-13), CH2Cl2, 40 °C.
b) silica gel, 6/3-45 (6%), 6/3-46 (63%).
c) TPAP, NMO, CH2Cl2, r.t., 61%.

Scheme 6/3.12. Use of pinene for domino metathesis reactions.

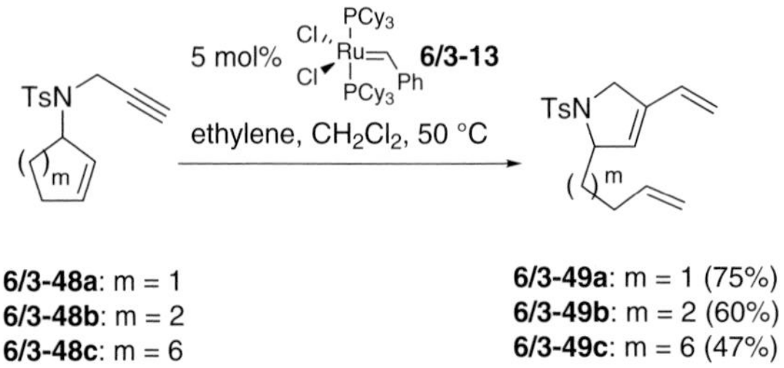
5 mol%
6/3-13
ethylene, CH2Cl2, 50 °C
TsN
6/3-48a: m = 1
6/3-48b: m = 2
6/3-48c: m = 6
6/3-49a: m = 1 (75%)
6/3-49b: m = 2 (60%)
6/3-49c: m = 6 (47%)

Scheme 6/3.13. Synthesis of substituted pyrrolines.

transformed into the enone **6/3-47** (Scheme 6/3.12). In addition to terpenes, carbohydrates have also been employed as substrates for domino metathesis; however, these reactions will be described later as they belong to the ene-yne metatheses.

One problem in the combination of metathesis transformations using alkenes is the fact that they are equilibrium reactions. In contrast, metathesis reactions of ene-ynes are irreversible as they give 1,3-butadienes, which are usually inert under the reaction conditions. Thus, the combination of a RCM and a ROM of ene-ynes of type **6/3-48** in the presence of an alkene (e. g., ethylene) led to **6/3-49** in good yield (Scheme 6/3.13) [242]. In these transformations the terminal triple bond reacts first. The process is not suitable for the formation of six-ring heterocycles.

A slightly different process takes place using substrates containing an internal triple bond, such as **6/3-50**. Here, the first reaction occurs at the terminal double bond to give **6/3-51** as the final product, as shown by Barrett and coworkers [243].

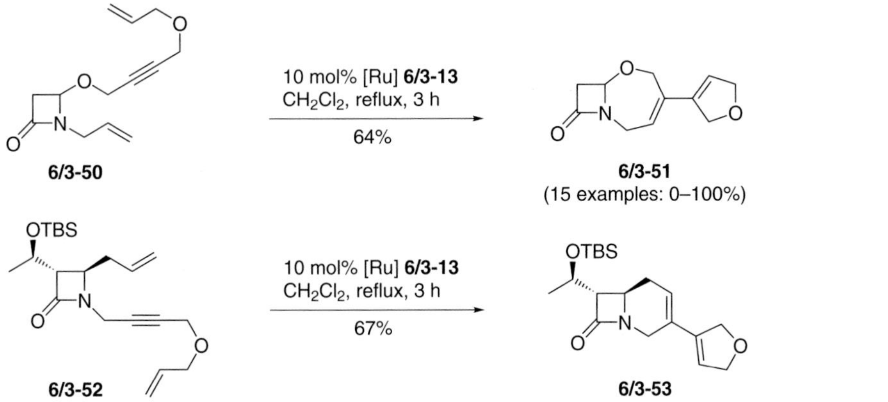

Scheme 6/3.14. Synthesis of bicyclic β-lactams.

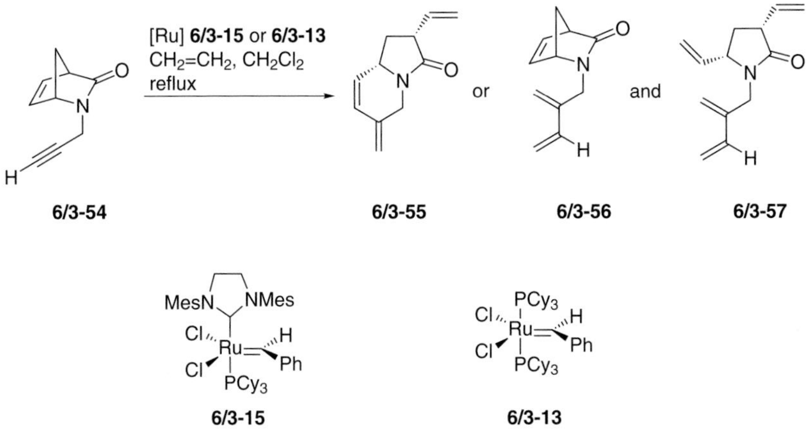

Scheme 6/3.15. Domino metathesis process of the enyne **6/3-54**.

Two new rings are formed in the reactions. The procedure also allows the synthesis of novel condensed carbocyclic β-lactams as **6/3-53** from **6/3-52** (Scheme 6/3.14).

A domino metathesis process using the enyne **6/3-54** and ethylene as substrates was developed by Arjona, Plumet and coworkers (Scheme 6/3.15) [244]. Interestingly, by using catalyst **6/3-15** (Grubbs II), the pyrrolidone **6/3-55** was obtained in 98% yield as the only product, whereas with catalyst **6/3-13** (Grubbs I), compounds **6/3-56** (60%) and **6/3-57** (25%) were formed.

Enantiopure 1-azabicyclic compounds of different ring size have also been prepared using the described method [245].

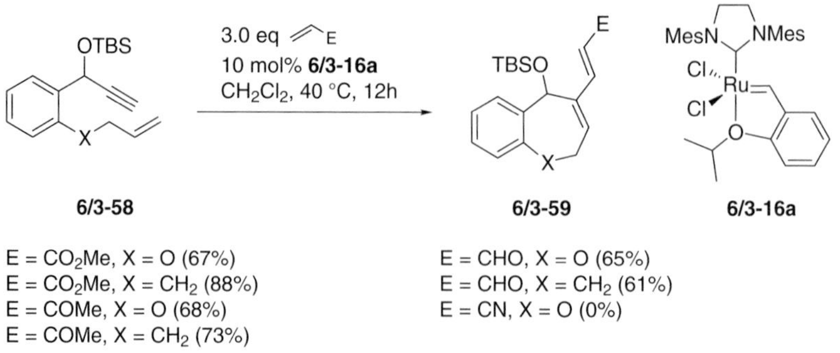
OTBS
3.0 eq E
10 mol% 6/3-16a
CH_2Cl_2, 40 °C, 12h
X
TBSO
E
X
MesN
NMes
Cl
Ru
Cl
O
6/3-58
6/3-59
6/3-16a
E = CO_2Me, X = O (67%)
E = CO_2Me, X = CH_2 (88%)
E = COMe, X = O (68%)
E = COMe, X = CH_2 (73%)
E = CHO, X = O (65%)
E = CHO, X = CH_2 (61%)
E = CN, X = O (0%)

Scheme 6/3.16. Domino metathesis process of enynes **6/3-58**.

Grimaud and coworkers [246] developed another domino RCM/CM process reacting the eneyne **6/3-58** with acrylate, acrolein and methylvinyl ketone in the presence of the Ru-catalyst **6/3-16a** to give **6/3-59** in good yield (Scheme 6/3.16).

A CM followed by a RCM was observed by Diver and coworkers in the reaction of alkynes **6/3-61** and 1,5-hexadienes **6/3-60** to give a mixture of cyclic and acyclic products **6/3-62** and **6/3-63** (Scheme 6/3.17) [247].

Another example of a twofold metathesis reaction of Undheim's group [248] is the Ru-catalyzed transformation of **6/3-64** to give **6/3-65** in 97% yield (Scheme 6/3.18).

Kaliappan and coworkers [249] prepared dioxa-triquinanes **6/3-68** together with the spiro compound **6/3-67** by a domino enyne RCM of the ene-yne **6/3-66** (Scheme 6/3.19).

A domino RCM of an ene-yne was also used by Granja and coworkers [250] for their synthesis of the B-bishomo-steroid analogue **6/3-70**. Reaction of the substrate **6/3-69** with the ruthenium catalyst **6/3-13** led to **6/3-70** in 48% yield as a 6.5:1-mixture of the two C-10-epimers (Scheme 6/3.20). The aim of this study was to prepare haptenes for the production of catalytic monoclonal antibodies that could be used to study the mechanism of the physiologically important transformation of previtamin D_3 into vitamin D_3 [251].

Another example of an efficient domino RCM is the synthesis of the highly functionalized tricyclic ring system **6/3-72** by Hanna and coworkers [252], which is the core structure of the diterpene guanacastepene A (**6/3-73**) (Scheme 6/3.21) [253]. Reaction of **6/3-71** in the presence of 10 mol% of Grubbs II catalyst **6/3-15** led to **6/3-372** in 93% yield. Interestingly, the first-generation Ru-catalyst **6/3-13** did not allow any transformation.

In addition to terpenes (as described above), carbohydrates have also been used as substrates in domino metathesis reactions, the aim being to synthesize enantiopure polyhydroxylated carbocyclic rings. These structures are components of several biologically active compounds such as aminoglycoside antibiotics [254], inositol phosphates [255], and carbanucleosides [256]. An efficient entry to this skeleton was developed by Madsen's group using a domino RCM/CM of the carbohy-

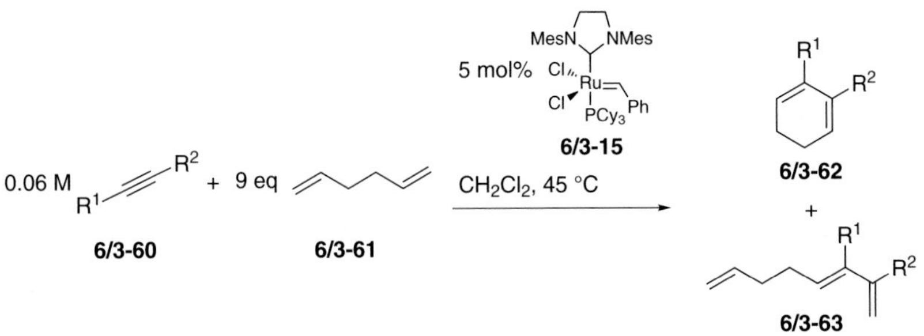

Entry	Substrates	Products	Time [h]	Yield [%] [6/3-62/6/3-63 (GC)]
1	(≡–OBz)		1.0	99 (1:1.2)
2	(≡–OBn)		2.0	99 (1:1.5)
3	(≡–N(Ts)C₄H₉)		2.0	99 (1:1.3)
4	(≡–CH(CH₂Ph)OAc)		2.0	99 (1:1.2)
5	(≡–Ph)		0.5	99 (1:1.4)
6	(≡–CH(CH₂Ph)OH)		6.0	48 (1:1)
7	(AcO–≡–OAc)		3.0	68 (1:1.9)

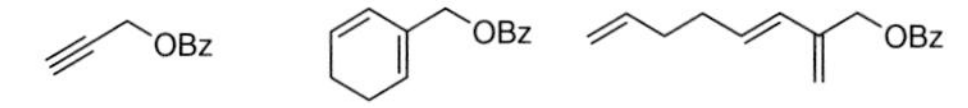

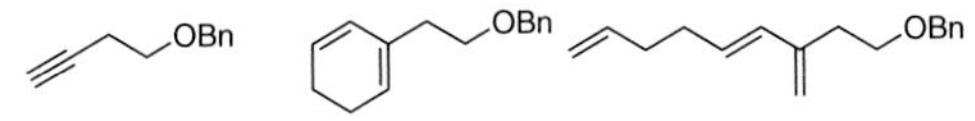

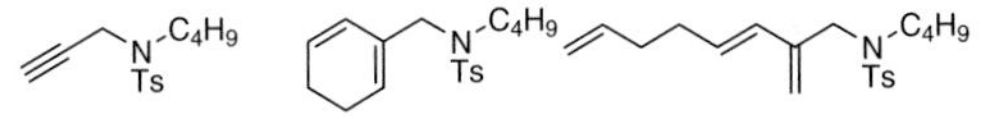

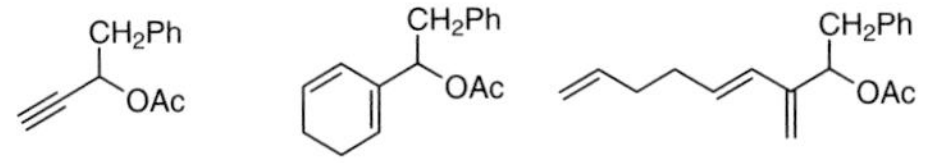

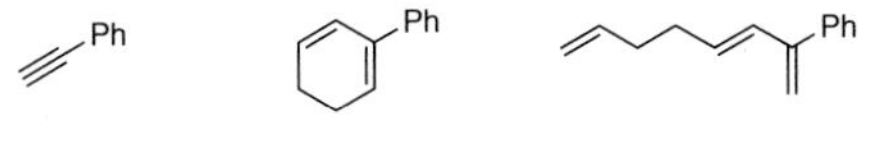

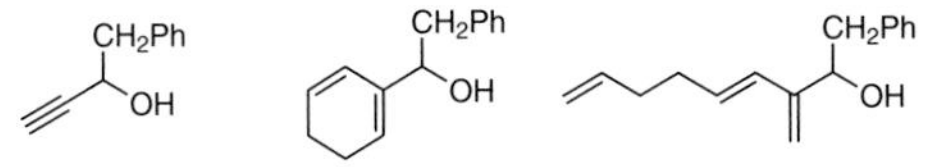

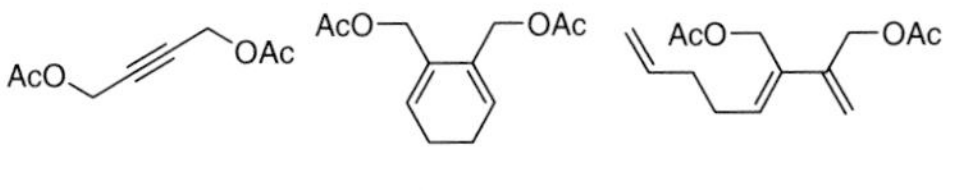

Scheme 6/3.17. Domino cross-metathesis/ring-closing-metathesis process.

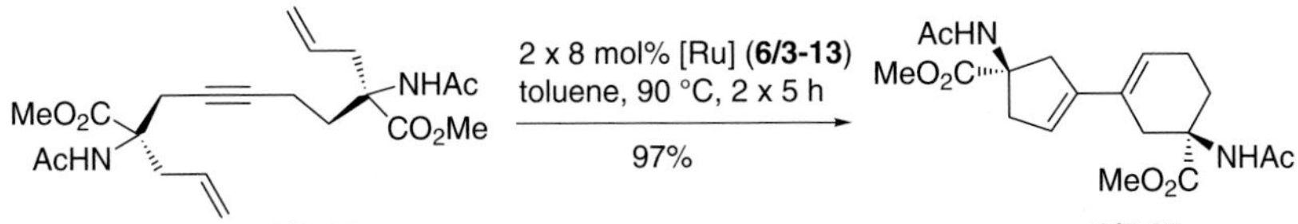

Scheme 6/3.18. Twofold metathesis process.

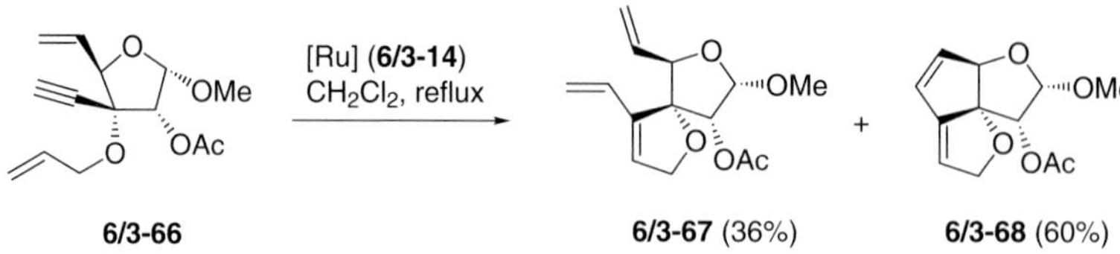

Scheme 6/3.19. Simple and twofold ring-closing metathesis.

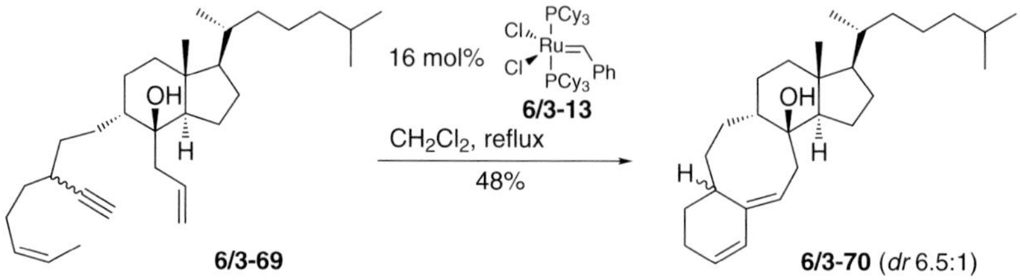

Scheme 6/3.20. Synthesis of a β-bishomo-steroid.

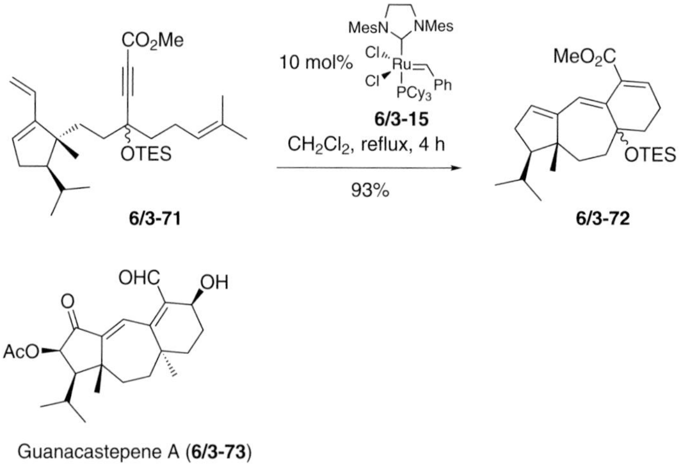

Guanacastepene A (**6/3-73**)

Scheme 6/3.21. Synthesis of the core structure of the diterpene guanacastepene A (**6/3-73**).

drate ene-yne derivative **6/3-76** to give the cyclohexene **6/3-77** [257]. **6/3-76** was obtained from methyl 5-deoxy-5-iodopentafuranoside (**6/3-74**) by treatment with Zn, followed by the dropwise addition of propargyl bromide and acetylation in a 9:1 diastereomeric ratio with 73% yield (Scheme 6/3.22).

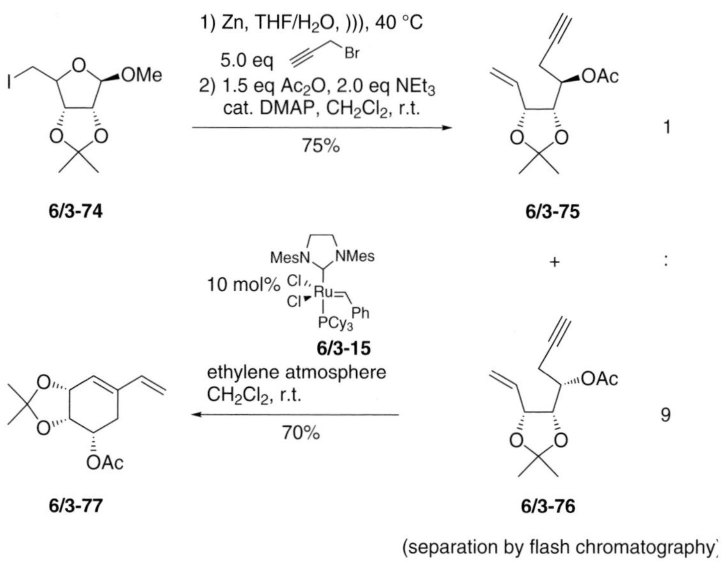

Scheme 6/3.22. Synthesis of polyhydroxylated carbocyclic compounds.

Hanna and coworkers [258] also used carbohydrate ene-yne derivatives for a domino metathesis reaction. The twofold intramolecular RCM of the glucose and ribose derivatives **6/3-78a,b** and **6/3-81**, respectively, using the second-generation Grubbs' catalyst **6/3-15**, gave the desired bicyclic cycloheptene and cyclooctene derivates **6/3-79a,b** and **6/3-82**, respectively, in high yield. The monocyclic products **6/3-80** and **6/3-83** were not formed. In contrast, when using the first-generation Grubbs' catalyst **6/3-13** in the reaction of **6/3-81** only **6/3-83** was obtained, in 85 % yield (Scheme 6/3.23).

6.3.1.2 Metathesis/Heck Reaction/Pericyclic Reaction/Hydrogenation

Metathesis can also be combined with other transition metal-catalyzed transformations and with pericyclic reactions, though the number of these domino processes remains quite small. Grigg and coworkers [259] combined a RCM with an intramolecular Heck reaction to allow the synthesis of bridged ring systems in good yield. A major problem in these reactions was poisoning of the ruthenium catalyst that was necessary for the metathesis reaction, by the palladium species used for the Heck reaction. This was especially pronounced when rings with more than five members were formed in the metathesis. The problem could be overcome by using a polystyrene-bound palladium catalyst [260], though in some cases a fluorous biphasic system also proved to be useful [261].

Thus, by using a mixture of the Grubbs' catalyst **6/3-13** and Pd(OAc)₂/PPh₃, **6/3-84a** was transformed into **6/3-85a** in 65 % yield. With the polystyrene-bound palladium catalyst 71 % yield was obtained; in contrast, the use of a biphasic system

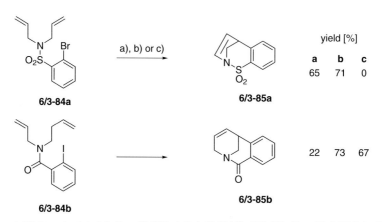

Dienyne	Catalyst, Time	Product: (Yield)	
6/3-78a: R = H	**6/3-15** (10%), 6 h	**6/3-79a**: (95%)	**6/3-80a**: –
6/3-78b: R = TES	**6/3-15** (10%), 6 h	**6/3-79b**: (96%)	**6/3-80b**: –
6/3-81	**6/3-13** (20%), 28 h	–	**6/3-83c**: (85%)
6/3-81	**6/3-15** (10%), 24 h	**6/3-82**: (84%)	–

Scheme 6/3.23. Ring-closing metathesis of diene-ynes **6/3-78** in refluxing CH$_2$Cl$_2$.

		yield [%]
		a b c
6/3-84a	a), b) or c) → **6/3-85a**	65 71 0
6/3-84b	**6/3-85b**	22 73 67

a: Mixture of catalyst: 1–5 mol% (PCy$_3$)$_2$Ru(=CHPh)Cl$_2$ (**6/3-13**), 10 mol% Pd(OAc)$_2$, 20 mol% PPh$_3$, 2 eq Tl$_2$CO$_3$, toluene, 1–8 h at rt then 16 h at 110 °C.
b: As in **a** with 10 mol% polystyrene bound palladium catalyst instead of Pd(OAc)$_2$/PPh$_3$.

c: As in **a** with 20 mol% of P–(⟨⟩–C$_6$F$_{13}$)$_3$.

Scheme 6/3.24. Domino metathesis/Heck reaction under different conditions.

was not successful. However, for the reaction of **6/3-84b**, the first procedure gave **6/3-85b** in only 22% yield, whereas using the two other methods **6/3-85b** was obtained in 73% and 67% yields, respectively (Scheme 6/3.24).

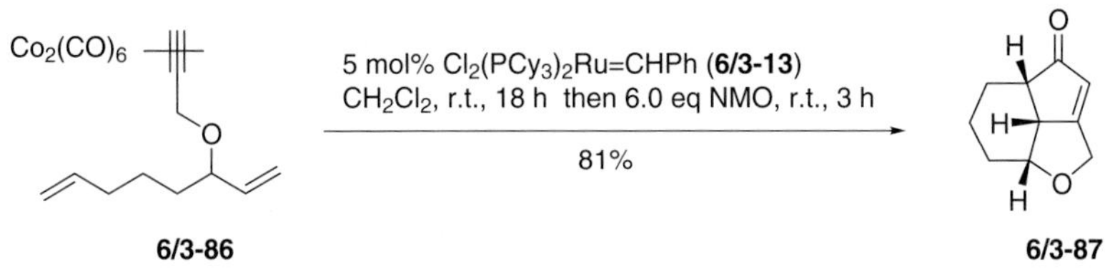

Scheme 6/3.25. Domino metathesis/Pauson–Khand reaction.

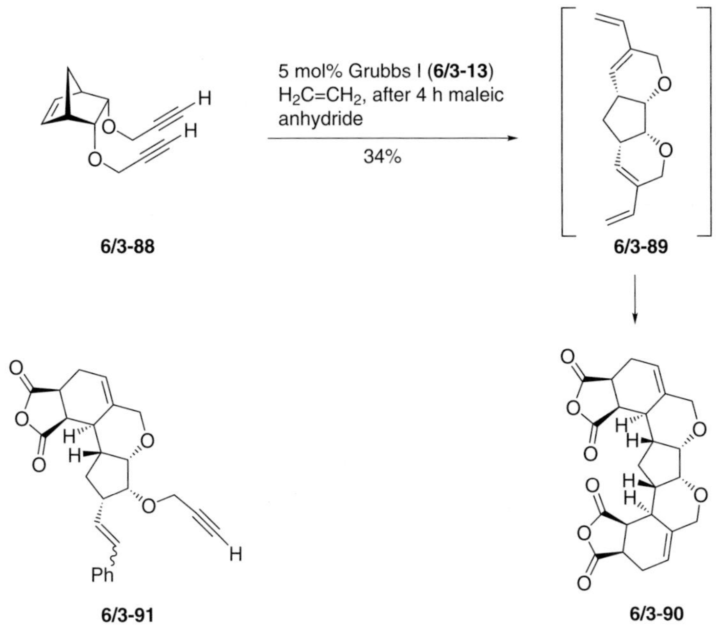

Scheme 6/3.26. Sequential metathesis/Diels–Alder reaction.

A combination of a metathesis and a Pauson–Khand reaction, which leads to tricyclic compounds starting from diene-ynes, has been described by Pérez-Castells and colleagues [262]. Treatment of the Co-complex **6/3-86**, obtained from the corresponding alkyne in 75 % yield, with 5 mol% of the Ru-catalyst **6/3-13** for 18 h, followed by addition of an *N*-oxide as trimethylamine-*N*-oxide (TMANO) or NMO as copromoters, gave **6/3-87** in 81 % yield.

A combination of a metathesis and a Diels–Alder reaction was published by North and coworkers [263]. However, this is not a true domino reaction, as the dienophile (e. g., maleic anhydride) was added after the *in situ* formation of the *bis*-butadiene **6/3-89** from the *bis*-alkyne **6/3-88** and ethylene. The final product is the *bis*-cycloadduct **6/3-90**, which was obtained in 34 % yield. Using styrene as an unsymmetrical alkene instead of ethylene, the mono-cycloadduct **6/3-91** was formed as a mixture of double-bond isomers, in 38 % yield (Scheme 6/3.26).

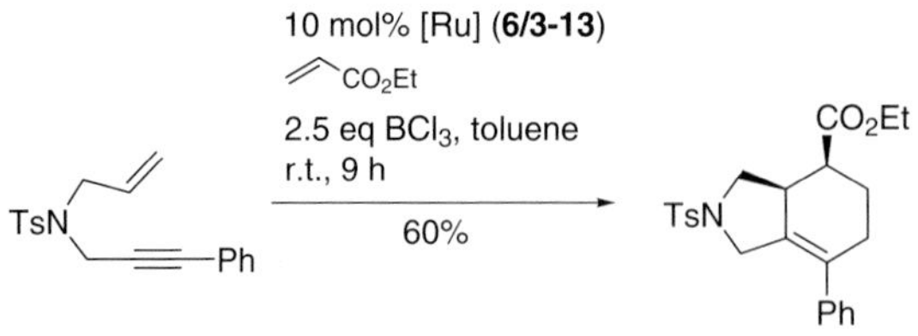

6/3-92 **6/3-93**

Scheme 6/3.27. Domino metathesis/Diels–Alder reaction.

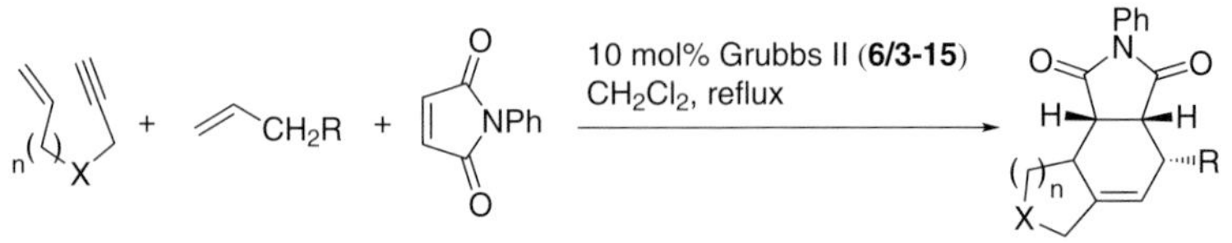

6/3-94 **6/3-95** **6/3-96** **6/3-97**

9 examples: 71–83%

X = NTs or $C(CO_2Et)_2$ (addition after formation of the diene)

Scheme 6/3.28. Synthesis of the tricyclic compounds **6/3-97**.

Although the metathesis of ene-ynes is a valuable method for the preparation of 1,3-butadienes, and may be used for Diels–Alder reactions, a problem arises from the need to employ either a high temperature or a Lewis acid to accelerate the cycloaddition, which is usually not feasible with the Grubbs' catalyst. Therefore, the combination of metathesis and cycloaddition is usually performed in sequential fashion (as just shown, and highlighted earlier) [264]. However, Laschat and coworkers [265] have shown the Lewis acid BCl_3 to be compatible with the Grubbs I catalyst (**6/3-13**). Reaction of **6/3-92** and ethyl acrylate using a mixture of 2.5 equiv. of the Lewis acid and 10 mol% of **6/3-13** led to **6/3-93** in 60% yield (Scheme 6/3.27).

Another intramolecular ene-yne metathesis followed by an intermolecular metathesis with an alkene to give a butadiene which is intercepted by a Diels–Alder reaction was used for the synthesis of condensed tricyclic compounds, as described by Lee and coworkers [266]. However, as mentioned above, the dienophile had to be added after the domino metathesis reaction was completed; otherwise, the main product was the cycloadduct from the primarily formed diene. Keeping this in mind, the three-component one-pot reaction of ene-yne **6/3-94**, alkene **6/3-95** and *N*-phenylmaleimide **6/3-96** in the presence of the Grubbs II catalyst **6/3-15** gave the tricyclic products **6/3-97** in high yield (Scheme 6/3.28).

Murakami and coworkers [267] described a combination of an intermolecular ene-yne metathesis followed by a disrotatory 6π-electrocyclic reaction to give six-membered cyclic dienes. Thus, reaction of **6/3-98** and of styrene (**6/3-99**) in the presence of a ruthenium catalyst at 100 °C led to the condensed cyclohexadienes **6/3-100**, in good yield (Scheme 6/3.29).

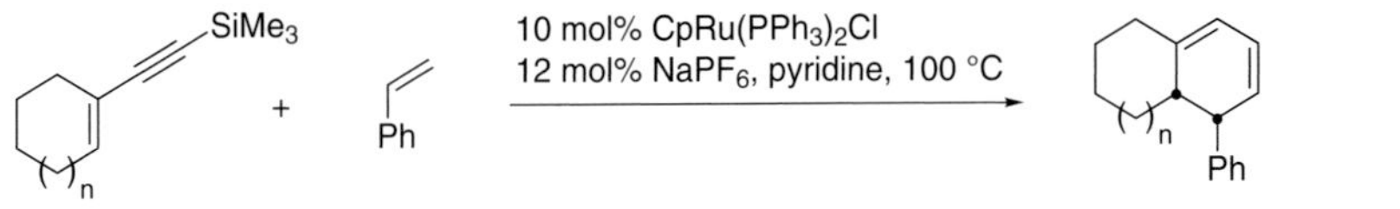

6/3-98a: n = 0   **6/3-99**   **6/3-100a:** n = 0, 60%
6/3-98b: n = 1   **6/3-100b:** n = 1, 70%
6/3-98c: n = 2   **6/3-100c:** n = 2, 67%

Scheme 6/3.29. Synthesis of condensed carbocycles.

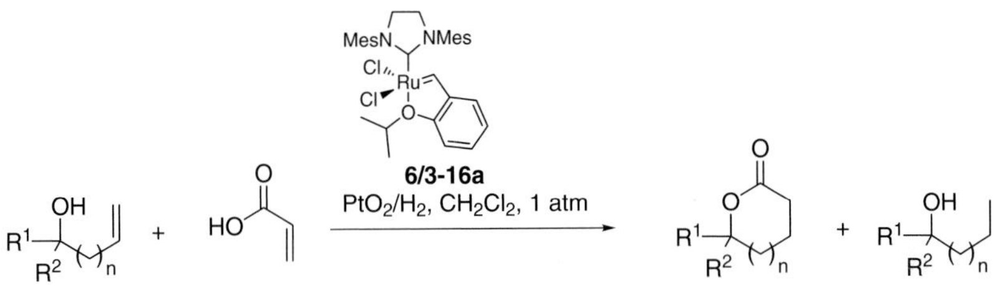

6/3-101   **6/3-102**   **6/3-103**   **6/3-104**

acrylic acid: 11 examples (10–70% yield of **6/3-103**)
acrolein: 9 examples (11–70% yield of the corresponding lactols)

Scheme 6/3.30. Synthesis of lactones and lactols.

A novel lactone and lactol synthesis was achieved by Cossy and coworkers [268], using a CM followed by a hydrogenation and a ring closure. In a typical procedure, a solution of acrylic acid or acrolein and an allylic or homoallylic alcohol is stirred at room temperature under 1 atm of H_2 in the presence of the ruthenium catalyst **6/3-16a** and PtO_2. Under these conditions, the homoallylic alcohols **6/3-101** (n = 1) and acrylic acid **6/3-102** led to the lactones **6/3-103** and the reduced alcohol **6/3-104**; with acrolein, the corresponding lactols were obtained, together with **6/3-104** (Scheme 6/3.30).

6.3.2
Other Ruthenium-Catalyzed Transformations

To date, very few other ruthenium-catalyzed domino processes have been identified in addition to the metathesis reactions. An intriguing example is the reaction of substituted propargylic alcohols containing a nucleophilic moiety, such as a second hydroxyl group with allylic alcohols in the presence of catalytic amounts of the ruthenium complex $CpRu(Ph_3P)_2Cl$. Thus, heating of a neat mixture of **6/3-105** and **6/3-106** containing 10 mol% NH_4PF_6 led to **6/3-107** in 57% yield. The substrate **6/3-105** is easily accessible by addition of acetylide to the corresponding cyclohexane carbaldehyde. The catalytic cycle, as outlined by Trost and coworkers [269]

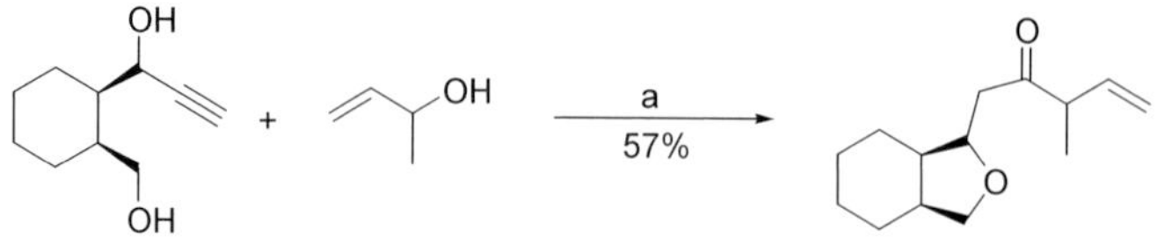

6/3-105   **6/3-106**   **6/3-107**

a) 10 mol% Cp(PPh₃)₂, RuCp(PPh₃)₂Cl, 20 mol% NH₄PF₆, 100 °C, 8 h.

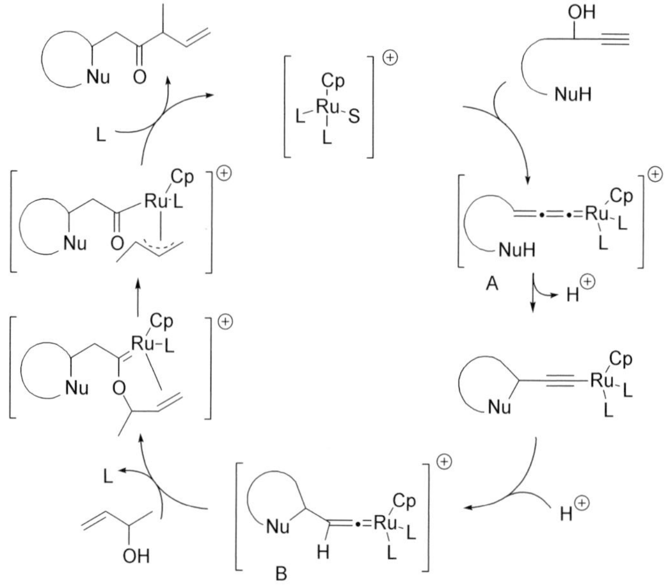

Scheme 6/3.31. Cyclization/reconstitutive addition process involving allenylidene ruthenium complexes as reactive intermediates.

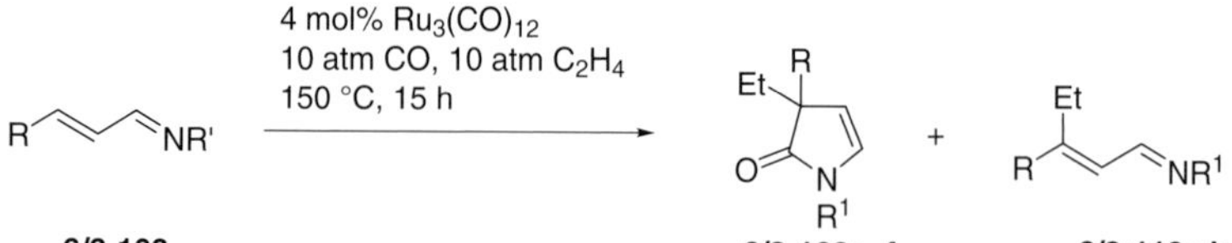

6/3-108   **6/3-109a–f**   **6/3-110a,b**

R	R¹	Yield [%] **6/3-109**		Yield [%] **6/3-110**
Ph	4-CF₃C₆H₄	38	a	58 (Z:E = 1:2.5)
Ph	Ph	49	b	41 (Z:E = 1:1)
Ph	4-tBuC₆H₄	15	c	–
Ph	tBu	44	d	–
Ph	Cy	93	e	–
Me	Cy	95	f	–

Scheme 6/3.32. Synthesis of unsaturated γ-lactams.

(see Scheme 6/3.31) involves a rapid formation and reaction of the first-formed allenylidene intermediate A. This then undergoes an intramolecular addition of the nucleophilic moiety in the molecule to give the vinylidene complex B; reaction of complex B with allyl alcohol then leads to the product.

Imhof and coworkers [270] described a Ru-catalyzed process starting from 1-aza-1,3-butadienes 6/3-108 under CO and C_2H_4 atmosphere to give 1,3-dihydropyrrol-2-ones (γ-lactams) 6/3-109. It can be assumed that, primarily, an insertion of CO into the C–H-bond at C-3 of the 1-aza-1,3-diene 6/3-108 takes place, followed by ring closure to give the pyrrolone. In a second process, one molecule of ethylene is inserted into the C–H bond adjacent to the newly formed carbonyl group. In some of these reactions the ethylazadiene 6/3-110 is formed as a byproduct (Scheme 6/3.32).

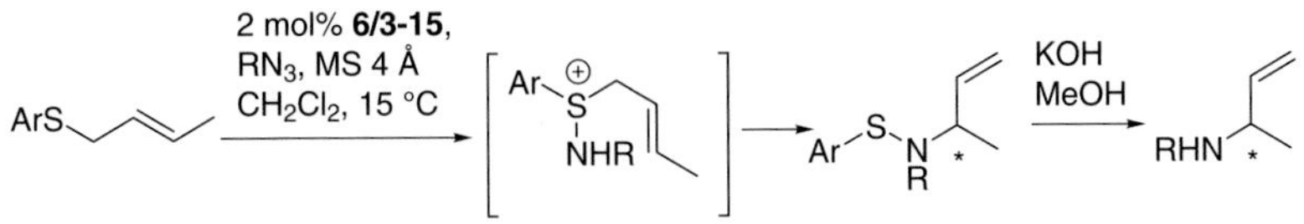

Entry	Ar	R	E/Z[a]	ee [%][b]	Config.[c]	Yield [%]
1	C_6H_5[d]	pMeC$_6$H$_4$SO$_2$	81:19	78	R	82
2	C_6H_5	pMeC$_6$H$_4$SO$_2$	81:19	84	R	65
3	C_6H_5[e]	pMeC$_6$H$_4$SO$_2$	81:19	86	R	16
4	p(O$_2$N)C$_6$H$_4$	pMeC$_6$H$_4$SO$_2$	76:24	81	R	47
5	pMeOC$_6$H$_4$	pMeC$_6$H$_4$SO$_2$	85:15	80	R	79
6	pBrC$_6$H$_4$	pMeC$_6$H$_4$SO$_2$	83:17	82	R	79
7	oBrC$_6$H$_4$	pMeC$_6$H$_4$SO$_2$	85:15	81	R	22
8	2-C$_{10}$H$_7$	pMeC$_6$H$_4$SO$_2$	83:17	78	R	88
9	C_6H_5	pMeOC$_6$H$_4$SO$_2$	81:19	83	–	57
10	C_6H_5	pBrC$_6$H$_4$SO$_2$	81:19	83	–	69

[a] E/Z-isomer ratio of the starting material.
[b] Determined by HPLC analysis using DAICEL CHIRACEL AD-H (hexane/2-propanol 9:1).
[c] Absolute configuration was determined by comparison of the sign of specific optical rotation with the reported one.
[d] Reaction was carried out at room temperature.
[e] Reaction was carried out at 0 °C.

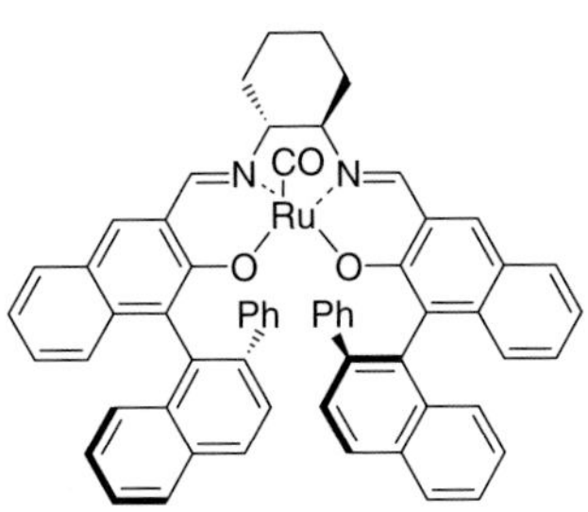

6/3-115

Scheme 6/3.33. Asymmetric conversion of crotyl aryl sulfides into *N*-allyl-arylsulfonamides using a chiral Ru-catalyst.

Finally, Katsuki and coworkers [271] described an enantioselective Ru-catalyzed domino reaction, which includes a sulfamidation of an aryl allyl sulfide **6/3-111** using the chiral Ru(salen)-complex **6/3-115**, followed by a 2,3-sigmatropic rearrangement of the formed **6/3-112** to give *N*-allyl-*N*-arylthiotoluenesulfonamides **6/3-113**. On hydrolysis, **6/3-113** yielded *N*-allyltoluenesulfonamides **6/3-114** (Scheme 6/3.33). The enantioselectivity ranged from 78 to 83 % *ee*.

6.4
Transition Metal-Catalyzed Transformations other than Pd, Rh, and Ru

Besides the already described Pd-, Rh- and Ru-catalyzed transformations, many other transition metals have also been used in domino processes, albeit to a lesser extent. Co- and also Ni-catalyzed transformations constitute the largest group in this section, though other examples include Cu, W, Mo, Fe, Ti, Cr, Au, Pt, Zr, and Lanthanide-catalyzed reactions.

6.4.1
Cobalt-Induced Transformations

Co-catalyzed transformations are concerned mainly with the [2+2+2] cycloadditions of three alkyne groups to give arenes. Another important reaction is the [2+2+1] cycloaddition of alkynes, alkenes and CO to give cyclopentenones, which is the well-known as Pauson–Khand reaction [272].

The forerunner in the Co-catalyzed [2+2+2] cycloaddition domino processes was that identified by Vollhardt and colleagues [273], with their excellent synthesis of steroids. Reaction of **6/4-1** with [CpCo(CO)$_2$] gave compound **6/4-3** with an aromatic ring B *via* the intermediate **6/4-2**. In this process, trimerization of the three alkyne moieties first takes place, and this is followed by an electrocyclic ring opening of the formed cyclobutene to give *o*-quinodimethane. This then undergoes a Diels–Alder reaction to provide the steroid **6/4-3** (Scheme 6/4.1).

Malacria and coworkers [274] used an intermolecular trimerization of alkynes to gain efficient access to the skeleton of the phyllocladane family. Thus, the Co-catalyzed reaction of the polyunsaturated precursor **6/4-4** gave **6/4-5** in 42 % yield. Here, six new carbon–carbon bonds and four stereogenic centers are formed. The first step is formation of the cyclopentane derivative **6/4-6** by a Co-catalyzed Coniaene-type reaction [275] which, on addition of *bis*(trimethylsilyl)ethyne (btmse), led to the benzocyclobutenes **6/4-7** (Scheme 6/4.2). The reaction is terminated by the addition of dppe and heating to reflux in decane to give the desired products **6/4-5** by an electrocyclic ring opening, followed by [4+2] cycloaddition.

As shown in the two examples described here, formation of the benzene nucleus by trimerization of alkynes is usually catalyzed by a Co-complex. However, Undheim and coworkers [276] have recently shown that a Ru^{II}-complex can also be used. Reaction of the triyne **6/4-9**, which was prepared from Schöllkopf's bislactim ether **6/4-8** [277] with Grubbs I catalyst **6/3-13**, led to **6/4-10** in an excellent yield of 90 %. Hydrolysis of **6/4-10** gave the desired as-indacene-bridged bis(α-amino acid) derivative **6/4-11** (Scheme 6/4.3).

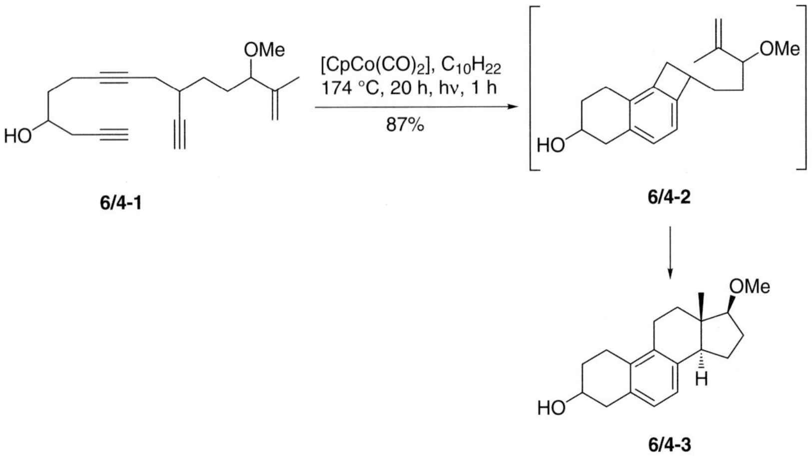

Scheme 6/4.1. Synthesis of steroids.

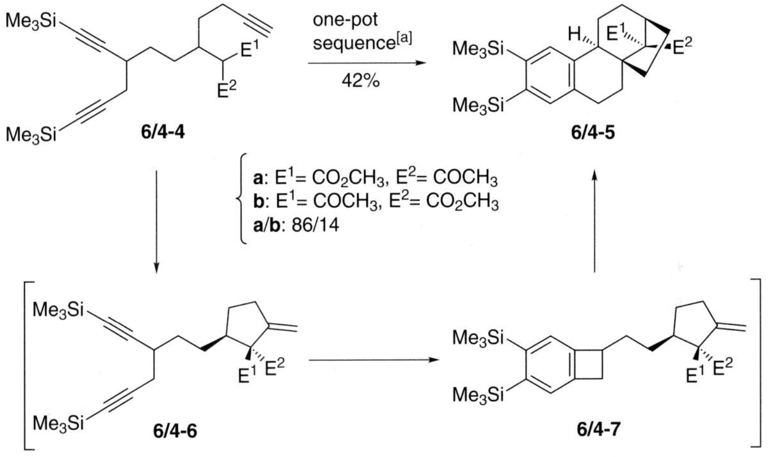

a: E¹= CO₂CH₃, E²= COCH₃
b: E¹= COCH₃, E²= CO₂CH₃
a/b: 86/14

[a] Reaction conditions: (1) 5% CpCo(CO₂), hv, 80 °C, 8 h; (2) btmse, 136 °C, hv, 15 min; (3) 5% dppe, decane, 175 °C, 12 h.

Scheme 6/4.2. Synthesis of the phyllocladane skeleton.

The Pauson–Khand reaction is the Co-induced formation of cyclopentenones from ene-ynes and CO. One impressive example of a domino Pauson–Khand process is the synthesis of fenestrane 6/4-15, as reported by Keese and colleagues [278]. The transformation is initiated by a double Grignard reaction of 4-pentynoic acid 6/4-12, followed by protection of the formed tertiary hydroxyl group to give 6/4-13. The Co-induced polycyclization of 6/4-13 led directly to the fenestrane 6/4-15

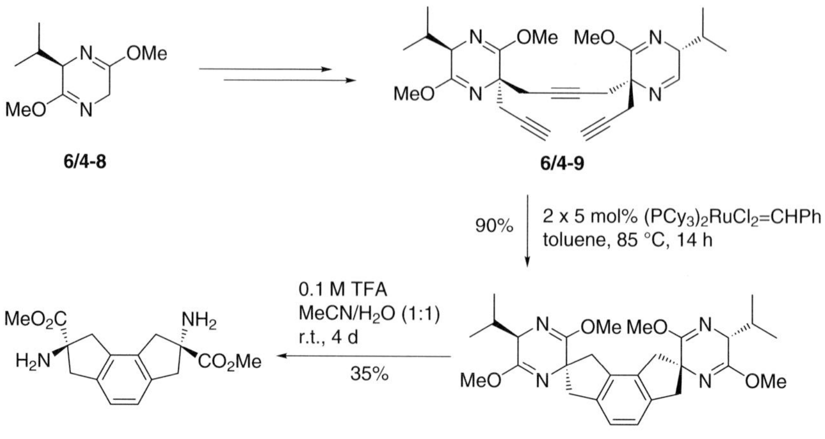

Scheme 6/4.3. Ru-catalyzed cyclotrimerization of alkynes.

via the proposed intermediate **6/4-14** (Scheme 6/4.4). Another interesting domino Pauson–Khand reaction was presented by the group of Cook [279], who generated six carbon–carbon bonds in a one-pot process in the synthesis of dicyclopenta[*a,e*]pentalene derivatives.

Chung and coworkers [280] combined a [2+2+1] with a [2+2+2] cycloaddition for the synthesis of multi-ring skeletons, angular triquinanes, and fenestranes. For the preparation of tetracyclic compounds such a **6/4-17**, these authors used diynes as **6/4-16** and CO as substrates (Scheme 6/4.5). Fully substituted alkynes gave low yields, and 1,5- as well as 1,7-dialkynes, did not react.

The same authors also combined the [2+2+1] cycloaddition with a Diels–Alder reaction [281]. For this multicyclization, the dienediyne **6/4-18** was treated with 5 mol% CO$_2$(CO)$_8$ under 30 atm CO at 130 °C for 18 h to give the tetracyclic compound **6/4-19** (Scheme 6/4.6).

It is not quite clear which step takes place first – the Co-catalyzed [2+2+1] cycloaddition of the outer alkyne moiety, or the Diels–Alder reaction of the diene with the inner alkyne to form a 1,4-cyclohexadiene, which then undergoes a Pauson–Khand reaction with the remaining alkyne. Recently, it has been shown that a domino reaction can also be performed using 1 mol of a 1,7-diphenyl-1,6-diyne **6/4-20** and a 1,3-diene **6/4-21** in the presence of Co/C at 150 °C under 30 atm CO, to give the polycyclic compounds **6/4-22** as sole product (Scheme 6/4.7) [282].

The Pauson–Khand reaction can be facilitated by preparing the necessary eneyne *in situ* by an allylic substitution of an alkyne with allylic acetate using a Pd^0- and Rh-catalyst. The yield of the cyclization product **6/4-24** ranges from 0% with X = O **(6/4-24a)** to 92% with X = NTs, as well as X = C(CO$_2$Et)$_2$ **(6/4-24c)** (Scheme 6/4.8) [283].

A Co-mediated domino [5+1]/[2+2+1] cycloaddition to give tricyclic δ-lactones **6/4-26** in good yields was observed by Liu and coworkers (Scheme 6/4.9) [284].

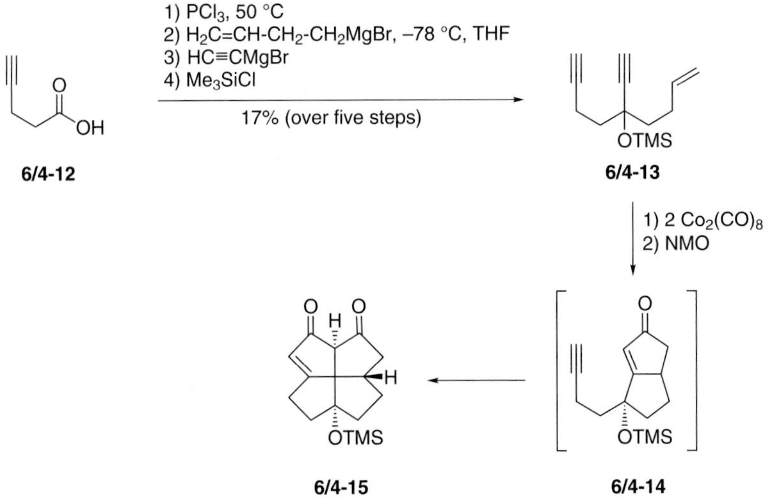

1) PCl$_3$, 50 °C
2) H$_2$C=CH-CH$_2$-CH$_2$MgBr, −78 °C, THF
3) HC≡CMgBr
4) Me$_3$SiCl

17% (over five steps)

6/4-12

6/4-13

1) 2 Co$_2$(CO)$_8$
2) NMO

6/4-15

6/4-14

Scheme 6/4.4. Synthesis of fenestrane.

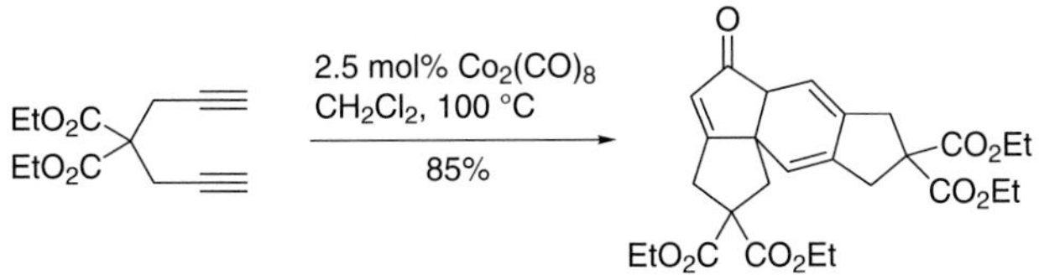

EtO$_2$C.
EtO$_2$C

2.5 mol% Co$_2$(CO)$_8$
CH$_2$Cl$_2$, 100 °C

85%

6/4-16

-CO$_2$Et
CO$_2$Et

EtO$_2$C CO$_2$Et

6/4-17

6 examples: 19–89% yield

Scheme 6/4.5. Synthesis of multiring skeletons.

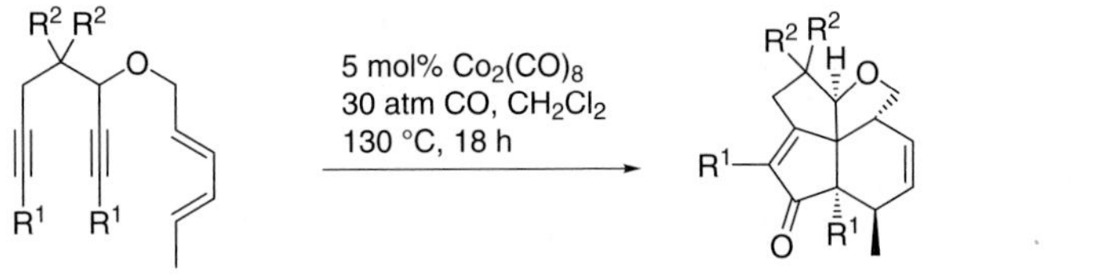

R^2 R^2

5 mol% Co$_2$(CO)$_8$
30 atm CO, CH$_2$Cl$_2$
130 °C, 18 h

R^1 R^1

6/4-18

6/4-19a: R^1= Ph, R^2= Me (84%)
6/4-19b: R^1= Ph, R^2= H (74%)
6/4-19c: R^1= R^2= H (51%)

Scheme 6/4.6. Domino Pauson–Khand/Diels–Alder reaction.

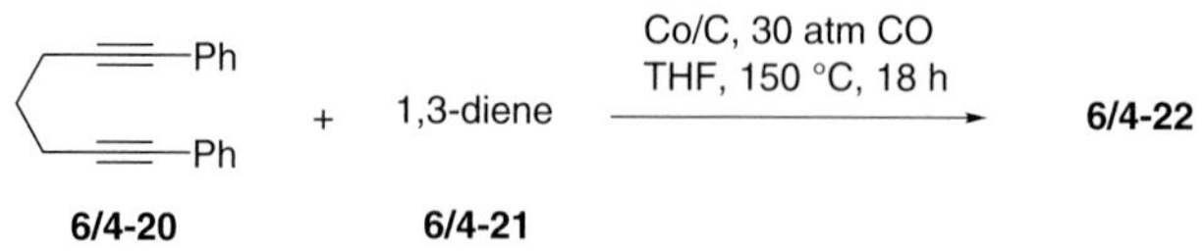
Ph
Ph
+ 1,3-diene
Co/C, 30 atm CO
THF, 150 °C, 18 h
6/4-22
6/4-20
6/4-21

Entry	1,3-Diene (6/4-21)	Product (6/4-22)	Yield [%]
1	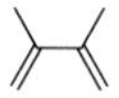	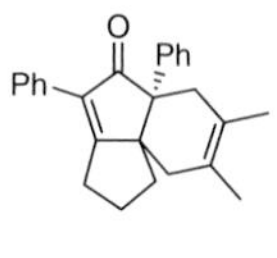	87
2	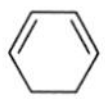	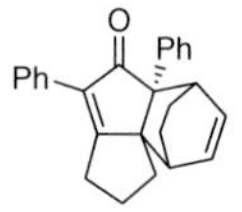	62
3		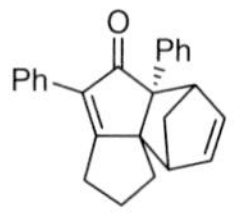	77

Scheme 6/4.7. Co-catalyzed domino reactions of a 1,7-diyne with a 1,3-diene

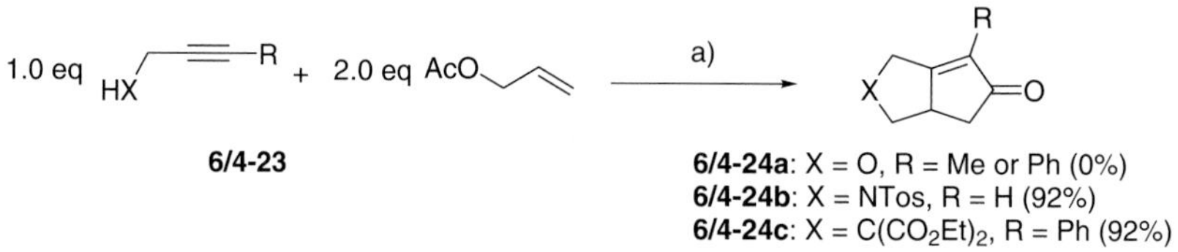
1.0 eq
HX
R
+ 2.0 eq AcO
a)
R
X
O
6/4-23
6/4-24a: X = O, R = Me or Ph (0%)
6/4-24b: X = NTos, R = H (92%)
6/4-24c: X = C(CO2Et)2, R = Ph (92%)

a) $Pd_2(dba)_3$•$CHCl_3$, $[RhCl(CO)(dppp)]_2$, dppp, BSA, 1 atm CO, toluene, r.t. → 110 °C.

Scheme 6/4.8. Combination of a nucleophilic substitution and a Pauson–Khand reaction.

Thus, reaction of **6/4-25a** gave **6/4-26a** as a single product by applying a CO pressure of 50 psi; under a nitrogen atmosphere, the isomer **6/4-27** is the main product. The procedure can also be used for the synthesis of *O*- and *N*-tricyclic lactones as **6/4-26b** and **6/4-26c** with the epoxides **6/4-25b** and **6/4-25c** as substrates.

As described previously, the Co-mediated carbonylative Co-cyclization of an alkyne and an alkene, is a very powerful procedure in the preparation of cyclopentenones [268]. However, depending on the reaction conditions it also allows the preparation of 1,3-dienes, which may be intercepted by a Diels–Alder reaction, as described by Carretero and coworkers [285]. As expected, reaction of **6/4-28** with $Co_2(CO)_8$ in refluxing acetonitrile led exclusively to the diastereomeric cyclopentenones **6/4-29** and **6/4-30** as a 59:41 mixture. However, using trimethylamine-*N*-

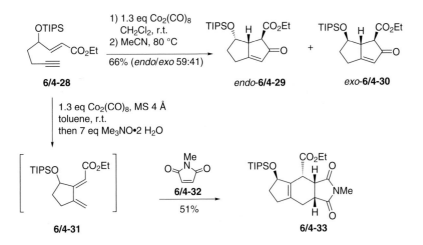

Gas	Temp [°C]	Yield [%]	
		6/4-26	6/4-27
CO (50 psi)	80	74	–
CO (1 atm)	75	38	32
N₂ (1 atm)	80	3	75

Scheme 6/4.9. Synthesis of δ-lactones.

Scheme 6/4.10. Co-mediated reaction of ene-ynes under different conditions leading either to cyclopentenones or 1,3-butadienes.

oxide as promoter in toluene **6/4-28** led to the 1,3-diene **6/4-31**, which at higher temperature in the presence of a dienophile such as **6/4-32** gave the cycloadduct **6/4-33** in 51% yield (Scheme 6/4.10).

A combination of Co-mediated amino-carbonylation and a Pauson–Khand reaction was described by Pericàs and colleagues [286], with the formation of five new bonds in a single operation. Reaction of 1-chloro-2-phenylacetylene **6/4-34** and dicobalt octacarbonyl gave the two cobalt complexes **6/4-36** and **6/4-37** *via* **6/4-35**, which were treated with an amine **6/4-38**. The final products of this domino process are azadi- and azatriquinanes **6/4-40** with **6/4-39** as an intermediate, which can also be isolated and separately transformed into **6/4-40** (Scheme 6/4.11).

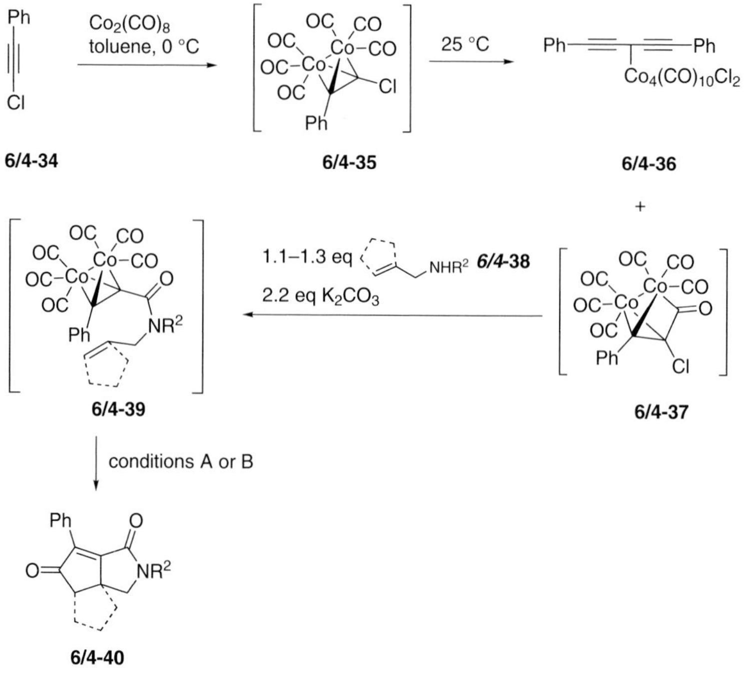

Entry	Allylamine	Conditions[b]	*dr*	Yield [%]	Yield [%]
1	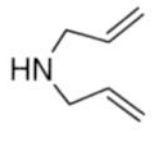 HN	A	–	–	48 (0)[a]
2	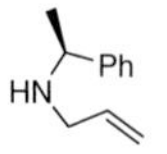 HN—Ph	A	1.2:1	89	55 (63)[a]
3	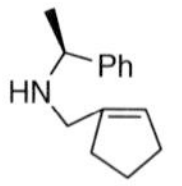 HN—Ph	B	1.4:1	84	(64)

[a] Values in parentheses: yield based on isolated **6/4-39**.
[b] A: 4 h, 45 °C. B: NMO, 20 min, 0 °C.

Scheme 6/4.11. Synthesis of azadi- and azatriquinanes.

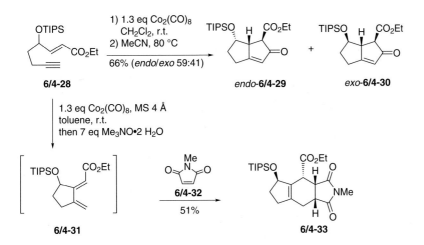

6/4-25a

6/4-26a

6/4-27

Gas	Temp [°C]	Yield [%]	
		6/4-26	6/4-27
CO (50 psi)	80	74	–
CO (1 atm)	75	38	32
N₂ (1 atm)	80	3	75

6/4-25b

6/4-26b (85%)

6/4-25c

6/4-26c (73%)

Scheme 6/4.9. Synthesis of δ-lactones.

6/4-28

endo-6/4-29

exo-6/4-30

66% (endo/exo 59:41)

1) 1.3 eq Co₂(CO)₈
CH₂Cl₂, r.t.
2) MeCN, 80 °C

1.3 eq Co₂(CO)₈, MS 4 Å
toluene, r.t.
then 7 eq Me₃NO•2 H₂O

6/4-31

6/4-32

6/4-33

51%

Scheme 6/4.10. Co-mediated reaction of ene-ynes under different conditions leading either to cyclopentenones or 1,3-butadienes.

oxide as promoter in toluene **6/4-28** led to the 1,3-diene **6/4-31**, which at higher temperature in the presence of a dienophile such as **6/4-32** gave the cycloadduct **6/4-33** in 51% yield (Scheme 6/4.10).

A combination of Co-mediated amino-carbonylation and a Pauson–Khand reaction was described by Pericàs and colleagues [286], with the formation of five new bonds in a single operation. Reaction of 1-chloro-2-phenylacetylene **6/4-34** and dicobalt octacarbonyl gave the two cobalt complexes **6/4-36** and **6/4-37** *via* **6/4-35**, which were treated with an amine **6/4-38**. The final products of this domino process are azadi- and azatriquinanes **6/4-40** with **6/4-39** as an intermediate, which can also be isolated and separately transformed into **6/4-40** (Scheme 6/4.11).

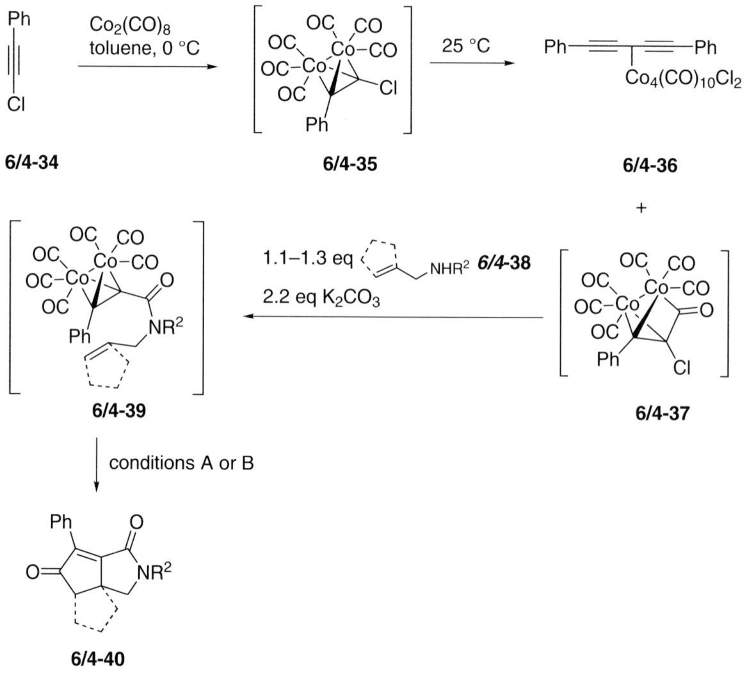

Entry	Allylamine	Conditions[b]	dr	Yield [%]	Yield [%]
1	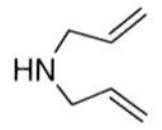 HN	A	–	–	48 (0)[a]
2	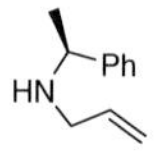 HN—Ph	A	1.2:1	89	55 (63)[a]
3	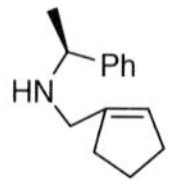 HN—Ph	B	1.4:1	84	(64)

[a] Values in parentheses: yield based on isolated **6/4-39**.
[b] A: 4 h, 45 °C. B: NMO, 20 min, 0 °C.

Scheme 6/4.11. Synthesis of azadi- and azatriquinanes.

6.4.2
Nickel-Induced Transformations

Most studies on nickel-catalyzed domino reactions have been performed by Ikeda and colleagues [287], who observed that alkenyl nickel species, obtained from alkynes **6/4-41** and a (π-allyl)nickel complex, can react with organometallics as **6/4-42**. If this reaction is carried out in the presence of enones **6/4-43** and TMSCl, then coupling products such as **6/4-44** are obtained. After hydrolysis, substituted ketones **6/4-45** are obtained (Scheme 6/4.12). With cyclic and β-substituted enones the use of pyridine is essential. Usually, the regioselectivity and stereoselectivity of the reactions is very high. On occasion, alkenes can be used instead of alkynes, though this is rather restricted as only norbornene gave reasonable results [288].

The addition of a Ni-catalyst to an alkyne as **6/4-47** and a cyclic enone as **6/4-46** in the presence of the chiral ligand **6/4-49** followed by a methyl transfer using Me$_2$Zn led to 3-substituted cycloalkanones as **6/4-48** in good to medium yield, with an *ee-*

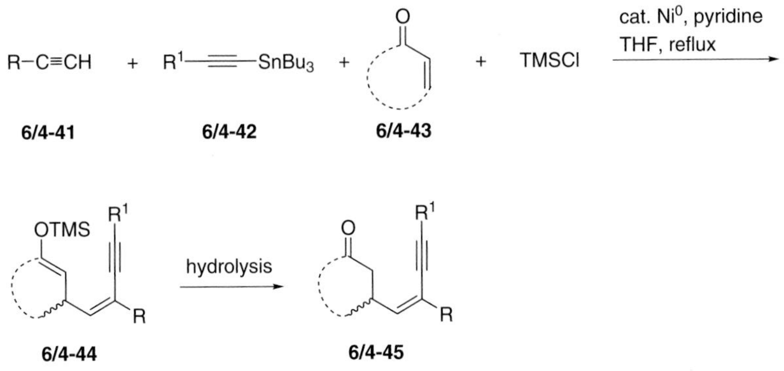

Scheme 6/4.12. Four component Ni^0-catalyzed reaction of alkynes, organometallics, enones and TMSCl.

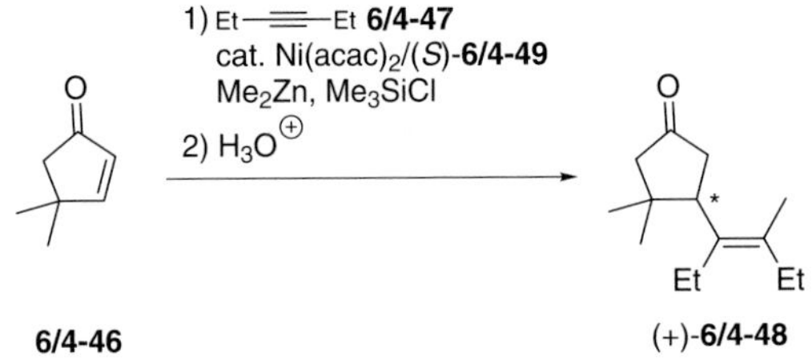

Solvent	*ee* [%]	Yield [%]
DME	59	73
diglyme	63	68
triglyme	81	78
tetraglyme	63	75

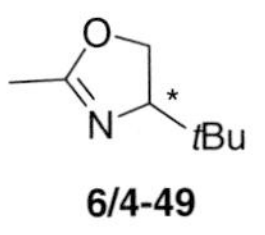

Scheme 6/4.13. Synthesis of 3-substituted cycloalkanones.

value of up to 81% (Scheme 6/4.13). The enantioselectivity is highly solvent-dependent; moreover, cyclopentenones gave better results than cyclohexenones [289].

A further development is the reaction of enones with ene-ynes in the presence of $ZnCl_2$ [290]. Reaction of **6/4-50a** with 2 equiv. of **6/4-51a** in the presence of equimolar amounts of $Ni(cod)_2$ and 1.5 mol% of $ZnCl_2$ in acetonitrile at room temperature for 6 h led to **6/4-52aa** with an (*E*)-configuration in 50% yield and 96% selectivity. In this transformation, **6/4-53aa** is formed as a minor isomer. Nickelacycles **6/4-54** or **6/4-55** can be assumed as a first intermediate (Scheme 6/4.14).

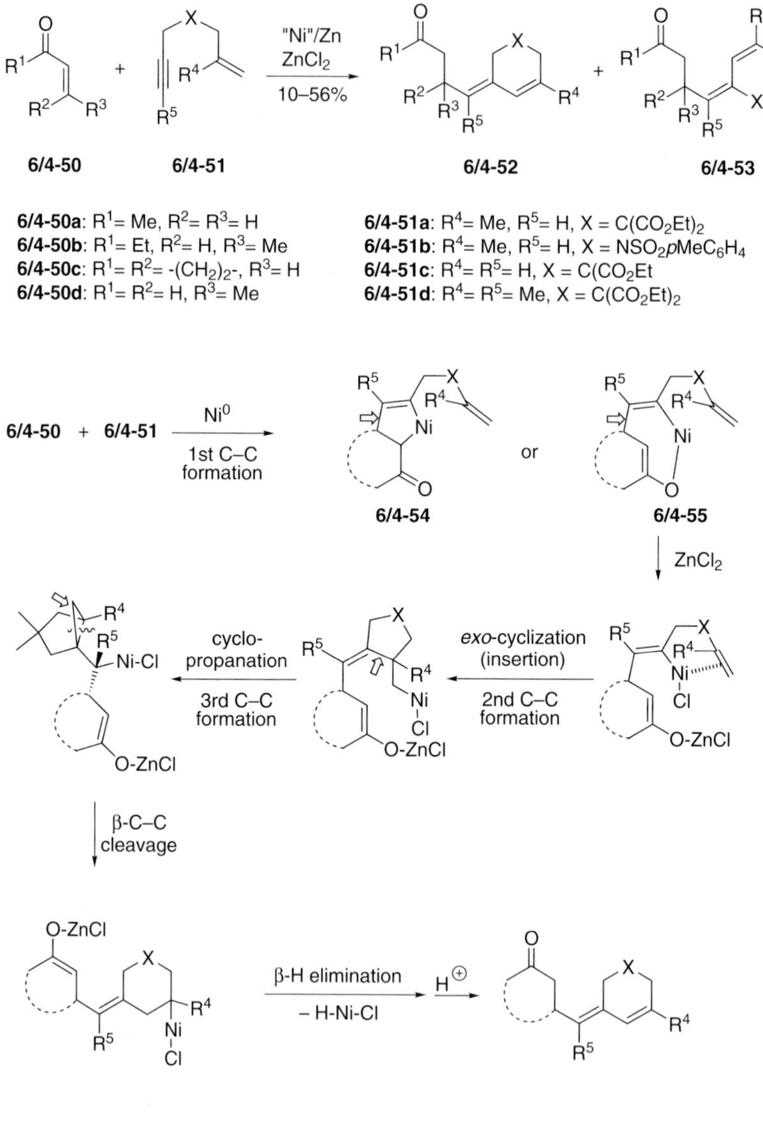
"Ni"/Zn
ZnCl2
10–56%
6/4-50
6/4-51
6/4-52
6/4-53
6/4-50a: R1= Me, R2= R3= H
6/4-50b: R1= Et, R2= H, R3= Me
6/4-50c: R1= R2= -(CH2)2-, R3= H
6/4-50d: R1= R2= H, R3= Me
6/4-51a: R4= Me, R5= H, X = C(CO2Et)2
6/4-51b: R4= Me, R5= H, X = NSO2pMeC6H4
6/4-51c: R4= R5= H, X = C(CO2Et
6/4-51d: R4= R5= Me, X = C(CO2Et)2
6/4-50 + 6/4-51
Ni0
1st C–C formation
or
6/4-54
6/4-55
ZnCl2
exo-cyclization (insertion)
2nd C–C formation
cyclo-propanation
3rd C–C formation
Ni-Cl
O-ZnCl
β-C–C cleavage
β-H elimination
– H-Ni-Cl
H⊕

Scheme 6/4.14. Nickel(0)-catalyzed reaction of enones and ene-ynes.

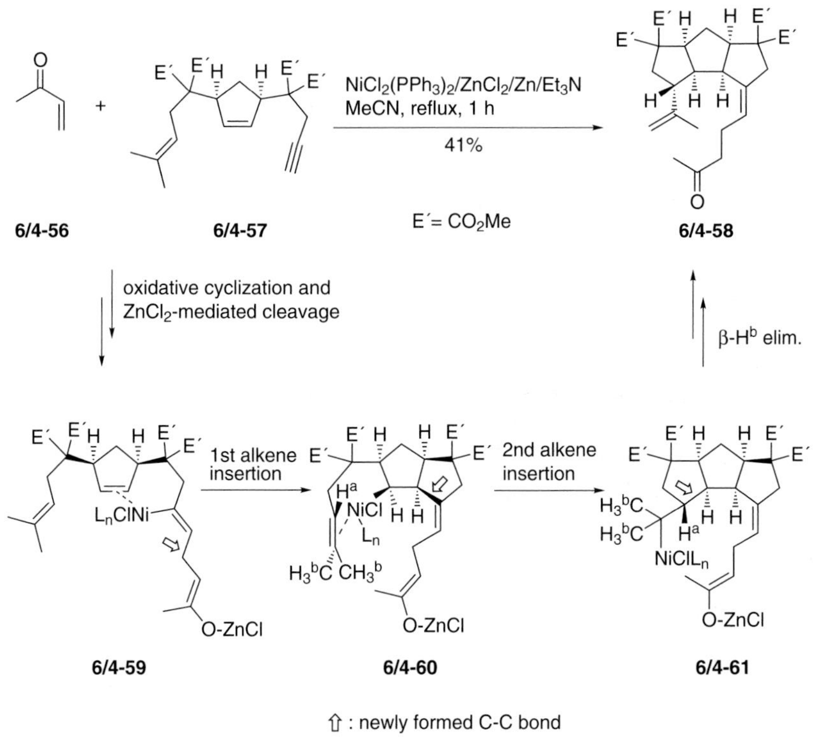

Scheme 6/4.15. Nickel-mediated formation of the triquinane skeletons.

When using this domino coupling, polycyclic compounds can also be obtained [291]. An additional prerequisite in this reaction, however, is inhibition of a premature β-hydrogen elimination. Reaction of **6/4-56** and **6/4-57** led to **6/4-58** with 41 % yield. Again, one can assume that first a Ni-complex **6/4-59** is formed, which gives the bicyclic **6/4-60** followed by formation of the triquinane skeleton **6/4-58** *via* **6/4-61** with a β-hydride elimination being the last step (Scheme 6/4.15).

Another approach to the rapid synthesis of complex polycycles from simple acyclic compounds is *via* the Ni(COD)$_2$-catalyzed reaction of an enone containing an alkyne moiety to give a carbo- or heterocyclic intermediate, which is quenched with an electrophile. A typical transformation as shown by Montgomery and his group [292] is the reaction of **6/4-62** with Ni(COD)$_2$, followed by addition of benzaldehyde to give **6/4-66** in 82 % yield as a single diastereomer (Scheme 6/4.16). The Ni-complexes **6/4-63**–**6/4-65** can be assumed as intermediates, while iodomethane, allyliodide, benzyliodide, formaldehyde, benzoylchloride and even α,β-unsaturated aldehydes such as acrolein can also be used as electrophiles. The latter compound undergoes a 1,4-addition. Moreover, the reaction can also be performed in an intramolecular mode; reaction of **6/4-67** gave **6/4-68** in 61 % yield.

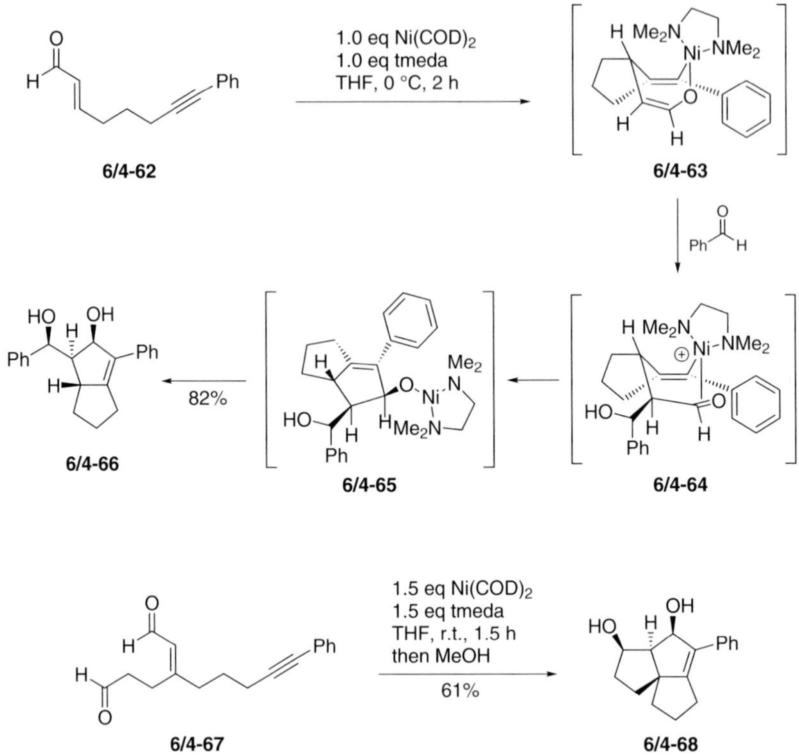

Scheme 6/4.16. Nickel(0)-mediated synthesis of polycycles.

The nickel-catalyzed reaction of arylhalides is a very appropriate method for the preparation of even highly hindered biaryls (Ullmann-type coupling). Reaction of **6/4-69** in the presence of $NiCl_2(PPh_3)_2/PPh_3$ led to the expected homo-coupled product. A different course however, was observed by Lin and coworkers [293] in the presence of the bidendate ligands dppf or BINAP, where the two phthalides **6/4-70** and **6/4-71** were obtained (Scheme 6/4.17). Using BINAP, an excellent facial selectivity was also found, especially in the transformation of **6/4-69b** to give **6/4-70b** in 71% yield and 98% *ee*. A methoxy group in para-position to the halide is necessary for high yields, where as a methoxy group in *meta*-position reduces the enantioselectivity. With regard to the mechanism, it can be assumed that first complex B is formed, which undergoes an insertion into the aldehyde moiety of a second molecule of **6/4-69**, with formation of complex D; this then leads to the products. Transmetalation of the arylnickel complex B to give complex C appears to be retarded using a bidentate ligand.

Phthalide derivatives are of pharmacological interest, as they show antitumor [294] and anticonvulsant [295] as well as antimicrobial activity [296]; they have also been found to have useful herbicidal activity [297].

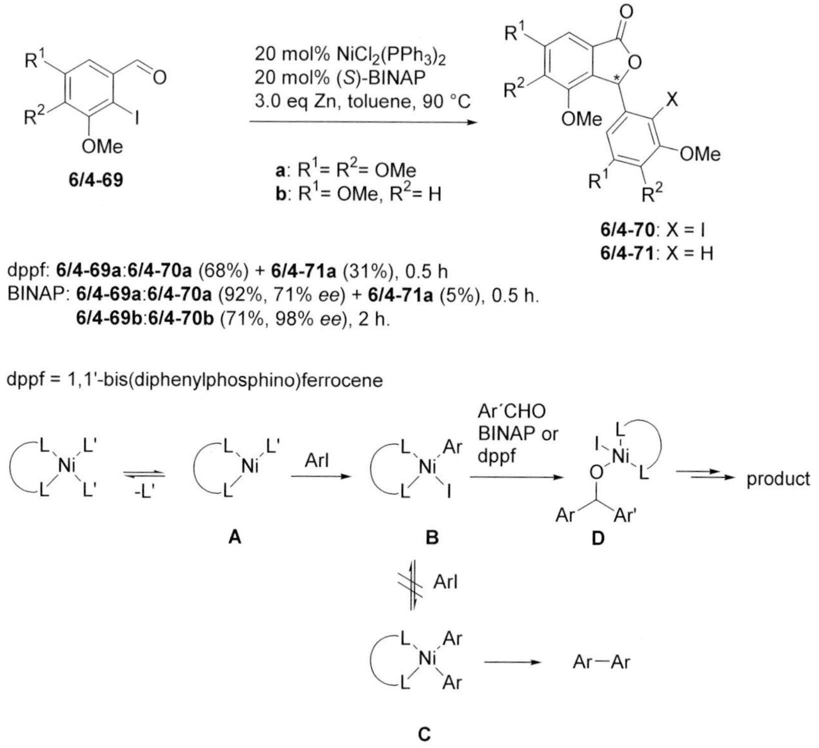
6/4-69
20 mol% $NiCl_2(PPh_3)_2$
20 mol% (S)-BINAP
3.0 eq Zn, toluene, 90 °C
a: R^1= R^2= OMe
b: R^1= OMe, R^2= H
6/4-70: X = I
6/4-71: X = H
dppf: 6/4-69a:6/4-70a (68%) + 6/4-71a (31%), 0.5 h
BINAP: 6/4-69a:6/4-70a (92%, 71% ee) + 6/4-71a (5%), 0.5 h.
6/4-69b:6/4-70b (71%, 98% ee), 2 h.
dppf = 1,1'-bis(diphenylphosphino)ferrocene
Ar´CHO
BINAP or
dppf
product
A
B
D
ArI
Ar—Ar
C

Scheme 6/4.17. Ni-catalyzed reaction of **6/4-69**.

A rather unusual combination of a Ni-catalyzed [2+2+2] cycloaddition of oxabenzonorbornadiene **6/4-72** with an alkyne **6/4-73** followed by a retro-electrocyclization to give an arene **6/4-75** and benzoisofuran **6/4-76** was described by Cheng and coworkers [298]. Under the reaction conditions, **6/4-76** reacted with another molecule of **6/4-72** to give **6/4-77** and **6/4-78** (Scheme 6/4.18). The best yields were obtained employing phenylacetylene with 98 % overall yield (58 % of **6/4-78** and 40 % of **6/4-77**). At a lower temperature (18 °C), intermediates of type **6/4-74** could be isolated. On occasion, azabenzonorbornadienes may also be used instead of **6/4-72**.

A $NiCl_2/CrCl_2$-mediated domino process of the chiral aldehyde **6/4-81** iodotriene **6/4-82** was used for the synthesis of (–)-ircinianin (**6/4-79**) and (+)-wistarin (**6/4-80**), as described by Uenishi and colleagues (Scheme 6/4.19) [299]. After the coupling to give **6/4-83**, an intramolecular Diels–Alder reaction occurred at room temperature, leading to the desired cyclic product **6/4-84** in 60 % yield.

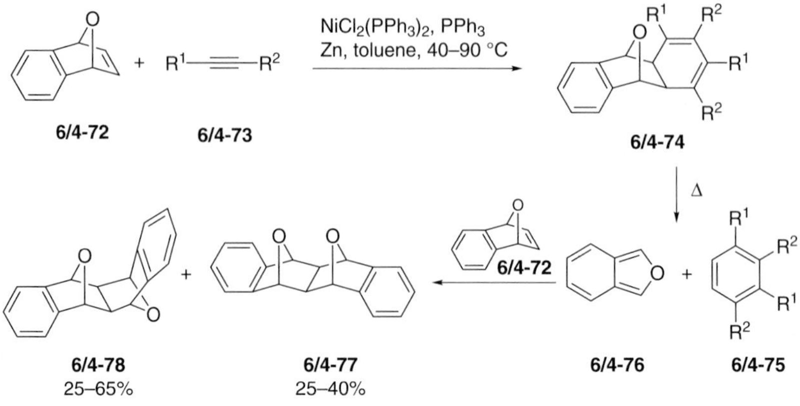

Scheme 6/4.18. Nickel-catalyzed reaction of oxabenzonornadiene with alkynes.

6.4.3
Copper-Induced Reactions

Cu-catalyzed domino reactions have been used for the synthesis of carbocycles, as well as for heterocycles such as indoles, benzoxazoles, and quinoxalines. A very useful process is also the combination of the formation of allyl vinyl ethers, followed by a Claisen rearrangement.

For the enantioselective allylic substitution with soft nucleophiles, Pd-, Mo- or Ir-transition metal catalysis using chiral ligands is most appropriate. However, for hard nucleophiles as alkyl groups, Cu is better suited. Alexakis and coworkers [300] developed a new chiral phosphoramidite ligand which allows the introduction of an alkyl group to cinnamylchloride with 92–96 % *ee*. If the employed Grignard reagent contains a C–C-double bond, a RCM can follow. Thus, reaction of **6/4-85** and the Grignard reagent **6/4-86** in the presence of copper thiophene-2-carboxylate (CuTC) and the chiral ligand **6/4-88**, followed by metathesis, gave **6/4-87** in good yields (Scheme 6/4.20).

A general method for the synthesis of highly substituted styrenes as **6/4-91**, vinyl-cyclohexadienes and related compounds was developed by Xi, Takahashi and coworkers [301] by reacting an intermediately formed five-membered zirconacycle **6/4-89** with propargyl derivatives **6/4-90** or allyl bishalides in the presence of CuCl (Scheme 6/4.21).

The formation of the indole moiety has found immense attention, since it exists in many bioactive compounds such as the indole alkaloids [302]. Whilst the Fischer indole synthesis remains the most important procedure, during the past few years several transition metal-catalyzed syntheses have been developed. Recently, a Cu^{II}-catalyzed cyclization of anilines containing an ortho-alkynyl group was published by Hiroya and coworkers [303], which allows a double cyclization in domino fashion to provide annulated indoles. Thus, reaction of **6/4-92** in the presence of

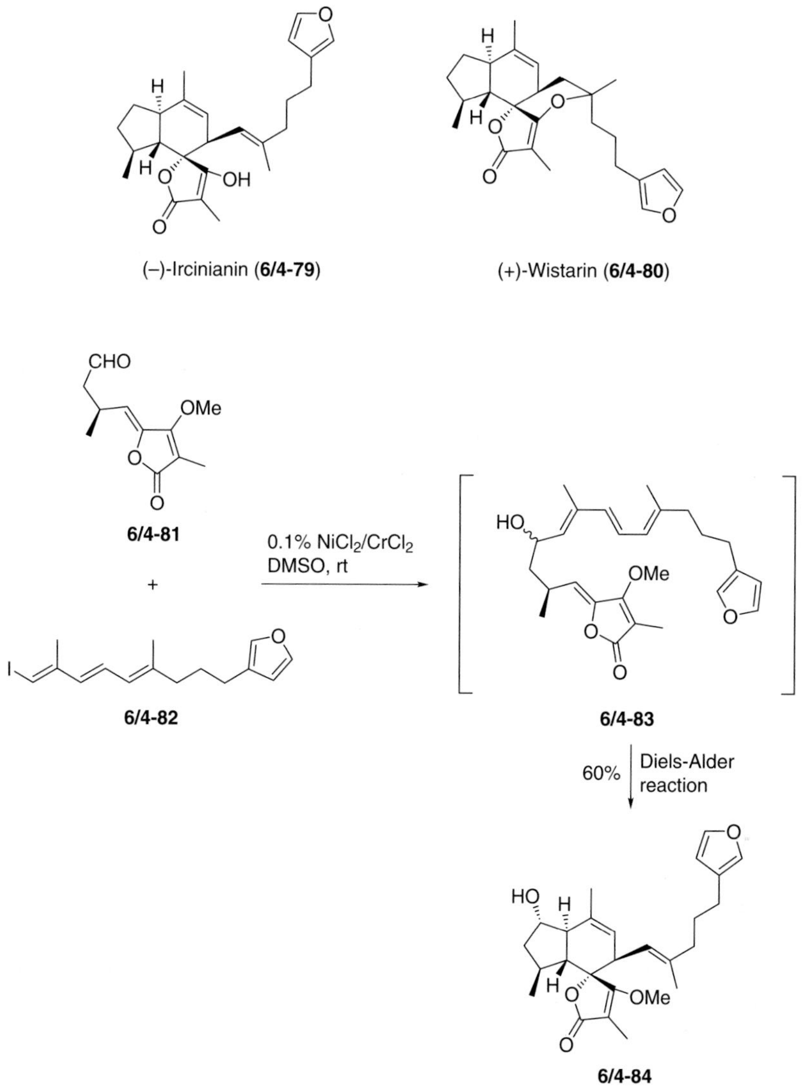

H
OH
(–)-Ircinianin (6/4-79)
(+)-Wistarin (6/4-80)
CHO
OMe
6/4-81
+
I
6/4-82
0.1% NiCl2/CrCl2
DMSO, rt
HO
OMe
6/4-83
60%
Diels-Alder
reaction
HO
OMe
6/4-84

Scheme 6/4.19. Synthesis of (–)-ircinianin (**6/4-79**) and (+)-wistarin (**6/4-80**).

50 mol% of $Cu(OAc)_2$ after formation of the amide anion using KH led to **6/4-93** in reasonable yield; in some cases, a small amount of the monoalkylated compound **6/4-94** is also found (Scheme 6/4.22).

Glorius and colleagues [304] described a useful Cu^{I}-catalyzed preparation of 2-phenyl benzoxazoles **6/4-97** and related compounds from 1,2-dihaloaromatic substrates **6/4-95** and benzamide **6/4-96** (Scheme 6/4.23).

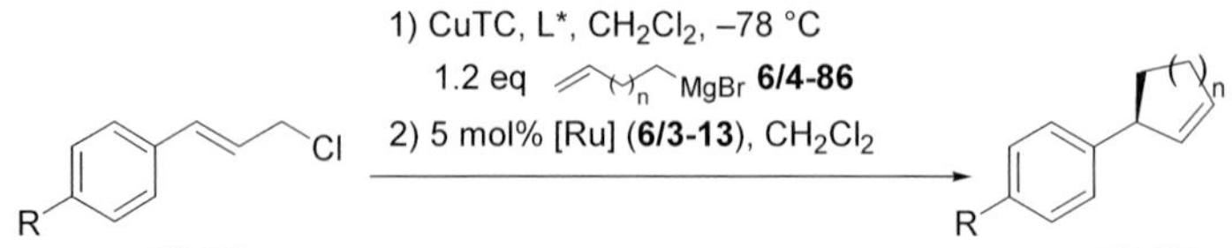
1) CuTC, L*, CH_2Cl_2, –78 °C
1.2 eq MgBr 6/4-86
2) 5 mol% [Ru] (6/3-13), CH_2Cl_2
Cl
R
n

6/4-85 6/4-87

L* =

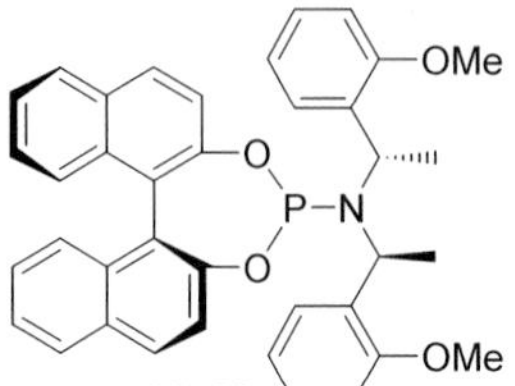
OMe
O
P–N
O
OMe

6/4-88

CuTC = copper thiophene-2-carboxylate

Entry	R	n	Isolated yield [%]	*ee* [%]
1	H	1	77	94
2	Me	1	79	93
3	H	2	69	96
4	Me	2	72	94

Scheme 6/4.20. Combination of a Cu-catalyzed allylic substitution with a metathesis.

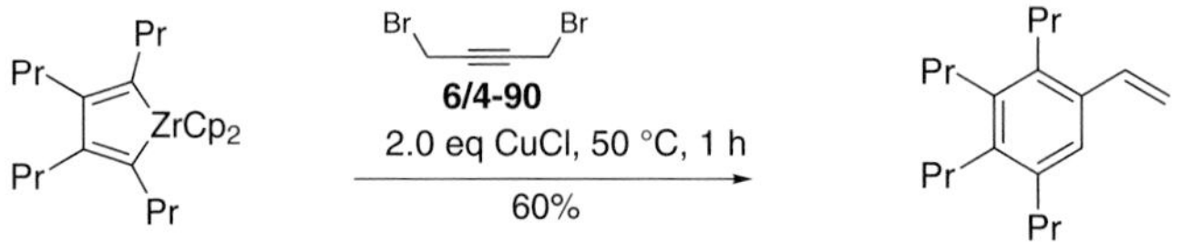
Pr
Pr
ZrCp2
Pr
Pr
Br
Br
6/4-90
2.0 eq CuCl, 50 °C, 1 h
60%
Pr
Pr
Pr
Pr

6/4-89 6/4-91

Scheme 6/4.21. Synthesis of styrenes.

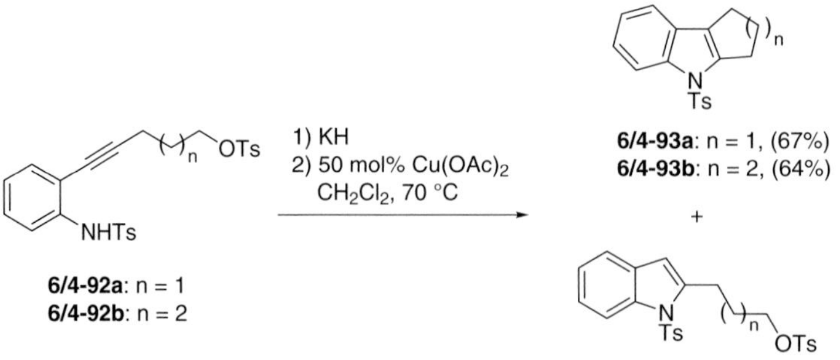
OTs
n
NHTs
1) KH
2) 50 mol% Cu(OAc)2
CH2Cl2, 70 °C
N
Ts
n
+
N
Ts
n
OTs

6/4-92a: n = 1
6/4-92b: n = 2

6/4-93a: n = 1, (67%)
6/4-93b: n = 2, (64%)

6/4-94a: n = 1, (0%)
6/4-94b: n = 2, (14%)

Scheme 6/4.22. Synthesis of indoles.

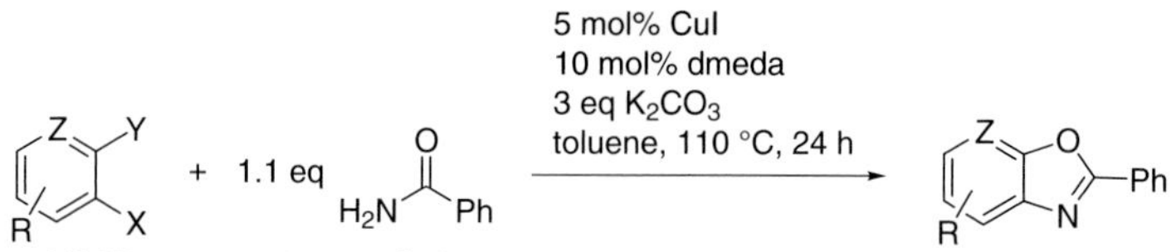

Scheme 6/4.23. Synthesis of benzoxazoles and related compounds.

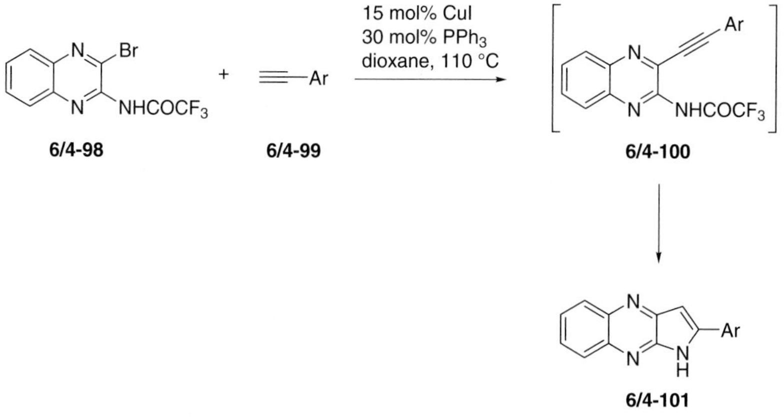

Entry	o-Dihalo substrate	Product	Isolated yield [%]
1	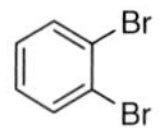	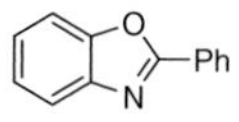	90
2	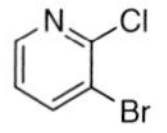	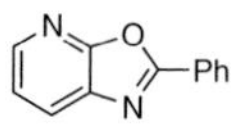	77
3	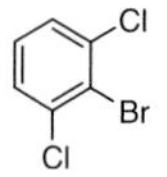	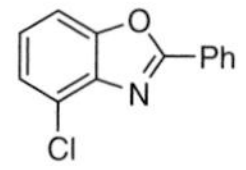	75
4	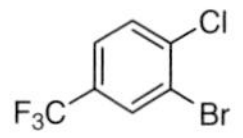	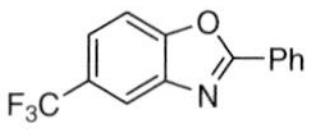	59
5	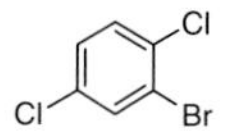	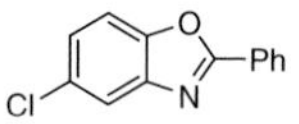	72

Scheme 6/4.24. Synthesis of pyrrolo[2,3-*b*]quinoxalines.

A Cu-catalyzed domino reaction has been developed by Cacchi and his group [305] for the synthesis of pyrrolo[2,3-*b*]quinoxalines **6/4-101** using a 2-bromo-3-tri-fluoroacetamido quinoxaline **6/4-98** and a tolane **6/4-99** containing electron-donating or -withdrawing groups as substrates, and with **6/4-100** as a proposed intermediate (Scheme 6/4.24).

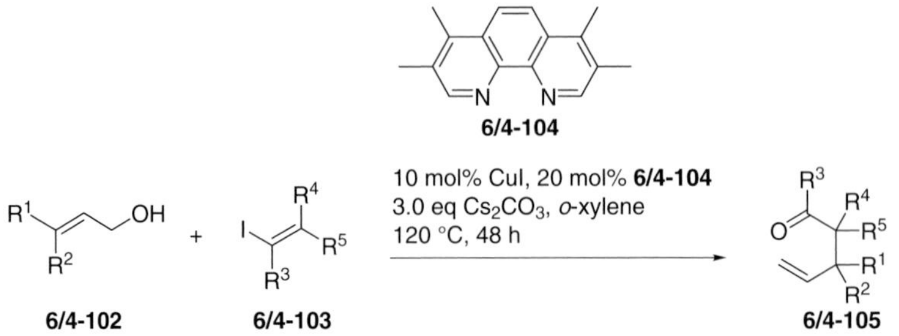

6/4-104

$$R^1 \diagdown OH \quad (R^2) \quad + \quad I \diagdown R^5 \ (R^3, R^4) \xrightarrow[\substack{3.0 \text{ eq } Cs_2CO_3, \text{ o-xylene} \\ 120\ °C,\ 48\ h}]{10\ \text{mol}\%\ CuI,\ 20\ \text{mol}\%\ \mathbf{6/4\text{-}104}} \quad \mathbf{6/4\text{-}105}$$

6/4-102   **6/4-103**   **6/4-105**

Entry	Alcohol (6/4-102)	Vinyl halide (6/4-103)	Product (6/4-105)	dr[a]	Yield [%]
1	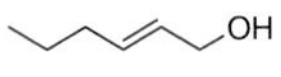	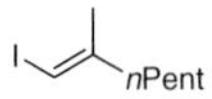	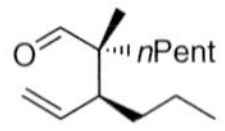	92:8	55
2	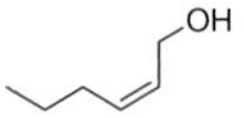	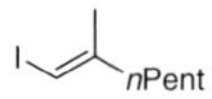	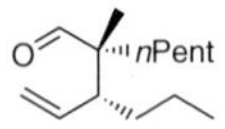	5:95	68
3	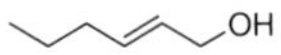	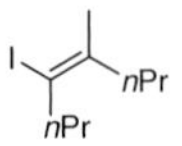	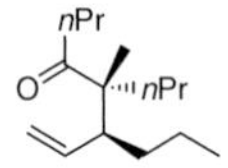	88:12[b]	59
4	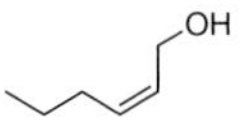	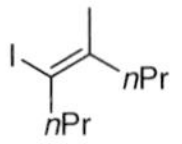	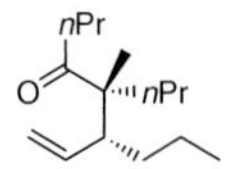	9:91[b]	–
5	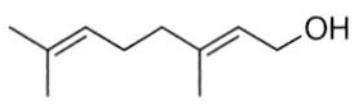	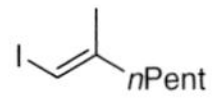	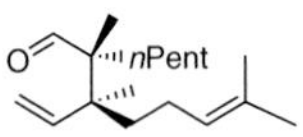	92:8	77
6	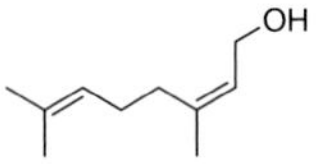	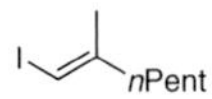	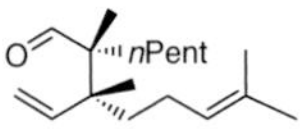	6:94	77
7	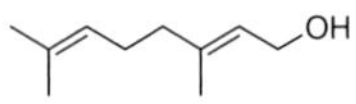	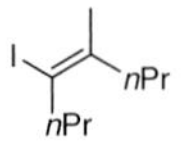	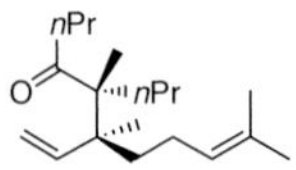	87:13	62
8	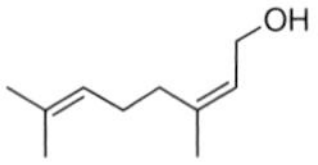	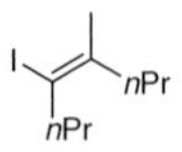	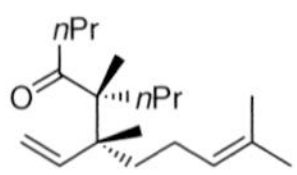	12:88	69

[a]Based on 300 or 500 MHz ^{1}H-NMR spectra of reaction mixture.
[b]Based on GC of the crude reaction mixture.

Scheme 6/4.25. Domino allyl yinyl ether formation/Claisen rearrangement.

As shown earlier in many examples, the Claisen rearrangement of allyl vinyl ethers also provides a very powerful method for carbon–carbon bond formation in domino processes. Usually, the necessary ethers are formed in a separate step. However, both steps can be combined in a novel domino reaction developed by Buchwald and Nordmann [306]. This starts from an allylic alcohol **6/4-102** and a vinyl iodide **6/4-103**, using copper iodide in the presence of the ligand **6/4-104** at 120 °C to give **6/4-105** (Scheme 6/4.25). The reaction even allows the stereoselective formation of two adjacent quaternary stereogenic centers in high yield.

6.4.4
Tungsten-Induced Reactions

Domino reactions involving tungsten are based throughout on tungsten-carbenes, which can undergo cyclization, followed by another cyclization. Barluenga and co-workers [307] prepared fluorene derivatives of type **6/4-108** using this procedure, in which a [4+2] cycloaddition of a 2-amino-1,3-butadiene **6/4-106** with a Fischer alkynyl carbene complex **6/4-107** is followed by a cyclopentannulation (Scheme 6/4.26). The same authors also performed a double [4+2] cycloaddition/cyclopentannulation.

A combination of cyclizations and a cycloaddition was observed by Aumann and coworkers [308] by reaction of a cycloalkenyl-ethynyl tungsten-carbene **6/4-109** with an alcohol to give **6/4-112** (Scheme 6/4.27). The last step in the sequence is a π-cyclization with insertion of CO of the intermediate cycloadduct **6/4-111** formed from **6/4-110** and **6/4-109**.

When the o-ethynylphenyl isopropyl ketone **6/4-113** was treated with 1 equiv. of $W(CO)_5 \cdot THF$ for 2 h in the presence of 4 equiv. of 1,1-diethoxyethylene, the bridged compound **6/4-118** was obtained in 73 % yield, and not the expected 3-ethoxy-1-isopropylnaphthalene (Scheme 6/4.28). Iwasawa and coworkers [309] proposed a mechanism for their new domino reaction, which includes a [3+2] cycloaddition followed by insertion of the resulting tungsten carbene moiety into the neighboring C–H bond, with **6/4-114**–**6/4-117** as possible intermediates. These authors have further shown that 10–20 mol% of $W(CO)_5 \cdot THF$ is sufficient, and

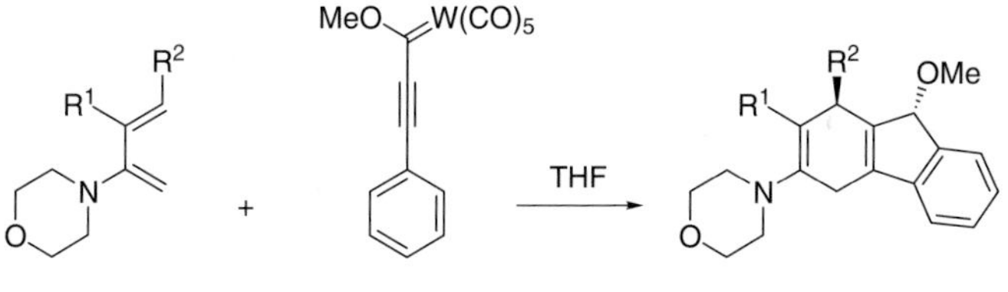

6/4-106   **6/4-107**

6/4-108a: R^1= Me, R^2= H: 95% yield
6/4-108b: R^1= Me, R^2= CH_2OMe: 95% yield
6/4-108c: R^1, R^2= $CH_2(CH_2)_2CH_3$: 92% yield

Scheme 6/4.26. Synthesis of fluorene derivatives.

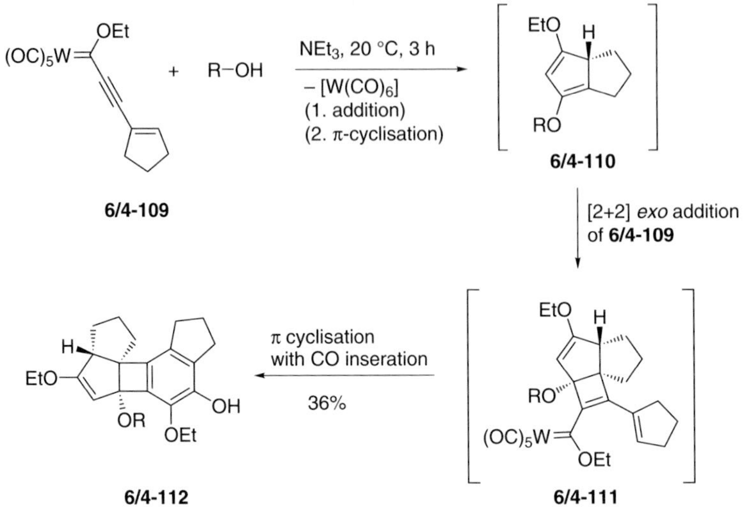

Scheme 6/4.27. Reaction of a cycloalkenylethenyl carbene.

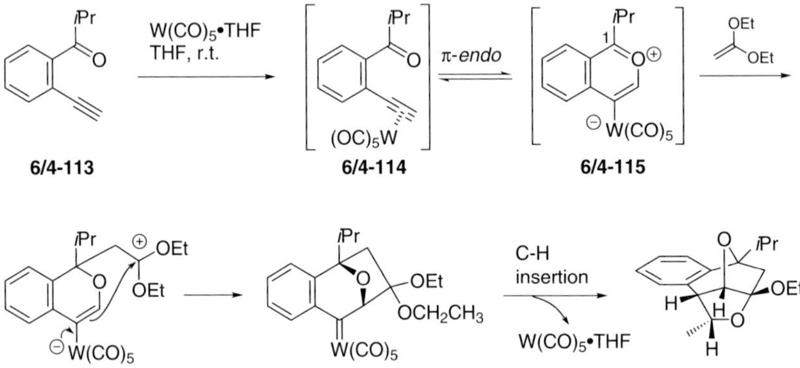

Scheme 6/4.28. Synthesis of bridged compounds from *o*-ethenylphenyl isopropyl ketone.

that other ketones, ketene acetals and even vinyl ethers can be used with 50 to 94 % yield.

6.4.5
Molybdenum-Induced Reactions

As discussed earlier in detail, metathesis reactions have been used *in extensio* in domino processes. Almost all published examples employ Grubbs I and II catalysts

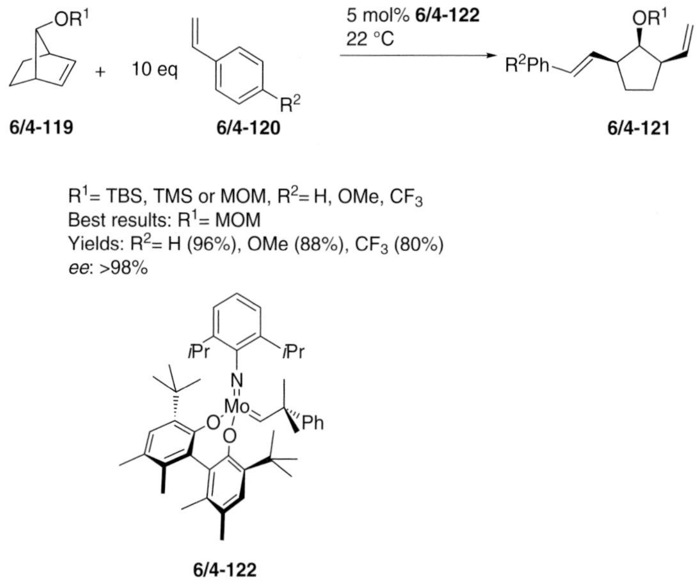

6/4-119 + 10 eq 6/4-120 → 6/4-121

5 mol% **6/4-122**
22 °C

R^1= TBS, TMS or MOM, R^2= H, OMe, CF$_3$
Best results: R^1= MOM
Yields: R^2= H (96%), OMe (88%), CF$_3$ (80%)
ee: >98%

6/4-122

Scheme 6/4.29. Enantioselective domino ring-opening/ring-closing metathesis with a Mo-catalyst.

(**6/3-13-6/3-16**) due to their higher stability and better handling. However, one example of an enantioselective domino process using a chiral Schrock type Mo-complex has been reported by Hoveyda and colleagues [310]. Domino ROM/RCM of the norbornene derivatives **6/4-119** and styrenes **6/4-120** in the presence of catalytic amounts of the biphen-based Mo-complex **6/4-122** gave the cyclopentanes **6/4-121** in high yield and enantioselectivity (Scheme 6/4.29). The best results were obtained using a MOM-protecting group, with up to 96 % yield and over 98 % *ee*.

Usually, the goal of domino reactions is an increase of complexity of the products in comparison to the substrates. However, a domino-type cleavage of bonds might also be advantageous if the starting material is easily accessible and the product difficult to obtain by other methods.

Tam and coworkers [311] developed a method for the synthesis of 1,3-disubstituted cyclopentanes **6/4-124** and cyclopentenes. Thus, reaction of the condensed isoxazolines **6/4-123**, easily obtainable by a 1,3-dipolar cycloaddition, gave **6/4-124** in good yields using Mo(CO)$_6$ (Scheme 6/4.30).

6.4.6
Titanium-Induced Reactions

Beller and coworkers [312] have recently published a very useful domino reaction using a Ti-based Lewis acid as catalyst to prepare indoles. The new procedure consists of a titanium-catalyzed amination of a chloroalkylalkyne to give an aryl hydra-

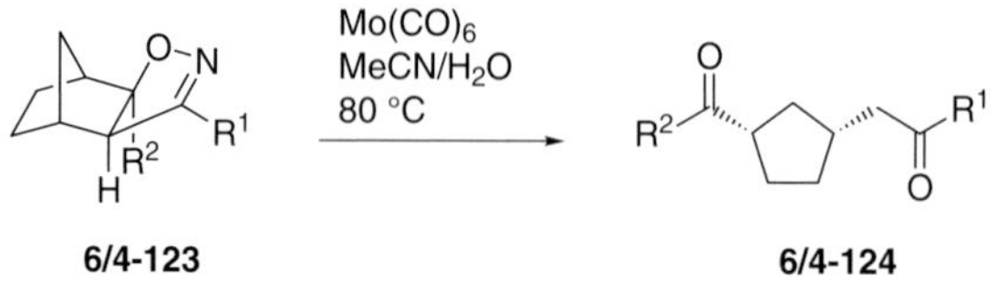

6/4-123 6/4-124

Entry	R^1	R^2	Yield [%]
1	Ph	nC_6H_{13}	75
2	Me	nC_6H_{13}	82
3	Ph	$SiMe_3$	65
4	Me	$SiMe_3$	84

Scheme 6/4.30. Domino ring-cleavage reactions.

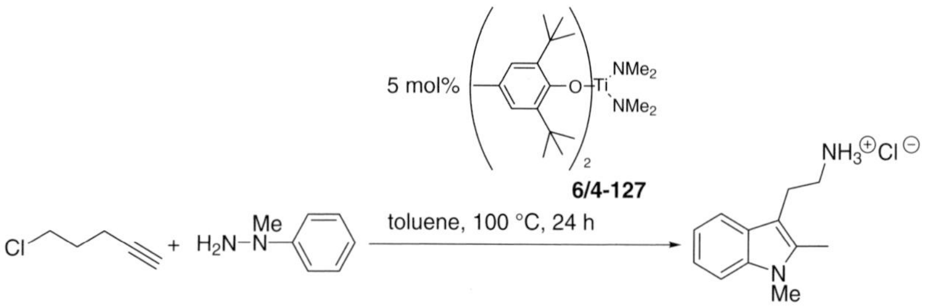

6/4-125 6/4-126 6/4-128

Scheme 6/4.31. Domino process to tryptamines.

zone; this then undergoes a Fischer-type indole formation, followed by a nucleophilic substitution of the alkyl chloride, with ammonia being liberated in the reaction. A multitude of different tryptamine derivatives and analogous have been prepared in this way. As an example, the reaction of **6/4-125** and **6/4-126** in the presence of catalyst **6/4-127** led to **6/4-128**, in 86% yield (Scheme 6/4.31).

Another Ti-catalyzed domino reaction has been developed by Rück-Braun and coworkers [313] using β-cyclopentadienyl (dicarbonyl)iron-substituted enals **6/4-129** and primary amines **6/4-130** to give dihydropyrrolones **6/4-131** (Scheme 6/4.32).

A methylenation of cyclic carbonates such as **6/4-132** using dimethyltitanocene to give a ketene acetal, followed by a subsequent Claisen rearrangement, allowed the synthesis of medium-ring lactones such as **6/4-133** in good yields; these are otherwise difficult to obtain. In this transformation, **6/4-133** is formed as a 1:1-mixture of the two atropisomers **6/4-133a** and **6/4-133b** (Scheme 6/4.33). The substrate

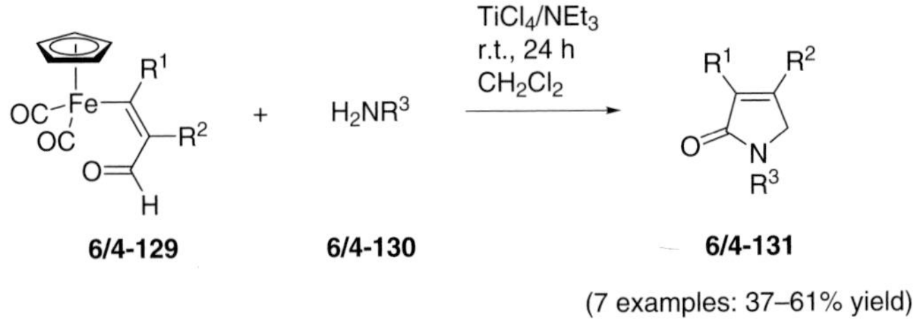

6/4-129   **6/4-130**   **6/4-131**

(7 examples: 37–61% yield)

Scheme 6/4.32. Synthesis of dihydropyrrolones.

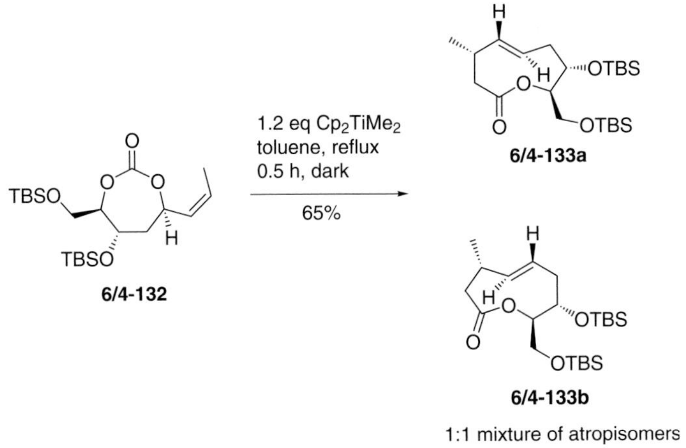

Scheme 6/4.33. Reaction of cyclic carbonates with Cp$_2$TiMe$_2$.

for this transformation, developed by Burton, Holmes and coworkers [314], was obtained from sugars.

6.4.7
Chromium-Induced Transformations

Rather complex structures are obtained by a novel chromium(0)-mediated three-component domino [6π+2π] cycloaddition described by Rigby and coworkers [315]. Irradiation of a mixture of the chromium complex **6/4-134** and the tethered diyne **6/4-135** with a Pyrex filter at 0 °C gave the polycyclic compounds **6/4-136** in medium to good yield (Scheme 6/4.34).

Another unprecedented domino cycloaddition process of a chromium complex, namely a [2+2+1]/[2+1] cycloaddition, was observed by Barluenga and coworkers [316]. These authors treated norbornene **6/4-137** with the Fischer alkynyl Cr carbene **6/4-138** and obtained, as the main product, not the expected cyclopropane derivative **6/4-139**, but compound **6/4-140** (Scheme 6/4.35).

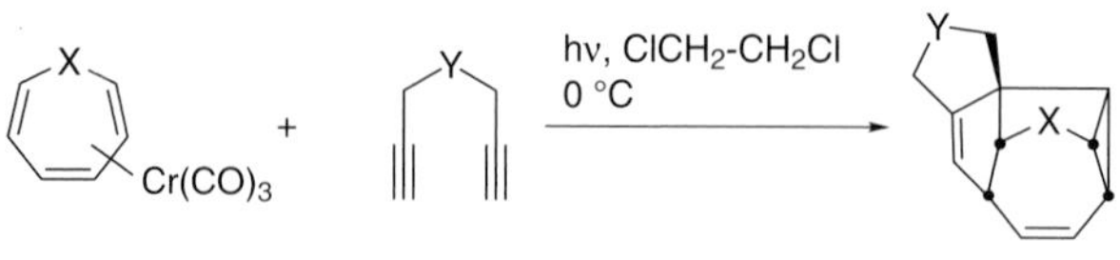

Scheme 6/4.34. The domino [6π+2π], [6π+2π] cycloaddition process.

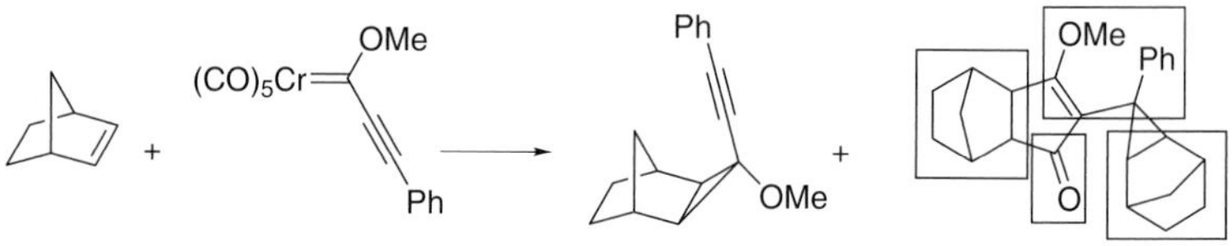

Entry	X	Y	Run time [h]	Yield [%]
1	CH$_2$	(CH$_2$)$_2$	1.5	76
2	CH$_2$	(CH$_2$)$_3$	2	69
3	CH$_2$	O	1.5	47
4	CH$_2$	C(CO$_2$Et)$_2$	1.5	77
5	CH$_2$	C(CH$_3$)$_2$	1	36
6	CH$_2$	C(Ph)CO$_2$Et	1.5	53
7	N(CO$_2$Et)	CH$_2$	0.5	40
8	SO$_2$	CH$_2$	2	57
9	SO$_2$	(CH$_2$)$_2$	2	32

Scheme 6/4.35. Unexpected formation of **6/4-140** by reaction of norbornene and an alkynyl Cr carbene.

6.4.8
Platinum- and Gold-Induced Reactions

Fürstner and coworkers developed a new Pt- and Au-catalyzed cycloisomerization of hydroxylated enynes **6/4-141** to give the bicylo[3.1.0]hexanone skeleton **6/4-143**, which is found in a large number of terpenes [317]. It can be assumed that, in the case of the Pt-catalysis, a platinum carbene **6/4-142** is formed, which triggers an irreversible 1,2-hydrogen shift. The complexity of the product/substrate relationship can be increased by using a mixture of an alkynal and an allyl silane in the presence of PtCl$_2$ to give **6/4-143** directly, in 55 % yield (Scheme 6/4.36).

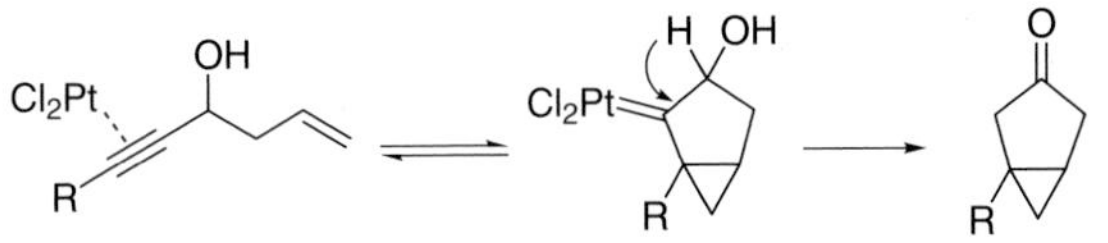

6/4-141 6/4-142 6/4-143

Substrate[a]	Product	Yield [%]
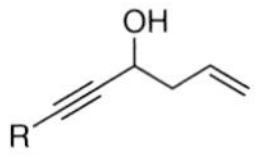	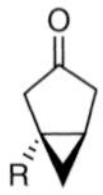	74/75[b] (R = Ph) 72 (R = C_5H_{11}) 81 (R = C_7H_{15})
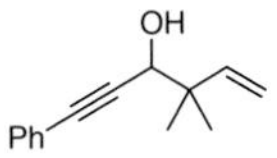	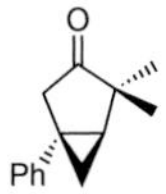	66
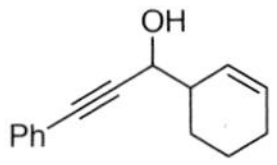	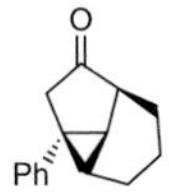	52
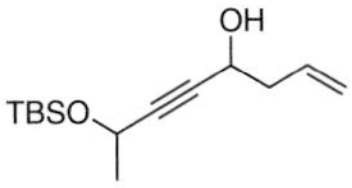	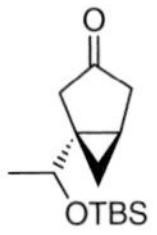	94[c]

[a] All reactions were performed in toluene at 60–80 °C with 5 mol% $PtCl_2$ unless stated otherwise. [b] Using 2 mol% $(PPh_3)AuCl/AgSbF_6$ in CH_2Cl_2 at 20 °C. [c] *dr* = 3:1.

Scheme 6/4.36. Pt- and Au-catalyzed cycloisomerization of hydroxylated ene-ynes.

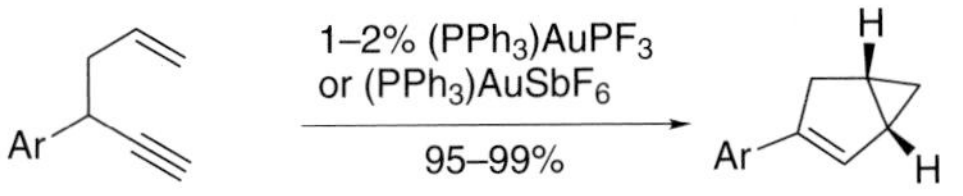

6/4-144 6/4-145

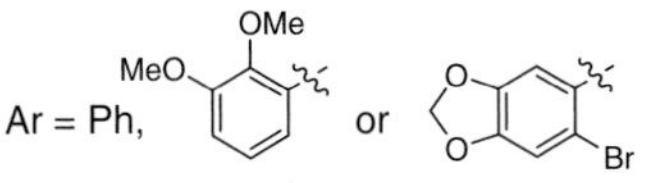

Scheme 6/4.37. Gold-catalyzed formation of bicyclohexenes from ene-ynes.

Another example of a catalytic isomerization of 1,5-ene-ynes as **6/4-144** to afford bicyclo[3.1.0]hexenes as **6/4-145** was described by Toste and coworkers (Scheme 6/4.37) [318]. In this reaction, Pd or Pt salts gave only <5 % yield; however, with Au salts the products could be obtained with excellent yield.

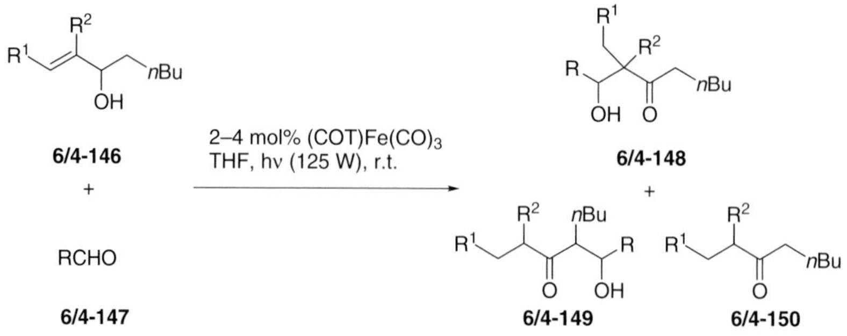

Scheme with structures 6/4-146 + RCHO (6/4-147) reacting under 2–4 mol% (COT)Fe(CO)₃, THF, hv (125 W), r.t. to give 6/4-148 + 6/4-149 and 6/4-150.

Entry	R	R¹	R²	Cat. [%]	Time [min]	Ratio: 6/4-148: syn/anti, 6/4-149: isomers	Yield [%] 6/4-148+ 6/4-149	Yield [%] 6/4-150
1	Ph	H	H	2	20	64/28/5/3	84	12
2	H	H	H	4	45–60	94/–/6	84	9
3	Me₂CHCH₂	H	H	3	35	63/27/5/5	72	19
4	Me₂CH	H	H	3	35	53/34/7/6	63	27
5	Cy	H	H	3	35	59/35/4/2	64	27
6	Et₂CH	H	H	3	35	65/33/2/–	36	42
7	Et₂CH=CH(CH₂)₂	H	H	4	180	55/36/5/4	78	6
8	pAcNHPh	H	H	5	180	57/36/4/3	54	4
9	Ph	Me	H	5	4 h		<20% conv.	
10	Ph	H	Me	5	4 h		<20% conv.	

Scheme 6/4.38. Domino isomerization/aldol reaction of allylic alcohols and aldehydes using (COT)Fe(CO)₃ as catalyst.

For the total synthesis of heliophenanthrone, the Dyker group has recently used a platinum-catalyzed cyclization of an *o*-alkynylbenzaldehyde leading to an isobenzopyrylium cation which underwent a subsequent intramolecular Diels–Alder reaction to give a diastereomeric mixture of the tricyclic structure [319].

A very new example for the combination of an Auᴵ-catalyzed [3,3]-rearrangement and a Nazarov reaction has been disclosed by Zhang and coworker. Thus, cyclopentenones could be easily achieved by converting en-ynyl acetates in the presence of AuCl(PPh₃)/AgSbF₆ [320].

6.4.9
Iron- and Zirconium-Induced Reactions

As shown in the proceeding sections, the aldol reaction – especially in a domino fashion – is one of the most suitable procedures for C–C-bond formation. Usually, one uses an enolate and a carbonyl moiety, but allylic alcohols can also be em-

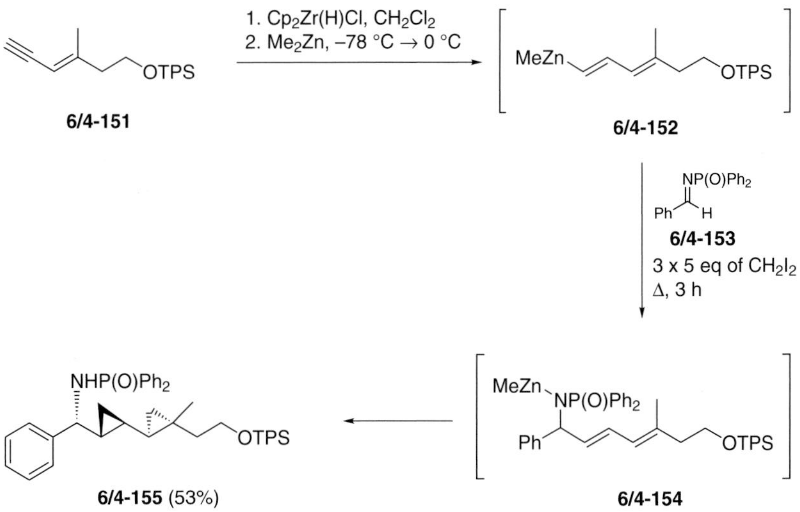

Scheme 6/4.39. Hydrozirconation of ene-ynes.

ployed, as these undergo an isomerization under transition metal catalysis [321]. Grée and coworkers [322] have shown that, for this purpose, iron complexes are most suitable. Thus, a mixture of the allylic alcohols **6/4-146** and the aldehydes **6/4-147**, under irradiation and in the presence of 2–4 mol% of [(COT)Fe(CO)₃], gave the aldol adducts **6/4-148** as the main products, together with the regioisomers **6/4-149** and the ketones **6/4-150** (Scheme 6/4.38). The process has also been used to produce chiral aldehydes [323].

A remarkable one-pot process which allows the stereoselective formation of five carbon–carbon bonds was developed by Wipf and coworkers [324], starting from ene-ynes such as **6/4-151** and Cp₂Zr(H)Cl and Me₂Zn. In the first step, hydrozirconation of the ene-yne **6/4-151** takes place, followed by *in situ* transmetalation to give the zinc compound **6/4-152**, which then reacts with the aldimine **6/4-153**. In the presence of CH₂I₂, the formed *N*-metalated allylic amide **6/4-154** is cyclopropanated, as well as the other C–C double bond, to give conjugated biscyclopropanes **6/4-155** (Scheme 6/4.39). In a similar way, several (14 examples) α-amido-substituted monocyclopropanes have also been prepared, starting from alkynes, in 45 to 91 % yield.

6.4.10
Lanthanide-Induced Reactions

Based on the novel lanthanide-mediated sequential intramolecular hydroamination/C–C-bond forming cyclization by Li and Marks [325], the Molander group [326] used this process for the synthesis of tricyclic and tetracyclic-aromatic nitro-

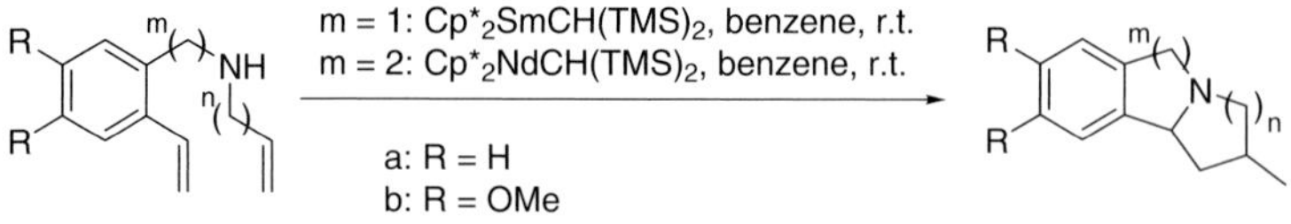

6/4-156: m = 1, n = 1
6/4-157: m = 1, n = 2
6/4-158: m = 2, n = 1
6/4-159: m = 2, n = 2

6/4-160: m = 1, n = 1, **a**: 79% (5.51:1), **b**: 84% (4.7:1)
6/4-161: m = 1, n = 2, **a,b**: 0%
6/4-162: m = 2, n = 1, **a**: 76% (1.7:1), **b**: 82% (1.4:1)
6/4-163: m = 2, n = 2, **a**: 73% (26:1), **b**: 69% (16:1)

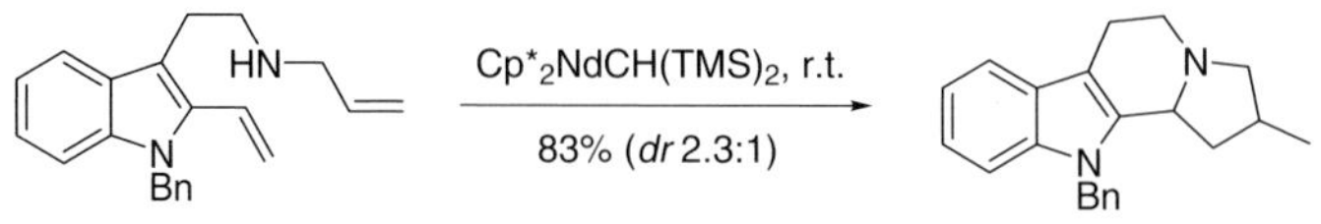

6/4-164 **6/4-165**

Scheme 6/4.40. Lanthanide-mediated cyclisations.

gen heterocycles. Reaction of the substrates **6/4-156–6/4-159**, as well as **6/4-164** with $Cp^{*}_{2}SmCH(TMS)_2$ or $Cp^{*}_{2}NdCH(TMS)_2$, led to pyrrolizidines **6/4-160**, indolizidines **6/4-162** and **6/4-165**, as well as quinolizidines **6/4-163** (Scheme 6/4.40). Astoundingly, the indolizidines **6/4-161** proved unusually unstable and could not be isolated. Although this is not a domino reaction in the strictest sense, it nonetheless provides a very useful transformation for preparation of the described heterocycles.

References

1 (a) L. F. Tietze, I. Hiriyakkanavar, H. P. Bell, *Chem. Rev.* **2004**, *104*, 3453–3516; (b) E. Negishi, *Handbook of Organopalladium Chemistry for Organic Synthesis*, John Wiley & Sons, Inc., Hoboken NJ, **2002**, *2*, 1689–1705; (c) M. Beller, C. Bolm, *Transition Metals for Organic Synthesis*, Wiley-VCH Weinheim, **1998**, Vol. 1 and Vol. 2.

2 (a) M. Oestreich, *Eur. J. Org. Chem.* **2005**, *5*, 783–792; (b) A. B. Dounay, L. E. Overman, *Chem. Rev.* **2003**, *103*, 2945–2963; (c) N. J. Whitcombe, K. K. Hii, S. E. Gibson, *Tetrahedron* **2001**, *57*, 7449–7476; (d) I. P. Beletskaya, A. V. Cheprakov, *Chem. Rev.* **2000**, *100*, 3009–3066.

3 (a) K. C. Nicolaou, P. G. Bulger, D. Sarlah, *Angew. Chem. Int. Ed.* **2005**, *44*, 4442–4489; (b) S. Schröter, C. Stock, T. Bach, *Tetrahedron* **2005**, *61*, 2245–2267; (c) U. Christmann, R. Vilar, *Angew. Chem. Int. Ed.* **2005**, *44*, 366–374; (d) P. Espinet, A. M. Echavarren, *Angew. Chem. Int. Ed.* **2004**, *43*, 4704–4734; (e) R. R. Tykwinski, *Angew. Chem. Int. Ed.* **2003**, *42*, 1566–1568.

4 E. Negishi, *Handbook of Organopalladium Chemistry for Organic Synthesis*, John Wiley & Sons, Inc., Hoboken NJ, **2002**, Vol. 2, pp. 1669–1687.

5 S. Ma, *Eur. J. Org. Chem.* **2004**, 1175–1183.

6 T. Morimoto, K. Kakiuchi, *Angew. Chem. Int. Ed.* **2004**, *43*, 5580–5588.

7 M. D. Charles, P. Schultz, S. L. Buchwald, *Organic Letters* **2005**, *7*, 3965–3968.

8 (a) T. Punniyamurthy, S. Velusamy, J. Iqbal, *Chem. Rev.* **2005**, 105, 2329–2363; (b) B. de Bruin, P. H. M. Budzelaar, A. W. Gal, *Angew. Chem. Int. Ed.* **2004**, 43, 4142–4157.

9 L. F. Tietze, K. M. Sommer, J. Zinngrebe, F. Stecker, *Angew. Chem. Int. Ed.* **2005**, 44, 257–259.

10 A. Padwa, *Helv. Chim. Acta* **2005**, 88, 1357–1374.

11 M. Beller, J. Seayad, A. Tillack, H. Jiao, *Angew. Chem. Int. Ed.* **2004**, 43, 3368–3398.

12 (a) K. C. Nicolaou, P. G. Bulger, D. Sarlah, *Angew. Chem. Int. Ed.* **2005**, 44, 4490–4527; (b) T. J. Katz, *Angew. Chem. Int. Ed.* **2005**, 44, 3010–3019; (c) D. J. Wallace, *Angew. Chem. Int. Ed.* **2005**, 44, 1912–1915; (d) J.-C. Wasilke, S. J. Obrey, R. Tom Baker, G. C. Bazan, *Chem. Rev.* **2005**, 105, 1001–1020; (e) A. Deiters, S. F. Martin, *Chem. Rev.* **2004**, 104, 2199–2238; (f) S. J. Connon, S. Blechert, *Angew. Chem. Int. Ed.* **2003**, 42, 1900–1923; (g) A. Fürstner, *Angew. Chem. Int. Ed.* **2000**, 39, 3012–3043.

13 (a) G. Poli, G. Giambastiani, *J. Org. Chem.* **2002**, 67, 9456–9459; (b) G. Poli, G. Giambastiani, A. Heumann, *Tetrahedron* **2000**, 56, 5959–5989.

14 (a) M. C. de la Torre, M. A. Sierra, *Angew. Chem. Int. Ed.* **2004**, 43, 160–181; (b) K. M. Koeller, C.-H. Wong, *Nature*, **2001**, 409, 232–240.

15 (a) L. F. Tietze, T. Nöbel, M. Spescha, *Angew. Chem. Int. Ed. Engl.* **1996**, 35, 2259–2261; (b) L. F. Tietze, T. Nöbel, M. Spescha, *J. Am. Chem. Soc.* **1998**, 35, 8971–8977.

16 N. E. Carpenter, D. J. Kucera, L. E. Overman, *J. Org. Chem.* **1989**, 54, 5846–5848; b) M. M. Abelman, L. E. Overman, *J. Am. Chem. Soc.* **1988**, 110, 2328–2329; c) Y. Zhang, E. Negishi, *J. Am. Chem. Soc.* **1989**, 111, 3454–3456; d) G. Wu, F. Lamaty, E. Negishi, *J. Org. Chem.* **1989**, 54, 2507–2508.

17 Y. Zhang, G. Wu, G. Agnel, E. Negishi, *J. Am. Chem. Soc.* **1990**, 112, 8590–8592.

18 (a) B. Burns, R. Grigg, V. Sridharan, T. Worakun, *Tetrahedron Lett.* **1988**, 29, 4325–4328; (b) B. Burns, R. Grigg, P. Ratananukul, V. Sridharan, P. Stevenson, T. Worakun, *Tetrahedron Lett.* **1988**, 29, 4329–4332.

19 (a) B. M. Trost, *Acc. Chem. Res.* **1990**, 23, 34–42; (b) B. M. Trost, Y. Shi, *J. Am. Chem. Soc.* **1992**, 114, 791–792; (c) B. M. Trost, Y. Shi, *J. Am. Chem. Soc.* **1991**, 113, 701–703; (d) B. M. Trost, M. Lautens, C. Chan, D. J. Jebaratnam, T. Mueller, *J. Am. Chem. Soc.* **1991**, 113, 636–644.

20 B. L. Feringa, W. F. Jager, B. de Lange, *Tetrahedron* **1993**, 49, 8267–8310.

21 L. F. Tietze, K. Kahle, T. Raschke, *Chem. Eur. J.* **2002**, 8, 401–407.

22 W. A. Herrmann, C. Broßmer, K. Öfele, C.-P. Reisinger, T. Priermeier, M. Beller, H. Fischer, *Angew. Chem. Int. Ed. Engl.* **1995**, 34, 1844–1848.

23 (a) L. F. Tietze, K. Heitmann, T. Raschke, *Synlett* **1997**, 35–37; (b) L. F. Tietze, T. Raschke, *Liebigs Ann.* **1996**, 1981–1987; (c) L. F. Tietze, T. Raschke, *Synlett* **1995**, 597–598; d) L. F. Tietze, R. Schimpf, *Angew. Chem. Int. Ed. Engl.* **1994**, 33, 1089–1091.

24 L. F. Tietze, R. Ferraccioli, *Synlett* **1998**, 145–146.

25 M. E. Fox, C. Li, J. P. Marino, Jr., L. E. Overman, *J. Am. Chem. Soc.* **1999**, 121, 5467–5480.

26 a) H. C. Brown, G. G. Pai, *J. Org. Chem.* **1985**, 50, 1384–1394; b) M. M. Midland, R. S. Graham, *Organic Syntheses*, Wiley: New York, **1990**, Vol. 7, pp. 402–406.

27 F. Miyazaki, K. Uotsu, M. Shibasaki, *Tetrahedron* **1998**, 54, 13073–13078.

28 S. P. Maddaford, N. G. Andersen, W. A. Cristofoli, B. A. Keay, *J. Am. Chem. Soc.* **1996**, 118, 10766–10773.

29 (a) M. J. Stocks, R. P. Harrison, S. J. Teague, *Tetrahedron Lett.* **1995**, 36, 6555–6558; (b) L. F. Tietze, K. M. Sommer, G. Schneider, P. Tapolcsanyi, J. Woelfling, P. Mueller, M. Noltemeyer, H. Terlau, *Synlett* **2003**, 1494–1496; (c) S. Ma, E. Negishi, *J. Am. Chem. Soc.* **1995**, 117, 6345–6357.

30 D. C. Harrowven, T. Woodcock, P. D. Howes, *Tetrahedron Lett.* **2002**, 43, 9327–9329.

31 G. Wu, A. L. Rheingold, S. J. Geib, R. F. Heck, *Organometallics* **1987**, 6, 1941–1946.

32 G. Dyker, A. Kellner, *Tetrahedron Lett.* **1994**, 35, 7633–7636.

33 S. Cacchi, M. Felici, B. Pietroni, *Tetrahedron Lett.* **1984**, 25, 3137–3140.

34 R. C. Larock, Q. Tian, *J. Org. Chem.* **2001**, 66, 7372–7379.

35 (a) G. Dyker, F. Nerenz, P. Siemsen, P. Bubenitschek, P. G. Jones, *Chem. Ber.* **1996**, *129*, 1265–1269; (b) G. Dyker, P. Siemsen, S. Sostman, A. Wiegand, I. Dix, P. G. Jones, *Chem. Ber.* **1997**, *130*, 261–265.

36 S. Schweizer, Z. Z. Song, F. E. Meyer, P. J. Parsons, A. de Meijere, *Angew. Chem. Int. Ed.* **1999**, *38*, 1452–1454.

37 R. Grigg, V. Loganathan, V. Sridharan, *Tetrahedron Lett.* **1996**, *37*, 3399–3402.

38 D. Brown, R. Grigg, V. Sridharan, V. Tambyrajah, *Tetrahedron Lett.* **1995**, *36*, 8137–8140.

39 R. Grigg, V. Sridharan, *Tetrahedron Lett.* **1992**, *33*, 7965–7968.

40 H. Ohno, K. Miyamura, Y. Takeoka, T. Tanaka, *Angew. Chem. Int. Ed.* **2003**, *42*, 2647–2650.

41 A. García, D. Rodríguez, L. Castedo, C. Saá, D. Domínguez, *Tetrahedron Lett.* **2001**, *42*, 1903–1905.

42 Y. Hu, J. Zhou, X. Long, J. Han, C. Zhu, Y. Pan, *Tetrahedron Lett.* **2003**, *44*, 5009–5010.

43 (a) S. P. Maddaford, N. G. Andersen, W. A. Cristofoli, B. A. Keay, *J. Am. Chem. Soc.* **1996**, *118*, 10766–10733; (b) S. Y. W. Lau, B. A. Keay, *Synlett* **1999**, 605–607.

44 (a) M. Catellani, G. P. Chiusoli, *J. Organomet. Chem.* **1983**, *250*, 509–515 and references therein; (b) M. Catellani, G. P. Chiusoli, A. Mari, *J. Organomet. Chem.* **1984**, *275*, 129–138; (c) M. Catellani, G. P. Chiusoli, S. Concari, *Tetrahedron* **1989**, *45*, 5263–5268; (d) R. C. Larock, S. S. Hershberger, K. Takagi, M. A. Mitchell, *J. Org. Chem.* **1986**, *51*, 2450–2457 and references therein; (e) R. C. Larock, P. L. Johnson, *J. Chem. Soc., Chem. Commun.* **1989**, 1368–1370; (f) S. Torii, H. Okumoto, H. Ozaki, S. Nakayasu, T. Kotani, *Tetrahedron Lett.* **1990**, *31*, 5319–5322; (g) S.-K. Kang, J.-S. Kim, S.-C. Choi, K.-H. Lim, *Synthesis* **1998**, 1249–1251; (h) M. Kosugi, H. Tamura, H. Sano, T. Migita, *Chem. Lett.* **1987**, 193; (i) M. Kosugi, H. Tamura, H. Sano, T. Migita, *Tetrahedron* **1989**, *45*, 961–967; (j) M. Kosugi, T. Kumura, H. Oda, T. Migita, *Bull. Chem. Soc. Jpn.* **1993**, *66*, 3522–3524; (k) H. Oda, K. Ito, M. Kosugi, T. Migita, *Chem. Lett.* **1994**, 1443–1444.

45 (a) F. E. Goodson, B. M. Novak, *Macromolecules* **1997**, *30*, 6047–6055; (b) K. M. Shaulis, B. L. Hoskins, J. R. Townsend, F. E. Goodson, *J. Org. Chem.* **2002**, *67*, 5860–5863.

46 S. Couty, B. Liégault, C. Meyer, J. Cossy, *Org. Lett.* **2004**, *6*, 2511–2514.

47 C.-W. Lee, K. S. Oh, K. S. Kim, K. H. Ahn, *Org. Lett.* **2000**, *2*, 1213–1216.

48 C. Zhou, R. C. Larock, *J. Org. Chem.* **2005**, *70*, 3765–3777.

49 B. Salem, J. Suffert, *Angew. Chem. Int. Ed.* **2004**, *43*, 2826–2830.

50 D. M. D'Souza, F. Rominger, T. J. J. Müller, *Angew. Chem. Int. Ed.* **2005**, *44*, 153–158.

51 D. Flubacher, G. Helmchen, *Tetrahedron Lett.* **1999**, *40*, 3867–3868.

52 (a) J. W. Daly, *J. Med. Chem.* **2003**, *46*, 445–452; (b) J. W. Daly, T. Kaneko, J. Wilham, H. M. Garraffo, T. F. Spande, A. Espinosa, *Proc. Natl. Acad. Sci. USA* **2002**, *99*, 13996–14001.

53 E. W. Dijk, L. Panella, P. Pinho, R. Naasz, A. Meetsma, A. J. Minnaard, B. L. Feringa, *Tetrahedron* **2004**, *60*, 9687–9693.

54 P. Pinho, A. J. Minnaard, B. L. Feringa, *Org. Lett.* **2003**, *5*, 259–261.

55 S. M. Verbitski, C. L. Mayne, R. A. Davis, G. P. Concepcion, C. M. Ireland, *J. Org. Chem.* **2002**, *67*, 7124–7126.

56 G. D. Artman III, S. M. Weinreb, *Org. Lett.* **2003**, *5*, 1523–1526.

57 C. Copéret, E. Negishi, *Org. Lett.* **1999**, *1*, 165–167.

58 P. Mauléon, I. Alonso, J. C. Carretero, *Angew. Chem. Int. Ed.* **2001**, *40*, 1291–1293.

59 J. Zhao, R. C. Larock, *Org. Lett.* **2005**, *7*, 701–704.

60 L. F. Tietze, F. Lotz, unpublished results

61 This type of reaction was first reported by Catellani and coworkers: (a) M. Catellani, M. C. Fagnola, *Angew. Chem. Int. Ed. Engl.* **1994**, *33*, 2421–2422; (b) M. Catellani, F. Frignani, A. Rangoni, *Angew. Chem. Int. Ed. Engl.* **1997**, *36*, 119–122; (c) M. Catellani, C. Mealli, E. Motti, P. Paoli, E. Perez-Carreño, P. S. Pregosin, *J. Am. Chem. Soc.* **2002**, *124*, 4336–4346; (d) M. Catellani, *Synlett* **2003**, 298–313.

62 S. Pache, M. Lautens, *Org. Lett.* **2003**, *5*, 4827–4830.

63 H. Nüske, S. Bräse, S. I. Kozhushkov, M. Noltemeyer, M. Es-Sayed, A. de Meijere, *Chem. Eur. J.* **2002**, *8*, 2350–2369.

64 S. Körbe, A. de Meijere, T. Labahn, *Helv. Chim. Acta* **2002**, *85*, 3161–3175.

65 M. Knoke, A. de Meijere, *Synlett* **2003**, *2*, 195–198.

66 J. Suffert, B. Salem, P. Klotz, *J. Am. Chem. Soc.* **2001**, *123*, 12107–12108.

67 C. Osterhage, R. Kaminsky, G. M. Kônig, A. D. Wright, *J. Org. Chem.* **2000**, *65*, 6412–6417.

68 T. Thiemann, M. Watanabe, S. Mataka, *New J. Chem.* **2001**, *25*, 1104–1107.

69 R. Grigg, B. Putnikovic, C. Urch, *Tetrahedron Lett.* **1997**, *35*, 6307–6308.

70 (a) U. Anwar, R. Grigg, M. Rasparini, V. Savic, V. Sriharan, *J. Chem. Soc., Chem. Commun.* **2000**, 645–646; (b) U. Anwar, R. Grigg, V. Sridharan, *J. Chem. Soc., Chem. Commun.* **2000**, 933–934.

71 R. Grigg, M. L. Millington, M. Thornton-Pett, *Tetrahedron Lett.* **2002**, *43*, 2605–2608.

72 (a) R. Grigg, V. Sridharan, M. York, *Tetrahedron Lett.* **1998**, *39*, 4132–4139; (b) R. Grigg, M. York, *Tetrahedron Lett.* **2000**, *41*, 7255–7258.

73 H. A. Dondas, G. Balme, B. Clique, R. Grigg, A. Hodgeson, J. Morris, V. Sridharan, *Tetrahedron Lett.* **2001**, *42*, 8673–8675.

74 R. Grigg, V. Sridharan, J. Zhang, *Tetrahedron Lett.* **1999**, *40*, 8277–8280.

75 R. Grigg, R. Savic, V. Tambyrajah, *Tetrahedron Lett.* **2000**, *41*, 3003–3006.

76 R. Grigg, A. Hadgson, J. Morris, V. Sridharan, *Tetrahedron Lett.* **2003**, 1023–1026.

77 R. Grigg, W. Maclachlan, M. Rasparini, *J. Chem. Soc. Chem. Commun.* **2000**, 2241–2242.

78 K. Yamazaki, Y. Nakamura, Y. Kondo, *J. Org. Chem.* **2003**, *68*, 6011–6019.

79 A. F. Littke, G. C. Fu, *J. Am. Chem. Soc.* **2001**, *123*, 6989–7000.

80 See also: S. K. Chattopadhyay, S. Maity, B. K. Pal, S. Panja, *Tetrahedron Lett.* **2002**, *43*, 5079–5081.

81 S. D. Edmondson, A. Mastracchio, E. R. Parmee, *Org. Lett.* **2000**, *2*, 1109–1112.

82 P. Vittoz, D. Bouyssi, C. Traversa, J. Goré, G. Balme, *Tetrahedron Lett.* **1994**, *35*, 1871–1874.

83 (a) G. Dyker, P. Grundt, *Tetrahedron Lett.* **1996**, *37*, 619–622; (b) G. Dyker, H. Markwitz, *Synthesis* **1998**, 1750–1754.

84 A. Bengtson, M. Larhed, A. Hallberg, *J. Org. Chem.* **2002**, *67*, 5854–5856.

85 G. Battistuzzi, S. Cacchi, G. Fabrizi, *Eur. J. Org. Chem.* **2002**, 2671–2681.

86 K. G. Dongol, B. Y. Tay, *Tetrahedron Lett.* **2006**, *47*, 927–930.

87 A. Kojima, T. Takemoto, M. Sodeoka, M. Shibasaki, *J. Org. Chem.* **1996**, *61*, 4876–4877.

88 A. Kojima, S. Honzawa, C. D. J. Boden, M. Shibasaki, *Tetrahedron Lett.* **1997**, *38*, 3455–3458.

89 H. A. Wegner, L. T. Scott, A. de Meijere, *J. Org. Chem.* **2003**, *68*, 883–887.

90 R. Sekizawa, S. Ikeno, H. Nakamura, H. Naganawa, S. Matsui, H. Iinuma, T. Takeuchi, *J. Nat. Prod.* **2002**, *65*, 1491–1493.

91 D. C. Swinney, *Drug Discovery Today* **2001**, *6*, 244–250.

92 J. E. Moses, L. Commeiras, J. E. Baldwin, R. M. Adlington, *Org. Lett.* **2003**, *5*, 2987–2988.

93 J. M. Humphrey, Y. Liao, A. Ali, T. Rein, Y.-L. Wong, H.-J. Chen, A. K. Courtney, S. F. Martin, *J. Am. Chem. Soc.* **2002**, *124*, 8584–8592.

94 S. Brückner, E. Abraham, P. Klotz, J. Suffert, *Org. Lett.* **2002**, *4*, 3391–3393.

95 (a) M. A. J. Duncton, G. Pattenden, *J. Chem. Soc. Perkin Trans. 1* **1999**, *10*, 1235–1246; (b) E. Marsault, P. Deslongchamps, *Org. Lett.* **2000**, *2*, 3317–3320.

96 (a) M. Murakami, K. Itami, Y. Ito, *J. Am. Chem. Soc.* **1997**, *119*, 7163–7164; (b) J. M. Takacs, F. Clement, J. Zhu, S. V. Chandramouli, X. Gong, *J. Am. Chem. Soc.* **1997**, *119*, 5805–5817.

97 R. Skoda-Földes, G. Jeges, L. Kollár, J. Horváth, Z. Tuba, *J. Org. Chem.* **1997**, *62*, 1326–1332.

98 R. Skoda-Földes, K. Vándor, L. Kollár, J. Horváth, Z. Tuba, *J. Org. Chem.* **1999**, *64*, 5921–5925.

99 C. M. Beaudry, D. Trauner, *Org. Lett.* **2002**, *4*, 2221–2224.

100 K. Kurosawa, K. Takahashi, E. Tsuda, *J. Antibiot.* **2001**, *54*, 541–547.

101 K. A. Parker, Y.-H. Lim, *J. Am. Chem. Soc.* **2004**, *126*, 15968–15969.

102 L. R. Pottier, J.-F. Peyrat, M. Alami, J.-D. Brion, *Synlett* **2004**, *9*, 1503–1508.

103 V. Fiandanese, D. Bottalico, G. Marchese, A. Punzi, *Tetrahedron* **2004**, *60*, 11421–11425.

104 B. A. Chauder, A. V. Kalinin, N. J. Taylor, V. Snieckus, *Angew. Chem. Int. Ed.* **1999**, *38*, 1435–1438.

105 (a) R. D. Stephens, C. E. Castro, *J. Org. Chem.* **1963**, *28*, 3313–3315; (b) C. E. Castro, E. J. Gaughan, D. C. Owsley, *J. Org. Chem.* **1966**, *31*, 4071–4078.

106 R. C. Larock, E. D. Yum, M. J. Doty, K. K. C. Sham, *J. Org. Chem.* **1995**, *60*, 3270–3271.

107 N. Rasool, A. Q. Khan, V. U. Ahmad, A. Malik, *Phytochemistry* **1991**, *30(8)*, 2803–2805.

108 B. Clique, S. Vassiliou, N. Monteiro, G. Balme, *Eur. J. Org. Chem.* **2002**, 1493–1499.

109 E. Bossharth, P. Desbordes, N. Monteiro, G. Balme, *Org. Lett.* **2003**, *5*, 2441–2444.

110 *o*-Iodophenols can also be used in this type of strategy to give benzofurans: J. H. Chaplin, B. L. Flynn, *Chem. Commun.* **2001**, 1594–1595.

111 M. Kotora, E. Negishi, *Synthesis* **1996**, 121–128.

112 M. Pal, V. Subramanian, V. R. Batchu, I. Dager, *Synlett* **2004**, *11*, 1965–1969.

113 N. Olivi, P. Spruyt, J.-F. Peyrat, M. Alami, J.-D. Brion, *Tetrahedron Lett.* **2004**, *45*, 2607–2610.

114 M. Gruber, S. Chouzier, K. Koehler, L. Djakovitch, *Applied Catalysis A: General* **2004**, *265*, 161–169.

115 A.-C. Carbonelle, J. Zhu, *Org. Lett.* **2000**, *2*, 3477–3480.

116 I. C. F. R. Ferreira, M.-J. R. P. Queiroz, G. Kirsch, *Tetrahedron Lett.* **2003**, *44*, 4327–4329.

117 S.-K. Kang, Y.-H. Ha, B.-S. Ko, Y. Lim, J. Jung, *Angew. Chem. Int. Ed.* **2002**, *41*, 343–345.

118 T. Hirashita, Y. Hayashi, K. Mitsui, S. Araki, *J. Org. Chem.* **2003**, *68*, 1309–1313.

119 W. Oppolzer, J.-M. Gaudin, *Helv. Chim. Acta* **1987**, *70*, 1477–1481.

120 L. F. Tietze, H. Schirok, *Angew. Chem. Int. Ed. Engl.* **1997**, *36*, 1124–1125.

121 L. F. Tietze, G. Nordmann, *Eur. J. Org. Chem.* **2001**, 3247–3253.

122 G. Poli, G. Giambastiani, *J. Org. Chem.* **2002**, *67*, 9456–9459.

123 (a) Y. Damayanthi, J. W. Lown, *Curr. Med. Chem.* **1998**, *5*, 205–252; (b) Y. Zhang, K.-H. Lee, *Chin. Pharm. J.* **1994**, *46*, 319–369; (c) A. C. Ramos, R. Peláéz-Lamamié de Clairac, M. Medarde, *Heterocycles* **1999**, *51*, 1443–1470; (d) R. S. Ward, *Chem. Soc. Rev.* **1982**, 75–125; (e) R. S. Ward, *Synthesis* **1992**, 719–730.

124 M. Szlosek-Pinaud, P. Diaz, J. Martinez, F. Lamaty, *Tetrahedron Lett.* **2003**, *44*, 8657–8659.

125 G. Zhu, Z. Zhang, *Org. Lett.* **2004**, *6*, 4041–4044.

126 W. Oppolzer, C. Robyr, *Tetrahedron* **1994**, *50*, 415–424.

127 M. Yoshida, H. Nemeto, M. Ihara, *Tetrahedron Lett.* **1999**, *40*, 8583–8586.

128 C. Jousse-Karinthi, C. Riche, A. Chiaroni, D. Desmaële, *Eur. J. Org. Chem.* **2001**, 3631–3640.

129 B. M. Trost, Y. Shi, *J. Am. Chem. Soc.* **1992**, *114*, 791–792.

130 T. Sugihara, C. Copéret, Z. Owczarczyk, L. S. Harring, E. Negishi, *J. Am. Chem. Soc.* **1994**, *116*, 7923–7924.

131 B. M. Trost, A. J. Frontier, *J. Am. Chem. Soc.* **2000**, *122*, 11727–11728.

132 C. W. Holzapfel, L. Marais, *Tetrahedron Lett.* **1997**, *38*, 8585–8586.

133 G. Balme, E. Bossharth, N. Monteiro, *Eur. J. Org. Chem.* **2003**, 4101–4111.

134 Y. Inoue, Y. Itoh, I. F. Yen, S. Imaizumi, *J. Mol. Cat.* **1990**, *60*, L1–L3.

135 M. Cavicchioli, E. Sixdenier, A. Derrey, D. Bouyssi, G. Balme, *Tetrahedron Lett.* **1997**, *38*, 1763–1766.

136 M. Bottex, M. Cavicchioli, B. Hartmann, N. Monteiro, G. Balme, *J. Org. Chem.* **2001**, *66*, 175–179.

137 G. Liu, X. Lu, *Tetrahedron Lett.* **2003**, *44*, 467–470.

138 S. Azoulay, N. Monteiro, G. Balme, *Tetrahedron Lett.* **2002**, *43*, 9311–9314.

139 S. Garçon, S. Vassiliou, M. Cavicchioli, B. Hartmann, N. Monteiro, G. Balme, *J. Org. Chem.* **2001**, *66*, 4069–4073.

140 A. Arcadi, S. Cacchi, G. Fabrizi, F. Marinelli, L. M. Parisi, *Tetrahedron* **2003**, *59*, 4661–4671.

141 L.-M. Wei, C.-F. Lin, M.-J. Wu, *Tetrahedron Lett.* **2000**, *41*, 1215–1218.

142 H. Nakamura, M. Ohtaka, Y. Yamamoto, *Tetrahedron Lett.* **2002**, *43*, 7631–7633.

143 S. Cacchi, G. Fabrizi, P. Pace, *J. Org. Chem.* **1998**, *63*, 1001–1011.

144 E. Bossharth, P. Desbordes, N. Monteiro, G. Balme, *Org. Lett.* **2003**, *5*, 2441–2444.

145 X. Xie, X. Lu, *Tetrahedron Lett.* **1999**, *40*, 8415–8418.

146 C. Gómez-Reino, C. Vitale, M. Maestro, A. Mouriño, *Org. Lett.* **2005**, *7*, 5885–5887.

147 P. H. Lee, Y. Kang, *J. Am. Chem. Soc.* **2006**, *128*, 1139–1146.

148 K. Okuro, H. Alper, *J. Org. Chem.* **1997**, *62*, 1566–1567.

149 R. Grigg, A. Liu, D. Shaw, S. Suganthan, D. E. Woodall, G. Yoganathan, *Tetrahedron Lett.* **2000**, *41*, 7125–7128.

150 (a) K. Grohe, *Chemistry in Britain* **1992**, 34–36; (b) M. P. Wentland, J. B. Cornett, *Annu. Rep. Med. Chem.* **1985**, *20*, 145–154.

151 F. E. Ward, D. L. Garling, R. T. Buckler, D. M. Lawer, D. P. Cummings, *J. Med. Chem.* **1981**, *24*, 1073–1077.

152 Y. Hayakawa, H. Yamamoto, N. Tsuge, H. Seto, *Tetrahedron Lett.* **1996**, *37*, 6363–6364.

153 R. Grigg, A. Liu, D. Shaw, S. Suganthan, M. L. Washington, D. E. Woodall, G. Yoganathan, *Tetrahedron Lett.* **2000**, *41*, 7129–7133.

154 R. D. Dghaym, R. Dhawan, B. A. Arndtsen, *Angew. Chem. Int. Ed.* **2001**, *40*, 3228–3230.

155 D. A. Evans, A. M. Golob, *J. Am. Chem. Soc.* **1975**, *97*, 4765–4766.

156 (a) T. T. Dabrah, H. J. Harwood, L. H. Huang, N. D. Jankovich, T. Kaneko, J.-C. Li, S. Lindsey, P. M. Moshier, T. A. Subashi, M. Therrien, P. C. Watts, *J. Antibiot.* **1997**, *50*, 1–7; (b) T. T. Dabrah, T. Kaneko, W. Jr. Massefski, E. B. Whipple, *J. Am. Chem. Soc.* **1997**, *119*, 1594–1598.

157 D. L. J. Clive, S. Sun, X. He, J. Zhang, V. Gagliardini, *Tetrahedron Lett.* **1999**, *40*, 4605–4609.

158 M. M. Bio, J. L. Leighton, *J. Org. Chem.* **2003**, *68*, 1693–1700.

159 (a) M. Beller, M. Eckert, *Angew. Chem. Int. Ed.* **2000**, *39*, 1010–1027; (b) M. Beller, M. Eckert, E. W. Holla, *J. Org. Chem.* **1998**, *63*, 5658–5661.

160 G. Cuny, M. Bois-Choussy, J. Zhu, *Angew. Chem. Int. Ed.* **2003**, *42*, 4774–4777.

161 G. Cuny, M. Bois-Choussy, J. Zhu, *J. Am. Chem. Soc.* **2004**, *126*, 14475–14484.

162 S. D. Edmondson, A. Mastracchio, E. R. Parmee, *Org. Lett.* **2000**, *2*, 1109–1112.

163 Y.-S. Lin, H. Alper, *Angew. Chem. Int. Ed.* **2001**, *40*, 779–781.

164 R. Mutter, I. B. Campbell, E. M. Martin de la Nava, A. T. Merritt, M. Wills, *J. Org. Chem.* **2001**, *66*, 3284–3290.

165 J.-H. Kim, R. J. Kulawiec, *Tetrahedron Lett.* **1998**, *39*, 3107–3110.

166 D. P. Curran, W. Du, *Org. Lett.* **2002**, *4*, 3215–3218.

167 J. G. Liehr, B. C. Giovanella, C. F. Verschraegen, Eds., *Ann. N. Y. Acad. Sci.* **2000**, 922.

168 A. Pirillo, L. Verotta, P. Gariboldi, E. Torregiani, E. Bombardelli, *J. Chem. Soc., Perkin Trans. I* **1995**, 583.

169 D. Bom, D. P. Curran, S. Kruszewski, S. G. Zimmer, J. Thompson Strode, G. Kohlhagen, W. Du, A. J. Chavan, K. A. Fraley, A. L. Bingcang, L. J. Latus, Y. Pommier, T. G. Burke, *J. Med. Chem.* **2000**, *43*, 3970–3980.

170 H. Josien, S.-B. Ko, D. Bom, D. P. Curran, *Chem. Eur. J.* **1998**, *4*, 67–83.

171 (a) R. M. Trend, Y. R. Ramtohul, E. M. Ferreira, B. M. Stoltz, *Angew. Chem. Int. Ed.* **2003**, *42*, 2892–2895; (b) M. A. Arai, M. Kuraishi, T. Arai, H. Sasai, *J. Am. Chem. Soc.* **2001**, *123*, 2907–2908.

172 L. F. Tietze, K. M. Sommer, J. Zinngrebe, F. Stecker, *Angew. Chem. Int. Ed.* **2005**, *44*, 257–259.

173 T. Netscher, *Chimia* **1996**, *50*, 563–567.

174 T. D. Nelson, A. I. Meyers, *J. Org. Chem.* **1994**, *59*, 2655–2658.

175 L. F. Tietze, K. Sommer, *German Patent* DE 102004011265A1, **2004**.

176 Y. Liao, M. Reitman, Y. Zhang, R. Fathi, Z. Yang, *Org. Lett.* **2002**, *4*, 2607–2609.

177 H. Nemeto, M. Yoshida, K. Fukumoto, M. Ihara, *Tetrahedron Lett.* **1999**, *40*, 907–910.

178 M. Toyota, M. Rudyanto, M. Ihara, *J. Org. Chem.* **2002**, *67*, 3374–3386.

179 J. H. Koh, M. R. Gagné, *Angew. Chem. Int. Ed.* **2004**, *43*, 3459–3461.

180 C. W. Holzapfel, L. Marais, *J. Chem. Res.* **1999**, 190–191.

181 G. L. J. Bar, G. C. Lloyd-Jones, K. I. Booker-Milburn, *J. Am. Chem. Soc.* **2005**, *127*, 7308–7309.

182 G. Liu, X. Lu, *Org. Lett.* **2001**, *3*, 3879–3882.

183 B. Alcaide, P. Almendros, C. Aragoncillo, *Chem. Eur. J.* **2002**, *8*, 1719–1729.

184 J.-H. Sohn, N. Waizumi, H. M. Zhong, V. H. Rawal, *J. Am. Chem. Soc.* **2005**, *127*, 7290–7291.

185 (a) A. Padwa, M. D. Weingarten, *Chem. Rev.* **1996**, *96*, 223–269; (b) A. Padwa, C. S. Straub, *J. Org. Chem.* **2003**, *68*, 227–239; (c) M. P. Doyle, M. A. McKervey, T. Ye, *Modern Catalytic Methods for Organic Synthesis with Diazo Compounds* **1998**, Chapter 7; (d) J. S. Clark; Nitrogen, *Oxy-*

gen and Sulfur Ylide Chemistry **2002**; (e) G. Mehta, S. Muthusamy, *Tetrahedron* **2002**, *58*, 9477–9504.

186 A. Padwa, A. T. Price, *J. Org. Chem.* **1998**, *63*, 556–565.

187 A. Padwa, L. Precedo, M. A. Semones, *J. Org. Chem.* **1999**, *64*, 4079–4088.

188 A. Padwa, Z. J. Zhang, L. Zhi, *J. Org. Chem.* **2000**, *65*, 5223–5232.

189 A. Padwa, C. S. Straub, *Org. Lett.* **2000**, *2*, 2093–2095.

190 S. Muthusamy, S. A. Babu, C. Gunanathan, E. Suresh, P. Dastidar, *Bull. Chem. Soc. Jpn.* **2002**, *75*, 801–811.

191 S. Muthusamy, S. A. Babu, C. Gunanathan, B. Ganguly, E. Suresh, P. Dastidar, *J. Org. Chem.* **2002**, *67*, 8019–8033.

192 T. Graening, W. Friedrichsen, J. Lex, H.-G. Schmalz, *Angew. Chem. Int. Ed.* **2002**, *41*, 1524–1526.

193 (a) B. Chen, R. Y. Y. Ko, M. S. M. Yuen, K.-F. Cheng, P. Chiu, *J. Org. Chem.* **2003**, *68*, 4195–4205; (b) P. Chiu, B. Chen, K. F. Cheng, *Org. Lett.* **2001**, *3*,1721–1724.

194 (a) B. N. Zhou, B. P. Ying, G. Q. Song, Z. X. Chen, J. Han, Y. F. Yan, *Planta Med.* **1983**, *47*, 35–38; (b) M. O. Hamburger, H. L. Shieh, B. N. Zhou, J. M. Pezzuto, G. A. Cordell, *Magn. Res. Chem.* **1989**, *27*, 1025–1030; (c) E. Li, A. M. Clark, C. D. Hufford, *J. Nat. Prod.* **1995**, *58*, 57–67; (d) D. J. Pan, Z. L. Li, C. Q. Hu, K. Chen, J. J. Chang, K. H. Lee, *Planta Med.* **1990**, *56*, 383–385.

195 D. M. Hodgson, A. H. Labande, F. Y. T. M. Pierard, M.Á. Expósito Castro, *J. Org. Chem.* **2003**, *68*, 6153–6159.

196 C.-Y. Zhou, P. W. H. Chan, W.-Y. Yu, C.-M. Che, *Synthesis* **2003**, *9*, 1403–1412.

197 G.-Y. Li, J. Chen, W.-Y. Yu, W. Hong, C.-M. Che, *Org. Lett.* **2003**, *5*, 2153–2156.

198 D. Muroni, A. Saba, N. Culeddu, *Tetrahedron Asymm.* **2004**, *15*, 2609–2614.

199 G. Maas, A. Müller, *Org. Lett.* **1999**, *1*, 219–221.

200 J. A. May, B. M. Stoltz, *J. Am. Chem. Soc.* **2002**, *124*, 12426–12427.

201 (a) H. M. L. Davies, J. J. Matasi, L. M. Hodges, N. J. S. Huby, C. Thornley, N. Kong, J. H. Houser, *J. Org. Chem.* **1997**, *62*, 1095–1105; (b) H. M. L. Davies, *Curr. Org. Chem.* **1998**, *2*, 463–488.

202 H. M. L. Davies, D. G. Stafford, B. D. Doan, J. H. Houser, *J. Am. Chem. Soc.* **1998**, *120*, 3326–3331.

203 H. M. L. Davies, B. D. Doan, *J. Org. Chem.* **1999**, *64*, 8501–8508.

204 (a) B. Breit, *Acc. Chem. Res.* **2003**, 36, 264–275; (b) O. Roelen, German Patent DE 849,548, 1938/1952; U. S. Patent 2,317,066, **1943**; *Chem. Abstr.* **1944**, *38*, 550.

205 C. Hollmann, P. Eilbracht, *Tetrahedron* **2000**, *56*, 1685–1692.

206 M. D. Keränen, P. Eilbracht, *Org. Biomol. Chem.* **2004**, *2*, 1688–1690.

207 B. Breit, S. K. Zahn, *Angew. Chem. Int. Ed.* **1999**, *38*, 969–971.

208 B. Breit, S. K. Zahn, *Angew. Chem. Int. Ed.* **2001**, *40*, 1910–1913.

209 (a) H. Chen, R. Tan, Z. L. Liu, Y. Zhang, L. Yang, *J. Nat. Prod.* **1996**, *59*, 668–670; (b) H. Kizu, N. Sugita, T. Tomimori, *Chem. Pharm. Bull.* **1998**, *46*, 988–1000.

210 R. E. Minto, C. A. Townsend, *Chem. Rev.* **1997**, *97*, 2537–2555.

211 R. Roggenbuck, A. Schmidt, P. Eilbracht, *Org. Lett.* **2002**, *4*, 289–291.

212 P. Köhling, A. M. Schmidt, P. Eilbracht, *Org. Lett.* **2003**, *5*, 3213–3216.

213 R. W. Hoffmann, D. Brückner, *New J. Chem.* **2001**, *25*, 369–373.

214 (a) G. D. Cuny, S. L. Buchwald, *J. Am. Chem. Soc.* **1993**, *115*, 2066–2068; (b) L. A. van der Veen, P. C. J. Kramer, P. W. N. M. van Leeuwen, *Angew. Chem. Int. Ed.* **1999**, *38*, 336–338.

215 R. Settambolo, G. Guazzelli, A. Mandoli, R. Lazzaroni, *Tetrahedron Asymm.* **2004**, *15*, 1821–1823.

216 Y. Fukuta, I. Matsuda, K. Itoh, *Tetrahedron Lett.* **1999**, *40*, 4703–4706.

217 (a) S. J. O'Malley, J. L. Leighton, *Angew. Chem. Int. Ed.* **2001**, *40*, 2915–2917; (b) M. J. Zacuto, S. J. O'Malley, J. L. Leighton, *J. Am. Chem. Soc.* **2002**, *124*, 7890–7891; (c) M. J. Zacuto, S. J. O'Malley, J. L. Leighton, *Tetrahedron* **2003**, *59*, 8889–8900.

218 S. D. Rychnovsky, *Chem. Rev.* **1995**, *95*, 2012–2040.

219 For a review, see: G. R. Jones, Y. Landais, *Tetrahedron* **1996**, *52*, 7599–7662.

220 D. R. Schmidt, P. K. Park, J. L. Leighton, *Org. Lett.* **2003**, *5*, 3535–3537.

221 P. A. Evans, J. E. Robinson, E. W. Baum, A. N. Fazal, *J. Am. Chem. Soc.* **2002**, *124*, 8782–8783.

222 P. A. Evans, J. E. Robinson, *J. Am. Chem. Soc.* **2001**, *123*, 4609–4610.

223 T. Muraoka, H. Asaji, Y. Yamamoto, I. Matsuda, K. Itoh, *Chem. Commun.* **2000**, 199–200.

224 A. L. Balch, M. M. Olmstead, *Chem. Rev.* **1998**, *98*, 2123–2165.

225 (a) S. J. Connon, S. Blechert, *Angew. Chem. Int. Ed.* **2003**, *42*, 1900–1923; (b) T. M. Trnka, R. H. Grubbs, *Acc. Chem. Res.* **2001**, *34*, 18–29; (c) A. Fürstner, *Angew. Chem. Int. Ed.* **2000**, *39*, 3012–3043; (d) R. Roy, S. K. Das, *Chem. Commun.* **2000**, 519–529; (e) M. L. Randall, M. L. Snapper, *J. Mol. Catal. A: Chemical* **1998**, *133*, 29–40; (f) S. E. Gibson, S. P. Keen, *Top. Organomet. Chem.* **1998**, *1*, 155–181; g) M. Mori, *Top. Organomet. Chem.* **1998**, *1*, 133–154.

226 W. J. Zuercher, M. Hashimoto, R. H. Grubbs, *J. Am. Chem. Soc.* **1996**, *118*, 6634–6640.

227 J. P. A. Harrity, D. S. La, D. R. Cefalo, M. S. Visser, A. H. Hoveyda, *J. Am. Chem. Soc.* **1998**, *120*, 2343–2351.

228 M. Schuster, S. Blechert, *Angew. Chem. Int. Ed.* **1997**, *36*, 2036–2055.

229 E. R. Freitas, C. R. Gum, *Chem. Eng. Prog.* **1979**, *75*, 73–76.

230 Philipps Petroleum Company, *Hydrocarbon Process* **1967**, *46*, 232.

231 R. Stragies, S. Blechert, *Tetrahedron* **1999**, *55*, 8179–8188.

232 H. Ovaa, C. Stapper, G. A. van der Marel, H. S. Overkleeft, J. H. van Boom, S. Blechert, *Tetrahedron* **2002**, *58*, 7503–7518.

233 N. Buschmann, A. Rückert, S. Blechert, *J. Org. Chem.* **2002**, *67*, 4325–4329.

234 M. Zaja, S. Blechert, *Tetrahedron* **2004**, *60*, 9629–9634.

235 J.-A. Funel, J. Prunet, *Synlett* **2005**, *2*, 235–238.

236 (a) M.-A. Virolleaud, O. Piva, *Synlett* **2004**, *12*, 2087–2090; (b) M.-A. Virolleaud, C. Bressy, O. Piva, *Tetrahedron Lett.* **2003**, *44*, 8081–8084.

237 A. Wrobleski, K. Sahasrabudhe, J. Aubé, *J. Am. Chem. Soc.* **2004**, *126*, 5475–5481.

238 T.-L. Choi, R. H. Grubbs, *Chem. Commun.* **2001**, 2648–2649.

239 K. C. Nicolaou, J. A. Vega, G. Vassilikogiannakis, *Angew. Chem. Int. Ed.* **2001**, *40*, 4441–4445.

240 (a) M. J. Bassindale, A. S. Edwards, P. Hamley, H. Adams, J. P. A. Harrity, *Chem. Commun.* **2000**, 1035–1036; (b) M. J. Bassindale, P. Hamley, A. Leitner,

J. P. A. Harrity, *Tetrahedron Lett.* **1999**, *40*, 3247–3250.

241 (a) G. Mehta, J. Nandakumar, *Tetrahedron Lett.* **2001**, *42*, 7667–7670; (b) G. Mehta, J. Nandakumar, *Tetrahedron Lett.* **2002**, *43*, 699–702.

242 A. Rückert, D. Eisele, S. Blechert, *Tetrahedron Lett.* **2001**, *42*, 5245–5247.

243 A. G. M. Barrett, S. P. D. Baugh, D. C. Braddock, K. Flack, V. C. Gibson, M. R. Giles, E. L. Marshall, P. A. Procopiou, A. J. P. White, D. J. Williams, *J. Org. Chem.* **1998**, *63*, 7893–7907.

244 O. Arjona, A. G. Csák–, V. León, R. Medel, J. Plumet, *Tetrahedron Lett.* **2004**, *45*, 565–567.

245 O. Arjona, A. G. Csák–, R. Medel, J. Plumet, *J. Org. Chem.* **2002**, *67*, 1380–1383.

246 F. Royer, C. Vilain, L. Elkaïm, L. Grimaud, *Org. Lett.* **2003**, *5*, 2007–2009.

247 J. A. Smulik, S. T. Diver, *Tetrahedron Lett.* **2001**, *42*, 171–174.

248 J. Efskind, C. Römming, K. Undheim, *J. Chem. Soc. Perkin Trans. 1*, **2001**, 2697–2703.

249 K. P. Kaliappan, R. S. Nandurdikar, *Chem. Comm.* **2004**, 2506–2507.

250 E. M. Codesido, L. Castedo, J. R. Granja, *Org. Lett.* **2001**, 1483–1486.

251 W. H. Okamura, M. M. Midland, M. W. Hammond, N. A. Rahman, M. C. Dormanen, I. Nemere, A. W. Norman, *J. Steroid Biochem. Mol. Biol.* **1995**, *53*, 603–613.

252 F.-D. Boyer, I. Hanna, *Tetrahedron Lett.* **2002**, *43*, 7469–7472.

253 (a) S. F. Brady, M. P. Singh, J. E. Janso, J. Clardy, *J. Am. Chem. Soc.* **2000**, *122*, 2116–2117; (b) S. F. Brady, S. M. Bondy, J. Clardy, *J. Am. Chem. Soc.* **2001**, *123*, 9900–9901.

254 M.-P. Mingeot-Leclercq, Y. Glupczynski, P. M. Tulkens, *Antimicrob. Agents Chemother.* **1999**, *43*, 727–737.

255 D. J. Jenkins, A. M. Riley, B. V. L. Potter, in: Y. Chapleur (Ed.), *Carbohydrate Mimics – Concepts and Methods*, Wiley VCH: Weinheim **1998**, p. 171.

256 M. T. Crimmins, *Tetrahedron* **1998**, *54*, 9229–9272.

257 C. Storm Poulsen, R. Madsen, *J. Org. Chem* **2002**, *67*, 4441–4449.

258 F.-D. Boyer, I. Hanna, L. Ricard, *Org. Lett.* **2001**, *3*, 3095–3098.

259 R. Grigg, M. York, *Tetrahedron Lett.* **2000**, *41*, 7255–7258.

260 C. Le Drian, I. Fenger, *Tetrahedron Lett.* **1998**, *39*, 4287–4290.

261 (a) I. Horvath, *Acc. Chem. Res.* **1998**, *31*, 641–650; (b) P. Bhattacharyya, D. Gudmunsen, E. G. Hope, R. D. Kemmitt, D. R. Paige, A. M. Stuart, *J. Chem. Soc., Perkin Trans. 1* **1997**, 3609–3612.

262 M. Rosillo, L. Casarrubios, G. Domínguez, J. Pérez-Castells, *Org. Biomol. Chem.* **2003**, *1*, 1450–1451.

263 D. Banti, M. North, *Tetrahedron Lett.* **2002**, *43*, 1561–1564.

264 S. C. Schürer, S. Blechert, *Chem. Commun.* **1999**, 1203–1204.

265 D. Bentz, S. Laschat, *Synthesis* **2000**, 1766–1773.

266 H.-Y. Lee, H. Y. Kim, H. Tae, B. G. Kim, J. Lee, *Org. Lett.* **2003**, *5*, 3439–3442.

267 M. Murakami, M. Ubukata, Y. Ito, *Chem. Lett.* **2002**, 294–295.

268 J. Cossy, F. Bargiggia, S. BouzBouz, *Org. Lett.* **2003**, *5*, 459–462.

269 B. M. Trost, J. A. Flygare, *J. Am. Chem. Soc.* **1992**, *114*, 5476–5477.

270 D. Berger, W. Imhof, *Chem. Commun.* **1999**, 1457–1458.

271 M. Murakami, T. Katsuki, *Tetrahedron Lett.* **2002**, *43*, 3947–3949.

272 (a) I. U. Khand, P. L. Pauson, *Heterocycles* **1978**, *11*, 59–67; (b) T. Sugihara, M. Yamaguchi, M. Nishizawa, *Chem. Eur. J.* **2001**, *7*, 1589–1595; (c) K. M. Brummond, J. L. Kent, *Tetrahedron* **2000**, *56*, 3263–3283; (d) Y. K. Chung, *Coord. Chem. Rev.* **1999**, *188*, 297–341; (e) N. E. Schore, *Org. React.* **1991**, *40*, 1–90; (f) N. E. Schore, *Chem. Rev.* **1988**, *88*, 1081–1119.

273 (a) S. H. Lecker, N. H. Nguyen, K. P. C. Vollhardt, *J. Am. Chem. Soc.* **1986**, *108*, 856–858; (b) J. Germanas, C. Aubert, K. P. C. Vollhardt, *J. Am. Chem. Soc.* **1991**, *113*, 4006–4008; (c) E. P. Johnson, K. P. C. Vollhardt, *J. Am. Chem. Soc.* **1991**, *113*, 381–382; (d) K. P. C. Vollhardt, *Lect. Heterocycl. Chem.* **1987**, *9*, 59; (e) K. P. C. Vollhardt, *Pure Appl. Chem.* **1985**, *57*, 1819–1826; (f) K. P. C. Vollhardt, *Angew. Chem. Int. Ed. Engl.* **1984**, *23*, 539–556.

274 (a) P. Cruciani, C. Aubert, M. Malacria, *J. Org. Chem.* **1995**, *60*, 2664–2665; (b) C. Aubert, O. Buisine, M. Petit, F. Slowinski, M. Malacria, *Pure Appl. Chem.* **1999**, *71*, 1463–1470.

275 P. Cruciani, R. Stammler, C. Aubert, M. Malacria, *J. Org. Chem.* **1996**, *61*, 2699–2708.

276 G. B. Hoven, J. Efskind, C. Rømming, K. Undheim, *J. Org. Chem.* **2002**, *67*, 2459–2463.

277 T. Beulshausen, U. Groth, M. Schöllkopf, *Liebigs Ann. Chem.* **1991**, 1207–1209.

278 M. Thommen, R. Keese, *Synlett* **1997**, 231–240.

279 S. G. van Ornum, J. M. Cook, *Tetrahedron Lett.* **1997**, *38*, 3657–3658.

280 (a) S. H. Hong, J. W. Kim, D. S. Choi, Y. K. Chung, S.-G. Lee, *Chem. Commun.* **1999**, 2099–2100; (b) S. U. Son, S.-J. Paik, S. I. Lee, Y. K. Chung, *J. Chem. Soc. Perkin Trans. 1* **2000**, 141–143; (c) S. U. Son, D. S. Choi, Y. K. Chung, S.-G. Lee, *Org. Lett.* **2000**, *2*, 2097–2100.

281 (a) S. U. Son, Y. K. Chung, S.-G. Lee, *J. Org. Chem.* **2000**, *65*, 6142–6144; (b) S. U. Son, Y. A. Yoon, D. S. Choi, J. K. Park, B. M. Kim, Y. K. Chung, *Org. Lett.* **2001**, *3*, 1065–1067; (c) D. H. Kim, S. U. Son, Y. K. Chung, S.-G. Lee, *Chem. Commun.* **2002**, 56–57.

282 S. I. Lee, S. U. Son, M. R. Choi, Y. K. Chung, S.-G. Lee, *Tetrahedron Lett.* **2003**, *44*, 4705–4709.

283 N. Jeong, B. Ki Sung, J. S. Kim, S. B. Park, S. D. Seo, J. Y. Shin, K. Y. In, Y. K. Choi, *Pure Appl. Chem.* **2002**, *74*, 85–91.

284 A. Odedra, C.-J. Wu, R. J. Madhushaw, S.-L. Wang, R.-S. Liu, *J. Am. Chem. Soc.* **2003**, *125*, 9610–9611.

285 (a) M. Rodríguez Rivero, J. Adrio, J. C. Carretero, *Eur. J. Org. Chem.* **2002**, 2881–2889; (b) M. Rodríguez Rivero, J. C. Carretero; *J. Org. Chem.* **2003**, *68*, 2975–2978; c) C. Aubert, O. Buisine, M. Malacria, *Chem. Rev.* **2002**, *102*, 813–834.

286 J. Balsells, A. Moyano, A. Riera, M. A. Pericàs, *Org. Lett.* **1999**, *1*, 1981–1984.

287 S. Ikeda, *Acc. Chem. Res.* **2000**, 511–519.

288 D.-M. Cui, H. Yamamoto, S. Ikeda, K. Hatano, Y. Sato, *J. Org. Chem.* **1998**, *63*, 2782–2784.

289 S. Ikeda, D.-M. Cui, Y. Sato, *J. Am. Chem. Soc.* **1999**, *121*, 4712–4713.

290 S. Ikeda, H. Miyashita, M. Taniguchi, H. Kondo, M. Okano, Y. Sato, K. Odashima, *J. Am. Chem. Soc.* **2002**, *124*, 12060–12061.

291 S. Ikeda, R. Sanuki, H. Miyachi, H. Miyashita, M. Taniguchi, and K. Odashima,

J. Am. Chem. Soc. **2004**, *126*, 10331–10338.

292 G. M. Mahandru, A. R. L. Skauge, S. K. Chowdhury, K. K. D. Amarasinghe, M. J. Heeg, J. Montgomery, *J. Am. Chem. Soc.* **2003**, *125*, 13481–13485.

293 (a) G.-Q. Lin, R. Hong, *J. Org. Chem.* **2001**, *62*, 2877–2880; (b) R. Hong, R. Hoen, J. Zhang, G.-Q. Lin, *Synlett* **2001**, 1527–1530; (c) J.-G. Lei, R. Hong, S.-G. Yuan, G.-Q. Lin, *Synlett* **2002**, 927–930.

294 (a) G. Q. Zheng, P. M. Kenny, J. Zhang, L. K. Lam, *Nutr. Cancer* **1993**, *19*, 77–86; (b) S. Kobayashi, Y. Mimura, K. Notoya, I. Kimura, M. Kimura, *Jpn. J. Pharmacol.* **1992**, *60*, 397–401.

295 S. Yu, S. You, H. Chen, *Yaoxue Xuebao* **1984**, *19*, 486, *Chem. Abstr.* **1984**, *101*, 22490c.

296 N. V. Purohit, S. N. Mukherjee, *J. Inst. Chemists (India)* **1997**, *69*, 149.

297 D. D. Wheeler, D. C. Young, *Chem. Abstr* **1961**, *55*, 2577.

298 T. Sambaiah, D.-J. Huang, C.-H. Cheng, *J. Chem. Soc. Perkin Trans. 1* **2000**, *1*, 195–203.

299 J. Uenishi, R. Kawahama, O. Yonemitsu, *J. Org. Chem.* **1997**, *62*, 1691–1701.

300 K. Tissot-Croset, D. Polet, S. Gille, C. Hawner, A. Alexakis, *Synthesis* **2004**, *15*, 2586–2590.

301 Z. Xi, Z. Li, C. Umeda, H. Guan, P. Li, M. Kotora, T. Takahashi, *Tetrahedron* **2002**, *58*, 1107–1117.

302 G. W. Gribble, *J. Chem. Soc. Perkin Trans. 1* **2000**, 1045–1075.

303 K. Hiroya, S. Itoh, M. Ozawa, Y. Kanamori, T. Sakamoto, *Tetrahedron Lett.* **2002**, *43*, 1277–1280.

304 G. Altenhoff, F. Glorius, *Adv. Synth. Catal.* **2004**, *346*, 1661–1664.

305 S. Cacchi, G. Fabrizi, L. M. Parisi, R. Bernini, *Synlett* **2004**, *2*, 287–290.

306 G. Nordmann, S. L. Buchwald, *J. Am. Chem. Soc.* **2003**, *125*, 4978–4979.

307 J. Barluenga, F. Aznar, S. Barluenga, M. Fernández, A. Martín, S. García-Granda, A. Piñera-Nicolás, *Chem. Eur. J.* **1998**, *4*, 2280–2298.

308 (a) H.-P. Wu, R. Aumann, R. Fröhlich, P. Saarenketo, *Chem. Eur. J.* **2001**, *7*, 700–

710; (b) H.-P. Wu, R. Aumann, R. Fröhlich, B. Wibbeling, *Eur. J. Org. Chem.* **2000**, 1183–1192.

309 N. Iwasawa, M. Shido, H. Kusama, *J. Am. Chem. Soc.* **2001**, *123*, 5814–5815.

310 D. S. La, J. G. Ford, E. S. Sattely, P. J. Bonitatebus, R. R. Schrock, A. H. Hoveyda, *J. Am. Chem. Soc.* **1999**, *121*, 11603–11604.

311 G. K. Tranmer, W. Tam, *Org. Lett.* **2002**, *4*, 4101–4104.

312 V. Khedkar, A. Tillack, M. Michalik, M. Beller, *Tetrahedron Lett.* **2004**, *45*, 3123–3126.

313 K. Rück-Braun, T. Martin, M. Mikulás, *Chem. Eur. J.* **1999**, *5*, 1028–1037.

314 E. A. Anderson, J. E. P. Davidson, J. R. Harrison, P. T. O'Sullivan, J. W. Burton, I. Collins, A. B. Holmes, *Tetrahedron* **2002**, *58*, 1943–1971.

315 J. H. Rigby, C. R. Heap, N. C. Warshakoon, *Tetrahedron* **2000**, *56*, 2305–2311.

316 J. Barluenga, M. A. Fernández-Rodríguez, F. Andina, E. Aguilar, *J. Am. Chem. Soc.* **2002**, *124*, 10978–10979.

317 (a) V. Mamane, T. Gress, H. Krause, A. Fürstner, *J. Am. Chem. Soc.* **2004**, *126*, 8654–8655; (b) A. Fürstner, F. Stelzer, H. Szillat, *J. Am. Chem. Soc.* **2001**, *123*, 11863–11869.

318 M. R. Luzung, J. P. Markham, F. D. Toste, *J. Am. Chem. Soc.* **2004**, *126*, 10858–10859.

319 G. Dyker, D. Hildebrandt, *J. Org. Chem.* **2005**, *70*, 6093–6096.

320 L. Zhang, S. Wang, *J. Am. Chem. Soc.* **2006**, *128*, 1442–1443.

321 J. C. Anderson, B. P. McDermott, E. J. Griffin, *Tetrahedron* **2000**, *56*, 8747–8767.

322 R. Uma, N. Gouault, C. Crévisy, R. Grée, *Tetrahedron Lett.* **2003**, *44*, 6187–6190.

323 D. Cuperly, C. Crévisy, R. Grée, *J. Org. Chem.* **2003**, *68*, 6392–6399.

324 P. Wipf, C. Kendall, C. R. J. Stephenson, *J. Am. Chem. Soc.* **2003**, *125*, 761–768.

325 Y. Li, T. J. Marks, *J. Am. Chem. Soc.* **1998**, *120*, 1757–1771.

326 G. A. Molander, S. K. Pack, *Tetrahedron* **2003**, *59*, 10581–10591.

7
Domino Reactions Initiated by Oxidation or Reduction

During the past few years, increasing numbers of reports have been published on the subject of domino reactions initiated by oxidation or reduction processes. This was in stark contrast to the period before our first comprehensive review of this topic was published in 1993 [1], when the use of this type of transformation was indeed rare. The benefits of employing oxidation or reduction processes in domino sequences are clear, as they offer easy access to reactive functionalities such as nucleophiles (e. g., alcohols and amines) or electrophiles (e. g., aldehydes or ketones), with their ability to participate in further reactions. For that reason, apart from combinations with photochemically induced, transition metal-catalyzed and enzymatically induced processes, all other possible constellations have been embedded in the concept of domino synthesis.

7.1
Oxidative or Reductive/Cationic Domino Processes

This chapter begins by classifying the combinations of oxidation/reduction processes with subsequent cationic transformations, though to date the details of only two examples have been published. The first example comprises an asymmetric epoxidation/ring expansion domino process of aryl-substituted cyclopropylidenes (e. g., 7-1) to provide chiral cyclobutanones 7-3 via 7-2, which was first described by Fukumoto and coworkers (Scheme 7.1) [2].

This procedure was used for the asymmetric total synthesis of the steroid (+)-equilenin (7-7) [3]. Cyclopropylidene derivates 7-4 could be converted into the cyclobutanones 7-5 in good yields by applying an asymmetric epoxidation using the chiral (salen)Mn^{III} complex 7-6 (Scheme 7.2) [4]. It is of interest that the demethoxylated substrate 7-4b led to 7-5b with a very high enantiomeric excess of 93%, whereas 7-4a gave 7-5a with only 78% ee.

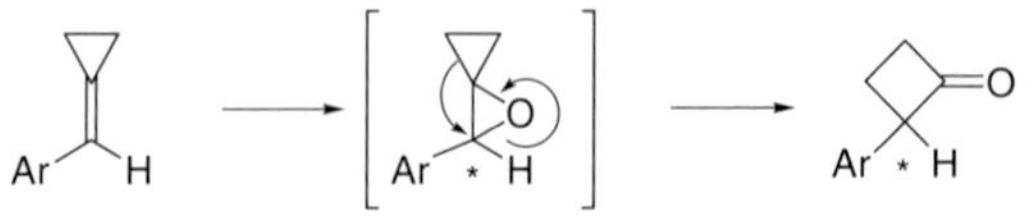

Scheme 7.1. Domino asymmetric epoxidation/ring expansion reaction.

Domino Reactions in Organic Synthesis. Lutz F. Tietze, Gordon Brasche, and Kersten M. Gericke
Copyright © 2006 WILEY-VCH Verlag GmbH & Co. KGaA, Weinheim
ISBN: 3-527-29060-5

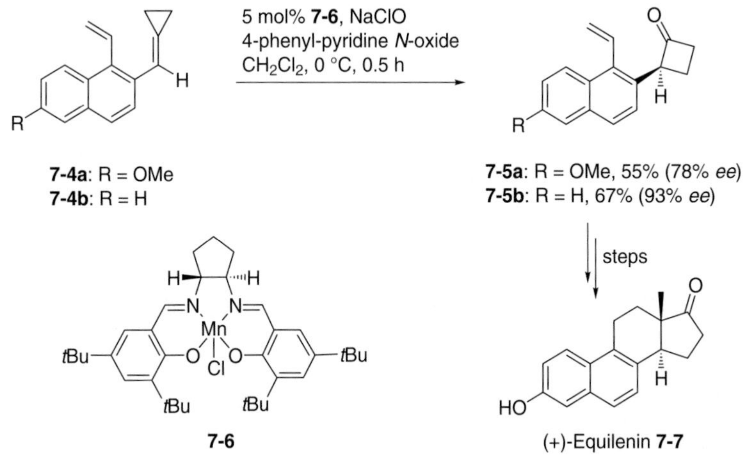
5 mol% 7-6, NaClO
4-phenyl-pyridine N-oxide
CH2Cl2, 0 °C, 0.5 h
7-4a: R = OMe
7-4b: R = H
7-5a: R = OMe, 55% (78% ee)
7-5b: R = H, 67% (93% ee)
steps
tBu
Mn
Cl
7-6
HO
(+)-Equilenin 7-7

Scheme 7.2. Asymmetric expoxidation/ring-expansion reaction using chiral (salen)Mn^{III} complex **7-6**.

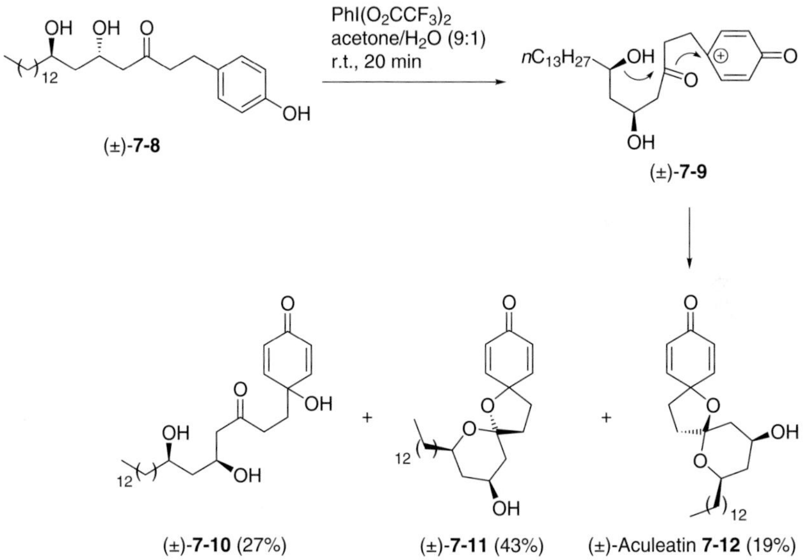
PhI(O2CCF3)2
acetone/H2O (9:1)
r.t., 20 min
(±)-7-8
nC13H27
(±)-7-9
(±)-7-10 (27%)
(±)-7-11 (43%)
(±)-Aculeatin 7-12 (19%)

Scheme 7.3. Biomimetic oxidative domino cyclization.

The second example was reported by Baldwin, Bulger and coworkers, and features the oxidation of a phenol initiating a cationic cyclization sequence to afford the natural product (±)-aculeatin D (**7-12**) [5]. Thus, when the dihydroxyketone **7-8** is treated with $PhI(O_2CCF_3)_2$, a formal two-electron oxidation of the phenol moiety takes place, triggering a twofold cationic-based cyclization to furnish the desired (±)-aculeatin D [(±)-**7-12**] in 19% yield, together with 43% of the isomer (±)-**7-11** and the side product (±)-**7-10** *via* the cation **7-9** (Scheme 7.3).

7.2
Oxidative or Reductive/Anionic Domino Processes

Anion-triggered reactions, as discussed earlier in Chapter 2, embody transformations in most cases, in which a nucleophile acts as the attacking species towards an electrophile. Since oxidation and reduction procedures are well established for providing nucleophilic or electrophilic functionalities, they can be combined with anionic process.

Taylor and coworkers described an *in situ* alcohol oxidation/Wittig (or Horner–Wadsworth–Emmons) domino reaction which is the only example of this type of domino reaction [6]. The main virtue of this approach is that there is no need to isolate the intermediate aldehydes, which are often volatile, toxic, and highly reactive. As seen in Schemes 7.4 and 7.5, it was possible to convert benzylic, allylic and propargylic alcohols with nonstabilized phosphonium salts and stabilized phosphonates directly into the desired alkenes, by using MnO_2 as oxidizing agent, a guanidine as base and, in some cases, titanium isopropoxide as a promoter, in good to excellent yield. However, the stereoselectivity is not always sufficient.

In the construction of C=N bond-containing compounds, such as nitrogen heterocycles, the aza-Wittig methodology has received increased attention as the method of choice [7]. Thus, an easy access to optically active (–)-vasicinone **(7-15)**, a pyrrolo[2,1-*b*]quinazoline alkaloid which is used in indigenous medicine [8], was

Entry	Alcohol	Conditions[a]	Product	Z:E	Yield [%]
1	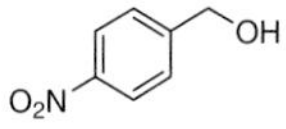 O_2N ... OH	Ph_3PCH_3Br 4 h	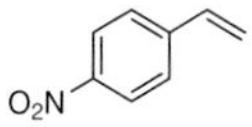 O_2N ...	–	91
2	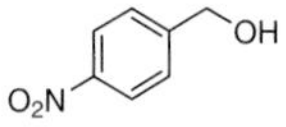 O_2N ... OH	Ph_3PCH_2PhCl 12 h	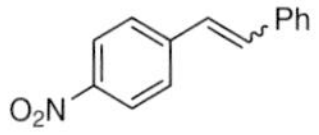 O_2N ... Ph	4:5	92
3	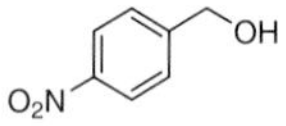 O_2N ... OH	Ph_3PCH_2PrBr $Ti(iPrO)_4$, 24 h	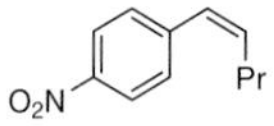 O_2N ... Pr	4:1	64
4	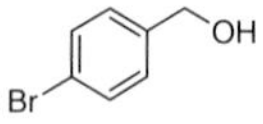 Br ... OH	Ph_3PCH_3Br $Ti(iPrO)_4$, 48 h	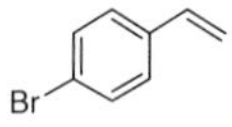 Br ...	–	88
5	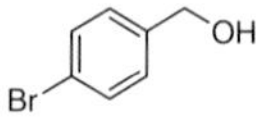 Br ... OH	Ph_3PCH_2PrBr $Ti(iPrO)_4$, 72 h	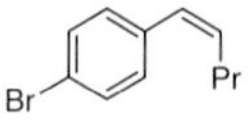 Br ... Pr	5:1	64
6	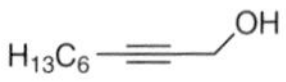 $H_{13}C_6$ ... OH	Ph_3PCH_2PhCl $Ti(iPrO)_4$, 24 h	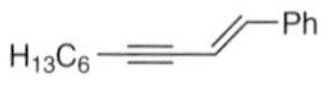 $H_{13}C_6$ ... Ph	1:2	62

[a] 10 eq MnO_2 in refluxing THF, 1.1 eq phosphonium salt, 2.2 eq guanidine 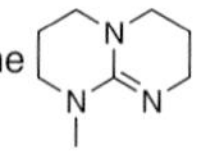 and in some cases 1.0 eq $Ti(iPrO)_4$.

Scheme 7.4. Oxidation/Wittig sequence I.

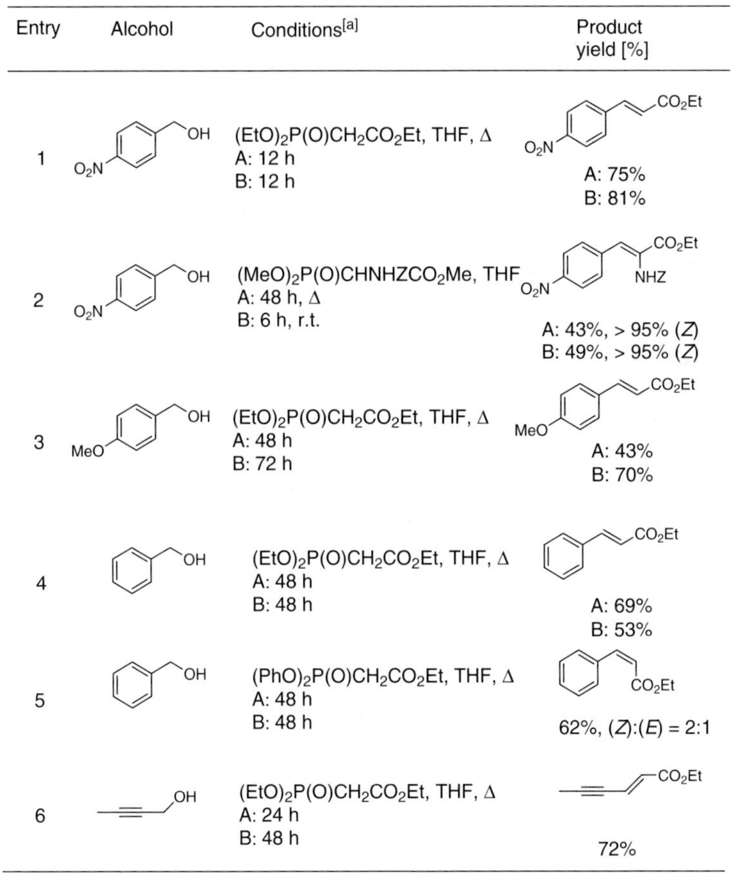

Entry	Alcohol	Conditions[a]	Product yield [%]
1	O$_2$N–C$_6$H$_4$–CH$_2$OH	(EtO)$_2$P(O)CH$_2$CO$_2$Et, THF, Δ A: 12 h B: 12 h	O$_2$N–C$_6$H$_4$–CH=CH–CO$_2$Et A: 75% B: 81%
2	O$_2$N–C$_6$H$_4$–CH$_2$OH	(MeO)$_2$P(O)CHNHZCO$_2$Me, THF A: 48 h, Δ B: 6 h, r.t.	O$_2$N–C$_6$H$_4$–C(=CH...)(CO$_2$Et)(NHZ) A: 43%, > 95% (Z) B: 49%, > 95% (Z)
3	MeO–C$_6$H$_4$–CH$_2$OH	(EtO)$_2$P(O)CH$_2$CO$_2$Et, THF, Δ A: 48 h B: 72 h	MeO–C$_6$H$_4$–CH=CH–CO$_2$Et A: 43% B: 70%
4	C$_6$H$_5$–CH$_2$OH	(EtO)$_2$P(O)CH$_2$CO$_2$Et, THF, Δ A: 48 h B: 48 h	C$_6$H$_5$–CH=CH–CO$_2$Et A: 69% B: 53%
5	C$_6$H$_5$–CH$_2$OH	(PhO)$_2$P(O)CH$_2$CO$_2$Et, THF, Δ A: 48 h B: 48 h	C$_6$H$_5$–CH=CH–CO$_2$Et 62%, (Z):(E) = 2:1
6	HC≡C–CH$_2$OH	(EtO)$_2$P(O)CH$_2$CO$_2$Et, THF, Δ A: 24 h B: 48 h	HC≡C–CH=CH–CO$_2$Et 72%

[a] 10 eq MnO$_2$ in refluxing THF, 1.2 eq phosphonate and
A) 2.2 eq guanidine.
B) 2.0 eq LiOH/MS 4 Å.

Scheme 7.5. Oxidation/Wittig sequence II.

achieved using a domino Staudinger reduction/intramolecular aza-Wittig process, as shown by the Eguchi group in 1996 (Scheme 7.6) [9]. Addition of tri-n-butyl-phosphine to o-azidobenzoic imide **7-13** led to a reduction of the azide group and additional formation of a phosphine imide species. The latter conducts an aza-Wittig reaction with the more suitable carbonyl group of the imide to give the tricycle **7-14** in 76% yield. Finally, cleavage of the silyl ether in **7-14** utilizing TBAF provided (–)-vasicinone (**7-15**) in excellent 82% yield and 97% *ee*.

Recently, the same procedure has been used by Williams and coworkers in their approach to the natural product (–)-stemonine (**7-18**) [10], an alkaloid which exhibits broad biological activity [11]. Conversion of the formyl azide **7-16** in a Staudinger reaction with ethyldiphenylphosphine first generated a phosphineimide which was used for the subsequent aza-Wittig process (Scheme 7.7). The furnished seven-

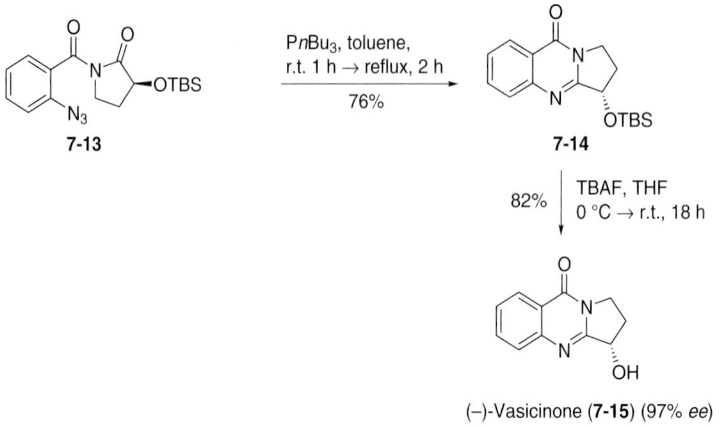

Scheme 7.6. Synthesis of (–)-vasicinone (**7-15**) by a domino Staudinger-reduction/aza-Wittig reaction.

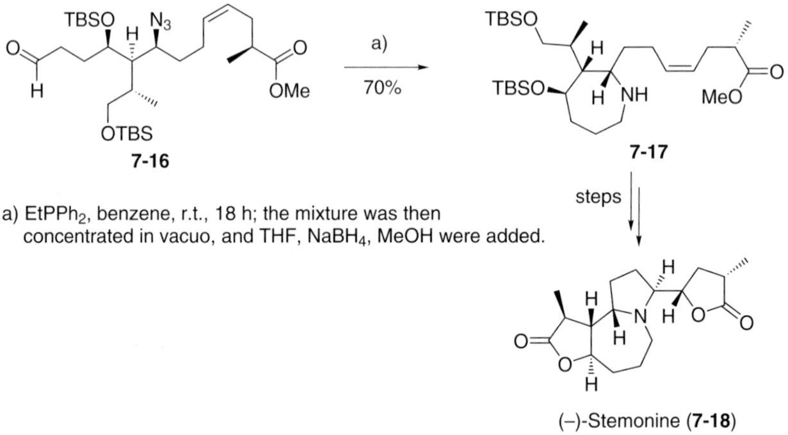

a) EtPPh$_2$, benzene, r.t., 18 h; the mixture was then
concentrated in vacuo, and THF, NaBH$_4$, MeOH were added.

Scheme 7.7. Synthesis of (–)-stemonine (**7-18**).

membered imine was immediately treated with NaBH$_4$ after changing the solvent from benzene to a mixture of THF and methanol to give the desired amine **7-17** in 70 % overall yield.

The Staudinger/aza-Wittig procedure has recently also been used by Mellet and Fernández for the synthesis of calystegine B$_2$, B$_3$, and B$_4$ analogues, emphasizing the growing importance of this methodology [12].

α-Amino acids constitute a valuable natural source of chiral substrates [13]. However, the use of dicarboxylic amino acids, such as aspartic or glutamic acid, is frequently complicated as a consequence of the attendance of two different carboxyl

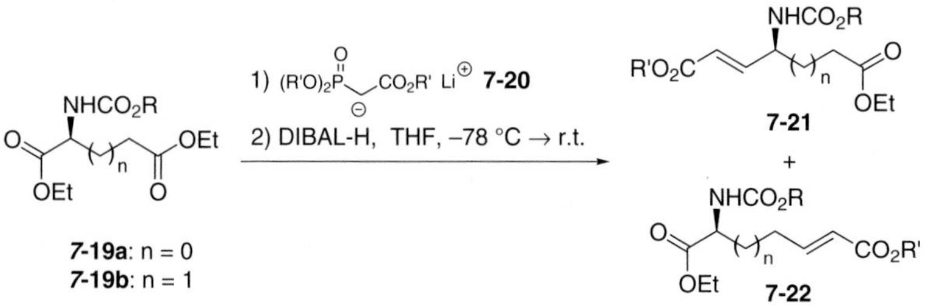

Scheme 7.8. Synthesis of dialkyl N-alkoxycarbonyl-α,β-unsaturated dicarboxylates.

Entry	7-19	7-20	Ratio 7-21/7-22	Yield [%] 7-21
1	n = 0, R = Me	R' = Et	100:0	62
2	n = 1, R = Me	R' = Et	> 40:1	74
3	n = 0, R = Me	R' = Me	> 20:1	58
4	n = 1, R = tBu	R' = Et	> 13:1	66

groups. Knaus and coworkers have developed a method for the differentiation of both acid functionalities in these compounds by applying a one-pot DIBAL-H reduction/Wittig–Horner reaction at the α-ester group [14]. When N-alkoxycarbonyl-protected L-diethyl aspartate 7-19a or L-diethyl glutamate 7-19b was treated with DIBAL-H at low temperatures, and in the presence of triethylphosphonoacetate lithium salt (7-20), γ-amino-α,β-unsaturated dicarboxylates 7-21 were obtained in good yields, high regioselectivity, and without loss of optical purity (Scheme 7.8). In some cases the side products 7-22 were also formed.

In a comparable approach, Roques and coworkers [15] converted the aspartic acid derivative 7-23, which bears a protected thiol in the β-position, into unsaturated compounds 7-27 and 7-28, respectively (Scheme 7.9).

It should be noted that by using the phosphonate 7-25 as well as the triphenylphosphane derivative 7-26, only the E-isomers were formed and a loss of the optical purity was not detectible. In contrast, complete racemization at both stereogenic centers was observed performing a Wittig–Horner reaction with the corresponding free aminoaldehyde. The observed result may be explained by the formation of a favored five-membered ring chelate 7-24 as intermediate using 7-23 as substrate, in which the aluminum coordinates to the amido- and the ester-group, leading to the observed preference of subsequent reduction.

Another domino process, designed by Polt and coworkers [16], deals with the consecutive transformation of an in situ-prepared aldehyde to give β-amino allylic alcohols 7-31 from β-amino acids. When the β-amino acid ester derivative 7-29 is sequentially treated with iBu$_5$Al$_2$H and vinyl magnesium bromide, a 3:2 mixture of the allylic alcohol derivatives 7-30 is obtained in 60% yield, which can be hydrolyzed to give 7-31 (Scheme 7.10).

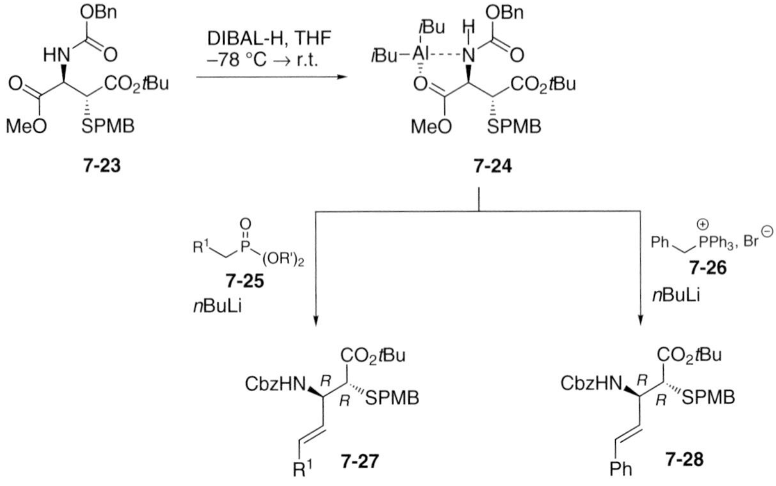

Scheme 7.9. Domino reduction/Wittig-Horner olefination process of aspartates.

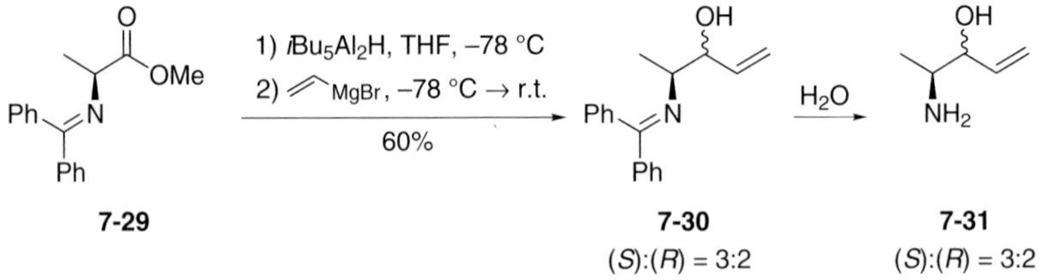

Entry	Reagent **7-25** and **7-26**	R^1	Product **7-27** and **7-28** yield [%]
1	$(Et_2O)_2P(O)CH_2SO_3CH_2tBu$	SO_3CH_2tBu	68
2	$(Et_2O)_2P(O)CH_2CO_2Bn$	CO_2Bn	58
3	$(Et_2O)_2P(O)CH_2CONHtBu$	$CONHtBu$	17
4	$(Et_2O)_2P(O)CH_2CON(CH_3)$-OMe	$CON(CH_3)$-OCH_3	72
5	$Ph_3P^{\oplus}CH_2Ph, Br^{\ominus}$	Ph	60

Scheme 7.10. Synthesis of β-amino allylic alcohols.

A facile synthesis of cyclic five-, six-, and seven-membered oxonitriles **7-35** by a combination of an ozonolysis and an aldol reaction has been described by the Fleming group [17]. As starting material the unsaturated oxonitriles **7-33** are used, which are easily accessible by reaction of unsaturated esters **7-32** with $LiCH_2CN$ (Scheme 7.11). It can be assumed that **7-34** acts as an intermediate.

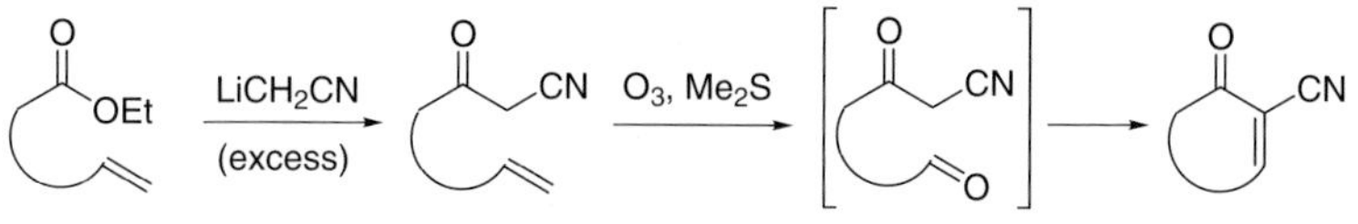

Entry	Ketonitrile	Yield [%]	Oxonitrile	Yield [%]
1	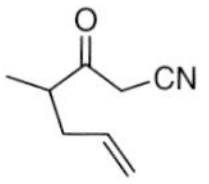	81	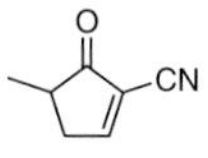	83[a]
2	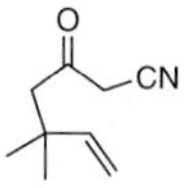	73	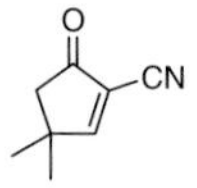	91[a]
3	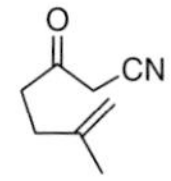	71	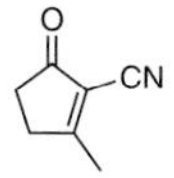	91[b]
4	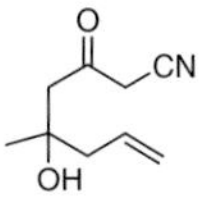	70	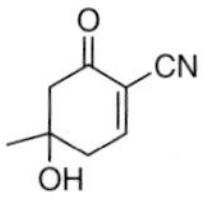	53
5	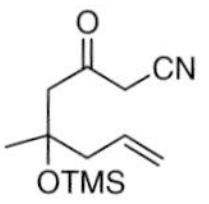	58	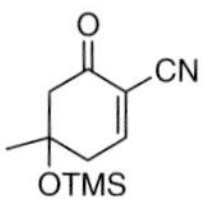	57

[a] The intermediate aldehydes cyclize during silica gel chromatography.
[b] The intermediate ketone is cyclized by sequential treatment with
   CaH_2 and aqueous NH_4Cl.

Scheme 7.11. Domino ozonolysis/aldol cyclization of unsaturated ketonitriles.

A similar approach in which alcohols are oxidized by IBX or DMP to give aldehydes that are directly attacked in an aldol-fashion by a tethered malonate moiety
has been published by Malacria and coworkers [18].

A convenient procedure for the lactonization of alkenols has been recently revealed by Borhan and coworkers [19]. This methodology was successfully applied in
the total synthesis of (+)-tanikolide (**7-37**), a natural product of marine origin which
exhibits antifungal activities. Thus, when alkenol **7-36** is subjected to soluble oxone
and a catalytic amount of OsO_4, a smooth domino oxidative cleavage/lactonization
process takes place which leads, after debenzylation, to the desired product in good
overall yield (Scheme 7.12).

The Bunce group disclosed a straightforward domino process for the construction of aryl-fused nitrogen heterocycles by employing a combination of a reduction
and a Michael addition [20]. The transformation involves an initial Fe-mediated reduction of nitroarenes **7-38**, furnishing an aniline which undergoes a subsequent

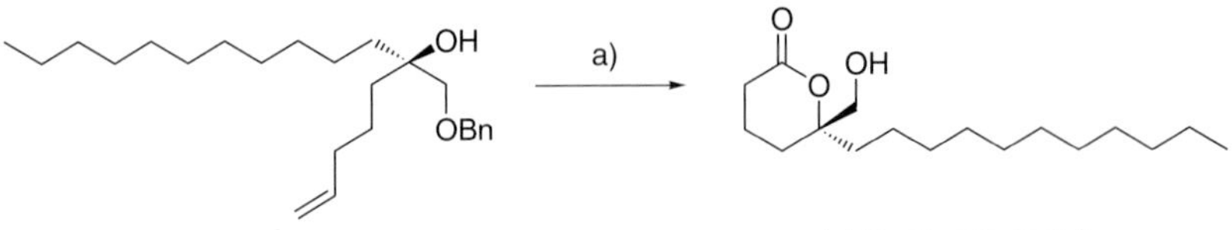

7-36

a)

(+)-Tanikolide (7-37)

a) 1) nBu₄NHSO₅/OsO₄ (cat), THF (73%)
   2) Pd(OH)₂/H₂, EtOAc, r.t. (87%).

Scheme 7.12. Oxidative cleavage/lactonization in the synthesis of (+)-tanikolide (**7-37**).

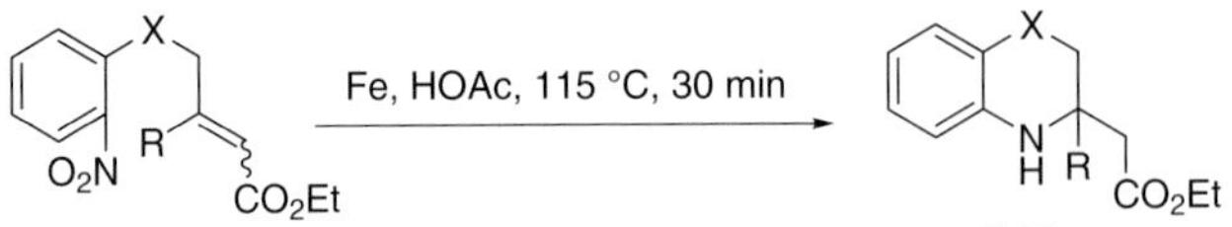

7-38   Fe, HOAc, 115 °C, 30 min   **7-39**

Entry	Substrate 7-38	X	R	Product 7-39	Yield [%]
1	a	CH₂	H	a	98
2	b	CH₂	Me	b	89
3	c E	O	H	c	94
4	d Z	O	H	c	98
5	e E	O	Me	d	88
6	f	NH	H	e	89
7	g	NH	Me	f	86

Scheme 7.13. Synthesis of aryl-fused nitrogen heterocycles.

Michael addition to give the azaheterocycles **7-39** through a favorable 6-*exo*-trig ring closure (Scheme 7.13). Even highly sensible substrates such as aryl-allyl ethers and amines could be transformed to the desired products in excellent yields.

Domino reduction/anionic processes have also been used for the synthesis of natural products. Such an approach was used by the Nicolaou group [21] in the total synthesis of diazonamide A (**7-43**), a substance which was of marine origin, has a highly strained and unprecedented structure, and exhibited cytotoxicity against several tumor cell lines. As one of the last steps, the *N,O*-acetal moiety in **7-43** was formed by reduction of the five-membered ring lactam in **7-40** with DIBAL-H to give an imine **7-41**; this is followed by a ring closure with the adjacent phenolic hydroxyl group to deliver **7-42** in 55 % overall yield (Scheme 7.14). The total synthesis of diazonamide A (**7-43**) was then accomplished by reductive Cbz group removal, followed by installation of the remaining side chain **7-44** in 82 % overall yield.

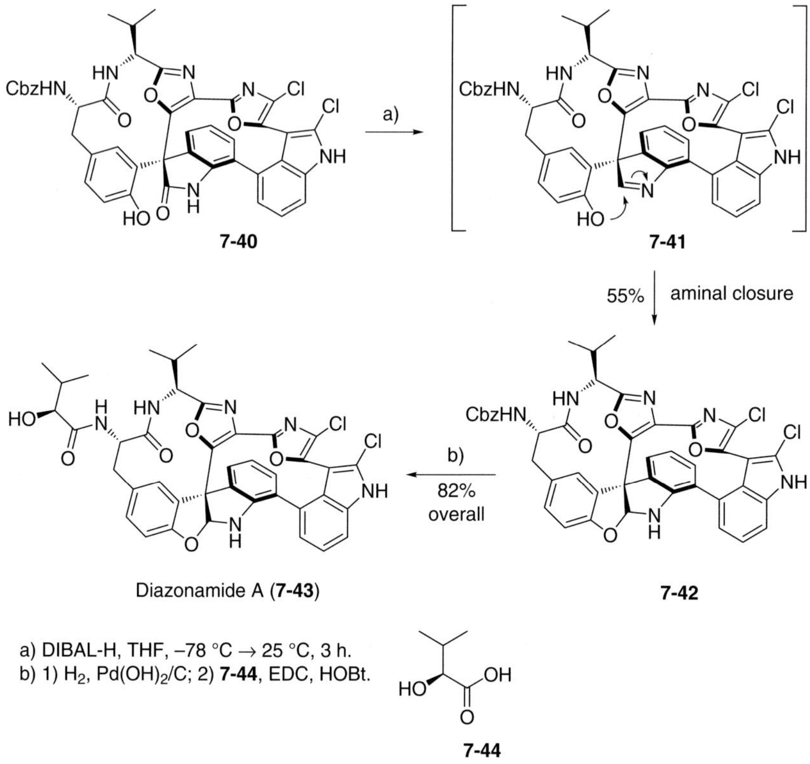

a) DIBAL-H, THF, −78 °C → 25 °C, 3 h.
b) 1) H$_2$, Pd(OH)$_2$/C; 2) **7-44**, EDC, HOBt.

Scheme 7.14. Domino reduction/nucleophilic cyclization procedure in the synthesis of diazonamide A (**7-43**).

7.2.1
Oxidative or Reductive/Anionic/Anionic Domino Processes

Domino oxidative/reductive anionic processes with more than two steps are also known. For example, in an effort to synthesize analogues of CC-1065, a potent anti-tumor antibiotic [22], the group of Boger [23] successfully designed a novel domino sequence consisting of an initial oxidation followed by several anionic transformations. Thus, MnO$_2$-mediated oxidative coupling of 2-(benzyloxy)ethylamine **7-47** with 5-hydroxyindole **7-45** led to the pyrrolo[3,2-*e*]benzoxazole **7-53** in 48 % yield (Scheme 7.15). **7-53** can be considered as a scaffold of a DNA-binding subunit. It can be assumed that, in the conversion of **7-45** to give **7-53**, an oxidation of **7-45** first takes place affording *p*-quinone monoimine **7-46**; this then undergoes a regioselective C-4 addition with the primary amine **7-47** to furnish **7-48**. Several more steps follow this *via* the intermediates **7-49** to **7-52**.

2-Isoxazolines are valuable substrates in organic synthesis as they can be transformed into useful building blocks such as γ-amino alcohols, β-hydroxy ketones,

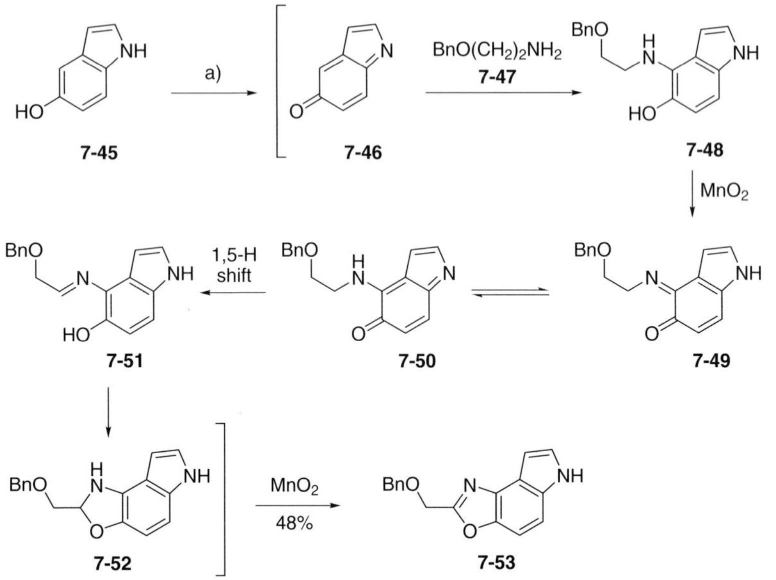

a) 30 wt eq MnO$_2$, 2 eq (benzyloxy)ethylamine (**7-45**), ethylenglycol dimethylether, 0 °C → r.t., 14 h.

Scheme 7.15. Synthesis of a DNA-binding subunit.

and β-hydroxy nitriles. 4-Hydroxy-2-isoxazolines have also been shown to be suitable intermediates for the synthesis of amino sugars and other biologically important molecules [24]. An established method for producing 2-isoxazolines is the reaction of nitroacetates and α-bromo enones or α-bromoaldehydes [25]. Recently, Righi and coworkers described a so-far unknown domino process towards 4-hydroxy-4,5-dihydroisoxazoles 2-oxides **7-56** using α-epoxyaldehydes **7-55** which were prepared *in situ* by the oxidation of epoxyalcohols such as **7-54** with (diacetoxy)iodobenzene (DAIB) and catalytic amounts of TEMPO (Scheme 7.16) [26]. Once formed, these were treated with α-bromoesters or amides, polymer-supported nitrite and diisopropylethylamine (DIPEA) to give the corresponding α-nitroesters or amides which induce the nitro-aldol reaction. There follows a nucleophilic ring closure with simultaneous opening of the epoxide to give the 4,5-dihydroisoxazoles 2-oxides **7-56** as an approximate 3:2 mixture of 4,5-*cis* and 4,5-*trans* isomers in excellent yields.

The isoxazoles can also be prepared in stepwise manner from the epoxyaldehydes **7-55** using α-nitroacetic acid derivatives and DIPEA, though the yields are much lower.

Another procedure for preparing 4-hydroxy-4,5-dihydroisoxazole 2-oxides derivatives in 86–100 % yield, and published by the same group [27], is based on an oxidative cleavage of D-mannitol-derivatives furnishing 2 equiv. of identical enantiopure (2R)-2-methanesulfonyloxyaldehydes which can react with α-nitroacetate in the usual manner.

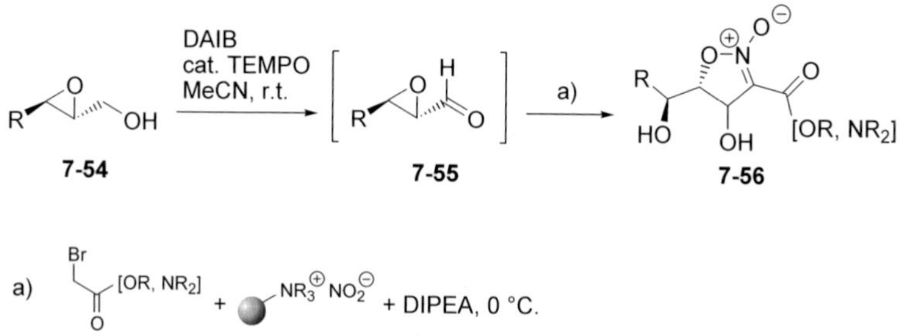

Scheme showing conversion of **7-54** → **7-55** → **7-56** with conditions DAIB, cat. TEMPO, MeCN, r.t.; reagent a).

a)

Br–CH(–C(=O)[OR, NR$_2$]) + —NR$_3^{\oplus}$ NO$_2^{\ominus}$ + DIPEA, 0 °C.

Entry	Isoxozoles **7-56**	Domino [%]	Two-step [%]
a	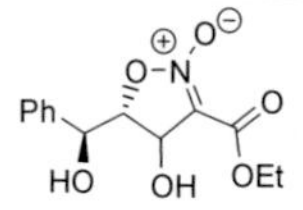 Ph, HO, OH, OEt	84	45
b	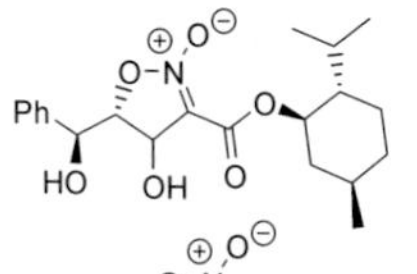 Ph, HO, OH (menthyl ester)	81	48
e	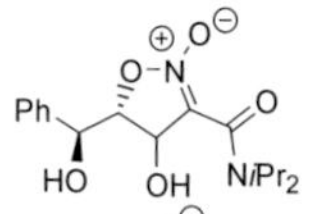 Ph, HO, OH, NiPr$_2$	74	20
f	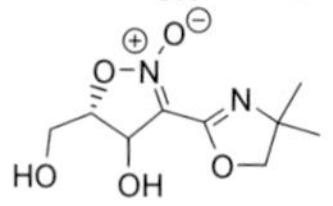 HO, OH (oxazoline)	89	33
g	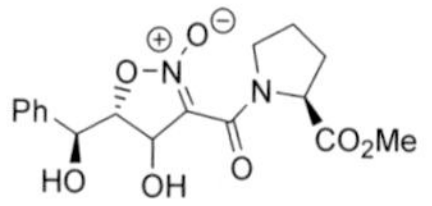 Ph, HO, OH, N (CO$_2$Me prolinate)	80	45

Scheme 7.16. Synthesis of 4-hydroxy-4,5-dihydroisoxazoles.

Another class of important heterocyclic compounds is represented by the piperazinones such as **7-61**, which have been often utilized in medicinal chemistry due to their structural similarity to constrained peptides. During the past few years, many examples of multiple bond-forming reactions between primary amines and doubly activated substrates to produce piperazinones have been developed [28]. Beshore and coworkers have discovered an effective and novel domino reaction towards these heterocycles, which is based on an reductive amination/transamidation/cyclization sequence by coupling an aldehyde such as **7-57** with an α-amino ester **7-58** in the presence of the reducing agent Na(AcO)$_3$BH (Scheme 7.17) [29]. The crucial and rate-limiting step within this procedure seems to be the intramolecular N,N'-acyl transfer converting intermediate **7-59** into **7-60**. The yields of the formed piperazinones **7-61** are very good, with the exception of compounds containing an aryl moiety in the α-position (R^1 = Ph, R^2 = H: 17% yield).

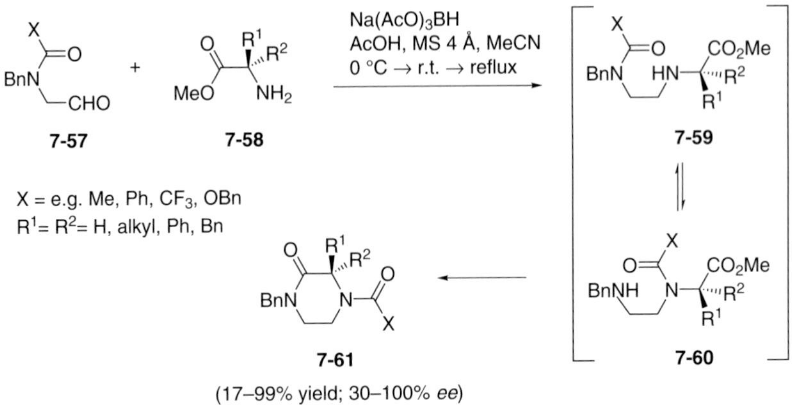

Scheme 7.17. Synthesis of piperazinone derivatives.

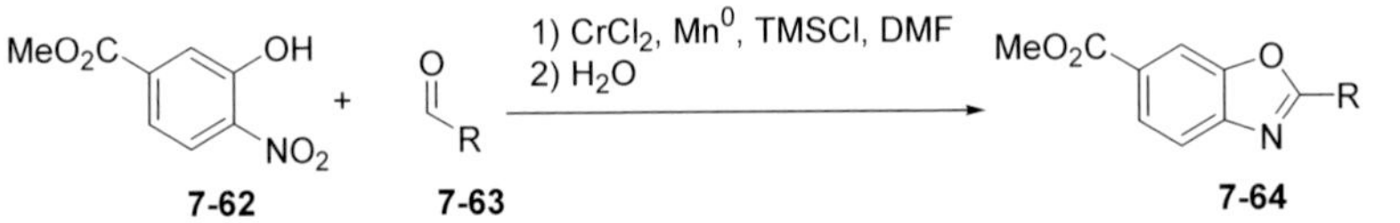

Entry	Aldehyde	Product	Yield [%]
1	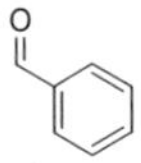	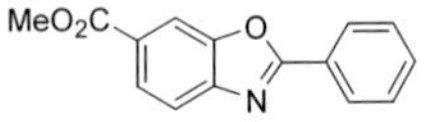	73
2	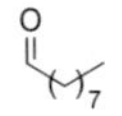	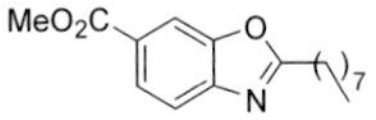	76
3	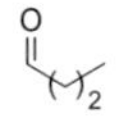	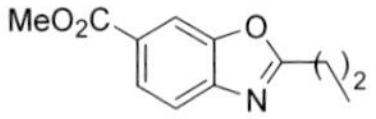	67
4	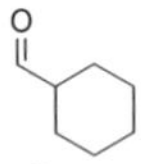	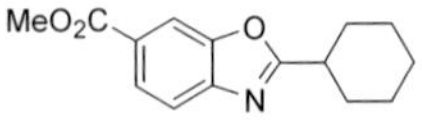	81
5	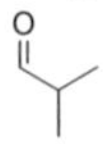	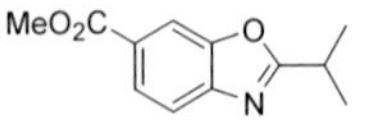	57
6	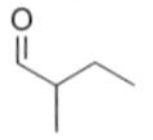	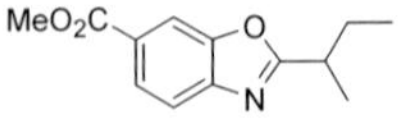	62
7	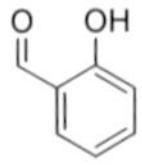	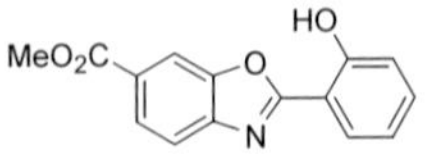	0

Scheme 7.18. Chromium-manganese redox couple-mediated domino process for the synthesis of benzoxazoles.

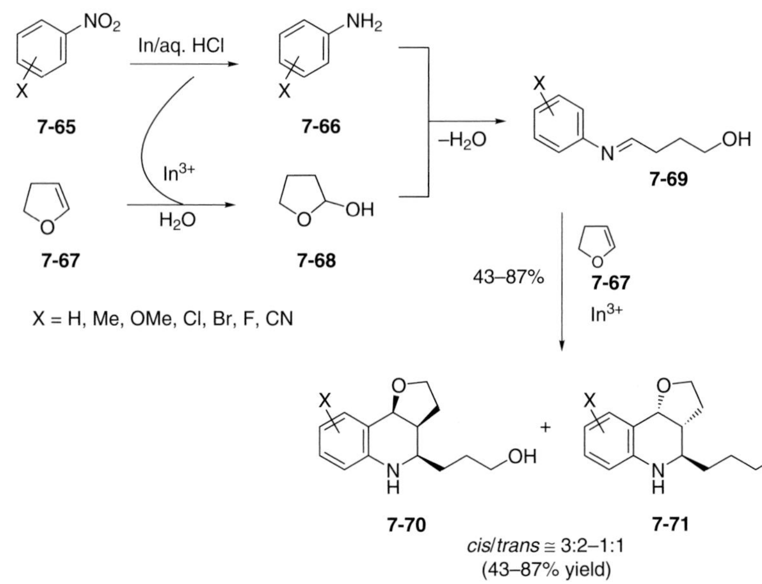

Scheme 7.19. Indium-mediated domino reaction for the synthesis of tetrahydroquinoline derivatives

It should be noted that, especially when using higher temperatures and longer reaction times, a considerable degree of racemization was observed. However, using
aldehydes **7-57** with X = CF$_3$, *ee*-values of up to 100% were found, together with
usually excellent yield.

A well-elaborated domino process for the synthesis of benzoxazoles **7-64** has
been published by the Miller group [30]. The strategy involves a chromium-manganese redox couple-mediated reduction of *o*-nitrophenol **7-62** to the corresponding
anilines (Scheme 7.18). Since TMSCl is needed as an additive, the formation of a
hexamethyldisilazane as intermediate is assumed. When the reaction was carried
out in the presence of an aldehyde such as **7-63**, instantaneous conversion to the
corresponding imine took place, which was then attacked by the phenolic hydroxy
group, leading to an aminal. Finally, a dehydrogenation reaction, which could be
caused by the *in situ*-formed Mn^{IV} species, furnished the desired benzoxazole
derivatives **7-64** in good yields. It is worth mentioning that this reaction only
proceeds in high yields when *o*-nitrophenols **7-62** are used which bear an electron-
withdrawing group *para* to the nitro group. Although this substitution pattern
decreases the ability of imine formation, it is proposed that the shift from the imine
to the aminal is facilitated.

A variety of natural products and pharmaceutical agents contain a tetrahydroquinoline moiety [31]. Recently, a simple and general access to these heterocycles by a so-far unknown domino reaction of aromatic nitro compounds **7-65** and
2,3-dihydrofuran mediated by indium in water has been described by Li and coworkers (Scheme 7.19) [32]. It is assumed that the process is initiated by reduction
of the nitro group in **7-65** to give the aniline **7-66** on treatment with indium in

aqueous HCl. The indium(III) ion thus generated catalyzes the hydration of 2,3-di-hydrofuran **7-67** to give hemiacetal **7-68**, which forms an imine **7-69** with **7-66**. There follows an indium(III)-promoted aza-Diels–Alder reaction with reversed electron-demand employing another equivalent of **7-67** as electron-rich dienophile, which leads to the tricycles **7-70** and **7-71** in almost 1:1 mixture. The obtained yields vary between 43 % and 87 %, but are generally higher when the substituent X has a low electron-withdrawing effect.

Due to its high complexity, the total synthesis of the *Strychnos* alkaloid (–)-strychnine (**7-76**) remains a major challenge [33]. A new approach for the enantioselective construction of this alkaloid was reported by Shibasaki and coworkers, in which the B- and D-rings are simultaneously assembled by an elegant domino reduction/biscyclization process (Scheme 7.20) [34]. After amination of **7-72** with **7-73**, the crude product **7-74** was treated with Zn in MeOH/aqueous NH_4Cl to provoke the formation of tetracycle **7-75** in 77 % yield. This domino process is thought to start with the reduction of the nitro group with zinc, generating an aniline. Subsequent 1,4-addition of the secondary amine to the enone and irreversible indole formation of the aniline moiety with the resulting ketone led to the desired tetracyclic compound **7-75** in 77 % over two steps. Remarkably, establishing this process made it possible to skip more than eight steps during the total synthesis, underlining once more the amazing potency of domino reactions.

The Hanna group developed a rapid route to (+)-calystegine B_2 (**7-79**) [35], a natural product with strong glycosidase enzyme-inhibiting properties [36]. As the key step, a triple domino procedure first published by Madsen and coworkers was em-

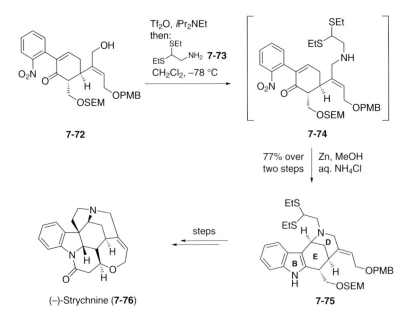

Scheme 7.20. Zn-mediated domino reduction/biscyclization process in the total synthesis of (–)-strychnine (**7-76**).

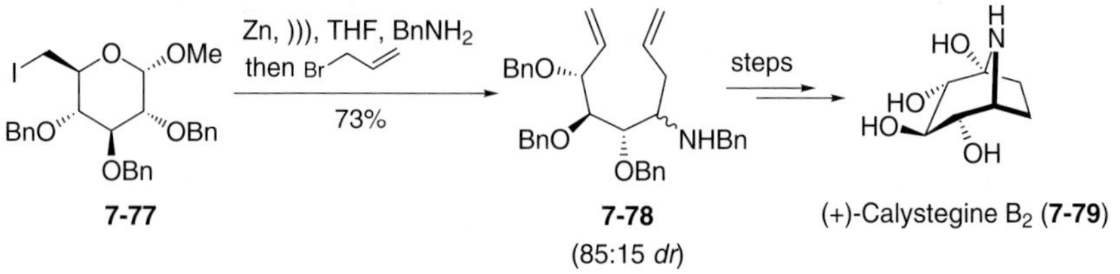

Scheme 7.21. Zinc-mediated domino reaction in the total synthesis of (+)-calystegine B₂ (**7-79**).

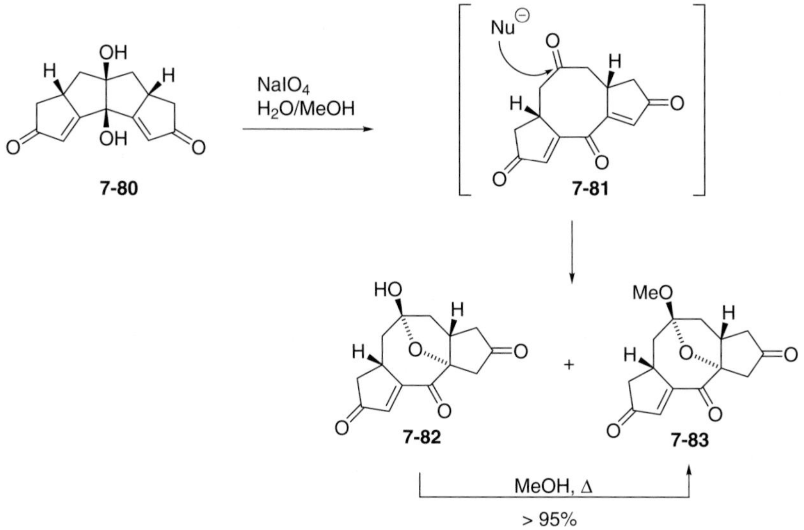

Scheme 7.22. Synthesis of a [5.8.5] ring system.

ployed [37]. Reaction of 6-iodoglucopyranose **7-77** with zinc dust in the presence of benzylamine, followed by addition of allyl bromide in THF under sonication, led to the amino diene **7-78** as a separable 85:15 mixture of diasteromers in 73 % yield (Scheme 7.21) [38, 39]. The reaction proceeds *via* an initially reductive fragmentation of **7-77** by zinc furnishing a 5,6-unsaturated aldehyde which is trapped *in situ* as the corresponding benzyl imine. The latter is then allylated by slow addition of allyl bromide, resulting in formation of the amino diene **7-78**. Metathesis and introduction of a hydroxy function afforded the desired natural product in a good overall yield (25 % over six steps).

The final example in this section deals with the potent synthesis of a [5.8.5] ring system, as achieved by Cook and coworkers [40]. When tetracyclic diol **7-80** is treated with NaIO₄, a tetraketone **7-81** is formed, presumably by an oxidative C–C bond cleavage (Scheme 7.22). There follows a nucleophilic attack on the nonconjugated carbonyl group by either water or methanol. The generated oxygen anion then reacts with the α,β-unsaturated ketone in a Michael fashion, furnishing the bridged hemiacetal **7-82** and acetal **7-83**. Evidence for the proposed mechanism was obtained from the conversion of the hemiacetal **7-82** into the acetal **7-83** in hot methanol. In this event, the tetraketone **7-81** is generated by undergoing a retro-Michael addition, followed by addition of methanol and reattack at the enone system to provide **7-83**.

7.2.2
Oxidative/Anionic/Pericyclic Domino Processes

The use of masked *o*-benzoquinones or *o*-benzoquinoid structures (MOB) as potential synthons for Diels–Alder reactions have furnished elegant and reliable approaches to a multitude of structurally diverse organic compounds [41, 42].

o-Benzoquinones can easily be formed by an oxidation of phenols using hypervalent iodine reagents such as bis(trifluoroacetoxy)-iodobenzene (BTIB), first reported by Wessely and coworkers during the 1950s [43]. However, due to their high tendency to dimerize, the handling of *o*-benzoquinones is rather difficult, and this reduces their usefulness as building blocks. A good solution to bypass this problem has been developed by Wood and coworkers, in which intramolecular trapping of the oxidation product by an appropriate tethered carboxylic acid moiety provides spirolactonedienones **7-85** which are relatively stable against dimerization (Scheme 7.23) [44].

When the oxidation of, for example, **7-84** was carried out in the presence of a suitable dienophile, the formed dienone is directly consumed within an intermolecular Diels–Alder reaction, leading to cycloadduct **7-89**, albeit in a yield below 40 %. In

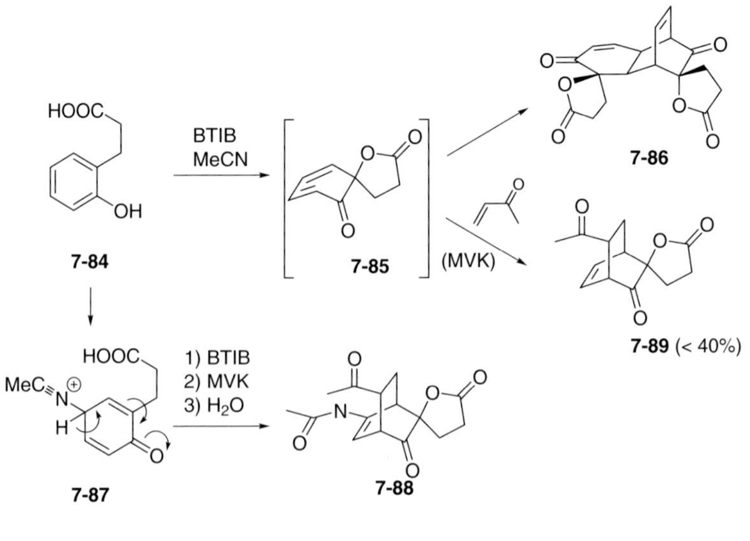

BTIB: bis(trifluoroacetoxy)-iodobenzene

Scheme 7.23. Domino oxidation/spirolactonization/Diels–Alder reaction sequence.

7-90 → **7-91**

| | | **7-91** Yield [%] | | |
Entry	Dienophile	R = I	R = Ac	R = CH$_3$
1	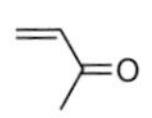	89	65	84
2	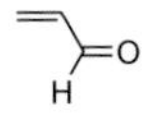	70	56	51
3	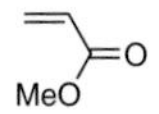	71	75	62
4	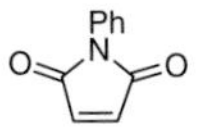	79	73	59

Scheme 7.24. Synthesis of spiro-compounds **7-91**.

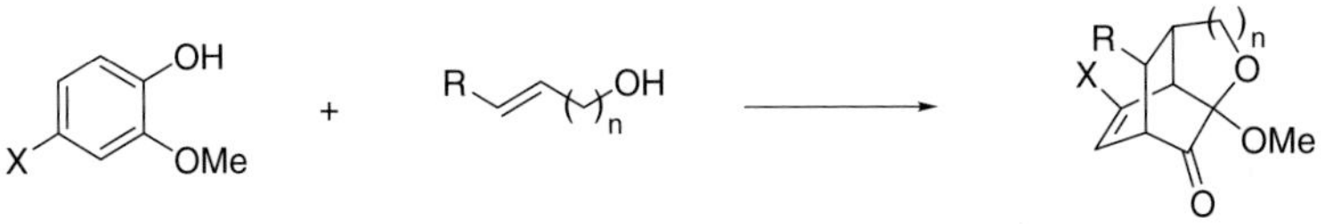

7-92
X = CO$_2$Me
X = Me

7-93
R = H, Me, Ph
n = 1 and 2

7-94
8 examples: 39–80% yield

Scheme 7.25. Acetalization/Diels–Alder sequence.

addition, the dimer **7-86** and cycloadduct **7-88**, which is formed *via* **7-87**, were obtained. The picture changed, however, when *para*-substituted aromatic phenols **7-90** in the presence of different dienophiles were used, which gave the desired compounds **7-91** in much better yield (Scheme 7.24).

Another synthesis using *in situ*-prepared *o*-benzoquinones has been reported by Liao and coworkers [45]. These authors observed that oxidation of 2-methoxy-phenols **7-92** with DAIB in the presence of unsaturated alcohols **7-93** furnish transient *o*-benzoquinone monoacetal intermediates, which easily undergo an intramolecular Diels–Alder reaction to provide bicyclo[2.2.2]octenones **7-94** with high regio- and stereoselectivity, as well as in acceptable yields (Scheme 7.25).

7.2.3

Oxidative or Reductive/Anionic/Oxidative or Reductive Domino Processes

It is well known that the reactivity of a double bond is highly influenced by its adjacent substituents. Electron-donating groups increase the reactivity towards an electrophilic addition, whereas electron-withdrawing groups favor a nucleophilic attack. A neat solution for performing a nucleophilic addition on electron-rich double bonds has recently been developed by the Williams group, featuring a reversible oxidation process [47]. This concept of temporarily oxidizing alcohols has been demonstrated to allow a Michael-addition of methylmalononitrile to an allylic alcohol **7-95** by a procedure involving a catalytic electronic activation of the substrate (Scheme 7.26). In this way, a domino Oppenauer-oxidation/Michael addition/Meerwein–Ponndorf–Verley reduction process led to substituted alcohols such as **7-96**.

An example of a highly chemo-, regio-, and stereoselective domino ester reduction/epoxide formation/reductive epoxide-opening process has been recently published by Huang and coworkers in their synthesis of pine sawfly sex pheromones [48]. Subjecting substrate (2*S*,3*R*)-**7-97** to LiAlH₄ triggers a reaction sequence, involving the reduction of the ester functionalities to afford **7-98**, S_Ni reaction with inversion of configuration leading to epoxide **7-99**, which then underwent a reductive epoxide-opening reaction at the terminal carbon giving the diol (2*S*,3*R*)-**7-100** in 79 % yield (Scheme 7.27).

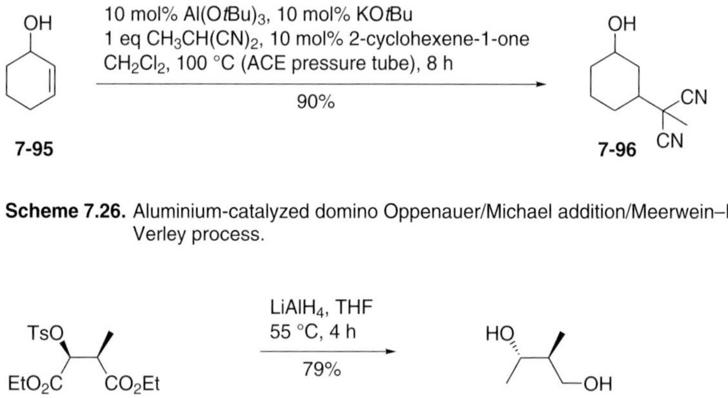

Scheme 7.26. Aluminium-catalyzed domino Oppenauer/Michael addition/Meerwein–Ponndorf–Verley process.

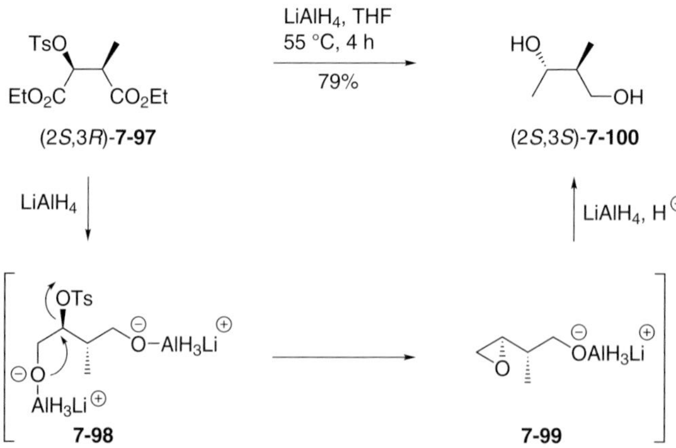

Scheme 7.27. Domino ester reduction/epoxide formation/reductive epoxide-opening reaction.

7.3
Oxidative or Reductive/Pericyclic Processes

Oxidative-pericyclic processes, and in particular the oxidative/Diels–Alder reaction, are quite common in nature. The so-called Diels–Alderase is usually an oxidizing enzyme, which induces, for example, the formation of a suitable dienophile such as an enone from an allylic alcohol [49].

Such reactions are also possible in vitro, as several mild oxidizing agents are at hand nowadays. Thus, the Dess–Martin periodinane (DMP) [50] has been proven to be a versatile and powerful reagent for the mild oxidation of alcohols to the corresponding carbonyl compounds. In this way, a series of new iodine(V)-mediated reactions has been developed which go far beyond simple alcohol oxidation [51]. Nicolaou and coworkers have developed an effective DMP-mediated domino polycyclization reaction for converting simple aryl amides, urethanes and ureas to complex phenoxazine-containing polycycles. For example, reaction of the *o*-hydroxy anilide **7-101** with DMP (2 equiv.) in refluxing benzene under exposure to air led to polycycle **7-103** *via* **7-102** in a yield of 35 % (Scheme 7.28) [52].

Interestingly, the oxidation of anilide **7-104** with 4 equiv. DMP led to **7-105** with the same tetracyclic core (Scheme 7.29).

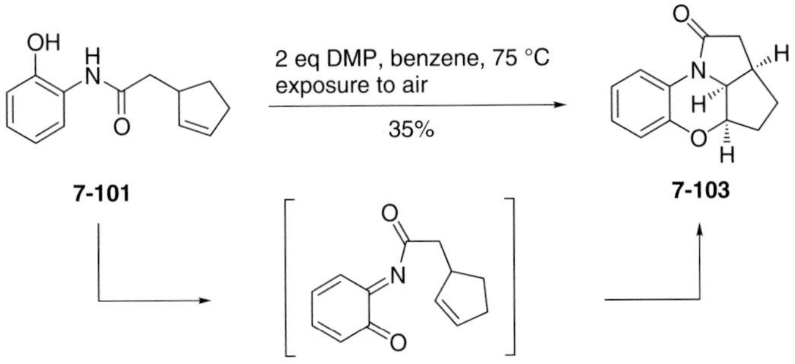

Scheme 7.28. DMP-induced domino oxidation/hetero-Diels–Alder sequence.

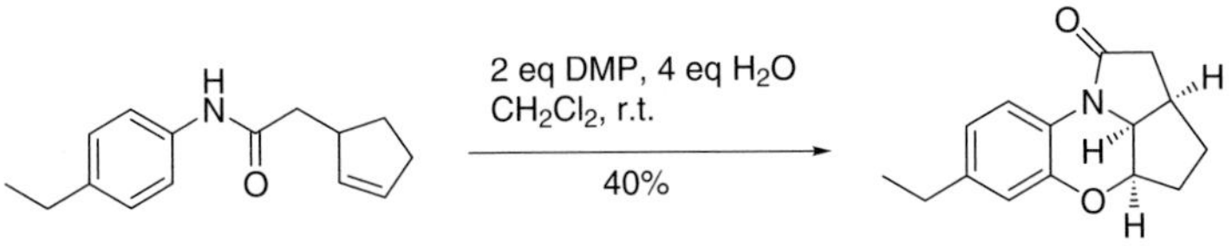

Scheme 7.29. Synthesis of tetracycle **7-105** from **7-104**.

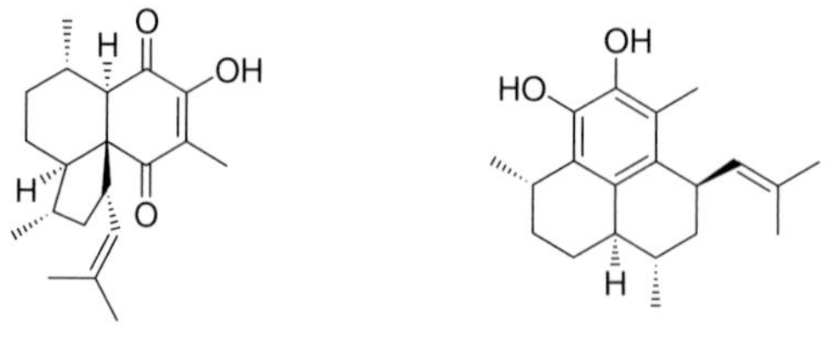

Elisabethin A (**7-106**)   Pseudopterosin A aglycone (**7-107**)

7-108   7-109   7-110

7-111   7-112   7-113

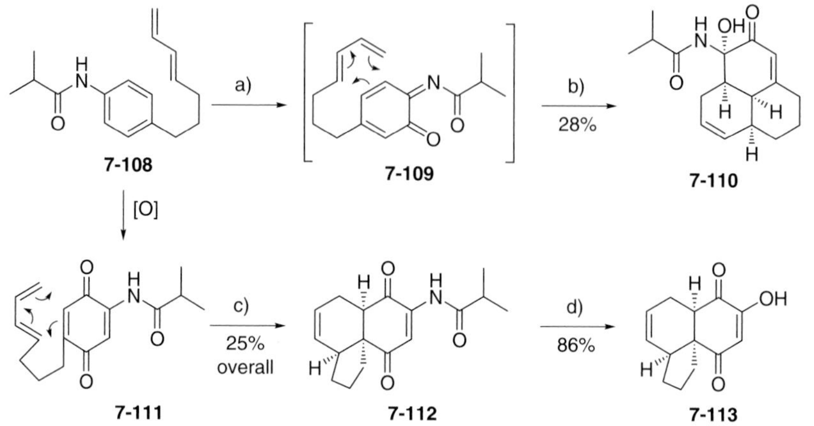

a) 4 eq DMP, 2 eq H$_2$O, CH$_2$Cl$_2$, 3 h, r.t.; b) 1. [4+2] cycloaddition; 2. + H$_2$O;
c) toluene, reflux, 12 h; [4+2] cycloaddition; d) K$_2$CO$_3$, MeOH.

Scheme 7.30. DMP-induced domino oxidation/hetero-Diels–Alder procedure for the
rapid entry into complex elisabethin and pseudopterosin structures.

The oxidative formation of *p*-benzoquinones from anilides such as **7-108** was
used for the synthesis of the core scaffold of the natural products elisabethin A (**7-106**) and pseudopterosin A aglycone (**7-107**) (Scheme 7.30). Exposure of anilide **7-108** to DMP [53] led to the formation of the *o*-imidoquinone **7-109**, which underwent an intramolecular Diels–Alder reaction to give **7-110** in 28% yield after hydration. In a competitive pathway, the *p*-quinone **7-111** is also formed from **7-108**, which on heating in toluene again underwent an intramolecular Diels–Alder reaction to give cycloadduct **7-112** in 25% overall yield. Hydrolysis of **7-112** furnished the carbocyclic skeleton **7-113** of elisabethin A (**7-106**).

En route to the total synthesis of tashironin (**7-114a**) and the debenzoylated compound **7-114b**, which shows an interesting promotion of neurite growth, Danishefsky and coworkers have developed a domino oxidative dearomatization/transannular Diels–Alder reaction [54]. In this line, treatment of **7-115** with phenyliodine(III) diacetate (PIDA) led to an intermediate **7-116**, which immediately underwent a transannular Diels–Alder reaction to furnish the complex cycloadducts **7-117** in good yields (Scheme 7.31).

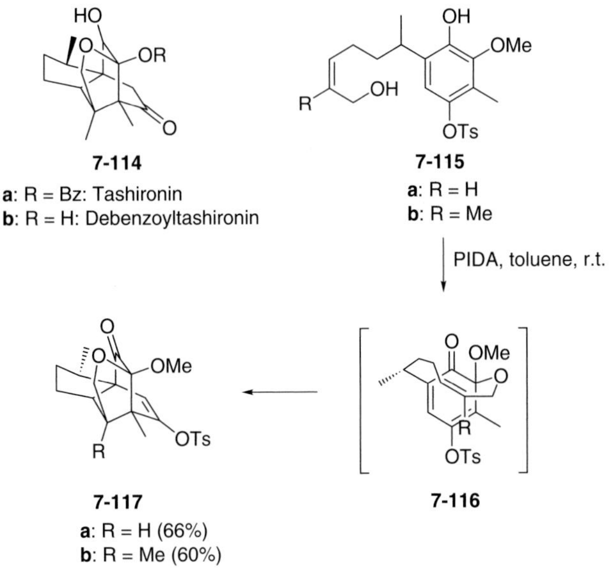

HO

7-114

a: R = Bz: Tashironin
b: R = H: Debenzoyltashironin

7-115

a: R = H
b: R = Me

PIDA, toluene, r.t.

7-117

a: R = H (66%)
b: R = Me (60%)

7-116

Scheme 7.31. Domino oxidative dearomatization/transannular-Diels–Alder reaction.

7.3.1
Oxidative/Pericyclic/Anionic Domino Processes

Lead tetraacetate is a well-established reagent for the oxidative cleavage of 1,2-diols, furnishing the corresponding aldehydes. An interesting aspect of this transformation is the design of domino processes by setting up the formed carbonyl functionalities in a way that further reactions can occur. As described by the group of Arseniyadis [55], a monocyclic unsaturated 1,2-diol such as **7-118** is cleaved to give the dialdehyde **7-119**, which can undergo an intramolecular hetero-Diels–Alder reaction (Scheme 7.32). The formed cyclic ene-acetal **7-120** suffers a rearrangement by an electrophilic attack of the metal on the electron-rich double bond in **7-120** *via* the temporary organolead intermediate **7-121** to give the carbocation **7-122**; this is trapped by an acetate group completing the synthesis of dioxa-bicyclo[2.2.2]octanes **7-123** which were obtained as a separable mixture of two diastereoisomers in 56% and 6% yield, respectively. It is important to mention that this reaction only proceeds when the conformational freedom of the hetero diene/dienophile motif is reduced by inserting appropriate groups (e. g. *gem*-dimethyl group).

The same group also published a similar procedure using unsaturated bicyclic diols **7-124** to yield the tricyclic ene-acetals **7-125**, using PhI(OAc)$_2$ as oxidizing agent. The products can be further transformed by treatment with Pb(OAc)$_4$, and this led to the tricyclic bisacetat **7-126** in remarkably good yield (Scheme 7.33) [56].

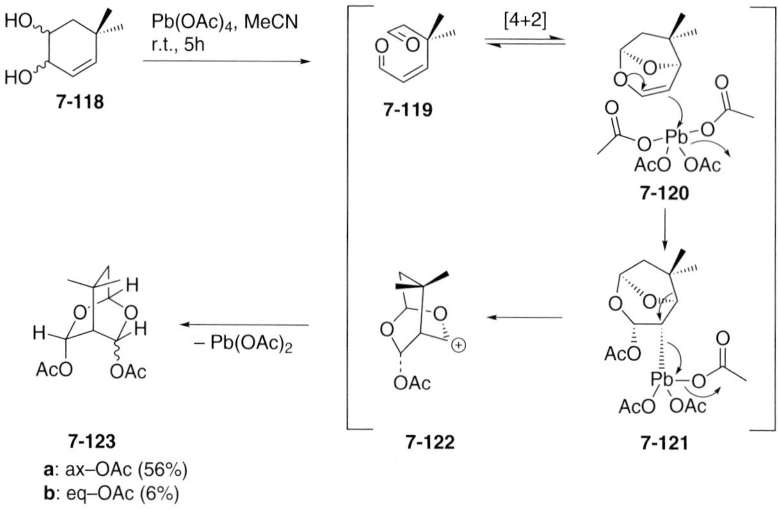

Scheme 7.32. Pb(OAc)₄-mediated synthesis of bicyclic **7-123**.

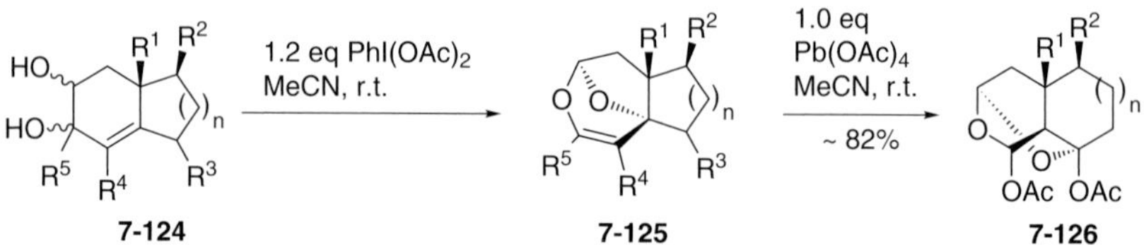

R^1 ⎫
R^2 ⎬ e.g. alkyl, alkenyl, alkoxy, free carbonyl, ketal

R^3= H, Me, iPr, OTMS
R^4= H, CO₂Me, Me
R^5= H, Me
n = 1,2

Scheme 7.33. General process of iodobenzene diacetate-mediated hetero-domino transformation.

Examples of the oxidative cleavage/bis-hetero-Diels–Alder domino sequence of compounds of type **7-124** to give acetals of type **7-125** are listed in Scheme 7.34. This shows that the reaction is high-yielding and can be carried out under rather mild conditions.

Moreover, compounds of type **7-127**, which were obtained from **7-125** by reaction with Pb(OAc)₄, can undergo a further domino process when treated with potassium carbonate in a mixture of water and methanol [57]. This includes saponification of the acetate moieties in **7-127** to provide **7-129** *via* the unstable cyclic hemiacetal **7-128** (Scheme 7.35). Retro-Claisen reaction and ring closure with the proposed intermediates **7-130** and **7-131** led to the bridged ring-system **7-132** as a mixture of diastereomers with preference of the β-isomer.

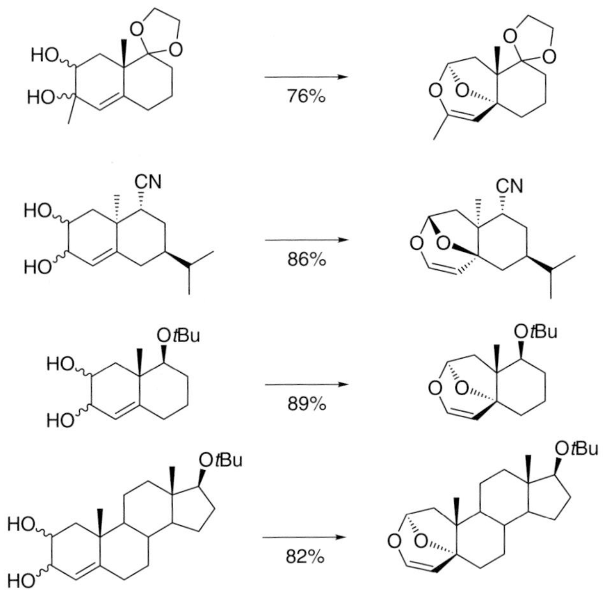

Reagents and conditions: 1.2 eq PhI(OAc)$_2$, MeCN, r.t., 12–24 h.

Scheme 7.34. Synthesis of endocyclic acetals of type **7-125**.

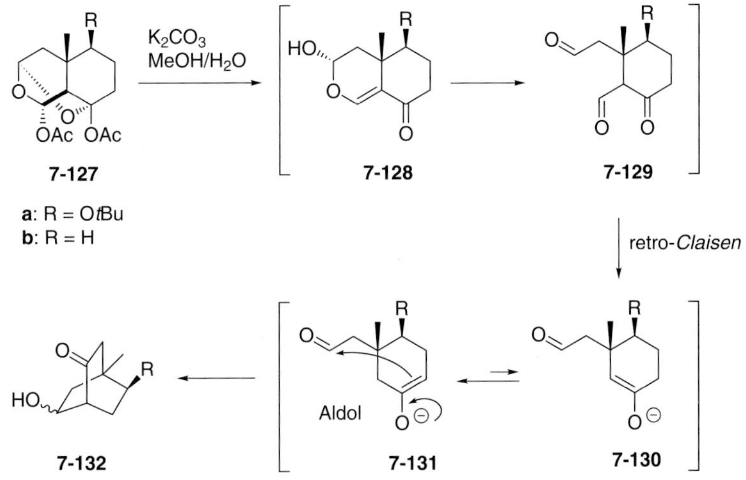

a: R = O*t*Bu, 68% (α:β = 1:10)
b: R = H, 61% (α:β = 1:8)

Scheme 7.35. Transformation of the bisacetate **7-127**.

7.3.2

Oxidative or Reductive/Pericyclic/Pericyclic Domino Processes

During their total synthesis of the cytotoxic natural product FR182877 (**7-133**) [58] and hexacyclinic acid (**7-134**) [59], the Evans group invented an impressive oxidation/twofold transannular Diels–Alder reaction process of the macrocycle **7-135** (Scheme 7.36) [60]. The sequence was initiated by introducing a conjugated double bond at C2–C3 of **7-135** using $Ph_2Se_2O_3$, SO_3/pyridine and NEt_3 in THF, leading to **7-136**. The setting was now prepared for a subsequent twofold cycloaddition, starting with an intramolecular Diels–Alder reaction with normal electron demand

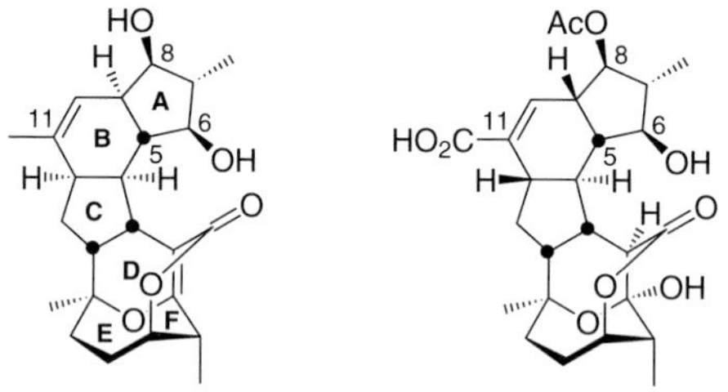

FR182877 (**7-133**)   Hexacyclinic acid (**7-134**)

7-135   **7-136**

7-138   63 %   **7-137**

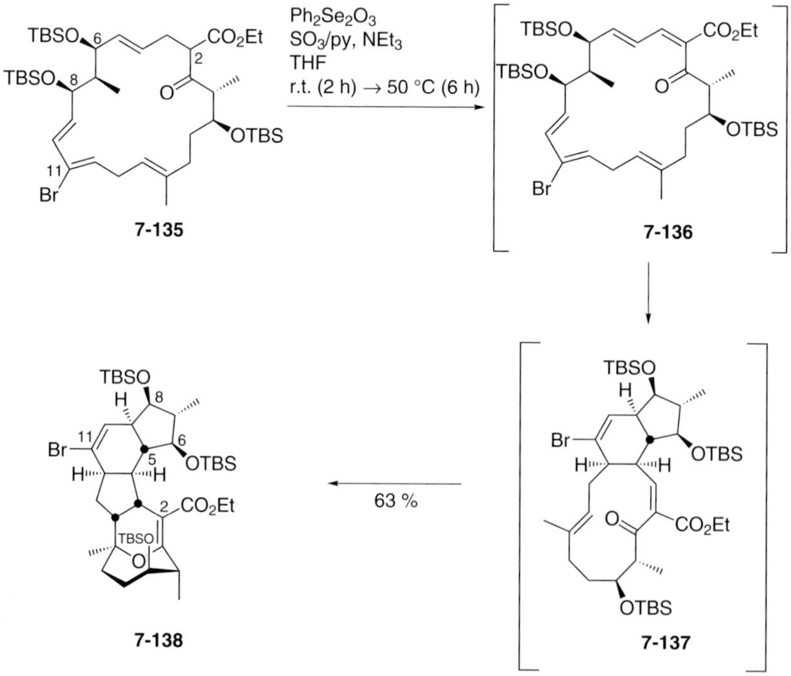

Scheme 7.36. Domino oxidation/twofold transannular-Diels–Alder reaction in the total synthesis of FR182877 (**7-133**).

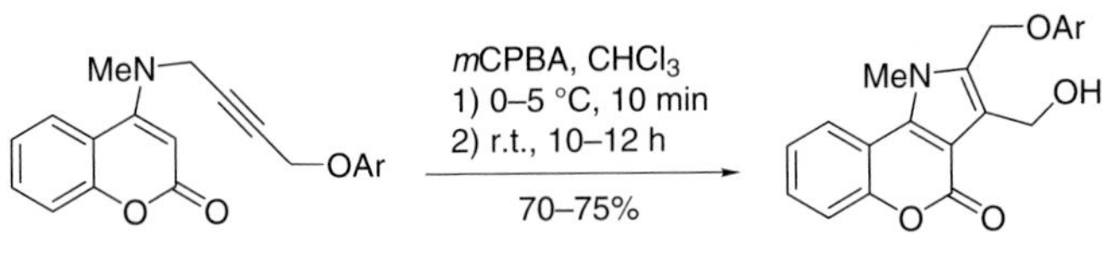

7-139

a: Ar = 2-ClC₆H₄
b: Ar = C₆H₅
c: Ar = 4-MeC₆H₄
d: Ar = 4-OMeC₆H₄
e: Ar = 2,4-Cl₂C₆H₃

7-140

Scheme 7.37. Oxidative twofold rearrangement sequence for the synthesis of coumarin derivatives.

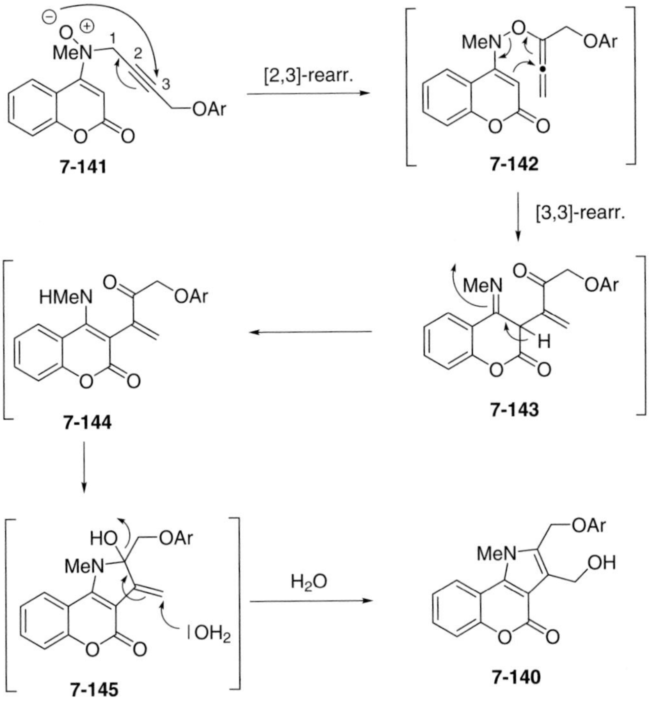

Scheme 7.38. Mechanism of the oxidative-twofold rearrangement sequence.

to give **7-137**, which was followed by an intramolecular hetero-Diels–Alder reaction with inverse electron demand to provide the pentacycle **7-138** as a single diastereoisomer in an excellent yield of 63 %.

Coumarins are known to have pronounced bioactivity. A rather new procedure for the synthesis of pyrrolo[3,2-c]coumarins uses a domino process in which an amine oxide rearrangement [61] is involved (Scheme 7.37) [62]. Reaction of the cou-

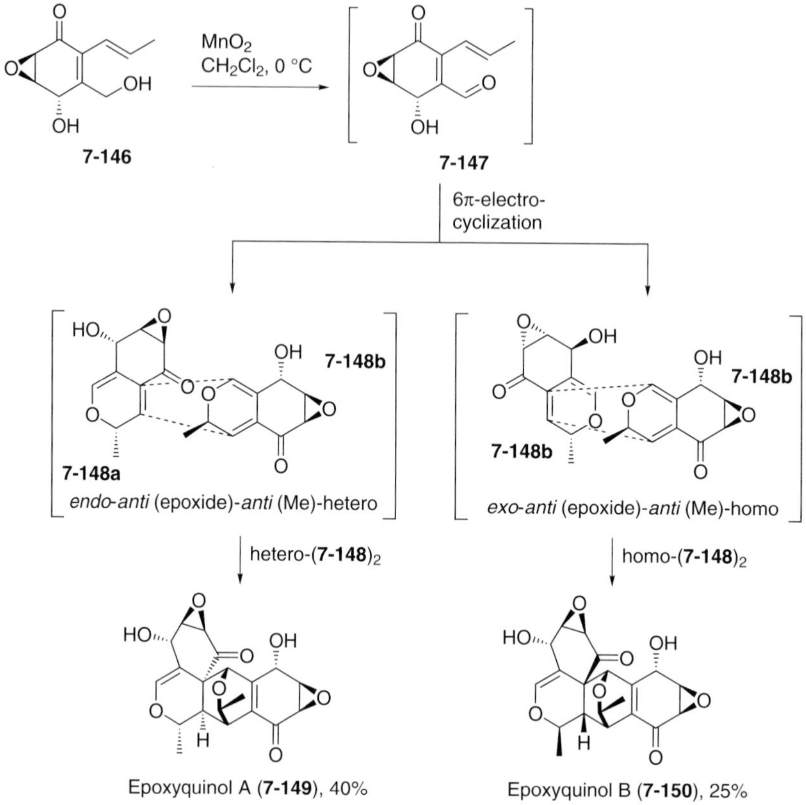

Scheme 7.39. Synthesis of epoxyquinol A (**7-149**) and B (**7-150**).

marin derivatives **7-139** with *m*CPBA provided **7-140** in a clean reaction, and good yields of 70–75 %.

A proposed mechanism for the transformation of **7-139** to **7-140** is depicted in scheme 7.38. Initially, the unstable *N*-oxide **7-141** is formed which undergoes a [2,3] sigmatropic rearrangement leading to allene **7-142**. There follows a [3,3] sigmatropic rearrangement providing imine **7-143**, which then tautomerizes to the enamine **7-144**. Addition of the amine moiety to the carbonyl group forms the hemiaminal, which reacts with water in a S_N2' displacement to yield the final products **7-140**.

An innovative approach to the epoxyquinols A (**7-149**) [63] and B (**7-150**) [64], two novel angiogenesis inhibitors with a heptacyclic ring system containing 12 stereogenic centers, has been developed by Hayashi and coworkers [65] using a biomimetic oxidation/6π-electrocyclization/(hetero)-Diels–Alder domino reaction. This type of transformation was first described by Porco, Jr. [66]. The reaction sequence is initiated by oxidation of the monomeric benzyl alcohol **7-146** to give aldehyde **7-147** using MnO₂ (Scheme 7.39). The latter undergoes a 6π-electrocyclization to generate the 2*H*-pyran derivatives **7-148a** and **7-148b**. Subsequent dimerization by a

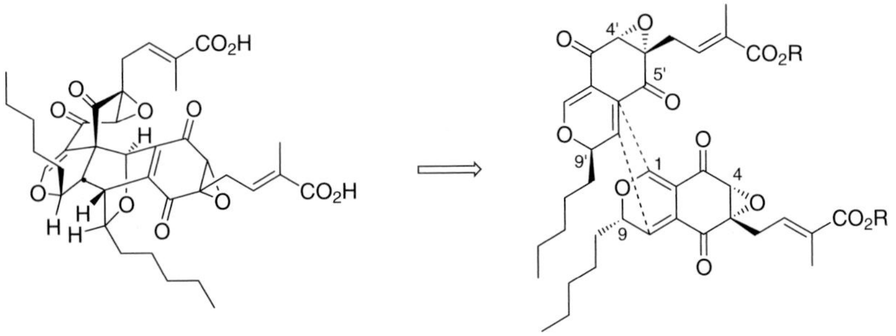

(±)-Torreyanic acid (**7-151**)　　　　　　　　**7-152**

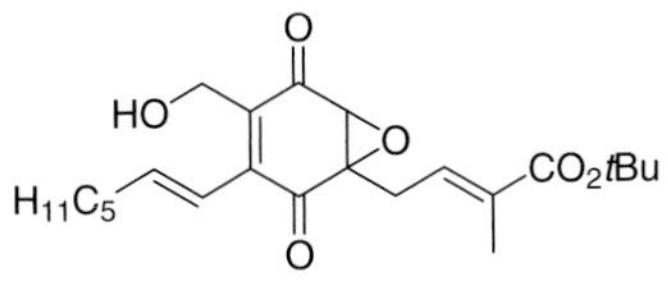

(±)-**7-153**

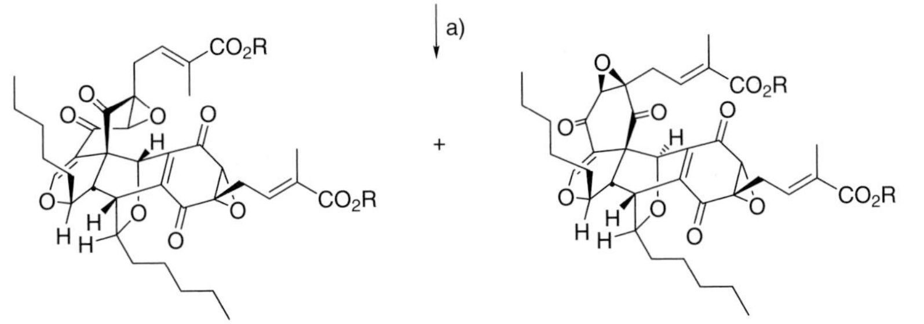

a)

+

(±)-**7-154** (39%)　　　　　　　　　　　(±)-**7-155** (41%)

b) ⌐ (±)-**7-154**: R = *t*Bu　　　　　　　b) ⌐ (±)-**7-155**: R = *t*Bu
　└→ (±)-**7-151**: R = H　　　　　　　　　└→ (±)-**7-156**: R = H

(±)-Torreyanic acid　　　　　　　　　　(±)-"*iso*"-Torreyanic acid

a) DMP, CH$_2$Cl$_2$, 1 h, SiO$_2$.
b) TFA/CH$_2$Cl$_2$ (25:75), 2 h, 100%.

Scheme 7.40. Synthesis of (±)-torreyanic acid (**7-151**).

Diels–Alder reaction of the same or different partners *via* the transition states het-ero-(**7-148**)$_2$ and homo-(**7-148**)$_2$ furnished epoxyquinol A (**7-149**) and B (**7-150**) in 40 % and 25 % yields, respectively. Remarkably, the reaction proceeds only *via* these two transition states, although there are 16 possible reaction modes.

Porco, Jr. and coworkers also used this biomimetic domino approach for the synthesis of the quinone epoxide dimer torreyanic acid **7-151** as racemic mixture [67]. This natural product, isolated from the fungus *Pestalotiopsis microspora*, shows a pronounced cytoxicity to tumor cells. Its retrosynthetic analysis leads to the 2*H*-pyran **7-152** (Scheme 7.40). As substrate for the domino process, the epoxide

7-153 was used, of which the hydroxyl group was oxidized with DMP to give the corresponding aldehyde. 6π-Electrocyclization afforded the desired pyran **7-152**, which underwent an intermolecular Diels–Alder reaction under the influence of silica gel, indicating acid catalysis in the dimerization. Since **7-153** was used as a racemic mixture the two racemic diastereomers **7-154** and **7-155** were obtained as products in nearly equal quantity with an overall yield of 80%. The final step in the synthesis of (±)-torreyanic acid (**7-151**) was the acid-catalyzed cleavage of the *t*-butyl ester in **7-154**. Under the same conditions, **7-156** was obtained from **7-155**.

7.4
Oxidative or Reductive/Oxidative or Reductive Processes

The domino oxidative polycyclization with rhenium(VII) reagents was first reported by Keinan and Sinha in 1995 [68]. Using this method, polyenic bis-homoallylic alcohols can be transformed in a single, highly efficient process into poly-THF products with excellent diastereoselectivity [69]. The procedure is particular beneficial for the asymmetric synthesis of polyether natural products, such as rollidecins C and D (**7-159a**) and (**7-159b**) (Scheme 7.41). These compounds belong to the family of *Annoaceous acetogenins*, and display a broad variety of biological activities [70]. Reaction of freshly prepared $CF_3CO_2ReO_3$ in CH_2Cl_2 with the alcohols **7-157a** and **7-157b** in

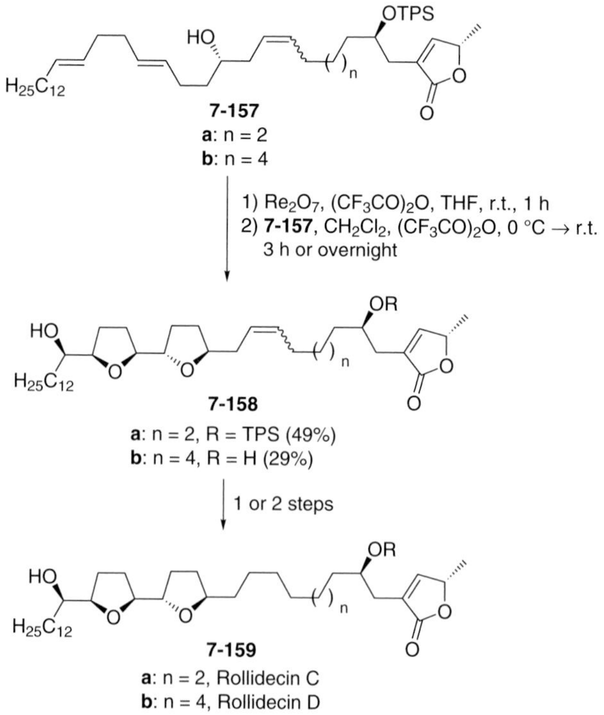

Scheme 7.41. Total synthesis of rollidecins C and D.

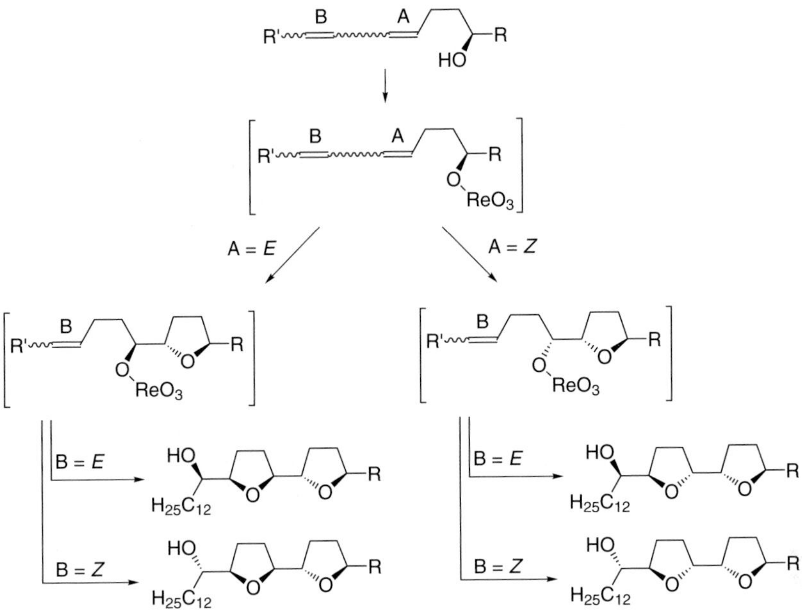

Scheme 7.42. Stereochemistry in oxidative polycyclization with trifluoracetylperrhenate.

the presence of TFAA, led to the desired polyethers **7-158a** and **7-158b** in 49% and 29% yields, respectively [71]. Hydrogenation and deprotection completed the total syntheses of **7-159a** and **7-159b**.

It should be noted that the stereochemistry of the products **7-159** depends not only on the configuration of the stereogenic centers but also on the double bond geometry in the substrates **7-157**. These correlations are indicated in Scheme 7.42 [72].

A useful and simple method for the one-pot preparation of highly functionalized, enantiomerically pure cyclopentanes from readily accessible carbohydrate precursors has been designed by Chiara and coworkers [73]. The procedure depends on a samarium(II) iodide-promoted reductive dealkoxyhalogenation of 6-desoxy-6-iodo-hexopyranosides such as **7-160** to produce a δ,ε-unsaturated aldehyde which, after reductive cyclization, is trapped by an added electrophile to furnish the final product. In the presence of acetic anhydride, the four products **7-161** to **7-164** were obtained from **7-160**.

Samarium(II) iodide also allows the reductive coupling of sulfur-substituted aromatic lactams such as **7-166** with carbonyl compounds to afford α-hydroxyalkylated lactams **7-167** with a high *anti*-selectivity [74]. The substituted lactams can easily be prepared from imides **7-165**. The reaction is initiated by a reductive desulfuration with samarium(II) iodide to give a radical, which can be intercepted by the added aldehyde to give the desired products **7-167**. Ketones can be used as the carbonyl moiety instead of aldehydes, with good – albeit slightly lower – yields.

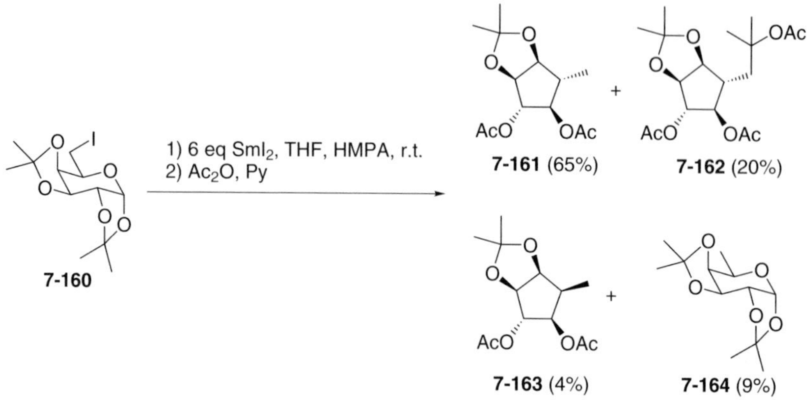

1) 6 eq SmI$_2$, THF, HMPA, r.t.
2) Ac$_2$O, Py

7-160

7-161 (65%) + **7-162** (20%)

7-163 (4%) + **7-164** (9%)

Scheme 7.43. Synthesis of highly functionalized enantiopure cyclopentanes.

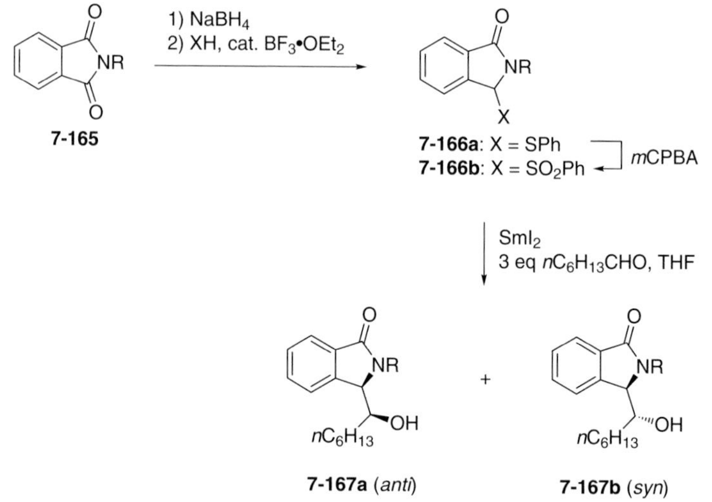

7-165

1) NaBH$_4$
2) XH, cat. BF$_3$•OEt$_2$

7-166a: X = SPh
7-166b: X = SO$_2$Ph ← *m*CPBA

SmI$_2$
3 eq *n*C$_6$H$_{13}$CHO, THF

7-167a (*anti*) + **7-167b** (*syn*)

Entry	X	R	SmI$_2$ [eq]	Temp [°C]	*anti/syn*	Yield [%]
1	SPh	Bn	3.0	r.t.	83/17	86
2	SO$_2$Ph	Bn	3.0	r.t.	79/21	80
3	SO$_2$Ph	Bn	5.0	r.t.	79/21	98
4	SPh	Me	3.0	0	82/18	71
5	SO$_2$Ph	Me	3.0	0	84/16	73
6	SPh	Me	3.0	r.t.	92/8	69
7	SO$_2$Ph	Me	3.0	r.t.	92/8	90

Scheme 7.44. SmI$_2$-promoted domino reductive desulfuration/reductive coupling process of lactams **7-166** with an aldehyde.

The final example in this section relates to the synthesis of the pharmaceutically interesting and structurally unusual CP molecules **7-168** and **7-169** (Scheme 7.45). Nicolaou and coworkers developed a neat domino Dess–Martin oxidation process for preparation of the CP core structure by transforming the 1,4-diol **7-170** into the bislactols **7-174** [75]. The reaction is initiated by a selective oxidation of the sterically hindered bridgehead alcohol in **7-170**, leading to formation of the isolable hemiacetal **7-171**. Further oxidation with the Dess–Martin reagent converts **7-171** to the keto aldehyde **7-173** *via* the ring-chain intermediate **7-172**. Subsequent reaction with a molecule of water furnished the stable diol **7-174** in an excellent yield of 82 %; this can be further oxidized, using TEMPO as an oxidizing agent, to give **7-175**.

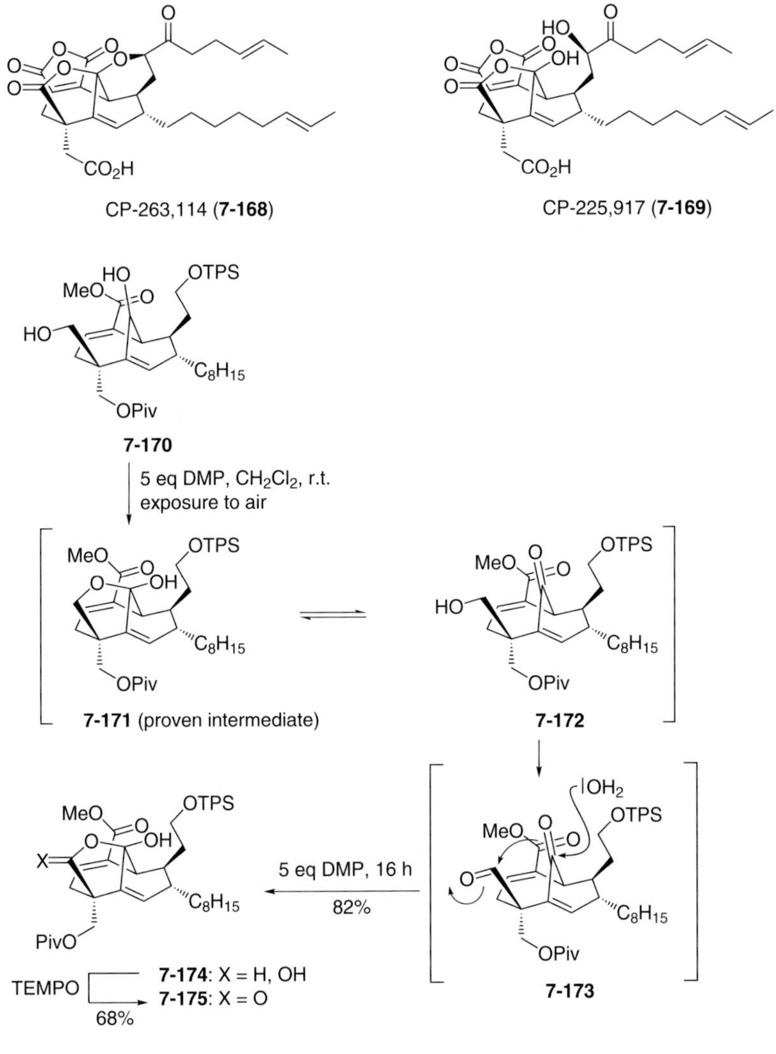

CP-263,114 (**7-168**)

CP-225,917 (**7-169**)

7-170

5 eq DMP, CH$_2$Cl$_2$, r.t.
exposure to air

7-171 (proven intermediate)

7-172

5 eq DMP, 16 h
82%

7-173

7-174: X = H, OH
7-175: X = O

TEMPO
68%

Scheme 7.45. Synthesis of the γ-hydroxy moiety of CP compounds.

References

1 L. F. Tietze, U. Beifuss, *Angew. Chem. Int. Ed. Engl.* **1993**, *32*, 137–170.
2 For a review, see: H. Nemoto, K. Fukumoto, *Synlett* **1997**, 863–875.
3 G. M. Anstead, K. E. Carlson, J. A. Katzenellenbogen, *Steroids* **1997**, *62*, 268–303.
4 M. Yoshida, M. A.-H. Ismail, H. Nemoto, M. Ihara, *J. Chem. Soc. Perkin Trans. 1* **2000**, 2629–2635.
5 J. E. Baldwin, R. M. Adlington, V. W.-W. Sham, R. Marquez, P. G. Bulger, *Tetrahedron* **2005**, *61*, 2353–2363.
6 L. Blackburn, C. Pei, R. J. K. Taylor, *Synlett* **2002**, 215–218.
7 (a) S. Eguchi, Y. Matsushita, K. Yamashita, *Org. Prep. Proced. Int.* **1992**, *24*, 209–244; (b) Y. G. Gololobov, L. F. Kasukhin, *Tetrahedron* **1992**, *48*, 1353–1406; (c) P. Molina, M. J. Vilaplana, *Synthesis* **1994**, 1197–1218; (d) H. Wamhoff, G. Rechardt, S. Stölben, *Advances in Heterocyclic Chemistry* **1996**, *Vol. 64*, pp. 159.
8 (a) A. H. Amin, D. R. Mehta, *Nature* **1959**, *184*, 1317; (b) D. R. Mehta, J. S. Naravane, R. M. Desai, *J. Org. Chem.* **1963**, *28*, 445–448.
9 S. Eguchi, T. Suzuki, T. Okawa, Y. Matsushita, E. Yashima, Y. Okamoto, *J. Org. Chem.* **1996**, *61*, 7316–7319.
10 D. R. Williams, K. Shamim, J. P. Reddy, G. S. Amato, S. M. Shaw, *Org. Lett.* **2003**, *5*, 3361–3364.
11 (a) K. Sakata, K. Aoki, C.-F. Chang, A. Sakurai, S. Tamura, S. Murakoshi, *Agric. Biol. Chem.* **1978**, *42*, 457–463; (b) H. Shinozaki, M. Ishida, *Brain Res.* **1985**, *334*, 33–40; (c) Y. Ye, G.-W. Qin, R.-S. Xu, *Phytochemistry* **1994**, *37*, 1205–1208; (d) R. A. Pilli, M. Ferriera de Oliviera, *Nat. Prod. Rep.* **2000**, *17*, 117–127.
12 M. I. García-Moreno, C. O. Mellet, J. M. García Fernández, *Eur. J. Org. Chem.* **2004**, 1803–1819.
13 J. W. Scott, in: *Asymmetric Synthesis*, (Ed.: J. D. Morrison), Academic Press, New York, **1984**.
14 (a) Z.-Y. Wei, E. E. Knaus, *Tetrahedron Lett.* **1994**, *35*, 2305–2308; (b) Z.-Y. Wei, E. E. Knaus, *Org. Prep. Proc. Int.* **1994**, *26*, 243–248.
15 C. David, L. Bischoff, B. P. Roques, M.-C. Fournié-Zaluski, *Tetrahedron* **2000**, *56*, 209–215.
16 H. Razavi, R. Polt, *Tetrahedron Lett.* **1998**, *39*, 3371–3374.
17 (a) F. F. Fleming, A. Huang, V. A. Sharief, Y. Pu, *J. Org. Chem.* **1999**, *64*, 2830–2834; (b) F. F. Fleming, A. Huang, V. A. Sharief, Y. Pu, *J. Org. Chem.* **1997**, *62*, 3036–3037.
18 S. Thorimbert, C. Taillier, S. Bareyt, D. Humilière, M. Malacria, *Tetrahedron Lett.* **2004**, *45*, 9123–9126.
19 J. M. Schomaker, B. Borhan, *Org. Biomol. Chem.* **2004**, *2*, 621–624.
20 R. A. Bunce, D. M. Herron, M. L. Ackerman, *J. Org. Chem.* **2000**, *65*, 2847–2850.
21 K. C. Nicolaou, M. Bella, D. Y.-K. Chen, X. Huang, T. Ling, S. A. Snyder, *Angew. Chem. Int. Ed.* **2002**, *41*, 3495–3499.
22 (a) L. H. Hurley, V. L. Reynolds, D. H. Swenson, G. L. Petzold, T. A. Scahill, *Science* **1984**, *226*, 843–844; (b) D. L. Boger, T. Ishizaki, H. Zarrinmayeh, P. A. Kitos, O. Suntornwat, *J. Org. Chem.* **1990**, *55*, 4499–4502; (c) D. L. Boger, T. Ishizaki, H. Zarrinmayeh, S. A. Munk, P. A. Kitos, O. Suntornwat, *J. Am. Chem. Soc.* **1990**, *112*, 8961–8971; (d) D. L. Boger, T. Ishizaki, H. Zarrinmayeh, *J. Am. Chem. Soc.* **1991**, *113*, 6645–6649; (e) D. L. Boger, W. Yun, *J. Am. Chem. Soc.* **1993**, *115*, 9872–9873; (f) D. L. Boger, W. Yun, S. Terashima, Y. Fukuda, K. Nakatani, P. A. Kitos, Q. Jin, *Bioorg. Med. Chem. Lett.* **1992**, *2*, 759–765.
23 D. L. Boger, W. Yun, H. Cai, N. Han, *Bioorg. Med. Chem.* **1995**, *3*, 761–775.
24 (a) V. Jäger, R. Schone, *Tetrahedron* **1984**, *40*, 2199–2210; (b) V. Jäger, I. Müller, R. Schone, M. Frey, R. Ehrler, B. Häfele, D. Schröter, *Lect. Heter. Chem.* **1985**, *6*, 79–98.
25 C. Galli, E. Marotta, P. Righi, G. Rosini, *J. Org. Chem.*, **1995**, *60*, 6624–6626.
26 N. Scardovi, A. Casalini, F. Peri, P. Righi, *Org. Lett.* **2002**, *4*, 965–968.
27 E. Marotta, M. Baravelli, L. Maini, P. Righi, G. Rosini, *J. Org. Chem.* **1998**, *63*, 8235–8246.
28 For a review, see: C. J. Dinsmore, D. C. Beshore, *Org. Prep. Proced. Int.* **2002**, *34*, 367–404.
29 D. C. Beshore, C. J. Dinsmore, *Org. Lett.* **2002**, *4*, 1201–1204.

30 A. Hari, C. Karan, W. C. Rodrigues, B. L. Miller, *J. Org. Chem.* **2001**, *66*, 991–996.

31 (a) G. D. Cuny, J. D. Hauske, M. Z. Hoemann, R. F. Rossi, R. L. Xie, PCT Int. Appl. WO 9967238, **1999**; *Chem. Abstr.* **1999**, *132*, 64182; (b) K. Hanada, K. Furuya, K. Inoguchi, M. Miyakawa, N. Nagata, PCT Int. Appl. WO 0127086, **2001**; *Chem. Abstr.* **2001**, *134*, 295752.

32 L. Chen, Z. Li, C.-J. Li, *Synlett* **2003**, *5*, 732–734.

33 For a representative review, see: J. Bonjoch, D. Solé, *Chem. Rev.* **2000**, *100*, 3455–3482.

34 T. Ohshima, Y. Xu, R. Takita, S. Shimizu, D. Zhong, M. Shibasaki, *J. Am. Chem. Soc.* **2002**, *124*, 14546–14547.

35 I. Hanna, L. Ricard, *Org. Lett.* **2000**, *2*, 2651–2654.

36 For an excellent review on calystegines, see: R. J. Molyneux, R. J. Nash, N. Asano, in: *Alkaloids: Chemical and Biological Perspectives*, *Vol. 11*, (Ed.: S. W. Pelletier), Elsevier Science, Oxford, **1996**, pp. 303–343.

37 For use of this method to synthesize aminocyclohexenes, see: (a) L. Hyldtof, C. S. Poulsen, R. Madsen, *Chem. Commun.* **1999**, 2101–2102; (b) L. Hyldtof, R. Madsen, *J. Am. Chem. Soc.* **2000**, *122*, 8444–8452.

38 F.-D. Boyer, I. Hanna, *Tetrahedron Lett.* **2001**, *42*, 1275–1277.

39 For a similar approach to calystegines, see: P. R. Skaanderup, R. Madsen, *Chem. Commun.* **2001**, 1106–1107.

40 H. Cao, S. R. Mundla, J. M. Cook, *Tetrahedron Lett.* **2003**, *44*, 6165–6168.

41 (a) P. Yates, N. K. Bhamare, T. Granger, T. S. Macas, *Can. J. Chem.* **1993**, *71*, 995–1001; (b) J. T. Hwang, C.-C. Liao, *Tetrahedron Lett.* **1991**, *32*, 6583–6586; (c) V. Singh, B. Thomas, *J. Org. Chem.* **1997**, *62*, 5310–5320; (d) P. Y. Hsu, Y. C. Lee, C.-C. Liao, *Tetrahedron Lett.* **1998**, *39*, 659–662; (e) W. C. Liu, C.-C. Liao, *J. Chem. Soc., Chem. Commun.* **1999**, 117–118.

42 S. Quideau, L. Pouysegu, *Org. Prep. Proced. Int.* **1999**, *31*, 617–680.

43 (a) F. Wessely, G. Lauterbach-Keil, F. Sinwel, *Monatsh. Chem.* **1950**, *81*, 811–818; (b) M. Metlesics, E. Schinzel, H. Vilsek, F. Wessely, *Monatsh. Chem.* **1957**, *88*, 1069–1076.

44 I. Drutu, J. T. Njardarson, J. J. Wood, *Org. Lett.* **2002**, *4*, 493–496.

45 (a) C.-S. Chu, T.-H. Lee, P. D. Rao, L.-D. Song, C.-C. Liao, *J. Org. Chem.* **1999**, *64*, 4111–4118; (b) K.-C. Lin, Y.-L. Shen, N. S. Kameswara Rao, C.-C. Liao, *J. Org. Chem.* **2002**, *67*, 8157–8165.

46 C.-S. Chu, C.-C. Liao, P. D. Rao, *Chem. Commun.* **1996**, 1537–1538.

47 P. J. Black, M. G. Edwards, J. M. J. Williams, *Tetrahedron* **2005**, *61*, 1363–1374.

48 P.-Q. Huang, H.-Q. Lan, X. Zheng, Y.-P. Ruan, *J. Org. Chem.* **2004**, *69*, 3964–3967.

49 (a) H. Oikawa, *Bull. Chem. Soc. Jpn.* **2005**, *78*, 537–554; (b) J. J. Agresti, B. T. Kelly, A. Jaschke, A. D. Griffiths, *Proc. Natl. Acad. Sci. USA* **2005**, *102*, 16170–16175; (c) V. Gouverneur, M. Reiter, *Chem. Eur. J.* **2005**, *11*, 5806–5815.

50 (a) D. B. Dess, J. C. Martin, *J. Org. Chem.* **1983**, *48*, 4155–4156; (b) D. B. Dess, J. C. Martin, *J. Am. Chem. Soc.* **1991**, *113*, 7277–7287; (c) S. D. Meyer, S. L. Schreiber, *J. Org. Chem.* **1994**, *59*, 7549–7552.

51 (a) K. C. Nicolaou, Y.-L. Zhong, P. S. Baran, *Angew. Chem. Int. Ed.* **2000**, *39*, 622–625; (b) K. C. Nicolaou, Y.-L. Zhong, P. S. Baran, *Angew. Chem. Int. Ed.* **2000**, *39*, 625–628; (c) K. C. Nicolaou, P. S. Baran, Y.-L. Zhong, J. A. Vega, *Angew. Chem. Int. Ed.* **2000**, *39*, 2525–2529; (d) K. C. Nicolaou, Y.-L. Zhong, P. S. Baran, *J. Am. Chem. Soc.* **2000**, *122*, 7596–7597; (e) K. C. Nicolaou, K. Sugita, P. S. Baran, Y.-L. Zhong, *Angew. Chem. Int. Ed.* **2001**, *40*, 207–210; (f) K. C. Nicolaou, P. S. Baran, R. Kranich, Y.-L. Zhong, K. Sugita, N. Zou, *Angew. Chem. Int. Ed.* **2001**, *40*, 202–206; (g) K. C. Nicolaou, P. S. Baran, Y.-L. Zhong, *J. Am. Chem. Soc.* **2001**, *123*, 3183–3185; (h) K. C. Nicolaou, Y.-L. Zhong, P. S. Baran, K. Sugita, *Angew. Chem. Int. Ed.* **2001**, *40*, 2145–2149.

52 K. C. Nicolaou, P. S. Baran, Y.-L. Zhong, K. Sugita, *J. Am. Chem. Soc.* **2002**, *124*, 2212–2220.

53 K. C. Nicolaou, K. Sugita, P. S. Baran, Y.-L. Zhong, *J. Am. Chem. Soc.* **2002**, *124*, 2221–2232.

54 S. P. Cook, C. Gaul, S. J. Danishefsky, *Tetrahedron Lett.* **2005**, *46*, 843–847.

55 J. I. Candela Lena, E. Altinel, N. Birlirakis, S. Arseniyadis, *Tetrahedron Lett.* **2002**, *43*, 2505–2509.

56 J. I. Candela Lena, E. Altinel, N. Birlirakis, S. Arseniyadis, *Tetrahedron Lett.* **2002**, *43*, 1409–1412.

57 L. Finet, J. I. Candela Lena, T. Kaoudi, N. Birlirakis, S. Arseniyadis, *Chem. Eur. J.* **2003**, *9*, 3813–3820; see also: E. M. Sanchez Fernandez, J. I. Candela Lena, E. Altinel, N. Birlirakis, A. F. Barrero, S. Arseniydis, *Tetrahedron: Asymm.* **2003**, *14*, 2277–2290.

58 B. Sato, H. Muramatsu, M. Miyauchi, Y. Hori, S. Takese, M. Mino, S. Hashimoto, H. Terano, *J. Antibiot.* **2000**, *53*, 123–130.

59 R. Höfs, M. Walker, A. Zeeck, *Angew. Chem. Int. Ed.* **2000**, *39*, 3258–3261.

60 D. E. Evans, J. T. Starr, *Angew. Chem. Int. Ed.* **2002**, *41*, 1787–1790.

61 (a) K. C. Majumdar, S. K. Chattopadhyay, *J. Chem. Soc., Chem. Commun.* **1987**, 524–525; (b) K. C. Majumdar, S. K. Chattopadhyay, A. T. Khan, *J. Chem. Soc. Perkin Trans. 1* **1989**, 1285–1288; (c) K. C. Majumdar, S. K. Ghosh, *J. Chem. Soc. Perkin Trans. 1* **1994**, 2889–2894; (d) K. C. Majumdar, U. Das, N. K. Jana, *J. Org. Chem.* **1998**, 63, 3550–3553.

62 K. C. Majumdar, S. K. Samanta, *Tetrahedron Lett.* **2002**, *43*, 2119–2121.

63 H. Kakeya, R. Onose, H. Koshino, A. Yoshida, K. Kobayashi, S.-I. Kageyama, H. Osada, *J. Am. Chem. Soc.*, **2002**, *124*, 3496–3497.

64 H. Kakeya, R. Onose, A. Yoshida, H. Koshino, H. Osada, *J. Antibiot.* **2002**, *55*, 829–831.

65 M. Shoji, S. Kishida, Y. Kodera, I. Shiina, H. Kakeya, H. Osada, Y. Hayashi, *Tetrahedron Lett.* **2003**, *44*, 7205–7207.

66 C. Li, S. Bardhan, E. A. Pace, M.-C. Liang, T. D. Gilmore, J. A. Porco, Jr., *Org. Lett.* **2002**, *4*, 3267–3270.

67 C. Li, E. Lobkovsky, J. A. Porco, Jr., *J. Am. Chem. Soc.* **2000**, *122*, 10484–10485.

68 S. C. Sinha, A. Sinha-Bagchi, E. Keinan, *J. Am. Chem. Soc.* **1995**, *117*, 1447–1448.

69 L. Zeng, Q. Ye, N. H. Oberlies, G. Shi, Z.-M. Gu, K. He, J. L. McLaughlin, *Nat. Prod. Rep.* **1996**, *13*, 275–306.

70 L. J. D'Souza, S. C. Sinha, S.-F. Lu, E. Keinan, S. C. Sinha, *Tetrahedron* **2001**, *57*, 5255–5262.

71 (a) B. T. Towne, F. E. McDonald, *J. Am. Chem. Soc.* **1997**, *119*, 6022–6028; (b) S. C. Sinha, A. Sinha, S. C. Sinha, E. Keinan, *J. Am. Chem. Soc.* **1997**, *119*, 12014–12015.

72 S. C. Sinha, E. Keinan, S. C. Sinha *J. Am. Chem. Soc.* **1998**, *120*, 9076–9077.

73 J. L. Chiara, S. Martínez, M. Bernabé, *J. Org. Chem.* **1996**, *61*, 6488–6489.

74 H. Yoda, H. Ujihara, K. Takabe, *Tetrahedron Lett.* **2001**, *42*, 9225–9228.

75 K. C. Nicolaou, Y. He, K. C. Fong, W. H. Yoon, H.-S. Choi, Y.-L. Zhong, P. S. Baran, *Org. Lett.* **1999**, *1*, 63–66.

8
Enzymes in Domino Reactions

Over the years of evolution, Nature has developed enzymes which are able to catalyze a multitude of different transformations with amazing enhancements in rate [1]. Moreover, these enzyme proteins show a high specificity in most cases, allowing the enantioselective formation of chiral compounds. Therefore, it is not surprising that they have been used for decades as biocatalysts in the chemical synthesis in a flask. Besides their synthetic advantages, enzymes are also beneficial from an economical – and especially ecological – point of view, as they stand for renewable resources and biocompatible reaction conditions in most cases, which corresponds with the conception of Green Chemistry [2].

Nature, however, has performed more than simple stepwise transformations; using a combination of enzymes in so-called multienzyme complexes, it performs multistep synthetic processes. A well-known example in this context is the biosynthesis of fatty acids. Thus, Nature can be quoted as the "inventor" of domino reactions. Usually, as has been described earlier in this book, domino processes are initiated by the application of an organic or inorganic reagent, or by thermal or photochemical treatment. The use of enzymes in a flask for initiating a domino reaction is a rather new development. One of the first examples for this type of reaction dates back to 1981 [3], although it should be noted that in 1976 a bio-triggered domino reaction was observed as an undesired side reaction by serendipity [4].

In recent years, due to their increased availability, enzymes have been used in the development of domino reactions. This interesting field has recently been summarized by Faber and coworkers in an excellent review [5]. In this chapter, the different types of biocatalyzed domino reactions are discussed and classified (Table 8.1).

Enzyme-triggered domino reactions follow a common scheme. Initially, the enzyme modifies a so-called "trigger" group in the starting material, furnishing a reactive species that can undergo a subsequent domino reaction. In the examples discussed below, the reactive group created in the first step may be a diene or dienophile (the latter generated, for example, by the oxidation of an allylic alcohol), or a negatively charged atom produced by hydrolysis of an ester or epoxide, which increases the electron density of a π-electron system or acts as a nucleophile. Accordingly, the intermediates provided can undergo further transformations, which may consist of a nucleophilic substitution, a cyclization (e. g., Diels–Alder cycloaddition), a fragmentation, or a rearrangement.

Domino Reactions in Organic Synthesis. Lutz F. Tietze, Gordon Brasche, and Kersten M. Gericke
Copyright © 2006 WILEY-VCH Verlag GmbH & Co. KGaA, Weinheim
ISBN: 3-527-29060-5

Enzymatic trigger reaction	Effect of trigger reaction	Domino process
oxidation	formation of dienophile	intermolecular Diels-Alder
transesterification	formation of activated diene	intramolecular Diels-Alder
ester hydrolysis	electron-donating group liberated	retro-[2+2] cycloaddition/fragmentation
ester hydrolysis	electron-donating group liberated	fragmentation
ester hydrolysis	electron-donating group liberated	rearrangement
ester hydrolysis	nucleophile liberated ($-CO_2-$)	cyclization
ester hydrolysis	nucleophile liberated ($-OH$)	cyclization
epoxide hydrolysis	nucleophile liberated ($-OH$)	cyclization

Table 8.1. Types of biocatalyzed domino reactions.

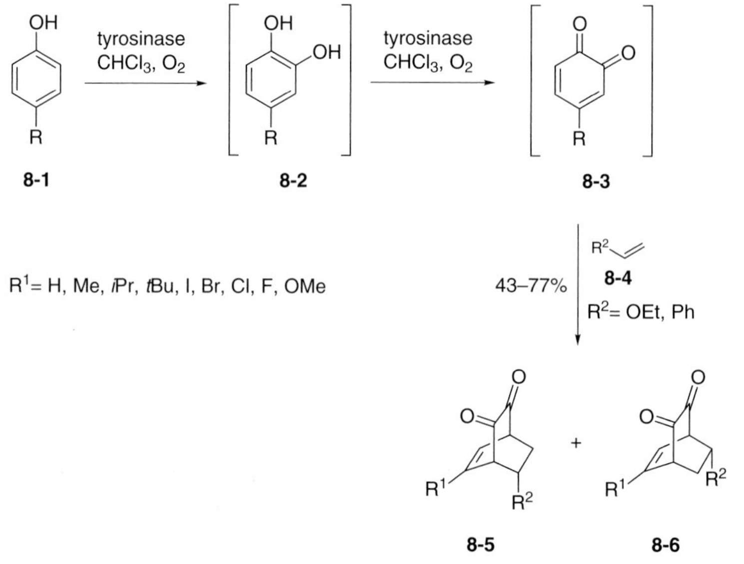

Scheme 8.1. Enzymatic oxidative generation of a diene followed by a Diels–Alder reaction

In 1996, the first successful combination of an enzymatic with a nonenzymatic transformation within a domino process was reported by Waldmann and co-workers [6]. These authors described a reaction in which functionalized bicy-clo[2.2.2]octenediones were produced by a tyrosinase (from *Agaricus bisporus*) -cata-lyzed oxidation of *para*-substituted phenols, followed by a Diels–Alder reaction with an alkene or enol ether as dienophile. Hence, treatment of phenols such as **8-1** and an electron-rich alkene **8-4** in chloroform with tyrosinase in the presence of oxygen led to the bicyclic cycloadducts **8-5** and **8-6** in moderate to good yield (Scheme 8.1). It can be assumed that, in the first step, the phenol **8-1** is hydroxylated by tyrosi-nase, generating the catechol intermediate **8-2**, which is then again oxidized enzy-

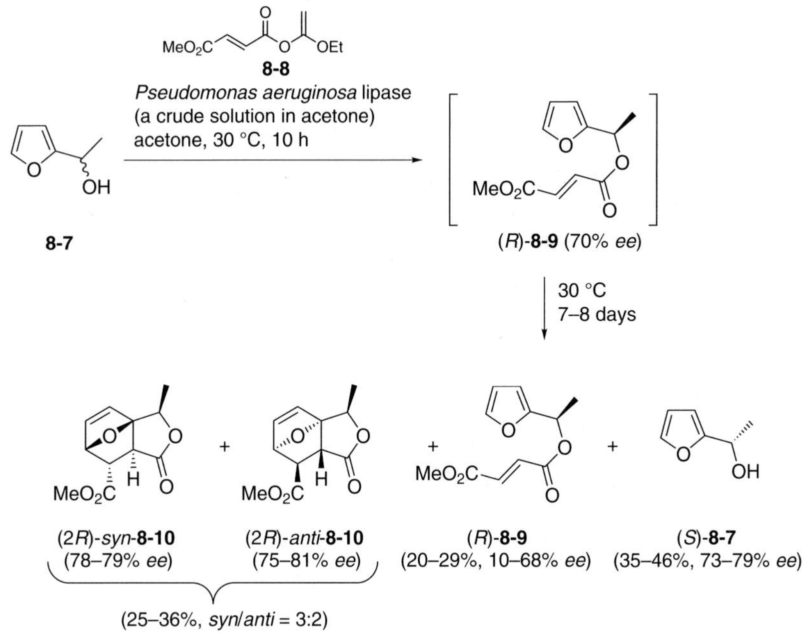

Scheme 8.2. Lipase-catalyzed domino transesterification/Diels–Alder reaction of (±)-**8-7** with **8-8**.

matically to produce the *o*-quinone **8-3** containing a highly reactive 1,3-butadiene moiety. This undergoes a Diels–Alder reaction with inverse electron demand with the present dienophile **8-4**. In principle, four different diastereoisomers can be formed in this chemoenzymatic domino process, but analysis of the products showed that mainly the *exo*-diastereoisomer **8-5** with a 1,3-substitution pattern was formed (in some cases as the only product). It is noteworthy to say that the overall process was rather slow and required up to three days for completion. Furthermore, no asymmetric induction was detected, since the addition seemed to proceed without the influence of the enzyme.

Another example for the combination of an enzymatic process with a cycloaddition is the lipase-catalyzed transformation of racemic alcohol **8-7** with fumaric acid ester **8-8** described by the Kita group (Scheme 8.2) [7]. As products, nonracemic 7-oxabicyclo[2.2.1]heptenes **8-10** were obtained, which are useful intermediates in the synthesis of various biologically important natural products [8]. An interesting aspect of this enzyme-triggered domino reaction using a lipase from *Pseudomonas aeruginosa* is the fact that during the transesterification a partial kinetic resolution occurred. As intermediate, the ester (*R*)-**8-9** was formed with 70 % *ee*, which can underwent an intramolecular cycloaddition to give the products (2*R*)-*syn*- and (2*R*)-*anti*-**8-10** with slightly increased *ee*-values compared to (*R*)-**8-9** in a 3:2 ratio and 25–36 % yield. A disadvantage of this reaction is the long reaction time of about 7 to 8 days. Besides the cycloadducts, nonracemic furfuryl alcohol (*S*)-**8-7** and the Diels–Alder precursor (*R*)-**8-9** were isolated in 35–46 % and 20–29 % yields, respectively.

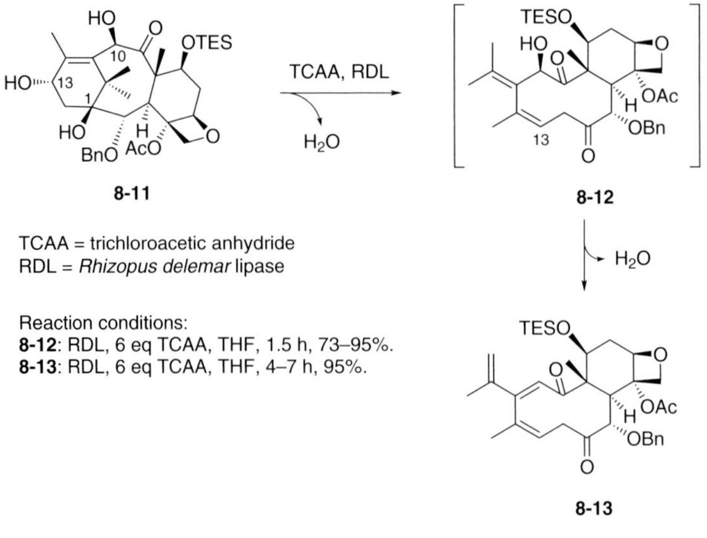

TCAA = trichloroacetic anhydride
RDL = *Rhizopus delemar* lipase

Reaction conditions:
8-12: RDL, 6 eq TCAA, THF, 1.5 h, 73–95%.
8-13: RDL, 6 eq TCAA, THF, 4–7 h, 95%.

Scheme 8.3. Enzymatic dehydration and rearrangement of a paclitaxel precursor.

Carbon skeleton rearrangements often lead to products of unrelated and exceptional structure; they are, therefore, of particular interest in organic synthesis. Such rearrangements as part of a domino process can also be triggered utilizing enzymes. During the development of a new strategy for the assembly of paclitaxel, a diterpene anticancer agent [9], an unexpected enzymatically induced selective dehydration and rearrangement has been observed by Kim and coworkers [10]. When the paclitaxel precursor **8-11** was treated with *Rhizopus delemar* lipase (RDL) in the presence of trichloroacetic anhydride, the tricyclic diterpene **8-12** was generated instead of the expected ester at the 13-hydroxyl group (Scheme 8.3). The authors assume that the elimination of the 13-hydroxyl group and the deprotonation of the 1-hydroxyl group occur in a concerted way, and thus a trichloroacetate intermediate is not involved. An evidence for the proposed mechanism is the stability of the separately prepared **8-11** containing a trichloroacetyl moiety at C-13 in the presence of RDL, trichloroacetic acid, and triethylamine in THF. Interestingly, the enzymatic reaction, when allowed to react for a longer time, furnished the new product **8-13**, which is formed by a second dehydration, in 95 % overall yield.

The application of enzymes for the mild hydrolysis of esters has been well established as a convenient method, and hence can be regarded as a standard methodology in organic synthesis. Accordingly, the expectation is met that such enzymatic ester cleavage reactions are included in domino processes. In this context, a number of publications describe enzyme-initiated fragmentation reactions.

In 1982, the Schaap group demonstrated that chemiluminescence can be induced by the addition of a base to dioxetanes bearing a phenolic substituent [11]. Herein, the same group presents a method utilizing aryl esterase to catalyze the cleavage of a naphthyl acetate-substituted dioxetane in aqueous buffer at ambient

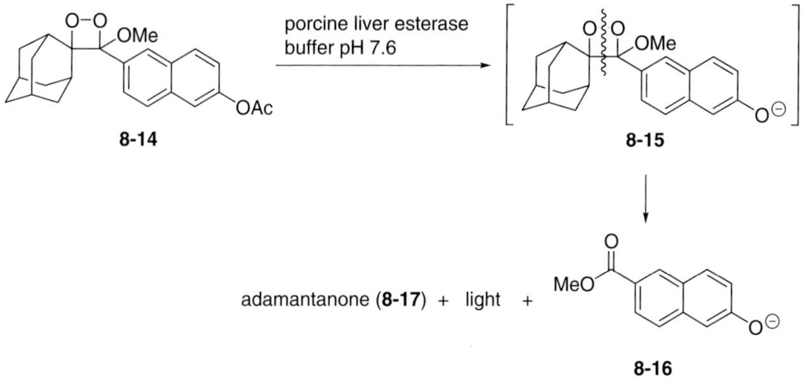

Scheme 8.4. Dioxetane fragmentation initiated by enzymatic ester hydrolysis.

temperature [12]. Hydrolysis of the acetate unit in the oxetane derivative **8-14** employing porcin liver esterase liberated the free naphtholate anion **8-15**, which underwent a subsequent fragmentation reaction generating naphthol methylester **8-16** and adamantanone with the concurrent chemiluminescence (Scheme 8.4).

The combination of an enzyme-catalyzed ester cleavage with a proximate fragmentation is of interest in protective group chemistry. According to this line, Waldmann and coworkers developed a new protecting group for amino, hydroxy and carboxy moieties containing a *p*-acetoxybenzyloxycarbonyl group during their synthesis of *N-Ras* lipopeptides [13]. After enzymatic cleavage of the acetate group of **8-18** using a lipase **8-19**, the resulting phenolate anion **8-21** induced (at an appropriate pH) a fragmentation leading to the formation of a quinone methide with liberation of the desired product **8-22** or **8-23** (Scheme 8.5). Thus, depending on the reactive group within the target molecule, liberation can occur with or without decarboxylation (amines, alcohols and carboxylic acids, respectively). Another beneficial feature of this strategy is the possibility of employing a phenylacetate ester instead of an acetate, which can be cleaved with high chemoselectivity using penicillin G acylase (**8-20**).

An ester hydrolysis is also the starting transformation for the domino reaction depicted in Scheme 8.6. Here, an unexpected rearrangement of achiral norbornene dicarboxylates **8-24** has been observed by Niwayama and coworkers to give a chiral bicyclo[3.1.0]hexene framework **8-27** by enzymatic hydrolysis using a pig liver esterase (PLE) [14]. This enzyme-triggered domino reaction proceeds *via* the formation of the monoester **8-25**, which immediately executes a Meinwald rearrangement providing **8-27** in near-quantitative yield *via* carbocation **8-26**. Though the enantiomeric excess turned out to be moderate, with 48 % *ee* for R = CH$_3$ and 65 % for R = Et, the reaction allows the formation of interesting molecules with three new stereogenic centers.

Besides fragmentation or rearrangement, the carboxylic acid anions, formed by an enzymatic hydrolysis, can also act as nucleophiles. Kuhn and Tamm used the asymmetric hydrolysis of *meso*-epoxy diester **8-28** with PLE to synthesize γ-lactone

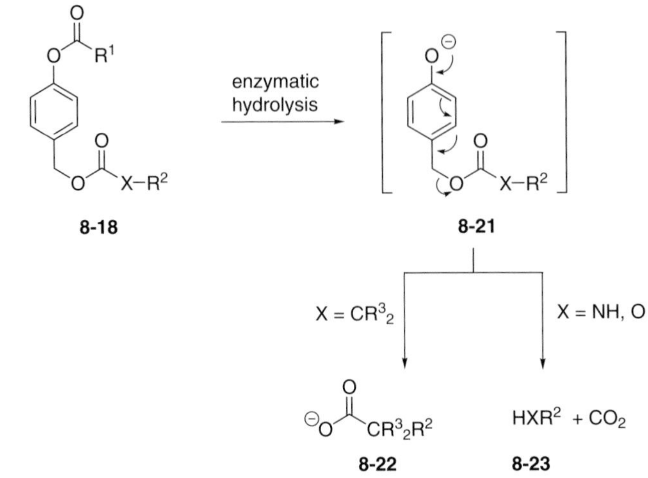

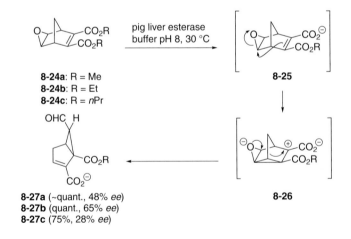

Scheme 8.5. Enzyme-triggered fragmentation of protective groups for amines, alcohols and carboxylic acids.

R¹ (trigger group)	Enzyme	X	R² (target molecule)
CH₃, nC₃H₇, nC₇H₁₅	Lipase (**8-19**)	O	Alcohol
		CR³₂	Carboxylic acid
CH₂Ph	Penicillin G Acylase (**8-20**)	NH	Amine

The R¹ subscripts: CH_3, nC_3H_7, nC_7H_{15}, CH_2Ph, and X entries O, CR^3_2, NH.

Scheme 8.6. Asymmetric ester hydrolysis followed by Meinwald rearrangement.

8-31, in which the formed carboxylic acid anion **8-29** in a S_N2-type fashion attacks the epoxide moiety (Scheme 8.7) [15]. It is of interest that, after formation of the lactone **8-30**, the remaining ester moiety is turned into a more accessible equatorial

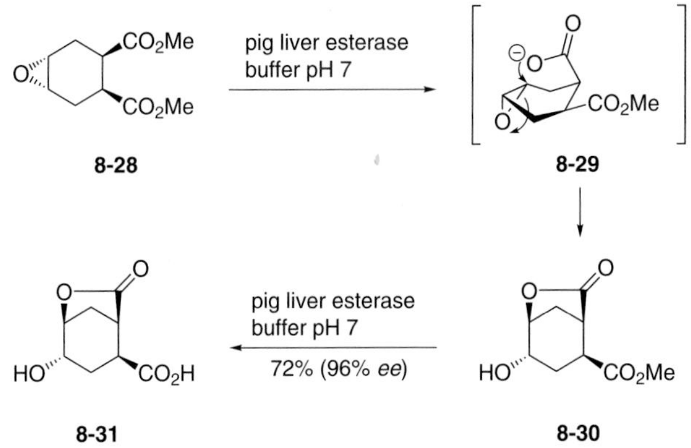

Scheme 8.7. γ-Lactone formation initiated by an enzymatically liberated Nu (– CO$_2^-$).

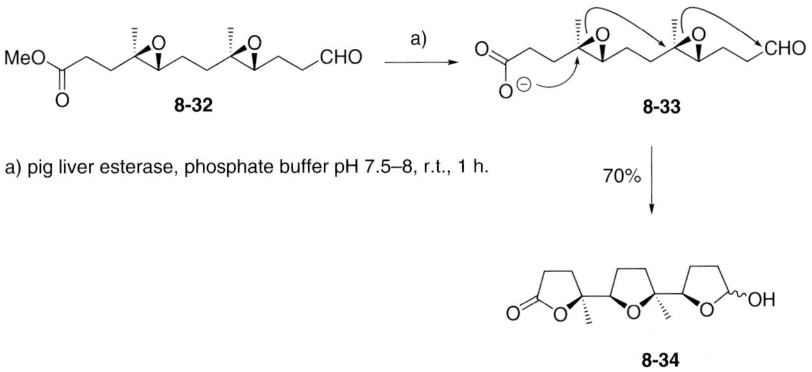

a) pig liver esterase, phosphate buffer pH 7.5–8, r.t., 1 h.

70%

Scheme 8.8. Synthesis of a tris-tetrahydrofuran derivative.

position, which also allows a hydrolysis by PLE to give **8-31** in 72 % yield and excellent enantiopurity of 96 % *ee*.

A similar domino process involving the opening of two epoxide moieties after an enzymatic ester hydrolysis has been described by Robinson and coworkers [16]. Treatment of **8-32** with PLE in an aqueous buffer solution at pH 7.5–8 led to **8-34** in 70 % yield after formation of **8-33** (Scheme 8.8).

In another approach, the alcohol moiety, formed by an enzymatic hydrolysis of an ester, can act as a nucleophile. In their synthesis of pityol (**8-37a**), a pheromone of the elm bark beetle, Faber and coworkers [17] used an enzyme-triggered reaction of the diastereomeric mixture of (±)-epoxy ester **8-35** employing an immobilized enzyme preparation (Novo SP 409) or whole lyophilized cells of *Rhodococcus erythropolis* NCIMB 11540 (Scheme 8.9). As an intermediate, the enantiopure alcohol **8-36** is formed *via* kinetic resolution as a mixture of diastereomers, which leads to the diastereomeric THF derivatives pityol (**8-37a**) and **8-37b** as a separable mixture with a

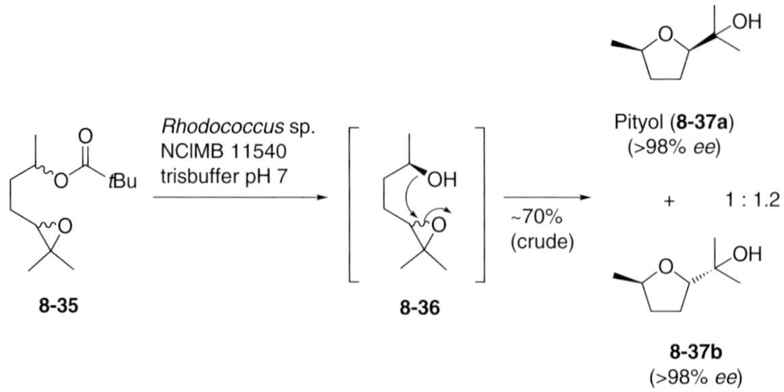

Scheme 8.9. Synthesis of tetrahydrofurans.

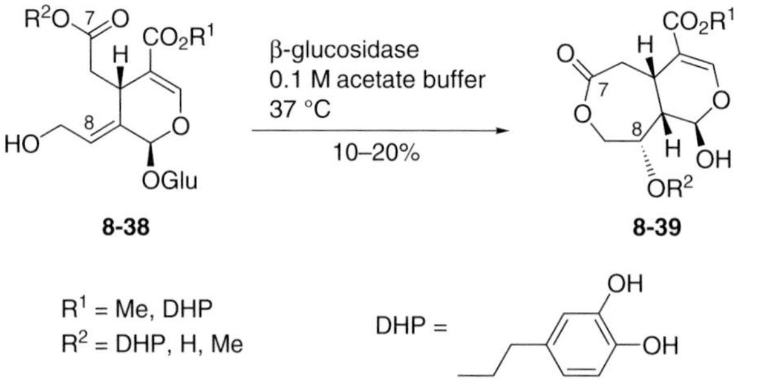

Scheme 8.10. Enzyme-triggered rearrangement of multifloroside **8-38**.

ratio of 1:1.2 in approximately 70% yield by nucleophilic opening of the epoxide moiety. Interestingly, there is evidence for a participation of the biocatalyst in the formation of the furan ring, since in the absence of the enzyme a reaction of iso-lated **8-36** to give **8-37** did not take place at pH 7, but only under basic conditions.

The Shen group has devised a domino process wherein after an enzymatic cleav-age of a glycoside a rearrangement sequence takes place [18]. Subjecting multi-floroside **8-38** to β-glucosidase in acetate buffer (Scheme 8.10) afforded jasmolac-tone analogues such as **8-39** in a rather low yield.

In addition to the enzymatic hydrolysis of esters, there also ample examples where an epoxide has been cleaved using a biocatalyst. As described by the Faber group [19], reaction of the (±)-2,3-disubstituted *cis*-chloroalkyl epoxide *rac*-**8-40** with a bacterial epoxide hydrolase (BEH), led to the formation of *vic*-diol (2R,3S)-**8-41** (Scheme 8.11). The latter underwent a spontaneous cyclization to give the desired product (2R,3R)-**8-42** in 92% *ee* and 76% yield. The same strategy was used with the homologous molecule *rac*-**8-43**, which afforded the THF derivative (2R,3R)-**8-45** in 86% *ee* and 79% yield.

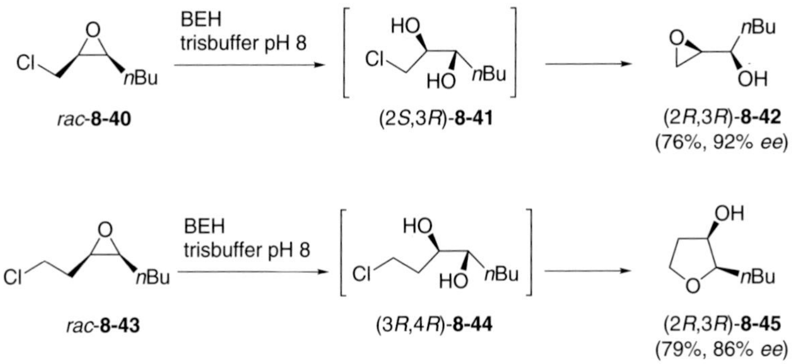

Scheme 8.11. Enzyme-initiated domino reaction of (±)-disubstituted chloroalklyl epoxides.

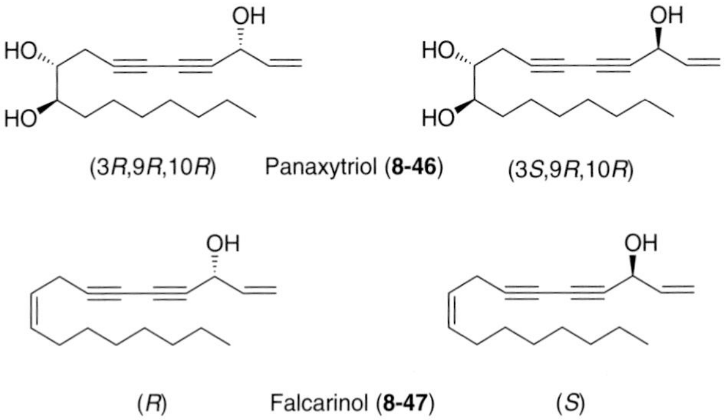

Scheme 8.12. Natural products panaxytriol (**8-46**) and falcarinol (**8-47**).

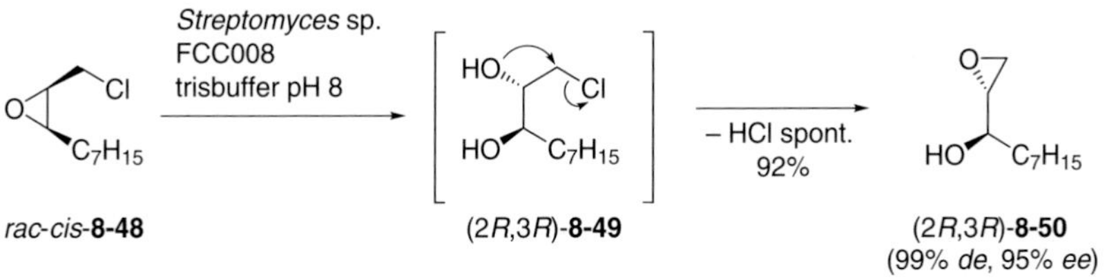

Scheme 8.13. Enzyme-assisted synthesis of the epoxide **8-50**.

The strategy has been used for the synthesis of the natural products panaxytriol (**8-46**) and falcarinol (**8-47**) [20], both of which exhibit strong neurotoxicity, antifungal, and cytotoxic activities [21] (Scheme 8.12).

Using strain *Streptomyces* sp. FCC008, *rac-cis-***8-48** could be converted into (2*R*,3*R*)-**8-50** in 92 % yield, 99 % *ee* and 95 % *de* (Scheme 8.13). **8-50** is one of the two building blocks required for the synthesis of panaxytriol (**8-46**).

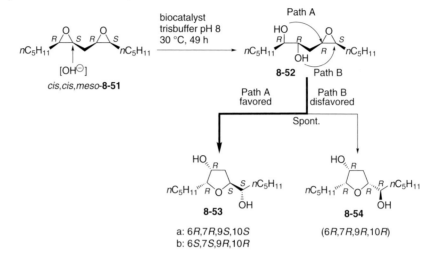

Scheme 8.14. Enzyme-catalyzed transformation of bis-epoxide *cis,cis,meso*-**8-51**.

Entry	Biocatalyst	Yield [%][a]	ee: **8-53** [%]	ee: **8-54** [%]	**8-53**:**8-54** [%]
1	*Rhodococcus ruber* DSM 44540	87	a: 51	–	>99:1
2	*Rhodococcus ruber* DSM 44539	56	a: 24	–	>99:1
3	*Rhodococcus equi* IFO 3730	40	b: 59	–	>99:1
4	*Mycobacterium paraffinicum* NCIMB 10420	38	b: 37	–	>99:1
5	*Rhodococcus* sp. CBS 717.73	84	a: 94	65	94:6

[a] Conversion after 49 h, calculated as the sum of the products.

In a similar manner, and as shown again by the Faber group, the catalyzed reaction of bis-epoxides led to THFs containing four stereocenters [22]. Thus, treatment of *cis,cis,meso*-**8-51** with the epoxide hydrolase *Rhodococcus* sp. CBS 717.73 predominantly yielded the THF derivative **8-53a** in 94 % *ee* and 89 % *de*, whereas the use of other biocatalysts has shown only low to moderate stereoselectivity (Scheme 8.14). As intermediate, the diol **8-52** can be assumed, whereby for the further transformation path **A** is always favored.

It should be noted that as early as 1993, Kurth and coworkers investigated the enzymatic transformation of bis-epoxides of type **8-51** using cytosolic epoxide hydrolase from rat liver. However, at that time the regio- and stereochemistry of the obtained THFs had not been investigated.

A combination of an enzymatic kinetic resolution and an intramolecular Diels–Alder has recently been described by Kita and coworkers [23]. In the first step of this domino process, the racemic alcohols (±)-**8-55** are esterified in the presence of a *Candida antarctica* lipase (CALB) by using the functionalized alkenyl ester **8-56** to give (*R*)-**8-57**, which in the subsequent Diels–Alder reaction led to **8-58** in high enantioselectivity of 95 and 91 % *ee*, respectively and 81 % yield (Scheme 8.15). In-

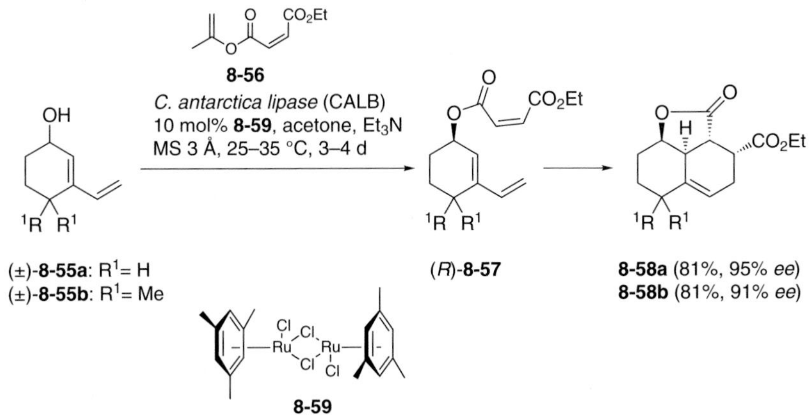

Scheme 8.15. Preparation of optically active polysubstituted decalines by a lipase-catalyzed domino esterification/Diels–Alder reaction.

stead of **8-56**, the methyl ester can also be used, with similar results. A special feature of this domino process is the possibility of performing a rapid racemization of the slow-reacting enantiomer (S)-**8-55** with the help of ruthenium catalyts **8-59**. In this way, a complete transformation of the racemic substrate (±)-**8-55** into almost enantiopure products in 81 % yield can be performed. Normally, in such a kinetic resolution process the yield is limited to <50%.

Beside the use of a single enzyme, a cocktail of different biocatalysts can also be used in performing a domino process, provided that the enzymes do not interfere one with another. This approach was used by Scott and coworkers in the synthesis of precorrin-5 (**8-61**) (Scheme 8.16) [24]. Starting from δ-amino levulinic acid (ALA) **8-60**, a mixture of eight different enzymes including the ALA-dehydratase to form porphobilinogen (PBG), as well as PBG deaminase and co-synthetase to furnish the tetracyclic uroporphyrinogen III (**8-62**) as intermediates, was employed to provide precorrin-5 (**8-61**) in 30% yield.

In a very new example, the Kita group has presented a neat combination of a lipase-catalyzed kinetic resolution of α-hydroxynitrones and a 1,3-dipolar cycloaddition, which was successfully applied in the asymmetric total synthesis of (−)-rosmarinecine [25].

Another very recent development in the field of enzymatic domino reactions is a biocatalytic hydrogen-transfer reduction of halo ketones into enantiopure epoxides, which has been developed by Faber, Bornscheuer and Kroutil. Interestingly, the reaction was carried out with whole lyophilized microbial cells at pH ca. 13. Investigations using isolated enzymes were not successful, as they lost their activity under these conditions [26].

The reactions presented in this chapter show clearly that enzyme-triggered domino reactions offer a great potential in asymmetric synthesis. It remains to be seen, whether this methodology becomes a general tool, since the design of such enzyme-induced domino processes is not trivial. Nonetheless, this emerging field obviously has great potential.

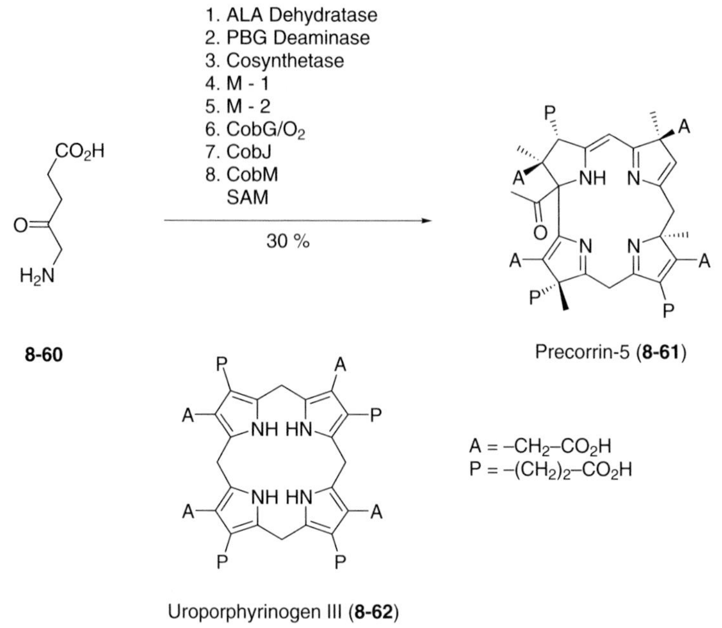

1. ALA Dehydratase
2. PBG Deaminase
3. Cosynthetase
4. M - 1
5. M - 2
6. CobG/O$_2$
7. CobJ
8. CobM
   SAM

30 %

CO$_2$H

O

H$_2$N

8-60

Precorrin-5 (**8-61**)

A = –CH$_2$–CO$_2$H
P = –(CH$_2$)$_2$–CO$_2$H

Uroporphyrinogen III (**8-62**)

Scheme 8.16. Multienzyme cocktail for the domino synthesis of precorrin-5 (**8-61**).

References

1 (a) *Enzyme Catalysis in Organic Synthesis*, (Eds.: K. Drauz, H. Waldmann), VCH, Weinheim, **1995**; (b) K. Faber, *Biotransformations in Organic Chemistry*, Springer, Berlin, **2004**.

2 (a) J. H. Clark, *Green Chemistry* **1999**, *1*, 1–8; (b) J. H. Clark, P. Smith, *Innovations in Pharmaceutical Technology* **2005**, *16*, 94–97.

3 G. Bellucci, G. Berti, M. Ferretti, F. Marioni, F. Re, *Biochem. Biophys. Res. Commun.* **1981**, *102*, 838–844.

4 K. Imai, S. Marumo, *Tetrahedron Lett.* **1976**, *15*, 1211–1214.

5 S. F. Mayer, W. Kroutil, K. Faber, *Chem. Soc. Rev.* **2001**, *30*, 332–339.

6 (a) G. H. Müller, H. Waldmann, *Tetrahedron Lett.* **1996**, *37*, 3833–3836; (b) G. H. Müller, A. Lang, D. R. Seithel, H. Waldmann, *Chem. Eur. J.* **1998**, *4*, 2513–2522.

7 (a) Y. Kita, T. Naka, M. Imanishi, S. Akai, Y. Takebe, M. Matsugi, *Chem. Commun.* **1998**, 1183–1184; (b) S. Akai, T. Naka, S.

Omura, K. Tanimoto, M. Imanishi, Y. Takebe, M. Matsugi, Y. Kita, *Chem. Eur. J.* **2002**, *8*, 4255–4264.

8 For reviews, see: (a) P. Vogel, J. Cossy, J. Plumet, O. Arjona, *Tetrahedron* **1999**, *55*, 13521–13642; (b) C. O. Kappe, S. S. Murphree, A. Padwa, *Tetrahedron* **1997**, *53*, 14179–14233; (c) M. Murakami, H. Igawa, *Chem. Commun.* **2002**, 390–391.

9 *Taxane Anticancer Agents: Basic Science and Current Status* (Eds.: G. I. Georg, T. T. Chen, I. Ojima, D. M. Vyas), American Chemical Society, Washington, DC, **1995**, ACS Symposium Series 538.

10 D. Lee, M.-J. Kim, *Org. Lett.* **1999**, *1*, 925–927.

11 A. P. Schaap, S. D. Gagnon, *J. Am. Chem. Soc.* **1982**, *104*, 3504–3506.

12 A. P. Schaap, R. S. Handley, B. P. Giri, *Tetrahedron Lett.* **1987**, *28*, 935–938.

13 E. Naegele, M. Schelhaas, N. Kuder, H. Waldmann, *J. Am. Chem. Soc.* **1998**, *120*, 6889–6902.

14 S. Niwayama, S. Kobayashi, M. Ohno, *J. Am. Chem. Soc.* **1994**, *116*, 3290–3295.

15 T. Kuhn, C. Tamm, *Tetrahedron Lett.* **1989**, *30*, 693–696.

16 S. T. Russell, J. A. Robinson, D. J. Williams, *J. Chem. Soc. Chem. Commun.* **1987**, 351–352.

17 M. Mischitz, A. Hackinger, I. Francesconi, K. Faber, *Tetrahedron* **1994**, *50*, 8661–8664.

18 Y.-C. Shen, C.-H. Chen, *Tetrahedron Lett.* **1993**, *34*, 1949–1950.

19 S. F. Mayer, A. Steinreiber, R. V. A. Orru, K. Faber, *Eur. J. Org. Chem.* **2001**, 4537–4542.

20 S. F. Mayer, A. Steinreiber, R. V. A. Orru, K. Faber, *J. Org. Chem.* **2002**, *67*, 9115–9121.

21 H. Matsunaga, M. Katano, T. Saita, H. Yamamoto, M. Mori, *Cancer Chemother. Pharmacol.* **1994**, *33*, 291–297.

22 S. M. Glueck, W. M. F. Fabian, K. Faber, S. F. Mayer, *Chem. Eur. J.* **2004**, *10*, 3467–3478.

23 S. Akai, K. Tanimoto, Y. Kita, *Angew. Chem. Int. Ed.* **2004**, *43*, 1407–1410.

24 A. I. Scott, *Synlett* **1994**, 871–883.

25 S. Akai, K. Tanimoto, Y. Kanao, S. Omura, Y. Kita, *Chem. Commun.* **2005**, 2369–2371.

26 T. M. Poessl, B. Kosjek, U. Ellmer, C. C. Gruber, K. Edegger, K. Faber, P. Hildebrandt, U. T. Bornscheuer, W. Kroutil, *Adv. Synth. Catal.* **2005**, *347*, 1827–1834.

9

Multicomponent Reactions

Chemical transformations utilizing more than two starting materials as the input for product formation are usually referred to as multicomponent reactions (MCRs). They can be considered as a subclass of domino processes as they are usually performed employing all substrates in one pot under similar reaction conditions, where the compounds undergo the transformations in a time-resolved mode, meaning "after each other". Since several substrates are put together, it is not only the molecular complexity that is built up very rapidly; indeed, there is also the possibility of generating manifold analogues. This diversity and easy accessibility to a large number of compounds, combined with high-throughput screening techniques, means that MCRs represent a very important tool in modern drug discovery processes [1].

For this reason, we have included an extra chapter in this book which deals with this topic. Quite recently, an excellent book on multicomponent transformations, as well as many highly informative reviews on the subject [1a,b,2] have been published. Therefore, only general aspects and the latest developments will be presented here, especially of isocyanide-based MCRs.

With regard to the classification of MCRs, three different types are often distinguished in the literature:
- Type I, when all the participating reactions are reversible.
- Type II, when the majority of the reactions are reversible, but the final product is formed irreversibly.
- Type III, when practically all of the reactions are irreversible.

Type I MCRs are usually reactions of amines, carbonyl compounds, and weak acids. Since all steps of the reaction are in equilibrium, the products are generally obtained in low purity and low yields. However, if one of the substrates is a bifunctional compound the primarily formed products can subsequently be transformed into, for example, heterocycles in an irreversible manner (type II MCRs). Because of this final irreversible step, the equilibrium is forced towards the product side. Such MCRs often give pure products in almost quantitative yields. Similarly, in MCRs employing isocyanides there is also an irreversible step, as the carbon of the isocyanide moiety is formally oxidized to C^{IV}. In the case of type III MCRs, only a few examples are known in preparative organic chemistry, whereas in Nature the majority of biochemical compounds are formed by such transformations [3].

Domino Reactions in Organic Synthesis. Lutz F. Tietze, Gordon Brasche, and Kersten M. Gericke
Copyright © 2006 WILEY-VCH Verlag GmbH & Co. KGaA, Weinheim
ISBN: 3-527-29060-5

Officially, the history of MCRs dates back to the year 1850, with the introduction of the Strecker reaction (S-3CR) describing the formation of α-aminocyanides from ammonia, carbonyl compounds, and hydrogen cyanide [4]. In 1882, the reaction progressed to the Hantzsch synthesis (H-4CR) of 1,4-dihydropyridines by the reaction of amines, aldehydes, and 1,3-dicarbonyl compounds [5]. Some 25 years later, in 1917, Robinson achieved the total synthesis of the alkaloid tropinone by using a three-component strategy based on Mannich-type reactions (M-3CR) [6]. In fact, this was the earliest application of MCRs in natural product synthesis [7].

The first MCR involving isocyanides (IMCR) was reported in 1921 with the Passerini reaction (P-3CR) [8], and over the years these reactions have become increasingly important and have been highlighted in several publications (for discussions, see below). Another older MCR which leads to (non-natural) α-amino acids is the Bucherer–Bergs reaction (BB-4CR), which was first reported in 1929 [9]. This type of transformation is closely related to the Strecker reaction, with CO_2 employed as a fourth component.

In recent years, many novel MCRs – including Michael addition-initiated three-component domino sequences [10], Knoevenagel/hetero-Diels–Alder-based MCRs [11], radical chain MCRs [12], transition metal-catalyzed Pauson–Khand MCRs [13], as well as Petasis MCRs [14], have been added to the chemist's armamentarium and successfully applied to all fields of organic synthesis.

Modern MCRs that involve isocyanides as starting materials are by far the most versatile reactions in terms of available scaffolds and numbers of accessible compounds. The oldest among these, the three-component Passerini MCR (P-3CR), involves the reaction between an aldehyde 9-1, an acid 9-2, and an isocyanide 9-3 to yield α-acyloxycarboxamides 9-6 in one step [8]. The reaction mechanism has long been a point of debate, but a present-day generally accepted rational assumption for the observed products and byproducts is presented in Scheme 9.1. The reaction starts with the formation of adduct 9-4 by interaction of the carbonyl compound 9-1 and the acid 9-2. This is immediately followed by an addition of the oxygen of the carboxylic acid moiety to the carbon of the isocyanide 9-3 and addition of this carbon to the aldehyde group, as depicted in TS 9-5 to give 9-5. The final product 9-6 is

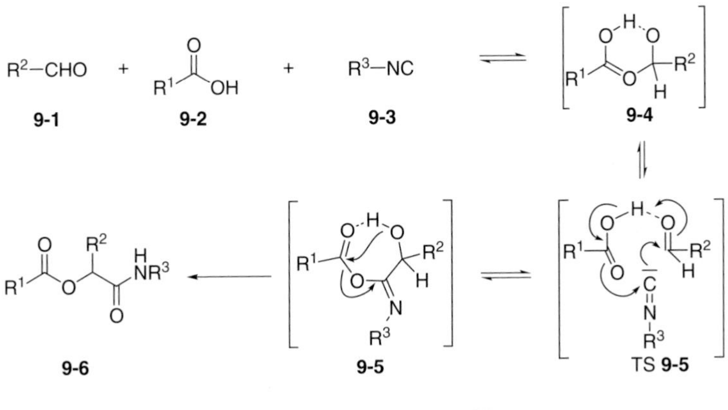

Scheme 9.1. Assumed mechanism of the Passerini-3CR.

Schme 9.2. Passerini condensation involving protected α-aminoaldehydes.

Entry	R¹	R²	R³	R⁴CO₂H	9-8	dr	Yield [%]
1	PhCH₂	H	Bn	PhCH₂CO₂H	a	67:33	81
2	Me	H	cHx	pMePhCO₂H	b	67:33	80
3	iPr	H	MeO₂CCH₂	(Z)-Gly	c	60:40	77
4	–(CH₂)₃–		tBu	PhCH₂CO₂H	d	65:35	94

then derived from **9-5** by a rearrangement [1a]. Consequently, catalytic asymmetric Passerini-3CRs have been recently developed on this basis [15].

The group of Banfi and Guanti has reported on a Passerini reaction employing *N*-protected α-aminoaldehydes of type **9-7**, though to date these have been used only rarely [16]. The general approach depicted in Scheme 9.2 involves the formation of a MCR product **9-8** followed by a one-pot Boc-deprotection and acyl migration under the release of peptide-like substances **9-9** possessing a central α-hydroxy-β-amino acid unit (Scheme 9.2). Interestingly, such compounds have been used in the synthesis of enzyme inhibitors [17]. In addition, the oxidized compounds **9-10** are even more attractive, as they are similar to a protease transition state [18].

The transformation has a broad scope, and higher oligopeptides can easily be obtained using higher functionalized carboxylic acids and/or isocyanides as starting materials [19]. The only disadvantage is the modest diastereoselectivity of approximately 2:1.

Dömling and colleagues have described interesting new variants of the Passerini-3CR. The reaction of arylglyoxals **9-11**, isocyanides **9-12** and α-phosphonatoalkanoic acids **9-13** led to the products **9-14**, which cyclized on treatment with LiBr and NEt₃ in a one-pot procedure. These authors obtained the substituted butenolides **9-15** in medium to high yield (Scheme 9.3) [20]. The butenolide moiety is found in several bioactive natural products, including the cardiotonic digitoxines [21] and the antifungal incrustoporin [22].

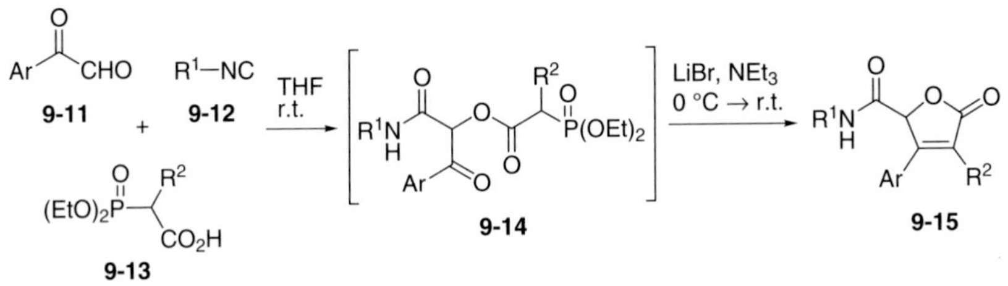

Entry	Isocyanide	Glyoxal	Phosphono acetic acid diethylester	Butenolide	Yield [%]
1	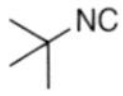	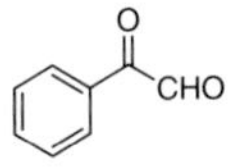	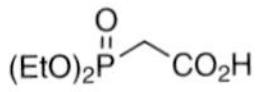	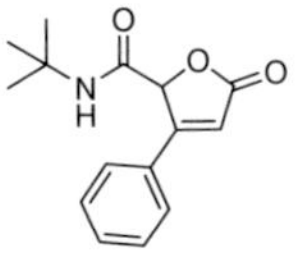	87
2	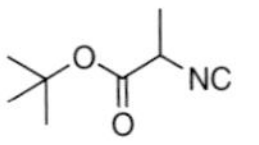	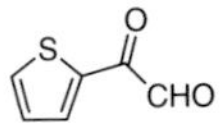	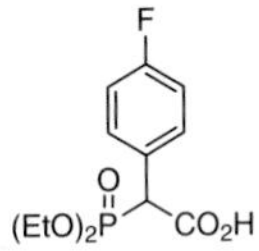	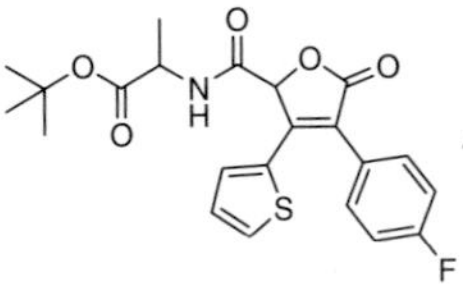	47
3	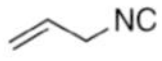	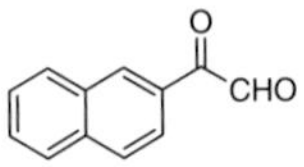	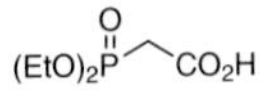	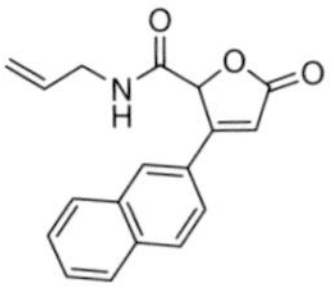	75
4	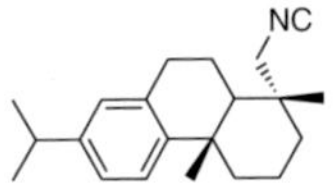	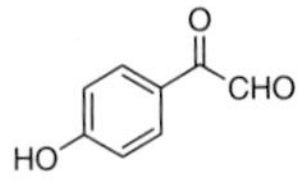	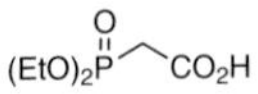	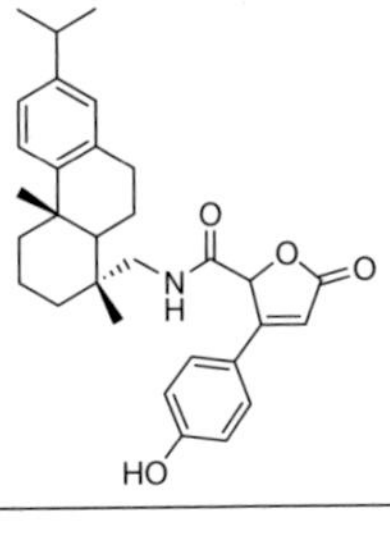	52

Scheme 9.3. Examples of the synthesis of butenolides by Passerini-3CR.

In a more recent approach, the same group synthesized macrocycles using a Passerini reaction followed by a ring-closing metathesis [23], but the final cycliza-tion gave only low yields.

It should be mentioned that the Passerini reaction has also been used by Marcac-cini's group to prepare β-lactams [24], oxazoles [25], and furanes [26]. Natural pro-ducts have also been accessed using this procedure as one of the key steps. The syn-theses of azinomycin by Armstrong [27] and eurystatin A by Schmidt [28] represent two good examples of this procedure.

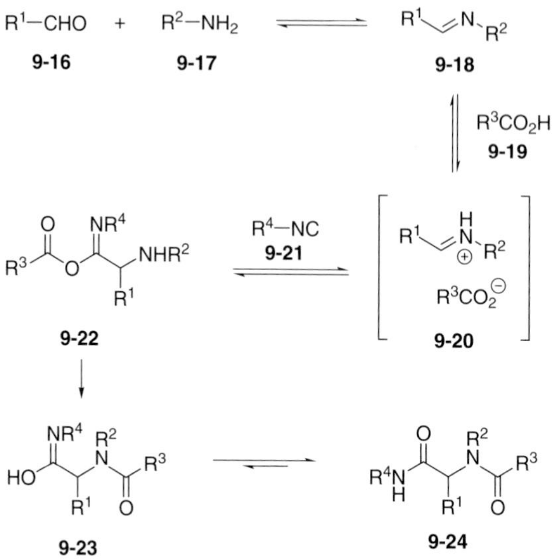

Scheme 9.4. Simplified mechanism of the Ugi-4CR involving a carboxylic acid

Perhaps the best-known MCR is the Ugi four-(or higher) component condensation (U-4CR) [29]. First reported in 1959, the Ugi-4CR describes the conversion of carbonyl compounds 9-16, amines 9-17, various types of acids 9-19 and isocyanides 9-21, the final product being peptide-like structures 9-24. A rather simplified mechanism of the Ugi-4CR is depicted in Scheme 9.4.

In the first step, condensation of the oxo-component 9-16 with the amine 9-17 takes place. Protonation of the obtained imine 9-18 by means of the acid 9-19 generates the highly electrophilic iminium salt 9-20, which in the following step adds to the isocyanide 9-21 to give the intermediate 9-22. This so-called α-adduct is a very strong acylating agent. Accordingly, the nitrogen of the former amine is immediately attacked to give the stable product 9-24 via 9-23.

The diversity of the Ugi-MCR mainly arises from the large number of available acids and amines, which can be used in this transformation. A special case is the reaction of an aldehyde 9-26 and an isocyanide 9-28 with an α-amino acid 9-25 in a nucleophilic solvent HX 9-30 (Scheme 9.5). Again, initially an iminium ion 9-27 is formed, which leads to the α-adduct 9-29. This does not undergo a rearrangement as usual, but the solvent HX 9-30 attacks the lactone moiety. Such a process can be used for the synthesis of aminodicarboxylic acid derivatives such as 9-31 [3, 30].

Moreover, Kim and coworkers have shown that α-amino-butyrolactones can be synthesized by a related process employing the amino acid homoserine with an unprotected hydroxy functionality [31]. In a more recent publication by the same research group, morpholin-2-one derivatives of type 9-37 have been prepared (Scheme 9.6) [32]. Herein, glycolaldehyde dimer 9-32 acts as a bifunctional compound, which first reacts with the α-amino acids 9-33 to give the iminium ions 9-34,

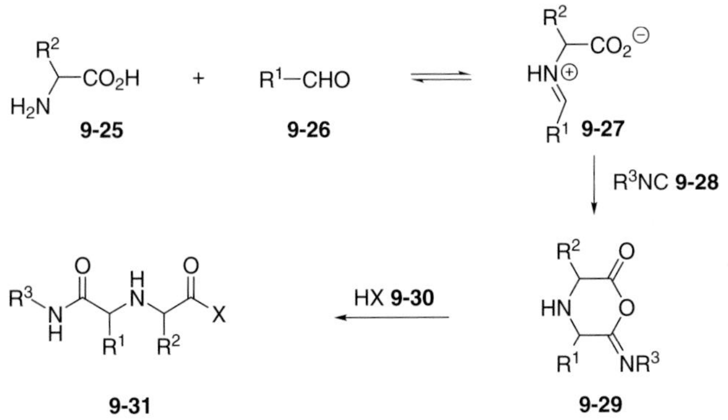

Scheme 9.5. Classical Ugi-4CR based on α-amino acids.

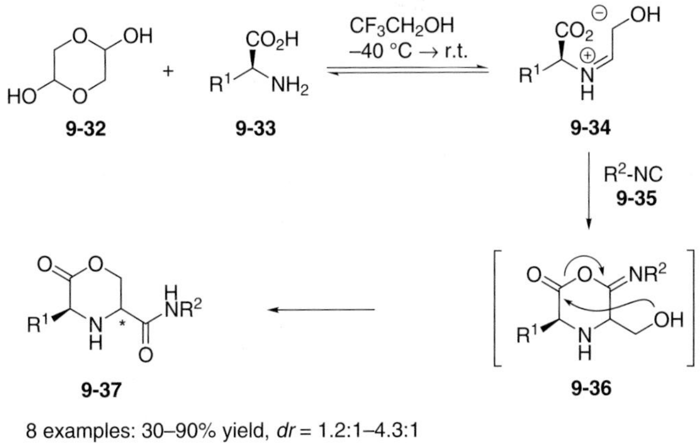

8 examples: 30–90% yield, dr = 1.2:1–4.3:1

Scheme 9.6. Synthesis of morpholine derivatives.

and then with the isocyanides **9-35** to afford the α-adducts **9-36**. Subsequent in-tramolecular acylation of the free hydroxy moiety led to the products **9-37**.

The process is not limited to acyclic amino acids, as cyclic compounds such as proline and pipecolic acid derivatives can also be employed, though in these cases more complex heterobicyclic substances are formed.

Ugi and Dömling have shown that the U-4CR can also be combined with other MCRs, thus creating sequences which involve up to nine different substrates [33]. An example of such an approach is the combination of an Ugi-4CR with the as-yet not mentioned Asinger reaction (A-3CR or A-4CR). The latter allows the formation of thiazolines from ammonia, carbonyl compounds and sulfides [34]. As shown in Scheme 9.7, a mixture of α-bromoisobutyraldehyde, isobutyraldehyde, sodium hy-drogensulfide and ammonia yields the imine **9-38** which, by reaction with t-butyl-isocyanide, methanol, and CO_2, led to the final product **9-39** [35].

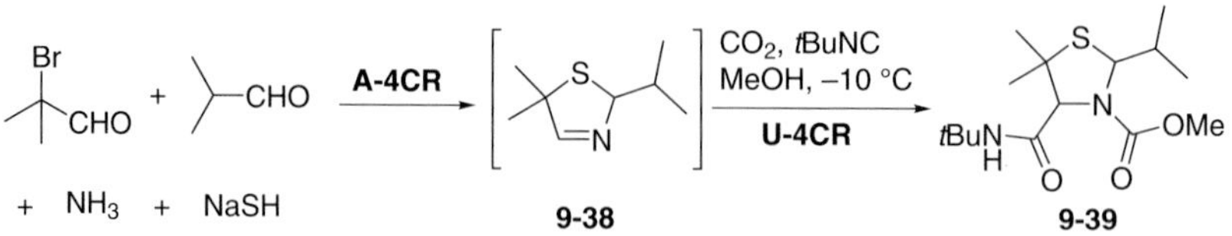

Scheme 9.7. Combination of an A-4CR and U-4CR corresponding to a 7CR.

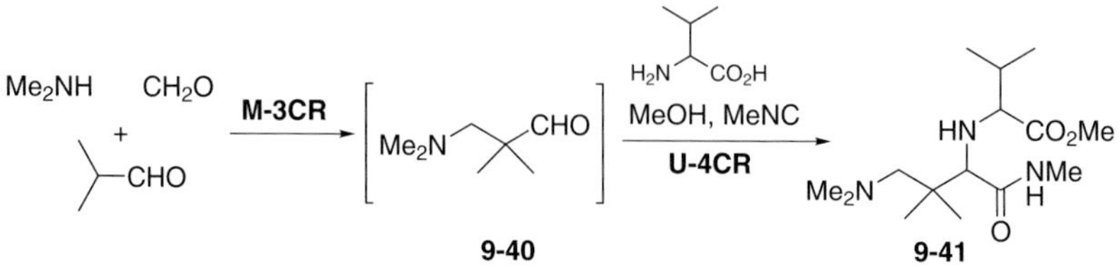

Scheme 9.8. Combination of a M-3CR and an U-4CR.

The combination of a Mannich-3CR with an Ugi-4CR results in a seven-component reaction (7CR), as shown in Scheme 9.8. Reaction of dimethylamine, formaldehyde and isobutyraldehyde led to the β-aminoaldehyde **9-40**, which was treated with valine, methyl isocyanide, and methanol to give the anticipated product **9-41** [3].

Other variations such as a combination of an Ugi-5C-4CR with an Ugi-4CR, creating an Ugi-9C-7CR, or a combination of an Ugi-4CR and a Passerini-3CR to form a 7C-6CR process, are also known [3, 36]. However, the limitations of these approaches are clear, as in "higher MCRs" competing reactions can occur more often such that the formation of unwanted side products is preprogrammed. In some cases the addition of a catalyst, which would accelerate single transformations in these domino processes, can be helpful [3].

Studies describing the combination of Ugi-type reactions and conventional transformations are also known. One report by Torroba and coworkers noted that quinoline-2-(1*H*)-ones could be obtained *via* a one-pot Ugi-4CR/intramolecular Knoevenagel condensation [37], while interesting heterocyclic compounds containing a β-lactam and a thiazole moiety were available using a highly efficient Ugi-3CR/Michael addition-initiated cyclization/elimination sequence discovered by the group of Dömling [38]. Another very new domino process combines a Staudinger reduction, an aza-Wittig reaction and an Ugi-3CR (SAWU-3CR) (Scheme 9.9) [39].

The latter sequence, as reported by Overkleeft, van Boom and coworkers, employs substrates of type **9-42** containing both azide and aldehyde functionalities. Treatment of **9-42** with PMe₃ in MeOH at room temperature forms a cyclic imine **9-43** *via* an intermediate phosphazene. Following the addition of an acid and an isocyanide **9-43**, products of type **9-44** are obtained. According to this scheme, enantiopure carbohydrate-derived azido aldehydes **9–45** and **9-48** led to the morpholino compounds **9-46** and **9-47**, as well as to the pipecolic acid scaffolds **9-49** and **9-50**, re-

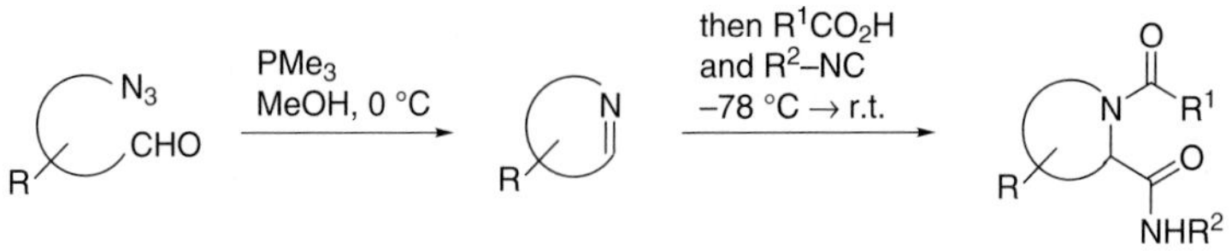

9-42 9-43 9-44

Scheme 9.9. Staudinger/aza-Wittig/Ugi (SAWU-3CR)-process producing N-heterocycles.

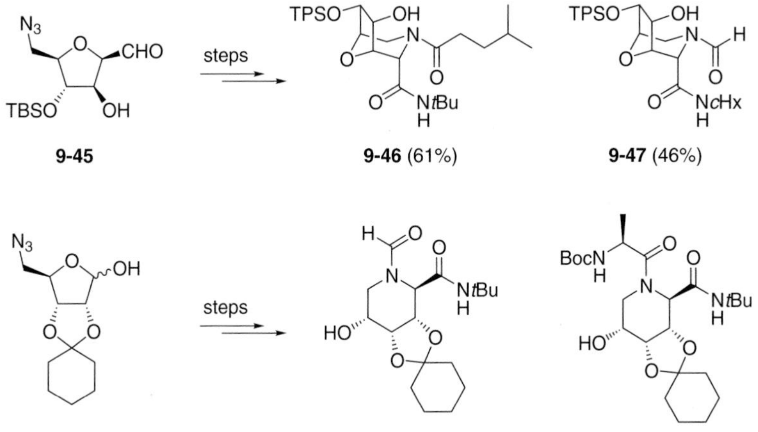

9-45 **9-46** (61%) **9-47** (46%)

9-48 **9-49** (78%) **9-50** (54%)

Scheme 9.10. Morpholine and pipecolic amide SAWU-3CR products.

spectively (Scheme 9.10). In the described cases the final products were isolated as sole diastereomers in yields ranging from 46% to a maximum of 78%.

It should be noted that the group of Martens has also produced pipecolic acid derivatives, although in their Ugi-3CR approach pre-formed cyclic imines were employed as the starting materials [40].

Other recent reports on Ugi multicomponent processes describe the generation of lactams of different ring size. To give some examples of this, Wessjohann and Ruijter [41] published some interesting results on an Ugi-based macrocycle assembly involving bifunctional starting materials, while Banfi's group described the synthesis of novel nine-membered lactams containing a C–C double bond with *cis* geometry [42]. The formation of these compounds is accomplished by utilizing unsaturated substrates in the Ugi reaction, which deliver an ideal ring-closing metathesis precursor for the subsequent step. According to a report by Hulme's group, it is also feasible to prepare γ-lactams using tethered *N*-Boc aldehydes in a Ugi-4CR followed by Boc removal and base-promoted cyclization (Ugi/De-Boc/Cyclize strategy; UDC) [43]. Kennedy and colleagues have used a somewhat related approach to access 2,5-diketopiperazines and 1,4-benzodiazepine-2,5-diones [44]. Such motifs are interesting from a pharmacological point of view, as they form part of several lead-generation libraries [45]. In their approach, Kennedy and colleagues used a so-called "resin-bound carbonate convertible" isocyanide 9-54 [43, 46] This was obtained from the oxazoline 9-51 by treatment with nBuLi to give 9-52 followed by acylation using the resin-bound chlorocarbonate 9-53. The procedure allows a convenient and safe preparation of otherwise hazardous and malodorously smelling isocyanides.

For the preparation of the 2,5-diketopiperazines 9-57 and 1,4-benzodiazepine-2,5-diones 9-58, respectively, the isocyanide 9-54 was either treated with an aldehyde and an amino acid, or with an aldehyde and an anthranilic acid, to give either 9-55 or 9-56, using the conditions depicted in Scheme 9.11. Further transformations include liberation from the resin with KOtBu forming N-acyloxazolidones and treatment with NaOMe to afford the corresponding esters, which are then cyclized to the desired products 9-57 and 9-58 under acidic conditions.

Sheehan and coworkers [47] have also prepared a novel isocyanide derivative, which was used in an Ugi-4CR to prepare noncovalent inhibitors of factor Xa, a key element in the coagulation process [48].

Furthermore, the groups of both Martens [49] and Dömling [50] have independently employed an Ugi-MCR as key reaction in the synthesis of so-called peptide

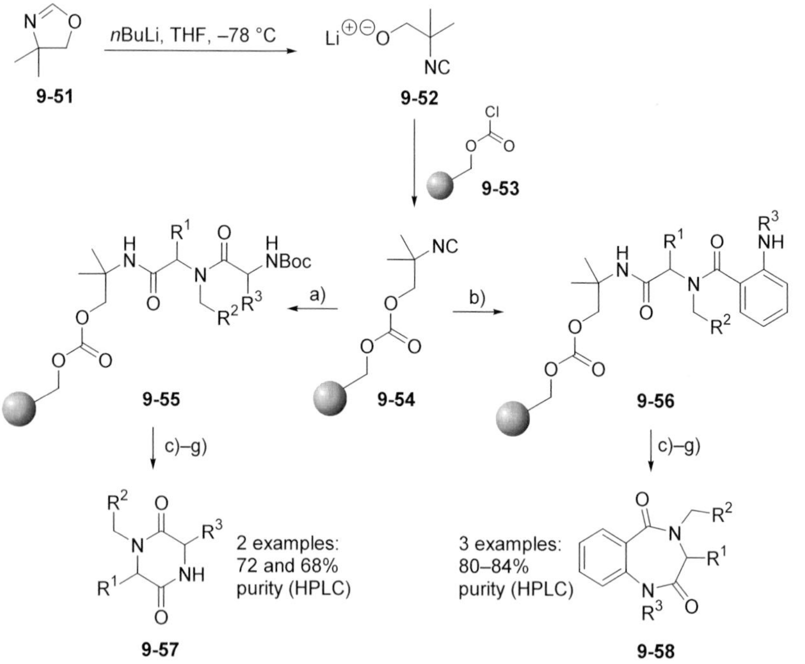

Reagents and conditions. a) 10 eq R^{2}CH$_2$NH$_2$, 10 eq R^{1}CHO, 10 eq Boc-D,L-amino acids, F$_3$CCH$_2$OH, MS 4 Å, CH$_2$Cl$_2$, r.t.; b) same as a) but with 10 eq NR^3-anthranilic acids in THF; c) 2 eq KOtBu, THF, r.t.; d) 1.2 eq NaOMe, THF, r.t.; e) 70/30 hexafluoroisopropanol, TFA, r.t.; f) Silicycle TMA-carbonate, THF; g) Silicycle isocyanate-3, THF, 16 h.

Scheme 9.11. Synthesis of 2,5-diketopiperazines 9-57 and 1,4-benzoazepine-2,5-diones 9-58 by an Ugi-4CR process.

nucleic acids (PNAs) [51], compounds that are of interest both as diagnostics and therapeutics [52]. An example of the approach of the latter research group is depicted in Scheme 9.12. Herein, the simple PNA **9-63** simply precipitated by stirring the four substrates **9-59**, **9-60**, **9-61**, and **9-62** in MeOH. However, this procedure also allows the synthesis of PNA polymers.

A straightforward application of an Ugi reaction in natural product synthesis has been elucidated by Bauer and Armstrong [53]. These authors prepared the intermediate **9-68** in the synthesis of the complex protein phosphatase inhibitor motuporin (**9-69**), by using an U-4CR process starting from the acid **9-64**, the aldehyde **9-65**, methylamine, and the isocyanide **9-66** via **9-67**.

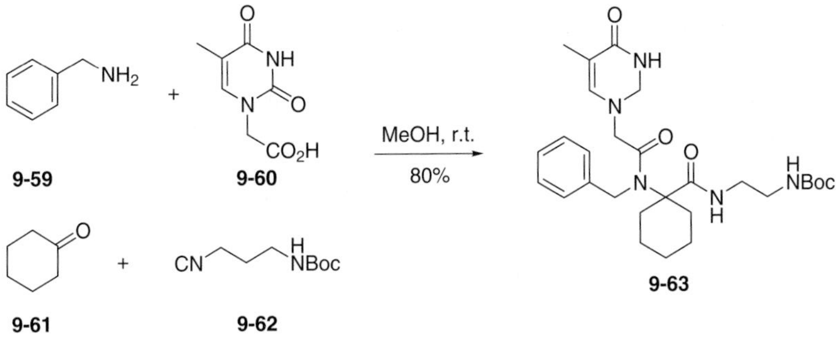

Scheme 9.12. Synthesis of a PNAs by an U-4CR process.

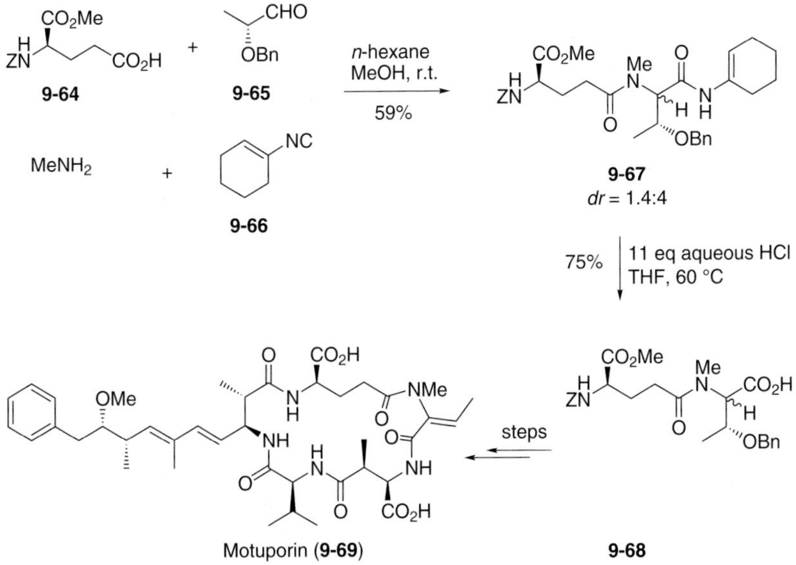

Scheme 9.13. Synthesis of a motuporin (**9-69**) by an U-4CR process.

Beside the Ugi-MCRs, several other novel multicomponent processes using isocyanides as central substrates are known. Kaime and coworkers have described a MCR between nitro compounds, isocyanides and an acylating agent such as acetic acid anhydride [54]. The α-oximinoamides of type **9-75** obtained are probably formed *via* the intermediates **9-72** to **9-74** (Scheme 9.14).

Acylation of the nitronate **9-71** leads to the iminium ion **9-72** which, by the addition of an isocyanide, forms the cation **9-73**. Following two acyl group migrations, the compound **9-75** is obtained *via* **9-74**. The best results were obtained when allylic nitro derivatives were used, as these can form the corresponding enolate in the presence of NEt₃. Aliphatic nitro compounds could also be employed, but in these cases it was necessary to use the more basic DBU.

Almost accidentally, Bienaymé and Bouzid discovered that heterocyclic amidines **9-76** as 2-amino-pyridines and 2-amino-pyrimidines can participate in an acid-catalyzed three-component reaction with aldehydes and isocyanides, providing 3-amino-imidazo[1,2-*a*]pyridines as well as the corresponding pyrimidines and related compounds **9-78** (Scheme 9.15) [55]. In this reaction, electron-rich or -poor (hetero)aromatic and even sterically hindered aliphatic aldehydes can be used with good results. A reasonable rationale for the formation of **9-78** involves a non-concerted [4+1] cycloaddition between the isocyanide and the intermediate iminium ion **9-77**, followed by a [1,3] hydride shift.

A whole set of related powerful isocyanide-based MCRs has been developed by Zhu's research group. By varying the structures of the starting materials and slightly changing the reaction conditions, the group was able to produce different hetereocyclic scaffolds selectively. For example, heating a methanolic mixture of an aldehyde, an amine and an isocyanoacetamide **9-80** allowed the clean formation of 5-amino-oxazoles **9-82** (Scheme 9.16) [56]. It can be assumed that, in the formation of the products **9-82**, the imines **9-79** as well as the adducts **9-81** act as intermediates.

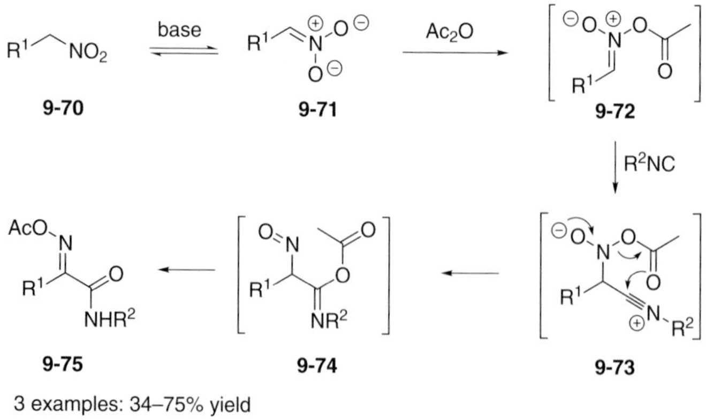

3 examples: 34–75% yield

Scheme 9.14. Formation of compounds **9-75** and proposed intermediates.

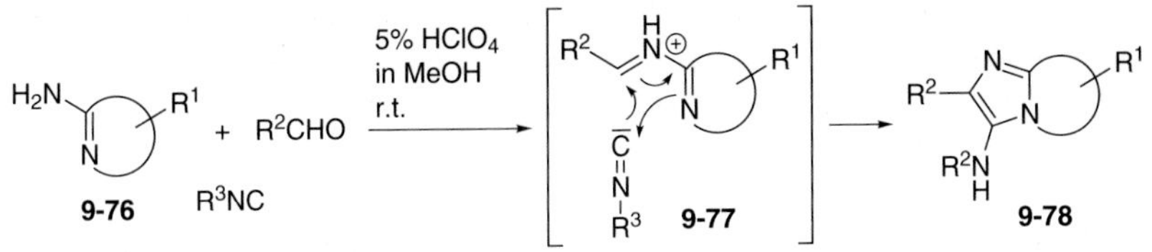

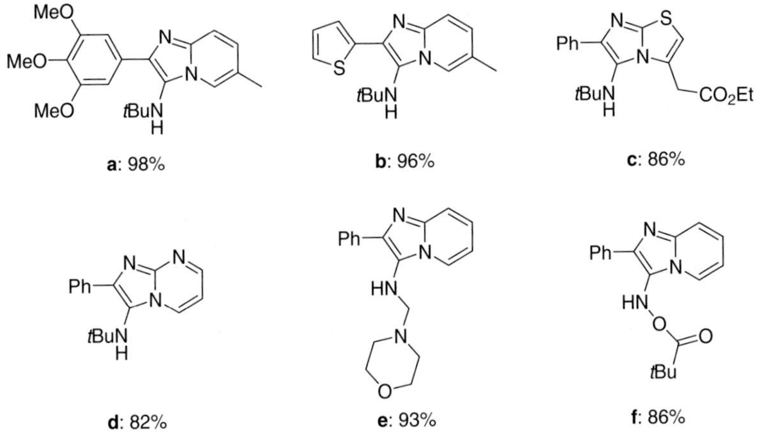

Scheme 9.15. MCR involving amidines as starting materials.

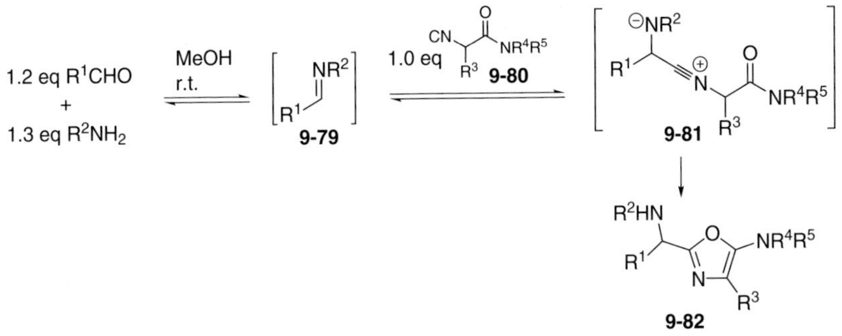

Examples of **9-78**:

a: 98%   **b**: 96%   **c**: 86%

d: 82%   **e**: 93%   **f**: 86%

Examples of **2-82**:

a: 68%   **b**: 60%   **c**: 90%

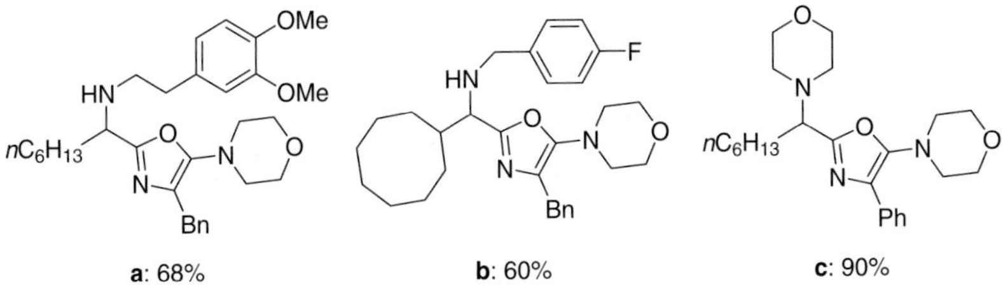

Scheme 9.16. Synthesis of oxazoles.

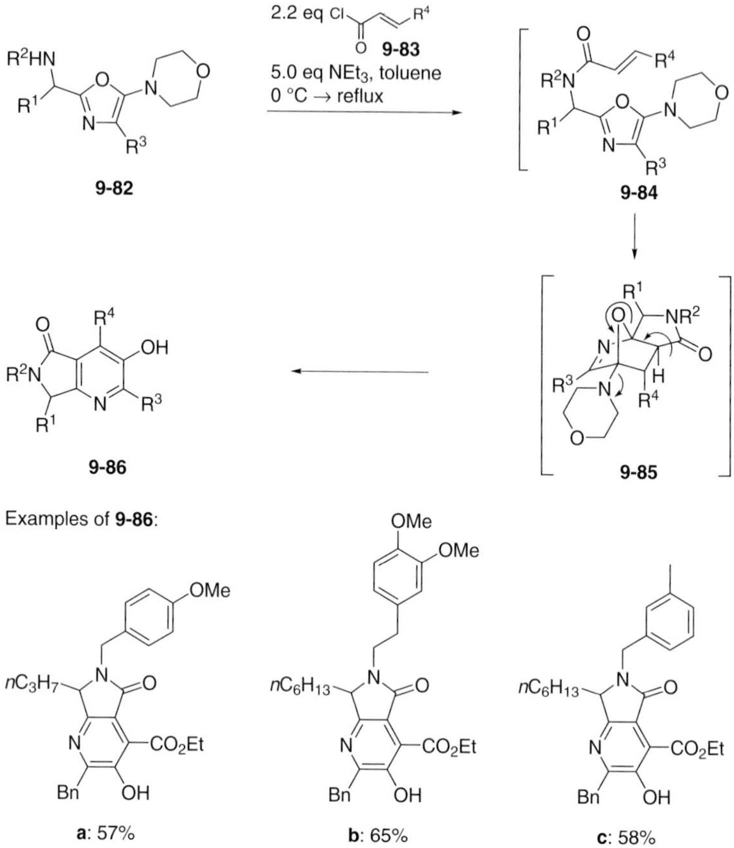

Scheme 9.17. Domino amide-formation/hetero-Diels–Alder reaction/Michael-cycloreversion producing pyrrolopyridines **9-86**.

Furthermore, oxazoles of type **9-82** bearing a secondary amino functionality can be converted into pyrrolo[3,4-*b*]pyridines **9-86** by reaction with appropriate acid chlorides **9-83** in a triple domino process consisting of amide formation/hetero Diels–Alder reaction and retro-Michael cycloreversion *via* **9-84** and **9-85** (Scheme 9.17). The pyrrolo[3,4-*b*]pyridines can be obtained in even higher yields when the whole sequence is carried out as a four-component synthesis in toluene. Here, 1.5 equiv. NH$_4$Cl must be added for the formation of the now intermediate oxazoles [56b].

In addition, oxa-bridged pyrrolopyrimidines can be prepared by employing electron-poor allyl amines, with aldehydes and isocyanoacetamides [57]. The products can be transformed into the pyrrolopyridines by adding TFA at −78 °C to the reaction mixture.

When aniline derivatives **9-87** bearing an alkyne moiety were used as the reaction input together with an aldehyde and **9-80**, furo[2,3-*c*]quinolines **9-88** were obtained (Scheme 9.18) [58]. Here, an intermediate oxazole is also assumed to occur, enter-

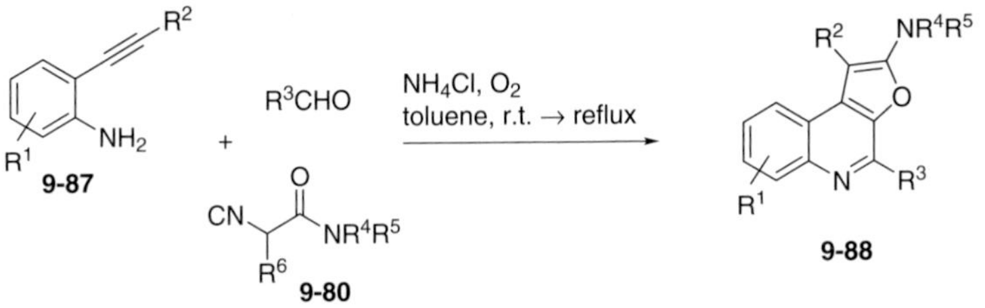

5 examples: 43–75% yield

Scheme 9.18. Three-component synthesis of furo[2,3-*c*]quinolines **9-88**.

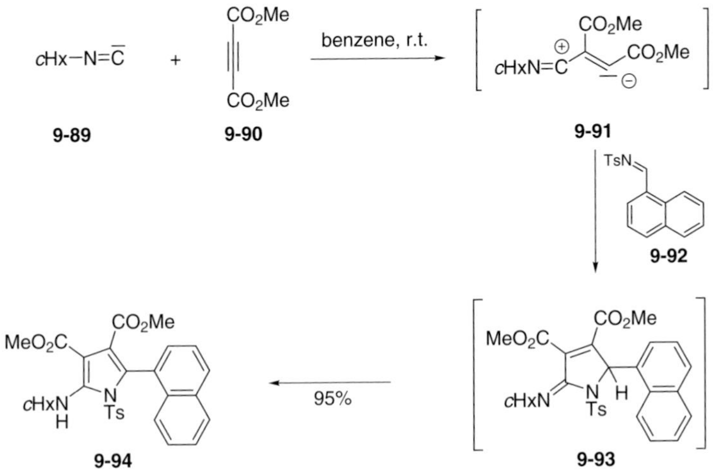

Scheme 9.19. Synthesis of aminopyrroles.

ing into a Diels–Alder/retro Diels–Alder reaction sequence followed by oxidation with atmospheric oxygen.

Other interesting multicomponent sequences utilizing isocyanides have been elaborated by Nair and coworkers. In a recent example, this group exploited the nucleophilic nature of the isocyanide carbon, which allows addition to the triple bond of dimethyl acetylenedicarboxylate (DMAD) (**9-90**) in a Michael-type reaction (Scheme 9.19) [59]. As a result, the 1,3-dipole **9-91** is formed, which reacts with *N*-tosylimines as **9-92** present in the reaction vessel to give the unstable iminolactam **9-93**. Subsequently, this undergoes a [1,5] hydride shift to yield the isolable aminopyrroles **9-94**. In addition to *N*-tosylimine **9-92** and cyclohexyl isocyanide (**9-89**), substituted phenyl tosylimines and *tert*-butyl isocyanide could also be used here.

Quinoneimines can also be used, but in this case iminolactams **9-93** are the final products because the aromatization process cannot take place [59]. A completely different mechanism is proposed when methylene active compounds such as 4-hy-

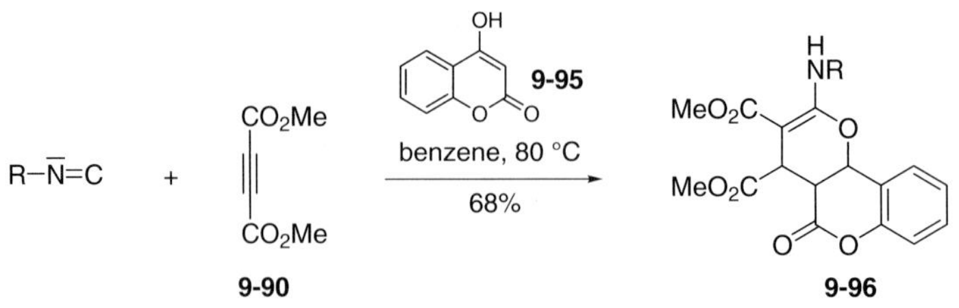

9 examples: R = *t*Bu: 65–98% yield
R = *c*Hx: 49–80% yield

Scheme 9.20. Synthesis of condensed dipyrano derivatives.

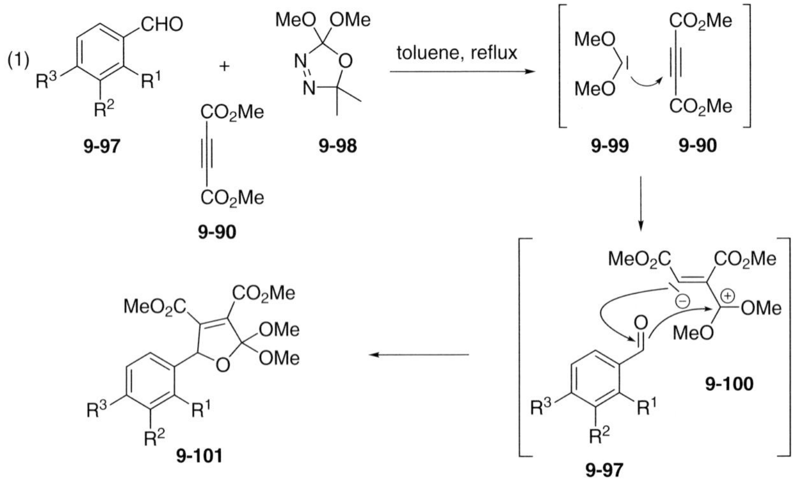

a: R^1= R^2= H, R^3= Me (85%)
b: R^1= R^2= benzo, R^3= H (90%)
c: R^1= R^3= H, R^2= Cl (75%)

Scheme 9.21. Synthesis of dihydrofurans **9-101**.

droxycoumarin (**9-95**) are utilized as the starting materials together with isocyanides and DMAD (**9-90**) (Scheme 9.20). The first steps are identical, but subsequently the formation of a ketenimine is proposed as intermediate; this finally cyclizes through its tautomer to give the condensed dipyran derivatives **9-96** [60].

Besides isocyanides, Nair and coworkers also used carbenes to add to alkynes such as DMAD (**9-90**) leading to 1,3-dipoles, which can be trapped in a formal 1,3-dipolar cycloaddition (Scheme 9.21) [61]. Thus, the dimethoxycarbene **9-99**, generated *in situ* through thermolysis of **9-98**, reacts with DMAD (**9-90**) to give the dipole **9-100**, which adds to an aldehyde **9-97** or a ketone. As the final product, dihydrofurans **9-101** are obtained in good yields.

Beller and coworkers have described a very useful 4-CR, which initially generates 1-acylamino-1,3-butadienes **9-104** as intermediates from two molecules of an alde-

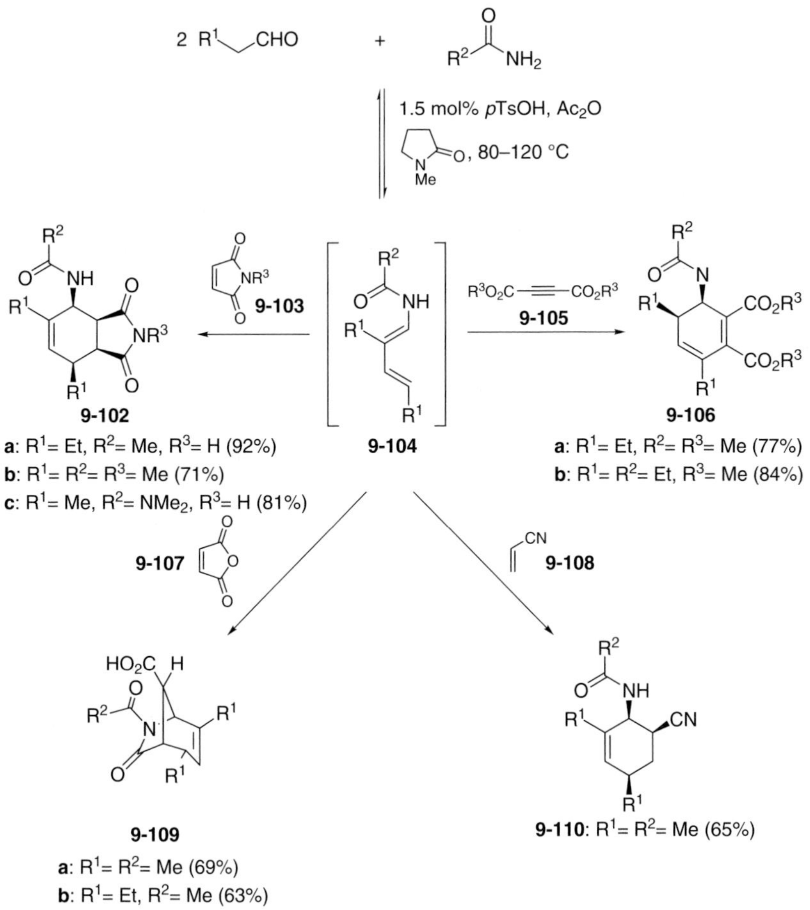

2 R^1 CHO + R^2-C(O)-NH$_2$

1.5 mol% pTsOH, Ac$_2$O
80–120 °C

9-102

a: R^1= Et, R^2= Me, R^3= H (92%)
b: R^1= R^2= R^3= Me (71%)
c: R^1= Me, R^2= NMe$_2$, R^3= H (81%)

9-103

9-104

9-105

9-106

a: R^1= Et, R^2= R^3= Me (77%)
b: R^1= R^2= Et, R^3= Me (84%)

9-107

9-108 CN

9-109

a: R^1= R^2= Me (69%)
b: R^1= Et, R^2= Me (63%)

9-110: R^1= R^2= Me (65%)

Scheme 9.22. Diels–Alder trapping of intermediate 1-arclamino-1,3-butadienes.

hyde and one molecule of an amide in the presence of Ac$_2$O and catalytic amounts of pTsOH·H$_2$O (Scheme 9.22) [62].

The formation of **9-104** proceeds through several equilibrium steps, but the process has been shown to be highly efficient when dienophiles such as maleimides **9-103**, acetylene dicarboxylates **9-105**, maleic anhydride (**9-107**) or acrylonitrile (**9-108**) are present in the reaction mixture. Thus, the formed butadienes **9-104** are trapped in a [4+2] cycloaddition and thereby the equilibria are shifted to the product side. The cycloadducts **9-102**, **9-106**, **9-109** and **9–110** are formed in good to excellent yields with high diastereoselectivity.

A classical non-isocyanide-based multicomponent process is the Biginelli dihydropyrimidine synthesis from β-keto esters, aldehydes and urea or thiourea [63]. The transformation was first reported in 1893 [64], but during the early part of the twentieth century it was somewhat neglected. Today, however, it is once more a "hot topic", and this is underlined by several current publications. One interest in this field is the design of novel catalysts which allow the process to be conducted under milder conditions and to deliver higher yields in comparison to the first protic acid-mediated version described by Biginelli. Besides the generally "normal" Lewis acids which are used as the triflates, Bi(OTf)$_3$ [65], Cu(OTf)$_2$ [66], Sr(OTf)$_3$ [67], Yb(OTf)$_3$ [68], the halides MgCl$_2$ [69], BiCl$_3$ [70], CuCl$_2 \cdot 6H_2O$ [71], FeCl$_3 \cdot 6H_2O$ [72], ZnCl$_2$ [73], CeCl$_3$ [74], RuCl$_3$ [75], VCl$_3$ [76], LaCl$_3 \cdot 7H_2O$ [77], LiBr [78], InCl$_3$ [79], InBr$_3$ [80], SmI$_2$ [81], TMSCl [82], TMSI [83] or other Lewis acid derivatives such as Bi(NO$_3$)$_3 \cdot 5H_2O$ [84], Mn(OAc)$_3 \cdot 5H_2O$ [85], LiClO$_4$ [86] somehow "exotic" catalyst-systems such as a polyaniline-bismoclite complex [87], silica aerogel-iron oxide nanocomposites [88] and ionic liquids [89] have been used. In addition, significant rate and yield enhancements for Biginelli reactions have been reported, for example by Kappe [90] and Stefani [91], as well as by Mirza-Aghayan [92], under the influence of microwave irradiation, while Jenner has recently studied the effect of high pressure on this multicomponent process [93].

In 1997, the controversial mechanism of the Biginelli reaction was reinvestigated by Kappe using NMR spectroscopy and trapping experiments [94], and the current generally accepted process was elucidated (see Scheme 9.23). The *N*-acyliminium ion **9-112** is proposed as key intermediate; this is formed by an acid-catalyzed reaction of an aldehyde with urea or thiourea *via* the semiaminal **9-111**. Interception of **9-112** by the enol form of the 1,3-dicarbonyl compound **9-113** produces the open-chain ureide **9-114**, which cyclizes to the hexahydropyrimidine **9-115**. There follows an elimination to give the final product **9-116**.

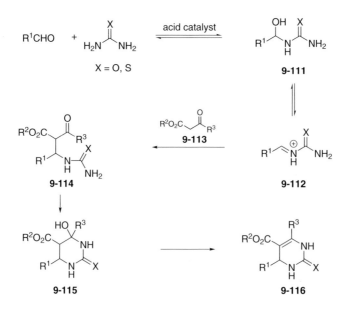

Scheme 9.23. Mechanism of the Biginelli multicomponent reaction.

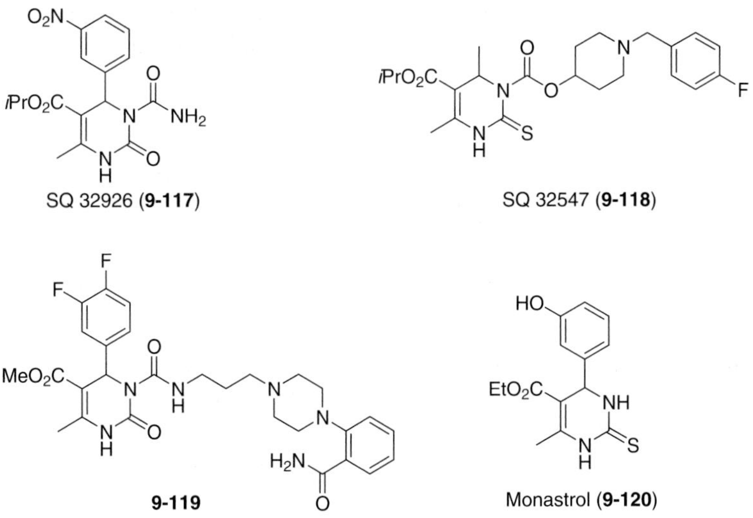

SQ 32926 (**9-117**)

SQ 32547 (**9-118**)

9-119

Monastrol (**9-120**)

Scheme 9.24. Examples of biologically active Biginelli products.

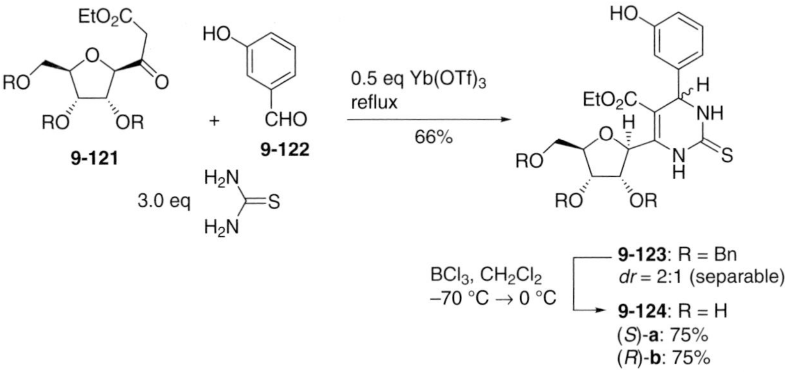

9-121

9-122

0.5 eq Yb(OTf)$_3$
reflux

66%

3.0 eq

BCl$_3$, CH$_2$Cl$_2$
−70 °C → 0 °C

9-123: R = Bn
dr = 2:1 (separable)

9-124: R = H
(S)-**a**: 75%
(R)-**b**: 75%

Scheme 9.25. Synthesis of a C-glycosylated monastrol derivative **9-124**.

The renaissance of the Biginelli MCR can be attributed to the obtained py-rimidine derivatives, which show remarkable pharmacological activity. A broad range of effects, including antiviral, antitumor, antibacterial, anti-inflammatory as well as antihypertensive activities has been ascribed to these partly reduced py-rimidine derivatives [96], such as **9-117** and **9-118** (antihypertensive agents) [97] and **9-119** (α_{1a}-adrenoceptor-selective antagonist) [98] (Scheme 9.24). Recently, the scope of this pharmacophore has been further increased by the identification of the 4-(3-hydroxyphenyl)-pyrimidin-2-thione derivative **9-120** known as monastrol [98], a novel cell-permeable lead molecule for the development of new anticancer drugs. Monastrol appears specifically to affect cell division (mitosis) by a new mechanism, which does not involve tubulin targeting. It has been established that the activity of **9-120** consists of the specific and reversible inhibition of the motility of mitotic kinesin Eg5, a promoter protein required for spindle bipolarity [98].

Not surprisingly, many newer reports are concerned with the synthesis of monastrol (**9-120**) and analogues thereof, as identified by the groups of Bose [99] and Dondoni [100]. The latter research group assembled an interesting C-glycosylated monastrol analogue **9-124** by reaction of the sugar derivative **9-121**, the aldehyde **9-122**, and thiourea in the presence of Yb(OTf)$_3$ (Scheme 9.25). In this way, the benzyl-protected **9-123** was obtained in 66 % yield as a separable 2:1-mixture of diastereomers. Subsequent debenzylation gave the desired compounds (*S*)-**9-124a** and (*R*)-**9-124b**, respectively. The advantage of these compounds is improved water solubility, while the sugar moiety might also positively influence the bioactivity [101]. These recent investigations have also shown that the configuration of the stereogenic center at the pyrimidine ring is important for the biochemical potency of monastrol [102].

The pyrimidine skeleton has also been found in several marine natural products with interesting biological activities [103]. Among the most notable of these are the crambescidin (**9-125**) [104] and the batzelladine (**9-126**) [105] alkaloids which show pronounced bioactivity (Scheme 9.26). Thus, batzelladine A and B are new leads for

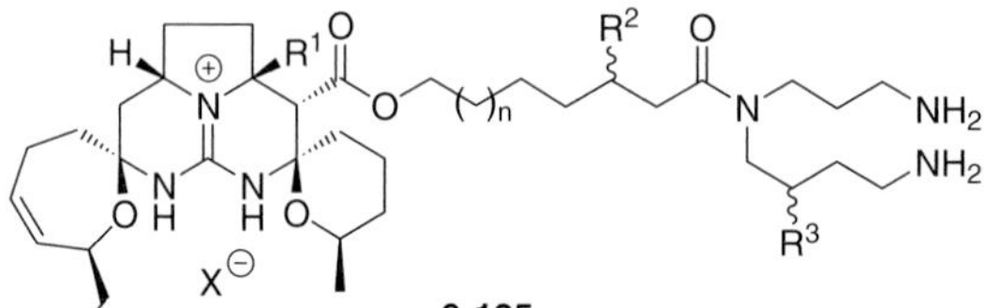

9-125

a: Ptilomycalin A: R^1= R^2= R^3= H, n = 10
b: Crambescidin 800: R^1= R^2= H, R^3= α-OH, n = 10
c: Crambescidin 816: R^1= OH, R^2= H, R^3= α-OH, n = 10
d: Crambescidin 844: R^1= R^3= OH, R^2= H, n = 13
e: Celeromycalin: R^1= R^3= H, R^2= β-OH, n = 10

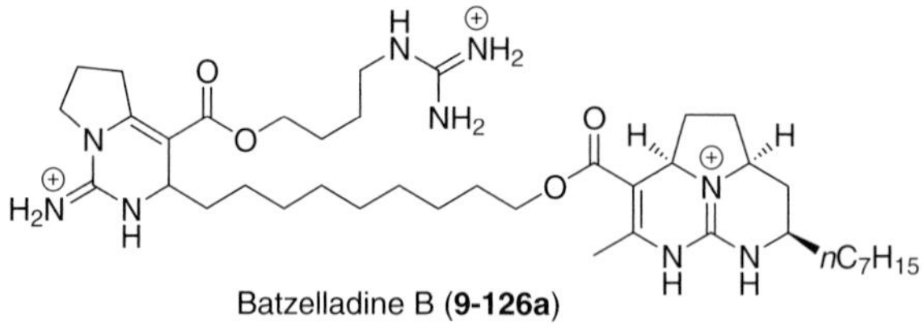

Batzelladine B (**9-126a**)

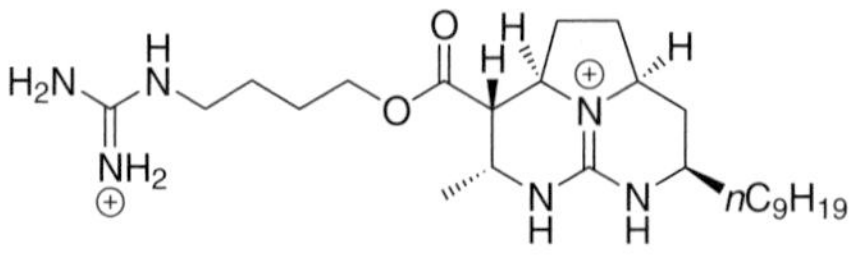

Batzelladine D (**9-126b**)

Scheme 9.26. Representative crambescidin and batzelladine alkaloids.

AIDS therapy, as they inhibit the binding of HIV envelope protein gp-120 to human CD4 cells [105a].

Overman and coworkers have synthesized crambescidin [106] and batzelladine alkaloids [107], as well as simplified analogues [108] using an intramolecular Biginelli condensation [109]. For the synthesis of (–)-dehydrobatzelladine C (**9-131**), as described recently [110], a Biginelli condensation of β-ketoester **9-127** with the enantiopure guanidine aminal **9-128** was used. **9-127** can be synthesized in 10 steps from commercially available materials [107a]. These authors obtained **9-129** with high stereoselectivity (>10:1), but the compound was difficult to separate from residual β-keto ester **9-127**. The mixture was therefore oxidized with CAN to give **9-130** in 33% yield after chromatographic purification, and this was then was transformed into the desired (–)-dehydrobatzelladine C (**9-131**).

A very important development in multicomponent domino reactions is the enantioselective approach using organocatalysts which has been recently discussed in an excellent review by Yus and Ramón [2c]. The latest great success in this field stem from Enders and coworkers, presence of an enantiopure proline derivative to give polyfunctionalized cyclohexenes with 99% ee [111].

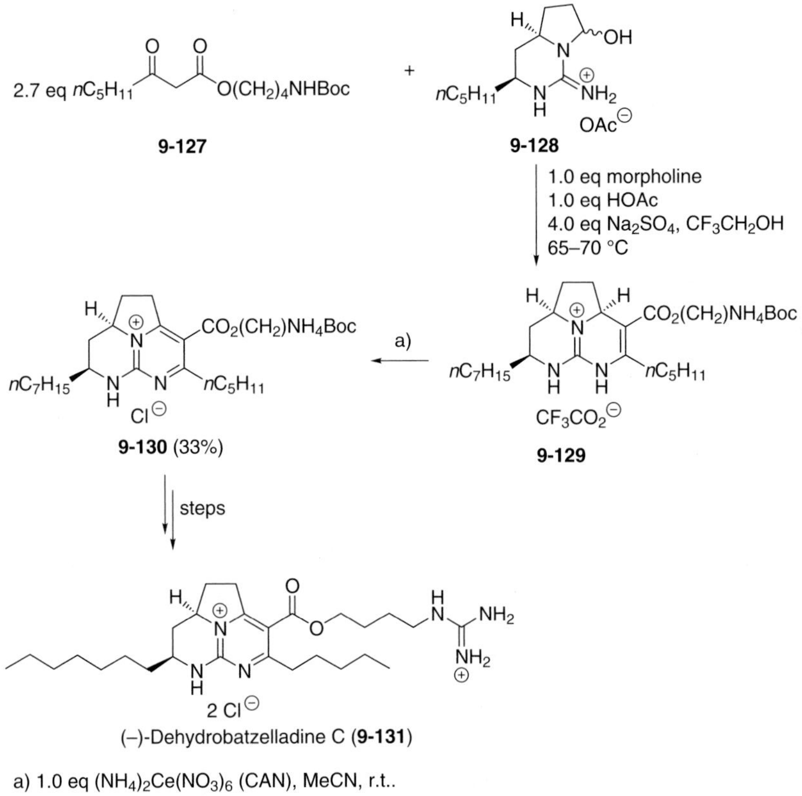

Scheme 9.27. Synthesis of (–)-dehydrobatzelladine C (**9-131**).

References

1 (a) A. Dömling, I. Ugi, *Angew. Chem. Int. Ed. Engl.* **2000**, *39*, 3168–3210; (b) H. Bienaymé, C. Hulme, G. Oddon, P. Schmitt, *Chem. Eur. J.* **2000**, *6*, 3321–3329; (c) L. Weber, *Drug Discovery Today* **2002**, *7*, 143–147; (d) A. Dömling, *Curr. Opin. Chem. Biol.* **2002**, *6*, 306–313; (e) A. Dömling, *Curr. Opin. Chem. Biol.* **2000**, *4*, 318–323; (f) C. Hulme, V. Gore, *Curr. Med. Chem.* **2003**, *10*, 51–80.

2 For recent reviews and monographs, see: (a) I. Ugi, *Pure Appl. Chem.* **2001**, *73*, 187–191; (b) J. Zhu, H. Bienaymé, *Multicomponent Reactions*, Wiley-VCH, Weinheim, **2005**; (c) D. J. Ramón, M. Yus, *Angew. Chem. Int. Ed.* **2005**, *44*, 1602–1634; (c) C. Simon, T. Constantieux, J. Rodriguez, *Eur. J. Org. Chem.* **2004**, 4957–4980; (d) I. Ugi, B. Werner, A. Dömling, *Molecules* **2003**, *8*, 53–66.

3 I. K. Ugi, B. Ebert, W. Hörl, *Chemosphere* **2001**, *43*, 75–81.

4 A. Strecker, *Liebigs Ann. Chem.* **1850**, *75*, 27–45.

5 A. Hantzsch, *Liebigs Ann. Chem.* **1882**, *215*, 1–82.

6 R. Robinson, *J. Chem. Soc.*, **1917**, *111*, 876–899.

7 M. Arend, B. Westermann, N. Risch, *Angew. Chem. Int. Ed.* **1998**, *37*, 1044–1070.

8 (a) M. Passerini, *Gazz. Chim. Ital.* **1921**, *51*, 126; (b) M. Passerini, *Gazz. Chim. Ital.* **1921**, *51*, 181.

9 (a) H. Bergs, DE-B 566,094, **1929**; (b) T. Bucherer, H. Barsch, *J. Prakt. Chem.* **1934**, *140*, 151.

10 For a detailed discussion, see Chapter 2.2 and the following chapters.

11 For detailed discussions, see: (a) Chapter 2.4 and Chapter 10; (b) L. F. Tietze, N. Rackelmann, *Pure Appl. Chem.* **2004**, *76*, 1967–1983.

12 For a recent example, see: K. Miura, M. Tojino, N. Fujisawa, A. Hosomi, I. Ryu, *Angew. Chem. Int. Ed.* **2004**, *43*, 2423–2425, and for a detailed discussion, see also Chapter 3.

13 For a detailed discussion, see Chapter 6.

14 For some recent examples, see: (a) N. A. Petasis, I. A. Zavialov, *J. Am. Chem. Soc.* **1997**, *119*, 445–446; (b) N. A. Petasis, A. Goodman, I. A. Zavialov, *Tetrahedron* **1997**, *48*, 16463–16470; (c) N. A. Petasis, Z. D. Patel, *Tetrahedron Lett.* **2000**, *41*, 9607–9611; (d) L. M. Harwood, G. S. Currie, M. G. B. Drew, R. W. A. Luke, *Chem. Commun.* **1996**, 1953–1954; (e) G. S. Currie, M. G. B. Drew, L. M. Harwood, D. J. Hughes, R. W. A. Luke, R. J. Vickers, *J. Chem. Soc. Perkin Trans. 1* **2000**, 2982–2990; (f) P. J. Pye, K. Rossen, S. A. Weissman, A. Maliakal, R. A. Reamer, R. Ball, N. N. Tsou, R. P. Volante, P. J. Reider, *Chem. Eur. J.* **2002**, *8*, 1372–1376; (g) N. A. Petasis, I. A. Zavialov, *J. Am. Chem. Soc.* **1998**, *120*, 11798–11799; (h) G. K. S. Prakash, M. Mandal, S. Schweizer, N. A. Petasis, G. A. Olah, *Org. Lett.* **2000**, *2*, 3173–3176; (i) G. K. S. Prakash, M. Mandal, S. Schweizer, N. A. Petasis, G. A. Olah, *J. Org. Chem.* **2002**, *67*, 3718–3723; Corrigendum: *J. Org. Chem.* **2002**, *67*, 6286; (j) T. Koolmeister, M. Södergren, M. Scobie, *Tetrahedron Lett.* **2002**, *43*, 5969–5970.

15 (a) P. R. Andreana, M. C. Liu, S. L. Schreiber, *Org. Lett.* **2004**, *6*, 4231–4233; (b) U. Kusebauch, B. Beck, K. Messer, E. Herdtweck, A. Dömling, *Org. Lett.* **2003**, *5*, 4021–4024; (c) S. E. Denmark, Y. Fan, *J. Am. Chem. Soc.* **2003**, *125*, 7825–7827.

16 L. Banfi, G. Guanti, R. Riva, *Chem. Commun.* **2000**, 985–986.

17 For examples, see: (a) B. E. Evans, K. E. Rittle, M. G. Bock, C. D. Bennett, R. M. DiPardo, J. Boger, M. Poe, E. H. Ulm, B. I. LaMont, E. H. Blaine, G. M. Fanelli, I. I. Stabilito, D. F. Veber, *J. Med. Chem.* **1985**, *28*, 1756–1759; (b) T. Mimoto, J. Imai, S. Tanaka, N. Hattori, S. Kisanuki, K. Akaji, Y. Kiso, *Chem. Pharm. Bull.* **1991**, *39*, 3088–3090.

18 For examples, see: (a) Z. Li, A.-C. Ortega-Vilain, G. S. Patil, D.-L. Chu, J. E. Foreman, D. D. Eveleth, J. C. Powers, *J. Med. Chem.* **1996**, *39*, 4089–4098; (b) J. Cacciola, R. S. Alexander, J. M. Fevig, P. F. W. Stouten, *Tetrahedron Lett.* **1997**, *38*, 5741–5744; (c) S. L. Harbeson, S. M. Abelleira, A. Akiyama, R. Barrett, III, R. M. Carroll, J. A. Straub, J. N. Tkacz, C. Wu, G. F. Musso, *J. Med. Chem.* **1994**, *37*, 2918–2929.

19 L. Banfi, G. Guanti, R. Riva, A. Basso, E. Calcagno, *Tetrahedron Lett.* **2002**, *43*, 4067–4069.

20 B. Beck, M. Magnin-Lachaux, E. Herdtweck, A. Dömling, *Org. Lett.* **2001**, *3*, 2875–2878.

21 K. R. H. Repke, R. Megges, J. Weiland, R. Schön, *Angew. Chem. Int. Ed. Engl.* **1995**, *34*, 282–294.

22 S. Zapf, T. Anke, O. Sterner, *Acta Chem. Scand.* **1995**, *49*, 233–234.

23 B. Beck, G. Larbig, B. Mejat, M. Magnin-Lachaux, A. Picard, E. Herdtweck, A. Dömling, *Org. Lett.* **2003**, *5*, 1047–1050.

24 R. Bossio, C. F. Marcos, S. Marcaccini, R. Pepino, *Tetrahedron Lett.* **1997**, *38*, 2519–2520.

25 R. Bossio, S. Marcaccini, R. Pepino, *Liebigs Ann. Chem.* **1991**, 1107–1108.

26 R. Bossio, S. Marcaccini, R. Pepino, T. Torroba, *Synthesis* **1993**, 783–785.

27 R. W. Armstrong, A. P. Combs, P. A. Tempest, S. D. Brown, T. A. Keating, *Acc. Chem. Res.* **1996**, *29*, 123–131.

28 U. Schmidt, S. Weinbrenner, *J. Chem. Soc. Chem. Commun.* **1994**, 1003–1004.

29 (a) I. Ugi, R. Meyr, U. Fetzer, C. Steinbrückner, *Angew. Chem.* **1959**, *71*, 386; (b) I. Ugi, C. Steinbrückner, *Angew. Chem.* **1960**, *72*, 267–268.

30 B. M. Ebert, I. K. Ugi, *Tetrahedron* **1998**, *54*, 11887–11898.

31 S. J. Park, G. Keum. S. B. Kang, H. Y. Koh, D. H. Lee, Y. Kim, *Tetrahedron Lett.* **1998**, *39*, 7109–7112.

32 Y. B. Kim, E. H. Choi, G. Keum, S. B. Kang, D. H. Lee, H. Y. Koh, Y. Kim, *Org. Lett.* **2001**, *3*, 4149–4152.

33 (a) I. Ugi, A. Dömling, W. Hörl, *Endeavour* **1994**, *18*, 115–122; (b) I. Ugi, A. Dömling, W. Hörl, *GIT Fachzeitschrift für das Laboratorium* **1994**, *38*, 430–437; (c) A. Dömling, *Combinatorial Chem. High Throughput Screening* **1998**, *1*, 1–22; (d) I. Ugi, A. Demharter, A. Hörl, T. Schmid, *Tetrahedron* **1996**, *52*, 11657–11664.

34 F. Asinger, *Angew. Chem.* **1956**, *68*, 413.

35 (a) A. Dömling, I. Ugi, *Angew. Chem. Int. Ed. Engl.* **1993**, *32*, 563–564; (b) A. Dömling, I. Ugi, E. Herdtweck, *Acta Chem. Scand.* **1998**, *52*, 107–113.

36 B. Ebert, Dissertation, Technical University of Munich, **1998**, p. 58.

37 S. Marcaccini, R. Pepino, M. C. Pozo, S. Basurto, M. García-Valverde, T. Torroba, *Tetrahedron Lett.* **2004**, *45*, 3999–4001.

38 J. Kolb, B. Beck, A. Dömling, *Tetrahedron Lett.* **2002**, *43*, 6897–6901.

39 M. S. M. Timmer, M. D. P. Risseeuw, M. Verdoes, D. V. Filippov, J. R. Plaisier, G. A. van der Marel, H. S. Overkleeft, J. H. van Boom, *Tetrahedron: Asymm.* **2005**, *16*, 177–185.

40 (a) W. Maison, A. Lützen, M. Kosten, I. Schlemminger, O. Westerhoff, J. Martens, *J. Chem. Soc. Perkin Trans. 1* **1999**, 3515–3525; (b) W. Maison, A. Lützen, M. Kosten, I. Schlemminger, O. Westerhoff, W. Saak, J. Martens, *J. Chem. Soc. Perkin Trans. 1* **2000**, 1867–1871.

41 A. Wessjohann, E. Ruijter, *Mol. Diversity* **2005**, *9*, 159–169.

42 L. Banfi, A. Basso, G. Guanti, R. Riva, *Tetrahedron Lett.* **2003**, *44*, 7655–7658.

43 C. Hulme, L. Ma, M.-P. Cherrier, J. J. Romano, G. Morton, C. Duquenne, J. Salvino, R. Labaudiniere, *Tetrahedron Lett.* **2000**, *41*, 1883–1887.

44 A. L. Kennedy, A. M. Fryer, J. A. Josey, *Org. Lett.* **2002**, *4*, 1167–1170.

45 (a) Y. Funabashi, T. Horiguchi, S. Iinuma, S. Tanida, S. Harada, *J. Antibiot.* **1994**, *47*, 1202–1218; (b) C.-B. Cui, H. Kakeya, H. Osada, *J. Antibiot.* **1996**, *49*, 534–540; (c) T. A. Keating, R. W. Armstrong, *J. Org. Chem.* **1996**, *61*, 8935–8939; (d) C. Hulme, M.-P. Cherrier, *Tetrahedron Lett.* **1999**, *40*, 5295–5299; (e) C. Hulme, M. M. Morrissette, F. A. Volz, C. J. Burns, *Tetrahedron Lett.* **1998**, *39*, 1113–1116.

46 For some other applications, see: (a) T. A. Keating, R. W. Armstrong, *J. Am. Chem. Soc.* **1995**, *117*, 7842–7843; (b) J. J. Chen, A. Golebiowski, J. McClenagan, S. R. Klopfenstein, L. West, *Tetrahedron Lett.* **2001**, *42*, 2269–2271; (c) C. Hulme, J. Peng, G. Morton, J. M. Salvino, T. Herpin, R. Labaudiniere, *Tetrahedron Lett.* **1998**, *39*, 7227–7230.

47 S. M. Sheehan, J. J. Masters, M. R. Wiley, S. C. Young, J. W. Liebeschuetz, S. D. Jones, C. W. Murray, J. B. Franciskovich, D. B. Engel, W. W. Weber II, J. Marimuthu, J. A. Kyle, J. K. Smallwood, M. W. Farmen, G. F. Smith, *Bioorg. Med. Chem. Lett.* **2003**, *13*, 2255–2259.

48 R. J. Leadly, Jr., *Curr. Top. Med. Chem.* **2001**, *1*, 151–159.

49 W. Maison, I. Schlemminger, O. Westerhoff, J. Martens, *Bioorg. Med. Chem. Lett.* **1999**, *9*, 581–584.

50 A. Dömling, K.-Z. Chi, M. Barrère, *Bioorg. Med. Chem. Lett.* **1999**, *9*, 2871–2874.

51 (a) P. E. Nielsen, M. Egholm, R. H. Berg, O. Buchardt, *Science* **1991**, *254*, 1497–1500; (b) E. Uhlmann, A. Peyman, *Chem. Rev.* **1990**, *90*, 543–584; (c) A. De Mesmaeker, R. Häner, P. Martin, H. E. Moser, *Acc. Chem. Rev.* **1995**, *28*, 366–374.

52 B. Hyrup, P. E. Nielsen, *Bioorg. Med. Chem.* **1996**, *4*, 5–23.

53 S. M. Bauer, R. W. Armstrong, *J. Am. Chem. Soc.* **1999**, *121*, 6355–6366.

54 P. Dumestre, L. El Kaim, A. Grégoire, *Chem. Commun.* **1999**, 775–776.

55 H. Bienaymé, K. Bouzid, *Angew. Chem. Int. Ed.* **1998**, *37*, 2234–2237.

56 (a) X. Sun, P. Janvier, G. Zhao, H. Bienaymé, J. Zhu, *Org. Lett.* **2001**, *3*, 877–880; (b) P. Janvier, X. Sun, H. Bienaymé, J. Zhu, *J. Am Chem. Soc.* **2002**, *124*, 2560–2567.

57 R. Gámez-Montańo, E. González-Zamora, P. Potier, J. Zhu, *Tetrahedron* **2002**, *58*, 6351–6351.

58 A. Fayol, J. Zhu, *Angew. Chem. Int. Ed.* **2002**, *41*, 3633–3635.

59 V. Nair, A. U. Vinod, C. Rajesh, *J. Org. Chem.* **2001**, *66*, 4427–4429.

60 V. Nair, A. U. Vinod, R. Ramesh, R. S. Menon, L. Varma, S. Mathew, A. Chiaroni, *Heterocycles* **2002**, *58*, 147–151.

61 V. Nair, S. Bindu, L. Balagopal, *Tetrahedron Lett.* **2001**, *42*, 2043–2044.

62 H. Neumann, A. Jacobi von Wangelin, D. Gördes, A. Spannenberg, M. Beller, *J. Am. Chem. Soc.* **2001**, *123*, 8398–8399.

63 For a recent review, see: C. O. Kappe, *Acc. Chem. Res.* **2000**, *33*, 879–888.

64 P. Biginelli, *Gazz. Chim. Ital.* **1893**, *23*, 360–413.

65 R. Varala, M. M. Alam, S. R. Adapa, *Synlett* **2003**, 67–70.

66 A. S. Paraskar, G. K. Dewkar, A. Sudalai, *Tetrahedron Lett.* **2003**, *44*, 3305–3308.

67 W. Su, J. Li, Z. Zheng, Y. Shen, *Tetrahedron Lett.* **2005**, *46*, 6037–6040.

68 (a) L. Wang, C. Qian, H. Tian, Y. Ma, *Synth. Commun.* **2003**, *33*, 1459–1468; (b) Y. Ma, C. Qian, L. Wang, M. Yang, *J. Org. Chem.* **2000**, *65*, 3864–3868.

69 G.-L. Zhang, X.-H. Cai, *Synth. Commun.* **2005**, *36*, 829–833.

70 K. Ramalinga, P. Vijayalakshmi, T. N. B. Kaimal, *Synlett* **2001**, 863–865.

71 M. Mukut, D. Prajapati, J. S. Sandhu, *Synlett* **2004**, 235–238.

72 J. Lu, H. R. Ma, *Synlett* **2000**, 63–64.

73 Q. Sun, Y. Wang, Z. Ge, T. Cheng, R. Li, *Synthesis* **2004**, 1047–1051.

74 D. S. Bose, L. Fatima, H. B. Mereyala, *J. Org. Chem.* **2003**, *68*, 587–590.

75 S. K. De, R. A. Gibbs, *Synthesis* **2005**, 1748–1750.

76 G. Sabitha, G. S. K. K. Reddy, K. B. Reddy, J. S. Yadav, *Tetrahedron Lett.* **2003**, *44*, 6497–6499.

77 J. Lu, Y. Bai, Z. Wang, B. Yang, H. Ma, *Tetrahedron Lett.* **2000**, *41*, 9075–9078.

78 (a) S. Rudrawar, *Synlett* **2005**, 1197–1198; (b) G. Maiti, P. Kundu, C. Guin, *Tetrahedron Lett.* **2003**, *44*, 2757–2758.

79 B. C. Ranu, A. Hajra, U. Jana, *J. Org. Chem.* **2000**, *65*, 6270–6272.

80 (a) M. A. P. Martins, M. V. M. Teixeira, W. Cunico, E. Scapin, R. Mayer, C. M. P. Pereira, N. Zanatta, H. G. Bonacorso, C. Peppe, Y.-F. Yuan, *Tetrahedron Lett.* **2004**, *45*, 8991–8994; (b) N.-Y. Fu, Y.-F. Yuan, Z. Cao, S.-W. Wang, J.-T. Wang, C. Peppe, *Tetrahedron* **2002**, *58*, 4801–4807.

81 X. Han, F. Xu, Y. Luo, Q. Shen, *Eur. J. Org. Chem.* **2005**, *8*, 1500–1503.

82 Y. Zhu, Y. Pan, S. Huang, *Synth. Commun.* **2004**, *34*, 3167–3174.

83 G. Sabitha, G. S. K. K. Reddy, C. S. Reddy, J. S. Yadav, *Synlett* **2003**, 858–860.

84 M. M. Khodaei, A. R. Khosropour, M. Beygzadeh, *Synth. Commun.* **2004**, *34*, 1551–1557.

85 K. A. Kumar, M. Kasthuraiah, C. S. Reddy, C. D. Reddy, *Tetrahedron Lett.* **2001**, *42*, 7873–7875.

86 J. S. Yadav, B. V. S. Reddy, R. Srinivas, C. Venugopal, T. Ramalingam, *Synthesis* **2001**, 1341–1345.

87 B. Gangadasu, S. Palaniappan, V. J. Rao, *Synlett* **2004**, 1285–1287.

88 S. Martinez, M. Meseguer, L. Casas, E. Rodriguez, E. Molins, M. Moreno-Manas, A. Roig, R. M. Sebastian, A. Vallribera, *Tetrahedron* **2003**, *59*, 1553–1556.

89 J. Peng, Y. Deng, *Tetrahedron Lett.* **2001**, *42*, 5917–5919.

90 (a) A. Stadler, C. O. Kappe, *J. Chem. Soc. Perkin Trans. 1* **2000**, 1363–1368; (b) C. O. Kappe, D. Kumar, R. S. Varma, *Synthesis* **1999**, 1799–1803.

91 H. A. Stefani, P. M. Gatti, *Synth. Commun.* **2000**, *30*, 2165–2173.

92 M. Mirza-Aghayan, M. Bolourtchian, M. Hosseini, *Synth. Commun.* **2004**, *34*, 3335–3341.

93 G. Jenner, *Tetrahedron Lett.* **2004**, *45*, 6195–6198.

94 C. O. Kappe, *J. Org. Chem.* **1997**, *62*, 7201–7204.

95 (a) C. O. Kappe, *Tetrahedron* **1993**, *49*, 6937–6963; (b) D. Dallinger, C. O. Kappe, *Pure Appl. Chem.* **2005**, *77*, 155–161; (c) D. Dallinger, A. Stadler, C. O. Kappe, *Pure Appl. Chem.* **2004**, *76*, 1017–1024.

96 (a) K. S. Atwal, B. N. Swanson, S. E. Unger, D. M. Floyd, S. Moreland, A. Hedberg, B. C. O'Reilly, *J. Med. Chem.* **1991**, *34*, 806–811; (b) G. C. Rovnyak, K. S. Atwal, A. Hedberg, S. D. Kimball, S. Moreland, J. Z. Gougoutas, B. C. O'Reilly, J. Schwartz, M. F. Malley, *J. Med. Chem.* **1992**, *35*, 3254–3263; (c) G. J. Grover, S. Dzwonczyk, D. M. McMullen, D. E. Normandin, C. S. Parham, P. G. Sleph, S. Moreland, *J. Cardiovasc. Pharmacol.* **1995**, *26*, 289–294.

97 (a) D. Nagaratham, S. W. Miao, B. Lagu, G. Chiu, J. Fang, T. G. M. Dhar, J. Zhang, S. Tyagarajan, M. R. Marzabadi, F. Q. Zhang, W. C. Wong, W. Y. Sun. D. Tian, J. M. Wetzel, C. Forray, R. S. L. Chang, T. P. Broten, R. W. Ransom, T. W. Schorn, T. B. Chen, S. O'Malley, P. Kling, K. Schneck, R. Benedesky, C. M. Harrel, K. P. Vyas, C. Gluchowski, *J. Med. Chem.* **1999**, *42*, 4764–4777, and subsequent papers in the same issue; (b) J. C. Barrow, P. G. Nantermet, H. G. Selnick, K. L. Glass, K. E. Rittle, K. F. Gilbert, T. G. Steele, C. F. Homnick, R. M. Freidinger, R. W. Ransom, P. Kling, D. Reiss, T. P. Broten, T. W. Schorn, R. S. L. Chang, S. O'Malley, T. V. Olah, J. D. Ellis, A. Barrish, K. Kassahun, P. Leppert, D. Nagarathnam, C. Forray, *J. Med. Chem.* **2000**, *43*, 2703–2718.

98 (a) T. U. Mayer, T. M. Kapoor, S. J. Haggarty, R. W. King, S. L. Schreiber, T. J. Mitchison, *Science* **1999**, *286*, 971–974; (b) S. J. Haggarty, T. U. Mayer, D. T. Miyamoto, R. Fahti, R. W. King, T. J. Mitchison, S. L. Schreiber, *Chem. Biol.* **2000**, *7*, 275–286.

99 D. S. Bose, R. K. Kumar, L. Fatima, *Synthesis* **2004**, 279–282.

100 A. Dondoni, A. Massi, S. Sabbatini, *Tetrahedron Lett.* **2002**, *43*, 5913–5916.

101 A. Dondoni, A. Massi, S. Sabbatini, V. Bertolasi, *J. Org. Chem.* **2002**, *67*, 6979–6994.

102 C. O. Kappe, *Eur. Med. Chem.* **2000**, *35*, 1043–1052.

103 L. Heys, C. G. Moore, P. J. Murphy, *Chem. Soc. Rev.* **2000**, *29*, 57–67.

104 (a) Y. Kashman, S. Hirsh, O. J. McConnell, I. Ohtani, T. Kusumi, H. Kakisawa, *J. Am. Chem. Soc.* **1989**, *111*, 8925–8926; (b) E. A. Jares-Erijman, R. Sakai, K. L. Rinehart, *J. Org. Chem.* **1991**, *56*, 5712–5715.

105 (a) A. D. Patil, N. V. Kumar, W. C. Kokke, M. F. Bean, A. J. Freyer, C. Debrosse, S. Mai, A. Truneh, D. J. Faulkner, B. Carte, A. L. Breen, R. P. Hertzberg, R. K. Johnson, J. W. Westley, B. C. M. Potts, *J. Org. Chem.* **1995**, *60*, 1182–1188; (b) A. D. Patil, A. J. Freyer, P. B. Taylor, B. Carte, G. Zuber, R. K. Johnson, D. J. Faulkner, *J. Org. Chem.* **1997**, *62*, 1814–1819.

106 (a) L. E. Overman, M. H. Rabinowitz, P. A. Renhowe, *J. Am. Chem. Soc.* **1995**, *117*, 2657–2658; (b) L. E. Overman, M. H. Rabinowitz, *J. Org. Chem.* **1993**, *58*, 3235–3237; (c) D. S. Coffey, A. I. McDonald, L. E. Overman, M. H. Rabinowitz, P. A. Renhowe, *J. Am. Chem. Soc.* **2000**, *122*, 4893–4903; (d) D. S. Coffey, L. E. Overman, F. Stappenbeck, *J. Am. Chem. Soc.* **2000**, *122*, 4904–4914; (e) D. S. Coffey, A. I. McDonald, L. E. Overman, F. Stappenbeck, *J. Am. Chem. Soc.* **1999**, *121*, 6944–6945.

107 (a) A. S. Franklin, S. K. Ly, G. H. Mackin, L. E. Overman, A. J. Shaka, *J. Org. Chem.* **1999**, *64*, 1512–1519; (b) F. Cohen, L. E. Overman, S. K. Ly Sakata, *Org. Lett.* **1999**, *1*, 2169–2172; (c) F. Cohen, L. E. Overman, *J. Am. Chem. Soc.* **2001**, *123*, 10782–10783.

108 F. Cohen, S. K. Collins, L. E. Overman, *Org. Lett.* **2003**, *5*, 4485–4488.

109 For a recent review, see: Z. D. Aron, L. E. Overman, *Chem. Commun.* **2004**, 253–265.

110 S. K. Collins, A. I. McDonald, L. E. Overman, Y. H. Rhee, *Org. Lett.* **2004**, *6*, 1253–1255.

111 D. Enders, M. R. M: Hüttle, C. Grondal, G. Raabe, *Nuture* **2006**, in press.

10
Special Techniques in Domino Reactions

10.1
Domino Reactions under High Pressure

In organic synthesis, chemical transformation with a negative volume of activation ($\Delta V^{\ddagger}$) can be improved by application of high pressure, leading in many cases to an increase of the yield, an enhancement of the reaction rate, and an improvement of the selectivity (chemo-, regio- and stereoselectivity). This is especially true for cycloadditions with $\Delta V^{\ddagger}$ values of approximately $-50 \, \text{cm}^3 \, \text{mol}^{-1}$ [1].

Scheeren and coworkers used the pressure effect for a powerful domino process consisting of three cycloadditions [2], to form up to six bonds and eight stereogenic centers in a single operation. At 15 kbar and 50 °C, the reaction of a mixture of **10-1** (1 equiv.) and **10-2** (3 equiv.) led to tricyclic nitroso acetals **10-5, 10-6** and **10-7** in a 1:3:1 ratio *via* the first [4+2]-cycloadduct (±)-**10-3** and the second [4+2]-cycloadduct (±)-**10-4**. The final step in this sequence is a [3+2]cycloaddition (Scheme 10.1).

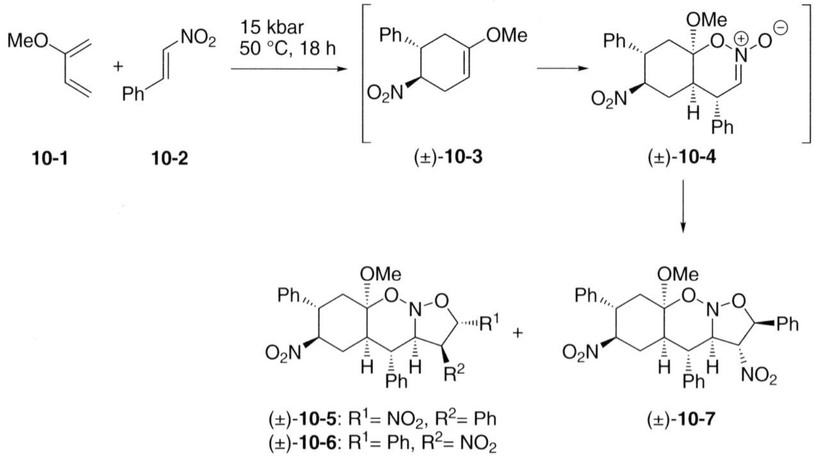

Scheme 10.1. Synthesis of tricyclic nitroso acetals under high pressure.

Domino Reactions in Organic Synthesis. Lutz F. Tietze, Gordon Brasche, and Kersten M. Gericke
Copyright © 2006 WILEY-VCH Verlag GmbH & Co. KGaA, Weinheim
ISBN: 3-527-29060-5

Moreover, the reaction can also be performed as a four-component domino process. Thus, treatment of a mixture of **10-1**, **10-2**, **10-8**, and **10-9** at 15 kbar gave the two tetracycles **10-10** and **10-11** in 83 % yield as a 65:35 mixture. It should be mentioned here that several other reports were published earlier by the same group, using a similar technique (Scheme 10.2) [3].

Another example of the application of high pressure in the field of domino transformations has been revealed by Brinza and Fallis, representing a carbonylation/cyclization procedure [4]. Thus, when hydrazones **10-12**, bearing a bromine atom, are subjected to standard radical conditions under a high-pressure atmosphere of carbon dioxide, cyclopentanones **10-16** and **10-17** are smoothly produced in good yields (Scheme 10.3).

It is assumed that the reaction is initiated by a radical bromine abstraction to give **10-13**, which after carbon monoxide insertion undergoes a rapid 5-*exo* cyclization onto the hydrazone moiety. The two diastereomeric hydrazinyl cyclopentanones **10-16** and **10-17** are formed with good yields, though with low stereoselectivity.

The final example of a domino process under high pressure, to be discussed in this chapter, is a combination of a Horner–Wittig–Emmons (HWE) reaction with a Michael addition developed by Reiser and coworkers [5]. Hence, reaction of a mixture of an aldehyde such as **10-18**, a phosphonate **10-19** and a nucleophile **10-20** in the presence of triethylamine at 8 kbar led to **10-21**. By this method, β-amino esters, β-thio esters and β-thio nitriles can be prepared in high yield (Scheme 10.4). Many of these transformations do not occur under standard conditions, thereby underlining the importance of high pressure in organic chemistry.

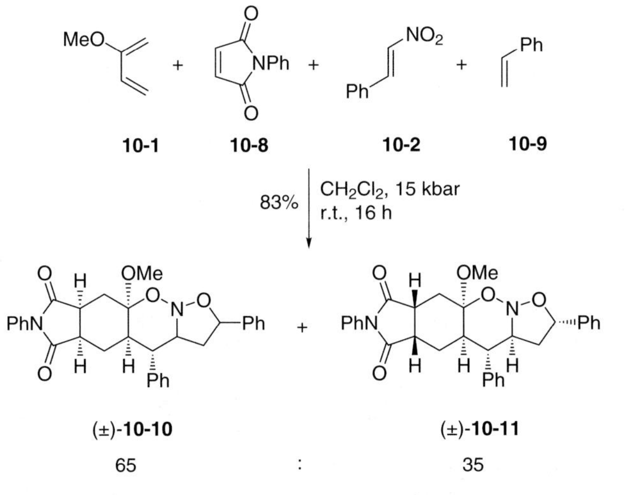

Scheme 10.2. Four-component domino process under high pressure.

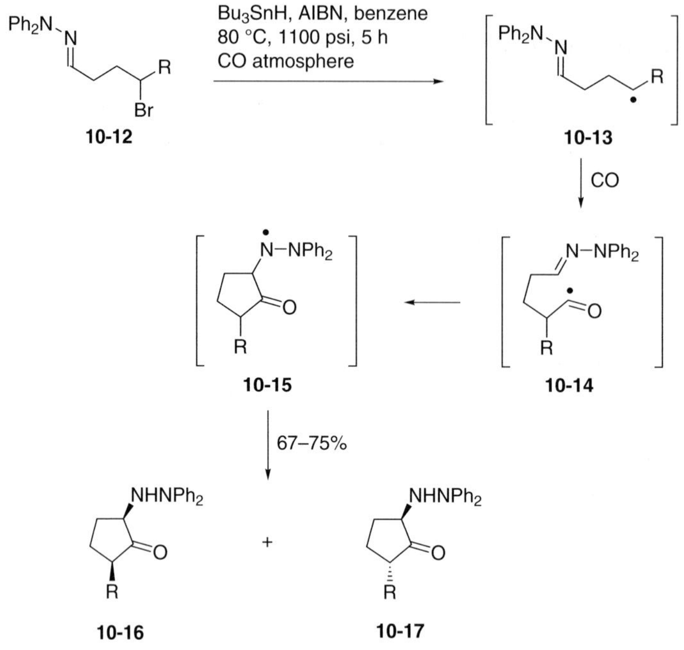

Scheme 10.3. Synthesis of hydrazinocyclopentanones.

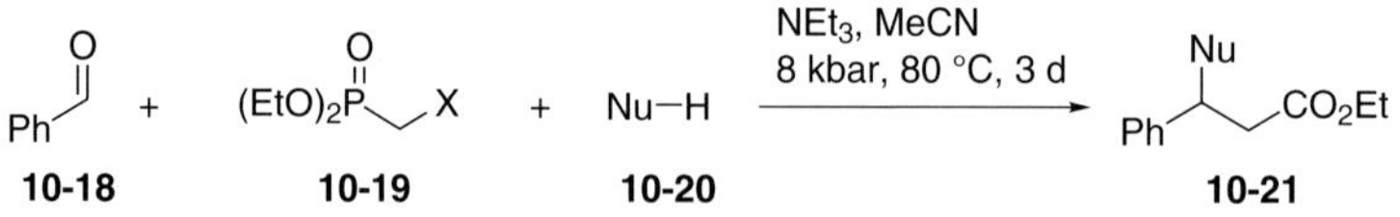

Entry	Substrate 10-12	R	Product 10-16/10-17	Ratio cis/trans	Yield [%]
1	a	H	a	–	69
2	b	Me	b	1:1	75
3	c	iPr	c	1:1.1	71
4	d	Cy	d	1:1.2	67

entry	Phosphonate	NuH	yield [%]
1	X = CO₂Et	piperidine	60
2	X = CO₂Et	PhSH	85
3	X = CN	PhSH	78

Scheme 10.4. Domino HWE/Michael reaction under high pressure.

10.2

Solid-Phase-Supported Domino Reactions

Today, multi-parallel synthesis lies at the forefront of organic and medicinal chemistry, and plays a major role in lead discovery and lead optimization programs in the pharmaceutical industry. The first solid-phase domino reactions were developed by Tietze and coworkers [6] using a domino Knoevenagel/hetero-Diels–Alder and a domino Knoevenagel/ene protocol. Reaction of solid-phase bound 1,3-dicarbonyl compounds such as **10-22** with aldehydes and enol ethers in the presence of piperidinium acetate led to the 1-oxa-1,3-butadiene **10-23**, which underwent an intermolecular hetero-Diels–Alder reaction with the enol ethers to give the resin-bound products **10-24**. Solvolysis with NaOMe afforded the desired dihydropyranes, **10-25** with over 90 % purity. Ene reactions have also been performed in a similar manner [7].

Combinatorial solid-phase synthetic methodologies have been used extensively in drug development [8]. A new solid-phase synthesis of 2-imidazolidones has been discovered by Goff, based on a domino aminoacylation/Michael addition reaction [9]. Thus, when immobilized amine **10-26** (HMPB-BHA resin) was treated with phenylisocyanate in the presence of triethylamine, a smooth formation of 2-imidazolidone took place. Acid-catalyzed removal from solid phase provided **10-27** in good yield (Scheme 10.6).

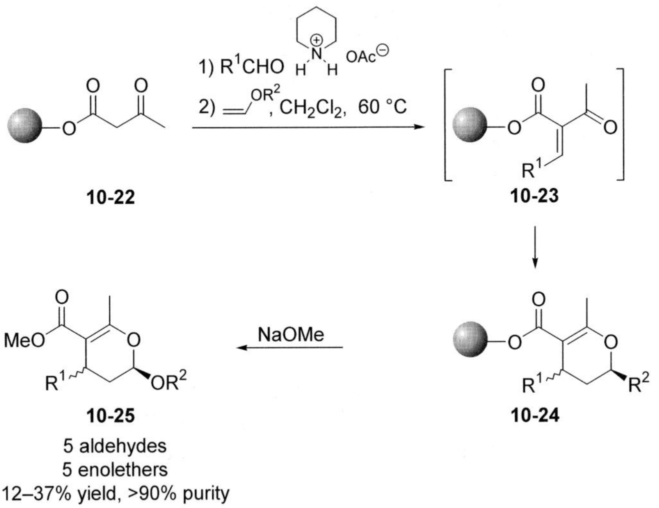

Scheme 10.5. Domino Knoevenagel/hetero-Diels–Alder reaction on solid phase.

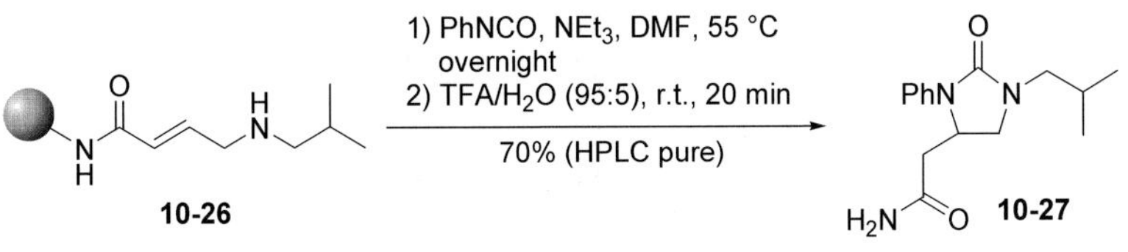

Scheme 10.6. Synthesis of 2-imidazolidone.

The concept for the synthesis of 4-hydroxy-4,5-dihydroisoxazoles by Righi and co-workers was discussed earlier, in Chapter 7. Here, an extension of this methodology by utilizing polymer-bound nitroacetate (hydroxylated Merrifield resin) is described [10]. Thus, the one-pot domino oxidation/nitroaldol cyclization of aziridine **10-28** with immobilized nitroacetate **10-29** furnished **10-30** which, after detachment from the resin, led to the desired product **10-31** in good yield and excellent *trans*-selectivity (Scheme 10.7).

An impressive diastereoselective domino Mannich/Michael condensation of polymer-bound (polystyrene) glycosylimines **10-32** with 1,3-butadiene **10-33** has been reported by Kunz and coworkers [11]. These authors obtained dihydropiperidinones **10-34** in good yield and high diastereoselectivity (Scheme 10.8).

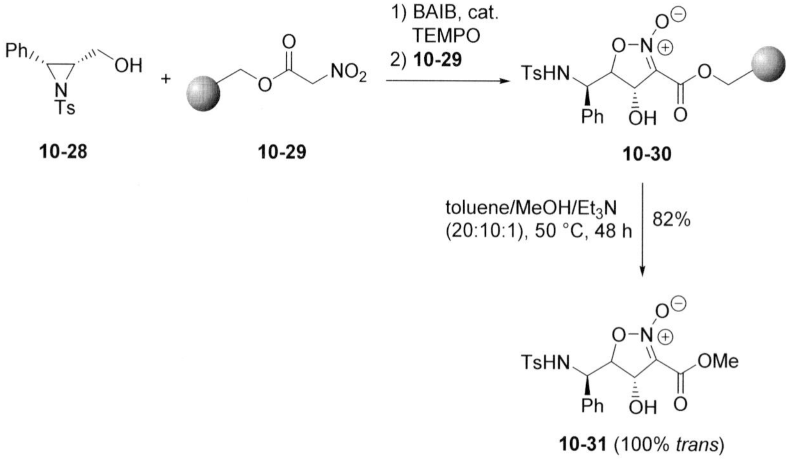

Scheme 10.7. Domino oxidation/nitroaldol/1,3-dipolar cycloaddition process.

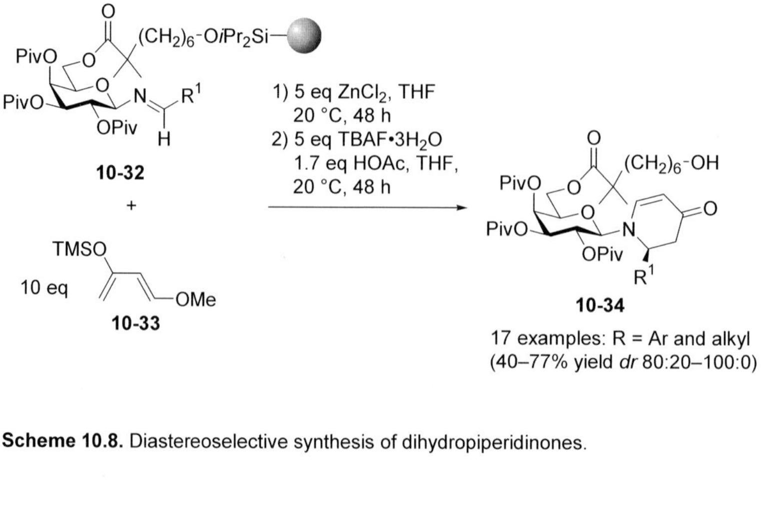

Scheme 10.8. Diastereoselective synthesis of dihydropiperidinones.

In a similar manner, diastereoselective Ugi reactions have been performed by utilizing immobilized amine **10-35** to give **10-36** (Scheme 10.9) [12].

A combination of a multicomponent Ugi transformation and an intramolecular Diels–Alder reaction has been developed by Paulvannan [13]. Hence, condensation of the resin-bound (acid-labile ArgoGel-Rink resin) amine **10-37** with a tenfold

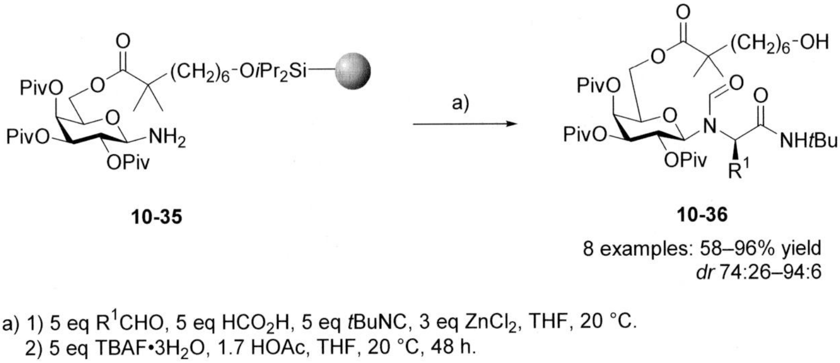

10-35 → a) → **10-36**

8 examples: 58–96% yield
dr 74:26–94:6

a) 1) 5 eq R^{1}CHO, 5 eq HCO$_2$H, 5 eq tBuNC, 3 eq ZnCl$_2$, THF, 20 °C.
2) 5 eq TBAF·3H$_2$O, 1.7 HOAc, THF, 20 °C, 48 h.

Scheme 10.9. Diastereoselective Ugi-reaction.

1) MeOH/CH$_2$Cl$_2$ (2:1), r.t., 36 h
2) TFA

10-37   **10-38**   **10-39**   **10-40**

10-42 ← **10-41**

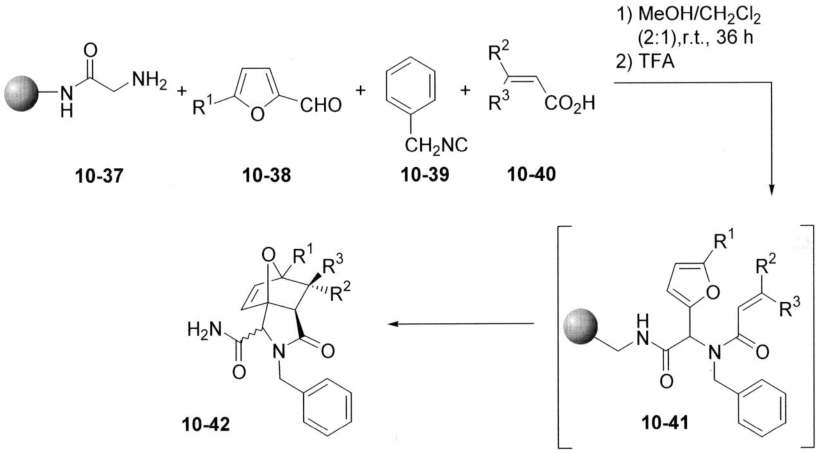

Entry	R^1	R^2	R^3	Isomeric ratio	Yield [%]
1	H	CO$_2$Et	H	91:09	95
2	H	H	CO$_2$Et	89:11	92
3	H	H	CONH–Bn	88:12	88
4	Me	CO$_2$Et	H	78:22	95
5	Me	H	CO$_2$Et	77:23	88
6	Me	H	CONH–Bn	88:12	92

Scheme 10.10. Domino Ugi/Diels–Alder process.

10-43

10-44a: R = Bn
10-44b: R = *p*-bromophenyl
10-44c: R = *p*-methoxyphenyl
10-44d: R = *m*-methoxyphenyl
10-44e: R = diphenylmethyl

1) MeOH, 50 °C, 48 h
2) TFA, CH$_2$Cl$_2$, r.t., 4 h

+ 20 eq R–NH$_2$ +

CH$_2$NC

10-45   **10-39**

10-46

Amine building blocks	Fumaric acid building blocks				
10-45	**10-44a**	**10-44b**	**10-44c**	**10-44d**	**10-44e**
	Product **10-46** [%]				
p-bromophenethylamine	41%	64%	49%	>99%	79%
3-aminopropanol	17%	21%	0%	>99%	35%
p-methoxybenzylamine	65%	73%	15%	79%	40%
n-hexylamine	32%	85%	45%	>99%	53%
n-decylamine	67%	84%	28%	>99%	40%
1-phenyl-2-propylamine	67%	34%	11%	>99%	71%
α-methylbenzylamine	98%	56%	9%	59%	71%
allylamine	45%	65%	7%	97%	85%
diphenylmethylamine	60%	63%	38%	>99%	99%
α-(2-Naphthyl)ethylamine	>99%	62%	54%	>99%	>99%
2-aminoethanol	29%	53%	0%	>99%	42%
p-chlorobenzylamine	63%	96%	46%	>99%	86%
p-methylbenzylamine	>99%	80%	63%	>99%	>99%
3-butoxypropylamine	>99%	64%	0%	>99%	70%
2,5-difluorobenzylamine	>99%	72%	81%	>99%	90%
1-adamantanamine	>99%	44%	63%	58%	0%
p-fluorobenzylamine	>99%	99%	60%	>99%	67%
o-methylbenzylamine	>99%	60%	59%	74%	92%
benzylamine	89%	74%	>99%	91%	69%
n-octylamine	90%	>99%	>99%	>99%	93%

Scheme 10.11. Domino Ugi/Diels–Alder process on a MPEG–O–CH$_2$-platform.

excess of furaldehyde **10-38**, benzyl isocyanide **10-39** and unsaturated acid **10-40** furnished the Ugi-product **10-41** which directly underwent an intramolecular Diels–Alder reaction to provide tricyclic lactams **10-42** after cleavage from the polymer using trifluoroacetic acid (TFA) (Scheme 10.10). Of note here was that all products were isolated in high yields and in reasonable diastereoselectivity.

Recently, Oikawa and coworkers also published a successful parallel synthesis using a domino Ugi/Diels–Alder process [14]. Thus, reaction of immobilized fur-

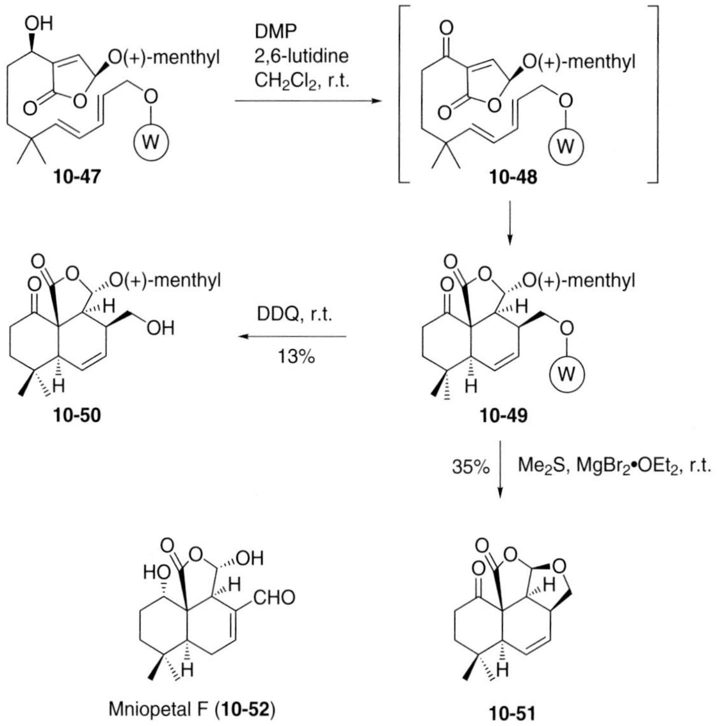

Scheme 10.12. Domino oxidation/hetero-Diels–Alder reaction.

fural **10-43** with five different fumaric acid building blocks **10-44** and twenty amines **10-45** in the presence of benzyl isocyanide **10-39** led to the formation of 96 products **10-46** in predominantly high yields (Scheme 10.11).

In this context it must be mentioned that the procedure by Kennedy and co-workers, in which a novel resin-bound isonitrile is applied in an Ugi multicomponent reaction for the synthesis of 2,5-diketopiperazines and 1,4-benzodiazepine-2,5-diones, was discussed earlier, in Chapter 9.

A solid-phase-supported synthesis of a building block for the synthesis of drimane sequiterpenes [15] such as mniopetals F (**10-52**), kuehneromycins and maras-manes using a domino process has been reported by Reiser and Jauch [16]. The sequence was initiated by oxidation of an immobilized allylic alcohol **10-47** onto Wang resin to give the corresponding α,β-unsaturated ketone which underwent an intramolecular Diels–Alder reaction to give **10-49** (Scheme 10.12). Removal of the polymer took place either by treatment with DDQ to give alcohol **10-50** in 13% yield, or by using Me$_2$S and MgBr$_2$·OEt$_2$, yielding the tetracycle **10-51** in 35%.

Since poly(ethylene glycols) (PEGs) are rather nontoxic resins, their application in the field of combinatorial chemistry is particularly attractive. A four-component domino process of immobilized aromatic amines **10-53** to give 1,2,3,4-tetrahydroquinolines **10-54** using this support has been developed by Benaglia and co-

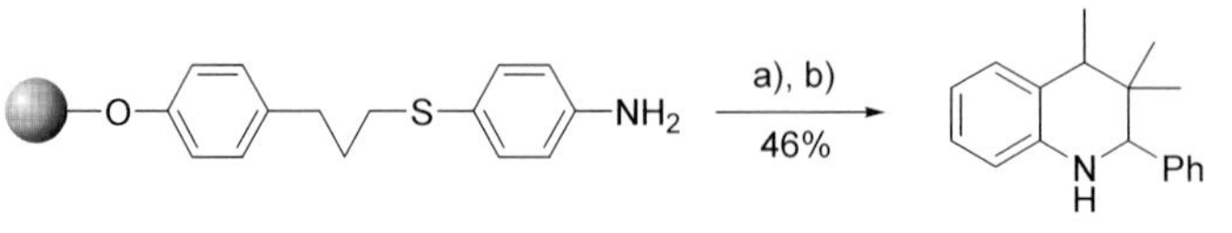
a), b)
46%
NH2
N
H
Ph

10-53 **10-54**

a) PhCHO, Me_2CHCHO, $Yb(OTf)_3$, MeOH, r.t., 96 h.
b) $NiCl_2{\cdot}6H_2O$, $NaBH_4MeOH/THF$ (3:1)r.t., 96 h.

Scheme 10.13. Poly(ethylene glycol)-supported domino synthesis of 1,2,3,4-tetrahydroquinolines.

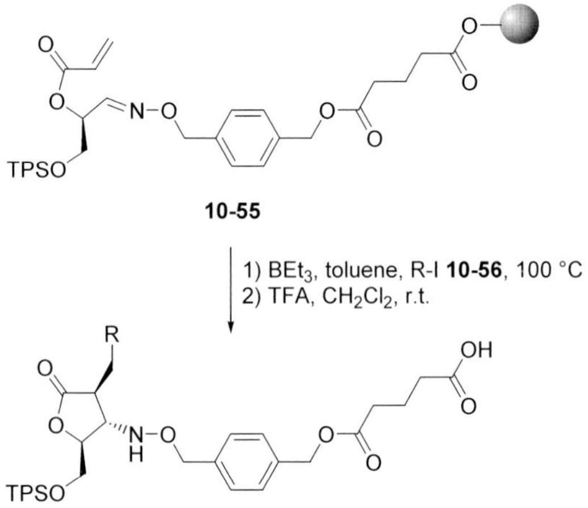
TPSO
10-55
1) BEt3, toluene, R-I 10-56, 100 °C
2) TFA, CH2Cl2, r.t.
R
OH
TPSO

10-57a: R = *i*Pr (57%, *ds* 8:1)
10-57b: R = cHx (50%, *ds* 8:1)
10-57c: R = Et (92%, *ds* 8:1)

Scheme 10.14. Domino radical addition-cyclization process of oxime ethers.

workers (Scheme 10.13) [17]. Removal of the resin was accomplished by a reductive desulfurization using $NiCl_2/NaBH_4$.

Domino radical transformations on solid support can also be used, as shown by Naito and coworkers [18]. Thus, immobilized oxime ether **10-55** (Wang resin) afforded the tetrahydrofuranones **10-57** after removal of the resin, when it was exposed to BEt_3 in the presence of different alkyl iodides **10-56** (Scheme 10.14).

10.3 Solvent-Free Domino Reactions

Today, in the field of organic chemistry, increasing attention is being given to "Green Chemistry" [19], using environmentally safe reagents and, in particular, solvent-free procedures. Reactions performed without the use of a solvent can indeed

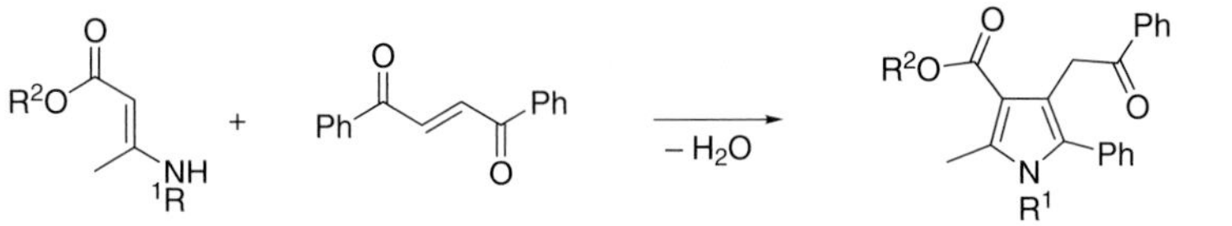

10-58   **10-59**   **10-60**

Entry	Product 10-60	R^1	R^2	Milling time [h] (T [°C])	Yield [%] in solution	solid state
1	a	H	Me	3 (25)	68	100
2	b	CH$_3$	Me	3 (25)	81	100
3	c	CH$_3$	C$_2$H$_5$	3 (−20)	78	100
4	d	CH$_2$Ph	C$_2$H$_5$	3 (0)	55	100

Scheme 10.15. Solid-state domino synthesis of pyrrole derivatives.

allow clean, eco-friendly and highly efficient transformations, with the additional advantage of a simplified work-up. In this way, liquid and solid substrates can be employed, and even reactions of solids with solids – so-called solid-state synthesis – can be performed. The latter transformations often profit from crystal-packing effects and are usually highly selective. Solid-state transformations take place in quantitative manner provided that three indispensable requirements are fulfilled: 1) In the phase rebuilding step, the molecules must be able to move long distances within the crystal; 2) the phase transformation step deals with unhindered crystallization of the product; and 3) disintegration of the new phase from the old one is essential in order to create a fresh surface such that the reaction is continued to completion [20].

By adhering to these conditions, Kaupp and coworkers performed a remarkable one-pot synthesis of highly substituted pyrroles [21]. When primary or secondary enamine esters **10-58** were ground with *trans*-1,2-dibenzoylethene **10-59** in a ball-mill, the pyrroles **10-60** were obtained in quantitative yield (Scheme 10.15). In contrast, when in solution the yields were much lower and the reaction also required higher reaction temperatures.

The following four examples deal with solid-phase transformations under microwave irradiation. Although microwave-assisted reactions are detailed in Section 10.4, the transformation will be discussed at this point under the heading of solid-phase transformations [22]. A recent access to flavonones of plant origin, which are of interest due to their use in medicine, has been published by Moghaddam and coworkers [23]. Their approach is based on a domino Fries rearrangement/Michael addition (Scheme 10.16). Under microwave irradiation, and in the presence of AlCl$_3$-ZnCl$_2$, supported on silica gel as promoter, the α,β-unsaturated esters **10-61** are transformed into the flavonones **10-63** *via* **10-62**.

1,3-Thiazines, a heterocyclic system which is found, for example, in antibiotics of the cephalosporin type, were synthesized using a solvent-free domino procedure by Yadav and coworkers [24]. Hence, reaction of *N*-acylglycines **10-64**, an aromatic

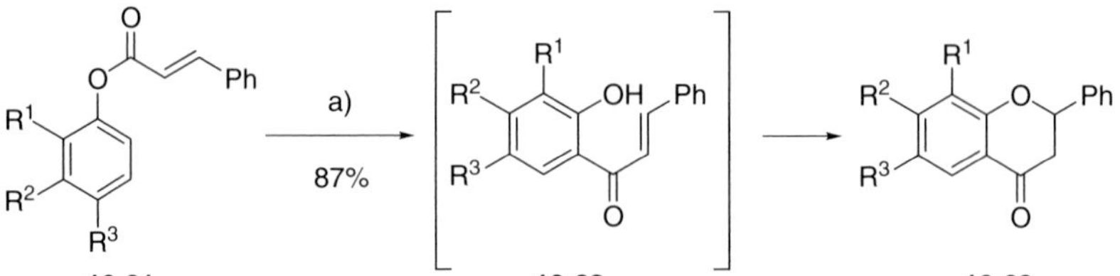

a) AlCl$_3$/ZnCl$_2$ mixture supported on silica gel, microwave irradiation, 7 min.

Entry	Substrate **10-61**	Product **10-63**	Yield [%]
1	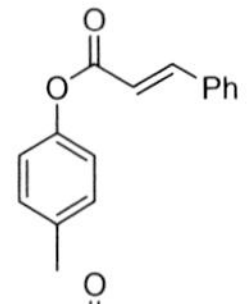	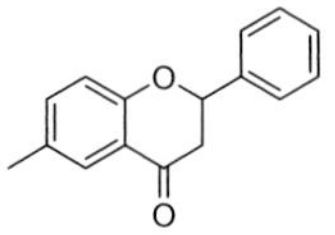	87
2	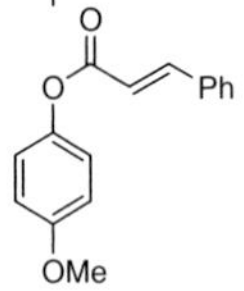	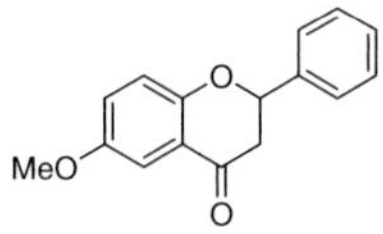	73
3	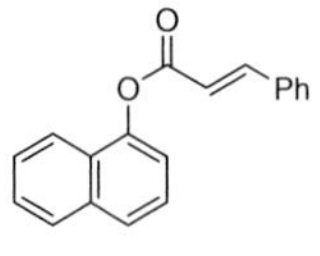	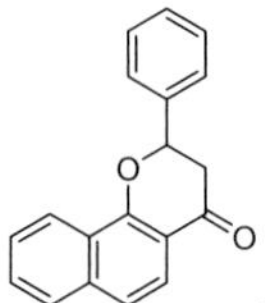	85
4	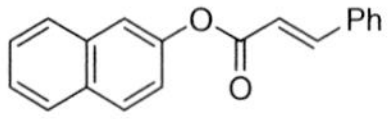	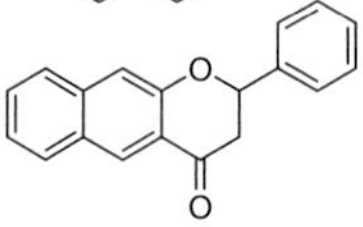	79

Scheme 10.16. Solvent-free domino synthesis of flavanones.

aldehyde **10-66** and ammonium *N*-aryldithiocarbamate **10-68** in the presence of acetic anhydride and anhydrous sodium acetate, gave the 1,3-thiazines **10-70** *via* **10-65**, **10-67** and **10-69** in high yield and diastereoselectivity (Scheme 10.17). Performing the reaction in an oil-bath gave much lower yield and diastereoselectivity.

A solvent-free synthesis of benzo[*b*]furan derivatives **10-79**, a class of compounds which is often found in physiologically active natural products, was described by Shanthan Rao and coworkers. These authors heated phosphorane **10-71** for 8 min in a microwave oven and obtained the benzo[*b*]furan **10-74** in 73% yield (Scheme 10.18) [25]. The sequence is initiated by an intramolecular Wittig reaction, providing alkyne **10-72**; this underwent a subsequent Claisen rearrangement to give the intermediate **10-73**. Also in this case, normal oil-bath heating gave much lower yields (5%) of the desired product; the authors hypothesize that the micro-

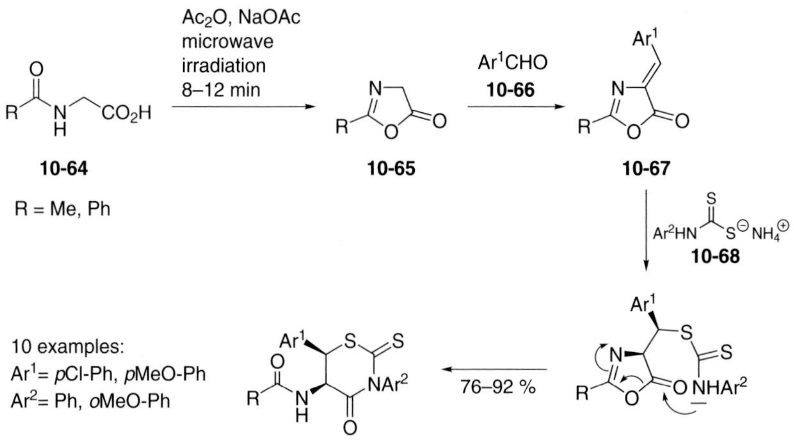

Scheme 10.17. Microwave-assisted solvent-free synthesis of 1,3-thiazines.

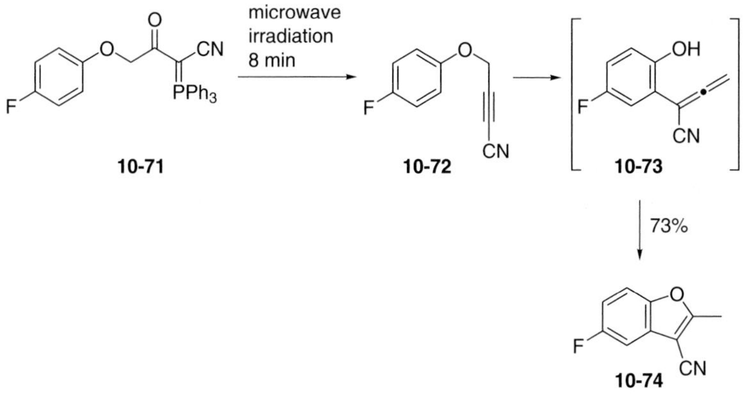

Scheme 10.18. Solvent-free domino Wittig–Claisen cyclization-process.

wave energy favors the dissociation of phenol **10-73** into the phenolate to initiate formation of the furane moiety.

Recently, the Texier-Boullet group [26] has prepared nitrocyclohexanols **10-77** by a twofold Michael addition/aldol reaction sequence (Scheme 10.19). Simply mixing chalcone **10-75** with nitromethane in the presence of a mixture of KF and Al$_2$O$_3$ under microwave irradiation gave **10-79** via the proposed intermediates **10-76**, **10-77** and **10-78** as a single diastereomer in 65 % yield. One possible explanation for the stereoselectivity of the transformation is fixation of the reactive species onto the solid KF/Al$_2$O$_3$, as depicted in **10-79**.

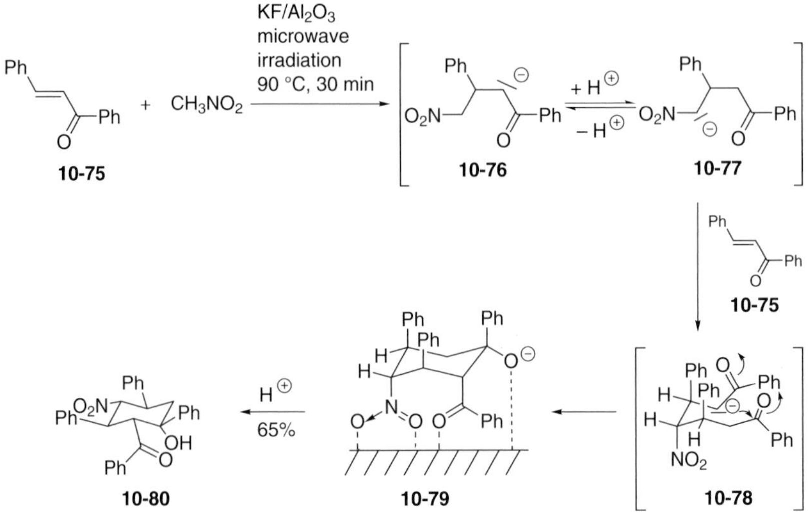

Scheme 10.19. Microwave-assisted solvent-free domino synthesis of nitrocyclohexanols.

10.4
Microwave-Assisted Domino Reactions

Microwave heating was first used for organic synthesis by Gedye in 1986 [27]. Since then, its use has grown in interest, due to shorter reaction times and reduced thermal strain of substances and products [28]. Microwave irradiation allows for a very rapid heating by direct transfer of energy to the molecules; this is in contrast to conventional heating, where energy transfer takes place *via* the walls of the vessel. Undheim and coworkers have shown that microwave irradiation can improve domino metathesis reactions [29]. For the transformation of **10-81** to give **10-82** in the presence of the catalyst **10-83**, heating at 85 °C for 14 h was necessary (Scheme 10.20) [30]. However, under microwave irradiation in toluene at 160 °C for 20 min, 100 % conversion was observed. It is of interest the Grubbs II catalyst **10-84** was much less effective in this respect.

In order to study any chemoselectivity influences of microwave irradiation on the domino Knoevenagel/hetero-Diels–Alder process (the so-called Tietze reaction), Raghunathan and coworkers [31a] investigated the transformation of 4-hydroxy coumarins (**10-85**) with benzaldehydes **10-86** in EtOH to afford pyrano[2,3-*c*]coumarin **10-87** and pyrano[2,3-*b*]chromone derivatives **10-88**. Normal heating of **10-85a** and **10-86a** at reflux for 4 h gave a 68:32 mixture of **10-87a** and **10-88** in 57 % yield, whereas under microwave irradiation a 97:7 mixture in 82 % yield was obtained. Similar results were found using the benzo-annulated substrates **10-85b** and **10-86b**.

1,2-Dihydro-2-quinolones were also re-investigated, and showed a higher yield and better selectivity under microwave irradiation compared to conventional heating [31b]. The resulting pyranoquinoline skeleton is found in several naturally oc-

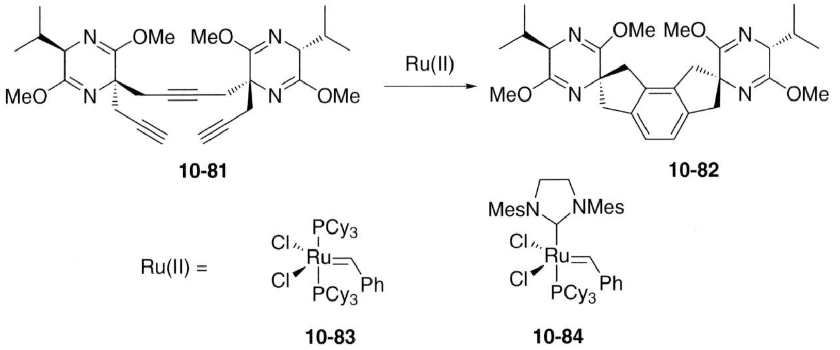

Scheme 10.20. Microwave-assisted domino metathesis reaction.

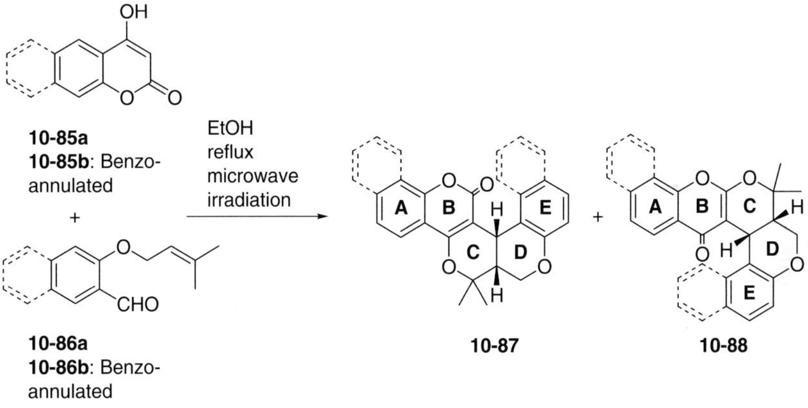

A
Conventional: 2 x 5 mol% **10-83**, toluene, 85 °C, 14 h, yield 90%.
Microwave heating: 8 mg **10-83**, 40 mg **10-81**, toluene, 160 °C, 20 min, conversion 100%.
B
Conventional: No transformation.
Microwave heating: 4 mg **10-84**, 50 mg **10-81**, toluene, 160 °C, 10 min, conversion 36%.

Scheme 10.21. Reactions of unsymmetrical 1,3-diones **10-85** with aromatic aldehydes **10-86**.

curring products, such as flindersine (**10-89**) and melicobisquinolone A (**10-90**) and B (**10-91**) (Scheme 10.22) [32]. All of these compounds have psychotropic, anti-allergenic, anti-inflammatory, antihistaminic and estrogenic activities [33].

A reductive amination/cyclization strategy was used by Surya Prakash Rao and coworkers to synthesize the annulated tricyclic compounds **10-93** [34]. These authors employed the microwave irradiation of a *cis/trans*-mixture of 1,5,9-triketone **10-92** with an excess of ammonium formate for 1 min in 87 % yield (Scheme 10.23).

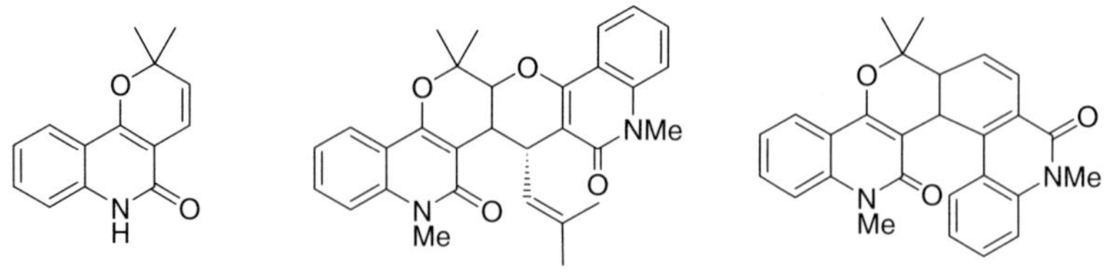
O
N
H
O
O
O
N
Me
O
NMe
O
O
N
Me
O
O
NMe

Flindersine (**10-89**) Melicobisquinolone A (**10-90**) Melicobisquinolone B (**10-91**)

Scheme 10.22. Selection of natural compounds possessing a pyranoquinolinone skeleton

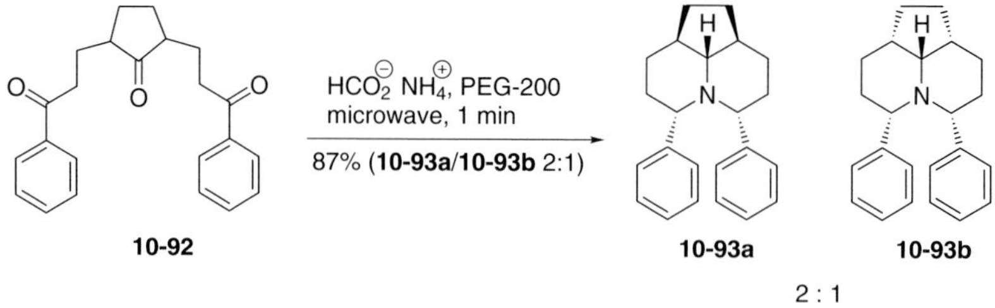
HCO2⁻ NH4⁺, PEG-200
microwave, 1 min
87% (10-93a/10-93b 2:1)
10-92
10-93a
10-93b
2 : 1

Scheme 10.23. Microwave-assisted synthesis of compounds **10-93**.

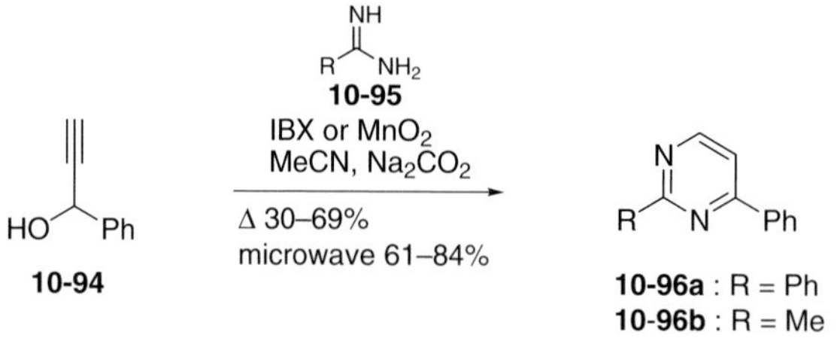
NH
R NH2
10-95
IBX or MnO2
MeCN, Na2CO2
Δ 30–69%
microwave 61–84%
HO Ph
10-94
R N Ph
10-96a : R = Ph
10-96b : R = Me

Scheme 10.24. Synthesis of pyrimidines

A new microwave-assisted domino reaction towards pyrimidines has been developed by the Bagley group, affecting up to four separate synthetic transformations in a single process [35]. Thus, irradiating a mixture of propargylic alcohol **10-94**, amidine hydrochloride **10-95** and MnO_2 or IBX as oxidant up to 120 °C gave **10-96** in good yield (Scheme 10.24). Under normal heating, the products were obtained in much lower yield. The reaction is proposed to proceed *via* the initial oxidation of the alcohol to the corresponding ketone with a subsequent Michael addition.

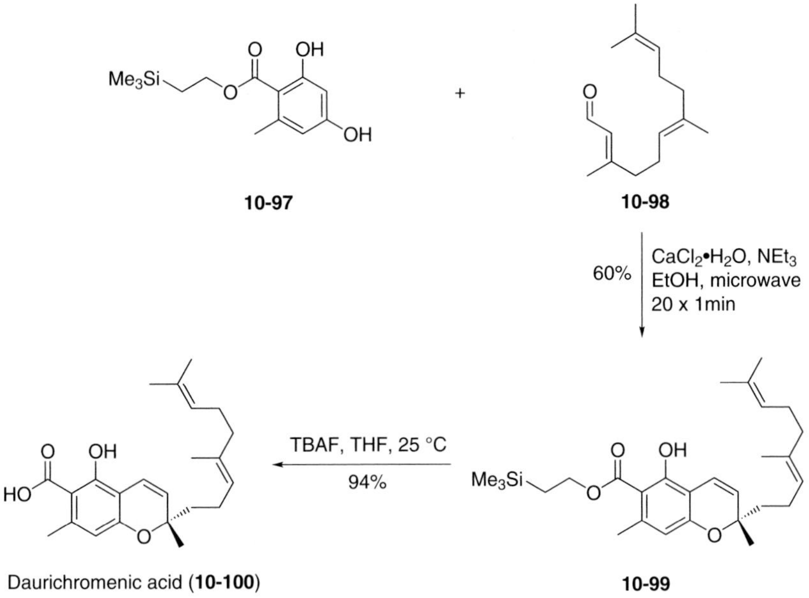

Scheme 10.25. Synthesis of daurichromenic acid (10-100).

In a similar manner, 2,4,6-trisubstituted 3-pyridinecarboxylic acid esters have been prepared using a mixture of α,β-ketoesters, propargylic alcohols, ammonium acetate and MnO_2. The reaction proceeds by oxidation of the propargylic alcohol to the corresponding ketone, which reacts with the enamine being formed from the β-ketoester and ammonium acetate. For this domino process, normal heating at 70 °C was used.

The highly potent anti-HIV natural product daurichromenic acid (10-100) was synthesized by Jin and coworkers [36] using a microwave-assisted reaction of the phenol derivative 10-97 and the aldehyde 10-98 (Scheme 10.25). Normal heating gave the desired benzo[b]pyran 10-99 by a domino condensation/intramolecular S_N2'-type cyclization reaction only in low yield. However, when the reaction mixture was irradiated twenty times in a microwave for 1-min intervals, 10-99 was obtained in 60 % yield. This compound was then transformed into 10-100 by cleavage of the ester moiety.

Functionalized 3-alkyltetronic acids can be obtained by a thermal domino Claisen/oxa-ene/ring-opening reaction of the corresponding allyl tetronates 10-101 [37]. However, under these conditions allyl tetronates with a trisalkylsubstituted alkene moiety such as 10-101 ($R^1 = R^2 = CH_3$) led to products of type 10-104 via 10-102 and 10-103, with two H-shifts as terminating transformations. Using microwave irradiation not only allows the reaction time to be shortened; rather, the reaction can also be halted at product 10-103 [38]. Moreover, by employing alkyl tetronates 10-101 with $R^1 = H$ and R^2 = chlorophenyl as well as tetranates, the reaction can be halted after the Claisen rearrangement under microwave irradiation.

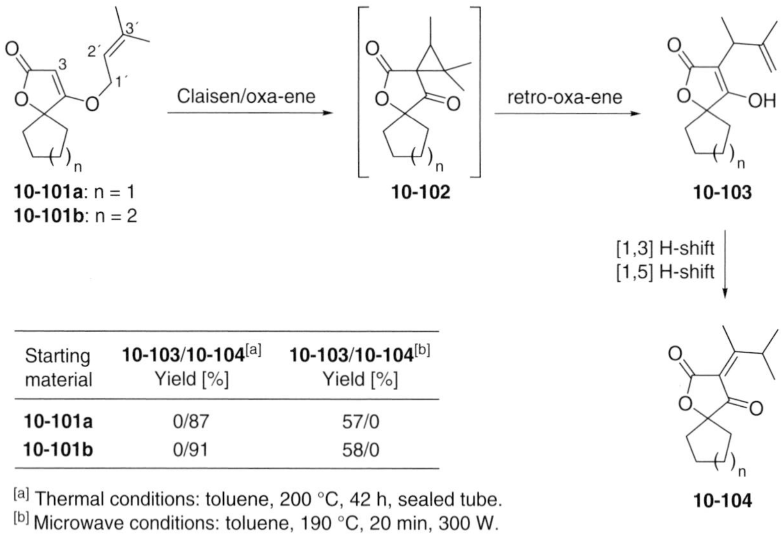

Starting material	**10-103/10-104**[a] Yield [%]	**10-103/10-104**[b] Yield [%]
10-101a	0/87	57/0
10-101b	0/91	58/0

[a] Thermal conditions: toluene, 200 °C, 42 h, sealed tube.
[b] Microwave conditions: toluene, 190 °C, 20 min, 300 W.

Scheme 10.26. Microwave-assisted domino Claisen/oxa-ene process.

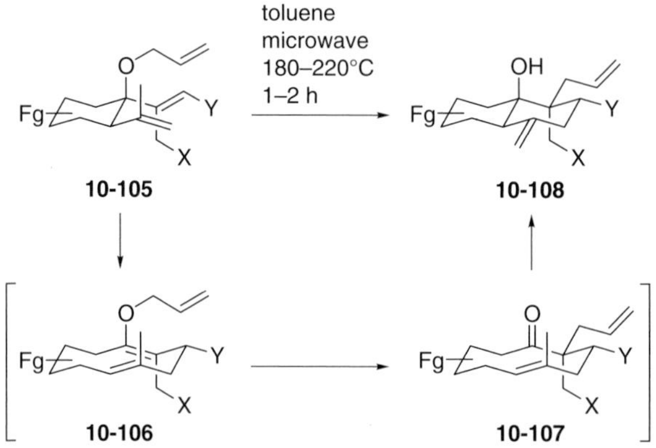

Scheme 10.27. Microwave-assisted domino oxy-Cope/Claisen/ene reaction

A combination of three highly stereoselective pericyclic reactions using microwave technology has been recently disclosed by Sauer and Barriault, leading to the formation of decalin skeletons possessing two contiguous quaternary centers [39]. The domino reaction is supposed to be initiated by an oxy-Cope rearrangement of precursor **10-105** to build up a ten-membered ring enol ether macrocycle **10-106** (Scheme 10.27). The latter immediately rearranges *via* a Claisen[3,3] shift to the corresponding (*E*)-cyclodec-6-en-1-one **10-107**, which spontaneously performs a cycli-

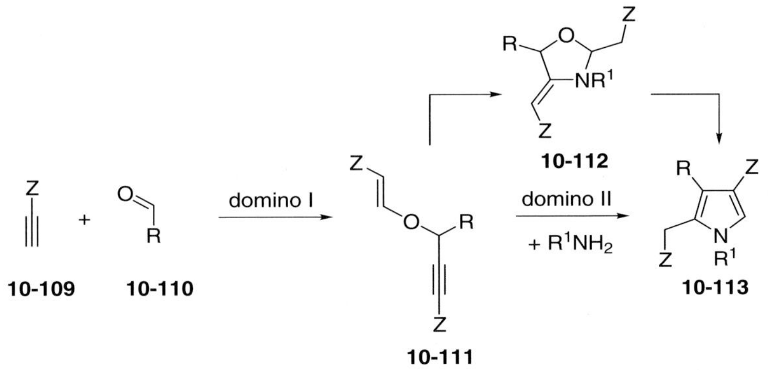

Scheme 10.28. Microwave-assisted domino processes to tetrasubstituted pyrroles.

Entry	Z	R	R¹	10-111 [%]	10-113	Domino II [%][a]	One-pot [%][b]
1	CO_2Me	Et	Bn	a (87)	a	74	44
2	CO_2Me	Et	Ph		b	38	
3	CO_2Me	Et	pOMe–Ph		c	56	
4	CO_2Et	Et	pOMe–Ph		d		53
5	CO_2Et	Et	Bn	e (88)	e	77	51
6	CO_2Me	Et	(S)PhCHMe		f	72	47
7	CO_2Me	Et	amino acid		g	69	42
8	CO_2Et	Me	Bn	h (73)	h	76	
9	CO_2Me	Hex	Bn	i (76)	i	61	
10	CO_2Et	Hex	Bn		j		47
11	CO_2Et	3-butenyl	Bn	k (79)	k	60	41
12	CO_2Et	cit	Bn	l (68)	l	65	49
13	CO_2Et	iPr	Bn	m (85)	m	63	46
14	CO_2Me	iPr	Bn	n (81)	n	58	
15	CO_2Et	cPr	Bn	o (72)	o	70	46
16	CO_2Et	$BnOCH_2$	Bn	p (41)	p	55	

[a] Conjugated alkynoate **10-111** (1 mmol) and the amine R^1-NH_2 (1.3 mmol) were absorbed on 1 g or 2 g of silica gel and irradiated at 900 W for 8 min; [b] (1) Aldehyde (1 mmol), alkyl propiolate (2 mmol), NEt_3 (0.5 mmol), 0 °C, 30 min. (2) silica gel (1 g), amine R^1-NH_2 (1.3 mmol). (3) μν-irradiation (900 W), 8 min.

zation *via* a transannular ene reaction to furnish decalin **10-108**, with high yield and excellent diastereoselectivity (>25:1).

It should be noted that microwave irradiation could also be useful for accelerating the reaction of aldoximes with dimedone. Thus, the corresponding N-hydroxylacridinediones were produced within a few minutes, in excellent yield [40].

The final example demonstrates that microwave irradiation allows a perfect fine-tuning of reaction conditions to obtain different products from the same starting materials. In the procedure developed by García-Tellado and coworkers [41], two domino processes were coupled. The first process consists of a high-yielding synthesis of enol-protected propargylic alcohols **10-111** starting from alkyne **10-109** and aldehyde **10-110** (Scheme 10.28). In the second process, transformation into tetrasubstituted pyrroles **10-113** takes place when enol-protected propargylic alcohols **10-111** adsorbed onto silica gel are irradiated in a microwave oven at 900 W for 5 min. It should be noted that the two domino processes could be performed as a one-pot procedure.

Interestingly, by conducting the second domino reaction at only 160 W for 90 min, it is also feasible to halt the process at the stage of the corresponding 1,3-oxazolidines, which are intermediates in the formation of pyrroles **10-113** [42].

It should be also noted that, in a very recent publication, Liu and coworkers were successful in applying microwave radiation within their domino approach towards the synthesis of pyrrolo[2,1-*b*]quinazoline alkaloids such as deoxyvasicinone, 8-hydroxydeoxyvasicinone, mackinazolinone and isaindigotone, which exhibit a promising broad spectrum of biological activities. In the case of isaindigotone, the authors were able to extend their strategy to a three-component procedure, which comprises the domino conversion of anthranilic acids and Boc-protected amino acids into the tricyclic core skeleton [43].

10.5
Rare Methods in Domino Synthesis

In addition to high pressure, solid-phase and microwave assistance, a number of less common and rarely occurring methods have been introduced into the field of domino reactions.

For example, McNab and coworkers have discovered that flash-vacuum pyrolysis (FVP) (1000 °C, 0.01 Torr) of pyrrole **10-114** led to the formation of pyrrolo[2,1-*a*]isoindol-5-one **10-117** in 79 % (Scheme 10.29) [44]. The transformation is proposed to proceed *via* an initial 1,5-aryl shift to give intermediate **10-115**, which then undergoes an elimination of methanol. Finally, electrocyclization of the ketene **10-116** results in the formation of **10-117**.

Ionic liquids such as 1-butyl-3-methylimidazolium tetrafluoroborate ([bmim]BF$_4$) or hexafluorophosphate ([bmim]PF$_6$) process unique properties, and are therefore nowadays often used in organic chemistry. As they exhibit only a negligible vapor pressure, formation of a separate liquid phase both with water and with common organic solvents, a good capacity for solubilizing organic substrates, an ease of recycling and reuse and a lack of flammability, they fit nicely into the concept of "green" solvents [45].

Recently, Sasson and coworkers have shown that a Knoevenagel condensation can be coupled effectively with catalytic hydrogenation to develop a domino process which leads directly to functionalized saturated esters [46]. When the reaction of aldehyde **10-118** and the active methylene compound **10-119** was carried out in an ionic liquid solvent ([bmim]BF$_4$) in the presence of EDDA and Pd/C under a hydrogen atmosphere (25–90 °C, 300 kPa), the product **10-120** could be obtained by extraction with diethyl ether after 4–12 h in excellent yield (Scheme 10.30). Interestingly, using DMA resulted in significant competing hydrogenation of the aldehyde to the alcohol.

The field of on-chip solution-phase synthesis and analysis, termed μSYNTAS (miniaturized-SYNthesis and Total Analysis Systems) is a rather new field in or-

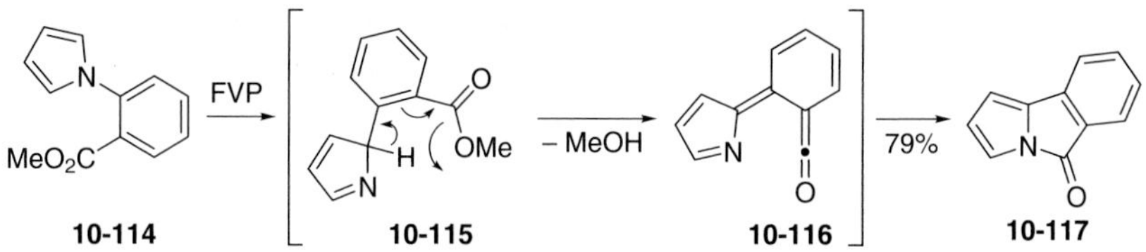

Scheme 10.29. Flash-vacuum pyrolysis (FVP).

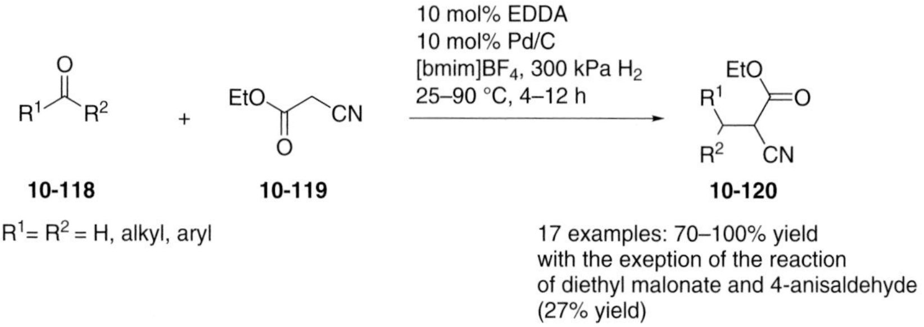

Scheme 10.30. Domino Knoevenagel condensation/catalytic hydrogenation reaction in ionic liquid.

ganic synthesis. The motivation behind the development of μSYNTAS is that a miniaturizing of processing systems leads to a significant enhancement in the efficiency of mixing and separation. Moreover, using extremely small amounts of materials in such systems makes them ideal for processing particularly valuable or hazardous reaction components. Furthermore, the large surface-to-volume ratios encountered in micro-reactor environments allow for rapid interruption of chemical processes, which may lead to higher reaction selectivity und therefore high-value products. De Mello and coworkers have applied the concept of μSYNTAS to the Ugi four-component condensation to give **10-126** starting from piperidine hydrochloride (**10-121**), formaldehyde (**10-122**) and cyclohexyl isocyanide (**10-124**) *via* **10-123** and **10-125** [47]. In the formation of **10-123**, water is formed which adds to **10-125** to give **10-126**. Interestingly, these authors were the first to confirm experimentally that the reaction proceeds by traversing a nitrilium cation species **10-125** (Scheme 10.31). Furthermore, these authors pointed out that the highly exothermic reaction requires no cooling when carried out in a micro-reactor.

Tietze and coworkers developed two new domino approaches in the field of combinatorial chemistry, which are of interest for the synthesis of bioactive compounds. Combinatorial chemistry can be performed either on solid phase or in solution using parallel synthesis. The former approach has the advantage that purification of the products is simple and an excess of reagents can be used. This is not possible for reactions in solution, but on the other hand all known transformations can be used. The Tietze group has now developed a protocol which combines the

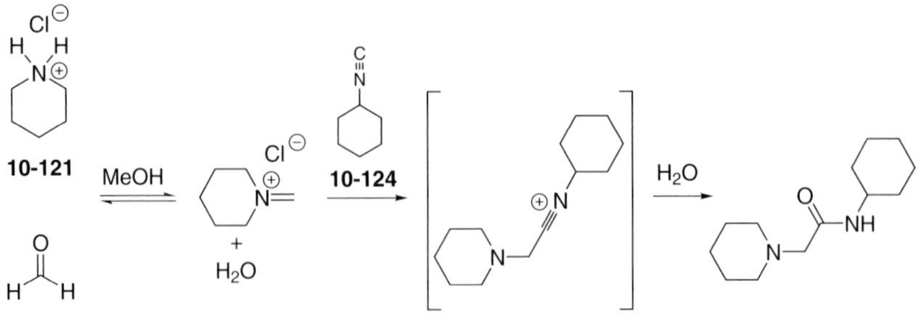

Scheme 10.31. Ugi-multicomponent reaction in a micro reactor.

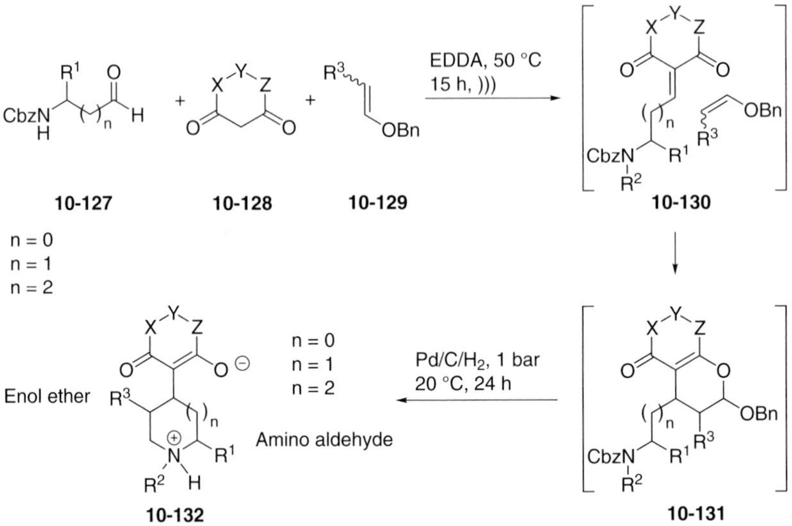

Scheme 10.32. General scheme for multicomponent domino Knoevenagel/hetero-Diels–Alder hydrogenation sequence.

advantages of solid- and solution-phase chemistry without having their disadvantages [48]. The main idea is to form salts which can be precipitated from the reaction mixture in excellent purity by simply adding diethyl ether (Scheme 10.32). Thus, reaction of *N*-Cbz-protected α-, β- or γ-aminoaldehydes **10-127** with a 1,3-dicarbonyl compound **10-128** in the presence of a benzyl enol ether **10-129** followed by hydrogenation led to substituted pyrrolidines, piperidines or azepanes **10-132** *via* the intermediates **10-130** and **10-131** in >95% chemical purity in nearly all cases as colorless solids after precipitation.

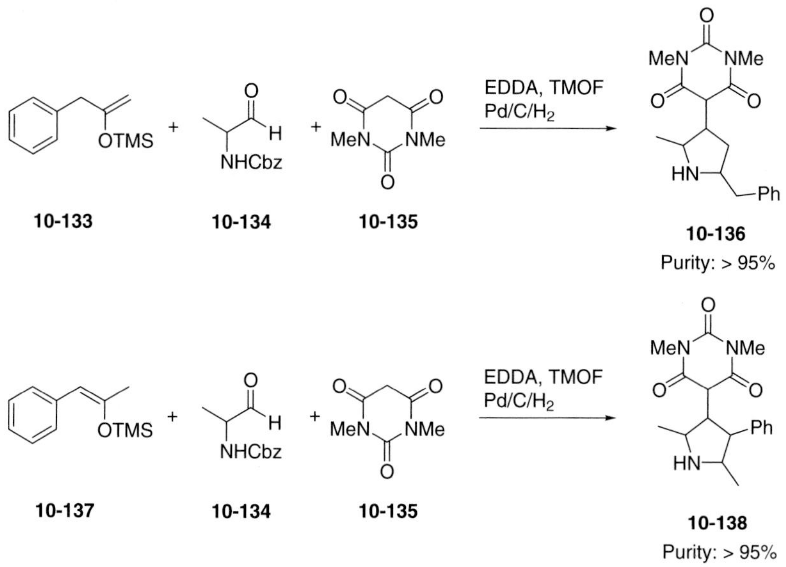

Scheme 10.33. Domino Knoevenagel/hetero-Diels–Alder hydrogenation sequence using TMS enol ethers as dienophiles.

The method can be further improved using trimethylsilyl (TMS) enol ethers, which can be prepared *in situ* from aldehydes and ketones [49]. TMS enol ethers of cyclic ketones are also suitable, and diversity can be enhanced by making either the kinetic or thermodynamic enol ether, as shown for benzyl methyl ketone. Thus, reaction of the "kinetic" TMS enol ether **10-133** with the amino aldehyde **10-134** and dimethylbarbituric acid **10-135** yielded **10-136**, whereas the "thermodynamic" TMS enol ether **10-137** led to **10-138**, again in excellent purity, simply by adding diethyl ether to the reaction mixture (Scheme 10.33).

To date, the main aim of combinatorial chemistry has been the preparation of a multitude of organic compounds with high constitutional diversity. Until now, stereochemical aspects have played only a minor role, although it is well known that the configuration of a molecule can have a dramatic affect on its biological activity. To address this problem, a new combinatorial strategy was developed by Tietze and coworkers [50], where stereogenic centers in a molecule are introduced by a catalyst-controlled transformation of a prostereogenic center. Using this approach in combination with a domino Knoevenagel/hetero-Diels–Alder reaction, 12 out of the 16 possible stereoisomers of emetine **10-149b** containing four stereogenic centers were synthesized. For this purpose, the two enantiomeric aldehydes **10-139** and *ent*-**10-139** obtained from the imine **10-141** by transfer hydrogenation with the ruthenium complexes (*R*,*R*)-**10-140** and (*S*,*S*)-**10-140** [51] irrespectively were used in the domino Knoevenagel/hetero-Diels–Alder reaction with Meldrum's acid **10-142** and the enol ether **10-143**, followed by solvolysis with methanol in the presence of potassium carbonate and hydrogenation.

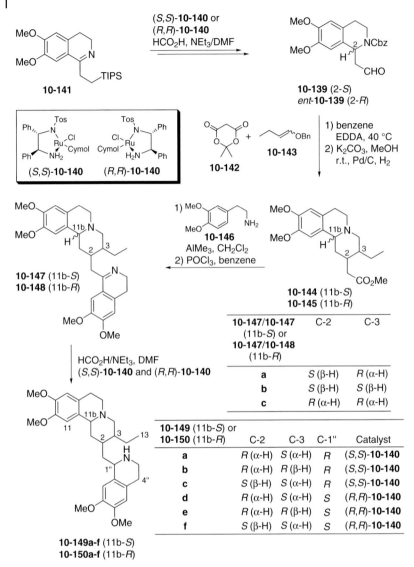

Scheme 10.34. Catalyst controlled synthesis of 12 stereoisomers of emetine.

Using the aldehyde **10-139** as substrate, the three diastereomers **10-144**, and using the aldehyde *ent-10-139*, the three diastereomers **10-145** were obtained. All six compounds were transformed independently into the imines **10-147a–c** and **10-148a–c**, which were then reduced independently again using the ruthenium complex (*R,R*)-**10-140** and (*S,S*)-**10-140**, respectively, to produce the desired 12 stereoisomers **10-149a–f** and **10-150a–f**.

References

1 For general overview about the concept, see: (a) L. F. Tietze, P. L. Steck, High Pressure in Organic Synthesis. Influence on Selectivity, in: *High Pressure Chemistry*, (Eds.: R. van Eldik, F.-G. Klärner), Wiley-VCH, Weinheim **2002**, pp. 239–283; (b) F.-G. Klärner, F. Wurche, *J. Prakt. Chem.* **2000**, *342*, 609–636.

2 L. W. A. van Berkom, G. J. T. Kuster, F. Kalmoua, R. de Gelder, H. W. Scheeren, *Tetrahedron Lett.* **2003**, *44*, 5091–5093.

3 (a) R. M. Uittenbogaard, J. P. G. Seerden, H. W. Scheeren, *Tetrahedron* **1997**, *53*, 11929–11936; (b) G. J. Kuster, H. W. Scheeren, *Tetrahedron Lett.* **1998**, *39*, 3613–3616.

4 I. M. Brinza, A. G. Fallis, *J. Org. Chem.* **1996**, *61*, 3580–3581.

5 S. Has-Becker, K. Bodmann, R. Kreuder, G. Santoni, T. Rein, O. Reiser, *Synlett* **2001**, *9*, 1395–1398.

6 L. F. Tietze, T. Hippe, A. Steinmetz, *Synlett* **1996**, 1043–1044.

7 L. F. Tietze, A. Steinmetz, *Angew. Chem. Int. Ed.* **1996**, *35*, 651–652.

8 (a) *Combinatorial Chemistry–Synthesis, Analysis, Screening* (Ed.: G. Jung), Wiley-VCH, Weinheim, **1999**; (b) *Solid-Phase Organic Synthesis* (Ed.: K. Burgess), Wiley, New York, **1999**; (c) F. Zaragoza Dörwald, *Organic Synthesis on Solid Phase*, Wiley-VCH, Weinheim, **2000**; (d) P. Seneci, *Solid-Phase Synthesis and Combinatorial Technologies*, Wiley, New York, **2000**; (e) *Solid-Phase Organic Syntheses*, Vol. 1 (Ed.: A. W. Czarnik), Wiley, New York, **2001**; (f) *Handbook of Combinatorial Chemistry* (Eds.: K. C. Nicolaou, R. Hanko, W. Hartwig), Wiley-VCH, Weinheim, **2002**.

9 D. Goff, *Tetrahedron Lett.* **1998**, *39*, 1477–1480.

10 P. Righi, N. Scardovi, E. Marotta, P. ten Holte, B. Zwanenburg, *Org. Lett.* **2002**, *4*, 497–500.

11 G. Zech, H. Kunz, *Angew. Chem. Int. Ed.* **2003**, *42*, 787–790.

12 G. Zech, H. Kunz, *Chem. Eur. J.* **2004**, *10*, 4136–4149.

13 K. Paulvannan, *Tetrahedron Lett.* **1999**, *40*, 1851–1854.

14 M. Oikawa, M. Ikoma, M. Sasaki, *Tetrahedron Lett.* **2005**, *46*, 415–418.

15 (a) A. Kuschel, T. Anke, R. Velten, D. Klostermeyer, W. Steglich, B. König, *J. Antibiot.* **1994**, *47*, 733–739; (b) R. Velten, D. Klostermeyer, B. Steffan, W. Steglich, A. Kuschel, T. Anke, *J. Antibiot.* **1994**, *47*, 1017–1024; (c) R. Velten, W. Steglich, T. Anke, *Tetrahedron: Asymmetry* **1994**, *5*, 1229–1232.

16 U. Reiser, J. Jauch, *Synlett* **2001**, *1*, 90–92.

17 R. Annunziata, M. Benaglia, M. Cinquini, F. Cozzi, *Eur. J. Org. Chem.* **2002**, 1184–1190.

18 H. Miyabe, K. Fujii, H. Tanaka, T. Naito, *Chem. Commun.* **2001**, 831–832.

19 (a) J. H. Clark, *Green Chemistry* **1999**, *1*, 1–8; (b) J. H. Clark, P. Smith, *Innovations in Pharmaceutical Technology* **2005**, *16*, 94–97.

20 G. Kaupp, J. Schmeyers, J. Boy, *Tetrahedron* **2000**, *56*, 6899–6911.

21 G. Kaupp, J. Schmeyers, A. Kuse, A. Atfeh, *Angew. Chem. Int. Ed.* **1999**, *38*, 2896–2899.

22 P. Spearing, G. Majetich, J. Bhattacharyya, *J. Nat. Prod.* **1997**, *60*, 399–400.

23 F. M. Moghaddam, M. Ghaffarzadeh, S. H. Abdi-Oskoui, *J. Chem. Res., Synop.* **1999**, 574–575.

24 L. D. S. Yadav, A. Singh, *Tetrahedron Lett.* **2003**, *44*, 5637–5640.

25 V. V. V. N.S. RamaRao, G. Venkat Reddy, D. Maitraie, S. Ravikanth, R. Yadla, B. Narsaiah, P. Shanthan Rao, *Tetrahedron Lett.* **2004**, *60*, 12231–12237.

26 O. Correc, K. Guillou, J. Hamelin, L. Paquin, F. Texier-Boullet, L. Toupet, *Tetrahedron Lett.* **2004**, *45*, 391–395.

27 R. Gedye, F. Smith, K. Westaway, A. Humera, L. Baldisera, L. Laberge, J. Rousell, *Tetrahedron Lett.* **1986**, *27*, 279–282.

28 For a review on microwave-assisted reactions in organic synthesis, see: (a) S. Caddick, *Tetrahedron* **1995**, *51*, 10403–10432; (b) S. A. Galema, *Chem. Soc. Rev.* **1997**, *26*, 233–238; (c) A. Loupy, A. Petit, J. Hamelin, F. Texier-Boullet, P. Jacquault, D. Mathé, *Synthesis* **1998**, 1213–1234; (d) G. Majetich, R. Hicks, *J. Microwave Power and Electromagnetic Energy* **1995**, *30*, 27–45; (e) A. Loupy, L. Perreux, *Tetrahedron* **2001**, *57*, 9199–9223; (f) P. Lidström, J.

Tierney, B. Wathey, J. Westman, *Tetrahedron* **2001**, *57*, 9225–9238; (g) M. Larhed, C. Moberg, A. Hallberg, *Acc. Chem. Res.* **2002**, *35*, 717–727.

29 J. Efskind, K. Undheim, *Tetrahedron Lett.* **2003**, *44*, 2837–2839.

30 (a) J. Efskind, C. Römming, K. Undheim, *J. Chem. Soc. Perkin Trans. 1* **2001**, 2697–2703; (b) K. Undheim, J. Efskind, *Tetrahedron* **2000**, *56*, 4847–4857.

31 (a) M. Shanmugasundaram, S. Manikandan, R. Raghunathan, *Tetrahedron* **2002**, *58*, 997–1003; (b) S. Manikandan, M. Shanmugasundaram, R. Raghunathan, *Tetrahedron* **2002**, *58*, 8957–8962.

32 (a) M. F. Grundon, *The Alkaloids: Quinoline Alkaloids Related to Anthranilic Acids*, Vol. 32, Academic, London, **1988**, p. 341; (b) J. P. Michael, *Nat. Prod. Rep.* **1995**, *12*, 77–89; (c) G. Brader, M. Bacher, H. Greger, O. Hofer, *Phytochemistry* **1996**, *42*, 881–884; (d) J. P. Michael, *Nat. Prod. Rep.* **1999**, *16*, 697–709.

33 (a) H. A. Abd El-Nabi, *Pharmazie* **1997**, *52*, 28–31; (b) I. S. Chen, S. J. Wu, I. J. Tsai, T. S. Wu, J. M. Pezzuto, M. C. Lu, H. Chai, N. Suh, C. M. Teng, *J. Nat. Prod.* **1994**, *57*, 1206–1211.

34 H. Surya Prakash Rao, K. Jeyalakshmi, S. P. Senthilkumar, *Tetrahedron* **2002**, *58*, 2189–2199.

35 M. C. Bagley, D. D. Hughes, H. M. Sabo, P. H. Taylor, X. Xiong, *Synlett* **2003**, *5*, 1443–1446.

36 Y. Kang, Y. Mei, Y. Du, Z. Jin, *Org. Lett.* **2003**, *5*, 4481–4484.

37 R. Schobert, S. Siegfried, G. J. Gordon, D. Mulholland, M. Nieuwenhuyzen, *Tetrahedron Lett.* **2001**, *42*, 4561–4564.

38 R. Schobert, G. J. Gordon, G. Mullen, R. Stehle, *Tetrahedron Lett.* **2004**, *45*, 1121–1124.

39 E. L. O. Sauer, L. Barriault, *J. Am. Chem. Soc.* **2004**, *126*, 8569–8575.

40 S. Tu, C. Miao, Y. Gao, F. Fang, Q. Zhuang, Y. Feng, D. Shi, *Synlett* **2004**, *2*, 255–258.

41 D. Tejedor, D. González-Cruz, F. García-Tellado, J. J. Marrero-Tellado, M. López Rodríguez, *J. Am. Chem. Soc.* **2004**, *126*, 8390–8391.

42 D. Tejedor, A. Santos-Expósito, D. González-Cruz, J. J. Marrero-Tellado, F. García-Tellado, *J. Org. Chem.* **2005**, *70*, 1042–1045.

43 J.-F. Liu, P. Ye, K. Sprague, K. Sargent, D. Yohannes, C. M. Baldino, C. J. Wilson, S.-C. Ng, *Org. Lett.* **2005**, *7*, 3363–3366.

44 H. McNab, S. Parsons, E. Stevenson, *J. Chem. Soc. Perkin Trans. 1* **1999**, 2047–2048.

45 S. Ce, Z. C. Chen, Q. G. Zheng, *Synthesis* **2003**, 555–559.

46 M. Baidossi, A. V. Joshi, S. Mukhopadhyay, Y. Sasson, *Tetrahedron Lett.* **2005**, *46*, 1885–1887.

47 (a) M. C. Mitchell, V. Spikmans, A. J. de Mello, *Analyst* **2001**, *126*, 24–27; (b) M. C. Mitchell, V. Spikmans, A. Manz, A. J. de Mello, *J. Chem. Soc. Perkin Trans, 1* **2001**, 514–518.

48 L. F. Tietze, H. Evers, E. Töpken, *Angew. Chem. Int. Ed.* **2001**, *40*, 903–905.

49 L. F. Tietze, H. Evers, E. Töpken, *Helv. Chim. Acta* **2002**, *85*, 4200–4205.

50 L. F. Tietze, N. Rackelmann, G. Sekar, *Angew. Chem. Int. Ed.* **2003**, *42*, 4245–4257.

51 N. Uematsu, A. Fujii, S. Hashiguchi, T. Ikariya, R. Noyori, *J. Am. Chem. Soc.* **1996**, *118*, 4916–4917.

Name Index

a
Abdel-Magid 194
Agami 186
Agosta 346
Ahn 371
Aizpurua 108
Alami 393, 395
Alcaide 184, 190, 314, 421
Alder 280
Alexakis 78, 470
Almendros 314
Alper 411, 414
Arcadi 191, 409
Arjona 447
Armstrong 545, 551
Arndtsen 412
Arseniyadis 515
Aubé 295, 443
Aumann 475
Avalos 298

b
Babiono 305
Bach 350
Bachi 270
Bagley 579
Bai 184
Bailey 132, 138
Baldwin 314, 495
Ballini 112, 127
Balme 191, 383, 395, 407, 409 ff.
Banfi 544, 549
Barbas III 93, 175
Barluenga 29, 295, 475, 479
Barrero 81
Barrett 446
Barriault 295, 315, 318, 582
Barton 350
Basavaiah 29
Battiste 285

Bauer 551
Behar 65
Beifuss 114, 181
Beller 95, 191, 412, 477, 556
Benaglia 573
Benetti 95, 133
Bennasar 273
Bernard 22
Beshore 505
Biehl 307
Bienaymé 552
Blechert 39, 121, 440 ff.
Bodwell 295
Boger 300, 503
Bonnet-Delpon 144
Booker-Milburn 420
Borhan 501
Bornscheuer 539
Bose 560
Bouzid 552
Bowman 158, 230, 236
Braish 71
Brandi 176, 331
Bräse 75
Braun 186
Breit 431
Brinza 567
Brown 319, 361
Brummond 185
Büchi 341
Buchwald 475
Bulger 495
Bunce 75, 501
Burger 188
Burton 479

c
Cacchi 367, 385, 409 ff., 473
Carreaux 302
Carreira 238

Carreño 50, 104, 287
Carretero 376, 462
Casey 110
Cassayre 232
Catellani 377
Chamberlin 127
Chatgilialoglu 261
Che 425
Chemla 110
Cheng 469
Chiara 523
Chiu 423
Chung 460
Clark 34
Clarke 93
Clive 265
Cohen 100, 329
Collignon 142, 175
Collin 51, 56
Cook 509
Cooke 186
Corey 3, 33
Cossy 371, 455,
Couty 186
Covarrubias-Zúñiga 69
Cram 283
Cravotto 168
Crews 196
Cuerva 246
Curran 254 ff., 417

d
Dai 41
Daïch 93, 325
Dailey 282
d'Angelo. 75
Danishefsky 71, 289, 514
Davies 429
Davis 89
de Groot 26

Domino Reactions in Organic Synthesis. Lutz F. Tietze, Gordon Brasche, and Kersten M. Gericke
Copyright © 2006 WILEY-VCH Verlag GmbH & Co. KGaA, Weinheim
ISBN: 3-527-29060-5

Subject Index

Domino Reactions in Organic Synthesis. Lutz F. Tietze, Gordon Brasche, and Kersten M. Gericke
Copyright © 2006 WILEY-VCH Verlag GmbH & Co. KGaA, Weinheim
ISBN: 3-527-29060-5

amine oxide rearrangement 519
α-amino acetals 28
amino acids 51,175
   – β-amino acids 50
β-amino allylic alcohols 499
β-amino carbonyl compounds 244
γ-amino dicarboxylates 499
β-amino esters 567
δ-amino levulinic acids 539
α-amino nitriles 31
amino sugars 504
aminoacylation/Michael addition
  domino reactions 569
aminoalcohols 42, 503
aminoaldehydes
   – α-aminoaldehydes 544, 546,
    586
   – β-aminoaldehydes 244 ff., 548,
    586
   – γ-aminoaldehydes 586
α-aminocyanides 543
aminodicarboxylic acids 546
α-aminoepoxides 42
aminoglycoside antibiotics 448
o-aminophenol 112
2-aminopyridine nucleosides 300
aminopyrroles 555
aminothiadiazol 180
amino-carbonylation 464
2-amino-4-cyclopentene-1,3-diol 341
α,ω-amino-1,3-dienes 374
N-amino-2-phenylaziridine 232
2-amino-1,3-propanediols 112
2-amino-pyridines 300, 552
2-amino-pyrimidines 92, 552
3-amino-imidazo[1,2-a]pyridines 552
5-amino-oxazoles 552
β-amyrin family 33
Anabaena flos aqua 96
anatoxin-α 96
angiogenesis 328
   – inhibitors 519
angular annulated tricyclopentanes
  354
anilines 377
anion capture 382
anionic domino processes 48 ff.
   – /anionic domino reactions 48
   – /oxidative processes 194 ff.
   – /pericyclic processes 160 ff.
   – /radical processes 156 ff
   – /reductive processes 194 ff.
   – /transition metal-catalyzed
    processes 191 ff.
Annonaceous acetogenins 16, 522

annulated indoles 470
   – lactams 447
3/5-annulated systems 78
3/6-annulated systems 78
annulation
   – [3+2] 114
   – [3+4] 116
   – [4+1] 240
   – [4+2] 240
   – [6+2] 148
anthracenes 65, 337
anthraquinones 314 f.
antiallergenic agents 579
antibacterial agents pyrimidines 559
antibiotics 395, 575
antibiotic(+)-CP-263,114 188
antibodies
   – catalytic monoclonal 448
anticancer agents 225, 268, 289, 291,
  313, 351, 399, 409, 417, 423, 468,
  502 ff., 518, 532, 537, 559
anticonvulsant agents 468
antifeedant natural products 292
antifungal antibiotics 280
antihistaminic agents 579
antimicrobial agents 468
antimycobacterial (+)-dihydropar-
  thenolide diols 96
antitumor agents, see anticancer
  agents
antiviral pyrimidines 559
anti-HIV agents 561, 581
anti-inflammatory agents 579
anti-influenza A virus agents 33
(+)-aphidicolin 291
aplysins 141
arcyriaflavin 385
ardeemins 179
areno annulated steroids 381
ArgoGel-Rink resin 570
aromatase 328
aromatic imines 57
aromatizations 291
(+)-arteannuin M 319
Artemisia annua L. 319
aryl imines 57
aryl tetralines 5
arylboronic acids 371
arylboronic esters 397
N-aryldithiocarbamate 576
arylglyoxals 544
α-aryliminium ions 181
arylnaphthalene lignans 89
aryloxiranes 73
1-arylthiobicyclo[4.1.0]heptane 223

QM LIBRARY
(MILE END)

WITHDRAWN
FROM STOCK
QMUL LIBRARY